METALLIC MULTILAYERS

METALLIC MULTILAYERS

Editors

A. Chamberod and J. Hillairet

Centre d'Etudes Nucléaires de Grenoble
Département de Recherche Fondamentale
Service de Physique - Groupe Métallurgie Physique
85X, F-38041 Grenoble Cédex, France

TRANS TECH PUBLICATIONS
Switzerland - Germany - UK - USA

ISBN 0-87849-609-2

Volumes 59 & 60 of
Materials Science Forum
ISSN 0255-5476

Distributed in the Americas by

Trans Tech Publications
Old Post Road
Brookfield VT 05036
USA

and worldwide by

Trans Tech Publications Ltd
Segantinistr. 216
CH-8049 Zürich
Switzerland

Scientific Committee

The school was organized with the support of the Formation Permanente du CNRS (12e circonscription, Grenoble).

This is the logo of the School. The external part represents at your convenience a C (for "couche" which is "layer" in French) or an E (for "école" which is "school" in French). Inside, you can easily distinguish the stack of two M for "metallic" and "multilayers". But that looks like two mountain peaks, as there are a number around Aussois. As a result, this logo is a subtile bilayer of science and poetry. Is not it?

Preface

This book brings together a series of lectures presented to the Summer School "Multicouches Métalliques" held in Aussois (France) in September 1989. The courses were designed to provide an in-depth review of the basic physics of the artificially multilayered metal-metal structures and the latest progress that has been made in the manufacturing, structural characterization, physical investigation and modeling of these novel materials.

Among the various topics covered are electronic and atomic transport, elastic and micromechanical behaviour, magnetic and superconducting properties. Emphasis is laid on the description and understanding of these properties on account of interlayer multiplicity, low dimensionality and nanometer periodicity. The course also offers insight into growth control, monolayer formation and related thermodynamics, together with a presentation of the most recent deposition and characterization techniques.

Hopefully this book will serve both as a stepping stone for those students interested in getting involved with research in this rapidly developing area and a guide for specialists in industry and the academic community.

We would like to thank both the lecturers and participants for having made the school a lively and enriching meeting.

A substantial grant was obtained from the departments "Mathématiques et Physique de base", and "Chimie" of the CNRS. Encouragement and financial aid came from ISA-RIBER France, THOMSON-CSF and CNET. All are gratefully acknowledged.

Grenoble, April 1990

A. Chamberod
J. Hillairet

TABLE OF CONTENTS

Materials Science Forum Vols. 59 & 60 (1990) pp. 1-40

FUNDAMENTAL PHENOMENA IN NUCLEATION AND GROWTH OF CRYSTALS
PART A - MACROSCOPIC CONCEPTS AND ENERGETICS

J.M. Bermond
CRMC2 - CNRS Campus Luminy - Case 913
F-13288 Marseille Cedex 09, France

The present chapter is an overview of the various phenomena that play a role in crystal growth and equilibrium . Covering the whole subject within the limited amount of space and time available for a Summer School is an impossible task . Besides , excellent comprehensive books or review papers exist [1 - 5] in which the interested reader can find detailed information . Therefore this chapter is merely aimed at introducing the physical quantities that are of importance to an experimentalist . Emphasis in put on distinguishing between (nearly) equilibrium situations and non-equilibrium ones .

Sections 1 to 5 are devoted to introducing the three basic equilibrium situations of crystallites on a surface and to relating them to the concept of surface tension. A macroscopic thermodynamical approach is used throughout. Non equilibrium situations are then presented, along with a brief account of surface diffusion, with the mere view to showing how surface diffusion can influence the growth shape in deposition experiments (sections 6 to 8). In section 9, the influence of epitaxy on the growth mode is examined. Since the binding energy of an atom to a surface is one of the basic parameters its experimental determination is described in section 10.

The nucleation of crystallites on surfaces is not dealt with. This is the subject of chapter 2 (by A. Masson) which was primarily intended to be the second half of the present chapter but was then separated for practical reasons. However chapters 1 and 2 were conceived in common and form the two pieces of a sequence about the same topic.

1 WHAT DOES A SOLID SURFACE LOOK LIKE ?

In this section we look at the contact between two bulk phases from an atomistic point of view , taking the solid-vacuum interface as an example .

1 . 1 Ideal surfaces . The geometrical approach

Let us take an infinite perfect crystal , draw a mathematical plane parallel to some (hkl) orientation , and remove all the atoms lying on one side of this dividing plane. The remaining piece of the crystal is bounded by an ideal (hkl) surface . For the moment the precise location of the dividing plane is unimportant [1] . It is enough that it runs through all points equivalent with respect to the adjacent matter . Several types of ideal (or perfect) surfaces can be obtained by using this "mental cleavage" process along various (hkl) orientations : flat , stepped and complex . Figure 1 shows one such complex surface.

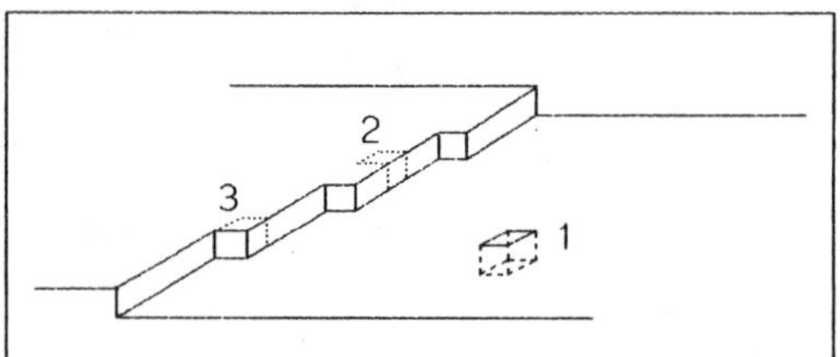

Fig. 1 - Schematic representation of an ideal complex surface . Each atom is pictured as a cube . The numbers show the atomic positions described in text .

This simple picture already aloows us to distinguish several possible atomic positions : The terrace site (1) where the atom is embedded in a lattice plane , the step atom (2) and the kink atom (3) . The kink position is the most important [1]. The number of neighbors of a kink atom is half that of an atom in the bulk and so is its binding energy (referred to that of an atom in the vapor).

1 . 2 Real surfaces

Real surfaces may considerably differ from ideal surfaces . Experimental evidence of this fact was first obtained by field ion microscopy and is now confirmed by scanning tunnelling microscopy .

The relative positions of the atoms in the topmost layers may not be those of the atoms in the infinite crystal . This is surface relaxation . One example is the Si(2x1) surface mesh .

The role of thermal activation is of major importance . It induces surface disorder and possibly facetting . A heated surface no longer achieves the perfect ordering of fig 1 but should be represented more realistically like on figure 2 (the TLK model) .

In addition to the previously mentionned atom positions there exist now adatoms and vacancies on (in) the terraces (4 and 5) , adatoms and vacancies at the steps (6 and 7) , which cause a roughening of the steps.

---

1 Although the structure of the surface may specifically depend on this position e. g. the structure of a NaCl (111) face .

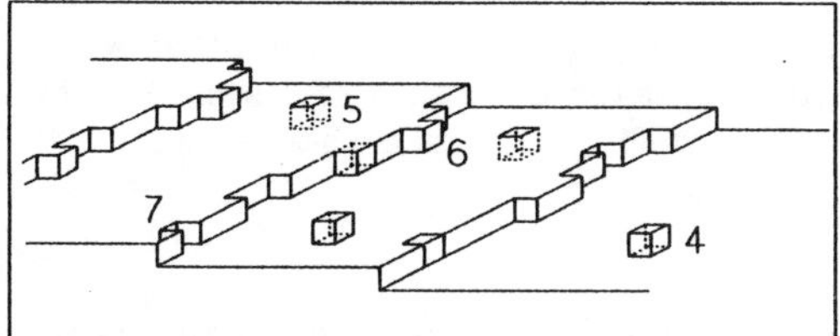

Fig. 2 - The TLK model of a surface.

Apart from their creation by thermal activation , adatoms may also be present for purely kinetic reasons (e. g. deposition from an external source) and contribute to the surface disorder or to the nucleation of new layers .

Adatoms and vacancies are mobile at temperatures where they can overcome the energy barriers that confine them on a terrace or at a step or a kink . This means, in particular , that surface diffusion occurs and possibly leads to a macroscopic change of shape of the crystal .

More extreme situations are possible : surface roughening and surface melting . Since these are treated at length in another chapter nothing more will be said here .

To conclude this brief section we generalize and point out that the junction between two three dimensional phases appears as a region of finite thickness — albeit possibly quite thin —which provides a transition between the two .

2 THE THREE BASIC EQUILIBRIUM SITUATIONS - EXPERIMENTAL FACTS

We describe here the three typical situations that can be met when a material , deposited on a foreign substrate , is in equilibrium with its own vapor phase , called here the "mother phase" .

2 . 1 The Volmer - Weber (VW) equilibrium

This equilibrium is characterized by the existence of three dimensional (3D) crystals in contact with the substrate while the rest of the substrate surface is devoid of any condensed phase . An example can be seen on figure 4 (Pb on a (00. 1) graphite substrate) . We use this example to continue the description .

All the crystallites have similar shapes except for a size factor [2] . Their external surface is composed of flat facets (the {111} and {100} for Pb) with smoothly rounded regions in-between .

Let us call T the temperature , p' and μ' the equilibrium pressure and chemical potential of the mother phase , p'_∞ and μ'_∞ those of the mother phase in equilibrium with

---

2 Thermodynamics requires that there should be only one crystal at equilibrium . Additionnally the equilibrium is stable only if the total volume and the mass of the system are kept constant [3] . However the exchange of matter between coexisting crystallites is generally slow compared to the displacement of atoms along the crystal surfaces . Consequently equilibrium of each crystallite is achieved locally before the equilibrium of the whole system is reached . Therefore we shall ignore these restrictions in sections 2 to 5 of this chapter .

the bulk (infinite) deposit . One observes that $p' > p'_{\infty}$ and $\mu' > \mu'_{\infty}$. In other words the chemical potential of a finite 3D crystallite is greater than that of its infinite bulk phase at the same temperature [3] . Therefore one defines the supersaturation

$$\Delta\mu = \mu' - \mu'_{\infty} = \mu_C - \mu_{C\infty} \quad (1)$$

where μ_C and $\mu_{C\infty}$ are the chemical potentials of the finite size crystallite and the infinite crystal respectively . So we take as an experimental characteristic of the VW equilibrium that $\Delta\mu$ be positive .

As we shall see (section 3.6) the VW equilibrium is typical for weak adhesion of the deposit on the substrate .

In many cases there exists a definite relationship between the lattices of the substrate and the crystallite . This phenomenon is called " epitaxy ". In the VW equilibrium the epitaxial relationship only depends on the substrate-deposit couple . It is independant of the external conditions : (T , $\Delta\mu$, . . .).

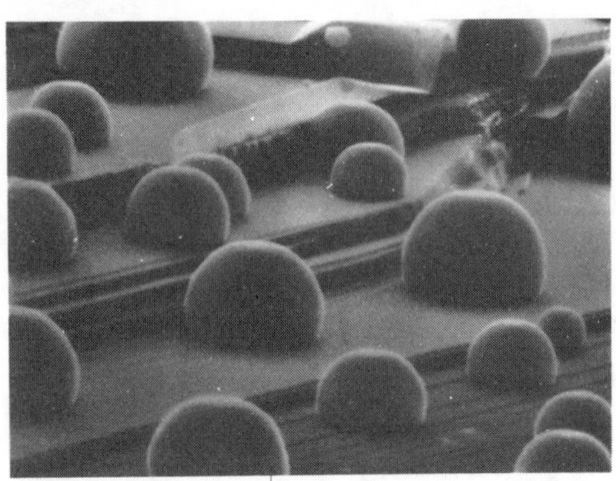

Fig. 3 - An example of the VW equilibrium : Pb on graphite .

2 . 2 The Frank - van der Merwe equilibrium

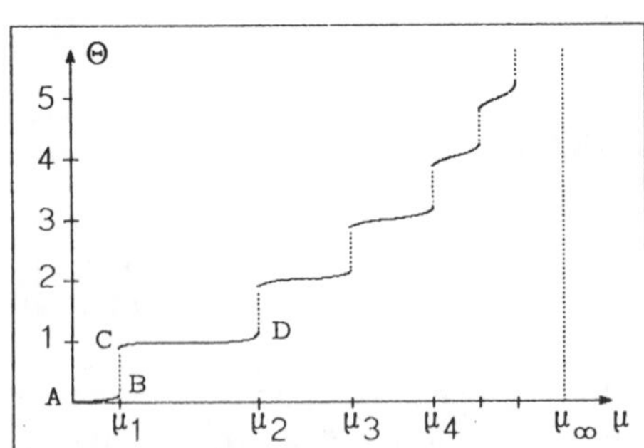

Fig. 4 - Schematic diagram of an isotherm of Xe on graphite .

In the Frank- van der Merwe (FM) equilibrium the material can cover the whole substrate surface in a stack of layers . The best examples are found in the physisorption of gases . Figure 4 shows a schematized isotherm of Xe on graphite . Θ (coverage) is defined here as the ratio of the number of Xe atoms to that of the atoms in a complete Xe layer (per unit area) . μ is the chemical potential of the Xe vapor – equivalently that of the Xe layer at equilibrium . μ_{∞} is the chemical potential of bulk Xe in equilibrium with its own vapor phase . We define $\Delta\mu$ as in eq. 1 . Starting

---

3 The same is true for a 3D crystallite .

from low coverage (point A) one first obtains a 2D random dilute phase of Xe on the substrate(2D gas) . At $\mu = \mu_1$ a 2D condensed phase appears (point B), in equilibrium with the 2D gas along BC . If $\mu > \mu_1$ a jump-wise change in Θ occurs . The condensed 2D phase now covers the whole substrate and gradually increases its density until a new 2D gas forms on top (CD). At $\mu = \mu_2$ a new condensed phase is formed in the second layer , etc . . .

In other words layers are condensed successively , each at a specific $\Delta\mu_i = \mu_i - \mu_\infty$. 2D layers can only be stable at $\Delta\mu < 0$. The structrure of the layers may be different. Structural changes within one layer are possible for a given substrate-deposit couple depending on T and $\Delta\mu$. Epitaxial relationships are also observed . They may change upon building a new layer on top .

2 . 3 The Stranski - Krastanov equilibrium

The Stanski-Krastanov (SK) case is intermediate between the VW and the FM equilibrium. Starting from a bare substrate a limited number of 2D layers can be formed upon increasing the chemical potential μ of the vapor . In this regime $\Delta\mu$ is negative and the same facts as for the FM equilibrium are observed . The finite number of layers is a characteristic of the deposit-substrate couple . A further increase of the amount of deposited material is only possible if $\Delta\mu$ becomes positive . Then 3D crystals are formed " on top " of the preexisting layers and the characteristics become those of the VW equlibrium .

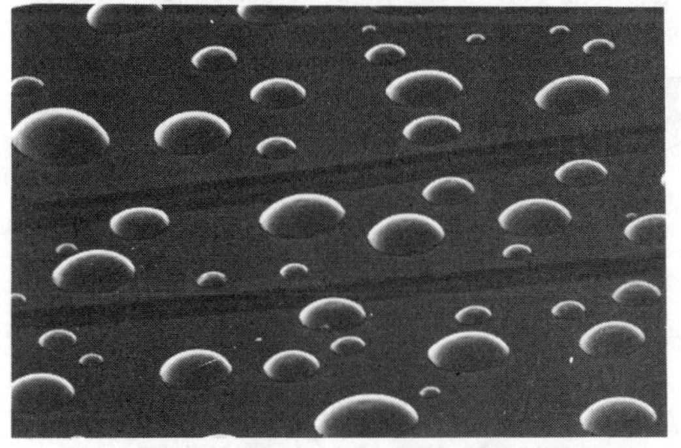

Fig . 5 - Example of SK equilibrium . Pb/Ge(111) . AES shows that one 2D condensed layer of Pb exists between the crystallites .

A wide variety of epitaxial relationships may be observed during the formation of the 2d layers . Many metal/metal or metal/semi-conductor systems display this behaviour . An example is shown by figure 5 (Pb / Ge(111)) . This allows us to compare the properties of one and the same deposit in VW and SK equilibria . The result [6] is that the 3D cristallytes of Pb have a shape similar to that observed for Pb on graphite (VW equilibrium) however with a different height above the substrate [6] .

2 . 4 Surface phases and 2D equilibria

The VW and the FM equilibria show us two extreme cases for the state of a deposit on a foreign substrate .

For 3D crystals the transition region (the interface) is thin enough , compared to the crystal thickness , for the properties of the crystallite to be fairly close to those of a bulk phase . Therefore the properties of the interface can be characterized by specific thermodynamic functions (section 3. 1) . We shall speak of a surface phase between two bulk phases .

On the contrary 2D layers are strongly influenced by the substrate and thus have structures and properties quite different from those of the bulk deposit . Moreover 2D layers can exist on a substrate even though no mother phase is present . This is the case for many metal/metal systems at temperatures where the deposit has a negligible vapor pressure . Within the domain of existence of a layer a 2D phase diagram can be constructed [7, 8] . Therefore we can speak of specific 2D phases on a substrate and of phase eqilibria within a surface layer .

Recent atomistic theories have led to a detailed understanding of the properties of 2D phases in terms of wetting . We refer the interested reader to [8] and to the chapter by J. Villain in the present book . For the specific purposes of the present chapter we shall use a more macroscopic language : that of the capillarity theory [5, 9, 10, 11] . The physical limits of this approximation will be stressed when necessary .

3 THE SURFACE PHASE . SURFACE TENSION AND LEDGE TENSION

3 . 1 Contact between 3D phases . Surface properties as excess quantities

We consider two bulk phases (labelled by the superscripts ' and ") separated by a macroscopically flat transition region . Several components (labelled i) are present. Figure 6a shows the thermodynamic system under study . It has a volume V , a free energy F , a total number of particles N , etc. . . Its properties only depend on the coordinate z normal to the interface . Sufficiently far from the latter they are identical to those of the infinite bulk phases . There the concentrations of component i *are* c'_i and c''_i and there exist homogeneous volume densities of energy, etc . . .

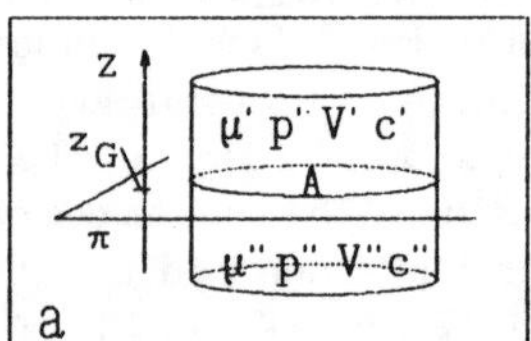

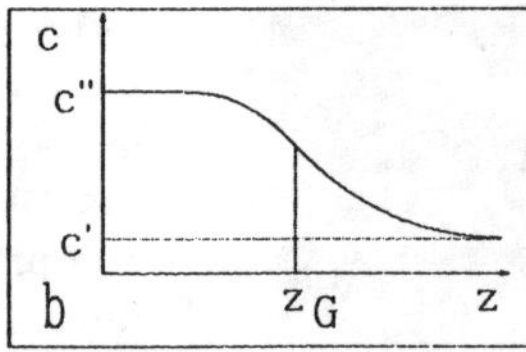

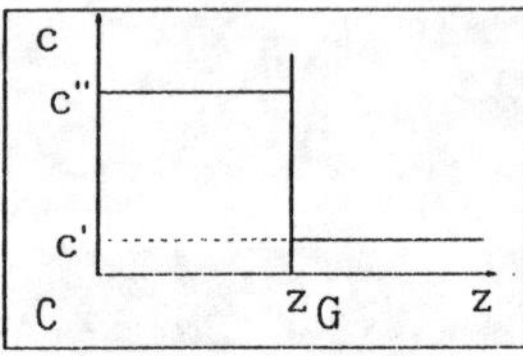

Fig. 6 - a (left) The Gibbs thermodynamic system - A is the base area , π is the dividing plane , z_G is its coordinate .
- b (center) Concentration as a function of z in the real system .
c (right) Concentration as a function of z in the Gibbs system .
For clarity one component only is assumed .

Using the Gibbs model consits in remplacing the real system by the following idealized one : Phases ' and " keep the properties of the bulk phases up to a mathematical plane located somewhere in the physical transition region (a convenient position will be choosen later on) . This plane is called the dividing surface . It defines two volumes V' and V" . The total Gibbs system has the same N , N_i 's , V , etc. . . as the real one . Hence one must write for any component

$$N_i = N' + N'' + N_i^\sigma$$

where N_i^σ is the number of atoms of i in the real system in excess with respect to the Gibbs system . This is in fact an implicit definition of N_i^σ . Since $V = V' + V''$, N_i^σ atoms must be placed in the dividing surface of the Gibbs system . This defines the excess number of atoms of i per unit surface area

$$\Gamma_i = \frac{N_i^\sigma}{A} \tag{2}$$

which is also called the adsorption of component i . In other words we have just defined a surface phase , " concentrated " in the dividing plane , whose composition is specified by the Γ_i 's .

Any extensive quantity of this surface phase can be similarily defined as an excess quantity . In particular one has

$$F = F' + F'' + F^\sigma \tag{3}$$

However **there is one exception** :

$$V = V' + V'' = 0 \tag{4}$$

since the surface phase has zero volume .

Next we define the specific thermodynamic functions of the surface phase as the surface densities of E^σ , F^σ , etc. . . In particular the specific surface free energy is

$$f^\sigma = F^\sigma/A \tag{5}$$

Finaly we calculate the chemical potentials of the various components in the surface phase : At equilibrium

$$\mu_i^\sigma = \mu_i' = \mu_i'' = \left.\frac{\partial F'}{\partial N_i'}\right)_{T,V',N_j'} = \left.\frac{\partial F''}{\partial N_i''}\right)_{T,V'',N''_j} \tag{6}$$

Strictly speaking all these definitions implicitely assume that the volume densities of free energy are uniform throughout the bulk phases . This is not always true for solids in which the state of strain may vary from point to point . Fortunately the elastic contribution to f' or f" is relatively small [10] so that we ignore this point in the following .

The Gibbs convention . Among the different possible locations of the dividing plane [10] one often chooses that defined by Gibbs . It consists in locating the dividing plane so that one of the Γ_i 's is nil . Generally the principal component is selected .

3 . 2 Contact between 3D phases . The surface tension

We now proceed by defining a quantity (the surface tension γ) which can be independent of the location of the dividing plane . This , along with the properties mentioned in section 3. 4 make it the quantity of experimental relevance . The grand potential of the surface phase is

$$\Omega^\sigma = F^\sigma - \sum_i \mu_i^\sigma \Gamma_i \tag{7}$$

We define the surface tension γ as the surface density of Ω , i. e.

$$\Omega^\sigma = \gamma A \tag{8}$$

or
$$\gamma = f^\sigma - \sum_i \mu_i^\sigma \Gamma_i \tag{9}$$

It can be shown that γ is independent on the position of the dividing plane (unlike the N_i^σ 's and f^σ) if isotropic pressures can be defined in the bulk phases . .

3 . 3 Contact between 2D phases . The ledge tension

What has just been done for 3D phases can be repeated for 2D phases . The only difference is that we are now in two dimensions and the boundary between phases is now a line . The two-dimensional equivalent to γ is defined in the same manner . It is called the ledge tension γ_l . It has the same characteristic properties as γ .

3 . 4 Thermodynamic relations [4]

3. 4. 1 The Gibbs-Duhem equation

Since f^σ is a surface density it is an intensive quantity . Then, thermodynamics tells us that it only depends on T and the Γ_i 's :

$$df^\sigma = - s^\sigma dT + \sum \mu_i^\sigma d\Gamma_i \tag{10}$$

By differentiating eq. 9 and combining with eq. 10 , one gets

$$d\gamma = - s^\sigma dT - \sum \Gamma_i \, d\mu_i^\sigma \tag{11}$$

which is the Gibbs-Duhem equation for the surface phase . It tells us that γ depends on T and the μ_i^σ 's , i. e. also on the chemical potentials in any one of the bulk phases (eq. 6) . In particular, at constant temperature, one obtains the Gibbs adsorption isotherm

$$d\gamma = - \sum \Gamma_i \, d\mu_i^\sigma = - \sum \Gamma_i \, d\mu_i \tag{12}$$

3. 4. 2 γ as a reversible work of creation .

Differentiating the relation $F^\sigma = Af^\sigma$ and combining with eq. 11 and 3 yields , for the whole system (summation over all phases and components) ,

$$dF = - S\, dT - p'\, dV' - p''\, dV'' + \gamma\, dA + \sum \mu_i \, dN_i \tag{13}$$

Hence

$$\gamma = \left.\frac{\partial F}{\partial A}\right)_{T,V',V'',N_j} \tag{14}$$

---

4 Unless otherwise indicated , all the summations in this section are over the components . Therefore we omit the subscript i for brevity .

Eq. 13 or 14 express γ as a reversible work per unit surface . It is the reversible work done on the system by the external forces to create a unit surface area at constant T , V' , V'' and numbers of particles . This fundamental property is the basis of the calculation of γ by " mentally cleaving " a crystal .

3 . 4 . 3 Simple examples .

Let us take a c. f. c. simple metal (say Ag) of lattice constant a and assume pair-wise additive atomic interactions limited to nearest neighbors . We only consider a perfect surface and ignore all entropic terms i. e. we calculate at T = OK . Let Φ be the enthalpy per bond . We first cleave between two adjacent (100) planes . We thus create two (100) surfaces of area $a^2/2$ per atom . Four bonds are broken per atom . The reversible work spent in this process per unit area is thus $W = 2\gamma = 8\Phi/a^2$. Hence $\gamma_{100} = 4\Phi/a^2$. One finds similarly $\gamma_{111} = 2\sqrt{3}\,\Phi/a^2$. Φ can then be evaluated from the sublimation enthalpy of the crystal : $\Delta H = Z\Phi/2$, where Z is the number of nearest neighbors per atom . Despite the crudeness of the model it is now obvious that γ is an anisotropic quantity . The anisotropy is grossly exagerated but the tendancy is shown : The most close-packed surfaces have the smallest γ .

The results just obtained may be misleading . It is a contradiction with the thermodynamic definitions of sections 3-1 and 3-2 that γ be determined by the properties of one phase only . In fact we have just calculated γ for a solid-vacuum interface . It is a sufficient approximation for the solid-gas interface . For the contact between two bulk condensed phases , yet , a full knowledge of the structures and interactions inside the phases and across the interface is required . In this case the interfacial tension is best obtained (or interpreted) by considering the adhesion work (section 3.6) .

3 . 4 . 4 Extremal properties of $\int\gamma dA$.

We consider now a cristalline phase " bounded by surfaces of several types (labelled k) in equilibrium with a fluid phase ' . We assume that we can define an isotropic pressure p" . Then

$$\Omega^\sigma = \sum \gamma^k A^k \qquad (15)$$

or

$$\Omega^\sigma = \int \gamma \, dA \,, \qquad (16)$$

a more concise form which includes the case where some parts of the surface are continuously curved . For the whole system

$$\Omega = -p'V' - p''V'' + \int \gamma \, dA \qquad (17)$$

and we know that Ω is the thermodynamic potential at constant T , V' , V'' and μ_i 's . Hence $\int\gamma\, dA$ must be an extremum at equilibrium with respect to all virtual changes at constant T , V' , V'' and μ_i 's (a minimum for stable equilibrium) . This requirement is the basis for the Wulff theorem and the generalized Gibbs-Tompson equation (section 5) .

3 . 5 The case of a continuously curved surface

So far only a dividing plane has been used . Figure 3 shows us that curved dividing surfaces must also be considered . In this case the chemical potential of the surface phase entails a curvature dependant term [11, 12] . Let M be a point on the surface where the external normal is $\vec{n}$ and the principal radii of curvature are R_1 and R_2. The local coordinates at M in the tangent plane are u and v in the principal section planes . If $\gamma(\vec{n})$ is a twice differentiable function of u and v , the equilibrium chemical potential μ^σ of an atom at M differs from its value μ^σ_∞ for a plane surface . One has , with v" the atomic volume in the solid phase ,

$$\mu^\sigma - \mu^\sigma_\infty = v''\left[\gamma\left(\frac{1}{R_1} + \frac{1}{R_1}\right) + \frac{1}{R_1}\frac{\partial^2\gamma}{\partial u^2} + \frac{1}{R_2}\frac{\partial^2\gamma}{\partial v^2}\right] \qquad (18)$$

So the chemical potential of a surface atom is higher the smaller are the local radii of curvature . At equilibrium the crystal so adjusts its curvature that equality of its chemical potential is ensured over its surface . On the other hand if the crystal deviates from its equilibrium shape there exists a gradient of chemical potential along its surface , which provides a " driving force " to restore equilibrium (in general by surface diffusion) . In most cases the anisotropy of γ is weak and eq. 18 is simplified by dropping the last two terms .

3 . 6 Contact between two bulk phases . The adhesion work

Let A anb B be two condensed in contact and γ^* be their interfacial tension . We separate them along the interface . The latter disappears and two surfaces of A and B are created . **By definition** , the reversible work spent is this process per unit area is

$$\beta = \gamma^A + \gamma^B - \gamma^* \qquad (19)$$

Hence $$\gamma^* = \gamma^A + \gamma^B - \beta \qquad (20)$$

β is called the adhesion work or equivalently the work of separation . Although eq. 19 is no of thermodynamic origin , β is an intresting quantity . It can be calculated if one knows the interaction between the atoms of the two species A and B . Hence we again recognize from eq. 20 that γ cannot be obtained from the mere properties of the individual phases . A realistic estimate of β should include the rearrangement of atoms at the created interfaces after separation [5] . This involves in particular the elastic energy in the interface if A and B are not isomorphous .

Later on we shall examine the situation where a deposit A lies on a substrate B . Let us assume $\gamma^A < \gamma^B$. Then the crucial quantity will be seen to be

$$\Delta\gamma = \gamma^A - \gamma^B + \gamma^* \qquad \text{or} \qquad (21)$$

$$\Delta\gamma = 2\gamma^A - \beta \qquad (22)$$

If $\beta > 2\gamma^A$, $\Delta\gamma < 0$. One says that phase A wets the substrate " more than perfectly " . The physical meaning is clear if one considers the surfaces of A and B separately as solid-vacuum interfaces . The adhesion work is greater than the cohesion (free) energy of A . So the system tends to form 2D layers of A on B . In our simple approach the various wetting conditions may be summarized as follows :

$\beta > 2\gamma^A$	$\Delta\gamma < 0$	More than perfect wetting.	2D layers are formed
$\beta = 2\gamma^A$	$\Delta\gamma = 0$	Perfect wetting .	2D layers are formed
$0 < \beta < 2\gamma^A$	$\Delta\gamma > 0$	Imperfect wetting .	3D crystallites are formed
$\beta = 0$	$\Delta\gamma = 2\gamma^A$	No wetting .	No interface is formed

3 . 7 The physical limits of the capillary approximation

Are γ and β thickness dependant ? Let us take γ as typical . If γ is calculated with a nearest neighbor interaction , only one atomic layer has to be taken into account in each bulk phase . This immediately points to the limit of the capillary approximation . γ is thickness independant if the number of atomic layers on each side of the dividing surface can be considered as larger than the range of atomic interactions. Only then can γ be defined as a characteristic of an interface . Experimental studies show that the condition is fulfilled to a good approximation for 3D crystals , either isolated or lying on a substrate . On the contrary 2D condensed phases are not suited to defining γ in the manner of section 3. 2 .

3 . 7 The surface tension of a 2D phase on a substrate

For a thermodynamical treatment , we consider the 2D phase as infinitely flat (zero volume) , we call its area A and we define γ off-handedly as the conjugate quantity of the extensive state variable A , i. e. we write the basic thermodynamic relation

$$dE^\sigma = T\,dS^\sigma + \gamma\,dA + \mu^\sigma dN^\sigma \qquad (23)$$

where we have assumed one component only for brevity . Then by defining $F^\sigma = E^\sigma - TS^\sigma$ $G^\sigma = F^\sigma - \gamma A$, $\Omega^\sigma = F^\sigma - \mu^\sigma N^\sigma$ the same relations as in section 3.2 are obtained , in particular

$$dF^\sigma = -S^\sigma dT + \gamma\,dA + \mu^\sigma dN^\sigma \qquad (24)$$

and $$d\gamma = -s^\sigma dT - \Gamma d\mu^\sigma \qquad (25)$$

where $\Gamma = N/A$ is now the coverage of the 2D phase . Eq. also makes γ the reversible work of creation of the unit surface phase area . Thus γ can again be calculated by using a cleavage process [5, 13] . The difference is that γ depends on the surface chemical potential even though a single component is present . It is easy to calculate the change in γ as a function of $\Delta\mu$ [5] . Starting from the bare substrate of surface tension γ^O , integration of eq. 25 along an isotherm (figure 7) yields

$$\Delta\gamma = \gamma(\mu') - \gamma^O = -\int_{\mu_O}^{\mu'} \Gamma\,d\mu \qquad (26)$$

where μ_O is the chemical potential of th 2D gas on the substrate at infinite dilution . Here it is convenient to define $\Delta\gamma_\infty$ by

$$\Delta\gamma_\infty = -\int_{\mu_O}^{\mu_\infty} \Gamma\,d\mu \approx -\int_{\mu_e}^{\mu_\infty} \Gamma\,d\mu \qquad (27)$$

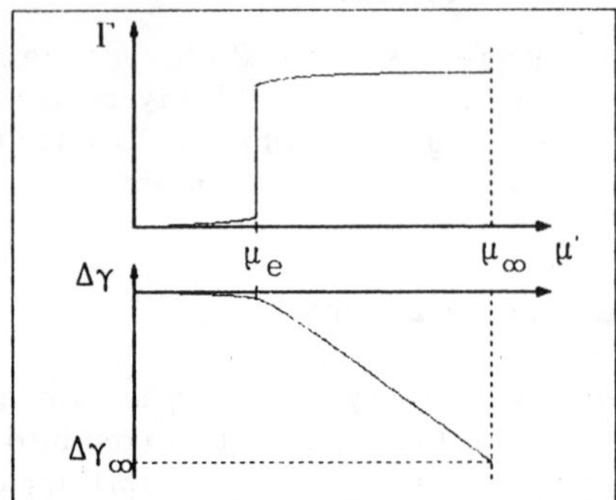

Fig 7 - Integration of eq. 25 along an isotherm . Graphs of Γ (top) and of Δγ (bottom) versus μ' .

Generalising to n layers is obvious . Then from eq. 26 , naming a^2 the area per atom of the full layer , one gets

$$a^2 (\gamma - \gamma^0) = a^2 \Delta\gamma_\infty - a^2 \int_{\mu_\infty}^{\mu'} \Gamma \, d\mu \approx a^2 \Delta\gamma_\infty - (\mu' - \mu_\infty) \qquad (28)$$

We now seek the thermodynamic stability of a substrate of area A_0 fully covered by a 2D condensed phase , at constant T , total volume V' and μ' of the mother phase , assuming that the state of the bulk substrate is unaltered . Starting from the bare substrate as the initial state of the system the change in the thermodynamic potential of the system is easily seen to be

$$\Delta\Omega = A_0 [\gamma (T , \mu') - \gamma^0] \qquad (29)$$

For the equilibrium to be stable ΔΩ must be negative , hence $\gamma - \gamma^0$ too . From eq. 28 one must have at negative Δμ

$$a^2 \Delta\gamma_\infty < \Delta\mu < 0 \qquad (30)$$

4 PROPERTIES OF THE SURFACE TENSION γ

In this section we consider γ as a size-independent characteristic of an interface . The properties of its 2D equivalent γ_1 are similar .

4 . 1 The anisotropy of γ

This is perhaps the most important property of γ. The anisotropy can be measured on the equilibrium shape of a crystal by using the inverse Wulff construction . Figure 8 shows the results obtained forPb [14] . The anisotropy is weak . It decreases upon increasing the temperature . The equilibrium shape is composed of the {111} and {100} facets with rounded regions inbetween . Figure 8 shows that these facets correspond to minimal values of γ , the most closely packed (the {111}) having the smallest γ value . This is a general rule . Other cases are known (Ni) where , in addition , the {110} belong to the equilibrium shape . These also present a (relative)minimum of γ .

Instead of a diagram like that of figure 8 on can draw a polar plot of γ as a function of the surface orientation (figure 9) . This " γ-plot " exhibits cusps in the

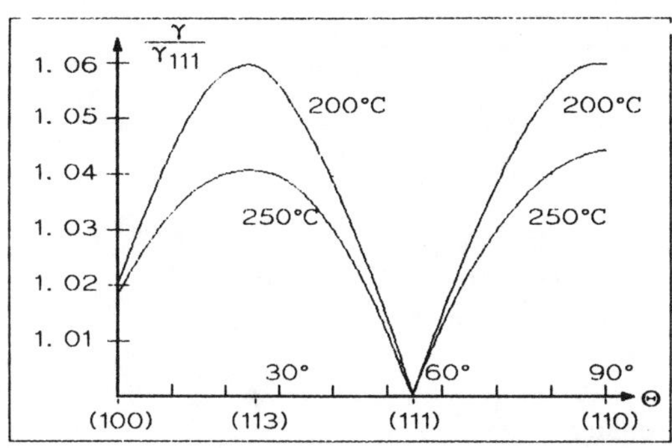

Fig. 8 - Anisotropy of the surface tension of Pb at two temperatures (from ref. [14]).

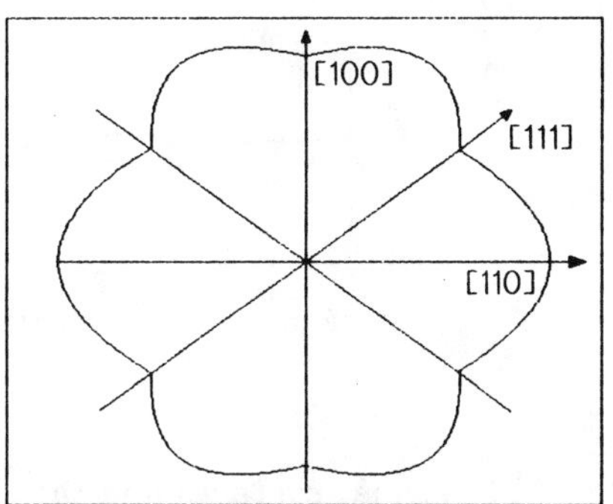

Fig. 9 - Schematic representation of The γ-plot of Pb . The anisotropy is greatly exagerated for clarity.

directions normal to the plane facets . The derivatives of γ are discontinuous at a cusp . Although the anisotropy is small (at least for metals) the cusps are responsible for the existence of flat facets on the equilibrium shapes [3] .

4 . 2 The absolute value of γ . Some experimental results

Unlike the anisotropy of γ , its absolute value is difficult to measure for solids . A review of the experimental techniques is given in [15] . Theoretical calculations are also useful . We merely quote here a few examples of absolute values of surface tensions to give the reader a feeling for the orders of magnitude .

Metal	γ (erg. cm^{-2})	T (°K)	Conditions	Reference
Pb	610	590	Vacuum or Ar	[16]
	534	300	Calculated	[17]
	610	601 (M. P.)	Calculated	[17]
Ag	1180	1073	N_2	[16]
	1500	1193	N_2	[16]
	1302	300	Calculated	[17]
	1046	1323 (M. P.)	Calculated	[17]
	1300		Calculated	[18]
W	2900	2000	Vacuum	[16]
	2800	2300	Vacuum	[16]

Table I
Examples of measured and calculated values of γ for the solid-vacuum interfaces of Pb , Ag and W (M. P. Melting point)

The solid-liquid interface tensions are in general smaller than the solid- gas or solid-vacuum surface tensions. This is shown by a comparison of the values in Table I and in Table II .

Metal	γ solid-liquid (erg. cm^2)	T (°K)	Ref.
Pb	40	600	[15]
Ag	126	1234	[15]

Table II

Examples of experimental values of the solid-liquid interfacial tensions of Pb and Ag .

4 . 2 Vicinal surfaces and the step tension

Vicinal surfaces are those whose orientations lie close to a cusp in the γ-plot , for example, the (10 1 0), (810), (610) in the f. c. c. lattice . Out of geometrical necessity ideal vicinal surfaces are composed of terraces of equal width separated by (equidistant) monoatomic steps . Figure 10 shows a cross section of such a vicinal . We note γ_s, the ledge tension of a terrace plane . It is called the " step tension ".

Let Θ be the polar angle defining the (vicinal) surface orientation referred to the terrace orientation . Since Θ is small the γ-plot can be approximated , close to the cusp , by the tangent circle at the cusp that runs through the Wulff point , i. e. we write

$$\gamma \approx \gamma_o \cos \Theta + \gamma_m \sin \Theta \qquad (31)$$

where γ_m is a constant and γ_o the surface tension of the terrace orientation . The more vicinal the surface, the better the approximation . Inspection of figure 10 immediatly shows that γ_m must be identified with γ_s / a , where a is the interlayer distance . Hence for a vicinal surface

$$\gamma \approx \cos \Theta (\gamma_o + \gamma_s \tan \Theta / a) \qquad (32)$$

Now let D be the step density projected on to the terrace orientation . From fig. 10

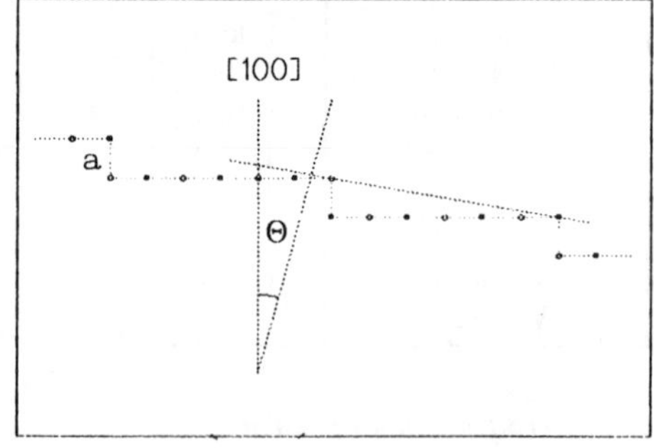

Fig 10 - Schematic cross section of an ideal (810) surface . The terraces are (100) planes . The steps are [001] rows normal to the plane of the figure . Different symbols are employed to represent the atoms since every second row is shifted by a/2 .

$D = \tan \Theta / a$. Hence

$$\gamma \approx \cos \Theta (\gamma_o + D \gamma_s) \qquad (33)$$

The step contribution to γ is proportionnal to the step density . The physical meaning of the approximation is clear . If Θ is small enough the steps are so widely separated that each can be treated as the individual ledge of an infinite terrace of ledge tension γ_S . (This the 2D equivalent to considering γ as thickness independant) . When the present approximation breaks down the steps are said to interact (see chapter by J. Villain) .

V CONSEQUENCES . INTERPRETATION OF THE EXPERIMENTAL FACTS

5 . 1 The equilibrium shape of an isolated crystal

In this section we consider a bulk fluid phase surrounding a crystallite of finite (small) dimensions . The notations are those of section 3 .

5 . 1 . 1 Thermomechanical equilibrium . The Wulff theorem .

In many instances a crystal can change its shape at constant volume without being necessarily in thermodynamic equilibrium with the surrounding medium . This happens for many metals in vacuum at temperatures where the evaporation rate is negligible compared to the rate of surface diffusion . The crystal is in thermomechanical equilibrium but its chemical potential is not specified by the mother phase . What is the equilibrium shape under these conditions ? The answer is obtained [10]from section 3-4-4 , condition (17) . The solution to condition (17) is expressed by the Wulff theorem [12 , 19] :

At thermomechanical equilibrium , there exists one single point in the crystal whose normal distances h_i to the various surface planes satisfy the relations

$$\frac{\gamma_1}{h_1} = \frac{\gamma_2}{h_2} = \ldots = \frac{\gamma_i}{h_i} = \ldots \qquad (34)$$

Note that the common value of the ratios γ_i / h_i is not specified by the theorem . One of the h_i 's may be chosen arbitrarily, which fixes the others . The size of the crystal is undetermined in thermomechanical equilibrium .

Implicit for the validity of the Wulff theorem is the hypothesis that all the γ_i 's are size independent [20] .

Figure 11b illustrate the Wulff theorem in the (rather particular) case where only flat facets belong to the equilibrium shape. In fact eqs. 34 are a tangential definition of the crystal surface . Therefore this shape can be deduced from the γ-plot by the Wulff construction (figure 11b) . Take the origin of the γ-plot as the Wulff point , draw the normal $\vec{n}$ to the (hkl) orientation , find its intersection P with the γ-plot and construct the plane through P mormal to $\vec{n}$. The equilibrium shape is homothetical to the innermost volume envelopped by this plane when $\vec{n}$ is varied . Note the important fact that inward cusps in the γ-plot are responsible for the existence of flat facets on the equilibrium shape . Nothing more will be said here since this subject is exhaustively treated elsewhere [3] .

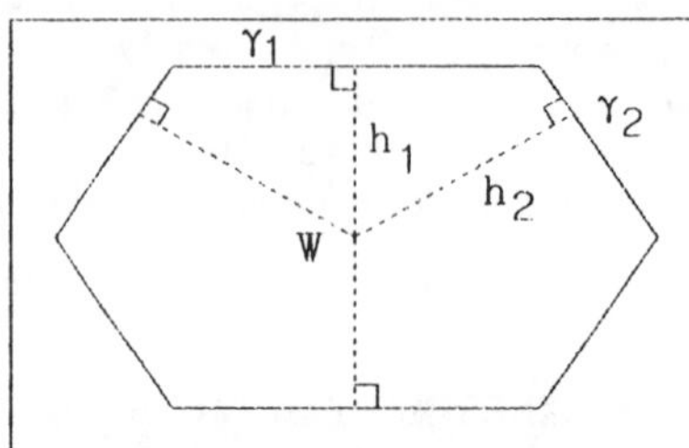

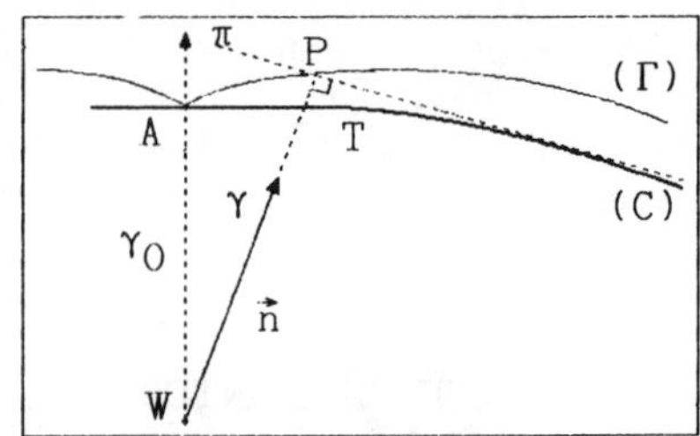

*Fig. 11 - a (left) The Wulff theorem applied to an hypothetical centro-symmetric crystal. W: Wulff's point. The h_i 's are the **normal** distances from W to the faces .*
*- b (right) The Wulff construction . (Γ) is the γ-plot . The crystal (C) is the enveloppe of plane π when P is varied . The **normal** distance from W to π is WP . The cusp at A produces a flat facet normal to WA . Only a part of the facet (AT) is drawn on the figure .*

5 . 1 . 2 Thermodynamic equilibrium .

If the crystal is in thermodynamic equilibrium with its mother phase its turns out that its shape is again determined by the Wulff theorem [10] . However its chemical potential is now equal to that of the fluid phase . Consequently the common value of the γ_i / h_i ratios is now specified . One arrives [10] at the generalized Gibbs-Thompson equation

$$\frac{\gamma_1}{h_1} = \frac{\gamma_2}{h_2} = \ldots = \frac{\gamma_i}{h_i} = \frac{\Delta\mu}{2\, v''} \qquad (35)$$

where $\Delta\mu = \mu' - \mu'_\infty$.

v" is the atomic volume of the crystal and μ'_∞ is the chemical potential of the fluid phase in equilibrium with an infinite crystal ($h_i \to \infty$) .

The Wulff theorem is again expressed by the (i-1) left hand equalities in eqs. 35 . It ensures that the various surface orientations all have the same chemical potential. Now, the latter is equal to that of the fluid phase (μ') . The size of the crystal is now specified by $\Delta\mu$. It decreases if $\Delta\mu$ increases .

Since the γ_i's and h_i's are positive $\Delta\mu$ is positive . The equilibrium chemical potential is higher for a finite crystallite than for an infinite crystal phase . $\Delta\mu$ in eq. 35 is expressed as an excess in the chemical potential of the fluid phase . It is called the supersaturation of this phase .

Implicit for the validity of eq. 35 are the same hypotheses as in section 5.1.1 plus that of a negligible crystal compressibility .

5 . 2 A 3D crystal on a foreign substrate . The VW equilibrium

The same considerations apply in this case . One also obtain the Wulff theorem and the generalized Gibbs-Thompson equation , however in a slightly different form:

$$\frac{\gamma_1}{h_1} = \frac{\gamma_2}{h_2} = \ldots = \frac{\gamma^* - \gamma^S}{h'} = \frac{\Delta\mu}{2v''} \quad (36)$$

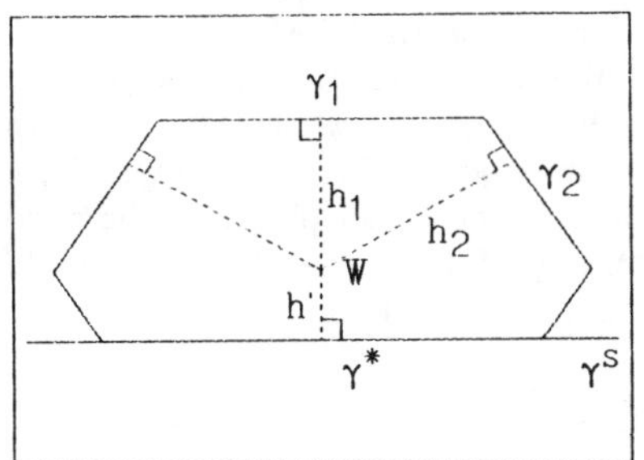

Fig. 12 - The Wulff theorem applied to a crystallite on a foreign substrate .

Figure 12 explains the notations . $\Delta\mu$ is positive again . In the VW equilibrium too , the crystallites are in equilibrium with a supersaturated mother phase .

A comparison of eq. 35 and 36 shows that the equilibrium chemical potential (equivalently the vapor pressure of the crystallite) is independent of the substrate and interfacial tensions . At equal normal distances of the faces in contact with the mother phase , a crystallite on a substrate has the same vapor pressure as an isolated one . This immediatly gives the reason why the epitaxial relationship is independant of $\Delta\mu$. $\Delta\mu$ only specifies the size of the crystallite whereas the selection of the epitaxy relationship is determined by the minimization of $\int\gamma\, dA$ (thermomechanical equilibrium). Full evidence of this statement is obtained from the nucleation theory [5] .

The particular model of a box-shaped crystal with its top facet parallel to the substrate is often used in epitaxy problems (figure 13) . Eq. 36 is easily transformed:

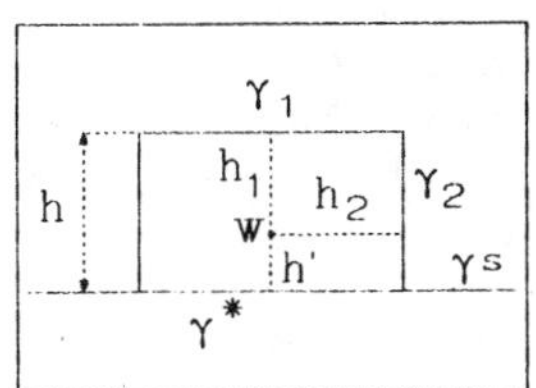

Fig. 13 - Model case of a box-shaped crystallite on a foreign substrate .

$$\frac{\gamma_1}{h_1} = \frac{\gamma_2}{h_2} = \frac{\gamma_1 + \gamma^* - \gamma^S}{h} = \frac{\Delta\mu}{2v''} \quad (37)$$

where $h = h_1 + h'$ is the total height of the crystallite (figure 13) .

Since h must be positive for a 3D crystal to exist , one must have

$$\Delta\gamma = \gamma_1 + \gamma^* - \gamma^S > 0 \quad (38)$$

Within the limits of the capillary approximation this provides us with a criterion for the existence of 3D crystals on a substrate [20] . Using eq. 20 and combining with eq.38 yields

$$\Delta\gamma = \gamma_1 + \gamma^* - \gamma^S = 2\gamma_1 - \beta > 0 \quad (39)$$

where β is the adhesion work between the substrate and a surface of type 1 . So the VW equilibrium is achieved when the substrate is imperfectly wet by the deposit . Conversely , if $\Delta\gamma < 0$, 3D crystals cannot exist on the substrate : the FM or SK equilibrium is thermodynamically favored .

5. 3 Beyond the capillary approximation .

If the crystal is so thin that the range of atomic interactions exceeds its thickness the capillary approximation breaks down and so does the Wulff theorem [20] . A full treatment of the adsorbate-adsorbate and adsorbate-substrate interactions is necessary . A clear account of the physical basis of the calculations can be found in [21] . More sophisticated methods of statistical mechanics are described in another chapter , to which we refer the reader . Here we only state that , when 3D crystallites are stable on a foreign substrate , their shape tends to be flatter the longer the range of interactions . Moreover they are no more homothetical when the size of the crystallite increases . An simpler approach has been used by Mutaftschiev [13] . Although restrictive hypothesis are made (in particular the isomorphism of the deposit-substrate couple) it lends itself to the understanding of the FM and SK equilibria in a simple manner and to the use of a criterion formally identical to condition 39 . Therefore we use this approach in the next section .

5 . 4 The FM and SK equilibria

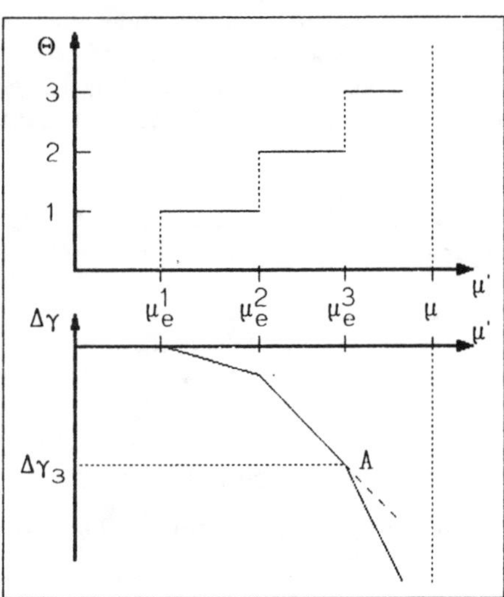

Fig. 14 - Idealization of a multilayer isotherm after Mutaftschiev [13].

Let us idealize the case of a layer by layer growth by remplacing the real isotherm by a sequence of abrupt transitions separated by perfecly flat steps (Figure 14) . This is equivalent to neglecting the roughness due to vacancies in each 2D condensed phase or to adatoms on top of them . Recalling that γ is the reversible work of creation of the unit surface it is possible to express γ_i for a stack of i layers in terms of interaction energies . Similarly one can express β_i , the work of separation of the stack from the substrate . Mutaftschiev has shown [13] that

$$a^2 (2\gamma_i - \beta_i) = \sum_k \Delta\mu_e^k - i\,\Delta\mu_e^i \qquad (40)$$

where $\Delta\mu_e^i = \mu_e^i - \mu_\infty$ (figure 14) and a^2 is the area per atom . Let us now consider the situation where the ith layer has just formed (e. g. point A in fig. 14) and let $\Delta\gamma_i$ be the value of $\Delta\gamma$ then achieved. ($\Delta\gamma_3$ in fig. 14) . It can then be established that $a^2\Delta\gamma_i$ is equal to the right-hand term in eq. 40 . Hence

$$\Delta\gamma_i = 2\gamma_i - \beta_i \qquad (41)$$

For the stack of i layers to be stable at $\Delta\mu_i^e < 0$ it is necessary that $\Delta\gamma_i < 0$ or , from eq. 44

$$2\gamma_i - \beta_i < 0 \qquad (42)$$

Inequality 42 is analogous to 39 . It expresses the stability of the stack in terms of wetting . However γ_i and β_i are now thickness dependent, equivalently $\Delta\mu$ dependent, since the range of atomic interactions now exceeds the thickness of the stack. Although eq. 42 was established under rather restrictive conditions we take it as a convenient stability criterion to explain the FM and SK equilibria .

As long as the stack of layers wets the substrate more than perfectly the condensed 2D phase can cover the substrate . Then the FM equilibrium occurs . Cases are known (Ag/Au , Au/Ag) where a very large number of layers can be formed .

Conversely the SK equilibrium is observed when condition 42 breaks down after the completion of a finite number of layers . Predicting when this exactly occurs seems beyond the possibilities of the existing theories at least for metallic systems . However some influences have been recognized . In particular the strain energy must be included in γ_i and β_i when the deposit is not exactly isomorphous to the substrate . The existence of a strain energy invariably tends to favor 3D crystallites [21, 22] .

6 EXAMPLES OF NON EQUILIBRIUM SITUATIONS

Only equilibrium situations have been considered so far . We have obtained the conditions for the existence of 3D crystals or 2D condensed layers on a substrate at equilibrium . In contrast, growth experiments are non equilibrium situations . In a typical vapor growth experiment a beam of atoms is condensed on to a substrate over which the adatoms can diffuse . If the adatom population is high enough 2D or 3D clusters nucleate and start growing . Morphologies quite different from those predicted by the equilibrium theory can be created .

6.1 The kinetically induced layer by layer growth mode

As an example we take the growth of Ag on a Mo(110) surface [23] . If deposition is made at room temperature a quasi-perfect layer by layer growth of Ag is obtained . Up to 8 monolayers (ML) have been detected (here the density of a monolayer is very close to 1 Ag atom per Mo atom) . So one could think of the Ag/Mo(1110) system as belonging to the FM class . However upon heating this stack of layers above 400 K 3D Ag crystallites are formed on top of the first two layers . Experiments under equilibrium conditions confirm the result [24] . Only the first two ML are stable at equilibrium . The system is in fact an SK one .

This shows that a pseudo FM growth may be observed at room temperature (possibly higher , depending on the system) even though the system belongs to the VW or SK classes. The reason is to be found in the easy surface diffusion of the deposited atoms. They have enough thermal energy to migrate either on the substrate or on the existing Ag monolayers . The layers spread out through the trapping of diffusing adatoms at their edges . On the other hand the temperature is too low for the edge atoms to escape from their positions . Therefore the 2D layers have no possibility of being reorganized into 3D crystals .

6.2 Kinetic shapes versus equilibrium shapes

Figure 15 shows the shapes of Pb crystallites formed on a Ge(111) surface [6] . A comparison with fig. 5 shows striking differences . The heights and shapes of the crystallites are not homothetical any more . Crystallites grown on the disordered portions of the substrate (top and bottom parts of fig. 15) are thicker than those grown on a flat portion (middle part of fig. 15) . The latter are bounded by nearly perfect

Fig. 15 -Example of growth shapes : Pb/Ge(111) .
The Ge surface was obtained by cleavage at room temperature under ultra high vacuum . The top and bottom parts of the figure show perturbed regions of the cleaved surface .
Deposition temperature 250°C .
The deposited mass of Pb is the equivalent to 600 ML .
From ref. [6] (with permission).

(111) facets while the former are somewhat more irregular . In contrast the equilibrium shape (fig. 5) displays one (111) facet parallel to the substrate and rounded regions that make a definite contact angle with the Ge surface ($\alpha = 58°$) . However the same epitaxial relationship is observed on both equilibrium and growth shapes .

A similar behaviour is observed for many other systems (see for example [25] and references therein) . Although a wide variety of growth shapes may be produced some qualitative rules are the following , in cases where the VW or the SK modes are thermodynamically favored

a) The higher the temperature and the smaller the impingement rate , the thicker are the growing crystals (though their density on the substrate is in general not constant) .

b) Annealing without flux tends to reshape the crystallites and drives them toward the equilibrium shape. Achieving the latter may be a very slow process , though .

c) However the thinner crystals ,that are presumably the most perfect , show a lesser tendancy to round themselves off .

d) One often observes the coexistence of very thin and perfect crystals and of thicker , quite differently shaped ones . In many cases this can be related to the defect density , either on the substrate itself or in the growing crystallite .

Points a) and b) show that the growth shape results from a competition between kinetic factors (impingement rate , surface diffusion and incorporation of adatoms into the growing crystal) and the thermodynamic " driving force " toward the equilibrium shape . Points c) and d) point to the importance of defects for the incorporation of diffusing atoms into the lattice . (However the detailed interpretation of point c) may be more subtle in relation with surface melting) .

7 SURFACE DIFFUSION . A BRIEF ACCOUNT

Since surface diffusion plays a major role in growth we examine here its most important features very briefly . A detailed account can be found in [26] . We only treat surface diffusion of adatoms since we expect that the high adatom population present during deposition will greatly reduce the contribution of surface vacancies .

An adatom on a perfect lattice plane is subject to the periodic field of the substrate . This field is itself perturbed by defects like steps or kinks . Figure 16 shows

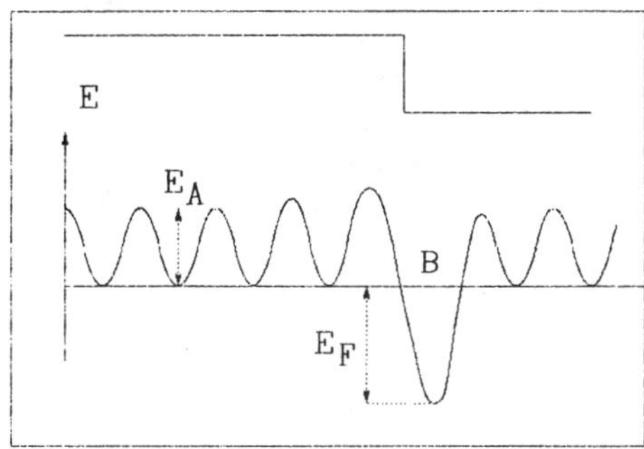

Fig. 16 - Schematic potential energy of an adatom in the substrate field . Top : A step between two terraces . Bottom : The related energy diagram . E_A : Activation energy of diffusion on the terrace - E_f : Formation energy of an adatom from the step - B : Step site .

the (schematic) potential energy of an adatom in the middle of a terrace and near a step . The energy diagram near a kink would look the same except for a deeper well at B . Energy barriers impede the displacements of the adatoms along the surface and must be overcome by thermal activation . Moreover steps and kinks act either as extra barriers or as traps for the the adatoms(they are preferential sites of nucleation — see chapter by A. Masson).

7 . 1 Surface migration

In the middle of a terrace an adatom can jump at random between adjacents sites a distance a apart . This random walk is characterized by an intrinsic diffusion coefficient

$$D_i = a^2 \nu \exp - \frac{E_A}{kT} = D_o \exp - \frac{E_A}{kT} \qquad (43)$$

where ν is a jump frequency . D_o and E_A can be measured if one is able to follow the motion of an adatom at the atomic scale . Such measurements are best performed by Field Ion Microscopy . For a description of the technique and for detailed experimental results we refer the interested reader to [27] and the references therein . Table III only displays a few experimental values quoted from [27] , with the main purpose of showing that surface migration may occur at rather low temperatures .

Diffusing species	E_A eV/atom	D_0 $cm^{-2}\ s^{-1}$	Temperature range (K)
W	.83	7.7×10^{-3}	264 - 319
Mo	.71	2.0×10^{-3}	235 - 286
Re	.67	6.1×10^{-3}	180 - 220

Table III

Surface migration data for adatoms on the (211) surface of tungsten . Unidimensional diffusion along the [$\bar{1}$ 1 1] channels .

7 . 2 Surface diffusion in a chemical potential gradient . Matter transport [27]

If a chemical potential gradient exists along the surface , atoms tend to migrate toward the regions of relatively smaller chemical potential . The case of interest here is that of a crystallite that has not achieved its equilibrium shape . The gradient of chemical potential is determined by the local departure from the equilibrium shape . Even though no atoms are exchanged with an external mother phase , matter is transported along the surface by surface diffusion until the whole system has reached its equilibrium state , i. e. until the crystallite has achieved its equilibrium shape at constant volume in our case . Since we expect the number of kinks to be much greater than the number of atoms along the steps the key phenomenon is now the extraction of an atom from a kink site by thermal activation followed by its migration and trapping by an another kink elsewhere .

Let us assume that there is only one diffusing species (namely the adatoms) in surface concentration n . Then , in one dimension , their drift velocity is

$$v = -\frac{D_i}{kT}\frac{\partial \mu}{\partial x}$$

where D_i is the intrinsic diffusion coefficient in the direction of interest . The diffusion flux is

$$J = n v = -\frac{n D_i}{kT}\frac{\partial \mu}{\partial x} = -\frac{N_o D_s}{kT}\frac{\partial \mu}{\partial x}$$

with $D_s = \frac{n}{N_o} D_i$

where N_o is the number of kink sites per unit surface area . For a complex or curved surface N_o is very close to the number of atoms per unit surface area .

D_s is called the mass transfer diffusion coefficient .

Assuming local equilibrium between the adatoms and the kinks one has

$$n = N_o A \exp -\frac{E_f}{kT}$$

where E_f is the formation energy of an adatom from a kink (figure 16) and A is a constant involving the formation entropy . Hence

$$D_s = A D_o \exp -\frac{E_f + E_A}{kT} \qquad (44)$$

Putting

$$E^{*} = E_f + E_A \, , \qquad (45)$$

the activation energy for matter transport , eq. (44) is generally written in concise form

$$D_s = D_0 \exp - \frac{E^{*}}{k\,T} \qquad (46)$$

where the same notations as in eq. 43 are used but where D_0 has another expression and numerical value .

Just as in section 7.1 D_0 and E^* are experimentally measurable quantities , however with different techniques [26] . Table IV shows some experimental results .

Diffusing species	E^* eV/atom	D_0 $cm^{-2} s^{-1}$	Temperature range (K)	Reference
W	3. 1 [a]	≈ 1	1200 - 1500	[29]
Ni	1. 96 [b]	$4.4 \; 10^{2}$	970 - 1570	[26]
	. 76 [c]	$4.3 \; 10^{-3}$	770 - 1070	[26]
Pb	. 75	≈ 10	330 - 550	[29]

Table IV

Examples of experimental values of E^* and D_0 for mass transport diffusion coefficients . a : blunting of field emission tips . b : transport on (110) in the [001] direction . c : transport on (110) in the [1 $\bar{1}$ 0] direction .

Tables III and IV illustrate some important features .

- Surface diffusion with mass transport generally occurs at higher temperatures than surface migration . The reason is that the adatom creation from a kink requires a higher activation energy than mere surface migration . The value of D_0 is also different .

– Surface diffusion is anisotropic . In measurement methods that use changes in shape to determine D_s the measured value is weighted by the slowest step in atomic displacement mechanisms .

To conclude this subsection we must mention that the values of D_0 and E^* depend on adsorption . For example the mass transport of W on a tungsten surface covered by 1 ML of Ni occurs with $D_0 = 3 \,.\, 10^{-3}\ cm^{-2}.\ s^{-1}$ and $E^* = 1.9$ eV/atom . Compared to clean tungsten (table IV) this increases the value of D_s by a factor 160 at 1200K and 20 at 1500 K [28] . The adsorption of carbon or oxygen has the opposite influence .

7 . 3 The time scale

Consider two crystals 1 and 2 homothetical in shape with a ratio λ , i. e. at equivalent positions the radii of curvature are $R_2 = \lambda R_1$. Away from equilibrium surface diffusion drives the crystal shapes toward the equilibrium shapes . What is the ratio of the times Δt_1 and Δt_2 needed to achieve a specified shape change ? The answer is given by Herring's scaling laws [30] . At corresponding points on the two crystal surfaces the chemical potentials and the diffusion fluxes are in the ratios $\mu_2/\mu_1 = 1/\lambda$ and $J_2/J_1 = 1/\lambda^2$. The ratio of the amounts of material to be transported is λ^3 and the total rates of transport along the surfaces (i. e. $\rho_i = L_i J_i$, where L is a length) are in the ratio $\rho_2/\rho_1 = L_2J_2/L_1J_1$. Hence $\Delta t_2/\Delta t_1 = \lambda^4$. A similar dimensional analysis can be carried out for different transport mechanisms . The results [30] are

Transport by surface diffusion	$\Delta t_2 = \lambda^4 \Delta t_1$	(47)
Transport by volume diffusion	$\Delta t_2 = \lambda^3 \Delta t_1$	
Transport by evaporation and condensation	$\Delta t_2 = \lambda^2 \Delta t_1$	

So the consideration of the time dependance of some linear dimension of a specimen allows the mass transport mechanism to be determined . It turns out that surface diffusion is the predominant mechanism in shape changes of crystals of linear dimensions less than 10 µm . The reason is that surface diffusion is much more rapid than volume diffusion and that the high Surface /Volume ratio favors the former .

A review of some applications of Herring's laws can be found in [31] . For our present purposes we shall only draw one consequence from equation 47 : A small crystal will reorganize itself by surface diffusion far more rapidly than a bigger one . For example if a crystal of , say 10 nm in mean diameter , is able to reorganize in 1 s , another one of 100 nm will need 10^4 s under the same conditions to undergo the same transformation . In other words the size of a crystal may place a severe kinetic limitation to the achievement of its equilibrium shape in a finite time .

8 CONSEQUENCES

8 . 1 Thin platelets versus thick crystals

From the few examples of section 7. 2 is it easy to understand that the actual growth shape results from a competition between (a) the arrival of adatoms at the crystal , either by direct impingement or by diffusion along the substrate ; (b) subsequent surface diffusion of the adatoms over the various faces of the growing crystal followed by their incorporation into the bulk at kink sites ; (c) the tendancy for the crystal to achieve its equilibrium shape by emission of adatoms from kinks and their subsequent trapping elsewhere . We note that surface diffusion alone is insufficient to explain growth or the return to the equilibrium form . There must also be enough kink sites to incorporate the adatoms . Additionally atoms can be returned to the mother phase by re-evaporation either from the substrate or from the crystal . In the following we consider only the case of complete condensation which seems to us more suitable to extract the fundamental parameters .

In treating this problem severe difficulties are found in relating the local surface diffusion flux to the macroscopic parameters which describe the deviation from the equilibrium shape . If the γ-plot is devoid of cusps the usual approach is the expansion of the crystal shape and of the γ-plot itself in spherical or cylindrical harmonics (see [25] and references therein) . Continuity seems to be a prerequisite for using these methods . Besides , an n-fold symmetry is necessary for applying them . Therefore a comprehensive treatment for a 3D crystal on a substrate seems beyond the range of such a theory . On the other hand the interplay of the fundamental phenomena can be understood by considering a fairly simple model : that of a box-shaped crystallite with a square top and bottom and vertical side facets [25] . Let the side length be 2r and the height h . The form factor is defined as $\lambda = h/r$. Its equilibrium value , required by the Wulff theorem , is named λ_0 . Adatoms are supplied to the crystal either by direct impingement (flux R) or by surface diffusion inside a square " collecting area " (side $2r_1$) . The densities of kinks on top and sides are treated as independant variables . Then the "driving force "toward the equilibrium shape can be explicitely taken into account . In fact, one can express the free energy of the system as a shape depending quantity through a function of λ and λ_0 . Different D_s are considered on the top and on the sides . Although rather crude , the model allows important qualitative conclusions to be drawn :

The kink density on the side facets is no major parameter provided it is greater than 10^{-5} ML (in the present model).

If there are enough kinks on top the important parameter is the ratio

$$u = \frac{1}{2}\,\frac{R r_1^2}{N_0 D_s}$$

where N_0 is the surface atomic density . u is in fact the ratio of the number of atoms supplied per site to the mass transfer diffusion coefficient D_s on the top surface . If $u \ll 1$ the " driving force " can compete with the arrival rate and the growth shapes approximate the equilibrium shape fairly closely . If $u \gg 1$ the "driving force " is outweighed by the arrival of adatoms . The crystallite flattens its shape even though growth normal to the substrate is still effective .

A similar outweighing of the " driving force " is produced when the kink density on top is decreased at constant R and constant D_s . In this case, growth mormal to the substrate is impeded by the lack of kinks available for incorporation . Flat and much thinner islands are grown . In the limit of a very small kink density ($< 10^{-9}$ M L) normal growth is practically forbidden and extremely thin platelets are conceivable .

In fact 2D nucleation occurs on top of the flat islands and sets a lower limit to the kink density (which roughly equals the number of critical clusters in equilibrium with the adatom population on the top) . New layers are built up very rapidly and the growth shape is less kinetically biased .

8 . 2 Surface defects and kinetic roughness

Even though the FM equilibrium may be thermodynamically stable , a kinetic roughness may exist due to the random nucleation of layers , combined with an incomplete spreading of each of them over the substrate . In this subsection we focus attention on the role of steps (of various origins) which may be fairly subtle . On the one hand, they are preferential sites for the nucleation and subsequent growth of crystalline layers . Direct experimental evidence was recently obtained by the new tchnique of Low Energy Electron Microscopy [32] . On the other hand, the upper edge of the step is a barrier to surface migration (section 7).

The development of a kinetic roughness (directly observed by electron microscopy) and its influence on Auger signals was recently demonstrated by Meinel et al. on the

Ag - Au system [33] . On a perfect Au substrate , the growth of Ag layers occurs by the nucleation of 2D clusters whose edges are steps of monoatomic height . Even though surface migration is possible the temperature may be too low for the migrating atoms to overcome the diffusion barrier at the steps . Consequently an increasing population of adatoms is progressively built up on top of each freshly created terrace until the nucleation of new clusters again occurs on it . The result of this sequence is the formation of hillocks by piling up incomplete 2D layers . This kinetically induced roughness is detectable if the AES signal of the adsorbate is plotted against the total deposited amount of material . If a perfect layer by layer growth occured one would expect a change in slope to be visible in the AES curves upon the completion of each layer . In reality these " breaks " are blurred by the kinetic roughness and may be barely noticeable [33] . Thus a certain amount of thermal energy must be retained by the adatoms in order to ensure a perfect layer by layer growth .

On the other hand, regularly spaced steps such as those found on vicinal surfaces may help develop a quasi-perfect stacking of layers . This behaviour was also examplified on the Au - Ag system [33] . If the migrating adatoms have enough thermal energy they can overcome the step barrier and are trapped at the step bottom . Here the key phenomenon is the preferential nucleation of new layers at steps . Each layer starts growing at the bottom of a substrate step and spreads out by trapping the incoming adatoms at its edge . The surface roughness may be practically suppressed . Steps thus favor (and possibly orient) the layer by layer growth . In this case the Auger curves show very sharp " breaks " after the completion of each 2D layer [33] . Exactly the same phenomenon has been invoked by Flynn [34] to interpret experimental observations of the growth of metallic superlattices . A minimum substrate temperature is required (given the deposition rate) to suppress the RHEED intensity oscillations during growth and to cancel the layer roughness . According to Flynn this cancellation is due to the adatom population being kept low by the ease of overcoming the step barriers and by the subsequent trapping of adatoms at the step bottoms . The optimum temperature range can be estimated from the knowledge of growth conditions (impinging flux , mean step spacing on the substrate) and from the temperature dependance of D_i . Good agreement with the experimental data is obtained .

9 SOME STRUCTURAL PROBLEMS - EPITAXY

9 . 1 The epitaxial relationship . Some examples

Within the interfacial region (for a VW system) or across the layer stacking (for a FM one) one has to accomodate two materials of different parameters and possibly different symmetries . Intuition tells us and experiments confirm that the atomic arrangements will in general differ from those of the bulk materials close to the interface whereas they will approach them at increasing distances . In addition an orientation relationship is found in many cases between the lattices of the two materials . This is " epitaxy ", to use the word coined by Royer to describe an order (taxis) which takes place in the overgrowth (epi) .

The epitaxy relationship is essentially a orientation order :

There is a preferential parallelism between a lattice plane of the bulk substrate and one of the bulk deposit (the "contact planes ") .

There is a preferential parallelism between some sets of rows that belong to the contact planes of the bulk materials (azimutal orientation) .

The few examples quoted hereafter show the most typical features of epitaxy .

Au on (00. 1) Graphite [35] . This a VW system . The relative orientations of the substrate and (3D) crystallites lattices are

{ 111 } Au // (00. 1) Graphite and ‹ $11\bar{2}$ › Au // [01. 0] Graphite

However the spacing between the ‹ $11\bar{2}$ › rows of Au is 1.5% larger than the lattice para-parameter of graphite .

Pb on Cu (111) [36] . This system is an SK one . At equilibrium 3D crystallites are formed on top of one monolayer (ML) of Pb .

In the ML the orientation relationship is

{ 111 } Pb // (111)Cu and ‹ $1\bar{1}0$ › Pb // [$1\bar{1}0$] Cu

Additionnaly, a p (4X4) periodicity (referred to Cu) is found . This means that the Pb atoms have the structure of a Pb (111) plane compressed by 2.7 % .

In the islands the same orientational order is observed but the interatomic distances have reached those of bulk Pb .

Cu on Mo (110) [23] . This , too, is an SK system . However the symmetries of the contact planes are different . 2 Cu layers are built up prior to the formation of 3D crystallites .

The first ML is pseudomorphous to the Mo (110) plane .

In the second ML the observed epitaxy is

(111) Cu // (110) Mo and [$1\bar{1}0$] Cu // [001] Mo

with the spacing between the Cu rows matched to that of the [110] Mo (2. 227 Å instead of 2. 213 in bulk Cu) . This is an example of the Nishiyama-Wasserman (NW_x) orientation . The Cu (111) plane is expanded by 0. 6% in the Mo [$\bar{1}10$] direction .

The 3D crystals obey the same orientation relationship but the spacing is now that in bulk Cu .

Ag on Mo (110) [23] . Up to 2 ML of Ag can exist on Mo prior to the appearance of 3D crystallites (equilibrium situation . More 2D layers can be grown by room température deposition). The layers form a distorted Ag (111) plane with a angle $\beta \approx 113°$ in the rhomboedral unit cell instead of 120 ° . The epitaxy is

(111) Ag // (110) Mo and [110] Ag //[111] Mo

which is an example of the Kurdjumov-Sachs (KS) orientation .

Rutile (TiO_2) on hematite (α-Fe_2O_3) . This is a VW system . Rutile is tetragonal and hematite trigonal . The contact occurs between the basal plane (00. 1) of hematite and the (100) plane of rutile . The [1 1 0] rows of the hematite substrate are parallel to the [0 1 0] rows of the rutile crystallites (figure 17) . If one compares the lattice planes just mentioned in the bulk crystals they must be in the relative orientations shown in fig. 17 to ensure the parallelism . Strictly speaking no coincidence between the bulk lattices exists . However one can find multiple surface meshes in the contact planes that are in reasonable metric coincidence . Misfits exist between these meshes ,

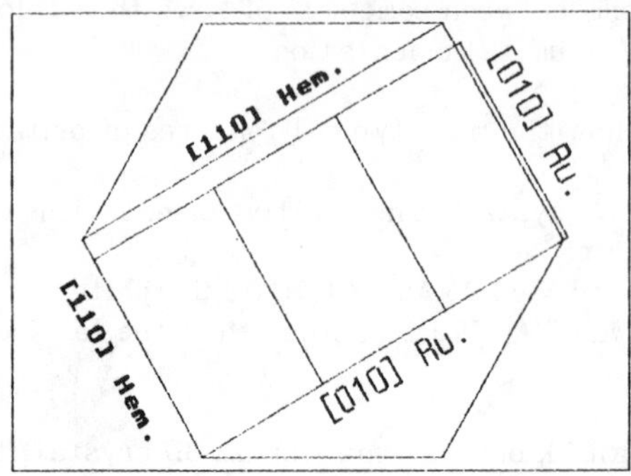

Fig. 17 - Epitaxy of rutile on hematite.

different with the direction . The (3x1) multiple mesh of TiO_2 is longer than the multiple (1x√3) of Fe_2O_3 in the [100] direction (+1. 4%) and shorter in the [001] direction (−10%) .

These few examples show the basic phenomena that must be taken into account in explaining or predicting epitaxy in a specified substrate-deposit system .

- Epitaxy is basically an orientationnal order characterized by the parallelism of some sets of rows . However, cases are known where pseudomorphism is observed within the very inner layers of the interface . Upon increasing the overlayer thickness pseudomorphism is relaxed and only the less severe requirement of row parallelism is satisfied .

- Some mismatch can be tolerated between the lattices of the bulk elements of the couple .

- Row parallelism alone requires a distorsion of the atomic arrangements in the interface region , i. e. a strain of the bulk lattices . In other words one part of the price to be paid for accomodating the materials is the storage of an elastic energy in the system .

Consequently two typical questions are raised : (a) Does the strain energy influence the growth mode ? (b) How does orientationnal order (possibly pseudomorphism) comply with the lattice mismatch ? Obviously the parameters that determine the answer are the mismatch itself (either purely metric or angular) and the interactions in the couples : substrate-deposit , substrate-substrate and deposit-deposit . No general solution is known at the moment . However approximate models can be constructed that help unravel the competition between these parameters .

9 . 2 Energy minimum considerations

There are basically two classes of models . (a) Models that use molecular dynamics and computer simulations [3, 38] . They differ in the potentials that are used to describe the interactions between the two materials in contact . They seem well suited to study the influence of mismatch on the growth mode . (b) Analytical models in which the total energy of the system is expressed as a function of strain and subsequently minimized to find the equilibrium atomic configuration . An analytical solution is found in general at the expense of some simplifying assumptions [39, 40, 41] . In the following we consider only the latter class of models since they seem to give the most comprehensive understanding of the basic phenomena and to have a fairly broad (though qualitative) predicting power . We also restrict ourselves to Metal/Metal systems although the principles are general .

In close-packed metals the row spacing is determined by the atomic diameter. Therefore we first define the natural misfit between the substrate A and the deposit B , of diameters a and b , by

$$f = \frac{|b - a|}{a} , \qquad \text{or equivalently by}$$
$$r = \frac{b}{a} , \qquad f = |r - 1|$$

Other " misfits " between various types of rows can be defined further and expressed as functions of f and of the angles in the surface unit meshes of A and B [40] . If the deposit is so homogeneously strained that an effective atomic diameter $\bar{b}$ can be defined an effective misfit $\bar{f} = |(\bar{b} - a)/a|$ or an effective ratio $\bar{r}$ are sometimes used [39] .

Since the substrate A is in general thicker than the deposit B all the models consider a strictly rigid substrate . The A–B interaction is modeled by a 2D periodic potential with the periodicity of the bulk substrate : in concise form

$$V_{2D} = E_0 + V(x,y) \qquad (48)$$

where E_0 is identified with the binding energy of an isolated atom of B on A and V defines the 2D modulation of the interaction . An overlayer of B (generally 1 ML) with initially the structure of the bulk is brought to interact with A , i. e. to be strained by the 2D potential . The B-B interaction is taken into account by expressing the elastic energy E as a function of strain and the elastic constants of the overlayer . The total energy of the system is then the sum of E and of the interaction energy $E_0 + V$. The equilibrium configuration is found by minimizing the total energy , i. e. in fact by minimizing E + V .

In such a model it is relatively easy to assign a physical significance to the parameters . For example one may write the modulation potential as follows for a square lattice of parameter a [22, 42]

$$V(x,y) = \frac{1}{2} W \left[2 - \cos 2\pi \frac{x}{a} - \cos 2\pi \frac{y}{a} \right] \qquad (49)$$

with W a constant . W is the height of the modulation . Hence it is also the activation energy for surface migration E_A (section 7. 1) . Besides , W may be empirically related to the binding energy E_0 (equivalently to the activation energy of desorption) by the rule of thumb $W = \chi E_0$ with $\chi \approx 0.1 - 0.3$. By so doing one makes W an estimate of the B-A interaction energy . On the other hand the B-B interaction is taken indirectly into account through the elastic constants of B . Since these constants appear in the minimizing conditions along with W, the competition between A-A and A-B interactions can be analysed . For example , in the various versions of the van der Merwe model [42, 40] , the significant quantities are found to be r and the lumped parameter l defined by

$$l^2 = \frac{v'' S}{W r^2} \qquad (50)$$

$$S = \frac{c_{11}^2 - c_{12}^2}{2\, c_{11}} \qquad (51)$$

where v" is the atomic volume and c_{11} and c_{12} are the stiffness constants of B .

It is now appropriate to mention the three basic atomic arrangements by which the energy E + V of the system is minimized [40] . Since we are only looking for a qualitative picture we consider a unidimensional representation of a ML (figure 18). If the B-B interaction is weak compared to the A-B one, the monolayer may be strained to match the rigid substrate completely ; pseudomorphism is then achieved (figure 18a). However there exists a critical misfit f_c above which the pseudomorphic layer is unstable . If $f > f_c$ the energy is minimized by an array of misfit dislocations with a strained layer in-between (figure 18b) . At constant misfit the same result is obtained upon increasing the strength of the B-B interaction with respect to A-B . Eventually when the latter (A-B) is weak compared to B-B the overlayer is only homogeneously strained and the misfit is accomodated by a misfit vernier (figure 18c) .

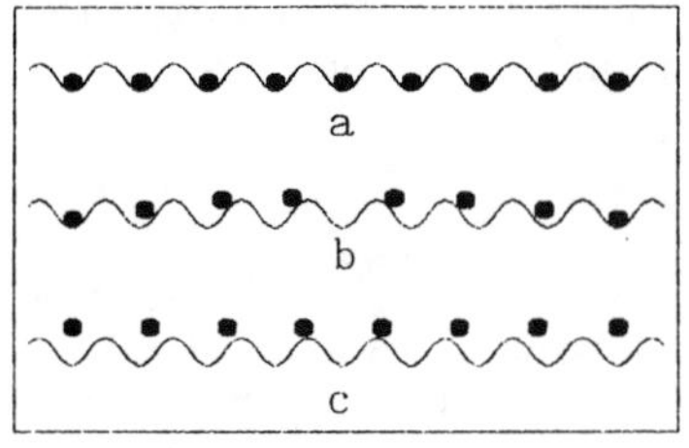

Fig. 18 - The three basic modes of misfit accomodation. The modulation of the A-B interaction potential is represented by the solid lines .
a - pseudomorphism
b - misfit dislocation
c - misfit vernier

To make contact with thermodynamics we may point out that the approach just explained is in fact the calculation of the elastic energy term in the adhesion work of the overlayer on the substrate . Temperature is implicitely considered in assuming that the equilibrium configuration is achieved whatsoever . However all entropic terms are ignored . Rewriting eq. 39 or 42

$$\Delta\gamma = 2\gamma^B - \beta \qquad (52)$$

(with obvious notations) shows that the wetting condition has now become strain (misfit) dependent . Minimizing the energy E + V is equivalent to maximizing the adhesion work .

9 . 3 Influence of misfit on the growth mode .

If the deposit were perfectly isomorphous to the substrate no strain would be required . Let β_0 be the adhesion work of the overlayer in this case . β_0 involves the chemical interaction that binds B to A and can in principle be calculated from the knowledge of E_0 and the structures of Aand B . Since the mismatch leads to an increase of the (equilibrium) energy of the system the actual value of β is smaller than β_0 , i. e.

$$\beta = \beta_0 - \beta_{strain} \qquad (53)$$

hence

$$\Delta\gamma = 2\gamma^B - \beta_0 + \beta_{strain} \quad (54)$$

Therefore the elastic energy due to the mismatch tends to increase $\Delta\gamma$. This result was recently demonstrated by van der Merwe and Bauer [22] . The misfit and the associated strain energy always favor SK or VW modes . In fact a flaw of the present reasoning is that γ^B is considered as constant . This is not strictly true since the strain in the overlayer changes γ^B too . However molecular dynamics models confirm the validity of the statement [39] . Since β_{strain} increases with the overlayer thickness there may conceivably exist situations where $\Delta\gamma$ is initially negative (layer by layer mode) but reaches a positive value beyond a critical thickness . Simple analytical calculations [21] and more elaborate dynamical models [38] confirm this point of view . The storage of elastic energy due to misfit accomodation induces the switching from 2D to 3D mode which is observed in the SK equilibrium . When this exactly happens depends in a subtle manner on the interactions and misfit . A precise prediction of the behaviour of a specified system seems beyond our present possibilities . That a detailed analysis of the interaction is required is demonstrated by a mere comparison of the systems Ag/Mo(110) and Au/Mo(110) [23] . The atomic diameters of Ag and Au differ by only . 2% (2. 889 Å versus 2. 884 Å) , which makes the misfit almost the same in both systems . However one Au ML only precedes the appearance of 3D crystallites whereas up to three Ag ML are built up prior to 3D crystals .

9 . 4 The selection of the epitaxial relationship

The general principle that determines the epitaxial relationship is already known to us (section 5. 2) . To minimize the total free energy of the system the deposit " chooses " the contact planes and the azimuthal orientation that maximize β (equivalently $\Delta\gamma$). It is now obvious that this is equivalent to minimizing the total energy $E + V_{2D}$ in our present language .

If the contact planes are taken for granted , predicting the azimuthal orientation and the atomic configuration is in principle possible by using the method outlined in section 9. 2 . Rectangular or square meshes [41, 42] as well as rhombic ones [41] have been investigated . At the moment, the most successfull model is that of van der Merwe, which is excellently reviewed in [40] . In this model, the properties of the system can be expressed in terms of the ratio r (section 9. 2) and the parameter l (eq. 50). In its various versions [40, 42, 43] the model correctly predicts the experimentally observed epitaxies . Its most recent improvements are described in [43] with an application to the epitaxy of close-packed overlayers (f. c. c. or h. c. p.) on (110) b. c. c. substrates and to the growth of metallic superlattices . In this particular case regions are defined in the (r,l) plane where specific epitaxial orientations are stable . Within these regions, the type of misfit accomodation is also determined . Although worked out for a monolayer only , the model can easily be extrapolated to the case of thicker , homogeneously strained , overlayers (no strain gradient normal to the interface) . The usefulness of this model to explain the experimentally observed properties of epitaxial systems has been established [43] . In particular the transition from pseudomorphism to a mere row parallelism upon increasing the overlayer thickness is understood in principle .

Purely geometrical considerations are sometimes used to predict the epitaxy relationship , mostly when dealing with f. c. c. (111) and h. c. p. (00. 1) contact planes . Since the symmetries and angles are identical for the epitaxy couple several linear combinations of the base vectors of the deposit (bulk) surface unit mesh can always be found that approximate a multiple surface unit mesh of the (bulk) substrate plane within a reasonably small mismatch . Implicit here is the assumption of a rigid substrate . For the preferred orientation one then selects the one which ensures the minimum strain of the deposit . The physical meaning of this criterium is clear : The surface corrugation potential V is assumed to be negligible compared to the elastic strain energy of the deposit. This is probably an oversimplification as is shown by a mere inspection of the experimental results obtained with simple epitaxial systems . For example, in the Pb/Cu(111) system, one observes a compression of the Pb<110> rows by 2. 7% so that every third Pb atom coincides with every fourth Cu atom (p4x4 coincidence mesh - see section 9.1) . However a smaller but opposite strain would be obtained with 5 Pb atoms per 7 Cu (+ 2. 24%) and a negligibe one (- 0.4%) with 11 Pb per 15 Cu . Other examples are Pb/Ag and Ag/Cu [44] . So a full scale energy calculation is necessary in any case . Additionally , when very large coincidence meshes are found , kinetic difficulties and substrate imperferctions would probably preclude such a long range order being achieved .

9 . 5 A few examples of real systems

To illustrate the complexity of real system and to show how subtle the interplay of the various quantities can be we summarize here some experimental results .

9. 5. 1 Influence of the nature of the deposit : Ag/Mo(110) [23, 24] and Au/Mo(110) [44] .

These two systems have practically equal misfits . They differ in their deposit-substrate interactions . Both are SK with 2 ML beneath 2D crystallites . The structure of the Ag layers is basically the same over the whole range of coverage and temperatures studied . On the contrary a succession of structures is observed for Au . These structures are different at room temperature and at equilibrium .

9. 5. 2 Influence of the substrate orientation : Pb/W(110) and Pb/W(100) [45] .

Pb/W(110) is a simple SK system . The Pb<110> rows are matched to the W<111> . Along them the Pb atoms are gradually compressed until the layer density is 11% higher than that of a bulk Pb(111) plane . 3D crystals then appear .

Pb/W(100) is more complex . At room temperature the first ML has the surface periodicity and coverage as W(100) [Θ =1] . At high temperature (600°C) a microfacetting of the tungsten surface occurs which is maximum at Θ = 0. 5 and disappears when Θ reaches 1 . Thick 3D crystals appear on top .

10 THE ADSORPTION ENERGY AND ITS MEASUREMENT

The binding energy of an isolated particle on a surface is the difference in energy between the following two states of the system : " surface + adsorbed particle " and

" surface + infinitely distant particle " [46] . In both systems the particle is taken in its lowest energy state . If E_0 is the difference in potential energy between these two states and E_{Ovibr} the vibrational rest energy of the adparticle , the binding energy E_B is

$$E_B = E_0 - E_{0\,vibr} \qquad (56)$$

In general E_{Ovibr} is negligible compared to E_0 .

In general E_B is not directly measurable and must be deduced from other quantities that are experimentally obtainable . The latter are either heats of adsorption or the activation energy of desorption .

Eq. 56 defines the binding energy of one particle on an initially nude surface . If other particles are already adsorbed , one defines $E(\Theta)$ as the difference in energy between the two following states of the system: "partially covered surface + one newly adsorbed particle " and " partially covered surface + one infinitely distant particle ".

Then , if $\Theta \to 0$, $E(\Theta) \to E_B$.

10 . 1 Thermodynamic quantities . Integral and differential heats of adsorption

The integral heat of adsorption Q is the heat received by a closed system when N_a moles of component are adsorbed on an initially clean substrate at constant temperature [46] . Let e_a be the mean molar energy of the adsorbed phase and e_g be that of the gas phase at T . Then

$$Q = N_a\,(e_a - e_g) \qquad (57)$$

It depends on T , N_a and the surface area A .

The differential heat of adsorption is defined as

$$q_d = \left.\frac{\partial Q}{\partial N_a}\right)_{T,A} = e_a - e_g + N_a \left.\frac{\partial e_a}{\partial N_a}\right)_{T,A} \qquad (58)$$

This is a molar , isothermal heat which depends on T and the coverage Θ . Its fundamental property is a limit one (see [46] p. 22 and p. 113) :

If $\Theta \to 0$, then

$$q_d \to E_B - (E_{avibr} - E_{Ovibr}) - \Delta E_{cin} \qquad (59)$$

E_{avibr} : mean molar vibration energy of the adparticle
ΔE_{cin} : difference in mean molar kinetic energies between the adparticle and the gas phase particle .

In general the last three terms in eq. 59 are small compared to E_B . Consequently the value of q_d extrapolated to zero coverage is a good measure of the energy E_B .

In the study of adsorption equilibria, one of the most easily accessible quantity is the isosteric heat of adsorption q_Θ . It is defined as

$$q_\Theta = R \left.\frac{\partial \mathrm{Ln}\, p}{\partial \frac{1}{T}}\right)_\Theta = -R T^2 \left.\frac{\partial \mathrm{Ln}\, p}{\partial T}\right)_\Theta \qquad (60)$$

where p is the pressure of the gas phase in equilibrium with the adlayer at T and R is the ideal gas constant . Eq. 60 shows that q_Θ can be determined from an experimental set of isotherms . q_Θ has the same fundamental interest as q_d since its can be related to q_d in a simple manner [46] . For example , if the vapor phase is an ideal gas ,

$$q_\Theta = q_d - R T \qquad (61)$$

10 . 2 Kinetic quantities

In many experiments the desorption flux of adparticles from a surface is monitored as a function of time and (or) temperature with the initial coverage as a parameter . The data are then analysed in terms of a coverage dependant rate R (normalized to the unit surface area)

$$R = \Theta^x \nu(\Theta) \exp - \frac{\Delta E_D(\Theta)}{R T} \qquad (62)$$

where Θ is the coverage , ν a desorption frequency , x the (apparent) order of desorption and ΔE_D the (experimental) activation energy of desorption .

For metal/ metal systems the adsorption is considered as non activated . This suggest identifying the experimental value ΔE_D with $E(\Theta)$:

$$\Delta E_D = E(\Theta) \qquad (63)$$

However eq. 63 cannot be used without caution . One must first assess what the overall kinetics of the desorption process really is .

The theoretical grounds for eq. 62 are the transition state theory of kinetic processes [47] . The fundamental hypotheses of the latter must be recalled . One assumes

(i) internal equilibrium of the adlayer throughout the desorption process
(ii) equilibrium between the adlayer and the substrate throughout
(iii) temperature independent ΔE_D and ν
(iiii) coverage independent ΔE_D and ν .

Hypothesis (iiii) is the most important for the experimentalist in interpreting the experimental data in terms of " binding states " . It is equivalent to saying that the desorbing species desorbs from a unique binding state . If it turns out that the experimental data have to be described by coverage dependent ν and ΔE_D (eq. 62), as is often the case , this means that more than one binding state exists in the adlayer . Consequently some desorption process has to be modelled , possibly involving several desorption steps or paths to account for the observed coverage dependance .

10. 2. 1 Desorption order and desorption mechanisms

Conversely desorption models show that a coverage independence of ν and ΔE_D

does not warrant a single binding state being involved [4] . The apparent order of desorption in 62 must be considered to infer a mechanism . As an example let us examine the physical situations that may lead to an apparent order x = 1 in eq. 62 . This specific value is found when the number of desorbing particles is proportional to the number of occupied surface sites . A first case where this happens is the direct desorption of isolated atoms from their adsorption sites into the gas phase ("true " first order). ΔE_D is then identical to the desorption energy of the adatom and eq. 63 is true . However a similar situation may be met in another case (" apparent " first order) where the desorption process is completed in two steps : The adsorbed particle is first extracted from the adsorption site to a intermediate state on the surface where it can migrate before being evaporated to the gas phase in a second step . Activation energies ΔE_1 and ΔE_2 are required for the first and the second step respectively (ΔE_1 is the sum of the formation and migration energies of the migrating species) . Let us assume $\Delta E_1 > \Delta E_2$. The migrating particule has a mean lifetime $\tau = [\nu_2 \exp - (\Delta E_2/kT)]^{-1}$ during which it travels a distance $l \approx \sqrt{D_i \tau}$ from an adsorption site . If the mean distance between the adsorption sites is greater than l (the diffusion zones do not overlap) the adsorption sites cannot be replenished by the migrating species . Step 1 is kinetically limiting and desorption is therefore first order . However the apparent desorption energy $\Delta E_D = \Delta E_1$ instead of the true desorption energy from the binding site [4]. Eq. 63 is not valid . Conversely if l is greater than the mean distance between sites (the diffusion zones overlap) the former are so replenished that the only way to transfer particles to the gas phase is the direct desorption from them . The apparent desorption energy ΔE_D is identical to the true desorption energy and eq. 63 holds. However , in the early stages of the desorption experiment , the apparent desorption order is 0 since the number of occupied adsorption sites is practically constant . Desorption processes from 2D condensed adlayers or 3D crystals have been similarly analysed [4] using the physical ideas just outlined . Similar treatments can be applied to gas adsorption on surfaces but will not be developed here (see [48 , 49] and references therein). Summarizing the results obtained by using simple models [4] we merely state :

Order x = 0 is obtained for (a) desorption from a 3D phase . ΔE_D equal to the sublimation energy ; (b) indirect desorption from a 2D gas with overlapping of the diffusion zones . ΔE_D equal to the true desorption energy ; (c) desorption from islands of a 2D condensed phase in equilibrium with a 2D gas with a mean inter-island distance smaller than the diffusion length in the 2D gas . ΔE_D equal to the desorption energy of the ledge atom .

Order x = 1/2 is obtained for desorption from a 2D condensed phase in equilibrium with a 2D gas with a mean inter-island distance greater than the diffusion length in the 2D gas .

Order x = 1 is obtained for (a) direct desorption from isolated adatoms ; (b) for indirect desorption of adatoms of a 2D gas without overlapping of the diffusion zones .

10. 2. 2 Analysis in terms of effective lifetime

A perhaps simpler method is data analysis in terms of first order kinetics only by writing the desorption rate as

$$R = \Theta \, \nu(\Theta) \exp - \frac{\Delta E_D(\Theta)}{R\,T} \qquad (64)$$

$$R = V(\Theta)\ \exp - \frac{\Delta E_D(\Theta)}{RT} \qquad (65)$$

which is equivalent to considering an effective , coverage dependent , lifetime

$$\tau = \left[\nu(\Theta)\ \exp - \Delta E_D(\Theta)/RT\right]^{-1} \qquad (66)$$

and writing

$$R = -\frac{d\Theta}{dt} = \Theta/\tau \qquad (67)$$

This method was introduced by Bauer and al. [50] and developped to a high degree of perfection . This is a convenient manner of analysing the data since R and Θ can be determined straightforwardly by most of the present experimental methods (see section 10.3). From a theoretical point of view it leads to physically significant results if desorption occurs from independent " binding states " characterized by specific (possibly coverage dependent) lifetimes τ_i (parallel mechanism) . Then the apparent lifetime τ is given by

$$\frac{1}{\tau} = \frac{1}{\Theta}\sum_i \frac{\Theta_i}{\tau_i} \qquad (68)$$

In any cases , and whatever experimental technique is used for studying a system , it is now obvious that the data evaluation is made most conveniently by determining the value of R at constant coverages , i. e. by determining desorption isosters . The success of these methods rests on the negligible temperature dependence of ν and ΔE_D at constant Θ (hypothesis (iii)) . Precise modern data acquisition methods allow the determination of ν and ΔE_D to be made in various regions of the phase diagram of 2D systems [51, 52] .

10 . 3 Experimental methods - Kinetics

The signal to be monitored in an experimental set up can be related either to the desorption rate itself or to the surface coverage (Auger signal, work function change etc. . .). The modern sensitive mass spectrometers now available make the first type the most popular at the moment . Although the reader is assumed to be familiar with the experimental arrangement we describe it very briefly . More details can be found in [52] and references therein .

Basically the apparatus used to measure the desorption energy of a metal-metal system consists in

- a source of metal atoms (most frequently a Knudsen cell) that supplies a collimated beam impinging on the sample surface
- the sample : a single crystal (cut to expose the desired orientation) whose temperature is controlled and can be varied
- the mass spectrometer as a detector .

For practical reasons the positions of the mass spectrometer and the sample are kept fixed in general . Consequently one assumes that desorption is isotropic . Then , the mass spectrometer signal is taken as a measure of the desorption rate itself , after a suitable calibration [52] .

The coverage can be determined by integration of the desorption (or adsorption) signal . In otherwords , the raw experimental data R (t) and T (t) can be converted into the more informative R (Θ , T) , which allows in particular the determination of desorption isosters .

10. 3. 1 Temperature Programmed Desorption

The Temperature Programmed Desorption (TPD) also known as Thermal Desorption Spectrometry (TDS) consists in condensing the deposit on the substrate and then heating the latter with a **linear** temperature sweep . Plots of the desorption rate as a function of T are obtained , with the initial coverage Θ_0 as a parameter . They show in general one or more maximum at temperatures $T_p(\Theta_0)$. If the peaks are well separated the apparent order may be checked by a mere inspection of the dependance of T_p on Θ_0 . For 0th order all the peaks have the same leading edge and T_p moves to higher values upon increasing Θ_0 . For 1srt order T_p is independent of Θ_0 . Data evaluation is then made according to one of the methods of section 10. 2 . Since R (Θ) and Θ are readily obtained the method lends itself to an analysis in terms of effective lifetime (eq. 64). Let us first assume that a single peak is present . If eq. 64 holds the data must be fitted by one [ν , E] pair over the whole temperature range for a given value of Θ . If this criterion is satisfied for any Θ one obtains the plots of ν(Θ) and E(Θ) against Θ . Extrapolation to zero coverage yields E_B (assuming that the ad- and desorption processes are identical) . When more than one peak are present one may tentatively assume a parallel mechanism of desorption . The criterion of the fit of the data by a set of [$\nu_i(\Theta_i)$, $E_i(\Theta_i)$] pairs is again checked . Now the difficulty is the assessment of the Θ_i's i. e. the decomposition of overlapping peaks . Reasonable procedures have been devised [52] . However some "experience" is admittedly required in applying them [52] .

The method is relatively fast and accurate in acquiring the data . However the rate of increase of the temperature must be low enough for equilibrium to be achieved within the adlayer (hypothesis (i)) . When the desorption peaks are weakly overlapping the determination of the [ν , E] pairs seems fairly unambiguous and detailed information on the desorption process can be obtained . On the contrary , if the peaks are too much superimposed their decomposition turns out to be impossible and only qualitative conclusions can be drawn [52] .

For practically all the metal-metal systems studied so far the overlap of the desorption peaks has been found reasonnably weak up to amounts of deposit equivalent about two monolayers . Then the assignment of the peak contributions to the desorption signal is possible . The parallel mechanism is found to be obeyed . It turns out that each adlayer gives rise to one desorption peak , with coverage dependent ν and E . Extrapolation of the E (Θ) plot to zero coverage yields E_B (if eq. 63 is assumed valid).

10. 3. 2 Adorption Desorption Transients

The Adsorption Desorption Transient (ADT) method allows isothermal adsorption and desorption to be made successively . As a " by product " equilibrium quantities (isosteric heats of adsorption) are also obtained (see section 10. 4) . The particle beam can be interrupted by means of a shutter . The temperature of the initially clean substrate is set to a definite stable value before the beam is allowed to impinge on the substrate. The sequence of events and the time dependence of the desorption signal are shown by figure 19 and explained in the following .

At $t = 0$ the shutter is opened ; isothermal adsorption occurs along part A of the curve . Initially all the impinging atoms are adsorbed . Upon building up the adlayer their sticking coefficient decreases , thus increasing R . At $t = t_1$ a steady state is achieved : the rate of desorption equals the impingment rate ($R = R_E$) and thus stays constant (part I of the curve) . At $t = t_2$ the shutter is closed , the beam is interrupted . Isothermal desorption occurs along part D of the plot . The coverages in the ad- or desorption processes can be determined by integration (see fig. 19) . By varying the temperature R (Θ , T) can be obtained during either ad- or desorption .

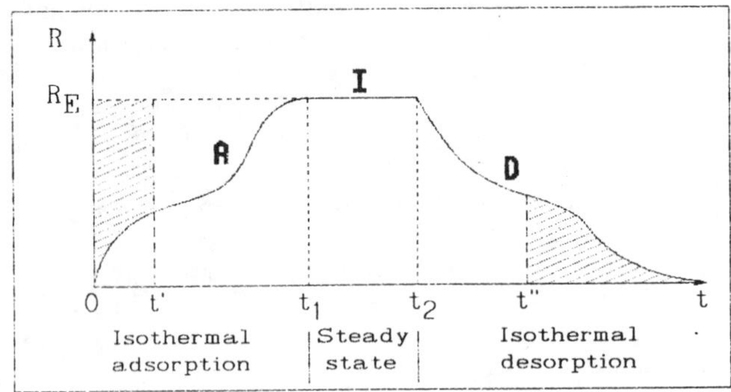

Fig. 19 - Schematic ADT desorption signal : Desorption rate R versus time in the case of complete accomodation . The coverages at times t' and t" are equal to the hatched areas .

The experimental data are then analysed with the same methods as in the TDS experiments .

Since the temperature of the substrate is kept constant throughout a run the method is devoid of the drawbacks of conventional isothermal desorption . It allows a comparison to be made between ad- and desorption phenomena . Data collection at low coverages is easier and more precise than for TDS . The sticking coefficient of the deposit may be measured as a function of coverage and temperature . In particular imcomplete accomodation of the deposit can be detected .

To compare the aforedescribed methods Bauer and al. have applied them to the Au/Mo and Cu/Mo systems [52, 53] . The conclusion of this consistency check is rather satisfactory . Both methods give equivalent information about the desorption energies and frequencies .

However an advantage of the ADT method is the possibility of monitoring the adsorption and desorption processes under isothermal conditions . A deeper insight on the phenemena can be gained from a comparison . In particular , the transition from 2D islands to 3D islands in the S. K. growth has been detected [53] .

10 . 4 Experimental methods - Equilibrium

The equilibrium methods have the fundamental interest of yielding quantities that can be unambiguously analysed in terms of thermodynamics and thus compared to statistical models . In particular the isosteric heat of adsorption can be determined if a set of isotherms is measured under equilibrium conditions . Unfortunately in the case of most metal/metal systems a true thermodynamic equilibrium is extremely difficult to established . Instead , a steady state can be readily obtained between the inpingement of particles on a substrate and their thermal desorption . Obviously the steady state which is reached during an ADT experiment can be used for the purpose of measuring the isotherms. For practical reasons it is easier to measure " isobars "by setting the substrate temperature and so varying the impingement rate that the coverage range of interest is covered . q_{Θ} is then readily determined .

AKNOWLEDGMENTS

We thank Drs. J. C. Heyraud and J. J. Métois for providing the photographs of Fig. 3, 5 and 15 . Both, along with Pr. R. Kern and Pr. M. Bienfait are also thanked for helpful indications and discussions in the preparation of this paper.

REFERENCES

1) Chernov A. A. " Modern Crystallography " Vol. 3 . Springer (Berlin 1984)
2) " Crystal Growth . An Introduction " P. Hartman ed. North Holland (1973)
3) Kern R. " The Equilibrium form of a crystal " in " Morphology of Crystals " I. Sunagawa ed. Terra Scientific Publ. Company , Tokyo and D. Reidel Publ. Company , Dordrecht (1987) p. 77
4) Kern R. , Le Lay G. and Métois J. J. " Basic Mechanisms in the Early Stages of Epitaxy " in " Current Topics in Material Science ", vol. 3 , E. Kaldis ed. North Holland (1979) p. 153
5) Mutaftshiev B. " Surface Thermodynamics " in " Interfacial Aspects of Phase Transformations " , B. Mutaftschiev ed. D. Reidel Publ. Company (1982) p. 63
6) Métois J. J. and Le Lay G. Surface Sci. 133 (1983) 422
7) Kolaczkiewicz J. and Bauer E. Surface Sci. 151 (1985) 333
8) Bienfait M. " Two Dimensional Phases Transitions " in " Current Topics in Material Science " vol 4 , E. Kaldis ed. North Holland (1980) p. 361
9) Pandit R. , Schick M. and Wortis M. Phys. Rev. B 26 (1982) 5112
10) Defay R. , Prigogine I. , Belleman A. and Everett D. H. " Surface tension and Adsorption " Longmans (1956)
11) Herring C. H. " Surface Tension as a Motivation for Sintering " in " The Physics of Powder Metallurgy " W. E. Kingston ed. McGraw Hill (1951) p. 143
12) Herring C. H. " The Use of Classical Macroscopic Concepts in Surface Energy Problems " in " Structure and Properties of Solid Surfaces " R. Gomer and C. S. Smith eds. Chicago Press (1953) p. 5
13) Mutaftschiev B. Surface Sci. 61 (1976) 93
14) Heyraud J. C. and Métois J. J. Surface Sci. 128 (1983) 334
15) Eustathopoulos N. and Joud J. C. in " Current Topics in Material Science " vol. 4 , E. Kaldis ed. North Holland (1980) p. 281
16) Tyson W. R. Can. Metall. Quart. 14 (1975) 307

17) Mezey L. Z. and Giber J. Jap. J. Appl. Phys. 21 (1982) 1569
18) Wynblatt P. Surface Sci. 136 (1984) L51
19) Burton W. K. , Cabrera N. and Franck F. C. Phil. Trans. Roy. Soc. London 243A (1951) 299
20) Bauer E. Zeit. für Kristall. 110 (1958) 372
21) Kern R. "Three and Two-dimensional Nucleation on Substrates " in " Interfacial Aspects of Phases Transformations " , B. Mutaftschiev ed. , D. Reidel Publ. Company , Dordrecht (1982) p. 287
22) van der Merwe J. H. and Bauer E. Phys. Rev. B 39 (1989) 3632
23) Bauer E. and Poppa H. Thin Solid Films 121 (1984) 159
24) Paunov M. and Bauer E. Surface Sci. 188 (1987) 123
25) Bermond J. M. and Venables J. A. J. Crystal Growth 64 (1983) 239
26) Bonzel H. P. " Mass Transport by Surface Self Diffusion " in " Surface Mobilities on Solid Materials ", Vu Thien Binh ed. Plenum , New York (1983) p. 195
27) Wang S. C. and Ehrlich G. Surface Sci. 206 (1988) 451
28) Roux H. , Piquet A. , Uzan R. and Drechsler M. Surface Sci. 59 (1976) 97
29) Morin R. and Drechsler M. Surface Sci. 111 (1981) 128
30) Herring C. H. J. Appl. Phys. 21 (1950) 301
31) Drechsler M. Jap. J. Appl. Phys. Suppl.2 Part 2 (1974) 25
32) Mundschau M. , Bauer E. , Telieps W. and Swiech W. Surface Sci. 213 (1989) 381
33) Meinel K. , Klaua M. and Bethge H. J. Crystal Growth 89 (1988) 447
34) Flynn C. P. J. Physics F 18 (1988) L195
35) Heyraud J. C. and Métois J. J. Surface Sci. 100 (1980) 519
36) Bauer E. Appl. Surface Sci. 11/12 (1982) 479
37) Dienes G. J. , Sieradzki K. , Paskin A. and Massoumzdeh B. Surface Sci. 144 (1984) 273
38) Grabow M. H. and Gilmer G. H. Surface Sci. 194 (1988) 333
39) Woltersdorf J. Appl. Surface Sci. 11/12 (1982) 495
40) van der Merwe J. H. " Recent Developments in the Theory of Epitaxy " in "Chemistry and Physics of Solid Surfaces " Vol. 4 R. Vanselow and R. Howe eds. Springer Berlin (1984) p. 365
41) Kato M. , Wada M. , Sato A. and Mori T. Acta Met. 37 (1989) 749
42) van der Merwe J. H. Phil. Mag. A 45 (1982) 127
43) Bauer E. and van der Merwe J. H. Phys. Rev. B 33 (1986) 3657
44) Pavlovska A. , Paunov M. and Bauer E. Thin Solid Films 126 (1985) 129
45) Bauer E. , Poppa H. and Todd G. Thin Solid Films 28 (1975) 19
46) Ross S. and Oliver J. P. "On Physical Adsorption" Interscience New York (1964)
47) Glasstone S. , Laidler K. J. and Eyring H. " Theory of Rate Processes " McGraw Hill New York (1941)
48) Menzel D. "Thermal Desorption " in "Chemistry and Physics of Solid Surfaces" Vol. 2 R. Vanselow and R. Howe eds. Springer Berlin (1982) p. 389
49) Cassuto A. and King D. A. Surface Sci. 102 (1981) 388
50) Bauer E. , Bonczek F. , Poppa H. and Todd G. Surface Sci. 53 (1975) 87
51) Kolaczkiewicz J. and Bauer E. Surface Sci. 175 (1986) 508
52) Pavlovska A. , Steffen H. and Bauer E. Surface Sci. 195 (1988) 207
53) Paunov M. and Bauer E. Appl. Phys. A 44 (1987) 201

Materials Science Forum Vols. 59 & 60 (1990) pp. 41-64

FUNDAMENTAL PHENOMENA IN NULEATION AND GROWTH OF CRYSTALS
PART B - STABILITY AND MOBILITY OF AGGREGATES ON SURFACES

A. Masson
Laboratoire de Physico-Chimie des Surfaces
Ecole de Chimie de Paris, 11, rue Pierre et Marie Curie
F-75005 Paris, France

In part A we have shown that a purely thermodynamic approach could describe the different growth processes according to the old Bauer's criteria [1]. This description only needs macroscopic parameters such as the specific surface free energy of a daughter phase condensed from a mother vapour phase, the surface free energy of the deposit and the interfacial free energy. Moreover, a macroscopie analysis of the interface by the way of the work of adhesion is necessary to take into account the elastic strain on the stability of the layer and multilayer growth [2].

In this part we want to stress the idea that the mathematical formalism developed to describe the nucleation and growth processes, and the new experiments of mass transport onto a surface, have shed new light on epitaxial phenomena in thin film growth.

In order to do that, we take aggregates onto a surface as a model, and we intend to study their stability, mobility and incidence of temperature on all the processes involved. Epitaxy is therefore described firstly as a nucleation phenomenon, then as a post-nucleation step. Finally with the introduction of new ionic-cluster beam techniques [3] it will be shown that new kinetic parameters must be studied.

I.1 Equilibrium description - Epitaxy as a nucleation process

The thermodynamic model of nucleation is well adapted to this process.

We start with an isotropic aggregate containing i atoms, with a spherical cap shape in equilibrium on a surface maintained at temperature T with a population of adsorbed monomers n_i coming from a mother vapour phase at pressure p.

The Young equation (1) is therefore written as $\gamma_s = \gamma_i + \gamma_e \cos\alpha$ (1)

γ_s, γ_c, γ_i being the specific free surface and interface energies of the system and α, a wetting angle * Fig. 1 [4].

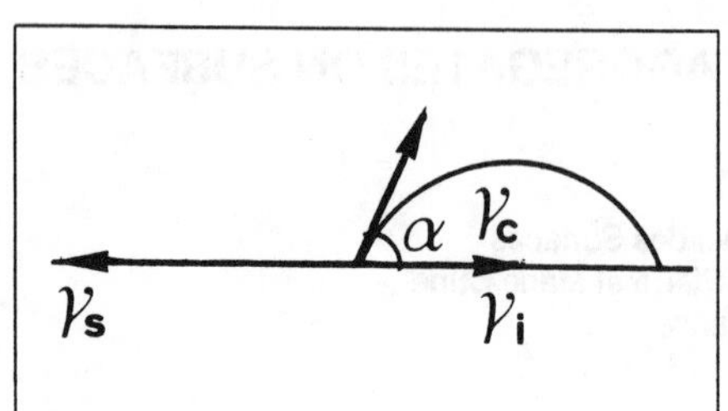

Fig. 1 Cap shape model for critical nucleus.

Within this consideration it is therefore possible to define the work of formation of such an aggregate and to discuss the importance of the concept of critical nucleus in the framework of the nucleation formalism so well described by the review papers of Volmer [5] Frenkel [6] and Hirth and Pound [7].*

We define here that the rate of nucleation is written without the non equilibrium factor correction as

$$J = \beta^* C_i^* \quad (2)$$

with C_i^*, the concentration of critical nuclei, and β^* a collision factor (collision of adatom on nuclei of critical radius r**.

We need also remind you that the supersaturation is written as = p/p_o (p_o the pressure of the bulk phase at temperature T).

The work of the formation of the aggregates is then

$$\Delta G_i = [-\Delta G_v V + \sum n_i S_i] F(\alpha) \quad (3)$$

$\Delta G_v = kT \ln p/po$ is the thermodynamic driving force at supersaturation and undersaturation when $p/po \neq 1$

---

* Different conclusions could be discussed when there are very strong interactions between the condensed phase and the substrate leading to a perfect wetting $\alpha = 0$.

* In part one we have already discussed the problem of anistropy of γ for different shape of the aggregates and the relations with Gibbs Wulff's construction needed for equilibrium description.

** See keynotes on heterogeneous nucleation, Appendix 1.

For an isotropic critical cluster the solution of equation (3) is

$$r^*=-2\gamma/\Delta G_v\ F(\alpha) \quad (4)$$

and $\Delta G^* = (16\ \pi\ \gamma^3/3\Delta G_v)F(\alpha)$ with $F(\alpha) = (1-\cos\alpha)^2(2+\cos\alpha)/4$ (5)

The rate of nucleation being defined as

$$J = \beta^* C_i^* \quad (6)$$

with $C_i^* = n_1 \exp(-\Delta G_i^*/kT)$ (7)

$$\beta^* = \nu\ n_o\ a\ r^*\ \sin\alpha\ \exp(-\Delta G_{diff}/kT) \quad (8)$$

n_o number of sites on the surface, ν lattice vibration frequency, r^* $\sin\alpha$ the capture line of critical nucleus, ΔG_{diff} the activation energy of surface diffusion, so the nucleation rate is rewritten as

$$J = n_i\ n_o\ a\ r^*\ \sin\alpha\ \exp\Delta G_{diff}/kT\ \exp(-16\pi\gamma^3/3RT\ \Delta G^2_v)\ F(\alpha) \quad (9)$$

We can now follow the description of Moazed [8] by writing J in the form:

$$J = A\ \sin\alpha\ \exp(-B(F(\alpha)) \quad (10)$$

with $\mathrm{Ln}\ J = \mathrm{Ln}A + \mathrm{Ln}\ \sin\alpha - BF(\alpha)$*** (10')

with A and B defined obviously by

$$A = n_i\ n_o a\ \nu\ n\ 2\gamma/\Delta G_v\ \exp(-\Delta G_{diff}/RT) \quad (11)$$

$$B = 16\ \pi\gamma^3/3RT\ \Delta G^2{}_v \quad (12)$$

The graphs of $F(\alpha)$ and LnA'= Ln A - LnJ as a function of are shown below in figures 2a and 2b for different values of taken from the work of Moazed [8].

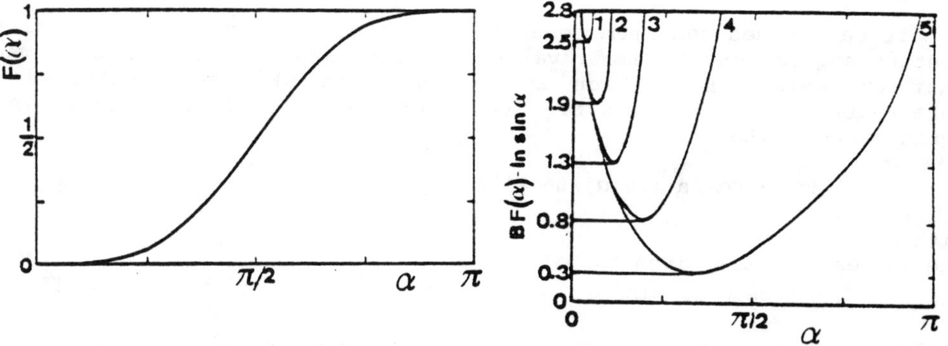

Fig.2a,2b Graph of F(a) and LnA'; in the Fig.2b, 1, 2, 3, 4 concern different values of supersaturation in B.

*** Only the single value of α is taken from equation (10').

I.2 Effect of supersaturation

So before discussing the role of macroscopic parameters in epitaxial growth, we need to know how the nucleation process starts with supersaturation. Still following the work of Moazed [8] which assumes that the variation of p has negligible effect on all terms of A and B factors, except the supersaturation p/po which appears in $\Delta Gv = -RT/W \ln p/po$ (W being the molar volume of the condensate).

It is obvious that when p/p0 is increasing A and B are decreasing but B is decreasing faster than A. Looking to the Fig. 2b it is easy to see, that the increase of supersaturation leads to a shift down and right of BF (a) and also a shift down for A' (which is representative of an horizontal line in the same figure), the discussion then centers on the intersection of these two curves corresponding to each critical supersaturation for nucleation related to tangential positions.

Only low supersaturations have to be taken into account for an epitaxial nucleation process. With this formalism it is easy to discuss the various components of supersaturation : temperature of substrate, and values of a which are of importance for the epitaxial nucleation process. We report here on the main conclusions, following the discussion of Moazed.

I.3 Wetting angle

If J_1 and J_2 are two nucleation rates for two populations of nuclei (1) and (2) having $\alpha 1$ and $\alpha 2$ wetting angles and if J_1 is an epitaxial orientation J_1 = Je the ratio J_1/J_2 >>1 is the measure of preferential epitaxial orientation. This ratio E_o (epitaxial orientation) can be written as

$$Je/J_2 = E_o = (\sin \alpha_1/\sin \alpha_2) \exp [B(F_1(\alpha_2)-F(\alpha_1)] \quad (13)$$

All other things being equal for the given experimental conditions in the case of epitaxy we still see that J_i has to be minimum which gives

$$\alpha_2 > \alpha_1 \rightarrow F(\alpha 2) > F(\alpha_1)$$

It can be deduced from equation (10) that very small and very large contact angles lead to small values of factors A and B. This also means that epitaxial orientation Eo is not favored. The conditions for enhancement or orientation are found for the zero of the derivative of equation (10') that is

$$d/d\alpha\ [B\ F\alpha - \mathrm{Ln}\ \sin \alpha]/\mathrm{Ln}\ A' = o \quad (14)$$

with A' = A/J
the values of a are given for

$$B = 4\cos \alpha/B3 \sin^4 \alpha \quad (15)$$

with a fixed value of B as in Fig. (2b) α is found as 5°C < α< 65°.

I.4 Temperature

The last parameter concerns the temperature. Obviously decreasing T gives a decrease of the pre-exponantial term in A. Because the rate of nucleation is written as equation (10) keeping J constant means that B must decrease. However, decreasing B decreases the ratio J_1/J_2 so a decrease in temperature always gives a decrease in the epitaxial orientation.

I.5 Nucleation epitaxy in the case of the atomistic model

It is easy to find with equation (4) for normal conditions of condensation that r* leads to atomic values when p/p_o is high (see table 1 in Appendix). This remark leads to the problem of the continuity of the capillary model just described, for the very small clusters. Walton [9] then Walton and Rhodin[10] have first developed the atomistic model, by a new calculation of the work of formation ΔG_i of clusters of i atoms They replace this work by an equivalent energy E_i of dissociation of a cluster of size i by writing $E_{ij} = iU_1 - U_{ij}$, with U_{ij} the energy of a cluster of size i in position J. E_{ij} represents the decrease of potential energy in forming the cluster i from i adatoms.

For a single state j the equilibrium relation of the nucleation formalism is written as

$$N_i/N_o = (n_1/N_o)_i \exp E_i/RT \tag{16}$$

with n_1 the concentration of adatoms and N_o the number of accessible sites in the substrate. Equation (16) is the Walton equation.

The analogy for J is also possible and we can take the notation of Rhodin (9) and write J as

$$J = R\ a\ 2\ n_o\ (R/\nu\ N_o) i^* \exp\ [(i^* + 1)\ E_a\ -\ E_d + E_i]/kT \tag{17}$$

R is the flux of impinging atoms, Ea and Ed respectively the activation energy for adsorption and diffusion of adatoms. Evidently α and γ have disappeared but the disadvantage of (17) is that we lose the possibility to have on explicit analytical expression for i* and Ei. These values are therefore obtained by experiment in this way, by changing the temperature of the substrate and by examining some particular structure for the very simple critical nuclei. By increasing the temperature, the dimers and then the trimers are no longer stable and we have to look for i*=2,3,4....

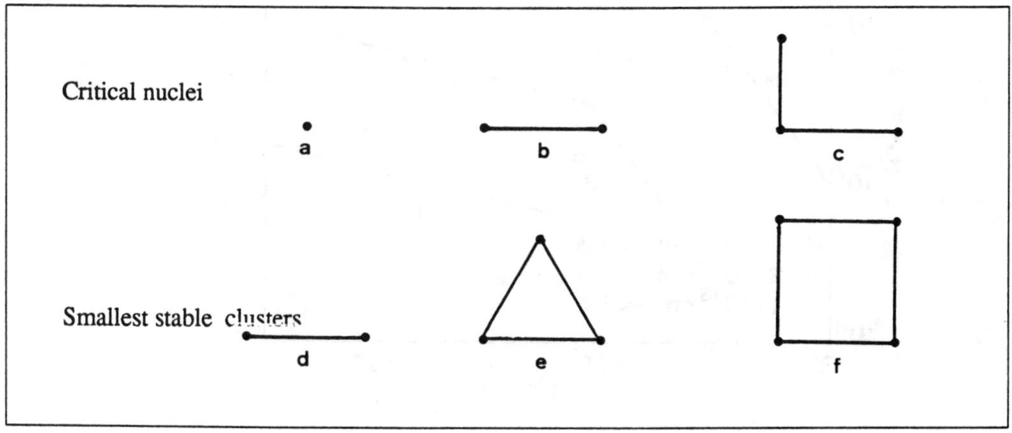

Fig. 3 Schematic atomic diagram of the smallest stable clusters and corresponding critical nuclei [9].

In the limit of a high supersaturation, the critical nucleus is a single atom and the rate J can be written as

$$J = R(R/\nu N_o) \exp [2E_a - E_d/kT]$$

As the substrate temperature is increased, a temperature is reached at which the pair is no longer stable. The next stable entity is one that has a minimum of two bonds per atom. Two possibilities are then found : the triangle and the square.

For the case b inFig.3 the corresponding nucleation rate is

$$J_2 = R(R/\nu N_o)2 \exp (3E_A + E_2 - E_d/kT) \quad (19)$$

and for the case c also in Fig.3

$$J_3 = R(R/\nu N_o)3 \exp (4E_A + E_3 - E_{diff}/kT) \quad (20)$$

with E_2 and E_3 the dissociation energy respectively of a dimer and a trimer.

The temperature T_1 and T_2 at which transitions occur can be obtained by equating J_1 and J_2, J_1 and J_3

as $T_1 = -(E_A + E_2)/kL_n(R/\nu N_o)$

and $T2 = -(E_a + 1/2E_3)/kL_n(R/\nu N_o)$

These transition temperatures are found experimentally by discontinuities in the graph J versus 1/T as shown on Fig.4.

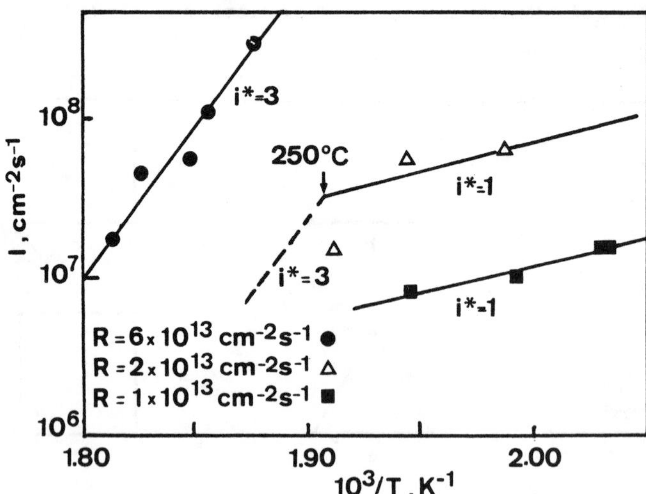

Fig.4 Experimental results of silver clusters on (100) NaCl for different R [9].

It is obvious that with this model some kinds of conclusion concerning the epitaxial nucleation started in the previous paragraph could be developed.Good agreement is found for example for silver on (100) NaCl surfaces. We need to have T > 250°C in order to observe a well oriented (100) film. Values of E_3, E_A and E_d have been deduced from experiments (Fig. 4).

This temperature may be related to the epitaxial temperature, since below this value a bad orientation is observed.

II MASS TRANSFER MOBILITY OF ATOMS AND AGGREGATES

II.1 Mobility of monomers

In the first part of this work, we have shown a diagram of a real surface as reported on Fig. 5, on which we can follow the different steps in mass transport at an atomic scale. To move matter accross a surface under a thermodynamic potential, one has first of all to detach an atom from a kink position, then to move it onto terraces of low index planes, to jump edges, and finally to position it on another kink site. All these mechanisms have been analysed generally by looking at the macroscopic shape transformation.

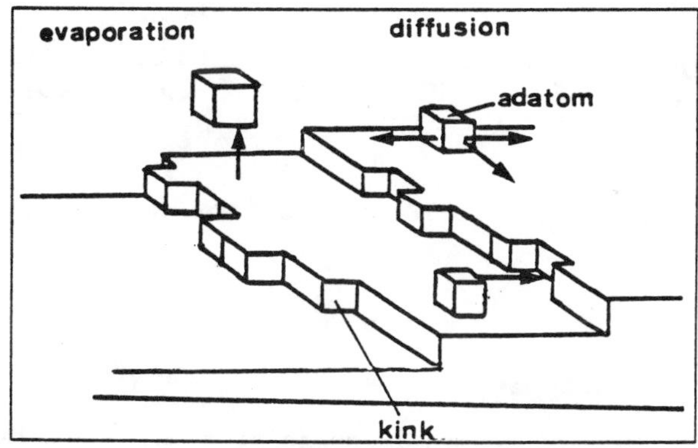

Fig. 5 Different mechanisms leading to mobility of adatoms on the surface.

It is the advent of the field ion microscope [12] that started the study at an atomic scale, since it became possible to visualize isolated adatoms by this technique. Binding energies for atom with different coordinations on a surface, as well as mean residence times and anisotropic diffusion have been also investigated. In fact the essential point is without doubt the possibility to measure atomic displacements on well characterized surfaces.The rate of shifting and the preferential directions are also described. Moreover, a migrating adatom is a probe of the variation of the interaction potential. Measurements of adatoms are shown in Fig.6 from the work of Erlich[15], which shows a sequence of atomic positions determined after movement obtained by annealing at high temperature.

If n is the number of jumps of length l, t the time of diffusion, then the atomic coefficient of diffusion can be written as

$$D = nl_2/2dt \qquad (21)$$

(d being a correction factor for diffusion along a line or in a plane).

The mean square displacement of the Brownian motion of an atom is then directly measured by experiment as

$$\langle R^2 \rangle = nl^2 \qquad (22)$$

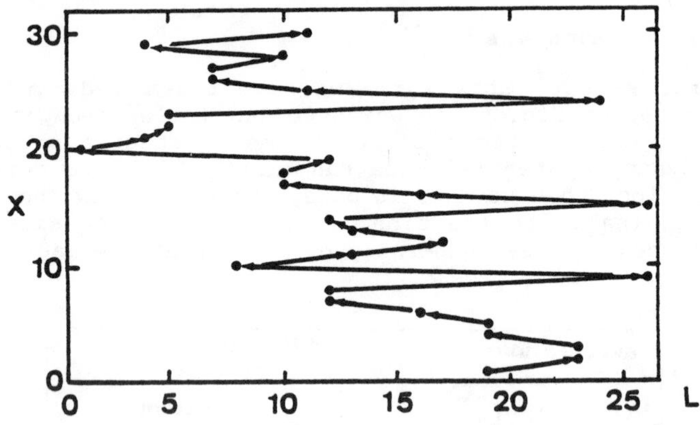

Fig. 6 Sequence of atomic positions taken from [16].

II.2 Mobility of polymers

The cluster formation and their correlated movements have been studied in details, by D.W. Basset [14] and G. Erlich [15] on tungsten, then with other foreign adatoms of rhodium, irridium and platinum also scattered on a tungsten surface.

On Fig.7 taken from Graham and Erlich [16] we can follow the variation of the diffusion coefficient for dimers and compare with the behaviour of monomers. The factor 2 obtained in binding energy of the dimer is due to correlation between adatoms when forming the dimer.

Later Basset [17] has also distinguished between the «worm» like shift of linear polymers and the very peculiar «random waltz» of big symmetric polymers [18].

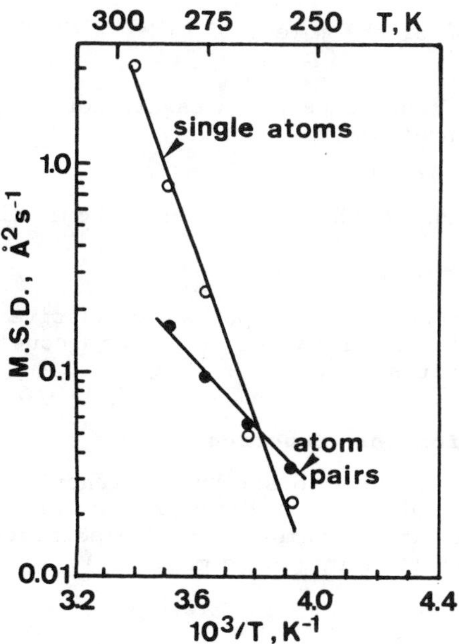

Fig. 7 Mean square displacement of monomers and dimers from [16].

II.3 Macroaggregates mobility

In the domain of discontinuous thin films, the pionner work of G.A. Basset [19] has to be mentioned. This work using in-situ electron microscopy has underlined the «liquid» behaviour of big gold aggregates scattered on a molybdenum sulfide cleavage surface. Huge translations and the disappearance of small clusters by coalescence have been described. At the same time morphologic transformations of spatial distributions of gold clusters deposited on SiO_2 surface have been studied by Skofronick [20] after annealing. The problem of intrinsic mobility of things others than isolated adatoms has to be deeply investigated, as well as the nucleation step in growth phenomena. The epitaxial relationship then becomes a post-nucleation phenomenon since aggregates can take new orientations during their displacements. Later in the eighties Poppa, Metois [21] and Honjo [22] have studied by ultra-high vacuum electron microscopy the rotation of such big clusters, avoiding the problem of impurities invoked in the first experiments. Before describing in some detail the different interface models, it is necessary to spend some time on the distribution functions which characterize these kinds of experiment.

II.3.1 Distribution functions

A collection of aggregates is well described by the following parameters:

- The mean size density ρo per unit surface of particles of size i $\rho_o = \Sigma_{i=1}\, n_i$.
- The histogram of size i, which is a statistical distribution of size, the moment of order 1 of this distribution being the mean values <d>.

- The positive square root of the moment of order 2 is the dispersion.

  $s = |\langle d^2\rangle - \langle d\rangle^2|^{1/2}$

- The radial distribution function P(r) is easily defined by measurement of all interparticle distances N(r,dr) so that

  $N(r,dr) = \rho_o \, P(r) \, 2\pi r \, dr$

- Finally the distribution of the nearest neighbour distances which is written as

  $W(r,dr) = 2 \rho_o \, e^{-\pi \rho_o r^2} \, dr$

Schmeisser [23] and Zanghi [24] have reported exhaustive investigations of all these functions and we have retained their main conclusions concerning the mobility of big aggregates.

II.3.2 Radial distribution function

Zanghi [24] has studied radial distribution functions of distributions of gold aggregates scattered on (100) cleavage surfaces of alkali halides before and after annealing at low temperature (temperature which is too low to invoke mechanisms other than intrinsic mobility). The results are shown in Fig. 8.

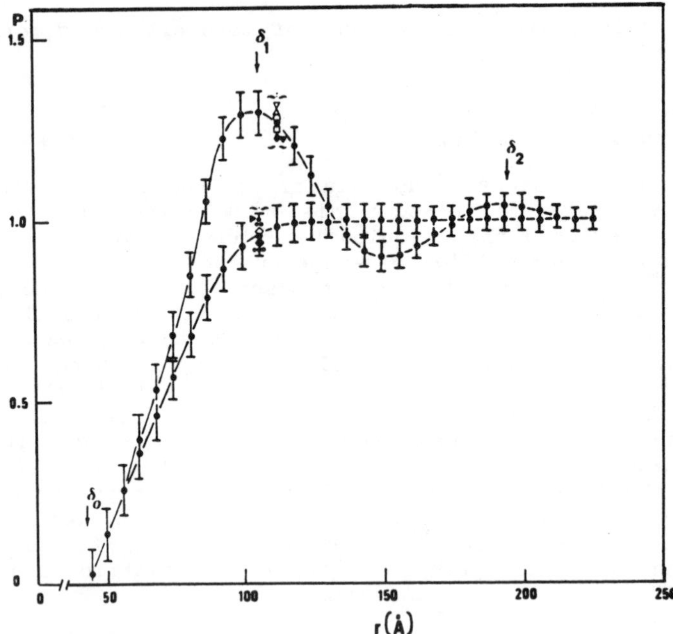

Fig. 8 Radial distribution function of gold clusters on KCl cleavage surfaces [24].

The oscillations obtained on curves (2) at distances r, ~ 100Å and r_2 ~ 200Å are given by an increase in the number of particles. This is due to intrinsic mobility, since the total mass is constant.

Other kinds of experiment needing intrinsic mobility have been studied by Schmeisser [25] and reported on Fig. 9. Non random distribution of particles is well observed by a systemac shift towards the big particles for the nearesst neighbour distance distributions.

Such a shift is well explained by taking in account intercorrelation particles and is interpreted by Schmeisser [25] as a result of migration of such particles after discussing the other possibility of depletion of monomers by a capture limited diffusion mechanism.

Finally the size distributions studied also by the same author are briefly reported to show that the exponential increase in different curves and the cut-off at a peculiar size are well explained by time dependent rate equations, in the formalism of Zinsmeister [26]. As shown on Fig. 10 after time t = 45 s there are big changes with the ρ_0 (shaded integral in Fig. 10) as the increasing of mean diameter <d> and the increase of dispersion s.

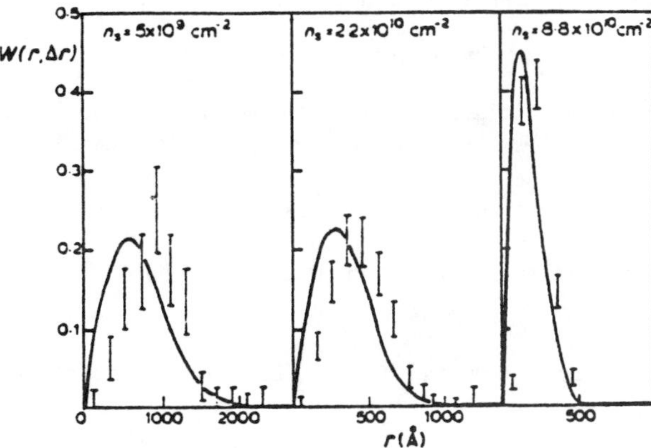

Fig. 9 Nearest neighbour distance distribution of gold particles on NaCl (100) surface from [25].

II.3.3 Size distribution

The simple distribution to follow with time is the size distribution. Taken also from Schmeisser, such a variation is shown on Fig. 10.

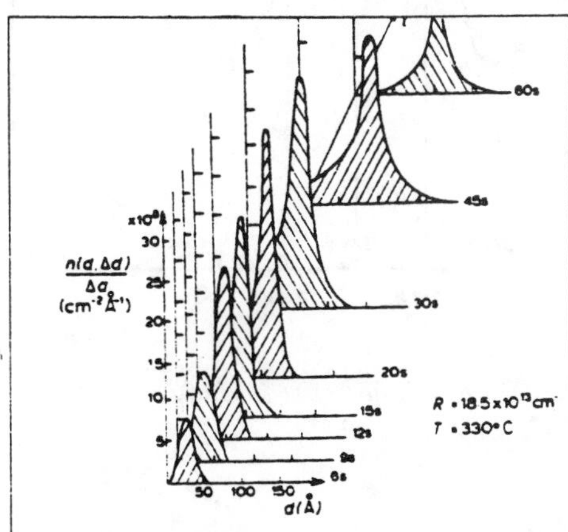

Fig. 10 Variation of cluster size-distribution with time for gold particles on NaCl substrate, from [25].

II.4 Interface models

In order to understand physically the translation and rotation of aggregates, two models will be described. The first is the model of Reiss [27] which is based on a description of binding energies of aggregates as a function of the misorientation angle. The second by Kern et al [28] is a description of interfaces by a thermodynamic model very close to grain boundary gliding. Let us now describe the two models.

II.4.1 - Reiss model

A two bidimensional island of quadratic symmetry is misoriented by an angle θ on a periodic substrate of the same symmetry with lattice parameter a (Fig. 11a). (2N+1)2 is the number of atoms in the cluster. For a single atom placed on a substrate, the potential is usually a sinusoidal function of its coordinates (x,y). If an atom in the cluster has coordinates (n,m) the difference between the coordinates of the atom (n,m) and the corresponding atom in the substrate can be written as $\Delta X(n,m)$ and $\Delta Y(n,m)$ in order that the potential seen by the atom can be described as

$$V_{nm} = V_1(1-\cos 2\pi/a\ \Delta X(n,m)\ \cos 2\pi/a\ \Delta Y(n,m)) + U \qquad (23)$$

U is a potential taking into account that the atom (n,m) is in the island. Reiss assumes U = 0, that is the island is frozen. Moreover for small θ

$$DX(n,m) \sim -ma\theta \quad \text{and} \quad \Delta Y(n,m) \sim na\theta \qquad \text{where } \theta < 0,5$$

in order that the V(n,m) for all (n,m) is written as

$$V\theta/V_1 = (2N)^2\ [1-\sin^2 \Theta/\Theta 2] \text{ with } \Theta = 2nN\theta \qquad (24)$$

This function is the complementary function of the diffraction function,as shown in Fig. 11b.

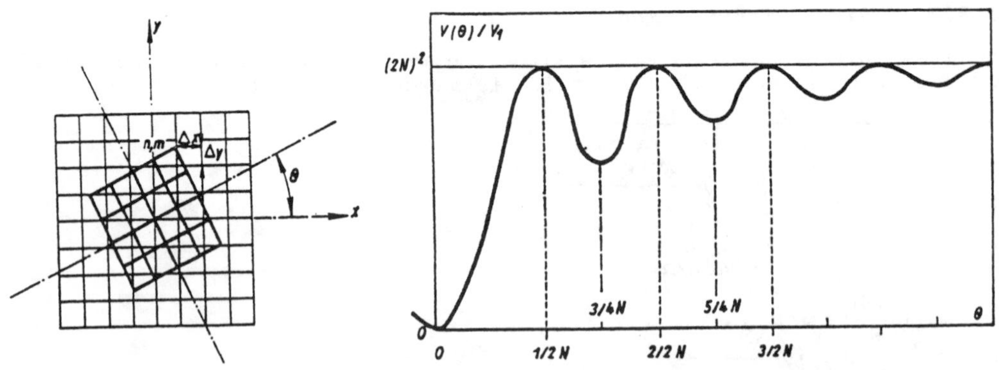

Fig. 11 a) Quadratic island model. b)Potential energy of the island of figure 11a, as a function of θ.

The lowest minima for $\theta = 0$ is obtained for epitaxial orientation. In order to move an island with $(2N + 1)^2$ atoms to its epitaxial orientation requires an energy of $(2N^2)V_1$. V1 being the activation energy of a single atom in the island. But on the other hand a misorientation $\theta_1 = 3/4N$ needs only an energy of $E = 1/25\ (2N^2)V_1$. This energy of rotation is close to a jump of an isolated atom if the cluster is no greater than 25 atoms. For bigger angles of disorientation, the same conditions could be found for bigger cluster. Reiss has also discussed the translation phenomenom with $\Delta X = ma\theta + u$ and $\Delta Y = na\theta + v$.

The relation of potential is then

$$V(\theta,u,v)V_1 = (2N^2)\ [1-\sin^2\Theta/\Theta^2 \cos 2nu/a \cos 2mv/a] \qquad (25)$$

The same kind of activation energy is also required for making translations with constant θ.

II.4.2 Kern et al. interface."Brownian gliding"

The Moiré description of interfaces is well adapted to this geometric interface as shown in Fig. 12.

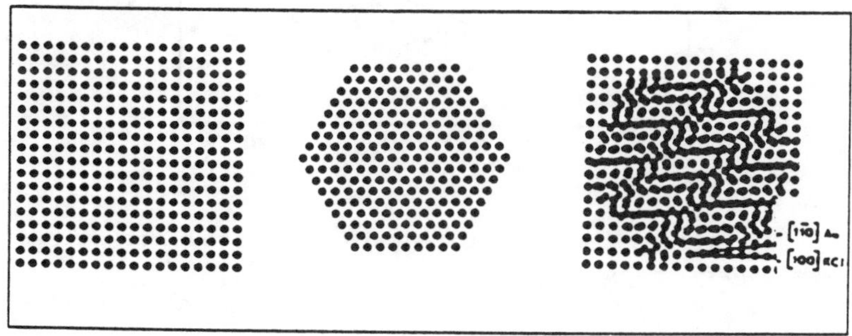

Fig. 12 Moiré model of interfaces.

In the model the interface is described by two kinds of zone. O ne where "black dot" atoms are in good match positions, where the work of adhesion is increasing. The other where the atoms are in a bad mutual position. This type of disorder can be assumed to look like a «liquid». The translation of the aggregate is then done by two processes. Firstly, a mutual fusion of the good interface at temperature T needing fusion enthalpy ΔHm and a fusion entropy ΔSm with an activation energy of translation written as

$$\Delta G_1 = \Delta H_m\ (1/\ - T/To) \qquad (26)$$

ΔH_m can be related to the atomic parameter Δh_m with d being the diameter of the aggregates and a^2 area of an atom by

$$\Delta H_m = \pi d^2\ \Delta h_m/4a^2 \qquad (27)$$

The second phenomenon needs a gliding of a «liquid» interface viewed as a hole diffusion needing a migration energy of

$$\Delta G2 = \Delta Ho - T\Delta So \qquad (28)$$

the gliding energy of the island being

$$\Delta G^{*} = \Delta G1 + \Delta G2 \qquad (29)$$

Assuming a Brownian gliding of a crystallite, following a kind of Langevin equation, [29] the diffusion coefficient is written as

$$D = 2kT/n\ a^2 \exp - \Delta G^{*}/kT \qquad (30)$$

$$\text{or with } D_d = 2\ a^2\ k\ T/h \exp \Delta G_2 \exp - \pi d^2\ \Delta hm/4a^2h\ (1/T - 1/T_o) \qquad (31)$$

Such a model has been studied by following log D versus 1/T as shown on Fig. 12 for different temperatures. By this way the temperature of fusion of the island and the pre-exponential term of equation (30) have been determinated.

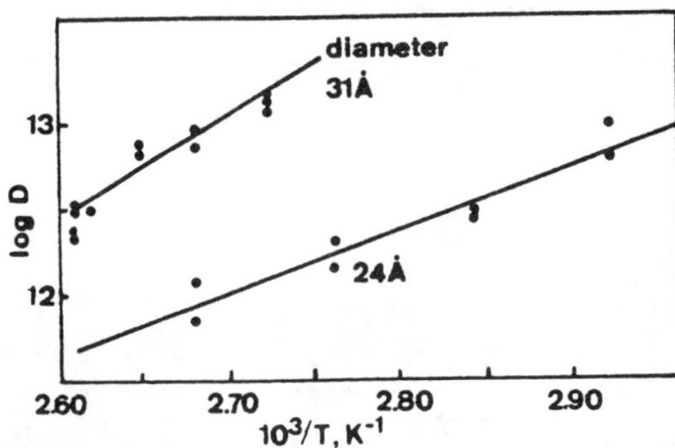

Fig. 13 Diffusion coefficients of crystallite versus temperature.

III FREE AGGREGATES

Each time a new atom or a molecule sticks on an aggregate, all the atoms in the cluster rearrange themselves, the cluster is reconstructed. If we intend to understand the crystalline growth at an atomic scale, it is necessary to know all the series of structures involved atom by atom. This reconstruction phenomenon (relaxation or contraction) takes place until a critical size is readed, for which the cluster is no longer able to change. A net cell is frozen in the cluster. During all these sequences one must also have in mind that electronic properties and chemical reactivity vary with size.

Experimentally this approach has been reserved for aggregates obtained by the condensation of gazeous materials, but is now extended to the most refractory solid materials.

This extension has been made possible by the development of new techniques such as lasers, sputtering or the liquid metal sources.

On Fig. 14a we can follow the different functions found, for instance, in the sputtering source of the Lausanne University.

We also show a spectrum of silver aggregates obtained after electron ionisation and size selection Fig. 14b. For our purpose here linked to growth process in thin films, we intend to develop only two special aspects.

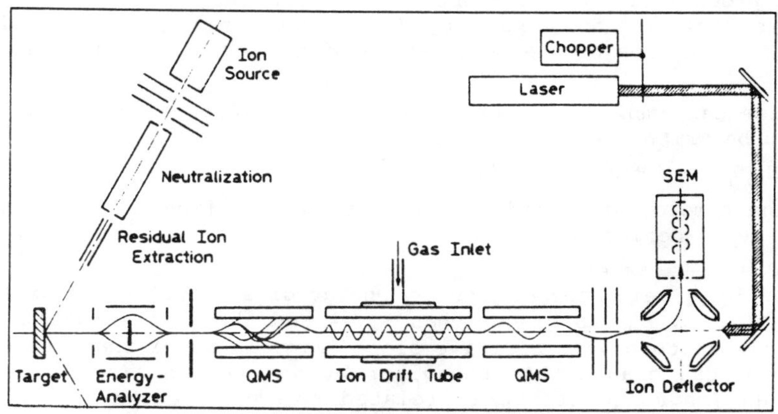

Fig.14a The sputtering source of Lausanne.

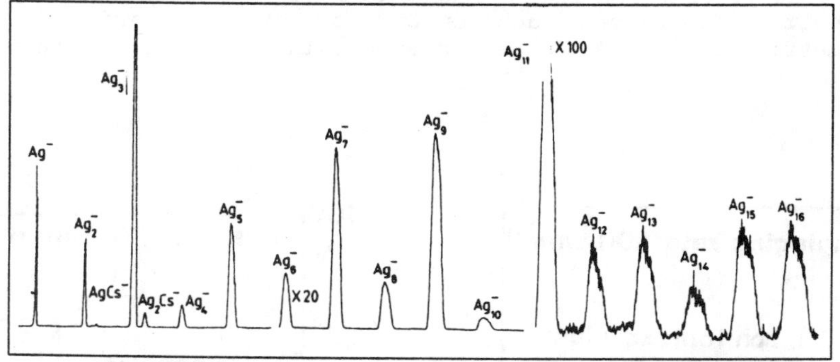

Fig.14b Mass spectrum of selected silver clusters.

III.1 Ionized aggregates on a surface

III.1.2 High energy interactions

Takagi [31] was the first to develop, in 1972, the possibility of growing thin films by interaction of a cluster beam with a surface. The main idea is to form clusters in the gas phase (homogeneous nucleation). The clusters are then ionised either by electronic impact or laser photoionisation after which it is then possible to accelerate them by an electric field. Takagi et al. have succeeded with clusters as big as a hundred thousand atoms per cluster. Following Takagi, the characteristics of the deposit can be of three kinds.

The first is a cleaning effect of the surface obtained by highly energetic cluster beams. For instance it has been possible to obtain epitaxial films of silicon on silicon surfaces at temperatures as low as 500°C.

The second and main effect is related to the bursting of clusters in the interaction with the surface. There is both an increase in the diffusion process of the atoms obtained and an increase in the nucleation rate. This effect can be seen by following the mass deposited on the surface as a function of the temperature of the substrate and the energy of the beam.

For a classical impinging flux R and J a rate nucleation, the deposited mass could be written as

$$M = R_t - NR/J \exp(-U/kT) \qquad (33)$$

with t, the time of evaporation, N the number of sites on the surface, the energy, U being defined as

$$U = E_{ad} - E_{diff}$$

E_a and E_{diff} being the activation energy of adsorption and diffusion of adatoms. Normally E_{ad} is bigger than E_{diff} and the linear variation of M against 1/T has to have a positive slope. Fig. 15 shows that the energy of acceleration can change this behaviour by an increase in the diffusion. Thirdly, an important effect is related to stoichiometry and has been observed with the deposition of complex semiconductors such as Cd Te and Mn Te. By changing the energy and the ionization of the products, for instance (Cd Te) + and neutral MnTe, it is possible to monitor the stoichiometry of the Cd_{1-x} Mn Te film Fig.16. This approach has also given epitaxial films of dielectric substances such as CaF_2 on Si(111) surfaces. Finally stoichiometric films of Al_2O_3 have been obtained at very low temperatures (100°C).

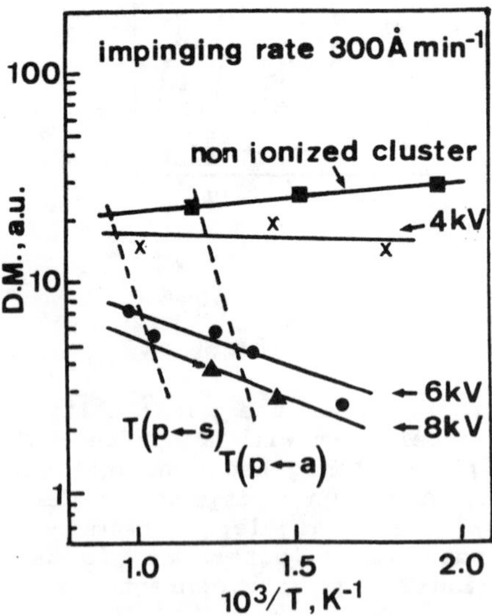

Fig. 15 Deposited mass versus reciprocal substrate temperature from [31].

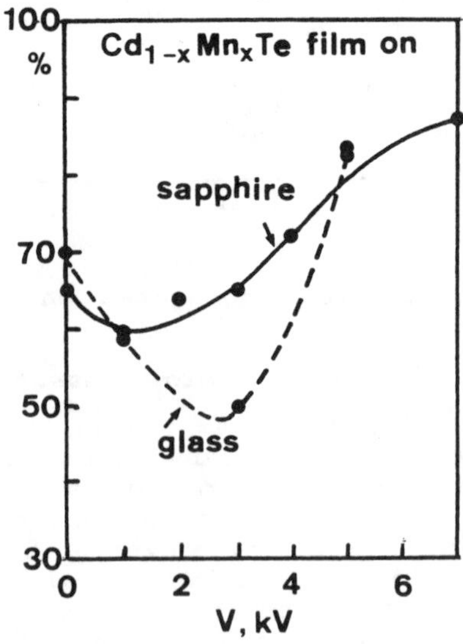

Fig. 16 Change of composition ratios of Cd_{1-x}, Mn_xTe films by controlling the acceleration voltage from [31].

III.2 Soft landing experiment

The major problem of heterogeneous nucleation is due to the fact that the classical method of creating condensed atomic beams on surfaces leads to multisize distributions. The reason is that the nucleation is generally dominated or controlled by nucleation on random distribution of defects surrounded by a surface area of capture in contact with neighbouring areas as described by Voronoi polygons [32].

In 1986 Di-Cenzo et al [33] considered that size selected aggregates could be «softed landed» on a surface. After ionisation, it is possible to make a size selection by time of flight spectrometry and to land the aggregates at zero energy. Such an experiment has been realised using gold aggregates on a graphite surface. In Fig. 17 it is possible to analyse the result of the stability of different sizes using photoelectron spectroscopic techniques. Selected size clusters, containing 5, 7, 27 and 33 atoms are obtained and a shoulder which appears on the high energy side of the 4f doublet that could be related to a possible fragmentation (this could well be related to the variation of cohesion energy with size).

This kind of experiment reproduced in other laboratories on other materials platinum or silver [35] opens the new way to control supercritical nuclei of known crystalline structure leading to epitaxial nucleation.

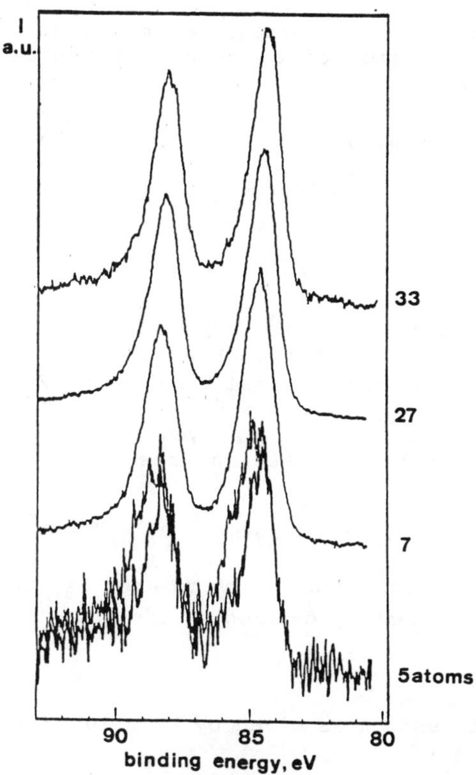

Fig.17 XPS spectra of size selected gold clusters on graphite, from [33].

APPENDIX 1

KEYNOTES FOR HETEROGENEOUS NUCLEATION

Thermodynamic model

Universally described after Volmer [5], the equilibrium of a population of adsorbed monomers of concentration C1, with a collection of polymers of size i, in concentration C_i is treated by the way of the law of conservation of mass of a chemical reaction.

$$C_i \longleftrightarrow {}_iC_1 \quad (1)$$

Given cluster of size i needing a work of formation ΔG_i, the Gibbs concept of a critical nucleus is related to a maximum in ΔG_i, for a minimum size i*. ΔGi is generally a sum of two terms, one related to the supersaturation of the mother phase, the other is a surface term so that

$$\Delta G_i = -n\ \Delta\mu + S\ \Sigma j \Delta S \quad (2)$$

with $\Delta\mu = kT/Ln\ p/po$

if with Volmer we take a «spherical cap model» it is possible to define the free energy of formation of a nucleus of radius r and to find the maximum value of ΔG for

$$\Delta G^* = 16/3\ \pi\ \gamma^3\ F(\alpha)/\Delta\mu^2 \quad (3)$$

F(α) being a function of the wetting angle α of the critical nucleus of radius

$$r^* = 2\gamma/\Delta\mu \quad (4)$$

The main advantage of this approach is that many different possibilities can be reached, for instance a disc shaped nucleus can be discribed by introducing a peripherical free energy 2prΣ.

Equation (2) can be solved for a polyedral nucleus A in contact with a surface B characterized by an adhesion free energy ß Formula (2) can be rewritten as

$$\Delta G = -n\ \Delta\mu + \Sigma_{i\neq j}\ (\gamma^A{}_i S_i) + (\gamma^{AB}{}_{i=j} - \beta)\ S_{i=j} \quad (5)$$

with j the contact face with a surface area S_j, the others faces being free. γ^{AB} is our excess quantity defined by the Dupre's relation as

$$\gamma^{AB} = \gamma^A + \gamma^B - \beta \quad (6)$$

An exhaustive discussion is given by Kern in [36]leading to an excess of activation energy ΔG;

$$\Delta G = [\Sigma_{i=j}\ \gamma_i\ ki + (\gamma_{AB} - \beta_j)\ k_j]\ \Delta\mu^2 \quad (7)$$

k_i and k_j being geometrical factors. As the free energy of adhesion b is positive, the surface always appears as a catalyst for nucleation since $D\Gamma$ becomes smaller when β increases. Following this idea, it is obvious that we can introduce other factors in (2) such as a surface roughness by changing $F(\alpha)$. Chakraverty and Pound [7] have discussed this point.

Finally, as has been discussed by Hirth [7]one can introduce impurity effects on and in the surface which change α and γ's of the crystal by the way of dissolved impurities. These impurities change the absolute values of the chemical potential between the daughter and the mother phase by kT $Ln[m(1-x)]$. Where x is the concentration of impurities and m the activity coefficient of the solvant.

In this special approach, we have shown in part A that we can also define the nucleus substrate interface mismatch (Cabrera) [37].

Time independant nucleation rate

The frequency of nucleation, is the rate at which critical nuclei become supercritical, and, by growing atom by atom reach the density of observable stable particles.

In supersaturated adatom populations, the driving force required to overcome the critical size i is a stochastic bimolecular reaction. Frenkel [6] has transformed this problem into a formally equivalent process of diffusion in i space (size space) through a diffusion of a particle of size i, under a force F derived from a potential ΔG_i.

The continuity equation is then written as

$$\delta C(i,t)/\delta t = \delta J/\delta_i \quad (8)$$

$$\text{with } J = D_i \text{ grad } C(i,t) - \mu\, C(i,t)F, \quad (9)$$

$\mu = D_i/kT$ is the mobility of particle of size i, D_i the diffusion coefficient,

$$F = \delta\Delta G_i/\delta i$$

C(i,t) is the number of particles of size i at time t.

The steady state solution of equation (8) is

$$J = Z\omega C_i^*. \quad (10)$$

Z is a non-equilibrium factor ignored here

C_i^* is easily written as

$$C_i^* = C_1 \exp -DG^*i/kT \quad (11)$$

by definition from our chemical bimolecular reaction, ΔGi being the work of formation of a cluster of size i*.

We need to define w in equation (10) which is a probability factor for atomic impact. This frequency term could be explicit if we know how to describe the way in which single adatoms join the nucleus. Here we consider only the surface diffusion mechanism, neglecting direct impingement and volume diffusion. From the kinetic theory of gases, the number of atoms impinging on a unit area of surface per unit time is $n_1 = p/(2\pi\ m\ kT)^{1/2}$ where p is the partial pressure of the vapour m its molecular mass, k Boltzmann's constant and T the absolute temperature. The probability per unit time that a monomer desorbs is given by $\nu \exp(-\Delta G_{des}/kT)$ where ν is a frequency term and ΔG_{des} the activation energy of desorption of an adatom.

With our cap shaped model of radius r and wetting angle α, the probability of capture of migrating atoms by a cluster is given by

$$\omega' = 2\ \pi\ a\ r\ \sin\ n\ \exp\ (\Delta G^{*}_{diff}/kT) \qquad (12)$$

ΔG_{diff} equals the activation energy for diffusion of an adatom. Finally the rate of nucleation can be written as

$$J = K\ \exp\ (\Delta G^{*}_{des} - \Delta G_{diff} - \Delta G^{*}/kT) \qquad (13)$$

$$\text{with } K = \nu\ 2\ \pi\ r^{*}\ a\ \sin\ \alpha C_1\ (2\pi\ mkT)^{-1/2} \qquad (14)$$

If we want to discuss the previous text in terms of macroscopic parameters p/po and α we can rewrite equation (10) as

$$J = \sin\ \alpha\ \exp\ (-BF(\alpha) \qquad (15)$$

which is the Moazed [9] approach, developed in the section I.

Time dependent nucleation

It is easy to see that with the previous thermodynamical formalism equation (4) the size of a critical nucleus is very close to an atom (see table 1).

Zinsmeister decided not to retain this assumption and to describe the nucleation process as a kinetic formalism of polymerisation through bimolecular collisions.

A series of differential equations is then written for the time-dependent concentration of polymers

$$dn_1/dt = R - n_i/\tau - \Sigma_{k>1}\ n_1\ w_{i,k}\ n_k$$

$$dn_2/dt = 1/2\ w_{1\ 1}\ h_1^2 - n_1\ w_{1\ 2}\ n_2$$

$$dn_k/dt = n_1\ w_{1,k-1}\ n_{k-1} - n_1\ w_{1-k}\ n_k$$

with ni the concentration of dimers, R the flux of impinging atoms, t the residence time of monomer, w1,k a collision factor of adatoms with a k mer.

Zinsmeister makes the following assumptions:

1) only monomers impinge, desorb or move on the surface.
2) direct impingement is negligible.
3) all aggregates of size i > 2 are stable and do not decompose by any mechanism.
4) spatial homogeneity is implicit. Many solutions have been obtained by changing n_1 and $w_{1,k}$ [38][39].

substrate temperature	kT Ln p/po	r* (Angström)
400°C	2 x 10	2.46
425°C	5 x 10^5	2.7
	5 x 10^4	3.26
475°C	1.46 x 10^4	3.44
	4.2 x 10^3	3.95
533°C	725	4.64
	138	6.2

Table 1 Calculation of the radius of the critical nucleus on the basis of classical nucleation theory.

An exhaustive discussion is presented in Lewis and Anderson [40]. With such an approach, Zinsmeister has been able to describe the behaviour of the monomer concentration and the cluster density as a function of time, with the collision factor constant or size dependent.

Finally Franck and Venables [41] have introduced cluster mobility and coalescence in the simplified kinetic equations. Space does not allow us discuss problems related to coalescence phenomena leading to thin continuous films.

For the static coalescence model, Oswald ripening has been described by Chakraverty [42], as for the dynamic approach the extended, Smolukosky colloid's model has been developed by Metois [43] for aggregates on a surface. Finally the different experimental methods for in situ study of thin films by molecular beam techniques have been exposed by Honjo [44] and Venables [45] using electron microscopy. As for the surface science approach, the reader is directed to the works of Bauer [46] and Masson [47].

REFERENCES

1) Bauer E., Z. Kristall 110, 305, (1958).

2) Bauer E. and Van der Merwe J.H., Phys. Rev., B33, 3657 (1986).

3) Takagi T. Yamada I. Kundri M. and Kokiyama S., in "2d int. conf. on Ions Sources" Vienne 750 (1972).

4) Young J., in "Theorie mecanique de la chaleur" Paris,Dupre. A .(1869).

5) Volmer M, in "Kinetic der Phasen bildung" Steinkopff, Dresden (1939).

6) Frenkel J., in "Kinetic. Theory of Liquids" Dover, New York, (1946).

7) Hirth J.P. and Pound G.M., in "Condensation and evaporation", Pergamon Press London (1963).

8) Moazed K.L. and Hirth J.P., Surf. Science 3, 49 (1965).

9) Walton D. J., Chem. Phys. 37, 1282 (1962).

10) Walton D. and Rhodin T., in «Single crystal film», Pergamon Oxford (1964).

11) Rhodin T. J., Chem. Phys. 38, 2698. (1963).

12) Muller E., in "Molecular Processes on Solid surface". M. Grow Hill, N.Y. 400 (1969).

13) Erlich G. and Ayraud F.G. J., Chem. Phys. 44, 1039 (1966).

14) Basset D.W. and Parsley M.J. J., Phys. D, 707 (1970).

15) Erlich E. and Ayraud G., J., Chem. Phys. 57, 1788 (1972).

16) Graham W. and Erlich G., Phys. Rev.Lett. 31, 1047 (1973).

17) Erlich G.,in "Studies of Surf. Science and Catalysis" 4, Bourdon ed. Elsevier (1980).

18) Basset D.W., Phil. Mag.4 ,52 (1980).

19) Basset G.A., in "Condensation and Evaporation of Solids" Gordon and Breach, N.Y. 599, (1962).

20) Skofronic J.G and Phillips W.B., J. App. Phys. 38, (1967).

21) Poppa H., in "Epitaxial growt" part. II, Acad. Press (1975).

22) Honjo G. Takayanagi K. Kobayashi K and Yagi K., Jap. J. Appl. Phys. Supp. 2 part 1, 13, 357, (1974).

23) Schmeisser H. in "nucleation and growth of thin film", Acad. Press (1978).

24) Zanghi A., Thèse Marseille (1975).

25) Schmeisser H., Thin Solid film, 22, 99 (1974-5).

26) Zinsmeister G., Thin solid film, 2, 497, (1968).

27) Reiss H. J., Appl. Phys. 39, 5045 (1968).

28) Kern R., Masson A. and Metois J.J., Surf. Science 27, 483 (1971).

29) Langevin P., in Œuvres ed. CNRS 301, (1950).

30) Metois J.J, Thèse, Marseille (1974).

31) Yamada Y., Kagka H., Usui H. and Tagagi T. J., Vac. Scienc. Tech. A4 3 722, (1986).

32) Elam W, Kerstein A.R. and Rehr J.J., Phys. Rev. Lett. 52, 1516, (1984).

33) Di Cenzo S,.Berry H.S., Phys. Rev. B.38, 8465 (1989).

34) Fayet P., Thèse Lausanne (1987).

35) Cox D., personal communication (1989).

36) Kern R. J., Min Crist.4,152(1978).

37) Cabrera .N., Edit. C.N.R.S.(1968) .

38) Zinsmeister G.in "Problem in thin physics", Vacuum, 16 (1966).

39) Zinsmeister G., Thin Solid Films, 4, 497 (1968).

40) Lewis B.and Anderson J.C., in "Nucleation and growth of thin Films" Acad Press (1978).

41) Venables J., in "Epitaxial Growth" J.Acad .Press. Part.I (1975).

42) Chakraverty B.K., in "North Holland Series in Crystal Growth.I" N.Holland.(1973).

43) Metois J.J., Zanghi J.C., Erre M. and Kern R., Thin Solid Films, 22,3, 331 (1976).

44) Honjo G., Yagy K., in "Current Topics in Material Science". North Holland (1984).

45) Venables J., Spiller G.D., T.in "Nato.Asi.Series.B.Physics", 86 (1983).

46) Bauer E., in "Techniques of Metal Research", Wyley, (1974).

47) Masson A, Mariot J.M., Roulet and Dufour G., J.Phys.C.14, 2531 (1981).

Materials Science Forum Vols. 59 & 60 (1990) pp. 65-92

STATISTICAL PHYSICS OF CLEAN SURFACES: ROUGHNESS AND SURFACE MELTING

J. Villain
Département de Recherche Fondamentale - Sph/MDN
C.E.N.G., 85X, F-38041 Grenoble Cédex, France

1 ROUGHNESS OF A SURFACE AT EQUILIBRIUM : THE ROUGHENING TRANSITION

When discussing the properties of surfaces, it is rather natural to begin with clean surfaces. One of the most typical properties is *roughness.* For a review, see Van Beijeren and Nolden (1989) or Nozières (1990).

1.1 Case of a highly symmetric surface

A highly symmetric surface (111 or 001) of a cubic crystal at equilibrium at low temperature $T \ll T_M$ is "flat" (or "smooth"). That is, the heights z of the surface at any distance R are equal with probability about $1-\exp(-u_A/T) -\exp(-u_V/T)$. u_A (resp u_V) is the energy of an extra-atom at the surface (resp. a surface vacancy, *Fig. 1.1*). The length unit is the interatomic distance and the energy unit is the Kelvin, throughout this lecture. It follows

$$G(\vec{R}) \equiv \left\langle [z(\vec{r})-z(\vec{r}+\vec{R})]^2 \right\rangle \simeq 2 \exp(-u_o/T) \qquad (1.1)$$

1.2 Case of a liquid surface

The surface free energy of an incompressible liquid of N atoms is

$$\mathcal{F}_{surf} = \int\int dx\, dy \left(\frac{1}{2}gz^2 + \sigma\sqrt{1+z'^2_x+z'^2_y}\right), \qquad (1.2)$$

where g is the gravity and σ is a constant (the *surface tension*). For small z'_x and z'_y an expansion to second order and a Fourier transformation yield

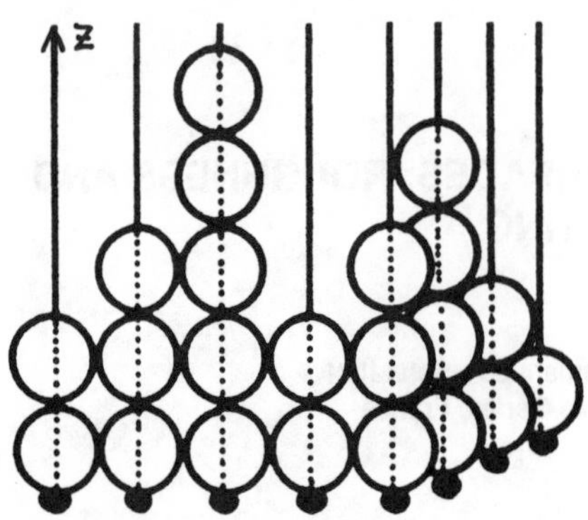

Fig. 1.1 The Solid on Solid (SOS) model is the simplest model to describe surface roughness. In this model, the atomic positions are fixed, and an atom is only allowed to be or not to be there. Each bond has an energy ϵ. For a simple cubic lattice $u_A=u_V=4\epsilon$ in eq. (1.1).

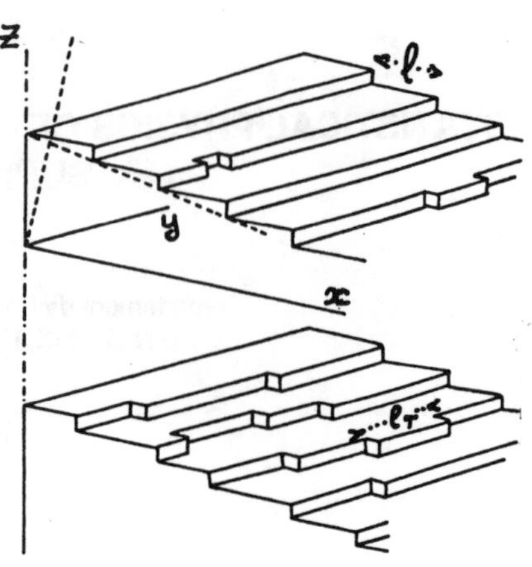

Fig. 1.2 Schematic representation of a stepped (or "vicinal") surface.

$$\mathcal{F} = \mu_o N + \sigma\Sigma + \frac{1}{2}\sum_q \left(g + \frac{\tilde{\sigma}}{2}(q_x^2 + q_y^2)\right) |z_q^2| \qquad (1.3)$$

where $\tilde{\sigma} = \sigma$. The notation $\tilde{\sigma}$ has been introduced because in a crystal, both quantities are different. N is the number of particles and Σ the area of the projection onto the xy plane. It results from Boltzmann statistics that

$$\langle |z_q^2| \rangle = \frac{T}{g + \tilde{\sigma}(q_x^2 + q_y^2)} . \qquad (1.4)$$

Hence $$\left\langle [z(\vec{r})-z(\vec{r}+\vec{R})]^2 \right\rangle = 2T \int_0^{1/a} dq_x\, dq_y \, \frac{1-\cos\,(\mathbf{q}.\mathbf{R})}{g + \tilde{\sigma}(q_x^2 + q_y^2)}$$

$$\approx \int_{1/R}^{1/a} 2\pi q dq \, \frac{2T}{g + \tilde{\sigma} q^2}$$

Neglecting gravity (and factors 2 and π!) this diverges :

$$G(\vec{R}) \equiv \left\langle [z(\vec{r})-z(\vec{r}+\vec{R})]^2 \right\rangle \approx (2\pi T/\tilde{\sigma}) \ln R \qquad (R << \sqrt{\sigma/g}) \qquad (1.5)$$

The difference with (1.1) is striking. A surface is said to be *rough* if $\left\langle [z(\vec{r})-z(\vec{r}+\vec{R})]^2 \right\rangle$ diverges as in (1.5), and *smooth* if it does not diverge as in (1.1). Is that a difference between a liquid and a solid surface? Not really. It will be seen below that a crystal surface may also be *rough*.

1.3 Vicinal surface of a crystal

"Vicinal" means (in latin) "neighbour" (of a highly symmetric orientation in the present case). For instance, the (115) face of Cu, studied by Fabre et al (1987) is a vicinal surface It can be regarded as a sequence of (001) faces separated by "ledges" (Fig. 1.2). Those ledges like to be straight. If they make a kink, it costs an energy W_o. On the other hand, ledges like to be at distance $7a/\sqrt{2}$. If they are nearer or farther, this costs an energy ϵ *per atomic distance*. W_o is of the order 1/10 of the cohesion energy, say 1000 K to 2000 K. The knowledge of ϵ is worse yet, but certainly $\epsilon \ll W_o$. We consider temperatures $T \ll W_o$. On an *isolated* ledge the average distance between two kinks would be

$$\ell_T = \exp(W_o/T). \tag{1.6}$$

Thus the degrees of freedom are not really each point of each ledge, but the points A_{mn} of ordinate $y_{mn} = n\ell_T$ on each ledge (labelled m, see Fig. 2.2, next Section). The free energy of an isolated ledge can be written as

$$\mathcal{F}_m = \sum_n T\,(x_{mn} - x_{m,n-1})^2 . \tag{1.7}$$

Actually, i) it should be a function of the nearest neighbour difference only if long-range interactions (e.g. elastic) are neglected. ii) Higher powers $(x_{mn} - x_{m,n-1})^{2p}$ can be neglected because the most probable value of $(x_{mn} - x_{m,n-1})$ are 0, 1 or -1, so that $(x_{mn} - x_{m,n-1})^{2p}$ has the same value for all integers p. iii) The factor T in (1.6) reproduces the correct correlation function of a single ledge, namely

$$\left\langle (x_{m,n+r} - x_{m,n})^2 \right\rangle = r .$$

The total free energy is now obtained by adding all free energies (1.7), and by adding to the result an interaction term which will also be assumed to be quadratic. The origin of the free energies is the free energy of the flat surface. Omitting an additive constant, one obtains

$$\mathcal{F} = \sum_{m,n} T\,(x_{mn} - x_{m,n-1})^2 + \epsilon\ell_T \sum_{m,n} (x_{mn} - x_{m-1,n})^2 \tag{1.8}$$

An important point is the factor ℓ_T which appears in order to take care of the ℓ_T points of each ledge which are near the points A_{mn} and are not explicitly taken into account in the summation. This procedure is called *rescaling*.

If $T \ll \epsilon\ell_T$, there are very few kinks because kinks at distance ℓ_T would cost a huge free energy $\epsilon\ell_T$. For $T \gg \epsilon\ell_T$ the interaction between ledges is not effective and the distance between kinks on the same step is really of order ℓ_T given by (1.6). Thus the local roughness changes in a temperature region given by

$$T/\epsilon \approx \exp(W_o/T) \tag{1.9}$$

This change in the local behaviour corresponds to a fairly steep decrease of the Bragg peak intensity (Villain et al 1985, Selke and Spilka 1986) described in Section 4. However, a change in the *local* properties does not necessarily imply the existence of a transition.

1.4 Roughening transition

As a matter of fact, there *is* a transition. A moderately complicated renormalisation group theory (Ohta and Kawasaki 1978, Wiegmann 1978) as well as exactly soluble models (van Beijeren 1977, Abraham 1983) show that there is a qualitative change in the long distance behaviour of $G(\vec{R})$. At low temperature it goes to a finite limit as in (1.1) while at high temperature it diverges logarithmically as in (1.5). At some intermediate "roughening temperature" T_R a transition occurs, which turns out to be *continuous*. The order of magnitude of the transition temperature is again given by (1.9) :

$$T_R/\epsilon \approx \exp (W_o/T_R) \quad . \tag{1.10}$$

Above T_R one can just forget the fact that the x_{mn} are integers. Then the free energy functional (1.8) looks very much (if Fourier transformed) like (1.3) except for the anisotropy. The equation analogous to (1.5) reads

$$\left\langle (x_{m'+m,n'+n} - x_{m',n'})^2 \right\rangle \approx \text{Const. ln} \left(m^2 \ell^{-2} + n^2 \ell_T^{-2}\right) \quad , \tag{1.11}$$

where ℓ is the average distance between ledges. The careful reader has noticed that in (1.5), z is the height, while x_{mn} is *not* a height. However, a fluctuation δx_{mn} of the position of the m'th ledge implies a height fluctuation $\ell^{-1} \delta x_{mn}$. Thus, height fluctuations of a crystal surface above its roughening transition are logarithmic, just as those of a liquid. Strangely, when ℓ goes to ∞, T_R goes to 0 because ϵ goes to 0. However, for $\ell=\infty$, there are no ledges and therefore no height fluctuations.

Highly symmetric faces (001) or (111) of some cubic materials (e.g. rare gases) undergo a roughening transition below the melting point, in other cases (e.g. Pb or In) they do not (Heyraud and Métois 1987). This is of crucial importance for crystal growth. In conclusion, the roughening transition separates (for a given orientation of the crystal surface) a low temperature region where the discrete nature of the lattice is essential, from a high temperature domain where the material may be treated as continuous.

1.5 Critical behaviour

It can be shown (Chui and Weeks 1976, Knops 1977) that the roughening transition has the same critical behaviour as the two-dimensional XY ferromagnet studied in a famous work of Kosterlitz and Thouless (1973). A difference, however, is that the high and low temperature behaviours are interchanged. The 2-D XY ferromagnet has the strange property to have a spin pair correlation function which vanishes like a power law, not only at the transition, but in the whole region *below* the transition. Rough surfaces have the equally unusual property to have a logarithmic height-height correlation function *above* the transition, as seen from (1.11).

This logarithmic divergence has been checked by Lapujoulade et al and by Conrad et al (1985) by atom beam diffraction. However, the height-height correlation function is not *directly* observable by diffraction. What is seen by atom diffraction (Blatter 1984, Levi 1984, Villain et al 1985) is (in some approximation) a measurement of $\langle \exp (ik(z(\vec{r}+\vec{R})-z(\vec{r}))) \rangle$ and it may be argued that this is not too badly approximated by $\exp\left[-k^2\left\langle [z(\vec{r}+\vec{R})-z(\vec{r})]^2\right\rangle/2\right]$. k is essentially the scattering vector. Inserting (1.11), the correlation function relevant for diffraction experiments is seen to be a power of the distance for a rough surface (as well as in a 2-D ferrromagnet below the transition).In both cases the exponents are known exactly at the transition. In other words, the constant in (1.11) has a universal value at the transition. This is very useful to determine the transition temperature (if one relies on theorists!... see § 1.7 below).

Other properties are not adequate to determine T_R. The specific heat has no *observable* singularity : the exponent α is $-\infty$. Also, the Bragg peak intensity vanishes at T_R with an exponent $\beta=\infty$ as a function of temperature. We shall come back in Section 3 to the experimental observation of the roughening transition.

1.6 Consequences for crystal growth

It is known that molecular beam epitaxy (MBE) can provide very smooth surfaces. We see from the above discussion that this is only possible for highly symmetric surfaces. Those can indeed be annealed at temperatures (say $2T_M/3$) where both following conditions are satisfied : i) the surface is below T_R. ii) The dynamics are reasonably fast.

Crystal growth in liquid phase does not lead to such smooth surfaces because it generates almost necessarily dislocations, which in fact play an essential part in the growth process. (Friedel 1956)

1.7 Recent developments and preroughening

As stated above, the existence of a roughening transition can be proved for certain models, in particular the *Solid on Solid* (SOS) model (Fig. 1). An essential feature of this model is that both sides of the interface have identical properties. In particular in (1.1) u_o was assumed to be the same for an extra atom or a lacking atom. This symmetry relation is satisfied for a domain wall, but not for the surface of a solid. Rys (1986) has investigated a more realistic, asymmetric model, and has found that the surface is *always smooth*. His argument is not easy to follow, but our interpretation is the following : in the rough phase (if there is one) each hill is within another (bigger) hill and also within another (deeper) valley. (This is related to the *self-similarity* property). It follows that an infinite sample contains arbitrary large hills and valleys. In Rys' model, hills and valleys have not the same energy; let hills be assumed to be preferred. Then, a "smooth" state with many little hills may be preferable to a rough state with very large hills. This smooth state will be smooth in this sense that $G(\vec{R})$ does not diverge, and therefore looks more like (1.1) than like (1.5). Such a smooth state with many little hills (and, possibly, little valleys) has been introduced by Den Nijs and Rommelse (1989). They have argued that this state is separated from the ordinary smooth state by a transition which they called *preroughening*.

According to Den Nijs (1989) the roughening transition is *not* affected by the asymmetric terms introduced by Rys. As a matter of fact, the concept of a roughening transitions seems to agree reasonably well with experiment as seen from the next sections (see subsection 4.2).

2 SURFACE THERMODYNAMICS : SURFACE TENSION AND CHEMICAL POTENTIAL

2.1 Surface tension : definitions

Definition : In the simplest cases (otherwise see Nozières 1990) the surface tension σ is the surface free energy per unit area. More precisely it can be defined by the following formula which gives the total free energy $\mathcal{F}$:

$$\mathcal{F} = \mu_o N + \iint \sigma(z'_x, z'_y)\, dS \qquad (2.1)$$

Here μ_o is a (not very interesting) constant and dS is the surface element. If the surface height $z(x,y)$ is a uniform function of the other two cartesian coordinates x,y, it is often preferable to introduce the notation

$$\varphi(z'_x, z'_y) = \sigma(z'_x, z'_y)\sqrt{1+ z'^2_x + z'^2_y} \qquad (2.2)$$

so that (2.1) is written as

$$\mathcal{F} = \mu_0 \iint z(x,y)\, dx\, dy + \iint \varphi(z'_x, z'_y)\, dxdy \qquad (2.3)$$

In these formulae, σ and φ have been assumed to be functions of the first derivatives z'_x and z'_y only. This is a rather generally accepted approximation, which is correct if the curvature is not exceedingly large.

In the case of an infinite crystal surface with a plane general orientation (chosen as z=Const) formula (1.3) can be derived from (2.1). Expanding σ and $dS/dxdy = \sqrt{1+z'^2+z'^2}$, (2.1) yields

$$\mathcal{F} = \mu_o N + \iint dxdy \left(\sigma+\sigma'_1 z'_x+\sigma'_2 z'_y+ \frac{1}{2}\sigma''_{11} z'^2_x+ \sigma''_{12} z'_x z'_y+ \frac{1}{2}\sigma''_{22} z'^2_y\right)$$

$$\left(1 + \frac{1}{2} z'^2_x + \frac{1}{2} z'^2_y\right)$$

with $\sigma'_1 = \partial\sigma/\partial x$, $\sigma''_{11} = \partial^2\sigma/\partial x^2$, etc. $\sigma''_{12} = \partial^2\sigma/\partial x\partial y$ can be set equal to 0 through an appropriate choice of the x and y directions. The integrals of z'_x and z'_y vanish since, for instance,

$$\int_{-L}^{L} z'_x(x,y)dx = z(L,y)-z(-L,y)$$

is a fluctuation term, negligible for a surface of large linear size 2L. For a surface of high symmetry (001 or 111) of a cubic crystal $\sigma''_{11} = \sigma''_{22} = \sigma''$ and one finds the following formula equivalent to (1.3).

$$\varphi(z'_x, z'_y) \simeq \tilde{\sigma}\left(z'^2_x+z'^2_y\right) , \qquad (2.4)$$

with $$\tilde{\sigma} = \sigma+\sigma''. \qquad (2.5)$$

The above calculation makes sense if second derivatives such as $\partial^2\sigma/\partial z'^2_x$ exist. They *are* expected to exist for $T>T_R$ where the discrete lattice is irrelevant. It will now be seen that they do not exist for $T<T_R$.

2.2 Singularities of the surface tension

Assume the orientation z=Const to be a highly symmetric orientation, e.g. (001), and below its roughening transition. We want to know σ or φ for a vicinal surface (Fig. 1.2) making a small angle θ with this orientation. The density of ledges per unit area of this orientation is tg θ. If γ is the free energy per unit length (called line tension) of a ledge, and if the interactions between ledges (addressed in § 2.4) are neglected, then

$$\varphi(\theta) \simeq \varphi(0) + \gamma\ |\mathrm{tg}\ \theta| \simeq \varphi(0) + \gamma\ \sqrt{z'^2_x + z'^2_y} \qquad (2.6).$$

For the sake of simplicity the line tension γ has been assumed to be isotropic. The right hand side of (2.4) has infinite second derivatives if $\vec{\nabla}z=0$. More generally if an orientation (z'_x, z'_y) of the surface is below its roughening transition the surface tension is a non-analytic function of z'_x and z'_y for this orientation. Such an orientation is called *singular*. The reason why the argument is wrong above T_R is that γ vanishes at T_R and is negative above. If a surface is not smooth, this is because its free energy is lowered if ledges are added.

To summarize, the deviations of φ resulting from weak deviations from the orientation z=0 are given by (2.4) for a singular orientation and by (2.6) above T_R. Generalisation to non-symmetric orientations (i.e. other than 001 or 111) or to a non-cubic crystal, is straightforward.

2.3 Chemical potential

Let the surface be locally modified (Fig. 2.1) so that the number of particles is changed by δN and the free energy by $\delta\mathcal{F}$. The system is assumed to be at *local equilibrium*. This means that $\delta\mathcal{F}$ is minimal within the constraint that δN is fixed. Then the quantity

$$\mu = (\delta\mathcal{F}/\delta N) \qquad (2.7)$$

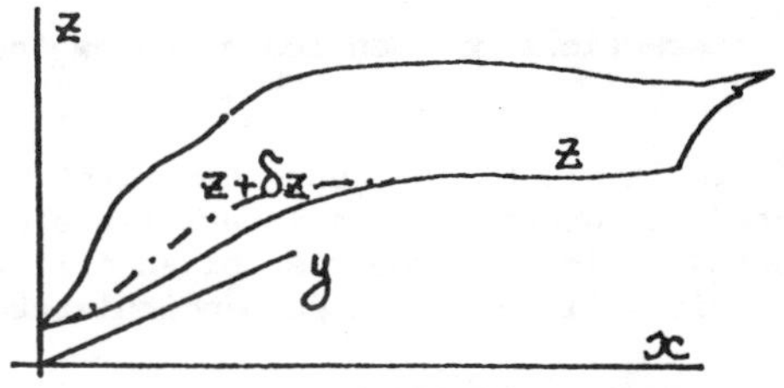

Fig. 2.1 Small deformation of a surface.

is well defined. At equilibrium it should be the same at each point of the surface since the equilibrium state is to minimize $\mathcal{F}$ (at given temperature and pressure); if μ were not uniform, then it would be lowered by removing an atom where it is high and putting it into a point where it is low. Thus, μ should be identified with the chemical potential since the latter satisfies (Landau, Lifshitz 1967) $\mu=(\partial\mathcal{F}/\partial N)_{P,T}$. μ is an important quantity because i) at equilibrium its uniformity determines the equilibrium shape. ii) away from equilibrium, the current of atoms is proportional to the gradient of chemical potential. μ will now be explicitly calculated, assuming in a first step the surface to have no singular orientation. If $z(x,y)$ is changed by $\delta z(x,y)$ the corresponding change of (2.7) is

$$\delta\mathcal{F} = \mu_0 \int\int \delta z(x,y)\ dx\ dy + \int\int \left(\frac{\partial\varphi}{\partial z'_x}\delta z'_x + \frac{\partial\varphi}{\partial z'_y}\delta z'_y\right) dx\ dy\ . \qquad (2.8)$$

Each term between brackets can be integrated by parts, provided the function φ is at least twice differentiable. For instance

$$\iint \left(\frac{\partial\varphi}{\partial z'_x}\,\delta z'_x\right) dx\,dy = \iint \left(\frac{\partial\varphi}{\partial z'_x}\,(\frac{\partial}{\partial x}\delta z)\right) dx\,dy$$

$$= \iint \frac{\partial}{\partial x}\left(\frac{\partial\varphi}{\partial z'_x}\,\delta z\right) dx\,dy - \iint \delta z\,\frac{\partial}{\partial x}\left(\frac{\partial\varphi}{\partial z'_x}\right) dx\,dy \quad .$$

We consider local variations δz for which the first term vanishes. The second term between brackets in (2.9) can be manipulated in the same way and the result is

$$\mathcal{F} = \iint \delta z(x,y) \left\{\mu_0 - \frac{\partial}{\partial x}\left(\frac{\partial\varphi}{\partial z'_x}\right) - \frac{\partial}{\partial y}\left(\frac{\partial\varphi}{\partial z'_y}\right)\right\} dx\,dy \ .$$

Since the variation δz is local the factor between brackets is constant and $\mathcal{F}$ is equal to this factor multiplied by the integral of δz, which is δN. Comparison with (2.7) yields

$$\mu = \mu_0 - \frac{\partial}{\partial x}\left(\frac{\partial\varphi}{\partial z'_x}\right) - \frac{\partial}{\partial y}\left(\frac{\partial\varphi}{\partial z'_y}\right) \quad . \tag{2.9}$$

This formula (Herring 1952, Mullins 1963) is valid when the surface has no singular orientation. The generalisation in the presence of a singular orientation has been given by Villain (1989).

2.4 Interactions between ledges below the roughening transition

Formula (2.6) is just the first term of an expansion. The next terms are expected to depend on the interactions between ledges. This interaction will be discussed below in the case of an orientation close to a singular orientation. The z axis will be chosen perpendicular to this orientation.

a) *short-range interactions*

The most obvious interaction is the impossibility for two ledges to cross themselves. At low temperatures this interaction is negligible with respect to elastic interactions. At higher temperatures this infinite contact interaction *energy* generates a *long range*, repulsive interaction *free* energy (Gruber and Mullins 1967, Pokrovskii and Talapov 1979). This free energy, as a matter of fact, is an entropy arising from the following mechanism : a single ledge has an entropy about ln 2 per length ℓ_T, i.e. $(\ln 2)/\ell_T$ per atomic length unit, if $\ell_T \approx \exp(W_o/T)$ is the average distance per kink. W_o is the energy of a kink and the number 2 means that on each length element ℓ_T the ledge has the choice between making a step to the left or to the right (Fig. 2.2). Now, if the ledge is surrounded with other ledges, this entropy is weaker. This is the origin of the interaction.

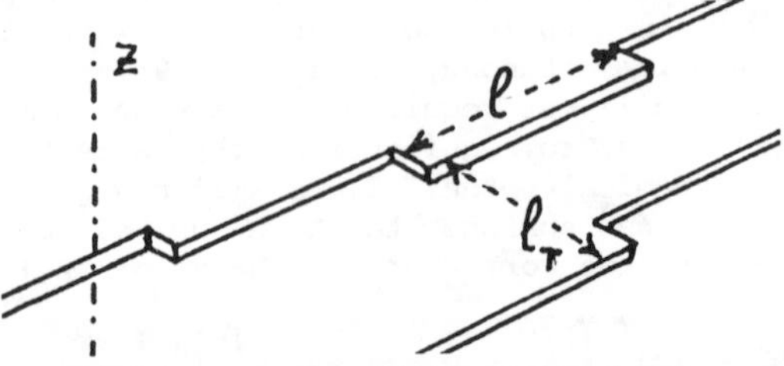

Fig. 2.2. Kinks on a ledge

It can be shown that this entropy loss is proportional, per unit length of each ledge, to the inverse square of the average distance $\ell \approx 1/\sqrt{z'^2_x+z'^2_y}$ between ledges. Since the density of ledges is $1/\ell$, the interaction term in the free energy density is

$$\varphi_{int} = G_3\ \ell^3 = G_3\ \left(z'^2_x+z'^2_y\right)^{3/2} \qquad (2.10)$$

b) *Elastic interactions.* A single ledge produces at distance $\vec{r}$ a strain $\epsilon(\vec{r})$ and the corresponding energy density is proportional to $\epsilon^2(\vec{r})$. A set of parallel ledges denoted i produces a strain $\sum_i \epsilon_i(\vec{r})$ which is the sum of the strain which would be produced by each ledge alone. This is because the equations of elasticity are linear. The total free energy is therefore the sum of the free energies of non-interacting ledges, plus the sum of pair interactions

$$W_{ij} = \int d^2r\ \epsilon_i(\vec{r})\epsilon_j(\vec{r})$$

For the sake of simplicity and because we want only a qualitative argument, the tensor nature of the strain has been disregarded. Now, the strain ϵ created by a single ledge is the gradient of the displacement $u(\vec{r})$. $u(\vec{r})$ should satisfy the equations of elasticity. A well-known solution is $u(\vec{r}) \sim \ln r$, which would have an infinite energy. Since this is not acceptable, the solution can be chosen to be the difference of two such solutions, $\ln|\vec{r}+\vec{a}| - \ln|\vec{r}-\vec{a}| \sim 1/r$. The corresponding strain is proportional to $1/r^2$ and the energy density is proportional to $1/r^4$. Then the interaction energy of two ledges of length L at distance ℓ results from the integration of an energy density of order $1/\ell^4$ integrated on a volume ℓ^2L. Finally the interaction per unit length is proportional to $1/\ell^2$ as in case (a). The problem is now : is this interaction repulsive or attractive? It is amusing to note that analogies are not very helpful in this particular case since the phonon mediated interaction between electrons (responsible for superconductivity) is attractive but the interaction between dislocations of identical sign is repulsive. The detailed calculation (Nozières 1990) indicates that the interaction between ledges of identical sign is repulsive. The case of ledges of different sign depends on the elastic constants because the displacements parallel and perpendicular to the surface act in opposite ways. In the case of a surface with an orientation close to the singular orientation z=0, the ledges have identical sign and the corresponding free energy density (obtained after multiplication by the density of ledges $1/\ell$) is again given by (2.10). However, elastic interactions are nonlocal, so that formula (2.1) is not stricly correct, though it is generally used in order to avoid extreme complications. There is no strong experimental confirmation of a $1/\ell^2$ interaction between ledges (Nozières 1990).

c) *Interactions through conduction electrons in a metal.* In a metal it may be possible to place the ledges at a distance such that an energy gap is created near the Fermi surface. Then the energy of occupied states is decreased and that of empty states is increased, so that the total energy is decreased by the presence of ledges. In contrast with (2.10) the interaction between ledges is no longer monotonous, but has a minimum for a well-defined value of the distance. This mechanism is related with a number of well-known phenomena (Jahn-Teller effect, Friedel oscillations, Kohn effect of phonons, Rudermann-Kittel-Kasuya-Yoshida interactions in magnetism).

If such interactions are present, they can produce a reconstruction of the surface. They are also important in the physics of adsorbed submonolayers. Although this topic is not within the scope of the present lecture, we give here data (table 1) summarizing work by Braun and Medvedev (1989). They provide rather convincing evidence of the presence of conduction electron-mediated interactions at the surface of W, Mo and Re. Table 1 gives a list of the superstructures which have been observed in submonolayers of various adsorbates on the (112) face of these metals. In all cases, adsorbates form rows perpendicular to the compact rows of the substrate surface (Fig. 2.2).

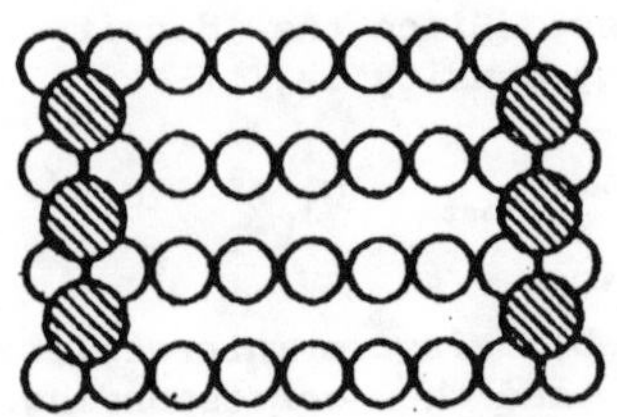

Fig. 2.3. The p(1x7) structure of a Sr submonolayer (hatched circles) adsorbed on W(112). Empty circles represent the surface atoms of the substrate.

Adsorbate *Substrate*	*W*	*Mo*	*Re*
Li	p(1x4),p(1x3),p(1x2)	p(1x4), p(1x2)	p(1x2), p(1x3/2)
Na	p(1x2)	p(1x4), p(1x2)	p(1x3),p(1x2),c(2x3)
K	c(2x2)	c(2x2)	c(2x2)
Cs	c(2x2)	c(2x2)	no data
Mg	p(1x7)	no data	p(1x3)
Sr	p(1x7)	p(1x9), p(1x5)	p(1x4)
Ba	c(2x2)	p(4x2)	c(2x2)
La	p(1x7), c(2x2)	c(2x2)	p(1x4),p(1x3),p(1x2)
Gd	p(1x7), c(2x2)	no data	
Dy	p(1x7), c(2x2)	no data	

Table 1. Submonolayer structures of various adsorbates on the (112) face of 3 metals. The frequent occurrence of the (1x7) structure (Fig. 2.3) on W testifies the existence of forces due to conduction electrons, which are oscillating functions of the distance between ledges.

3 SHAPE OF CRYSTALS

The investigation of crystal shapes is one of the major applications of the previous section. For the sake of simplicity all calculations will be done in the case when the height z of the surface depends only on the coordinate x.

3.1 **Equilibrium shape** (Wulff 1901)

For definiteness we consider a bar of a tetragonal material, the (xy) and (zy) orientations of which are below their roughening transition. The bar is parallel to the y axis (Fig. 3.1). Of course, a bar is never an equilibrium shape, but the cylindrical symmetry simplifies greatly the calculation! The system should be stable with respect to any perturbation which preserves the cylindrical symmetry. As in the case of true equilibrium, the chemical potential μ should be uniform. μ is given by the first term of the right hand side of (2.9), where φ is given according to (2.6) and (2.10) by

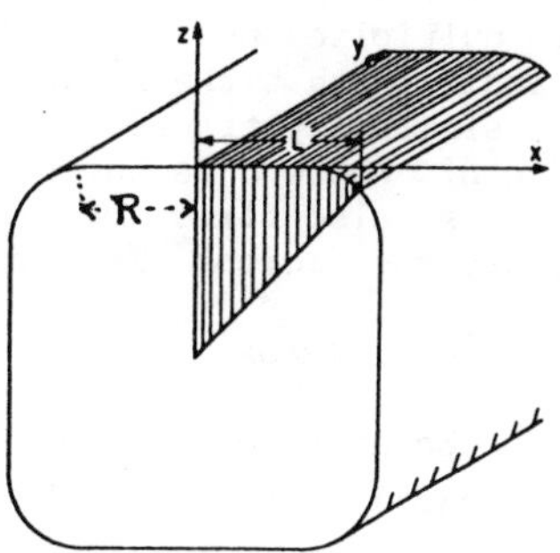

Fig. 3.1 Equilibrium shape of a bar.

$$\varphi(z') = \gamma|z'| + \frac{1}{3} G_3 |z'|^3 \quad . \qquad (3.1)$$

$z'=dz/dx$ has been assumed to be small. For finite, non-vanishing values of z' (2.9) yields the chemical potential

$$\mu = 2 G_3 |z'z''| \quad . \qquad (3.2)$$

The uninteresting constant μ_o has been set equal to 0. (3.2) should be constant. Integration yields $z'^2 \simeq \mu |\delta x|/G_3$, where δx is the distance along x to a line where $z'=0$. Introducing the analogous distance δz along z, a second integration yields

$$|\delta z| \simeq 2\mu |\delta x|^{3/2} /G_3 \qquad (3.3)$$

This singularity was noticed for the first time by Jayaprakash et al (1983) and is also present in the true equilibrium shape.

One might expect that the parts $z'=0$ are reduced to lines. This is the case above T_R, but not below (Andreev 1982) as will now be shown. The determination of the size 2R of the part $z'=0$ requires more complicated methods. (3.3) implies that z' and z'' tend respectively to 0 and to ∞ when δx goes to 0, so that (3.2) is no longer applicable. One has to minimize the free energy (2.5) with the constraint that the volume is fixed. In a region where z(x) is a uniform function of x, this implies that

$$\int dx \, [\varphi(x)+2\lambda z(x)] \qquad (3.4)$$

is stationary with respect to small deformations $\delta z(x)$ of the surface (Landau and Lifshitz 1963). λ is a Lagrange multiplier and the second term between brackets corresponds to the volume variation. This stationarity condition is the same that would be obtained by writing that the grand potential $\mathcal{F}-\mu N$ is stationary, and therefore -2λ can be identified with the chemical potential. Note, however, that in the present case the grand potential has an *extremum, but not a minimum* : a crystal smaller than a critical size decays, and a bigger crystal grows. Therefore the critical size corresponds to an unstable equilibrium. Equation (3.2) fixes the linear size p of the rounded part, which is inversely proportional to μ because (3.2) is invariant if all lengths are multiplied by a given constant and μ divided by this constant. This also implies that the shape of the rounded parts is independent of μ except for a similarity transformation. Therefore the volume of the rounded parts is $4\gamma p^2 L$, where L is the bar length. The surface free energy of the rounded part is also fixed by (3.2) and proportional to p. Let 4α be the proportionality coefficient. The free energy of the bar is

$$\mathcal{F} = 8RL\sigma_o + 4\alpha pL$$

where σ_o is the surface tension of the parts with $z'=0$ or ∞. Since $\mathcal{F}-\mu N = \mathcal{F}-\mu(4R^2L+8RpL+4\gamma p^2)$ must be stationary, one has

$$\partial(\mathcal{F}-\mu N)/\partial R = 8L(\sigma_o-\mu R-\mu p) = 0$$

Since p is proportional to $1/\mu$, $R=\sigma_o/\mu - p$ is also proportional to $1/\mu$ and has no reason to vanish at low temperature. If it vanishes at some temperature, the theory is no longer consistent above this temperature, and therefore this temperature should be the roughening temperature. Quite generally *a crystal exhibits facets parallel to the orientations which are below their roughening temperature* (Andreev 1982).

The calculation of the shape of a crystal which has not the shape of a bar is algebraically complicated. In the case of interactions of the Gruber-Mullins-Pokrovskii-Talapov type (§ 2.4 a) Akutsu et al (1988) have shown that the *gaussian curvature* (product of the two principal curvatures) is constant along a facet edge.

Heyraud and Métois (1986) tried to check the 3/2 singularity, and were moderately successful. This singularity is not a general property because electron-mediated interactions (§ 2.4 c) have been neglected. They can lead to a discontinuity of the slope at the facet edge. A similar discontinuity has been observed in Pb near the melting temperature (Heyraud and Métois 1989) and attributed (Nozières 1989) to surface melting (see section 7).

The present theory is a mean field theory. The effect of fluctuations has been investigated by Wolf and Villain (1989) and is very small.

3.2 Shape of a crystal surface out of equilibrium

A classical experiment (Mullins 1959, Yamashita et al 1981) consists of making grooves (Fig. 3.2) on a crystal surface and observing their decay. The dominant process is generally the diffusion of atoms or vacancies at the surface, and then the height z obeys the equation

$$\dot{z} = K\,\partial^2\mu/\partial t^2 \qquad (3.5)$$

where K is a transport coefficient. If no singular orientation is present, equations (3.5), (2.9) and (2.6) yield the linear equation

$$\dot{z} = -K\tilde{\sigma}\,\partial^4 z/\partial x^4 \qquad (3.6)$$

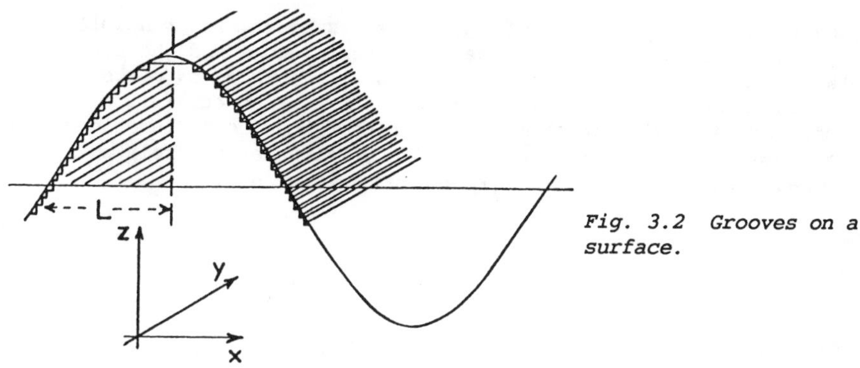

Fig. 3.2 Grooves on a surface.

which has solutions that are sinusoidal in space and exponential in time. If the average orientation is singular, (3.5) and (3.2) yield a non-linear equation. Note that the first term at the right hand side of (3.1) should be omitted when applying (2.9). The reason (apart from the fact that it would yield an unpleasant delta-function $\partial^2 |x|/\partial x^2$) is the following : the first term of (3.1) is the contribution of non-interacting ledges. The corresponding part μ_1 of the chemical potential is the ratio of the increases $\delta\mathcal{F}$ and δN in free energy and number of particles resulting from a displacement of ledges. But, for non-interacting ledges $\delta\mathcal{F}=0$. Now, since the chemical potential μ should be strictly positive and finite at the top of the profile, it follows that the top of the profile has a 3/2 singularity as in (3.3), but *without facet*. The experimental observation of facets (Yamashita et al 1981, Bonzel et al 1984) is presumably due to a miscut surface (Villain and Lançon 1989, Lançon and Villain 1989).

4 EXPERIMENTAL OBSERVATION OF THE ROUGHENING TRANSITION

4.1 Microscopic scales : diffraction

The roughness on a scale of about 100 Å can be observed by diffraction of particles which do not penetrate into the material. These particles can be

a) Atoms (e.g. He)
b) Electrons
c) X-rays (or, in principle, neutrons) at grazing incidence.

Note that X-rays (in contrast with visible light) undergo total reflection at low angle when they come from *outside* the material. The advantage of X-rays is the availability of powerful, tunable sources (synchrotron radiation) and an easy theoretical interpretation of the data because, at the atomic level, the scattering process is correctly described by the Born approximation (Robinson et al 1989, Held et al 1987, Mochrie 1987). Electron diffraction is, in our view, less appropriate than atom diffraction because electrons penetrate deeper into matter and the theory is more complicated. For the theory of atom diffraction, see e.g. Armand and Manson (1988).

In all cases, experiments yield a spectrum which describes the interference between beams diffracted by various parts of the surface. The spectrum of a smooth surface ($T<T_R$) shows Bragg peaks, i.e. delta-functions of the projection

$\vec{q}^{\perp}$ of the scattering vector $\vec{q}$. These Bragg peaks disappear, in principle, at T_R (Fig. 4.1). Far above T_R there are no coherent interferences and the spectrum is smooth. In the vicinity of T_R Bragg singularities are replaced by peaks (Villain et al 1985, Conrad et al 1986, Fabre et al 1987, Salanon et al 1988)) at the same location, but described by power laws rather than delta functions (Fig. 4.2). It is however difficult to detect any singularity related to the transition, because the singularities are very weak (Fig. 4.1).

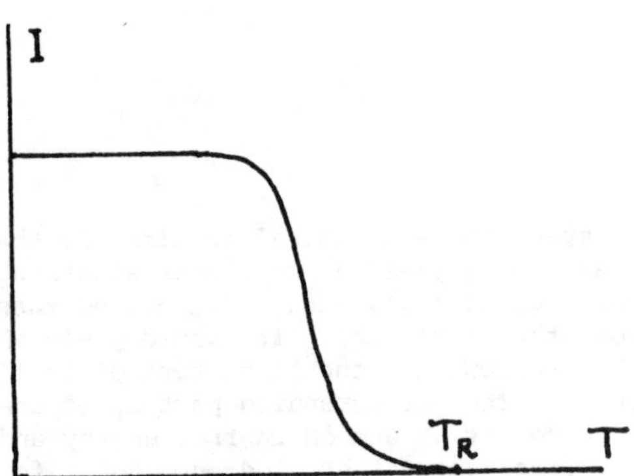

Fig. 4.1 Theoretical variation of the intensity of a typical Bragg peak with temperature.

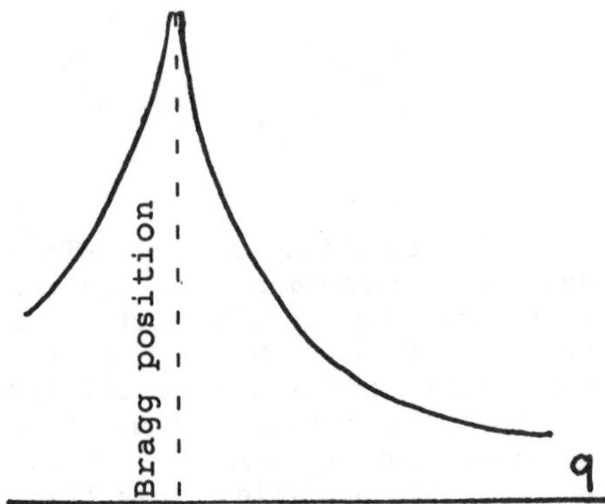

Fig. 4.2 Scattered intensity as a function of the scattering vector q at or slightly above T_R.

However, the experimentator who wants to determine the roughening temperature T_R can do it, if he has full confidence in the theory, because the exponents have universal values at T_R. Then the temperature where the observed scattering function (Fig. 4.1) is represented by the required exponent is T_R. Is the confidence in theory justified ? Mathematically, yes, but the calculations rely on models which may be questioned, as seen in Section 1.8.

Diffraction methods are mainly able to test the existence of a *local* roughness. Local roughening occurs well below T_R , as seen from Fig. 4.2. However, this local property is the one of practical interest, e.g. for crystal growth.

An experimental observation of the transition is generally to be expected from macroscopic methods like those described in the next subsection.

4.2 Equilibrium shape of single crystals

As seen in Section 3, the equilibrium shape of a crystal indicates whether any given orientation is below its roughening temperature (a facet parallel to this orientation exists) or above it (no facet). This method has been successfully applied by Heyraud and Métois (1987) to Pb and In.

Moreover the equilibrium shape of a crystal is a *measurement of the surface tension* σ through relations (2.9) and (2.2), since the chemical potential μ is the same on the whole surface and can be measured. The easiest system is solid ^{4}He in equilibrium with its superfluid, because the superfluid has a very high thermal conductivity and a weak latent heat. In addition, ^{4}He can be obtained at very high purity. Facets have been observed (Gallet et al 1987). In ^{3}He, they were not observed at equilibrium although, in principle, they should be (Rolley 1989). The reason is presumably that the ledge free energy is extremely low.

The equilibrium shape of small metal crystals in contact with their vapour has been observed by *in situ* electron microscopy (Heyraud and Métois 1987-8). One of the hardest problems to solve is metastability.

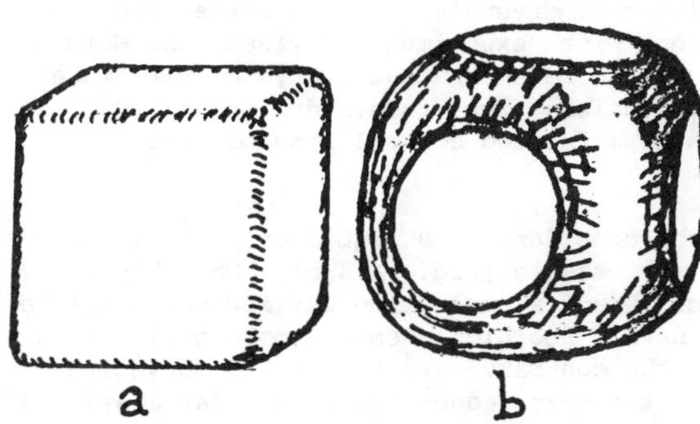

Fig. 4.3 a) A crystal grown at low temperature. Facets are too large and curved parts too small at higher temperature. b) A crystal prepared by cooling of a liquid droplet. In case (a) equilibrium cannot be reached because additional terraces can grow only through a slow nucleation process, which is absent in case (b).

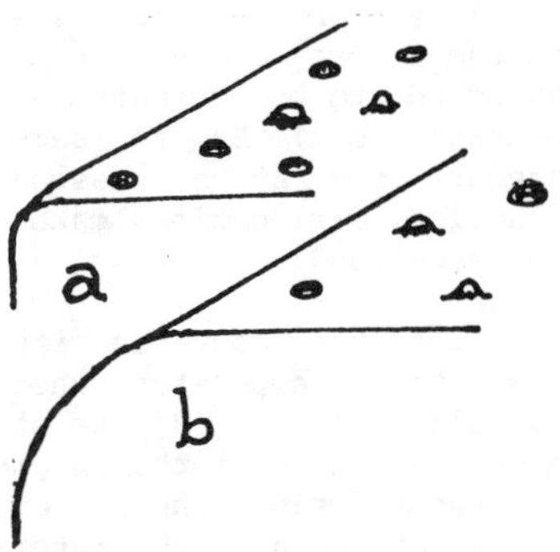

Fig. 4.4 The density of small surface defects is lower near a strongly curved region (a) than near a weakly curved part of the surface (b).

4.3 Metastability of crystal shapes (Burton et al 1951)

In this subsection the metastability of a crystal *free from dislocations* will be discussed. Consider first a crystal grown at low temperature (Fig. 4.3 a) and then heated at a higher temperature. The facets are too large, therefore the rounded parts are too small. If the total mass is fixed, the decay of facets requires the formation of additional terraces. If μ is the chemical potential (which is uniform at the surface of a metastable crystal) the grand potential of an additional terrace of radius ρ is

$$W(\rho) = 2\pi\gamma\rho - \pi\mu\rho^2 \tag{4.1}$$

where γ is the line tension of ledges. $W(\rho)$ is an increasing function of ρ if ρ is smaller than the nucleation radius

$$\rho_{nucl} = \gamma/\mu \tag{4.2}$$

The free energy barrier to overcome is

$$W_{nucl} = 2\pi\gamma\rho_{nucl} - \pi\mu\rho_{nucl}{}^2 = \pi\gamma^2/\mu \tag{4.3}$$

The problem is now : what is μ ? The answer is that the chemical potential should be consistent with the largest curvatures available on the crystal. Indeed, the chemical potential can be regarded as just a convenient measure of the density of extra atoms and vacancies at the surface (Fig. 4.4). In the vicinity of a highly curved region this density is high, and the extra excitations (atoms and vacancies) migrate into the rest of the crystal until the chemical potential is equal everywhere. In the case of Fig. 4.3 a one can write, according to (2.10),

$$\mu = G_3/R_1 \qquad (4.4)$$

where R_1 is the radius of curvature of the curved parts. Insertion of (4.4) into (4.3) yields (if for instance $G_3 \approx \gamma$) a large nucleation barrier for small terraces (say 50 Å). Thus, dislocation-free crystals with too large facets are expected to be metastable in agreement with experiment (Heyraud and Métois, private com.). By the same argument it can be shown that a crystal with facets cannot grow at a reasonable speed (unless the facets disappear). But, in practice, dislocations make the growth of a faceted crystal possible (see, e.g., Nozières 1990).

Let the converse situation be now considered : a liquid droplet is cooled down to a temperature where facets are stable (Fig. 4.3 b). What happens ? Formulae (4.1 to 3) are still applicable and a nucleation barrier can still be expected. However the regions which have a too high chemical potential are now the facets, since the curved parts, on the contrary, are too weakly curved. On a facet of radius R_f the interaction with the next ledges (which are far away) can be ignored and formula (2.7) yields

$$\mu = 2\pi\delta R_f/2\pi R_f\delta R_f = \gamma/R_f \qquad (4.5) \quad .$$

Now, insertion into (4.1) yields a nucleation radius equal to the facet radius, which means that the facet can decay without nucleation. Thus the equilibrium shape can be reached by cooling, not by heating. This is in agreement with experiment (Heyraud and Métois, private com.).

5 DYNAMICS (see Mullins 1963)

If a surface in equilibrium with its vapour has mountains and valleys as in Fig. 3.2, the equilibrium shape can only be reached if atoms go from mountains to valleys. 3 ways are possible.

5.1 Transport through the vapour

The work necessary to vaporize an atom at low temperature is by definition the cohesive energy E_o which, in transition metals for instance, is about 4 eV or 70 000 Kelvins. A given atom will vaporize on the average until a time of order exp (E_o/T) times a microscopic time τ_o which cannot be smaller than 10^{-13} seconds (the minimal time in solid state physics). Thus the air way is generally unfavourable. However, close to the melting temperature, it may be preferable for long distances, since at the low density of the vapour phase, the atoms are not scattered. Then the chemical potential μ_o of the vapour may be regarded as uniform and the height z of the surface obeys the equation

$$\dot{z}(x,y) = -B(\mu(x,y)-\mu_o) \quad . \tag{5.1}$$

where $B \approx (1/\tau_o)\exp(-E_o/T)$ and μ is the chemical potential at the surface, assumed to have μ_o as an average value so that the crystal does not grow or vaporize. If the surface has a sinusoidal profile (Fig. 3.2) without singular orientation, it follows from (2.7), (2.9) and (2.6) that (5.1) takes the form

$$\dot{z} = Bz'' \quad , \tag{5.2}$$

with $z''=\partial^2 z/\partial x^2$. Thus the decay is sinusoidal in space, exponential in time and the relaxation time τ is proportional to the square of the wavelength λ :

$$\tau \sim \lambda^2 \tag{5.3}$$

5.2 Surface diffusion

Point defects at the surface have a relatively weak energy in comparison with the cohesive energy E_o , and their mobility is rather high, so that the terrestrial way is the fastest one for short lengths. The equations have been written in Subsection 3.2 and it follows that (5.3) is replaced by

$$\tau \sim \lambda^4 \tag{5.4}$$

5.3 Volume diffusion

The problem is more difficult since one has to solve the diffusion equation in a three-dimensional, but semi-infinite medium. Qualitatively, the decay of the profile of Fig. 3.2 can be discussed as follows. If the height is h, the chemical potential variation is of order h/λ^2 and its gradient is of order h/λ^3 as well as the current density. Until now, everything is as in the case 5.2. However the flux from a maximum to a minimum is now obtained by multiplying by a surface of order $L\lambda$ (instead of a length L) where L is the length along the y direction. This flux is distributed (as in case 5.2) on an area $L\lambda$, so that $\dot{h} \approx -h/\lambda^3$ and (Mullins 1959)

$$\tau \sim \lambda^3 \tag{5.5}$$

Thus diffusion through the bulk can dominate surface diffusion over long distances at sufficiently high temperature, although defects have a higher energy and a lower mobility in the bulk.

5.4 Kinetics of a solid-liquid interface

At a solid-liquid interface, conservation of matter is important in the presence of impurities, and can give rise to instabilities in crystal growth. Similar instabilities can also arise in a pure system, as a consequence of energy conservation. They will be considered in the next section.

6 ROUGHNESS OF A GROWING SOLID

In the first 4 chapters we have described the roughness of a crystal surface at equilibrium. This roughness is due to thermal fluctuations. In this chapter a completely different type of roughness will be studied : roughness due to growth. The crystal is now far from equilibrium. Even after the end of the growth process it can remain practically indefinitely in its non-equilibrium state. Thus, this kind of roughness should be avoided when growing a crystal! We first describe a macroscopic model appropriate for crystal growth from the melt.

6.1 The Mullins-Sekerka instability

When growing a crystal one tries to reach a stationary regime. An appropriate device is a plate (Fig. 6.1) pulled at velocity V along the z direction between two ovens at temperatures T_1 and T_2 ($T_1<T_M<T_2$, T_M = melting temperature) and locations z_1 and $z_2=z_1+d$. One wishes of course to have a plane interface in order to avoid inhomogeneities of the impurity concentrations.

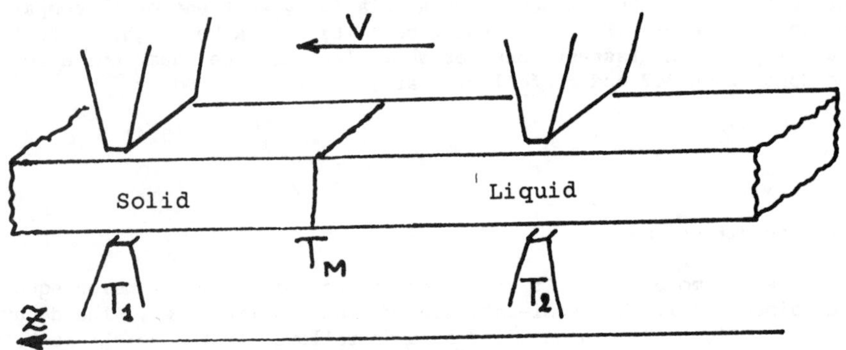

Fig. 6.1 Pulling a crystal.

It has been known for a long time (Mullins and Sekerka 1963) that *the interface can be plane only if the velocity V is lower than some threshold* V_{MS} which depends on the temperature gradient $(T_2-T_1)/d$. For $V>V_{MS}$ the surface forms wiggles (Fig. 6.2). This is called the *Mullins-Sekerka instability* (Langer 1980, Pelcé 1989, Caroli 1990, Pomeau 1990).

6.2 The mechanism of the Mullins-Sekerka instability in a pure material

The instability arises essentially from the evacuation of the latent heat which proceeds by diffusion because of energy conservation, and is therefore slow. In practice the instability is generally due to impurities which diffuse more slowly yet. The simplest method to investigate the stability of a plane interface is to assume that it is slightly distorted (Fig. 6.3) and to look whether the distortion will decay or be amplified. In the former case the plane interface is stable, in the latter case it is unstable. Now, the velocity of the interface is expected to be proportional to the local temperature gradient :

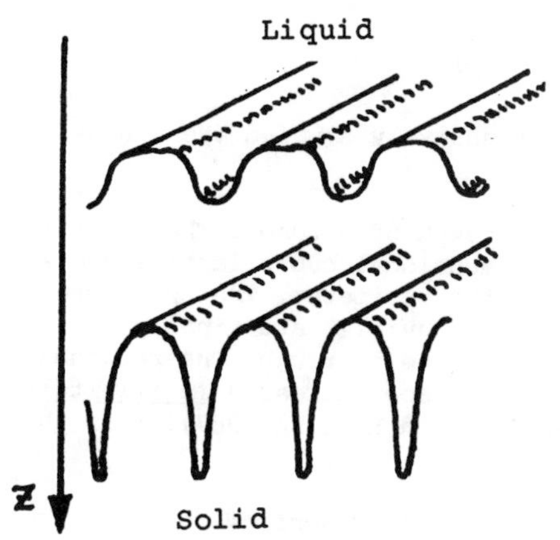

Fig. 6.2. Various types of solid-liquid interface shapes observed in the Mullins Sekerka instability.

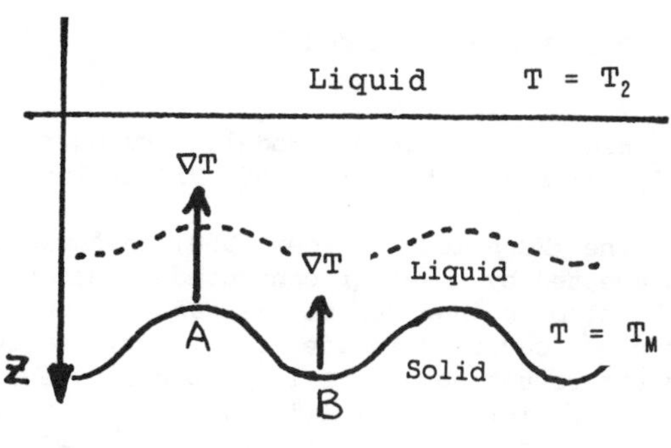

Fig. 6.3 Mechanism of the Mullins-Sekerka instability. If a wavy shape is given to the solid-liquid interface (lower solid line) the temperature gradient is stronger at the tips, which therefore grow faster, making the interface more wavy. The dashed line shows an intermediate isotherm surface.

$$V \sim \partial T/\partial z \tag{6.1}$$

Now, the isotherm surfaces should be approximately planes near $z=z_2$. On the surface, however, one has approximately

$$T_{surface} = T_M \tag{6.2}$$

so that the temperature is (approximately, as seen below) constant on the surface. It follows (Fig. 6.3) that the temperature gradient is stronger at the tips (A) than in the valleys (B) so that the tips proceed with a higher velocity and the plane interface is unstable. However, the above argument cannot be completely correct since the surface tension appears nowhere and should clearly have a stabilizing effect. Actually, equation (6.2) should be replaced by the so-called Gibbs-Thomson equation which expresses this stabilizing effect. Equation (6.1) should also be replaced by a more complicated equation (the Stefan equation) but this plays no essential role in the argument. The detailed calculation shows that the instability does not occur if the velocity is low enough, and this can be understood since then the energy has enough time to diffuse forward into the liquid, instead of being left behind.

The effect of concentration in a binary mixture or in an impure system can easily be included in the theory which has been briefly outlined above. On the other hand, the crystal nature of the solid is generally not taken into account.

Other types of instabilities can arise. For instance growth in a supersaturated solution generally gives rise to dendrites. Although a considerable amount of work has been devoted to the growth of needle crystals (Langer 1980, Pelcé 1989, Pomeau 1990) these do not really deserve the name of dendrites. The growth of snowflakes, which really look like trees (in greek δενδρον) is not yet precisely understood. Finally, a dynamical treatment of the roughening transition (Nozières and Gallet 1987), though quite different, should be mentioned.

6.3 Microscopic models

Microscopic growth models are currently extensively studied (Meakin et al 1987, Kim and Kosterlitz 1989, Viscek 1989).

The Eden model (Eden 1958) describes the growth of a tumor. The tumor is represented by a set of connected points on a d-dimensional cubic lattice and at each step a new point is added randomly at the perimeter. The "perimeter" includes cavities inside the tumor, so that these cavities are finally filled and the resulting object is compact. However the surface is rough. The roughness δh of an infinite, initially smooth, infinite surface increases indefinitely with time as a power h^{β} (Kardar et al 1986, Wolf and Kertesz 1987). h is the average thickness (proportional to time).

Diffusion limited aggregation (Meakin et al 1987) describes for instance metal deposition at the cathode in electrolysis. If M is the deposited mass and R the radius of the deposit, the average density M/R^d goes to 0 as $R^{-(d-D)}$ for large R. This is because the cations have to diffuse to the surface through a random walk, and have a very weak probability to reach the centre before being caught by the external branches. D is a constant ("fractal dimension").

If one wants to describe crystal growth from the vapour, models of *ballistic deposition*, in which the atoms follow a straight line instead of a random walk, are more appropriate (Edwards and Wilkinson 1982, Kariotis 1989, Ramanlal and Sander 1985, Meakin et al 1986, Kim and Kosterlitz 1989, Wolf 1989). However, these models disregard or underestimate the effect of diffusion in Molecular Beam Epitaxy. This effect is generally investigated in the annealing phase and without growth (Villain 1986). Note that ballistic growth models are believed to have the same roughness exponent as the Eden model (Kim and Kosterlitz 1989). The growth of amorphous materials or bad crystals by sputtering has been addressed by Bruinsma (1990).

7 SURFACE MELTING

At the surface of a solid, the atoms are expected to be less tightly bound than in the bulk. Thus one can wonder whether they can melt at lower temperature than the bulk. This phenomenon will be seen to be really predicted theoretically (Lipowski 1984) and observed experimentally (Frenken et al 1988, Bienfait et al 1988) and is called *surface melting*. For a review, see Dash (1989).

The concept of surface melting is only meaningful if there is a liquid layer between the solid and the vapour (Fig. 7.1) *and* if the thickness of this liquid layer is much larger than that of the solid-liquid interface.

A simple theoretical description will now be given in the framework of the Landau theory of phase transitions. This theory is based on a free energy functional of the form

$$\mathcal{F} = \int\int\int_{\text{Volume}} [\Phi(m(\vec{r}))+(\nabla m(\vec{r}))^2 - Hm(\vec{r})]d^3r - \int\int H_1 m(\vec{r})d^2r \qquad (7.1)$$

where the function Φ is represented (below the critical temperature) by Fig. 7.2. The "order parameter" $m(\vec{r})$ is related to the local density $\rho(\vec{r})$ by

$$m(\vec{r}) = \rho(\vec{r}) - \frac{1}{2}(\rho_L+\rho_S). \qquad (7.2)$$

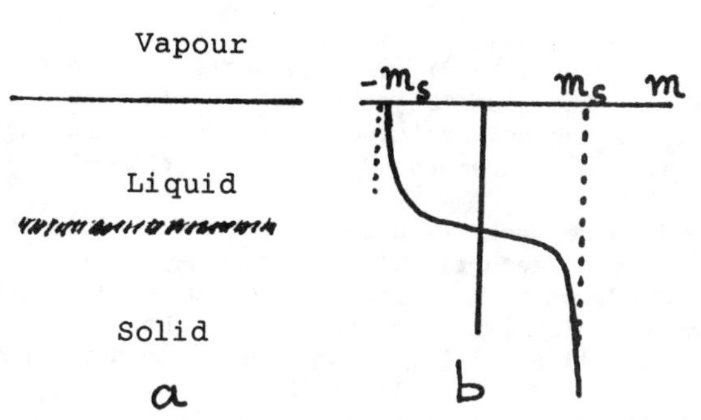

Fig. 7.1 a) Surface melting is characterized by a liquid layer separating the bulk solid from the surface. b) corresponding variation of the density with depth.

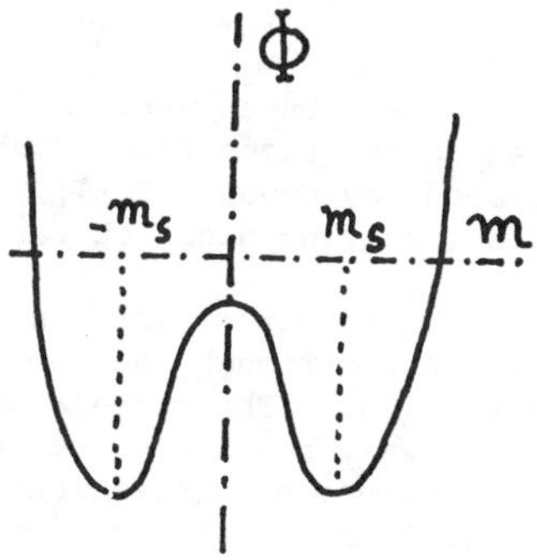

Fig. 7.2 The local free energy density $\Phi(m)$.

ρ_L and ρ_S are respectively the densities of the liquid and the solid phase. The reader may find it helpful to forget the solid-liquid transition and to think of an Ising ferromagnet below its critical point. m is now the local magnetization and H is the magnetic field. Both systems can be related through the lattice gas model. We want H_1 to favor the liquid phase, generally of lower density; then H_1 should be negative according to (7.2). The solid phase is stable when H is positive. The magnetic analogue of surface melting is a surface layer with negative magnetization separating the positive bulk from the surface.

Surface melting can actually be expected from the present model as will now be argued in the mean field approximation. In the volume, minimisation of (7.1) with respect to $m(\vec{r})$ yields

$$\nabla^2 m(r) = \Phi'(m(\vec{r})) - H \quad , \qquad (7.3)$$

where $\Phi'(m)=d\Phi/dm$.

(7.3) has of course a uniform solution (m=Const) but this solution is not expected to minimize (7.3) if $-H_1$ is much larger than H. The right solution has obviously the form $m(\vec{r})=f_o(z)$, where z is the direction perpendicular to the surface and f_o is approximately the profile of a Bloch wall. "approximately", because a Bloch wall generally corresponds to H=0. Neglecting this detail, the free energy, in the limit $-H_1 \gg H$, is found to be approximately,

$$\mathcal{F} \simeq W_w + H\ell - H_1 \, f_o(-\ell) \quad , \qquad (7.4)$$

where W_w is the free energy of a domain wall. For $-H_1 \gg H$, ℓ is large and it is sufficient to know the asymptotic behaviour of $f_o(\ell)$. As guessed from

Fig. 7.2, $\Phi(m) \simeq \frac{1}{2}\kappa^2(m+m_s)^2$, where κ is a constant and $m_s=f(\infty)$. (7.3) reads $m'' \simeq \kappa^2(m+m_s)-H$. The solution is $m(z) = f_o(-z) \simeq -m_s + \kappa^{-2}H + \text{Const} \times e^{-\kappa z}$. After insertion into (7.4) and minimisation with respect to ℓ one obtains $H/|H_1| \approx \kappa \exp(-\kappa\ell)$, i.e.

$$\ell \sim \ln(|H_1|/H) \tag{7.5}$$

ℓ is seen to vary *continuously* (Lipowski 1982). Thus, a surface can make a first order transition continuous! In the case of melting, $|H_1|/H$ should be replaced by Const/ (T_M-T), where T_M is the bulk melting temperature. Note that surface melting does not take place if the wall energy W_w is larger than $|H_1|m_s$.

The result (7.5) is in good agreement (Van der Veen et al 1990) with experiment although the above theory is a mean field one, which neglects local fluctuations. The correction is indeed weak in three dimensions because the domain wall roughness is only logarithmic according to (1.5). However, for large values of the thickness of the liquid film, long-range, Van der Waals interactions should be taken into account. Then the logarithmic result (7.5) is replaced by a power law $\ell \sim H^{-1/3}$ (if retardation effects are neglected).

Surface melting will not be discussed further since it can be in many respects considered as a particular case of the topic discussed in the next Section.

8 WETTING

Two reviews (with different and complementary points of view) have been written by DE GENNES (1985) and DIETRICH (1989).

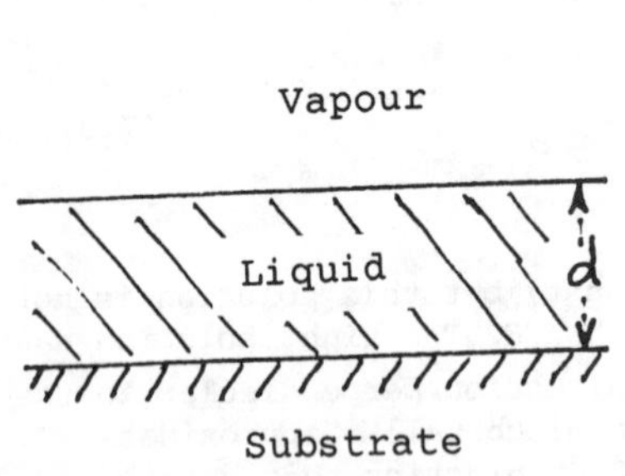

Fig. 8.1 An adsorbed liquid in equilibrium with its vapour. Note the analogy with Fig. 7.1 (surface melting). In surface melting as in the wetting transition the thickness of the intermediate liquid layer goes to infinity.

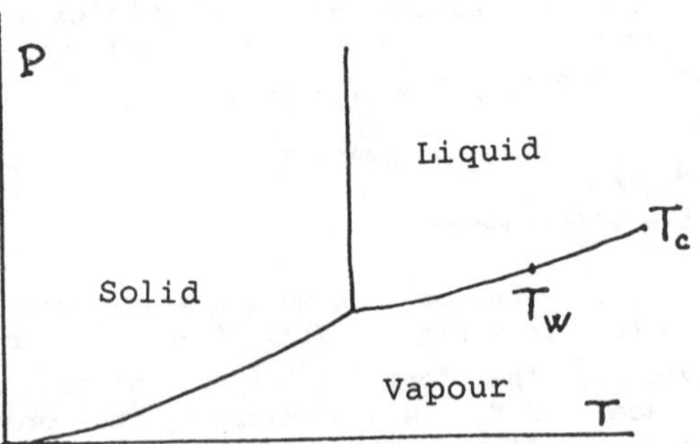

Fig. 8.2 A typical phase diagram of a material. Complete wetting occurs on the coexistence line between T_c and the wetting temperature T_W which depends on the substrate.

8.1 Definition

If a solid substrate is in contact with a saturating vapour, it usually happens that a part of the vapour condenses and forms a liquid film adsorbed by the substrate (Fig. 8.1). Changing temperature may increase the thickness to infinity. This new situation is called complete wetting and the transition to complete wetting is called *wetting transition.*

At equilibrium, this transition can only occur on the gas-liquid coexistence line of the phase diagram (Fig. 8.2) or on the gas-solid coexistence line. In the latter case the situation is richer since transitions corresponding to the formation of 1, 2, 3.... monolayers can sometimes be observed (Fig. 8.3).

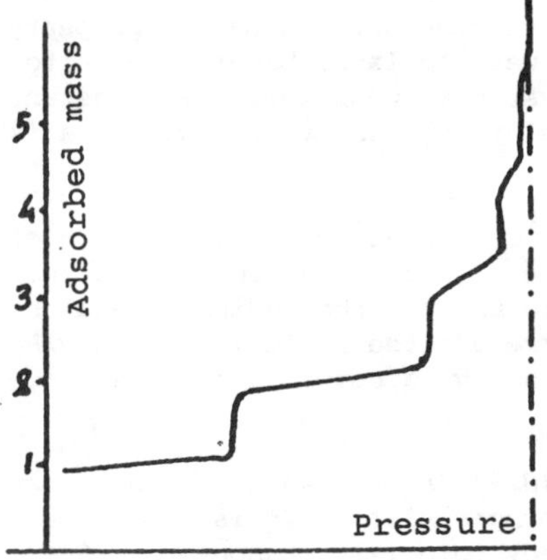

Fig. 8.3 *Krypton mass adsorbed on graphite at fixed temperature as a function of pressure (from Bouchdoug et al 1986)*

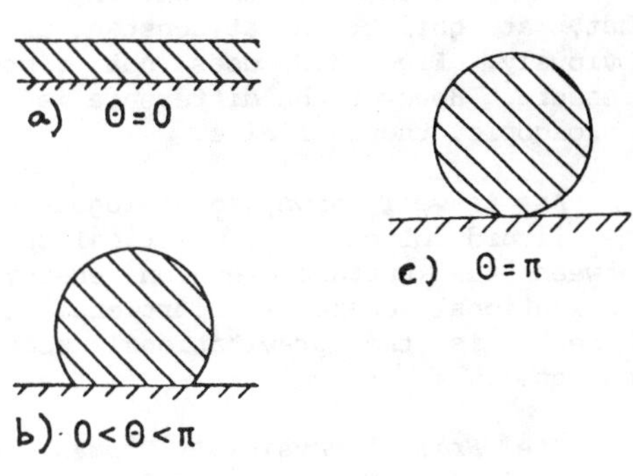

Fig. 8.4 *a) Complete wetting. The contact angle is $\theta=0$. b) A bubble with $0<\theta<\pi$. c) The case $\theta=\pi$.*

Wetting can be related to the substrate-liquid surface tension $\sigma_{s\ell}$, the substrate-vapour surface tension σ_{sg} and the vapour-liquid surface tension $\sigma_{g\ell}$. Complete wetting is only possible if the free energy $\sigma_{s\ell}+\sigma_{g\ell}$ with the liquid film is lower than σ_{sg}.

$$\sigma_{g\ell} < \sigma_{sg} - \sigma_{s\ell} \tag{8.1}$$

We have used the language appropriate to wetting by a fluid. Relation (8.1) also holds if the letter "ℓ" designates a solid.

Incomplete wetting can only be discussed if the interaction energy between both interfaces (Fig. 8.1) is taken into account. This will be done later. We now wish to relate the above definition of wetting to what is called "wetting" by the man is the street. The man in the street, not the man in a swimming pool! The latter is an uninteresting case since he is forced to get wet. The interesting situation is when wetting can occur or not.

8.2 The contact angle

If it rains, the man in the street does or does not get wet, depending on whether he wears waterproof clothes. A waterproof cloth is a substrate where water droplets remain as spherical as possible. This means that the contact angle θ (Fig. 8.4) should be as large as possible (i.e. as close to π as possible). θ was calculated by Young, who observed that the free energy variation if the contact line is slightly displaced by δx should vanish. The free energy variation is $\sigma_{s\ell}\delta x + \sigma_{g\ell}\delta x \cos\theta - \sigma_{sg}\delta x$ per unit length. Equating this to 0, one obtains Young's relation

$$\cos\theta = (\sigma_{sg} - \sigma_{s\ell}) / \sigma_{g\ell} \tag{8.2}$$

When relation (8.1) holds, any droplet is unstable : it spreads until it forms a microscopic film, the thickness of which is determined by the interaction between interfaces.

In Section 8.1, the wetting transition has been defined by the property that, at equilibrium at constant pressure, the wetting layer becomes infinite. Obviously, Fig. 8.3 does not represent a system at equilibrium at constant pressure. However the difference is not essential if the layer of Fig. 8.4 a is macroscopic, though finite.

As is well known, an analogous argument yields the ascensional height h of the liquid column in a vertical capillary tube. It results from the balance between the surface tension energy (proportional to the radius R) and the gravitational energy (proportional to the square of the radius). Thus $h \sim \sigma/Rg$ where g is the gravitational acceleration and σ a combination of surface tensions).

The British physicist Thomas Young (1773-1829) is well-known for his work on elasticity (Young's modulus) and optics (Young's slits). It is not so well known that he was also an outstanding egyptologist (Lacouture 1989). Jean-François Champollion was his younger rival before he supplanted him.

8.3 Relation between wetting and the critical point

In § 8.1 it was said that a wetting transition *can* occur when temperature T is changed. Must it occur? And does it occur when raising or lowering the temperature? A simple answer was given by Cahn (1977). Cahn's argument may be formulated as follows. Both sides of inequality (8.1) vanish at the critical point T_c (Fig. 8.2) but with *different* critical exponents. It turns out that the left hand side vanishes more rapidly. Therefore *there is always complete wetting on the coexistence line near T_c*. It may disappear at some *wetting temperature* $T_w < T_c$.

This statement can be justified in some detail. Let the right hand side of (1) be considered first. The fluid-substrate surface energy is mainly due to bonds between the substrate and the fluids, and to broken bonds in the fluid. The density of both is proportional to the density of matter. Thus $\sigma_{sg} - \sigma_{s\ell}$ should be proportional to the difference of the densities, $\rho_\ell - \rho_g$. Far from the surface, this is proportional to δT^β, with $\beta \approx 1/3$. However, at a surface, it is proportional to δT^{β_1}, where $\beta_1 \approx 0.8$ (Diehl 1986, Dietrich 1987). Now, the left hand side of (1) is analogous to the free energy of a domain wall per unit area in magnetism. This must be proportional to λf, where f is the typical energy density and λ is the domain wall thickness. The specific heat $C=(d/dT)(df/d\beta)$ is proportional to $\delta T^{-\alpha}$, therefore $f \sim \delta T^{2-\alpha}$. λ is proportional to $\delta T^{-\nu}$ (Widom 1985). It follows $\sigma_{\ell g} \sim \delta T^{2-\alpha-\nu} = \delta T^{(d-1)\nu} \approx \delta T^{1.3}$ (Jasnow 1984).

8.4 Thickness of a wetting layer and the Antonov rule

Coming back to Fig. 1, one can say that the double layer has an effective surface tension

$$\sigma_{sg}(d) = \sigma_{s\ell} + \sigma_{\ell g} + V(d) \quad , \qquad (8.3)$$

where V(d) is expected to vanish at ∞ since both interfaces do not "feel" each other. If V(d) is minimum for $d=\infty$ (Fig. 8.5 a) the surface is completely wet. If it is not completely wet, V(d) should therefore have a negative, absolute minimum at d_o (Fig. 8.5 b). d_o may be zero or not. At equilibrium with the saturating vapor, $d=d_o$ and therefore $\sigma_{sg} = \sigma_{sg}(d_o)$.

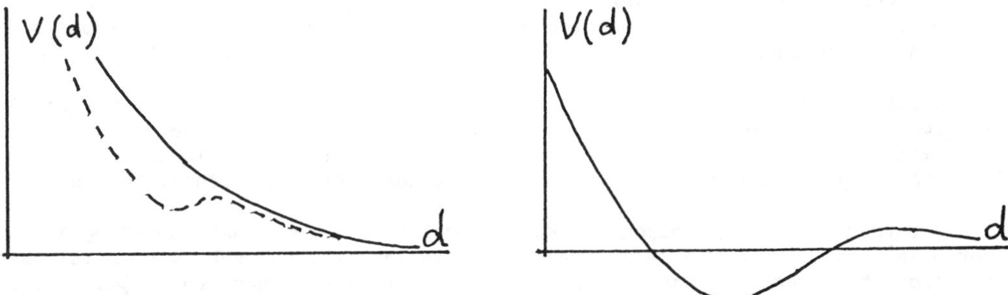

Fig. 8.5. a) The interaction V(d) between the interfaces of Fig. 8.1 in the case of complete wetting. b) The same in the case of partial wetting

For the case of *complete wetting*, d should, in principle, be infinite. In practice it is finite, but large, so that V(d) can be replaced by $V(\infty)=0$ in (8.3). Thus one can write

$$\sigma_{sg} = \sigma_{s\ell} + \sigma_{\ell g} \qquad \text{(Antonov's rule)} \qquad (8.4)$$

Thus, Young's relation (2) is still valid in practice at complete wetting. However, the meaning of σ_{sg} is not quite the same in (8.4) as in Cahn's argument summarized in § 8.3. We considered there a hypothetical solid-gas interface with no liquid film and therefore a fixed, but not necessarily realistic value of σ_{sg}. The correct argument is more complicated and left to the reader. On the other hand, the value of σ_{sg} to be inserted into (8.2) is $\sigma_{sg}(d_o)$ and thus takes into account the formation of a microscopic film.

In principle d_o can go to ∞ continuously when the wetting transition is approached, if this transition is continuous. Experimentally the transition is almost always of the first kind. In the case of a fluid of neutral molecules this can be understood as follows. The dominant interaction between molecules at large distances is a Van der Waals *attractive* interaction; however the resulting interaction between *interfaces* may be *repulsive*. It will be shown below that it behaves as d^{-2} at large distances, with a fixed coefficient. Since the minimum at d_o is due to competition with short range forces, the transition is of the first kind (Fig. 8.5). However it will be seen in the next subsection that the interaction between interfaces can be attractive. As a result, a continuous wetting transition is in principle allowed (Dietrich and Schick 1985, 1986).

8.5 Van der Waals interactions

Molecules or rare gas atoms interact at long distance through the so-called Van der Waals attraction (superimposed with a hard core repulsion at short distances). The origin of the Van der Waals interaction between two atoms at distance r is the quantum fluctuation of the dipole moment m. The dipole moment m_1 of the atom 1 induces a variation $\delta m_2 \sim m_1/r^3$ of the dipole moment of the atom 2, and the interaction energy is proportional to $-\delta m_2 . m_1/r^3 \sim 1/r^6$. The interaction energy of one atom with a layer (at distance z) parallel to the surface implies integration over x and y, and is therefore proportional to $-1/z^4$, where the minus sign indicates an attrractive interaction.

The energy of the layer of Fig. 1 contains four parts : substrate-adsorbate interaction, adsorbate-adsorbate interaction, substrate-vapour and fluid-vapour interaction.

Substrate-adsorbate interaction: The interaction energy of an atom at distance h from the surface implies integration of $-1/z^4$ over the depth of the substrate from h to infinity, and is proportional to $-1/h^3$. Integration on h from 1 to d yields a constant, plus a positive quantity proportional to $1/d^2$.

Substrate-vapour interaction: same calculation except that the integration over h is from d to ∞. The result is a negative (attractive) energy proportional to $-1/d^2$, but smaller than the previous contribution because the vapour has a lower density.

Vapour-adsorbate interaction: Same calculation as for the substrate-adsorbate interaction. Result proportional to $+1/d^2$.

Adsorbate-adsorbate interaction: The interaction energy between two layers at distance z is proportional to $-\mathcal{A}/z^4$, where $\mathcal{A}$ is the area. Integration at constant z yields now $-\mathcal{A}(d-z)/z^4$. Integration over z from 1 to d yields a negative energy proportional to d, plus an attractive contribution proportional to $1/d^2$.

The conclusion, when only Van der Waals interactions are present, is

$$V(d) \approx C/d^2 \qquad (d \to \infty) \tag{8.5}$$

where the so-called *Hamaker constant* C can be positive or negative.

The Van der Waals attraction induces an excess of density near the surface, which has not been taken into account.

The above calculation neglects the fact that electromagnetic interactions need some time to go from atom 1 to atom 2. Taking this retardation effect (Lifshitz 1956, Dzyaloshinskii et al 1961, Van Kampen et al 1968) into account changes the exponent 6 into 7 in (8.4) and the exponent 2 into 3 in (8.5).

REFERENCES

AKUTSU, Y., AKUTSU, N., YAMAMOTO, T. (1988) Phys. Rev. Lett. **61**,424
ANDREEV, A.F. (1981) Zh. Exp. Teor. Fiz. **80**, 2042 and Sov. Phys. J.E.T.P. **53**, 1063 (1982)
ARMAND, G., MANSON, J.R. (1988) Phys. Rev. B **37**, 4363
BALIBAR, S., GALLET, J.F., ROLLEY, E., WOLF, P.E. (1987) La Recherche, **18**, 1454
BIENFAIT, M., GAY, J.M., BLANK, H. (1988) Surface Sci **204**, 331
BLATTER, G. (1984) Surface Sci. **145**, 419
H.P. BONZEL, E. PREUSS, B. STEFFEN (1984) Appl. Phys. **A 35**, 1 and Surface Sci. **145** , 20 (1984).
BOUCHDOUG, M., MENAUCOURT, J., THOMY, A.(1986) J. Physique **47**, 1797
BRAUN, O.M., MEDVEDEV, V.K. (1989) Uspekhi fizicheskikh nauk **157**, 652
BRUINSMA, R. (1990) in "Kinetics of Ordering and Growth at Surfaces" ed. by M. Lagally (Plenum, New York)
BURTON, W.K., CABRERA, N. (1949) Discuss. Faraday Soc. **5**, 33
BURTON, W.K., CABRERA, N., FRANK, F.C. (1951) Phil. Trans. Roy. Soc. **243**, 299
CAHN, J.W. (1977) J. Chem. Phys. **66**, 3667
CAROLI, C. (1989) see GODRECHE (1989)
CHUI, S.T., WEEKS, J.D. (1976) Phys. Rev. B **14**, 4978
CONRAD, E.H., ATEN, R.M., KAUFMAN, D.S., ALLEN, L.R., ENGEL, T., den NIJS, M., RIEDEL, E.K. (1986) J. Chem. Phys. **84**, 1015
DASH, G. (1989) Contemporary Physics **30**, 89
DE GENNES, P.G. (1985) Rev. Mod. Phys. **57**, 827
DEN NIJS, M., ROMMELSE, K. (1989) Phys. Rev. B **40**, 4709.
DEN NIJS, M. (1989) Preprint
DIEHL, H.W. (1986) in *Phase Transitions and Critical Phenomena*, ed. by C. Domb and J. Lebowitz (Acad. Press, London) Vol. 10.
DIETRICH, S. (1989) in *Phase Transitions and Critical Phenomena*, ed. by C. Domb and J. Lebowitz (Acad. Press, London) Vol. 12.
DIETRICH, S., SCHICK, M. (1985) Phys. Rev. B **31**, 4718.
DIETRICH, S., SCHICK, M. (1986) Phys. Rev. B **33**, 4952.
DZYALOSHINSKII, I.E., LIFSHITZ, E.M., PITAEVSKII, L.P. (1961) Adv. Phys. **10**, 165
EDEN, M. (1958) in "Symposium on Information theory in Biology", ed. H.P. Yockey (Pergamon Press, New York) p. 359
EDWARDS, S.F., WILKINSON, D.R. (1982) Proc. Roy. Soc. **A 381**, 17
FABRE, F., GORSE, D., LAPUJOULADE, J., SALANON, B. (1987) Europhysics Lett. **3**, 737 and Surface Sci. **175**, L693
FRENKEN, J.W.M.., TOENNIES, J.P., WOLL, Ch. (1988) Phys. Rev. Lett. **60**, 1727
FRIEDEL, J. (1956) *Les dislocations* (Gauthier-Villars, Paris 1956) and English translation (1964, Pergamon, Oxford)
FRIEDEL, J. (1976) Ann. de Physique **1**, 257
GALLET, F., et al (1987) J. Physique **48**, 369
GODRECHE, C., editor (1990) Proceedings of the 1989 Saint-Pierre-Quiberon summer school
E.E. GRUBER, W.W. MULLINS (1967) J. Phys. Chem. Solids **28**, 875
HALPIN-HEALY, T. (1989) Phys. Rev. Lett. **62**, 442
HELD, G., JORDAN-SWEET, J.L., HORN, P.M., MAK, A., BIRGENEAU, R.J. (1987) Phys. Rev. Lett. **59**, 2075
HEYRAUD, J.C., METOIS, J.J. (1987) Surface Sci. **180**, 647
HEYRAUD, J.C., METOIS, J.J. (1988) J. Crystal Growth **82**, 269
HEYRAUD, J.C., METOIS, J.J. (1989) to be published
HERRING, C. (1952) chapter dans "Structure and Properties of Solid Sufaces", éd. R. GOMER et C.S. Smith (University of Chicago Press, Chicago 1952).
JAYAPRAKASH, C., SAAM, W.F., TEITEL, S. (1983) Phys. Rev. Lett. **50**, 2017
JASNOW, D (1984) Rep. Prog. Phys. **47**, 1059
JULLIEN, R., BOTET, R. (1985) J. Phys. A **18**, 2279
KARDAR, M., PARISI, G., ZHANG, Y.C. (1986) Phys. Rev. Lett. **56**, 889
KARIOTIS, R. (1989) J. Phys. A to be published

KIM, KOSTERLITZ, J.M. (1989) Phys. Rev. Lett. **62**, 2289
KNOPS, H.J.F. (1977) Phys. Rev. Lett. **39**, 766
KOSTERLITZ, J.M. (1974) J. Phys. C **7**, 1046
KOSTERLITZ, J.M., THOULESS, D.J. (1973) J. Phys. C **6**, 1181
LACOUTURE, J. (1989) *Champollion, une vie de lumières*, Grasset, Paris
LANDAU, L., LIFSHITZ, E. (1967) *Physique statistique*, Editions Mir, Moscow.
LANGER, J.S. (1980) Rev. Mod. Phys. **52**, 1
LEVI, A.C.,(1984) Surface Sci. **137**, 385
LIFSHITZ, E.M. (1956) Sov. Phys. J.E.T.Ph. **2**, 73
LIPOWSKI, R. (1982) Phys. Rev. Lett. **49**, 1575
LIPOWSKI, R. (1984) Phys. Rev. Lett. **52**, 1429
MEAKIN, P., RAMANLAL, P., SANDER, L., BALL, R.C. (1986) Phys.Rev. **A 34**, 5091
MEAKIN, P., BALL, R.C., RAMANLAL, P., SANDER, L. (1987) Phys.Rev. **A 35**, 5233
MOCHRIE, S.G.J. (1987) Phys. Rev. Lett. **59**, 304
MULLINS, W.W. (1957) J. Appl. Phys. **28**, 333
MULLINS, W.W. (1959) J. Appl. Phys. **28**
MULLINS, W.W. (1963) in "Metal Surfaces : Structure, Energetics and Kinetics", Am.Soc. Metals, Metals Park, Ohio, p. 17
MULLINS, W.W., SEKERKA (1963) J. Appl. Phys. **34**, 323 and **35**, **444** (1964)
NOZIERES, Ph., Gallet, F. (1987) J. Physique **48**, 353
NOZIERES, Ph. (1989) J. Physique **50**, 2541
NOZIERES, Ph. (1990) Cf GODRECHE (1990)
PAVLOVSKA, A. (1977) J. Crist. Growth **39**, 346
PELCE, P. (1988) "Dynamics of curved fronts" (Acad. press, New York)
POKROVSKII, V.L., TALAPOV, A.L. (1979) Phys. Rev. Lett. **42**, 65
POMEAU, Y. (1990) see GODRECHE (1990)
RAMANLAL, P., SANDER, L. (1985) Phys. Rev. Lett. 1822
RETTORI, A., VILLAIN, J. (1988) J. Physique France **49**, 257
ROBINSON, I.K., MAC DOWELL, A.A., ALTMAN, M.S., ESTRUP, P.J., EVANS-LUTTERODT, K., BROCK, J., BIRGENEAU, R.J. (1989) Phys. Rev; Lett. **62**, 1294
ROLLEY, E. (1989) Thèse de Doctorat, Paris 6
RYS, F. (1986) Phys. Rev. Lett. **56**, 624
SALANON, B., FABRE, F., LAPUJOULADE, J. (1988) Phys. Rev. B **33**, 7385
SELKE, W., SZPILKA, A.M.(1986) Z. Physik **B 62**, 381
VAN BEIJEREN, H. (1977) Phys. Rev. Lett. **38**, 993
VAN BEIJEREN, H., NOLDEN, I. (1987) in *Structure and Dynamics of Surfaces II*, W. Schommers and Von Blanckenhagen (eds) Springer, Heidelberg
VAN DER VEEN, J.F., PLUIS, B., DENIER VAN DER GON, A.W. (1990) in "Kinetics of Ordering and Growth at Surfaces" ed. by M. Lagally (Plenum, New York)
VAN KAMPEN, N.G., NIJBOER, B.R.A., SCHRAM, K. (1968) Phys. Lett. **26 A**
VILLAIN, J. (1986) Europhys. Lett. **2**, 531
VILLAIN, J., GREMPEL, D., LAPUJOULADE, J. (1985) J. Phys. F **15**, 809
VILLAIN, J. (1989) Comptes rendus de l'Académie des Sciences **309 II**, 517
VILLAIN, J., LANÇON, F. (1989) Comptes rendus de l'Académie des Sciences, **309** II, 647
VISCEK, T. (1989) "Fractal growth phenomena" (World Scientific, Singapore)
J.D. WEEKS (1980) in *Ordering in Strongly Fluctuating Condensed Matter Systems* Ed. T. Riste (Plenum, New York) p. 293
WIDOM, B. (1965) J. Chem. Phys. **43**, 3892
WIEGMAN, P. B. (1978) J..Phys C **11**, 1583
WOLF, D. (1990) in "Kinetics of Ordering and Growth at Surfaces" ed. by M. Lagally (Plenum, New York)
WOLF, D.E., KERTESZ (1987) J., Europhys. Lett. **4**, 651
WOLF, D., VILLAIN, J. (1989) Phys. Rev. to be published
WULFF, G. (1901) Z. Kristallogr. **34**, 449
K. YAMASHITA, H.P. BONZEL, H. IBACH (1981) Appl. Phys. **25**, 231

Materials Science Forum Vols. 59 & 60 (1990) pp. 93-140
Copyright Trans Tech Publications, Switzerland

X-RAY AND NEUTRON CHARACTERIZATION OF MULTILAYER SYSTEMS

M. Piecuch (a) and L. Nevot (b) *
(a) Laboratoire mixte CNRS Saint Gobain UMR 37
CRPAM BP 109
Pont à Mousson Cédex, France
(b) Institut d'Optique
Université Paris Sud, B.P. 43
Orsay Cédex, France

1 INTRODUCTION

The progress of materials science and particularly of deposition techniques have led to the synthesis of systems with controlled modulation of the chemical composition. The idea of obtaining chemically modulated films is usually attributed to the work of Dumond and Youtz [1, 2] who were the first to propose artificial multilayers for use as soft X-ray monochromators. However the present activity in the field of multilayer physics was initiated in the late sixties and early seventies by several workers in different areas:

i) The preparation of high quality semiconductor single-crystal multilayers which was initiated with a proposal in 1969-70 by Esaki and Tsu [3, 4] has produced enormous experimental activity and some major achievements in the field of quantum effects and electronic or optical properties (for more information see Esaki [5]).

ii) The development of X-UV mirrors [6, 7] which is related to several possible applications; high density thermonuclear plasma diagnostics, X-ray telescopes, optics for high flux synchrotron sources, soft X-ray monochromators...

iii) The early metallurgical studies initiated by Hilliard and coworkers [8, 9], which have led to an intense activity in diffusion experiments [10, 11...] and to the discovery of the so called supermodulus effect [12].

One of the problems associated with these various workers, operating in their various fields is that, according to their own needs and interests, they have used and defined their own tools, concepts and languages. Even for the very basic characterization by X-ray diffraction, the diffractometers, measurement methods and data analyses are different.

Structural studies on semiconductor superlattices have emphasized the utility of high resolution X-ray studies with sophisticated monochromators [13]. The goal of the multilayer manufacturer here is the crystalline

*Present address: Laboratoire d'électronique quantique
Université de Rennes I Beaulieu 35042 Rennes Cédex France

perfection, but the determination of interplanar distances is also crucial and relatively difficult since the various III/V crystals have nearly the same lattice parameters. To check this desired crystalline quality, the experimental spectra are recorded in a narrow angle interval around a Bragg peak of the single crystal (see, for instance, M. Quillec et al.[14]). Data analyses are made in the framework of classical X-ray kinematical theory adapted to a crystal with modulated interplanar spacing and (or) modulated scattering amplitudes [15, 16]. X-ray characterization gives information in the following areas: modulation period, amplitude of the modulation of interplanar spacing, crystal perfection as estimated from rocking curve widths, determination of the coherency strain imposed by the epitaxial growth, interplanar distances... This information is mainly obtained from the position of the reflections in reciprocal space. Few workers have used the limited intensity information to obtain an approximate model of chemical modulations. In that case, when one attempts to determine strain and stress with high precision, some groups have used the dynamical theory of X-ray diffraction as developed for instance by Tagaki [17] and Taupin [18] (see Tapfer and Ploog [19]).

People working on X-UV mirrors have another goal; obtaining high reflectivity. They require good interface sharpness and limited thickness variations in order to have the greatest possible contrast between low and high index components of the multilayer. But, as some elements in X-ray mirror multilayers are poorly crystallized and even amorphous, high precision diffraction around a Bragg peak cannot be used to characterize the samples. The investigators in this field have therefore developed small angle reflectometry [20, 21] with various experimental arrangements. Workers in small angle X-ray reflectometry come mainly from classical optics and they have used the formalisms and methods of the classical theory of visible light propagation in multilayers [22], only slightly modified for X-rays [23, 24, 25, 26]. The usual way for the determination of the multilayer parameters is to do some simulation and to compare the results with experimental spectra. The parameters deduced from X-ray characterization are the chemical composition of the layers, the interface perfection defined usually via the width of a smooth transition layer and the thickness variations.

Metallurgists have used the two different approaches of small angle reflectometry and high angle diffraction [27], small angle peak intensities being used for interdiffusion measurements in amorphous phases [28], and high angle intensities for stress determinations [8]. They have however mainly used the kinematical approximation for data interpretation even in the case of diffusion experiments where the measurements of diffusion coefficients are based on precise intensity determinations [8, 9, 11, 27, 28...]

To summarize, the state of the art is confusing and the published literature on diffraction is dispersed in many specialized journals.

Recently, the introduction of molecular beam epitaxy techniques in metal physics and the associated need for rather precise structural determinations have opened up a new branch of materials science: the metallurgy of multilayers. In this new field various questions remain unanswered:

i) Are we able to create new crystalline structures of elemental metals by epitaxy (preparation of pseudomorphic structures) and to characterize these structures?

ii) Are we able to choose new structures to obtain new unexpected (or expected) properties?

iii) Are we able to test the various predictions of the very impressive recent electronic structure calculations (see e.g. Moruzzi et al.[29]), and measure magnetic and electronic properties of new epitaxial phases on well defined samples, in order to achieve understandable comparisons with theory?

Several groups are now engaged in such a programme [30, 31, 32, 33, 34, 35, 36, 37...]. In all of these studies, the importance of using X-ray scattering as a routine probe of sample quality is stressed. Obviously, there is a need for careful structural determinations at high angles to measure precisely the crystal lattice parameters of the epitaxially grown phases and for in-situ control of growth by RHEED (Reflection High Energy Electron Diffraction) to have some idea of crystal growth and a knowledge of epitaxial mechanisms. But the workers will also gain some knowledge of interface perfection, a good understanding of interfacial diffusion and interfacial alloy formation... Then the use of X-ray (or neutron) diffraction should be made at small angles and around the Bragg peaks of the average lattice and the use of sophisticated data analysis and interpretation must be promoted. This requires measuring the intensities and positions of a great number of reflections so as to do complete structure determinations. Using the high fluxes and wavelength tunability of synchrotron sources also opens up new methods such as the use of the anomalous dispersion effect [38, 39].

At the same time, an intense activity continues in the domain of multilayer devices for X-ray or neutron systems [40, 41] and a lot of work is also being done on non-epitaxial magnetic multilayers [42, 43, 44...]. Concurrently, new fields of classical metallurgy have opened with the help of multilayers: amorphous phase formation by solid state reactions [45], ion beam mixing [46] etc... In all cases, small angle X-ray diffraction is one way of achieving structural characterization.

As the direct inversion of X-ray spectra to give the various atomic positions in the samples is not feasible, a lot of theoretical models have been developed to understand some features of the experimental spectra such as peak positions, peak intensities and in some cases peak shapes. The comparison of the X-ray or neutron spectra with predictions from theoretical results at various stages of sophistication and the qualitative understanding of the physics of diffraction by multilayers are the two goals of this paper. In the next section (part 2), we will introduce the essential phenomena of diffraction by multilayers by presenting two idealized models in the framework of simple kinematical theory (in other words, in the first Born approximation of scattering) and we will discuss the main experimental consequences of these models. Some intuitive feeling about diffraction in multilayers will also be developed. In part 3, in the same approximation, various kinds of disorder and imperfections will be discussed with the help of two specific models. In part 4, we shall give accounts of X-ray optics and of optical multilayer theory and some experimental spectra will be compared with calculated ones. Then, in part 5, we will give briefly some results of the dynamical X-ray theory (beyond the first Born approximation) and we will discuss some applications of this theory to multilayers. In part 6 two typical experimental devices will be briefly presented.

Several basic books have been used by the authors and can be useful to the reader to have more information on the various topics of this chapter. We have mainly used the classics by André Guinier [15], W.H. Zachariasen [47], or J. Cowley [48]. But the treatise of R.W. James [49] for X-rays and the principles of optics by M. Born and E. Wolf [50] can be also very profitable. The recent review papers by D.B. Mc Whan [51], E. Spiller [52], L. Nevot et al.[53] and B. Pardo et al.[54] among others, are also useful and the special issue of the Revue de Physique Appliquée edited by Pierre Dhez in 1988 [55] contains reviews on some of the topics of this work.

This chapter is mainly devoted to the classical X-ray range (wavelength between 0.05 to 0.5 nm). But a lot of work on multilayers was done in the soft X-ray region. Interested readers can find some specific information in the papers by Ph. Houdy et al.[56], Youn et al.[57], Ruterana et al.[58] and in the review articles by Henke [59] or Spiller [60]. The various optical constants in the soft X-ray region were computed by Henke et al.[61].

In a few cases neutrons have also been used [62, 63, 64, 65…]. They are obviously particularly useful in the case of magnetic structure determinations [32, 37] and certainly the chapters by Fert and by Gautier will come back to these points. Vergnat et al.[64] have used a specific advantage of neutrons : their isotopic sensitivity. They have demonstrated that one may use the difference in scattering length between hydrogen and deuterium to study SiH/SiD multilayers and to gain information about hydrogen positions in hydrogenated amorphous silicon. The same group [65] has used polarized neutrons to determine directly the interface width in Fe/Si magnetic multilayers using the fact that iron is fully magnetized in the centre of the layers while it is non-magnetic at the interfaces.

2 X-RAY DIFFRACTION FROM ONE DIMENSIONAL SUPERLATTICES

Throughout this chapter we will speak mainly in terms of scattering vectors rather that in terms of diffraction angles. The incident wave is a plane wave $\exp(-i\mathbf{k}_0\mathbf{r})$ of wave vector $\mathbf{k}_0$, $\mathbf{k}$ is the wave vector in the direction of observation and $\mathbf{q}$ is the scattering vector: $\mathbf{q} = \mathbf{k} - \mathbf{k}_0$. The geometry of the experiments will usually be the Bragg geometry ($\theta, 2\theta$ reflection, [47]): $\mathbf{q}$ is perpendicular to the film plane and parallel to the z direction of our coordinate system (unit vector $\mathbf{u}_z$). The modulus of the scattering vector is $q = 4\pi\sin\theta/\lambda$. It is obvious that one cannot obtain any information on the multilayer modulation in transmission geometry where $\mathbf{q}$ is in the film plane. However, in some cases, measurements with geometries different from the usual one can also give interesting structural information.

The kinematical theory of X-ray diffraction is equivalent to the first Born approximation of scattering theory. The diffracted wave $\mathbf{\Phi}(\mathbf{r})$ can be expressed as [48]

$$\mathbf{\Phi}(\mathbf{r}) = \exp(-i\mathbf{k}_0\mathbf{r}) + f(\mathbf{q})\,\frac{\exp(-ik_0 r)}{4\pi r} \qquad (1a)$$

with

$$f(\mathbf{q}) = \mu \int_V V(\mathbf{r})\exp(-i\mathbf{q}\mathbf{r})\, d^3r \qquad (1b)$$

$V(\mathbf{r})$ is the total scattering potential, μ a coupling constant and $f(\mathbf{q})$ the scattering amplitude. If $V(\mathbf{r})$ may be given by

$$V(\mathbf{r}) = \Sigma_i\, V_i(\mathbf{r}-\mathbf{r}_i) \qquad (1c)$$

where V_i is the potential of the ith atom, equation 1b becomes

$$f(\mathbf{q}) = \Sigma_i\, f_i \exp(-i\mathbf{q}\mathbf{r}_i) \qquad (2a)$$

f_i is the atomic scattering factor of the ith atom

$$f_i = \mu \int_V V_i(\mathbf{r})\, \exp(-i\mathbf{q}\mathbf{r})\, d^3r \qquad (2b)$$

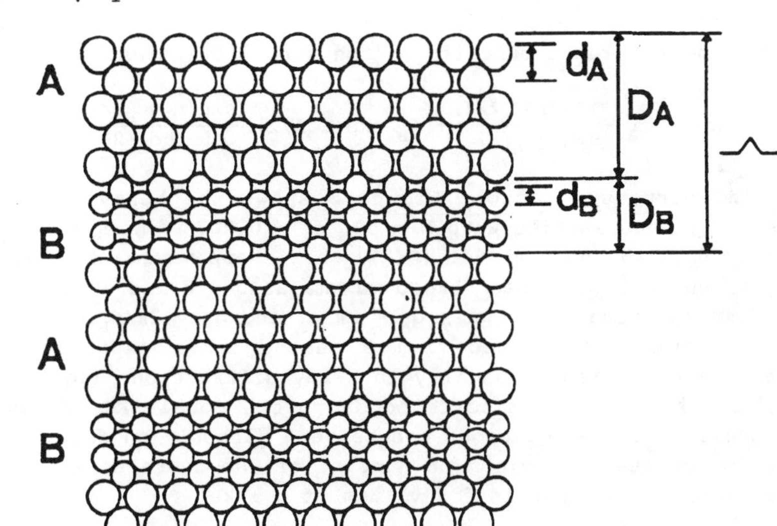

Fig. 1 Ideal phase-locked superlattice.

2.1 The ideal superlattice

To begin with an idealized situation, we suppose that we have a superlattice constituted of n_A planes of A atoms and n_B planes of B atoms.

The distance between two A planes is d_A, the distance between two B planes is d_B, the distance between two adjacent A and B planes is $(d_A+d_B)/2$, the scattering factor of the A atoms is f_A and the scattering factor of the B atoms is f_B (see figure 1). These planes being parallel to the film, it is the custom to call such a superlattice a phase-locked superlattice [66]. A non-phase-locked superlattice being a superlattice where the thickness of A or (and) B does not match exactly with an integer number of atomic layers. The calculation of the scattering amplitude by such a one dimensional system is relatively simple and gives

$$f(\mathbf{q}) = \frac{\sin(N\Lambda q/2)}{\sin(q\Lambda/2)} \left\{ f_A \frac{\sin(n_A q d_A/2)}{\sin(q d_A/2)} + f_B \exp(iq\Lambda/2) \frac{\sin(n_B q d_B/2)}{\sin(q d_B/2)} \right\} \tag{3}$$

where $\Lambda = n_A d_A + n_B d_B$ is the multilayer period, $\mathbf{q} = q\mathbf{u}_z$ and N is the number of bilayers. The intensity can be expressed as

$$I(\mathbf{q}) \propto |S(\mathbf{q})|^2 \; |F(\mathbf{q})|^2 \tag{4a}$$

$$S(\mathbf{q}) = \frac{\sin(Nq\Lambda/2)}{\sin(q\Lambda/2)} \tag{4b}$$

is the interference function of the superlattice and

$$F(\mathbf{q}) = f_A \frac{\sin(n_A q d_A/2)}{\sin(q d_A/2)} + f_B \exp(iq\Lambda/2) \frac{\sin(n_B q d_B/2)}{\sin(q d_B/2)} \tag{5}$$

is the structure factor. Figure 2 shows the behaviour of these two factors for a superlattice with $N = 50$, $d_A = 0.2075$ nm, $d_B = 0.2142$ nm, $n_A = 2$, $n_B = 2$, $f_A = 26$ and $f_B = 44$. $|S(\mathbf{q})|^2$ has sharp maxima at $q = 2m\pi/\Lambda$, with m integer, while $|F(\mathbf{q})|^2$ has broad maxima at $q = 0$ and at $q = mq_0$, where $q_0 \cong 29.47\ \mathrm{nm}^{-1}$ is situated between $q_A = 2\pi/d_A$ and $q_B = 2\pi/d_B$. This value of q_0 is an estimation calculated from a second order expansion from $|F(\mathbf{q})|^2$ (figure 2 is presented as usually in X-ray diffraction: $I(\mathbf{q})$ versus 2θ, $q = 4\pi\sin\theta/\lambda$ and $\lambda = 0.1789$ nm corresponds to the cobalt $K\alpha_1$ transition). The consequences of these equations can be seen directly on figure 2, $I(\mathbf{q})$ will have its more intense maximum on a point of the average reciprocal lattice not too far from q_0, that is for values of q like

$$q_M = \frac{2\pi}{d} \tag{6a}$$

where

$$d = \frac{n_A d_A + n_B d_B}{n_A + n_B} \tag{6b}$$

and secondary maxima or satellites occur around these main peaks at position $q_S = 2\pi/d \pm 2k\pi/\Lambda$, the widths of these peaks are mainly determined in this ideal model by the number of periods. Another description of the situation can be made if the superlattice is considered as a single crystal constituted by unit cells of n_A planes of A and n_B planes of B, and if the lattice of A is isomorphous to the lattice of B in the x and y directions. The reciprocal lattice vector in the z direction has a length $c^* = 2\pi/\Lambda$, but the structure factor enhances the peaks corresponding to $q_M = nc^*$ with $n = n_A + n_B$, and only some of the other peaks at $q_S = n'c^*$ with $n' = n \pm 1$ or $n' = n \pm 2$ are detectable. This discussion is only true if the structure factor has a single maximum at q_0. If the values of n_A and n_B are large, or (and), if the values of d_A and d_B are very different the interference between the various terms in $|F(\mathbf{q})|^2$ becomes destructive and the structure factor has two maxima which are close to $2\pi/d_A$ and $2\pi/d_B$ (it is obvious that if n_A and n_B tend to infinity we obtain the diffraction of two pure A and B crystals with peaks at $2\pi/d_A$ and $2\pi/d_B$).

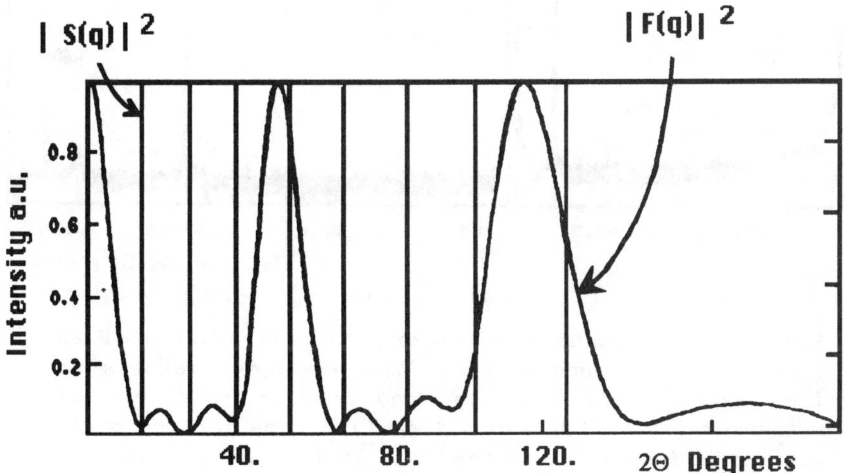

Fig. 2 Theoretical interference function and structure factor of an ideal phase-locked superlattice, $N = 50$, $d_A = 0.2075$ nm, $d_B = 0.2142$ nm, $n_A = 2$, $n_B = 2$, $f_A = 26$ and $f_B = 44$. $|S(\mathbf{q})|^2$ has sharp maxima at $q = 2m\pi/\Lambda$, with m integer, while $|F(\mathbf{q})|^2$ has several broad maxima at $q = 0$ and at $q = mq_0$, where $q_0 \cong 29.47\ nm^{-1}$ is situated between $q_A = 2\pi/d_A$ and $q_B = 2\pi/d_B$.

The physics of the diffraction by multilayers is essentially contained in equations 3 and 4 but the model which leads to these equations is too idealized and can only be used as a first step in order to determine n_A, d_A, n_B and d_B.

One experimental example of the possible use of this kind of consideration was made by Maurer et al.[34, 35]. Figure 3 shows the results of a high resolution X-ray diffraction experiment in the Bragg geometry (θ, 2θ reflections, X-ray intensity versus 2θ) on a Fe/Ru metallic superlattice. This superlattice is constituted by two layers of iron and two layers of ruthenium repeated 140 times. It is grown on a 20 nm ruthenium buffer deposited on $(11\bar{2}0)$ sapphire. The ruthenium (0001) plane is parallel to the substrate and RHEED has shown that iron is also in a hexagonal phase coherent with ruthenium.

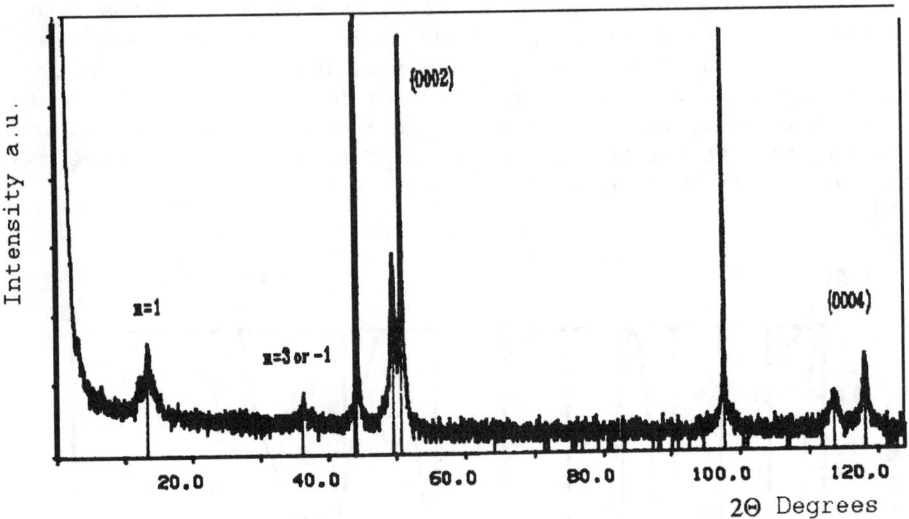

Fig. 3 High resolution X-ray diffraction profile of an iron ruthenium superlattice. The average thickness of iron layers is 0.4 nm and the average thickness of ruthenium layers is also 0.4 nm. The superlattice peaks are indexed by n or by the four indices of the average hexagonal lattice. The two intense non-indexed peaks are from the sapphire substrate while the two small peaks near superlattice (0002) or (0004) lines are those of the ruthenium buffer.

It can be seen on figure 3 that besides the two intense peaks of the substrate (at around 44° and 97°) there is an intense peak at 50.643° corresponding to $q = q_M = 30.043\ nm^{-1}$, a value of q which is approximately four times greater than the value corresponding to the first superlattice peak (this is only partially true because the real superlattice is not exactly phase locked). Other peaks are clearly seen in this case at 36.215° ($n' = 3$) and at 117.783° ($q = 2q_M$). The two other peaks on figure 3 correspond to the ruthenium buffer. The structure factor is important for the low q satellite ($n' = 3$) of the main peak but it is zero for the high q satellite which is then not visible on the figure. If such experiments are made on a great number of samples with different values of n_A and n_B, we are able to determine the values of d_A and d_B by using formula 6b [34]. We have plotted the values of d deduced from the X-ray diffraction measurements versus x ($x = n_{Fe}/(n_{Fe} + n_{Ru})$). The values of d are reasonably aligned versus x and extrapolate for x = 0 to 0.2142 nm, a value corresponding to pure HCP ruthenium. One may then believe that formula 6b deduce that the value obtained at x = 1 corresponds to the value of the spacing between the (0001) planes of iron in that hexagonal superlattice, d_{Fe} = 0.2075 nm.

In conclusion of this section, one can say that with the idealized multilayer model, one can understand the peak positions and thus one can deduce the multilayer period Λ and the average lattice spacing of the multilayer components. The interface sharpness and the multilayer perfection remain to be determined.

2.2 The diffuse multilayer

The above calculation corresponds to perfect coherent superlattices with sharp interfaces, the next step is to assume some interdiffusion. The lattice spacing and the atomic scattering factor are smooth periodic functions of z. These periodic functions are developed in a Fourier series [16]. To simplify the calculations, we have assumed that two Fourier coefficients are sufficient to obtain a good description of our samples:

$$d(z) = d_0\left(1 + \Delta_1 \cos\left(\frac{2\pi z}{\Lambda}\right) + \Delta_2 \cos\left(\frac{4\pi z}{\Lambda}\right)\right) \qquad (7a)$$

and

$$f(z) = f_0\left(1 + \phi_1 \cos\left(\frac{2\pi z}{\Lambda}\right) + \phi_2 \cos\left(\frac{4\pi z}{\Lambda}\right)\right) \qquad (7b)$$

$d_0 = (n_A d_A + n_B d_B)/(n_A + n_B)$ is the average lattice spacing and $f_0 = (n_A f_A + n_B f_B)/(n_A + n_B)$ the average structure factor. Δ_1, ϕ_1 and Δ_2 ,ϕ_2 are the amplitude of first and second harmonics for d and f respectively, and Λ is the modulation period. The preceding parameters have been evaluated by making the link with the ideal superlattice of the preceding section, but equivalent formulae may be obtained for a non phase-locked superlattice (n_A and n_B having non-integer values):

$$d_0 = \frac{D_A d_A + D_B d_B}{D_A + D_B} \quad \text{and} \quad f_0 = \frac{D_A f_A + D_B f_B}{D_A + D_B} \qquad (8)$$

D_A and D_B are the experimental thicknesses of A and B (see figure 1). In formula 7, all the variables are periodic and we are able to separate f(**q**) into S(**q**) and F(**q**). Again S(**q**) is given by formula 4b, S(**q**) = $\sin(Nq\Lambda/2)/\sin(q\Lambda/2)$. **q** is parallel to the direction of modulation (z axis, $\mathbf{q}=q\mathbf{u}_z$). The calculation of F(**q**) (or F(q)) is restricted to a single period $\Lambda = n_0 d_0$ and gives

$$F(q) = \sum_{n=1}^{n_0} f(z_n) \exp\left(-iq\left[z_n + \Lambda\left(\frac{\Delta_1}{2\pi}\sin\left(\frac{2\pi z_n}{\Lambda}\right) + \frac{\Delta_2}{4\pi}\sin\left(\frac{4\pi z_n}{\Lambda}\right)\right)\right]\right) \qquad (9a)$$

$$z_n = n d_0 \qquad (9b)$$

To obtain eq. 9, the z coordinate of the nth atomic layer is expressed as

$$z(n) = \Sigma_i d(z_i) = nd_0 + \Sigma_i d_0 (\Delta_1 \cos(\frac{2\pi z_i}{\Lambda}) + \Delta_2 \cos(\frac{4\pi z_i}{\Lambda}))$$

with $z_i = id_0$, i integer and we have approximated this sum by an integral. To obtain a more useful formula, $\exp[-iq\Lambda(\Delta_1/2\pi)\sin(2\pi z_n/\Lambda)]$ may be developed in a Fourier series with the help of a well known identity [67]:

$$\exp(-iq\Lambda\frac{\Delta_1}{2\pi}\sin(\frac{2\pi z_n}{\Lambda})) = \Sigma_p J_p(q\Lambda\frac{\Delta_1}{2\pi}) \exp(-i\frac{2\pi p z_n}{\Lambda}) \quad (10)$$

where $J_p(x)$ is the standard Bessel function of the first kind. The final expression of the structure factor is

$$F(q) = \Sigma_n \Sigma_{p,m} f_0 J_p(q\Lambda\frac{\Delta_1}{2\pi}) J_m(q\Lambda\frac{\Delta_2}{4p}) (1 + \phi_1 \cos(\frac{2\pi z_n}{\Lambda}) + \phi_2 \cos(\frac{4\pi z_n}{\Lambda}))$$

$$\exp(-iz_n(q + \frac{2\pi p}{\Lambda} + \frac{4\pi m}{\Lambda})) \quad (11)$$

It can be deduced from eq. 11 that F(q) will have a maximum when any one of the various arguments in the exponentials are simultaneously equal to 2π times an integer. Simplified expressions of F(q) have been obtained by neglecting the Bessel functions of order $p > 2$ [16]. For instance, the most intense peaks will be obtained for $q = mq_M$, with $q_M = 2\pi/d_0$ and m integer. This is almost the same conclusion as above, but we may also obtain an analytical expression of the intensities.

$$F(mq_M) = f_0 G_0 \{ J_0(mn_0\frac{\Delta_1}{2\pi}) J_0(mn_0\frac{\Delta_2}{4\pi}) - \phi_1 J_1(mn_0\frac{\Delta_1}{2\pi}) J_1(mn_0\frac{\Delta_2}{4\pi}) +$$

$$\phi_2 J_2(mn_0\frac{\Delta_1}{2\pi}) (J_2(mn_0\frac{\Delta_2}{4\pi}) + J_0(mn_0\frac{\Delta_2}{4\pi})) \} \quad (12a)$$

with

$$G_0 = \frac{\sin(\pi m n_0)}{\sin(\pi m)} \quad (12b)$$

The same kinds of complicated formulae are also obtained for the satellites at $q = mq_M \pm 2\pi/\Lambda$ [16], but the simplified forms of these various equations are more transparent. If it is assumed that the displacement wave amplitudes are small

$$\exp(-iq\Lambda\frac{\Delta_1}{2\pi}\sin(\frac{2\pi z_n}{\Lambda})) = 1 - iq\Lambda\frac{\Delta_1}{2\pi}\sin(\frac{2\pi z_n}{\Lambda}) \quad (13)$$

and

$$F(q) = \Sigma_n f_0 \exp(iqz_n) (1 + \phi_1 \cos(\frac{2\pi z_n}{\Lambda}) + \phi_2 \cos(\frac{4\pi z_n}{\Lambda})) F(z_n) \quad (14a)$$

$$F(z_n) = \left(1 - iq\Lambda\frac{\Delta_1}{2\pi}\sin(\frac{2\pi z_n}{\Lambda}) - iq\Lambda\frac{\Delta_2}{4\pi}\sin(\frac{4\pi z_n}{\Lambda}) \right) \quad (14b)$$

Following Guinier [15], one may express Δ_1, ϕ_1 and Δ_2, ϕ_2 with a single concentration wave

$$c(z) = c_0\left(1 + C_1\cos(\frac{2\pi z}{\Lambda}) + C_2\cos(\frac{4\pi z}{\Lambda}) \right) \quad (15)$$

and $\Delta_n = C_n\varepsilon$, $\phi_n = C_n\eta$ (the system is an interdiffused superlattice with components following Vegard's law). Here c_0 is the mean concentration of A, $\varepsilon = (d_A - d_B)/d_0$ and $\eta = (f_A - f_B)/f_0$. Finally, if in eq.14 the products of small quantities are neglected, one useful expression of F(q) is obtained

$$F(q) = f_0 \{ G(q) + \frac{1}{2} \{ (\eta + q\frac{\Lambda\varepsilon}{2\pi})C_1 G(q + \frac{2\pi}{\Lambda}) + (\eta - q\frac{\Lambda\varepsilon}{2\pi})C_1 G(q - \frac{2\pi}{\Lambda}) +$$
$$(\eta + q\frac{\Lambda\varepsilon}{4\pi})C_2 G(q + \frac{4\pi}{\Lambda}) + (\eta - q\frac{\Lambda\varepsilon}{4\pi})C_2 G(q - \frac{4\pi}{\Lambda}) \} \} \quad (16a)$$

where

$$G(q) = \frac{1 - \exp(-iqn_0d_0)}{1 - \exp(-iqd_0)} \quad (16b)$$

We can deduce from eq. 16 that the main maximum of I(q) at $q = mq_M$ will have an amplitude proportional to f_0n_0 while the amplitude of the first satellite at $q = mq_M + 2\pi/\Lambda$ will be proportional to $0.5C_1\ (\eta - (mn_0 + 1)\varepsilon)\ n_0f_0$ and the amplitude of the satellite at $q = mq_M - 2\pi/\Lambda$ will be proportional to $0.5C_1\ (\eta + (mn_0 - 1)\varepsilon)\ n_0f_0$. More generally, if we have a concentration wave with an infinite Fourier sum then

$$c(z) = c_0 + \Sigma_p\ C_p\ \cos(\frac{2\pi pz}{\Lambda}) \quad (17)$$

The amplitude of the high q pth satellite at $q = mq_M + 2\pi p/\Lambda$ will be: $0.5C_p\ (\eta - (mn_0/p + 1)\varepsilon)\ n_0f_0$ and the amplitude of the low q pth satellite at $q = mq_M - 2\pi p/\Lambda$ will be $0.5C_p\ (\eta + (mn_0/p - 1)\varepsilon)\ n_0f_0$. The amplitude asymmetry between low q and high q satellites is easily understandable by a simple argument [15]: the stronger satellite is determined by the more diffusing atom so if $f_A > f_B$ or $\eta > 0$ the A atom dominates, then if $d_A > d_B$, the low q peak will be the more intense and if $d_B > d_A$ the situation will be reversed. This is seen in figure 3 where the more diffusing atom, ruthenium, is also the biggest in size. The low q peak is therefore more intense than the high q one. The intensity formulae deduced from eq.16 have been used by some workers to determine composition waves [68, 69]. For instance Fleming et al.[68] have used this analysis to determine the change in the modulation of the composition with annealing at T = 860°C in a superlattice of $(GaAs)_{12}(AlAs)_9$. One can also use the intensity ratio between low q and high q satellites to determine ε.

Formula 16 is also useful to understand some of the features of low angle satellites. The intensity of the pth low angle satellite (m = 0) will be

$$I(\frac{2\pi p}{\Lambda}) = N^2 \ (C_p \ (\eta - \varepsilon) \ \frac{n_0 f_0}{2})^2 \quad (18)$$

The first qualitative consequence of this fact is that it is the density which is important to give some measurable contrast at low angles. If we have, for instance, a multilayer constituted from a rare earth and a late first transition series element, ε is of the order of 0.35 while η is of the order of 0.7 leading to a rather weaker contrast than expected simply from the atomic number argument (the scattering of electromagnetic radiation, like X-rays, by atoms depends mainly on the number of electrons ([48]p 76)). However as ε is usually small (a few percent), while η may be large, the main factor is often the atomic number difference. A second obvious consequence of formulae 16 and 18 is that the intensities of the different satellites are proportional to the square of the Fourier series coefficients. This is sometimes used as a qualitative estimation of interface sharpness; a "smooth" profile will have few non-zero Fourier coefficients (only one for a sinusoidal concentration wave for instance) while a " sharp" profile will have C_p smoothly decreasing with p; for example, for a rectangular concentration wave

$$C_p = \frac{2}{\pi p} \sin(\frac{\pi p D_A}{\Lambda}). \quad (19)$$

Then a great number of satellites at low angles is the first indication of good interface sharpness. This must not be taken too seriously however, as some other factors like the angular dependence of atomic structure factors, the importance of contrast $(\eta - \varepsilon)$ and the multiple scattering effects (dynamical effects) also play an important role in the observed satellite intensities at low angles. A small period multilayer will have a smaller number of satellites than a large period one with the same interface quality. A third interesting consequence of equations 18 and 19 has been used to determine the relative thickness of the components in a multilayer [70, 71] One can see that the sine in equation 19 will be zero if $pD_A/\Lambda = m$ where m is an integer, then if the pth peak at low angles is the first missing peak and if the (p + 1)th exists, one may deduce that $D_A = \Lambda/p$. If Λ is determined by the peak positions, one determines in this way the thickness of the A layer (This discussion is made in terms of kinematical variables and is not strictly true. To have exact quantitative estimations of relative thicknesses, one has to use the optical thicknesses i.e. $d_{opt} = nd$, where n is the optical index. A discussion of these considerations can be found in [72]).

The preceding discussion is based on formulae 16, 17, 18 and 19 which have been established for the case of coherent superlattices, but the behaviour of low-angle satellites is qualitatively the same in polycrystalline or even amorphous samples. The extension can be made in a straightforward way by replacing the sum in equation 14 by an integral:

$$F(q) = \int f_0(1 + \phi_1\cos(\frac{2\pi z}{\Lambda}) + \phi_2\cos(\frac{4\pi z}{\Lambda}))\exp(-iqz)\rho(z)\,dz \qquad (20)$$

where $\rho(z)$ is the probability of finding an atom between z and z + dz. If one takes $\rho(z) = \Sigma_n \delta(z - z(n))$, formula 20 is equivalent to formula 14. In the case of amorphous layers a good approximation to $\rho(z)$ is to assume that $\rho(z)$ is modulated like c so that

$$\rho(z) = \rho_0(1 + C_1\kappa\cos(\frac{2\pi z}{\Lambda}) + C_2\kappa\cos(\frac{4\pi z}{\Lambda})) \qquad (21)$$

where $\kappa = (\rho_A - \rho_B)/\rho_0$ and $\rho_0 = (D_A\rho_A + D_B\rho_B)/(D_A + D_B)$, the same kind of definitions as above (ρ_A and ρ_B are the mean values of ρ in A and B). With this expression of $\rho(z)$, F(q) is given by

$$F(q) = f_0\rho_0\{2G'(q) + (\eta + \kappa)C_1 G'(q + \frac{2\pi}{\Lambda}) + (\eta + \kappa)C_1 G'(q - \frac{2\pi}{\Lambda}) +$$
$$(\eta + \kappa)C_2 G'(q + \frac{4\pi}{\Lambda}) + (\eta + \kappa)C_2 G'(q - \frac{4\pi}{\Lambda})\} \qquad (22a)$$

where

$$G'(q) = \frac{\sin(qn_0d_0/2)}{q} \qquad (22b)$$

The main difference between 16 and 22 is the fact that F(q) has only a principal maximum at q = 0. There are then only low-angle satellites, but all the qualitative conclusions about the behaviour of the low-angle satellites remains true if one notes that the density difference κ and the difference of interplanar lattice spacing ε are opposite.

Figure 4 shows an example of an experimental illustration of the discussion we have just made. We see in figure 4 the low angle X-ray scattering profile of a tungsten carbon amorphous multilayer with Λ = 10.7 nm before and after annealing for 15 hours at 744°C [73]. Several of the qualitative points we have emphasized are observed in this experiment:

i) There is a relatively large number of peaks, this is partly due to a rather good interface quality, but also to the relatively high period and to the good contrast between amorphous carbon and polycrystalline tungsten.

ii) After annealing, Figure 4 shows an extinction of the fourth peak, while before annealing the third and sixth peaks do not exist, showing the diffusion of carbon in tungsten (the carbon layer thickness falling from 1/3 to 1/4 of the period).

iii) The seventh peak decreases markedly after annealing showing some interface smoothing which is presumably also due to the diffusion of the carbon atoms in tungsten.

To conclude this part, we have shown that some models of multilayer structure may be used to develop a good intuitive and qualitative feeling for multilayer quality and structural properties. An intuitive feeling is one of the first needs of a multilayer manufacturer. However, these simple models are only a first step to multilayer structural characterization. In the next part we will discuss, within the first Born approximation, some consequences of several imperfections like thickness variations, lack of coherency at the interfaces and strain.

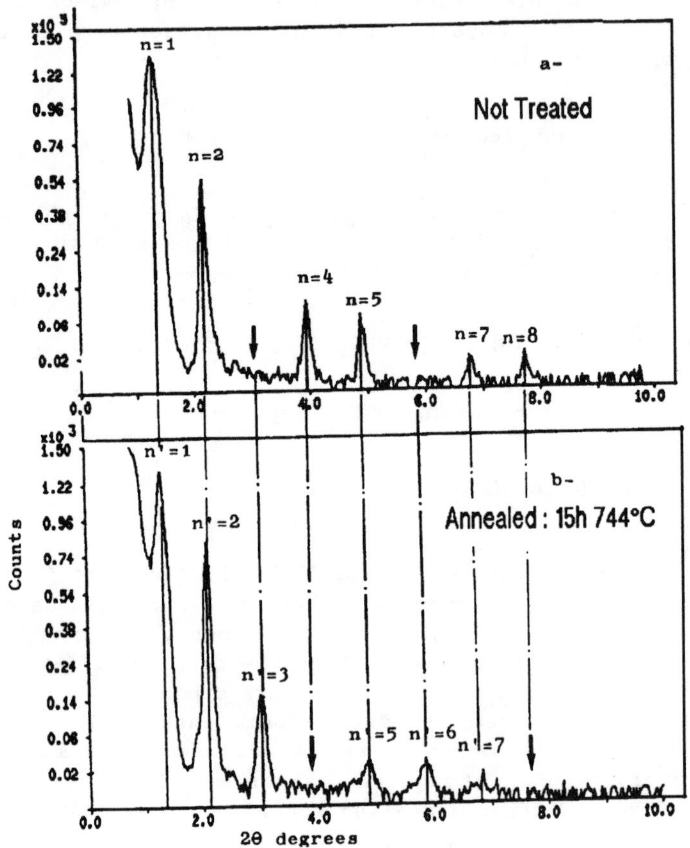

Fig. 4 Small angle scattering profile for a tungsten carbon multilayer showing the decrease of the carbon layer thickness after annealing.

3 EFFECT OF DISORDER AND IMPERFECTIONS IN MULTILAYER DIFFRACTION

In the preceding part, we assumed that the multilayer was perfect, and in a second approach we only introduced a coherent concentration wave. In real samples, we have to understand the effect of numerous imperfections.

The various fluctuations or defects encountered in multilayer systems are caused by different mechanisms such as imperfections in the deposition process, defects in the substrates, geometrical constraints between atoms of different size at the interfaces, or different crystalline symmetry between the components. These latter kinds of constraints are accommodated by interplane coherency strain or by the creation of misfit dislocations. The first mechanism occurs in multilayers with small [30] or medium [34] lattice mismatch, the second may occur in systems with large lattice mismatch [74,75]. A transition between the two regimes is observed in superlattices of small or medium lattice mismatch when the modulation wavelength increases.

In order to have some information about the type and magnitude of the imperfections which are present in our samples, the use of X-ray scattering and the comparison of the recorded spectra with some model is the most popular method in the multilayer community. However, realistic calculations of X-ray spectra with imperfections is a very difficult task. In the classical works on X-ray diffraction by imperfect crystals [15, 48], two main classes of defect are distinguished: those for which it is possible to define an average lattice and those for which it is not. It is customary to call the first class noncumulative disorder and the second cumulative disorder [76]. Noncumulative disorder means that the various atom positions fluctuate around a mean value (as is the case for thermal motion). Cumulative disorder means that random fluctuations in the position of one atom affect the position of the others. The various defects of superlattices are mainly in the second class. This kind of defect destroys the long-range order. Then, there are only a few results of analytical calculations for a rather limited number of typical defects: continuous or discrete fluctuations of the individual layer thickness [76, 78], fluctuations of modulation wavelength [74], variations of the number of atoms in a layer [77], lateral fluctuations in the layer thickness [79], strain [80], interface disorder [81]… The methods of calculation are not exactly the same in all of these works, but the overall principle can be simply shown [15, 48]. The next section, 3.1, will be devoted to the presentation of the method with the help of one single example [78], showing the drastic assumptions which are made to obtain analytical results. In paragraph 3.2 we will discuss one of the other specific models [81] which shows some interesting features and its relevance to the description of experiments. In paragraph 3.3 we will give some information on the results of simulations and other models and give some conclusions about the effects of imperfections on experimental spectra.

3.1 An example of analytical calculations of X-ray spectra in the presence of defects: the effect of the fluctuation in the number of atomic layers

The various methods of calculation of X-ray spectra in imperfect crystals with cumulative disorder [76] are mainly based on the averaging of the square of the scattering amplitude. That is, one has to calculate $<|f(\mathbf{q})|^2>$. Some workers [82] have used $|<f(\mathbf{q})>|^2$, but this gives only Bragg components of the intensity, and not Bragg and diffuse scattering, as pointed out by Mc Whan [83] (diffuse scattering is the difference between the average of the scattering function squared and the average of the intensities). One may note that noncumulative disorder does not give rise to diffuse scattering nor line broadening (like the usual Debye-Waller treatment of thermal motion). The first work on diffraction by disordered stacks is that of Hendricks and Teller on the scattering of X-rays by micas and other similar minerals [84]. Since then, a few calculations have been done, mainly on the problem of stacking faults [15, 48]. The more powerful method of calculation is the Patterson function method. We will recall briefly the principle of this method as given in standard textbooks ([48] p 139). Again the calculations are based on equation 2 and the z axis is perpendicular to the layer plane. The measured intensity, $I(\mathbf{q})$, is proportional to $|f(\mathbf{q})|^2$, the Patterson function $P(\mathbf{r})$ is defined as the Fourier transform of $|f(\mathbf{q})|^2$, that is, if equation 1 is used, as the convolution product of $V(\mathbf{r})$ and $V(-\mathbf{r})$:

$$P(\mathbf{r}) = \mu^2 V(\mathbf{r}) * V(-\mathbf{r}) = \mu^2 \int V(\mathbf{R}) V(\mathbf{r}+\mathbf{R}) \, d^3R \tag{23}$$

The utility of the Patterson function formalism is that all the averages due to disorder that must be done on $|f(\mathbf{q})|^2$ can be made on the Patterson function before performing the Fourier transform, then the true imperfect superlattice Patterson function can be expressed as

$$P_{eff}(\mathbf{r}) = \mu^2 \langle V(\mathbf{r}) * V(-\mathbf{r}) \rangle = \mu^2 \int \langle V(\mathbf{R}) V(\mathbf{r}+\mathbf{R}) \rangle \, d^3R \tag{24}$$

In some cases, $V(\mathbf{r})$ may be expressed as the convolution product of two functions $V(\mathbf{r}) = V_0(\mathbf{r}) * D(\mathbf{r})$ where $V_0(\mathbf{r})$ is the scattering potential of one period and $D(\mathbf{r})$ is a set of delta functions representing the reference positions of the various multilayers. Then the scattering intensity is the product of two functions: $|f(\mathbf{q})|^2 = |F(\mathbf{q})|^2 |S(\mathbf{q})|^2$ where $|F(\mathbf{q})|^2$ is the Fourier transform of the convolution product $\mu^2 V_0(\mathbf{r}) * V_0(-\mathbf{r})$ and $|S(\mathbf{q})|^2$ is the average Fourier transform of the correlation function $P_L(\mathbf{r}) = D(\mathbf{r}) * D(-\mathbf{r})$, the Patterson function of the multilayer skeleton [78]. In the examples of the preceding paragraph $|f(\mathbf{q})|^2$ is given by such formulae where $D(\mathbf{r})$ is periodic. In the general case $D(\mathbf{r}) * D(-\mathbf{r})$ can be calculated directly as

$$D(\mathbf{r}) * D(-\mathbf{r}) = \Sigma_n \Sigma_m \delta(z - (z_n - z_m)) \tag{25}$$

The effect of atomic layer number fluctuations was treated by Clemens and Gay [78]. They have taken a superlattice constituted by N layers of an amorphous block of thickness D_A (the A atoms are supposed to be non-diffracting) and n_B atomic layers of B, and they have supposed that D_A is a random variable. They have used a formalism based on the Patterson function method. The structure factor of one layer $|F(\mathbf{q})|^2$ is particularly simple and constant in that case since the fluctuating layers do not diffract. Two cases have been considered by Clemens and Gay [78]. In the first case they have assumed that the distribution of D_A is continuous and Gaussian of half-width σ about an average spacing $\langle D_A \rangle$. To evaluate the double sum in eq. 25, one notes that there are N terms in the sum where n and m are equal, but only N - 1 where $m = n \pm 1$, N - 2 where $m = n \pm 2$ and so on... Then one obtains:

$$D(\mathbf{r}) * D(-\mathbf{r}) = \sum_{n=-N+1}^{N-1} (N - |n|)\ \delta(z - nD_L) \tag{26a}$$

$$\text{where: } D_L = D_A + n_B d_B \tag{26b}$$

If the distribution function of distances between adjacent layers is called $d(D_L)$ ([15] p.603) and the distribution function of nth nearest layer distance, $d_n(nD_L)$, one obtains

$$\begin{aligned} \langle P_1(\mathbf{r})\rangle &= \langle D(\mathbf{r}) * D(-\mathbf{r})\rangle \\ &= \sum_{n=-N+1}^{N-1} (N - |n|) \int \delta(z - nD_L)\, d_n(nD_L)\, dD_L \end{aligned} \tag{27}$$

The calculation of $d_n(nD_L)$ can be made recursively as shown in refs [48, 15]

$$d_n(nD_L) = d_{n-1}((n-1)D_L) * d(D_L) \tag{28}$$

With a Gaussian distribution of first neighbours, eq.27 becomes [15]

$$\langle P_1(\mathbf{r})\rangle = \sum_{n=-N+1}^{N-1} (N - |n|) \left(\frac{1}{\sqrt{\pi |n| \sigma^2}}\ \delta(z - n\Lambda) * \exp(-\frac{z^2}{|n|\sigma^2}) \right) \tag{29a}$$

$$\Lambda = \langle D_A\rangle + n_B d_B \tag{29b}$$

is the average period of the multilayer. The Fourier transform of eq.29 gives ($\mathbf{q} = q\mathbf{u}_z$)

$$|S(\mathbf{q})|^2 = N + 2 \sum_{n=1}^{N-1} (N - n) \cos(n\Lambda q)\ \exp(-\frac{n\sigma^2 q^2}{4}) \tag{30}$$

We may also calculate $|F(\mathbf{q})|^2$

$$|F(\mathbf{q})|^2 = \left(\frac{f_B \sin(n_B q d_B/2)}{\sin(q d_B/2)} \right)^2 \tag{31}$$

Figure 5, presents X-ray spectra calculated with the help of formulae 30 and 31 [83]. The calculations have been done for a tungsten carbon multilayer with the assumption of completely (110) textured tungsten layers and amorphous carbon. The parameters used were: 10 sequences of 5 atomic layers of tungsten with interplanar lattice spacing d (d is chosen as the unit length), separated by carbon nonscattering layers of mean thickness $\langle D_A\rangle = 5.22$ d, the thicknesses of the carbon layers being distributed with a distribution function of width $\sigma = 0.5$ d [83]. In figure 5a, we show the total intensity $I(\mathbf{q})$ and in figure 5b the Fourier transform of the layer distribution function. In this case we have again the intensity given by formula 4a but that is only because we have made the simplifying assumption that the carbon layers are non diffracting. One notes that this kind of imperfection affects drastically the high angle part of the calculated spectra but has a relatively small influence on small angle diffraction.

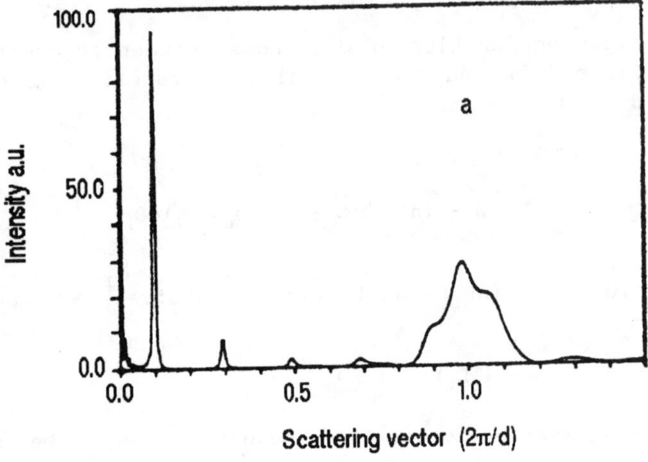

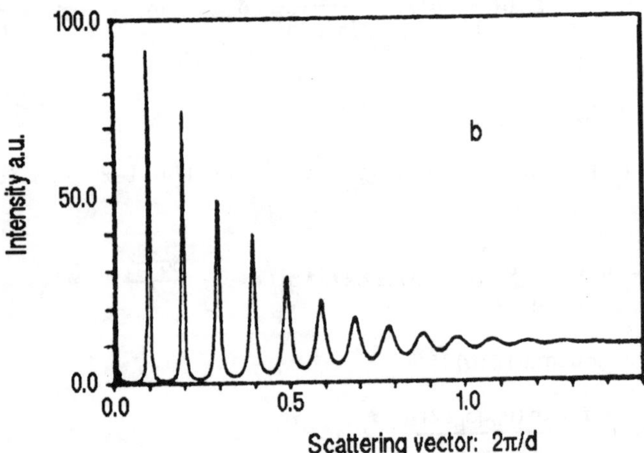

Fig. 5 Calculated intensity (a) and layer interference function (b) for a tungsten carbon multilayer [83]. The unit length, d, is the (110) tungsten interplanar spacing. There are 10 bilayers of 5 atomic planes of tungsten and random layers of carbon of an average thickness of 5.22 d and a mean square deviation of 0.5 d.

If we do a similar calculation near the main peak of the multilayer for two values of σ, 0.5 d and 0.2 d, the influence of the layer thickness distribution can be seen clearly, showing that, as the distribution becomes narrower, the peaks become narrower and more intense. The specific effect of the imperfection appears even in this very simplified model, causing degradation of the peak width. On the other hand, interdiffusion, as in 2.2, only affects the peak intensity. However the effect of discrete distributions is markedly different [78]. Clemens and Gay have shown that in that case the effect of layer thickness distribution on large angle peak width is considerably weaker, the disorder leading mainly to an intensity reduction [78]. To see the effect of other imperfections on line broadening it is useful to discuss another analytical model.

3.2 The effect of interface distortions

In order to understand the large linewidths obtained in some metallic superlattices like Nb/Cu [85, 86], Fe/Mg [74], Mo/Ni [68] or Fe/V [87], J.P. Loquet et al. [81] have tested the effect of lattice mismatch at the interface. In the preceding part we have either assumed a sharp coherent interface (in 2.1) with the interplane distance at the interface equal to $(d_A + d_B)/2$, or a smooth coherent transition between d_A and d_B as in paragraph 2.2. Here we assume that there are random imperfections at every interface like misfit dislocations or even a paracrystalline transition layer (See Ketterson et al.[88]) or random thickness errors leading to a non-integer number of layers. All these kinds of errors can be described, at least approximately, as a random interplanar distance at the interfaces. Following eq. 2 the scattering amplitude is expressed as $(\mathbf{q} = q\mathbf{u}_z)$

$$f(\mathbf{q}) = \Sigma_i f_i \exp(-i\mathbf{q}\mathbf{r}_i) = \Sigma_n f_A \exp(-iqnd_A) + \exp(-iq((n_A - 1)d_A + a_1))\Sigma_n f_B \exp(-iqnd_B) + \exp(-iq((n_A - 1)d_A + (n_B - 1)d_B + a_1 + a_2))\Sigma_n f_A \exp(-iqnd_A) + \ldots \quad (32)$$

The a_j are the random interface widths. Formula 32 can be compacted if one defines Δa_j:

$$\Delta a_j = a_j - (d_A + d_B)/2 \quad (33)$$

The average of the intensity becomes

$$\langle|f(\mathbf{q})|^2\rangle = A^2\langle S_1^2\rangle + B^2\langle S_2^2\rangle + AB[\langle S_1 S_2^*\rangle \exp(i(q\Lambda)/2) + \langle S_2 S_1^*\rangle \exp(-i(q\Lambda)/2)] \quad (34)$$

where we have defined

$$S_1 = 1 + \exp(-iq\Lambda)\exp(-iq(\Delta a_1 + \Delta a_2)) + \ldots + \exp(-iNq\Lambda)\exp(-iq(\Delta a_1 + \ldots + \Delta a_{2N-3} + \Delta a_{2N-2})) \quad (35)$$

and

$$A = f_A \frac{\sin(n_A q d_A/2)}{\sin(q d_A/2)} \quad (36)$$

S_2 and B being analogous to S_1 and A. The first result of this kind of disorder, as can be seen in formula 34, is that a structure factor and an interference function can no longer be defined, the two interferences inside the layers and between layers are mixed. The crucial quantities in this formula are the four averages: $<S_1^2>$, $<S_2^2>$, $<S_1S_2^*>$ and $<S_2S_1^*>$. Such products can be expressed, for example, as

$$<S_1^2> = <\sum_{n=-N+1}^{N-1} (N - |n|) \exp(-iqn\Lambda) \exp(-iq(\sum_{j=1}^{2n} \Delta a_j))> \quad (37)$$

These quantities are calculated with the same method as above with the help of distribution functions, $d_n(n\Delta a)$, the $d_n(n\Delta a)$ being defined and calculated as the $d_n(nD_L)$ in 3.1, if one assumes that the distribution of Δa, $d_1(\Delta a)$ is continuous and is Gaussian of width σ around 0 [81]:

$$\begin{aligned} I(\mathbf{q}) &= <|f(\mathbf{q})|^2> \\ &= N\left(A^2 + B^2 + 2AB\cos\left(\frac{q\Lambda}{2}\right)\exp\left(-\frac{\sigma^2q^2}{4}\right)\right) + 2\sum_{n=1}^{N-1}(N-n) \\ &\left\{(A^2+B^2)\exp\left(-\frac{2n\sigma^2q^2}{4}\right)\cos\left(\frac{2nq\Lambda}{2}\right) + AB\left[\cos\left(\frac{(2n+1)q\Lambda}{2}\right)\right.\right. \\ &\left.\left.\exp\left(-\frac{(2n+1)\sigma^2q^2}{4}\right) + \cos\left(\frac{(2n-1)q\Lambda}{2}\right)\exp\left(-\frac{(2n-1)\sigma^2q^2}{4}\right)\right]\right\} \end{aligned} \quad (38)$$

As pointed out before, in formula 38 we have a mixing between the interferences inside the layers and between the layers. J.P.Locquet et al.[81] have used formula 38 to analyse the lack of long range order in various superlattices. They have found that for small values of Λ, the value of the interfacial disorder, as deduced from σ, is close to the lattice mismatch for four different metallic systems. If Λ increases, the value of the interfacial disorder increases. This can be interpreted as due to the presence of interfacial dislocations. The width of the distribution of interfacial distances is then due to the combination of two types of disorder. The first one is strictly related to the difference in lattice parameter, while the second one is due to the dislocation density at the interface which grows as the strain energy, that is, with multilayer volume and wavelength.

Formula 38 presents also some interesting features for the interpretation of anomalous X-ray experiments, since it can be put in the form

$$I(\mathbf{q}) = f_A^2\, S_{AA}(q) + f_B^2\, S_{BB}(q) + 2f_Af_B\, S_{AB}(q) \quad (39)$$

with obvious definitions of $S_{AA}(q)$, $S_{BB}(q)$ and $S_{AB}(q)$. The variation of f_A or f_B with the help of the anomalous effect leads to direct determinations of $S_{AA}(q)$, $S_{BB}(q)$ and $S_{AB}(q)$ and of some of the multilayer parameters as the dispersion in interface width and the mean interface width if this theory is used for comparisons with experiments. The use of the anomalous dispersion can be essential; simple theories like those of 2.1 have four parameters:n_A, d_A, n_B, d_B, the diffused model has six parameters, the last model of this part has also six parameters, then the intensity variations measured with the anomalous dispersion are very useful to obtain some of these parameters.

One example of the application of formula 38 is shown on figure 6 where we have reported calculated and experimental X-ray spectra for an iron ruthenium superlattice of 50 periods with $n_{Fe} = 6$, $n_{Ru} = 13$, $d_{Fe} = 0.2085$ nm, $d_{Ru} = 0.21423$ nm and $\sigma = 0.02$ nm. With this rather large value of σ, one is able to reproduce the experimental linewidth, but the intensity of the satellite is too high. This is certainly a sign that there is not only interface disorder in this kind of superlattice but also a small interdiffusion (one or two layers of alloys).

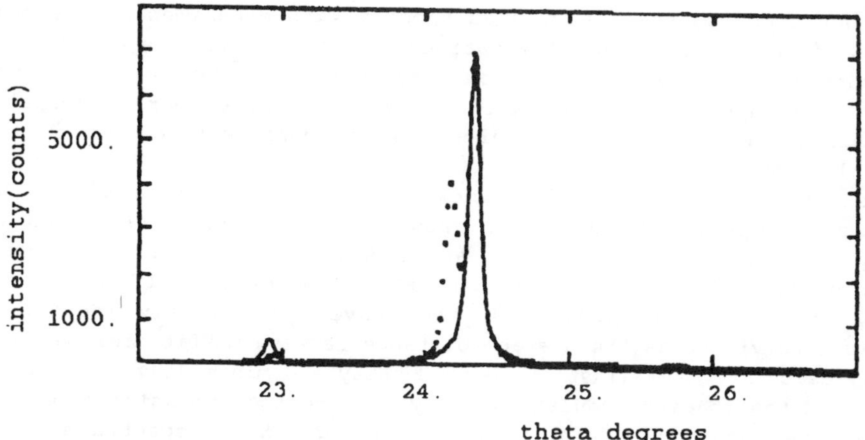

Fig. 6 Calculated X-ray spectrum for an iron ruthenium superlattice of 50 periods with $n_{Fe} = 6$, $n_{Ru} = 12$, $d_{Fe} = 0.2085$ nm, $d_{Ru} = 0.21423$ nm and $\sigma = 0.1$ nm. The calculated spectrum (continuous line) is compared with an experimental diffractogram taken on the same type of superlattice (points).

In these two paragraphs we have presented in some detail two analytical models of disordered multilayers and discussed their relevance for the interpretation of experimental X-ray spectra. The next section will discuss some other models of imperfections.

3.3 Some other theoretical results

Few studies have been performed in the domain of Monte Carlo simulations. One may however quote the results of Mitura and Mikołajczak [89]. They have studied the influence of grain size using different numbers of atomic planes, the influence of fluctuations in the number of A and B planes and of the interface sharpness. The calculations were performed for Bi/Sb and Pb/Ag multilayers. The grain size and the number of planes have been assumed to follow a Gaussian distribution, the interface sharpness is introduced via the size of a region where the atomic structure factor and the interplane distance vary linearly. The main conclusion of this work[89] is that both a small grain size and a large fluctuation in the number of planes broaden the diffraction peaks, but that a large interface thickness only weakens the satellite intensities. These kinds of conclusion are qualitatively in agreement with our preceding discussion. One other recent simulation has been proposed by M.B.Stearns [90]. She also uses a trapezoïdal composition profile and has assumed some mixture of crystallites of different orientations, namely in Co/Cr multilayers, the coexistence of (0002) and $(10\bar{1}0)$ Co hcp planes. These crystallites are assumed to be independent scatterers. She has also introduced Lorentzian broadened lines to obtain information on the multilayer coherence length. She has then compared some calculated spectra with experiments. The main conclusion of this work is that the various parameters used are essentially independent and that the peak positions may be used to determine the mean interplanar distance and the modulation wavelength. The peak intensities depend mainly on the interface thickness and the peak widths on the coherence length.

Ariosa et al.[80] have used a somewhat different point of view. They have assumed a partially coherent growth of the multilayers and an interplay between coherency strain and misfit dislocations. The main parameter of the model is thus the degree of coherency δ. If $\delta = \infty$, the coherency is perfect while if $\delta = 0$, they have a completely incoherent growth. δ can be viewed as the average distance between misfit dislocations. Then, they calculated the elastic energy density and minimising this energy they obtained the spacing modulation. They deduced the composition profile and stress free spacing modulation from low angle X-ray spectra and the actual spacing profile of an Mo/V superlattice was self-consistently determined by reproducing the high angle X-ray diffractograms. This work is one of few attempts to obtain coherent results from low and high angle X-ray patterns. The calculated spectra reproduce quite well the intensities but not really the peak widths.

Recent work in the field of semiconducting multilayers has shown the important role of other kinds of imperfections. For instance Neumann et al.[91] have emphasized the effect of substrate terracing. The terracing induces a tilt of the film with respect to the substrate and a transverse broadening of the superlattice peaks. This kind of terracing also produces dislocations at the various steps of the terraces.

Concluding part 3, we have reported some of the analytical calculations of the influence of multilayer imperfections on diffraction spectra. This kind of effect has also been studied experimentally during recent years with, however, few conclusive results. The main type of imperfection which has been treated with the help of some simplified models is the lack of constancy of the multilayer period. It was seen that this essentially produces line broadening. The two main types of imperfection in multilayers, interdiffused interfaces and period errors, act very differently. The former weakens the satellite intensities while the latter broadens all peaks and obviously also weakens their intensities.

Up to this point in the calculations of diffraction amplitudes it was tacitly assumed that the thicknesses of the films were small enough for the first Born approximation to be reasonable. Then the calculation of experimental reflectivity is done using standard methods: the experimental spectra are corrected for effects of polarization, Lorentz factor, Debye Waller factor and absorption (see e.g. James [49] for the details of the intensity measurements and for the calculation of the correction factors in X-ray diffraction). The proper application of all these corrections must be done carefully even though they are not always obvious in multilayer systems whose physical properties are unknown (for instance the Debye Waller factor and the absorption coefficient may be very different from those of usual bulk phases). These procedures are certainly correct for almost all the high angle studies on metallic systems, but in the small angle limit and in some semiconducting systems the samples fall outside the range where the kinematical theory is valid.

One useful criterion to see if we are in the dynamical regime is the use of the Darwin width Δq_{Darwin} which is given by (in Å^{-1}) [83]

$$\Delta q_{Darwin} = 50.2 r_e \frac{|F|}{Vq} \qquad (40)$$

where V is the unit cell volume and r_e is the classical electron radius: 2.82 10^{-5} Å. This formula has been simplified by assuming that the beam is linearly polarized and that the atomic structure factor is real.

When the experimental line width approaches the Darwin width the dynamical effects become important and one has to be cautious with the direct use of intensities. The Darwin width is typically in the 0.001 degree range for high angles but may be only in the 0.01 degree range at small angles. Another criterion was used by Bartels et al.[92]. They use the criterion that the absolute values of reflectivity must be smaller than 10% to use the first Born approximation. This again is correct in many metallic systems at large angles but is not always respected in the small angle region.

In addition to the multiple scattering effects, at small angles there also are optical effects. The X-ray refractive index is no longer one and the phenomenon of total reflection and shifting of the peak positions must be considered. All these points will be discussed in the next two parts.

4 OPTICAL THEORY OF MULTILAYERS

Two main theoretical approaches have been used to describe low angle X-ray spectra and to calculate reflectivity curves in the $\theta, 2\theta$ mode. The first type of theory was mainly used by workers coming from the area of visible light optics. The atomic structure is neglected and the various sublayers of A and B are considered as homogeneous media of constant electronic density. The multilayer system is then a periodic stack of two layers, and each layer has its own optical properties. The systematic study of the propagation of an electromagnetic wave in a stratified medium was made by Abelés [22]. His work deals with transparent materials but can be extended to any medium where one can define complex dielectric constants (or optical indices). In 1931 Kiessig [93] used a similar type of approach to explain the interference fringes observed in X-ray reflection at grazing angles on thin film coatings. Since then, a large amount of work has been done in the field of small angle X-ray reflectometry mainly initiated by the pioneering work of the Orsay group [94, 95, 96] and two principal types of calculation have been developed: the recursion method [23, 26], and the matrix method [25, 97, 98].

The second type of theory developed in the early days of X-rays by Darwin [99], Prins [100], Laue [101] and Ewald [102] is now called under the generic name of dynamical theory [47, 48]. A useful description of the method is proposed in a review article by Ewald [103]. The details of the calculations can be found in W.H. Zachariasen [47]. This theory was initially developed for perfect crystals, but can be extended to multilayers relatively easily if the main periodicity of electron density is only considered and in a first approach, if the crystalline periodicity is neglected [104]. In the following paragraph (4.1), we briefly present X-ray optics. The principle of the calculation of low angle X-ray spectra by the recursive method is presented in 4.2, and a brief description of the matrix method in 4.3. Then in 4.4 we will show how to take into account various types of imperfection and discuss some experimental results. The dynamical theory and some of its results will be discussed in part 5.

4.1 X-ray optics

4.1.1 Optical constants

The optical indices for X-ray wavelengths can be derived in two ways. The first one is to recall that X-rays are electromagnetic waves and then will behave like visible light and the dielectric constant or optical indices may be calculated by standard methods. The second one which can also be used to define the optical indices of neutrons is more direct and is presented briefly in appendix A.

As given by formula A4 the optical index of X-rays is

$$n = 1 - 2\pi N\, r_e \frac{f}{k_0^2} \tag{41a}$$

or, in a more usual way

$$n = 1 - N\, f\, r_e \frac{\lambda^2}{2\pi} \tag{41b}$$

In formula 41, f is the complex atomic structure factor, $f = f_0 + f' + if''$, where we have, as usual, separated the atomic structure factor in a normal part f_0 and an anomalous one, $f' + if''$, r_e is the classical electron radius

$$r_e = \frac{e^2}{4\pi\varepsilon_0\, mc^2} \tag{41c}$$

and N is the atomic density. The index may be written

$$n = 1 - \delta - i\beta \tag{42a}$$

where

$$\delta = N\,(f_0 + f')\,r_e\,\frac{\lambda^2}{2\pi}$$

and

$$\beta = Nf''\,r_e\,\frac{\lambda^2}{2\pi} \tag{42b}$$

More precisely, if there are different atomic species in our sample eq. 42b becomes

$$\delta = \Sigma_i\,N_i\,(f_0 + f')_i\,r_e\,\frac{\lambda^2}{2\pi} \tag{42c}$$

and

$$\beta = \Sigma_i\,N_i\,f''_i\,r_e\,\frac{\lambda^2}{2\pi} \tag{42d}$$

where the sums are over all the components of the sample.

In the classical X-ray range ($\lambda \cong 0.1 - 1$ nm) and far from the absorption edges, $f_0 + f'$ is practically equivalent to Z, the atomic number of the element and β is directly related to the linear absorption coefficient μ through the relation

$$\beta = \frac{\mu\lambda}{4\pi} \tag{43}$$

Typical values of δ are in the 10^{-5} range, while β is in the 10^{-6} range for classical X-ray wavelengths. The values used for the calculations are often those published by Cromer and Liberman [105], leading to an estimated accuracy of 1%.

4.1.2 Reflection at a single boundary

The coefficient of reflectivity r_{12} at a boundary between medium 1 and medium 2 is defined as the ratio of the reflected amplitude to the incident amplitude. It is calculated by expressing the usual continuity of tangential components of the electric and magnetic field and one obtains the well known Fresnel formula [50]

$$r_{12} = \frac{\xi_1 - \xi_2}{\xi_1 + \xi_2} \tag{44a}$$

with

$$\xi_i = \sqrt{n_i^2 - \cos^2\theta} \tag{44b}$$

for "s" polarization (Transverse electric wave or electric field perpendicular to the plane of incidence) or

$$\xi_i = \sqrt{\frac{n_i^2 - \cos^2\theta}{n_i^4}} \tag{44c}$$

for "p" polarization (Transverse magnetic wave or electric field in the plane of incidence). θ is defined, as usual in X-rays, as the angle between the X-ray direction in vacuum and the interface plane and so the various formulae differ from that of visible light optics.

The Snell Descartes law can be used to calculate the angle θ_i between the interface plane and the direction of propagation in the ith medium. With our definition of the angles, it is expressed as

$$n_i \cos\theta_i = \cos\theta \tag{45}$$

The first consequence of 44 and 45 is the existence of a critical angle below which the reflectivity is 100 % (total reflection) for a non-absorbing medium. The critical angle is obtained if one puts $\cos\theta_i = 1$ in formula 45, leading to $\cos\theta_c = n$, or

$$\sin\theta_c = \sqrt{2\delta} \tag{46}$$

if the imaginary part of the optical index is neglected. Formula 44 together with the analog formula for the transmitted amplitude

$$t_{12} = \frac{2\xi_1}{\xi_1 + \xi_2} \tag{47}$$

are the basis of the calculations of the meflective power of the multilayer.

4.2 Multilayer theory by the recursive method

4.2.1 Description of the method

The recursive method was developed initially by Parrat [106]. A recent account of this method has been given by Underwood and Barbee [26] and Pardo et al.[54]. One starts with the calculation of the ratio between the reflected wave and the incident wave amplitude, R_{12} of a single film of medium 2 of thickness d_2 situated between medium 1 and 3. The same method as above (continuity of tangential components of electric and magnetic field) leads to:

$$R_{12} = r_{12} + \frac{R_{23}t_{12}t_{21}\exp(2i\phi_2)}{1 + r_{12}R_{23}\exp(2i\phi_2)} \tag{48a}$$

where $\phi_2 = 2\pi d_2\xi_2/\lambda$ is the amplitude factor for transmission through the layer thickness and R_{23} is the reflected amplitude at the bottom of the layer. If the conservation of energy is used, $t_{12}t_{21} + r_{12}^2 = 1$, one obtains:

$$R_{12} = \frac{r_{12} + R_{23}\exp(2i\phi_2)}{1 + r_{12}R_{23}\exp(2i\phi_2)} \tag{48b}$$

These expressions of reflectivity coefficients mean that the ratio R_{12} of the reflected and incident field at the 1-2 interface can be expressed with the help of the various Fresnel coefficients as a function of the ratio of the incident and reflected field R_{23} at the 2-3 interface. More generally in a stack like that of figure 1, one has a relation between the ratio of the incident and reflected fields at interface j-1j, R_{j-1j} and the ratio of the incident and reflected fields at interface jj+1, R_{jj+1}

$$R_{j-1j} = \frac{r_{j-1j} + R_{jj+1}\exp(2i\phi_j)}{1 + r_{j-1j}R_{jj+1}\exp(2i\phi_j)} \tag{49}$$

Then, the computational procedure starts from the bottom of the stack where the substrate is supposed to be infinite leading to $R_{nn+1} = 0$ and to $R_{n-1n} = r_{n-1n}$ and, with successive applications of eq. 49, R_{12} and the reflectivity $R = R_{12}R_{12}^*$ can be calculated. One interest of this method is that we need no assumptions on the periodicity, the layer number and so on... So such computations may be used for disordered stacks, stacks with thickness errors, stacks with composition gradients (one then uses smaller sublayers each with different refractive indices)... The computer program is relatively easy to write but the computation time becomes long for a number of layers greater than 100.

4.2.2 Example of calculations

Figure 7 shows an example of calculations made with the help of such a computer program [53]. The theoretical results are compared with an experimental reflectivity curve obtained with CuKα_1 radiation for a W/C stack prepared by T.W. Barbee and containing 23 layers deposited on a silicon wafer by magnetron sputtering.

Some of the main features of the low angle reflectivity spectra are clearly seen on figure 7. There are 3 main peaks which are the Bragg peaks of the multilayer, but their positions are not exactly at Bragg angles like $2\Lambda\sin\theta = n\lambda$, but according to the expression for ϕ given above at positions where we have $\phi_w + \phi_c = 2n\pi$ or

$$2d_w\sqrt{n_w^2 - \cos^2\theta} + 2d_c\sqrt{n_c^2 - \cos^2\theta} = n\lambda \tag{50}$$

Here $d_w = 0.8$ nm and $d_c = 2.575$ nm, a small layer of silicon oxide has also been detected between the silicon substrate and the first carbon layer. The calculation leading to the theoretical curve in figure 7 has assumed an index profile at each interface given by the error function

$$n(z) = n_1 + (n_1 - n_2)\mathrm{erf}(z,\sigma) \tag{51a}$$

with

$$\mathrm{erf}(z,\sigma) = \frac{1}{\sigma\sqrt{2\pi}} \int_{-\infty}^{z} \exp(-\frac{y^2}{2\sigma^2})\,dy \tag{51b}$$

In formula 51 σ is an estimation of the interface roughness. In the calculations shown in figure 7, σ is different for the C/W interface and the W/C interface. σ(C/W) = 0.8 nm and σ(W/C) = 0.325 nm. This can be interpreted as due to islands of tungsten which do not completely coalesce. Some other interesting points are visible on figure 7.

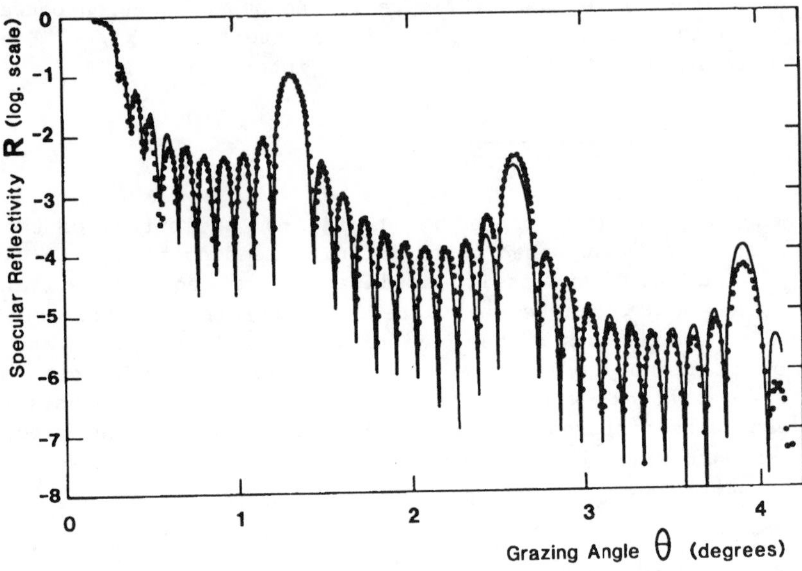

Fig. 7 Measured (dots) and calculated (solid line) X-ray reflectivity spectra. The experimental reflectivity curve is obtained with CuKα_1 radiation for a W/C stack prepared by T.W. Barbee and containing 23 layers deposited on asilicon wafer by magnetron sputtering. The multilayer period is 3.375 nm. The parameters of the calculation are discussed in the text.

The total reflection limit is seen at the inflexion point of $R(\theta)$ and can be used to determine the mean value of δ and to deduce the average value of the density of the multilayer. Between the 3 Bragg peaks, there are the 9 secondary peaks expected if there are 11 periods like in this stack. This structure can be directly deduced from the kinematical formula. In the case of a small number of periods, there are secondary minima of $|S(\mathbf{q})|^2 = (\sin(Nq\Lambda/2)/\sin(q\Lambda/2))^2$. However the positions of these minima are not at the angles deduced from that formula. These remarks show the great power of the method which is able to detect fine detail of the multilayer structure like concentration profile asymmetry.

4.3 The matrix method

In the following, the coordinate system is defined as follows. The z axis is perpendicular to the interface plane, the y axis is in the interface plane and in the plane of incidence and the x axis is in the interface plane and perpendicular to the plane of incidence. The amplitude of the x component of the electric field (in "s" polarization) or of the magnetic field (in "p" polarization) is called U and the amplitude of the y component of the magnetic field (in "s" polarization) or minus the y component of the electric field (in "p" polarization) is called V.

Abelés [22] has shown that there exists a relation between the values of U and V at z and the values of U and V at z'

$$\begin{pmatrix} U(z) \\ V(z) \end{pmatrix} = G(z,z') \begin{pmatrix} U(z') \\ V(z') \end{pmatrix} \tag{52}$$

G(z,z') is the propagator of the medium between z and z', and can be expressed as

$$G(z,z') = G(z).G^{-1}(z') \tag{53}$$

with

$$G(z) = \begin{pmatrix} \frac{i}{\xi}\cos(k_z z) & \sin(k_z z) \\ i\xi\sin(k_z z) & \cos(k_z z) \end{pmatrix} \tag{54}$$

where ξ is defined as above (eq.44) and $k_z = 2\pi\xi/\lambda$. Another form of matrix formalism has been developed by Lee [107] and Megademini and Pardo [108]. One defines two waves, the first one the T wave propagates in the positive z direction and the other one, the R wave in the opposite direction. This kind of representation doesn't allow the continuity relations to be easily expressed, but allows the direct calculation of reflectivity: $R = |R/T|^2$. In this representation G is diagonal

$$G(z) = \begin{pmatrix} \exp(ik_z z) & 0 \\ 0 & \exp(-ik_z z) \end{pmatrix} \tag{55}$$

Finally we need the matrix which transforms T and R at the interface between two different media 1 and 2, that is F_{12}. F_{12} can be calculated if the transformation between UV and TR representations [54] is used:

$$F_{12} = \frac{1}{2} \begin{pmatrix} 1 + \frac{\xi_2}{\xi_1} & 1 - \frac{\xi_2}{\xi_1} \\ 1 - \frac{\xi_2}{\xi_1} & 1 + \frac{\xi_2}{\xi_1} \end{pmatrix} \tag{56a}$$

or

$$F_{12} = \frac{1}{t_{12}} \begin{pmatrix} 1 & r_{12} \\ r_{12} & 1 \end{pmatrix} \tag{56b}$$

The calculation of the matrix of a bilayer uses the product of two G and F matrices and is given by

$$M_{bi} = F_{21}\, G(d_1)\, F_{12}\, G(d_2) \tag{57a}$$

$$M_{bi} = \frac{1}{1 - r_{12}^2} \begin{pmatrix} \exp(i\Phi) - r_{12}^2 \exp(-i\Delta) & r_{12}(\exp(i\Delta) - \exp(-i\Phi)) \\ r_{12}(\exp(-i\Delta) - \exp(i\Phi)) & \exp(-i\Phi) - r_{12}^2 \exp(i\Delta) \end{pmatrix} \tag{57b}$$

with $\Phi = \phi_1 + \phi_2$, $\Delta = \phi_1 - \phi_2$ and $\phi_i = 2\pi d_i \xi_i / \lambda$. The matrix of the whole medium can thus be calculated as

$$M = F_{01}\, F_{21}^{-1}\, M_{bi}^{N}\, F_{2s} \tag{58}$$

F_{01} is the interface matrix between the first layer and vacuum, F_{2s} is the interface matrix between the second layer and the substrate, F_{21}^{-1} has been introduced because F_{21} is counted N times in the power of M_{bi}. Several analytical forms of M have been proposed [22,54], for instance if M_{bi} is diagonalized, M may be computed with the help of eq. 58. Finally the reflectivity is deduced from M by assuming there are no R waves emerging from the substrate, so we have $T_0 = M_{11}T_s$ and $R_0 = M_{21}T_s$. The reflectivity is then

$$R = \left|\frac{M_{21}}{M_{11}}\right|^2 \tag{59}$$

The matrix method is more convenient for a perfect stack where the direct diagonalisation of M_{bi} allows economies in computing time. The matrix method can also be used to treat analytically some of the imperfections, as we will see in the next paragraph.

4.4 Imperfection effects

The various kinds of imperfections we have to treat are essentially of three different types at low angles:

i) Thickness errors or variations as above.

ii) Diffuse interfaces or interfacial roughness.

iii) Various kinds of inhomogeneities in the samples such as porosity, impurities, local fluctuations of composition.

We will treat mainly the effects of the thickness errors and interfacial roughness.

4.4.1 Thickness errors

To calculate analytically the influences of thickness errors, one cannot use directly the models we have developed, but if one notes that for angles a little larger than θ_c the Fresnel reflection coefficients are weak, equation 48 can be written as

$$R_{12} = r_{12} + R_{23}\exp(2i\phi_2) \quad (60)$$

where we have neglected in the denominator $r_{12} R_{23}\exp(2i\phi_2)$ which is much smaller than one in many cases. Equation 60 is often a reasonable approximation to calculate reflectivity spectra. Some useful applications of this kind of method have been made by Rieutord et al.[109] in the case of X-ray characterization of Langmuir Blodgett films. We have then a kind of kinematical approximation but with true phases. For a bilayer eq. 60 becomes

$$R_f = r_{21} + r_{12}\exp(2i\phi_1) + R_i\exp(2i(\phi_2 + \phi_1)) \quad (61)$$

where R_f and R_i are the amplitude ratios of the reflected and the incident waves before and after the bilayer. For a perfect stack, the amplitude ratio of the multilayer is obtained as

$$R = r_{21}(1 - \exp(2i\phi_1)) \frac{1 - \exp(2iN\phi_L)}{1 - \exp(2i\phi_L)} \quad (62)$$

with

$$\phi_L = \phi_2 + \phi_1$$

Formula 62 is a good approximation to calculate, for example, the peak positions in a perfect multilayer and to see again the condition of peak extinction ($\phi_1 = p\pi$). To show the influence of thickness errors, we rewrite eq. 62 in a slightly different form

$$R = r_{21} + r_{12}\exp(2i(\phi_{10}+ \Delta\phi_1)) + (r_{21} + r_{12}\exp(2i(\phi_{10} + \Delta\phi_3))) \exp(2i(\phi_{Lo} + \Delta\phi_2 + \Delta\phi_1)) + \ldots \quad (63)$$

ϕ_{Lo} is the mean phase change during a bilayer travel and ϕ_{10} is the mean phase change during the first layer crossing while the $\Delta\phi_i$ values are random errors. The equation 63 can be written as

$$R = r_{21} S_1 + r_{12} S_2 \exp(2i\phi_{10}) \quad (64a)$$

with

$$S_1 = 1 + \exp(2i(\phi_{Lo} + \Delta\phi_2 + \Delta\phi_1)) + \ldots \quad (64b)$$

$$S_2 = \exp(2i\Delta\phi_1) + \exp(2i(\phi_{Lo} + \Delta\phi_3 + \Delta\phi_2 + \Delta\phi_1)) + \ldots \quad (64c)$$

If the intensity is calculated, an expression like in 3.2 (equation 38) is obtained:

$$\langle R\rangle = R_0 N + 2R_1 \sum_{n=1}^{N-1} (N - n) \exp(-4n\langle\Delta\phi^2\rangle)\cos(2n\phi_{Lo})) \quad (65)$$

where

$$R_0 = 2r_{21}^2 \ (1 - \cos(2\phi_{10})\exp(-2<\Delta\phi^2>))$$

$$R_1 = 2r_{21}^2 \ (1 - \cos(2\phi_{10})\exp(+2<\Delta\phi^2>))$$

$<\Delta\phi^2>$ is the variance in ϕ due to thickness errors. This kind of calculation has not been used in the small angle region, but the calculation of the effect of thickness errors has been made mainly with non-cumulative errors [109, 110] leading to a formula notably different from eq. 65

$$R(\theta_P) = 4r_{21}^2 \ N \ (1 - \exp(-4<\Delta\phi^2>)) + 4r_{21}^2 \ N^2 \ \sin^2\phi_{10}\exp(-4<\Delta\phi^2>) \quad (66)$$

θ_P is the peak angle. Formula 66 was used in ref.[110] to analyse calculated curves, and to compare fitted $<\Delta\phi^2>$ with those introduced in MonteCarlo calculations. Surprisingly, the agreement between the calculated $<\Delta\phi^2>$ and the imposed one was reasonable. Direct comparisons with experiments remain to be made.

4.4.2 Interfacial roughness

The influence of interfacial roughness on the reflectivity of single interfaces has received some attention in visible light optics. Beckmann and Spizzichino [111] have given a detailed presentation of the methods and of the assumptions used:

i) The slope of the surface irregularities are small enough to neglect the polarization due to local variations in the angle of incidence.

ii) The radius of curvature of any surface irregularity is large compared to λ.

iii) Multiple scattering and shadowing between irregularities at the same interface are not considered.

iv) Scattering in the bulk media bounding the interface is negligible.

The main mathematical quantities which are included in the theory are the variance σ of the distribution of the interface positions z (this distribution is usually assumed to be Gaussian) and the coherence length ζ of the variations of the interface positions. L is the characteristic dimension of the sample on which the experiments are done or with more precision the dimension of the first Fresnel zone. The selfcorrelation function of the interface position is defined as (x is one of the coordinates parallel to the interface)

$$C(x) = \frac{1}{L} \int_0^L z(s)z(x+s)\, ds \quad (67a)$$

and assumed to behave like an exponential

$$C(x) = \exp(-\frac{|x|}{\zeta}) \quad (67b)$$

The calculations of Beckmann and Spizzichino [111] show in that case that the specular reflectivity is given by a modified Fresnel coefficient r' which is expressed as

$$r' = r \exp(-\frac{k_{z1}^2 \sigma^2}{2}) \quad (68)$$

k_{z1} is the z component of **k** in the first medium. In the other directions θ (outside of specular reflection θ_s where specular means that the reflected flux obeys the laws of classical optics for plane interfaces) one has a diffuse incoherent flux whose intensity is given by

$$I_D = \frac{I_0}{1 + y^2} \quad (69a)$$

and

$$y = \frac{2\pi\zeta}{\lambda}(\cos\theta_s - \cos\theta) \, , \; I_0 \, \alpha \, \frac{\zeta}{L} \quad (69b)$$

This function of cosθ is a Lorentzian of width $\lambda/\pi\zeta$ and its magnitude depends of the ratio between the coherence length ζ and the Fresnel zone size L. The extension of this formalism to X-rays is not so obvious.

There is some confusion about the term of interfacial roughness in small angle X-ray studies. Strictly speaking, interface roughness must induce scattered radiation in the whole space as in formula 69. The majority of workers in small angle X-ray reflectometry assume howewer that the total integrated scattered radiation remains localized near the specular direction, and thus expect that the main consequence of interfacial roughness is only to reduce the specular reflectance, being formally equivalent to a coherent, smooth concentration wave (non-rectangular concentration wave). This is equivalent in terms of Beckmann's formalism of assuming a coherence length smaller than the characteristic dimensions of the first Fresnel zone and bigger than the X-ray wavelength.

The first Fresnel zone is approximately elliptic. The calculation of a and b the length of the principal axes of the ellipse can be easily made and leads to

$$a = \sqrt{\frac{\lambda R_S R_D}{R_S + R_D}} \quad (70a)$$

and

$$b = \frac{a}{\sin\theta} \quad (70b)$$

R_S and R_D are the distances of the sample from source and detector respectively. The small axis a is of the order of 10^{-2} mm so the first condition may be easily respected, but the second one is a little more restrictive.

One recent work [112] has attempted to go beyond this kind of consideration, but the calculations of D.G. Stearns [112] are restricted to weak scattering within the first Born approximation. One may however quote a significant result of this work: the scattered amplitude depends only on specular reflection of the average interface profile, then in the weak scattering case the assumptions made by the majority of workers are valid. However some problems still remain and one has to be cautious with simple applications of this kind of method.

The theory of Beckmann has been extended to multilayers by Eastman [113] and some similar calculations in the X-ray region have been proposed by Megademini and Pardo [108] and Vidal and Vincent [25]. For instance Vidal and Vincent have shown that one cannot exactly propose simple formulae like 68 and that the whole transmission matrix of the two media and interfaces is modified. The extension of the theory to multilayers also gives complicated results [113] and only numerical procedures are practicable.

The situation is however not so bad, as some workers have proposed a lot of approximate methods which seem about right. For instance in the case of short wavelength variations of interface width ($\lambda_i < 1$ µm), Croce and Nevot [114] have proposed that the specular reflectivity of a rough interface with a Gaussian profile is

$$R(\theta) = R_0(\theta)\ \exp(-4k_{z1}\ k_{z2}\ \sigma^2) \tag{71}$$

where k_{z1}, k_{z2} are the normal components of the propagating wave vector in the two media, σ is the variance of the height distribution at the interface and $R_0(\theta)$ is the expected reflectivity of a flat interface. For long wavelengths they give

$$R(\theta) = R_0(\theta)\ \exp(-4k_{z1}^2\ \sigma^2) \tag{72}$$

For a multilayer they propose to modify in formulae 48 and 49 the various Fresnel coefficients r by

$$r' = r\ \exp(-2k_{z1}^2\ \sigma^2) \tag{73a}$$

or

$$r' = r\ \exp(-2k_{z1}\ k_{z2}\ \sigma^2) \tag{73b}$$

A remark can be made here, if one uses eq. 73 in 48a, one can take into account some losses of intensity, if one calculates the transmission coefficients directly with aq.47. But, if one applies aq. 48b, the energy losses are neglected. The last method seems to give the best results. This may be explained by the small diffuse intensity in the X-ray wavelength region.

A simple explanation of these latter formulae can be given if instead of the true roughness, an intermediate layer of refractive index $n_i = g(n_A, n_B)$, g being some appropriate function, and thickness 2σ is used. Formulae 56 and 55 with three layers, give a product of matrices

$$M = G_1(d_A - \sigma)\ F_{1i}\ G_i(2\sigma)\ F_{i2}\ G_2(d_B - \sigma) \tag{74}$$

instead of $G_1(d_A)\ F_{12}\ G_2(d_B)$ and then one may define a new effective interface matrix F'_{12} as

$$F'_{12} = G_1^{-1}(d_A)\ G_1(d_A - \sigma)\ F_{1i}\ G_i(2\sigma)\ F_{i2}\ G_2(d_B - \sigma)\ G_2^{-1}(d_B) \qquad (75a)$$

or

$$F'_{12} = G_1(-\sigma)\ F_{1i}\ G_i(2\sigma)\ F_{i2}\ G_2(-\sigma) \qquad (75b)$$

The evaluation of eq.75 can be made if the various exponentials are developed up to second order and if one takes $n_i^2 = (n_A^2 + n_B^2)/2$ (mean of dielectric constants). Then one obtains

$$F'_{12} = C(1-2) \begin{pmatrix} 1 & r_{12}(1 - 2k_{z1}k_{z2}\sigma^2) \\ r_{12}(1 - 2k_{z1}k_{z2}\sigma^2) & 1 \end{pmatrix} \qquad (76)$$

where

$$C(1-2) = \frac{1 - 0.5(k_{z1} - k_{z2})^2\ \sigma^2}{t_{12}} \qquad (76b)$$

If one use

$$r_{12}\ (1 - 2k_{z1}k_{z2}\sigma^2) \equiv r'_{12} = r_{12}\ \exp(-2k_{z1}\ k_{z2}\ \sigma^2) \qquad (77a)$$

in eq. 76b, and if t_{12} is augmented in the same approximation like

$$t'_{12} = t_{12}\ \exp(+0.5(k_{z1} - k_{z2})^2\ \sigma^2) \qquad (77b)$$

one obtains an expression of F'_{12} which is analogous to eq.56b. The various Fresnel coefficients are given by eq.77. which is in qualitative agreement with eq.73.

Figures 8 and 9 present some other comparisons of theoretical results with experiments. In figure 8, the sample is a stack of 40 tungsten and carbon layers, deposited on float glass by Ph. Houdy. the measurement were performed at the Institut d'Optique (Orsay). The significant feature of the experimental spectra is the asymmetry of the Bragg peaks. A noticeable broadening is observed on the small angle side, together with an increase of the secondary peaks. This can only be reproduced if thickness variations of 0.01 nm per period are assumed, the average thickness $\langle d_c \rangle$ and $\langle d_w \rangle$ being equal to 1.49 and 2.69 nm respectively. The continuous curve in figure 8 corresponds to a gradual change of $\sigma_{w/c}$ from 0.2 to 0.39 nm at the surface, the $\sigma_{c/w}$ being constant and equal to 0.9 nm. Figure 9 shows the comparison between calculated and experimental spectra on a Fe/Ru HCP superlattice [34], with 1.2 nm of iron and 4 nm of ruthenium. An interface roughness of 0.1 nm has been introduced, but the main originality of these calculations is the introduction of random thickness errors by Monte-Carlo methods. A rather high value of the mean thickness error has to be introduced in order to reproduce the experimental results.

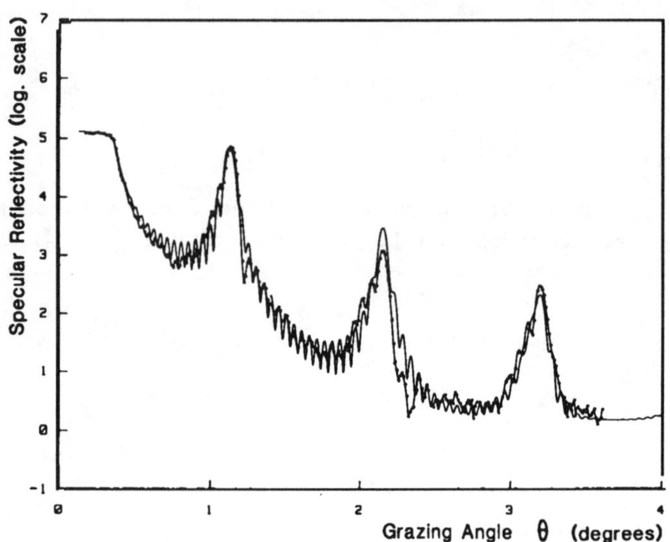

Fig. 8 Measured (dots) and calculated (solid line) reflectivity curve for a C/W stack containing 40 layers on a float glass substrate. The mean period is 4.18 nm.The calculation assumes a linear decrease of the period from the bottom to the top and a correlated interface roughness increase. Note that the reflectivity is not in absolute units.

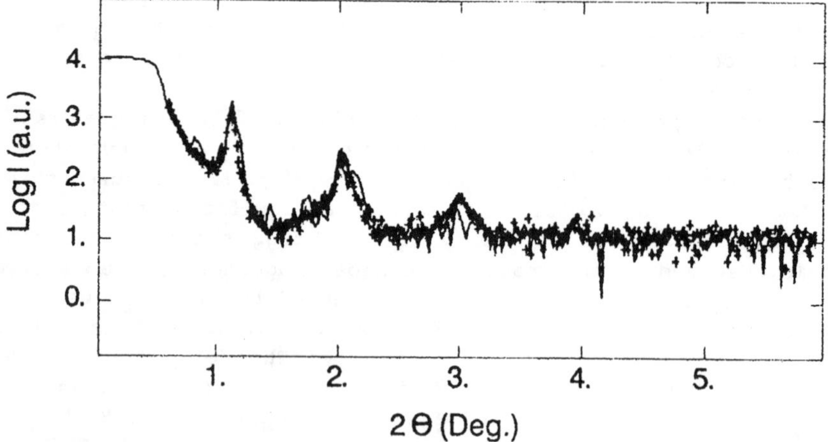

Fig 9: Calculated (solid line) and experimental (crosses) small angle diffraction profiles of a Fe/Ru superlattice with 50 periods of 1.2 nm of iron and 4 nm of ruthenium, the calculation includes an interfacial roughness of 0.1nm and cumulative thickness errors of 0.2 nm.

But this value (0.2 nm) is understandable if one imagines that the main imperfections in this non lattice-matched superlattice are monatomic steps.

These last two examples show that the best actual method to understand clearly the experimental spectra is to do simulations. The major inconvenience of this way of analysis is the computing time.

5 DYNAMICAL THEORY OF X-RAYS

In this part we will give briefly some principles and several results of X-ray dynamical theory. The general principle of the dynamical theory can be presented simply; one has to find the solutions of Maxwell equations with locally varying dielectric constants. The solutions of these equations can be found either by direct resolution of these differential equations or by summing the Born series [47, 48].

Recent discussion of the use of dynamical theory in the small angle X-ray region was made by Kadin and Keem [115] and Hollanders and Thijsse [116]. We have attempted here to present a simplified version of the theory and some of its simple applications to multilayers.

5.1 General results

The basic ingredient is the susceptibility χ, which is given by

$$\chi(\mathbf{r}) = -r_e \rho_e(\mathbf{r}) \frac{\lambda^2}{\pi} \tag{78}$$

where $\rho_e(\mathbf{r})$ is the density of electrons. In Fourier space χ can be expressed as

$$\chi(\mathbf{q}) = -r_e \rho_e(\mathbf{q}) \frac{\lambda^2}{\pi} \tag{79a}$$

with

$$\begin{aligned} \rho_e(\mathbf{q}) &= \frac{1}{V} \int_V \rho_e(\mathbf{r}) \exp(-i\mathbf{qr}) \, d^3r \\ &= \frac{1}{V} \int_V \rho(\mathbf{r}) \, f(\mathbf{r}) \exp(-i\mathbf{qr}) \, d^3r \end{aligned} \tag{79b}$$

The optical index n is related directly to $\chi(\mathbf{q}=0)$

$$n = 1 + \frac{1}{2} \chi(0) \tag{80}$$

The Maxwell equations in $\mathbf{r}$ space are

$$\Delta\mathbf{D} + \mathrm{curlcurl}(\chi\mathbf{D}) + \mathbf{k}^2 \mathbf{D} = 0 \tag{81}$$

where $\mathbf{D}$ is the electric displacement field of a monochromatic electromagnetic wave of wave vector $\mathbf{k}$. If one seeks $\mathbf{D}$ as a linear combination of Bloch wave corresponding to the eigenfunctions in the periodic medium under study:

$$\mathbf{D} = \Sigma_h \mathbf{D}_h \exp(-i(\mathbf{k}_i + \mathbf{q}_h)\mathbf{r}) \tag{82}$$

where $\mathbf{q}_h$ is a reciprocal space vector of the periodic medium. An infinite set of linear differential equations is obtained from equation 81. The most useful approximation in order to solve this system is to reduce this infinite set of waves to only two, (two-wave approximation), retaining in the development of **D** the components of wave vector closest to the incident and reflected wave (in the Bragg case). The incident wave is

$$\mathbf{D}_i = \mathbf{D}_{0i} \exp(-i\mathbf{k}_i\mathbf{r}) \tag{83a}$$

of amplitude $\mathbf{D}_{0i}$ and wave vector $\mathbf{k}_i$ in the medium (**k** and $\mathbf{D}_0$ in vacuum) and the reflected wave is

$$\mathbf{D}_r = \mathbf{D}_{rh} \exp(-i(\mathbf{k}_i + \mathbf{q}_h)\mathbf{r}) \tag{83b}$$

of amplitude $\mathbf{D}_{rh}$ and wave vector $\mathbf{k}_h = \mathbf{k}_i + \mathbf{q}_h$. The total wave is

$$\mathbf{D} = \mathbf{D}_i + \mathbf{D}_r = \mathbf{D}_{0i} \exp(-i\mathbf{k}_i\mathbf{r}) + \mathbf{D}_{rh} \exp(-i(\mathbf{k}_i + \mathbf{q}_h)\mathbf{r}) \tag{84}$$

The insertion of this kind of wave in eq.81 gives a system of two homogeneous equations for the amplitudes

$$\begin{aligned} &(k_i^2(1 - \chi(0)) - k^2)\ D_{0i} - k_i^2 C\chi(-\mathbf{q}_h) D_{rh} = 0 \\ &-k_h^2 C\chi(\mathbf{q}_h) D_{0i} + (k_h^2(1 - \chi(0)) - k^2)\ D_{rh} = 0 \end{aligned} \tag{85}$$

The secular equation of this homogeneous system is

$$(k_i^2(1 - \chi(0)) - k^2)\ (k_h^2(1 - \chi(0)) - k^2) = k_i^2 k_h^2 C^2 \chi(\mathbf{q}_h)\chi(-\mathbf{q}_h) \tag{86}$$

and the non-trivial solutions are related by

$$\frac{D_{rh}}{D_{0i}} = \frac{k_i^2(1 - \chi(0)) - k^2}{C\ k_i^2\ \chi(-\mathbf{q}_h)} = x \tag{87}$$

C is a polarization factor; C=1, "s" polarization and C=|cos2θ|, "p" polarization [47]. Equations 86 and 87 can be combined as a quadratic equation in x namely

$$C\ \chi(-\mathbf{q}_h)\ x^2 + x\ (2\chi(0) - \alpha) + C\ \chi(\mathbf{q}_h) = 0 \tag{88}$$

α is given by

$$\alpha = \frac{\mathbf{q}_h^2 + 2\mathbf{q}_h.\mathbf{k}}{k^2} \tag{89}$$

Equation 88 has two solutions, x' and x", which give two solutions for $\mathbf{k}_i$ and $\mathbf{D}_{0i}$ and for $\mathbf{k}_h$ and $\mathbf{D}_{rh}$. Then $\mathbf{D}_i$ and $\mathbf{D}_r$ can be written as

$$\mathbf{D}_i = \exp(-i\mathbf{kr})(\mathbf{D}'_{0i}\exp(-i\Delta k'_{iz}z) + \mathbf{D}''_{0i}\exp(-i\Delta k''_{iz}z)) \quad (90a)$$

$$\mathbf{D}_r = \exp(-i(\mathbf{k} + \mathbf{q}_h)\mathbf{r})(\mathbf{D}'_{0i}\,x'\exp(-i\Delta k'_{iz}z) + \mathbf{D}''_{0i}\,x''\exp(-i\Delta k''_{iz}z)) \quad (90b)$$

where we have used the continuity of the wave vector to write

$$\mathbf{k}_i = \mathbf{k} + \Delta k_{iz}\mathbf{u}_z \quad (90c)$$

Δk_{iz} is calculated with equation 86

$$2\Delta k_{iz}\sin\theta = k(\chi(0) + C\,x\,\chi(\mathbf{q}_h)) \quad (90d)$$

In the Bragg case the appropriate boundary conditions are

$$\mathbf{D}'_{0i} + \mathbf{D}''_{0i} = \mathbf{D}_0 \quad (91a)$$

continuity at the z=0 plane and

$$\mathbf{D}'_{0i}x'\exp(-i\Delta k'_{iz}t) + \mathbf{D}''_{0i}x''\exp(-i\Delta k''_{iz}t) = 0 \quad (91b)$$

infinite substrate, t is the sample thickness. The reflecting power is then the square of the ratio of the two amplitudes at z=0

$$R_q = \left|\frac{D_{rh}}{D_{0i}}\right|^2 = \left|\frac{D'_{0i}x' + D''_{0i}x''}{D_0}\right|^2 \quad (92a)$$

or

$$R_q = \left|x'x''\frac{\exp(-i\Delta k'_{iz}t) - \exp(-i\Delta k''_{iz}t)}{x'\exp(-i\Delta k'_{iz}t) - x''\exp(-i\Delta k''_{iz}t)}\right|^2 \quad (92b)$$

A rather complicated equation can be derived from eq.92 in the general case of absorbing crystals. One defines

$$L = \sqrt{y^2 - C^2\,\chi(\mathbf{q}_h)\,\chi(-\mathbf{q}_h)} \quad (93a)$$

with

$$y = \frac{2\chi(0) - \alpha}{2}$$

or

$$y = (\sin\theta - \sin\theta_B)\sin 2\theta_B + \chi(0) \quad (93b)$$

Here θ_B is the usual Bragg angle corresponding to $\mathbf{q}_h$

$$\mathbf{q}_h = \frac{4\pi}{\lambda}\sin\theta_B\,\mathbf{u}_z \quad (93c)$$

Then

$$\frac{I(\theta)}{I(0)} = |\, C\, \chi(\mathbf{q}_h)\,|^2 \, \frac{|\sin(ANL)|^2}{Den} \tag{94a}$$

with

$$Den = |\, y \sin(ANL) + i\, L \cos(ANL)\, |^2 \tag{94b}$$

N is the number of diffracting planes. $t = N\Lambda$ and $A = \dfrac{4\pi^2\Lambda}{\lambda^2 q_h}$.

5.2 Application to multilayers

Equation 94 can be applied relatively easily to the case of a perfect multilayer, and has some interesting features. Firstly, one obtains a shift of the peak position which, if absorption is neglected, is given by

$$\sin\theta = \sin\theta_B + \frac{\langle\delta\rangle}{\sin\theta_B} \tag{95}$$

θ_B is determined by the usual Bragg formula: $2\Lambda\sin\theta_B = n\lambda$, $\langle\delta\rangle$ is the average value of δ in the multilayer. This is nearly the same position as that obtained in eq.50. This can be used at small angles to determine the true value of θ_B and then the period of the superlattice. One way is to plot $\sin\theta_n/n$, where θ_n is the measured position of the nth order small angles peak, versus $1/n^2$. The slope of the straight line obtained gives $2\Lambda\langle\delta\rangle/\lambda$, and the intercept with the y axis gives $\lambda/2\Lambda$. We have then a direct determination of Λ, the multilayer period and of $\langle\delta\rangle$ which gives the average optical index.

Approximate expressions of eq.94 can be obtained in two different limits. In the case of weak absorption, one has [47]

$$\frac{I(\theta)}{I(0)} = C^2 \, \frac{|\chi(\mathbf{q}_h)|^2}{y^2 + L^2 \cot g^2(ANL)} \tag{96}$$

Formula 96 may be used to calculate peak intensities:

$$\frac{I(\theta_p)}{I(0)} = \tanh^2(NA\chi(\mathbf{q}_h)) \tag{97}$$

In this case it is very difficult to simply determine the Fourier coefficients of the multilayer. The kinematical limit is only obtained for $N\chi(\mathbf{q}_h) \ll 1$.

If the absorption becomes important one has

$$\frac{I(\theta)}{I(0)} = C^2 \, \frac{|\chi(\mathbf{q}_h)|^2}{B + \sqrt{B^2 - |\chi(\mathbf{q}_h)|^4}} \tag{98}$$

B is given by $B = |L|^2 + |y|^2$. The peak intensity is obtained as

$$\frac{I(\theta_p)}{I(0)} = C^2 \frac{\pi^2 |\chi(\mathbf{q}_h)|^2}{\lambda^2 \mu^2} \qquad (99)$$

Formula 99 has two interesting features, the intensity is independent of N and is proportional to $|\chi(\mathbf{q}_h)|^2$. Then it can be used to determine approximate values of Fourier coefficients.

A slightly different form of the two-wave dynamical theory has been used in reference [112]. The incident wave and the detected wave amplitudes S and T are calculated recurrently by a matrix method. Conclusions of this work are equivalent to our preceding results, but maybe the formulae obtained are more easily programmable with a desk computer.

The same type of equations has been used by Tapfer and Ploog for the analysis of high-angle diffraction on strained semiconducting superlattices [19].

Equation 99 was used recently for a non conventional experiment by Simon et al.[38]. They have studied the behaviour of Nd/Fe multilayers at small angle and in the anomalous regime. Absorption being important in this case, the use of eq.99 can be made. $\chi(\mathbf{q}_h)$ for the pth Bragg peak is expressed as

$$\chi(\mathbf{q}_h) = -r_e \left(\frac{f_{Fe}}{V_{Fe}} - \frac{f_{Nd}}{V_{Nd}} \right) \frac{\lambda^2}{\pi\Lambda} C_p \qquad (100)$$

In this formula, we have explicitly shown the dependence of X-ray scattering from electronic density, f_{Fe} or f_{Nd} being an effective number of electrons and C_p is the pth Fourier coefficient of the volume concentration of iron.

Thus as f_{Fe} and f_{Nd} vary near the absorption edges, the peak intensitie variation can be calculated as a function of the atomic volume ratio V_{Fe}/V_{Nd} which is then deduced [38].

6 SOME EXPERIMENTAL DEVICES

Two main types of apparatuses will be briefly described, the standard reflectometer of the Institut d'Optique (Orsay) which is one of the best devices to do small angle reflectometry and the high resolution diffractometer which is used at Pont à Mousson.

The reflectometer at the Institut d'Optique is limited to a range in theta from 0 to 5°. The source and detector arms are 51.25 cm long, the angular mechanical accuracy is 1 second of arc. Arm rotations are accomplished with stepping motors. The X-ray source is a sealed tube with a copper anticathode with a linear focus (0.1 x 10 mm) operating at 40 kV and 10 mA with a standard generator. The cooling water is regulated both in flow and temperature. The $K\alpha_1$, $K\alpha_2$ doublet is selected by a plane quartz crystal $(10\bar{1}1)$ monochromator. The $K\alpha_2$ radiation is then eliminated with a variable width slit (60 µm in most cases) located 35 cm from the monochromator.

The sample is maintained on the object holder by vacuum. The surface alignment is achieved using a retractable knife edge or a preadjusted telescopic sight.

During the analysis, the illuminated area varies as $1/\sin\theta$. In the vicinity of the total reflection angle θ_c, it is frequently of the order of 1 cm^2 , the use of sufficiently large samples must then be promoted. It is also important to control the surface flatness in order to avoid geometrical effects which may perturb the measurements. A thick(> 6 mm) float glass substrate often offers good flatness and has a good roughness. The various types of monocrystalline wafers used for obtaining monocrystalline superlattices (silicon, gallium arsenide, sapphire,...) however show facets and figure deviations which may perturb absolute reflectivity measurements. An object holder with mechanical deformations (obtained with a set of screws) was useful in some cases to compensate these deviations, while controlling with visible light, the interference colours generated by the air gap between the wafer and the reference optical surface.

The detection of the reflected photons is performed with the help of a proportional counter. A turret equipped with calibrated attenuation screens is used to keep the counting rate smaller than 10^3 counts per second. The slit located in front of the counter has a variable width from 10 to 200 μm, which enables compromise to be made between high resolution (at low grazing angles) and high counting rate (at higher grazing angles). The entire experimental set up is driven by a computer.

The goniometer used at Pont à Mousson is devoted mainly to high-angle, high-resolution experiments, but may reasonably be used at low angles and it is thus a good compromise. The X-ray beam is produced by a cobalt anticathode and passes through a four crystal Ge(220) monochromator [13] leading to a good selection of the $K\alpha_1$ line and to a very weak intrinsic angular divergence of about 3×10^{-2} mrad. The main advantage of the goniometer is then a linear position sensitive detector coupled with a microcomputer which simultaneously drives the goniometer. The results collected are then transfered to a larger computer which does all the analysis.

The linear position sensitive detector has an angular resolution of 0.03°. Concerning displacements of the detector, the entrance window stays tangent to the 2θ platform and may be kept fixed in position during the θ-scan of the multilayer or moved at a chosen angular speed. Discussions on the performance of the experimental setup are given in ref. [117].

Some examples of the use of these two systems have been shown previously: figures 3, 4, 6 and 9 have been recorded at Pont à Mousson showing the main interest of our experimental device: small angle and high angle experiments, but also one of its weaknesses, the small number of counts leading to low statistics. Figures 7 and 8 have been obtained at the Institut d'Optique and show the high quality of the recorded spectra with very good statistics.

Some other types of experiments can also be made on our diffractometer, for instance rocking curves have been recorded on some superlattices and experiments at different orientations of q are also possible [34], then in conclusion of this part, one can see that there are different possibilities of experimental set up , one has to make the choice as always between high specificity with high quality or some versatility but at the cost of some loss in quality.

7 CONCLUSION

In this chapter we have built up a picture of the structure of the multilayer starting from the more perfect one, the monocrystalline superlattices, and showing some of the imperfections we may encounter. We have then tried to show the usefulness of diffraction experiments to characterize these structures and their various defects.

We have shown how the comparison between experimental X-ray spectra of and analytical models or computer simulations can give useful information on the structure and the main defects of multilayers. This method is actually one of the best to characterize multilayers. Other techniques of characterization are also useful, but they are either destructive (transmission electron microscopy, various types of profiling...), or have not the same spatial resolution (Rutherford backscattering).

We hope that our main goal which was to develop some qualitative understanding of the main phenomena encountered in diffractions by multilayers has been achieved. We hope also that the few detailed calculations we have presented have shown that much work remains to be done both in the theoretical and the experimental sides of diffraction by multilayers. It is rather amusing that the references to theoretical results are either somewhat old or very recent showing once more the interplay between theory and experiment which, in this field, has only recently become possible.

APPENDIX A: calculation of the optical index [118]

If one writes the total wave function diffused by a sample of thickness t as

$$\Phi(\mathbf{r}) = \exp(-ik_0 r) + \Phi_{diff}(r) \qquad (A1)$$

where $\Phi_{diff}(\mathbf{r})$ the diffused wave function is given here by the integration of the amplitude diffused individually by each atom

$$\Phi_{diff}(\mathbf{r}) = \int f(\mathbf{x})\, r_e\, N(\mathbf{x}) \frac{\exp(ik_0\mathbf{R})}{R} d^3x \qquad (A2)$$

with $\mathbf{r} = \mathbf{R} + \mathbf{x}$, x being the coordinate inside the diffracting sample, R the distance from the sample point to the detector and r_e is the classical electron radius and $f(\mathbf{x})$ the atomic scattering factor at $\mathbf{x}$. If one has $r >> x$, one gets

$$\Phi_{diff}(\mathbf{r}) = -i2\pi N\, r_e\, f\, t \frac{\exp(-i\mathbf{k}_0\mathbf{r})}{k_0} \qquad (A3)$$

Then $\Phi(\mathbf{r})$ is not too far from $\exp(-i\mathbf{k}_0\mathbf{r} + ik_0\, t\, (n-1))$, if one defines n the usual optical index as

$$n = 1 - \frac{2\pi N\, r_e\, f}{k_0^2} \qquad \text{(A4)}$$

In the preceding calculations, N is the average atomic density and f the average atomic structure factor. One can see that formula (A4) is also valid for neutrons if one replaces fr_e by b.

REFERENCES

1) Dumond D.W.M. and Youtz J.P.: Phys. Rev., 1935, 48 ,703
2) Dumond D.W.M. and Youtz J.P.: J. Appl. Phys., 1940, 11, 357
3) Esaki L.and Tsu R.: IBM Res. Note, 1969, RC2418
4) Esaki L.and Tsu R.: IBM J. Res. Develop., 1970, 14 , 61
5) Esaki L.: Synthetic modulated structures, Chang Leroy L. and Giessen B.C. editors, Academic Press, 1985, p.3
6) Spiller E.: Appl. Optics, 1976, 15, 2333
7) Dhez P.: SPIE proceedings, 1986, 733, 308
8) Cook H.E. and Hilliard J.E. : J. Appl. Phys., 1969, 40, 2191
9) Philofsky E.M. and Hilliard J.E. : J. Appl. Phys., 1969, 40, 2198
10) Greer A.L. and Spaepen F.: Synthetic modulated structures, Chang Leroy L. and Giessen B.C. editors, Academic Press 1985, p 419
11) Piecuch M.: Rev. Phys. Appl., 1988, 23, 1727
12) Thaller B.J., Ketterson J.B. and Hilliard J.E.: Phys. Rev. Lett. , 1978, 41 , 336
13) Bartels W.J.: J. Vac. Sci. Techno., 1983, B1, 338
14) Quillec M., Goldstein L., Le Roux G., Burgeat J. and Primot J.: J. Appl. Phys, 1984, 55, 2904
15) Guinier A.: X-ray diffraction in crystals, imperfect crystals and amorphous bodies, W.H. Freeman and co San Francisco, 1963 and Guinier A.: Théorie et technique de la radiocristallographie, third edition Dunod Paris 1964.
16) Segmüller A. and Blakeslee A.E.: J. Appl. Cryst., 1973, 6 , 19
17) Tagaki S.: Acta Cryst., 1962, 15, 1311
18) Taupin D.: Bull. Soc. Fr. Min. Cristall., 1964, 87 , 469
19) Tapfer L. and Ploog K.: Phys. Rev. B , 1986, 33 , 5565
20) Nevot L. and Croce P.: Rev. Phys. Appl., 1980, 15, 761
21) Spiller E.: Rev. Phys. Appl., 1988, 23, 1687
22) Abélés F.: Ann. Phys. Fr., 1950, 12, 596
23) Nevot L. and Croce P.: J. Appl. Cryst., 1975, 8, 304
24) Pardo B., Nevot L. and André J.M.: SPIE Proc., 1988, 984, 855
25) Vidal B. and Vincent P.: Appl. Opt., 1984, 23, 1794
26) Underwood J.H. and Barbee T.W.: Appl.Opt., 1981, 20, 3027
27) Cook H.E., deFontaine D. and Hilliard J.E.: Acta Met., 1969, 17, 765
28) Bruson A., Marchal G. and Piecuch M.: J. Appl. Phys., 1985, 58, 1229
29) Moruzzi V.L. and Marcus P.M.: Phys. Rev. B, 1988, 38, 1613
30) Durbin S.M., Cunningham J.E. and Flynn C.P.: J. of Phys. F, 1982,

12, 75

31) Cunningham J.E. and Flynn C.P.: J. of Phys. F, 1985, 15, L221

32) Kwo J., Gyorgy E.M., McWhan D.B., Hong M., Di Salvo F.J., Vettier C. and Bower J.E.: Phys. Rev. Lett., 1985, 54, 2700

33) Jonker P.T., Walker E., Kisker E., Prinz G.A. and Carbone C.: Phys. Rev. Lett., 1986, 57, 142

34) Maurer M., Ousset J.C., Ravet M.F. and Piecuch M.: Europhysics Lett., 1989, 9, 803

35) Maurer M. , Ousset J.C., Ravet M.F., Piecuch M and Sanchez J.P.: MRS Proc., 1989, in press

36) Etienne P., Creuset G., Friedrich A., Nguyen Van Dau F., Fert A. and Massies J.: Appl. Phys. Lett., 1988, 53, 162

37) Rhyne J.J., Erwin R.W., Borchers J., Salamon M.B., Du D. and Flynn C.P.: Metallic multilayers and epitaxy edited by Hong M., Wolf F. and Gubser D.C., 1988, Metalurgical Society, Warrendale, p11

38) Simon J.P., Lyon O., Bruson A., Marchal G. and Piecuch M.: J. Appl. Cryst, 1988, 21, 317

39) De Andres A., De Santis M., Raoux P., Maurer M., Ravet M.F. and Piecuch M. conference on utilisation of synchrotron radiation Roma 1989

40) Chauvineau J.P.: Rev Phys Appl., 1988, 23, 1645

41) Mezei F.: Comm. Phys., 1976, 1 , 81

42) Piecuch M., Baczewski L.T., Durand J., Marchal G., Delcroix P. and Nabli H.: J. de Phys., 1988, 49, C8-1755

43) Shinjo T., Mibu K., Ogawa S., and Hosoito N.: MRS Proceedings, 1989, in press

44) Krishnan R., Porte M. and Tessier M.: J. de Phys. 1988, 49, C8-1751

45) Schwarz R.B. and Johnson W.L.: Phys. Rev. Lett.,1983,51, 415

46) Le Boité M.G., Traverse A., Névot L., Pardo B. and Corno J.: Nucl. Inst. and Meth. Phys. Res., 1988, B29, 653

47) Zachariasen W.H.: Theory of X-ray diffraction in crystals, 1967, Dover, New York

48) Cowley J.: Diffraction physics, second edition 1981, Elsevier New York

49) James R.W.: The optical principles of the diffraction of X-rays, second edition 1982, Ox Box Woodbrige Connecticut

50)Born M. and Wolf E.: Principles of optics sixth edition , 1984, Pergamon New York

51) Mc Whan D.B.: Synthetic modulated structures, Chang Leroy L. and Giessen B.C. editors, Academic Press 1985 p 43

52) Spiller E.: Rev Phys Appl., 1988, 23, 1687

53) Nevot L., Pardo B. and Corno J.: Rev. Phys. Appl., 1988, 23, 1675

54) Pardo B., Megademini T. and André J.M.: Rev Phys Appl., 1988, 23, 1579

55) Dhez P. editor: Rev Phys Appl., 1988, 23

56) Houdy P., Bodart V., Hily C., Ruterana P., Nevot L., Arbaoui M., Alehyane N., and Barchewitz R.: SPIE proc., 1986, 733, 389

57) Youn K.B., Sella C., Barchewitz R., Arbaoui M., and Alehyane N.: SPIE proc., 1986, 733, 316

58) Ruterana P., Chevalier J.P. and Houdy P.: J. Appl. Phys., 1989, 65, 3907

59) Henke B.L.: A.I.P. Proc., 75, Low energy X-ray diagnostic, 1981, edited by Atwood D.T. and Henke B.L., 85

60) Spiller E.: A.I.P. Proc., 75, Low energy X-ray diagnostic, 1981, edited by Atwood D.T. and Henke B.L., 124

61) Henke B.L., Lee P., Tanaka T.J., Shimabukuro R.L. and Fujikawa B.K.: Atomic and nuclear data tables , 1982, 27, 1

62) Janot C., Roth M., Marchal G., Piecuch M. and Bruson A.: J. Non Cryst. Solids, 1986, 47, 1751

63) Cebollada A., Martinez J.L., Callego J.M., de Miguel J.J., Miranda R., Ferrer S., Batallan F., Fillion G. and Rebouillat J.P.: Phys. Rev. B, 1989, 39, 9726

64) Vergnat M., Houssaini S., Dufour C., Bruson A., Marchal G. Mangin P., Erwin R., Rhyne J.J. and Vettier C.: Phys. Rev. B, 1989, 40, to appear.

65) Dufour C., Bruson A., George B., Marchal G., Mangin P., Vettier C., Erwin R. and Rhyne J.J.: Solid state comm., 1989, 10, 963

66) Barbee T.W.: Physics, fabrication and applications of multilayered structures,1988, Dhez P. and Weisbuch C. editors, North Holland Amsterdam

67) Courant R. and Hilbert D.: Methods of Mathematical Physics, Interscience New York 1960

68) Fleming R.M., Mc Whan D.B., Gossard A.C., Wiegmann W. and Logan R.A.: J. Appl. Phys., 1980, 51, 557

69) Dernier P.D., Moncton D.E., McWhan D.B., Gossard A.C., and Wiegmann W.: Bull. of Am. Phys. Soc., 1977, 22, 293

70) Lepetre Y.: Thesis , 1986, Marseille

71) Chauvineau J.P., Corno J., Naccache D., Nevot L., Pardo B.and Valiergue L.: J. Opt. (Paris), 1985, 15, 265

72) Nevot L., Pardo B., Chauvineau J.P., Corno J. and Valiergue L.: Le vide, les couches minces, 1986, supplement 233, 63

73) Dupuis V., Ravet M.F., Tête C. and Piecuch M.: SPIE Proc., 1989, in press

74) Fuji Y., Ohnishi T., Ishihara T., Yamada Y., Kawaguchi K., Nakayama N. and Shinjo T.: J. Phys. Soc. Jap., 1986, 55, 251

75) Kahn M.R., Chun C.S.L., Felcher G.P., Grimsditch M., Kueny A., Falco C.M. and Schüller I.K.: Phys. Rev. B, 1983, 27, 7186

76) Sevenhans W., Gijs M., Bruynseraede Y., Homma H. and Schüller I.K.: Phys. Rev. B, 1986, 34, 5955

77) Nakayama N., Takahashi K., Shinjo T., Takada T. and Ichinose H.: Jap. Jour. Appl. Phys., 1986, 25, 552

78) Clemens B.M. and Gay J.G.: Phys. Rev. B, 1987, 35, 9337

79) Chzran D. and Dutta P.: J. Appl. Phys., 1986, 59, 1504

80)Ariosa D., Fischer O., Karkut M.G. and Triscone J.M.: Phys. Rev. B, 1988, 37, 2415

81) Locquet J.P., Neerinck D.,Stockman L., Bruynseraede Y. and Schüller I.K.: Phys. Rev. B, 1988, 38, 3572

82) McWhan D.B., Gürvitch M., Rowell J.M. and Walker L.R.: J. Appl. Phys., 1983, 54, 3886

83) McWhan D.B.: Physics, fabrication and applications of multilayered structures,1988, Dhez P.and Weisbuch C. editors, North Holland Amsterdam

84) Hendricks S. and Teller E.: Jour. Chem. Phys., 1942, 10, 147

85) Schüller I.K.: Phys Rev. Lett., 1980, 44, 1597

86) Lowe W.P., Barbee T.W., Geballe T.H. and McWhan D.B.: Phys. Rev. B, 1981, 24, 6193

87) Endoh Y., Kawaguchi K., Hosoito N., Shinjo T., Takada T., Fuji Y. and Onishi T.: J. Phys. Soc. Jap., 1984, 53, 3481

88) Shiroishi Y., Sellers C., Hilliard J.E. and Ketterson J.B.: J. Appl. Phys., 1986, 62, 3694

89) Mitura Z. and Mikołajczak P.: J. Phys. F, 1988, 18, 183

90) Stearns M.B.: Phys. Rev. B, 1988, 38, 8109

91) Neumann D.A., Zabel H. and Morkoç H.: J. Appl. Phys., 1988, 64, 3024

92) Bartels W.J., Hornstra J. and Lobcek D.J.W.: Acta Cryst., 1986, A42, 539

93) Kiessig H.: Ann. Phys., 1931, 10, 769

94) Croce P. and Nevot L.: J. App. Cryst., 1974, 7, 125

95) Nevot L.: Thesis, Orsay, 1978

96) Croce P., Nevot L. and Pardo B.: Nouv. Rev. Opt. Appl., 1972, 3, 37

97) Lee P.: Opt. Comm., 1981, 37, 159

98) Megademini T.: Thesis, Orsay, 1984

99) Darwin C.G.: Phil. Mag., 1914, 27, 315 and 675

100) Prins J.A.: Zeitschr. Phys., 1930, 63, 477

101) Laue M. von,: Ergeb. Exackt Naturwiss., 1931, 10, 133

102) Ewald P.P.: Ann. Phys. , 1917, 54, 519 and 577

103) Ewald P.P.: Rev. Mod. Phys., 1965, 37, 46

104) Vinogradov A.V. and Zeldovich B.Y.: Appl. Opt., 1977, 16, 89

105) Cromer D.T. and Liberman D.: J.Chem. Phys., 1970, 53, 1891

106) Parrat L.G.: Phys.Rev., 1954, 95, 359

107) Lee P.: Appl. Opt., 1983, 22, 1241

108) Megademini T. and Pardo B.: J. Opt., 1985, 16, 289

109) Rieutord F., Benattar J.J., Bosio L., Robin P., Blot C. and de Kouchkovsky R.: J. de Phys. (Paris), 1987, 48, 679

110) Spiller E. and Rosenbluth A.E.: Opt. Eng., 1986, 25, 954

111) Beckmann P. and Spizzichino A.: The scattering of electromagnetic waves from rough surfaces, Pergamon New York, 1963

112) Stearns D.G.: J. Appl. Phys., 1989, 65, 491

113) Eastman J.M.: Phys. Thin Films, 1978, 10, 167

114) Croce P. and Nevot L.: Rev. Phys. Appl., 1976, 11, 113

115) Kadin A.M. and Keem J.E.: Script. Met., 1986, 20, 443

116) Hollanders M.A. and Thijsse B.J.: Less Comm. Met., 1988, 140, 33

117) Dhez P., Megtert S., Ravet M.F. and Ziegler E.: SPIE Proceedings, 1988, 984, 89

118) Levy-Leblond J.M. and Balibar F.: Quantique, Interedition Paris , 1984

Materials Science Forum Vols. 59 & 60 (1990) pp. 141-160

STRUCTURAL STUDIES OF METALLIC MULTILAYERS USING TRANSMISSION ELECTRON MICROSCOPY

J-P. Chevalier
Centre d'Etudes de Chimie Métallurgique
UPR A2801 - C.N.R.S.
15, rue Georges Urbain
F-94407 Vitry Cédex, France

1 INTRODUCTION

This lecture aims to provide a succinct presentation of the essential concepts of electron microscopy and of the use of diffraction and various forms of imaging to study metallic multilayer structures. Neither the whole of the field of microscopy nor the theories of dynamical diffraction or image contrast can be covered here. The reader can find these in a number of well known texts [1,2,3,4,5,6,7]. As we will see subsequently, the microscope allows both magnified images and electron diffraction patterns to be recorded. The electron diffraction patterns enable the structural identification of the phases present as well as a measure of the multilayer periodicity. Imaging brings information on the spatial distribution of the phases, the regularity of the layers, variations in layer thickness, roughness and abruptness of interfaces, size of crystallites within layers, presence and eventually nature of interfacial phases and on the size and extent of defects such as voids. In fact, it can be considered that, apart from specimen preparation, the techniques used in the microscopy of multilayers are the standard tools of electron microscopy in the field of materials science, and there are few, if any, techniques specific to this particular problem. The reader can thus apply what he may find in other reviews, notably those relating to the use of electron microscopy in the study of semiconductor interfaces [8, 9, 10, 11], disordered and amorphous materials [12, 13, 14] and poly-crystalline materials (this is more in the line of very conventional studies on materials, e.g. [1, 2 and 4]).

Since this presentation is given at a summer school, this paper will essentially deal with the microscope as a tool, with emphasis on understanding the principles and limitations of this technique, and no attempt will be made to review the field of the microscopy of metallic multilayers or the closely related subjects of solid state amorphisation, silicide formation and other phase changes in multilayers. Here, after treating rapidly the concept of the microscope and of image formation, the problems of specimen preparation will be examined. Finally, the use of electron diffraction and image contrast will be discussed and illustrated with examples of results on metallic multilayers.

2 THE MICROSCOPE

The electron microscope consists of an evacuated column (with vacuum in the 10^{-7} torr range for most commercial machines) with an electron gun at one end, a thin specimen somewhere in the middle, and some kind of recording device (usually photographic emulsion) at the other end. The electron gun produces a virtually monochromatic beam of electrons of energy commonly ranging from 100 keV to 400 keV, corresponding to electron wavelengths from 0.0370 Å to 0.0164 Å, respectively, and the electrons have to be treated as relativistic particles. Optically the column can be summarised as consisting of a set of condenser lenses, the objective lens and a set of intermediate and projector lenses. All these lenses are magnetic, electrostatic lenses being usually used in the gun, notably for field emission guns. The condenser lenses control the geometry of the electron beam impinging on the specimen. Modern systems, with three or four lenses, make it possible to vary independently both the size (from several tens of μm to the order of 10 Å) of the illuminated region and the convergence (from 10^{-4} to a few 10^{-2} rad) of the beam. The objective lens is the crucial lens of the microscope and its key role in image formation will be described later. Its design governs the ultimate performance of the microscope, and lens performance can, at least in an initial approach, be gauged by the value of the spherical aberration coefficient (C_s). Modern microscopes now have values of C_s below 1 mm, which yields values for the point resolution of the microscope of the order of 2 Å and below, at 200 keV. The specimen is immersed in the magnetic field of this lens, usually on a goniometric holder, with tilt angles approximately proportional to the attainable resolution (i.e. the better the resolution, the smaller the tilt angle). The projector and intermediate lenses are quite numerous (5-6) and are simply used to magnify either the image or diffraction plane of the objective lens to the conjugate observation plane. With the advent of microprocessor controlled microscopes it is now possible to program the excitation of these lenses so as to substantially reduce the rotation between object and final image, even up to very high magnification (typically up to x10^6), as well as make optically complex microscopes useable. In comparison with a light optical microscope, the optical elements are equivalent, but it is important to realise that in the electron microscope, the magnetic lenses do not move with respect to each other, but instead all have variable focal lengths, through variations in lens current.

The theory of image formation was first proposed by Ernst Abbe in the 19th. century, for the optical microscope. The theory is described in detail in most texts on optics, such as Born and Wolf [15], and can be applied readily to image formation in the electron microscope. It describes the relation between object plane, back focal plane and image plane of the objective lens. Abbe shows, that for coherent parallel illumination, there exists a Fourier transform relation between the wave function at the object plane and the wave function at the back focal plane. Furthermore, a second Fourier transform relates the wave function at the back focal plane to that of the image plane. For a perfect lens, the object and image wave fields are identical except for a magnification factor.

It is however, essential to realise that these Fourier transforms operate on both the amplitudes and phases of the wave function. The phase information is thus preserved in the image formation process, and this is true not only for the objective lens but also all the other image forming lenses. The phase information is lost only on recording either the image intensity or the intensity of the diffraction pattern. The use of the phase information is what differentiates a microscope from various scattering experiments which measure only the diffracted intensities. The phase contains the information concerning the spatial extent of the object wave field, as well as the phase shift due to the scattering process itself. The former enables us to obtain an image of the object, whilst the latter means that it is possible to obtain the projected atomic structure of the specimen.

This, in fact, is only true, in principle, since the aberrations of the objective lens introduce extra phase shifts in the waves (formally, these are readily introduced by multiplication of the diffracted wave field by the lens phase shifts at the level of the diffraction, or back focal, plane). These extra phase shifts make it more difficult to interpret the image in terms of atomic structure. It is now necessary to take into account the phase contrast transfer function of the microscope, and this is the field of high resolution imaging, particularly well treated by Spence [16].

There is another vital point to realise, which is implicit in the Abbe theory, and that is that we start with a three dimensional object, albeit a thin object, and that the image has necessarily two dimensions. Therefore structural information parallel to the electron beam has been lost in this projection, which is in general non-linear. Even for perfect crystals with no variation in structure parallel to the beam, this projection poses problems. At best (very thin weakly scattering specimens) the image is related to a linear projection of the potential. For thicker specimens or for specimens scattering more strongly, the projection is no longer linear (due to dynamical effects) and full image simulation is required to interpret, by comparison, the experimental images. For crystals with either structure or microstructure varying in the beam direction it is then very difficult indeed to unravel the effects of projection. For microstructural features on a scale > 10Å, stereo microscopy, where through the use of different projections the lost information is partly recovered, can be useful. Of course this problem is at its most acute in the study of amorphous materials at high resolution (e.g. [12, 13, 14]).

The Abbe theory relies on the conditions for Fraunhoffer diffraction, which imply elastic scattering. The theories for the elastic interaction of electrons with crystals are now well known (see e.g. [1]). The kinematical theory, valid for the scattering of X-rays, is both a weak and single scattering model, valid for light to medium **Z** elements for thin specimens (<< 100 Å). The various forms of dynamical theory (e.g [1]) take into account the strong nature of the scattering and also include multiple scattering. For high resolution imaging, the phase object theories (the physical optics approach, described by Cowley [17]) are more frequently used.

All these theories treat inelastic effects as a loss of electrons ('absorption'), that is, electrons which do not contribute to the image contrast. However, electrons which have undergone inelastic collisions lead to signals which can be characteristic of the chemistry of the specimen, either if we consider the energy loss spectrum (EELS) [18] (typically from 0 to 2 keV loss) or the X-ray photons emitted by atoms excited by such interactions (e.g. [19]). This is the field of X-ray microanalysis and coupled with electron diffraction, can enable complete phase identification to be readily carried out on a sub 1000 Å scale.

Finally, it has to be realised that the interaction (both elastic and inelastic) of electrons with matter is very strong, and this poses a severe constraint for the specimen: this has to be very thin, of the order of 100 to 1000 Å. This is really a very severe problem since it is not trivial to obtain such specimen thickness without creating grave structural artefacts. Furthermore, the total specimen volume examined is then very small and this obviously poses sampling problems in the case of heterogeneous samples (which are often of most interest!).

To conclude we can state that the electron microscope can give images (with a resolution better than 2 Å now), diffraction patterns (from areas ranging from several μm to about 10 Å in diameter), and using analytical signals produced by inelastic interactions, chemical micro-analysis on the ~100 Å scale.

3 SPECIMEN PREPARATION

For microscopy on any material, specimen preparation overwhelmingly controls the quality and reliability of the results which may be obtained. This is even more the case for materials, such as metallic multilayers, which are, by design, heterogeneous. Before considering the specimen preparation techniques as-such, it is first necessary to think about the specimen geometry. Metallic multilayers are generally prepared by deposition techniques on some substrate material. We can then wish to look through the multilayers, which can be very useful, say, after a reaction has taken place, since, on polycrystalline samples, this leads to electron diffraction patterns with more complete Debye rings, simply by improved sampling; this is either called a 'plan view' or 'through foil' microscopy. Often, especially if we are looking for information concerning interfaces, or microstructure within a single layer, then we have to look in the direction parallel to the layers; this is called 'cross-sectional' microscopy.

Specimen preparation for 'plan-view' geometry is close to that usually used for the preparation of thin foils of metals, alloys and semiconductors. Depending on the total thickness of the multilayers, the substrate/multilayer assembly is reduced in thickness by grinding either the substrate or both sides, to a thickness of 20 μm or so. Final thinning is generally then carried out using a ion beam thinner (e.g. [20]). Some care has to be taken since this may lead to phase transformations in the multilayer, either by ion beam mixing, or simply by heating. Often plan view geometry is used for 'sandwich' films rather than multilayers with a large number of layers. In that case, the layers can be deposited on a substrate which can be removed by dissolving it in the appropriate solvent (e.g. single crystal NaCl and water). The layer can then be recovered directly using a standard microscope grid, and no further thinning is required (see e.g. [21, 22, 23]).

Cross-sections are generally harder to prepare, and currently three different techniques are used. Lepêtre, who initiated a number of studies on multilayers for soft X-ray application, has developed a micro-cleavage technique which has the advantage of being rapid, clean and of not introducing artefacts in the electron thin sample (e.g. [24, 25, 26]). The main drawback of this method is that the total observable area is rather small. For multilayers deposited on crystalline Si or GaAs substrates, the micro-cleavage technique has been further developed by other authors [27, 28]. This technique is particularly useful when used in conjunction with ion beam thinning, notably as a means of evaluating the structural changes induced by the ion beam [27].

The cross-section specimen preparation technique much used for the study of interfaces in semiconducting materials (e.g. [29]) has been transposed to multilayer specimens. The substrates (of thickness ~ 0.5-1 mm) and the multilayers are glued either face to face (e.g. [30, 31]) or face to back [20], mechanically ground down to about 20 μm thickness and then ion beam thinned. This method can yield very considerable electron thinned areas (figure 1), thus leading to good sampling statistics. The ion beam can produce a number of modifications to the multilayer, ranging from inducing crystallisation in amorphous layers (it is also possible to amorphise crystalline layers), to mixing at the layer interface and to layer coalescence through sputtering away of the lighter element; these modifications and the related artefacts in the images have been discussed by Ruterana [27, 30]. This author has also considered in some detail the specimen modifications induced by the high energy electron beam in the microscope.

Finally, it is important to mention the use of ultramicrotomy in preparing cross-sections of multilayers (e.g. see figure 3). This technique is well suited to cases where the substrate material is appropriate (e.g. 'Kapton', see [23]) or where the multilayer has been floated off a NaCl substrate (e.g. [32, 33]). In both cases,the multilayers are embedded in resin before cutting. Ultramicrotomy is increasingly used in the preparation of metallic specimens for electron microscopy (e.g. [32, 33, 34, 35, 36, 37]) and the specimen modifications induced (see [33, 36, 37]) have been characterised.

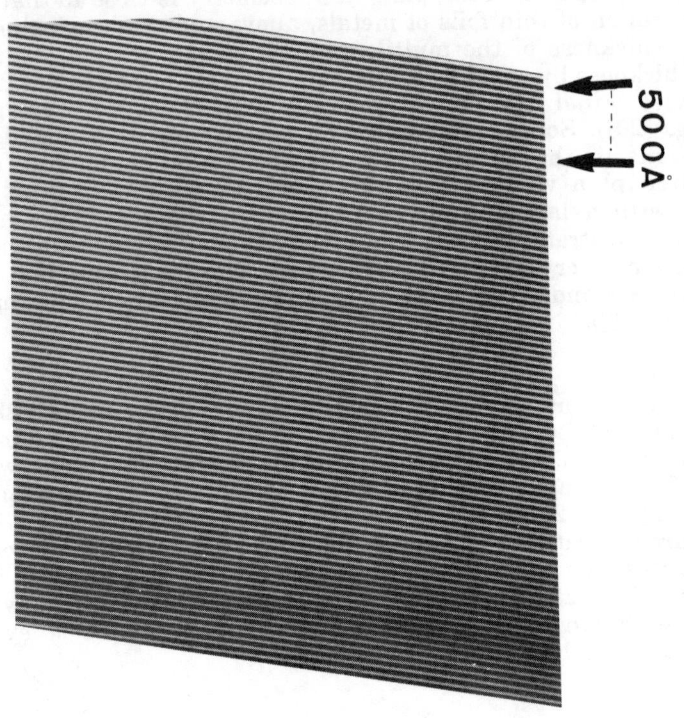

Fig. 1 A 90 period W/C multilayer on a (111) Si substrate, showing all the layers. The cross-section is prepared by ion beam thinning [27, 30]. Micrograph courtesy of P. Ruterana.

4 DIFFRACTION

Using the intermediate and projector lenses of the microscope, either the back focal or diffraction plane of the objective lens can be made conjugate to the observation screen. In the latter case, the diffraction pattern from a selected (either using the "selected area" mode [1] or by focussing the incident electron beam to a probe) region of the specimen can then be observed and recorded. For the study of multilayers, diffraction yields important information on the structure and phases within the layers and the layer spacings, depending on the specimen geometry (plan view or cross-section) and the size of the area selected.

The measurement of multi-layer periodicity can be readily obtained by diffraction with specimens in the cross-section geometry. Formally, we can describe a multi-layer consisting of atoms A and B as a fluctuating potential V_A-V_B, superposed on a constant potential V_B ($V_A > V_B$). Electrons will be diffracted by the fluctuating potential, whilst the constant potential simply gives rise to a refraction term (see [1]).

We now write

$$V_A - V_B = V(\mathbf{r}) \quad (1)$$

The term $V(\mathbf{r})$ can be decomposed into three terms :

1 - a function $T(\mathbf{r})$, with

$T(\mathbf{r}) = 1$ ($\mathbf{r}$ within the thickness of the multilayer)

and $T(\mathbf{r}) = 0$ ($\mathbf{r}$ outside) (2)

This is a usual form for the function which describes the spatial extent of the diffracting potential.

2 - the periodicity of the multilayer given by $D(\mathbf{r})$, with

$$D(\mathbf{r}) = \Sigma_i\, \delta(\mathbf{r}-\mathbf{r}_i) \quad (3)$$

$$\mathbf{r}_i = i\, .\, \mathbf{a} \quad (4)$$

where $\mathbf{a}$ is the multilayer repeat distance. This represents an infinite sequence of regularly spaced delta functions.

3 - the profile of the regular modulations of V_A-V_B described by a function $P(\mathbf{r})$, assuming that each layer has the same profile. This can range from the 'top-hat' form for layers with abrupt interfaces, to gaussian like profiles, for either diffuse or rough interfaces.

From this, the scattering potential $V(\mathbf{r})$ can be written as

$$V(\mathbf{r}) = T(\mathbf{r}) \times \{D(\mathbf{r}) * P(\mathbf{r})\} \quad (5)$$

where $*$ signifies a convolution.

If we now suppose a simple Fourier transform relation between scattering amplitude and diffracted amplitude (generally not true quantitatively in the case of electrons) then the diffracted amplitude, $\mathcal{v}(\mathbf{q})$, will be given by

$$\mathcal{v}(\mathbf{q}) = \mathcal{F}\, V(\mathbf{r}) \tag{6}$$

where $\mathcal{F}$ signifies a Fourier transform.

Then

$$\mathcal{v}(\mathbf{q}) = \mathcal{F}\, [T(\mathbf{r}) \times \{D(\mathbf{r}) * P(\mathbf{r})\}] \tag{7}$$

$$\mathcal{v}(\mathbf{q}) = \mathcal{F}\, T(\mathbf{r}) * \{\mathcal{F}\, D(\mathbf{r}) \times \mathcal{F}\, P(\mathbf{r})\} \tag{8}$$

$$\mathcal{v}(\mathbf{q}) = \mathcal{t}(\mathbf{q}) * \{\mathcal{d}(\mathbf{q}) \times \mathcal{p}(\mathbf{q})\} \tag{9}$$

with $\mathcal{t}(\mathbf{q})$, $\mathcal{d}(\mathbf{q})$ and $\mathcal{p}(\mathbf{q})$ being the Fourier transforms of $T(\mathbf{r})$, $D(\mathbf{r})$ and $P(\mathbf{r})$.

The function $\mathcal{d}(\mathbf{q})$ is simply an infinite sequence of regularly spaced delta functions with a period proportional to 1/a. The amplitudes of these peaks are modulated by the function $\mathcal{p}(\mathbf{q})$, which is the Fourier transform of the profile of the varying part of the multilayer potential $V(\mathbf{r})$. The peaks themselves have a fine structure (rarely visible) given by the function $\mathcal{t}(\mathbf{q})$.

Thus, from this equation, the multilayer period can be obtained from the electron diffraction pattern through the inverse of the spacing of $\mathcal{d}(\mathbf{q})$. This function appears as satellite peaks around the zero beam. An accurate measurement, however, generally requires in situ calibration of the microscope camera length. This is much easier for multilayers deposited on oriented single crystal substrates (e.g. Si or GaAs [27, 30]), where such internal calibration can be readily obtained (e.g. see figure 2).

The use of diffraction to identify the structures and phases present is very conventional and need not be treated here. The diffraction patterns can be obtained in selected area mode down to specimen areas of the order of the μm, with parallel illumination (which yields fine spots in the diffraction pattern, well suited for identification of structures and measurement of lattice parameters). There are numerous examples where this has been used for phase identification of both unreacted and reacted metallic multilayers (e.g. [20, 21, 22]). Diffraction also gives the information concerning eventual epitaxy between the layers or between the layers and the substrate. Notably Holloway and Sinclair [20] have shown that the Ti polycrystals in a Ti/Si multilayer are textured with $\{01\bar{1}0\}$ Ti planes predominantly parallel to the layer interfaces.

Diffraction patterns from smaller regions can be obtained in modern microscopes with a probe focussed by both the condenser lenses and the strong pre-field of the so-called condenser-objective lens. In these cases probe sizes down to about 10 Å or so can be obtained, but with conventional thermionic sources the brightness is very low. Furthermore, since the probe is focussed, the illumination is rather convergent, yielding discs instead of spots in the diffraction pattern. In some cases this may make structure identification more difficult. Finally mention should be made of the technique of convergent beam electron diffraction essentially developed by Steeds and coworkers at Bristol (e.g. [38]). Here the incident probe is strongly convergent and through a study of the fine structure within the diffracted discs it is possible to measure specimen thickness, lattice parameters and to determine the space group of the crystal [39]. These techniques have yet to be used in the context of metallic multilayers, but they are certainly a very powerful means of structure identification, especially when associated with quantitative X-ray microanalysis (e.g. [19]) in the microscope.

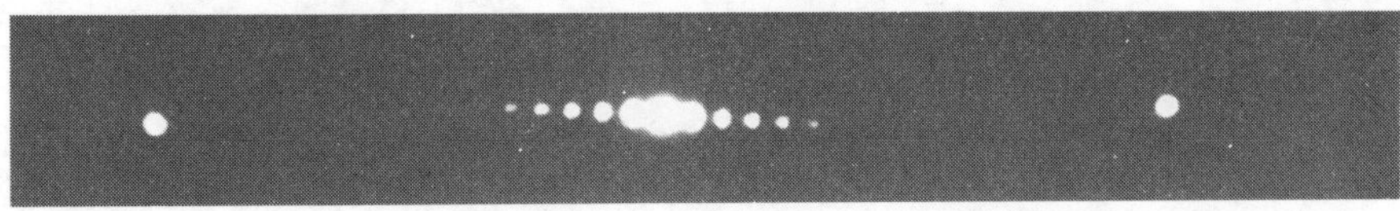

Fig. 2 Small angle region of a selected area diffraction pattern from the 90 period W/C multilayer shown in figure 1. The period of the function $\mathcal{L}(\mathbf{q})$ is that of the spots around the centre. The two separate spots at much larger angles are {111} reflections from the Si substrate, and allow a very accurate calibration of diffraction camera length to be carried out. Micrograph courtesy of P. Ruterana [27, 30].

5 IMAGE CONTRAST

It is essential that the reader realises that the image contrast obtained is the result not only of the structure and microstructure of the specimen and of the electron optics of the microscope but also of the choices, in terms of microscope operation, specimen diffraction conditions and geometry, made by the microscopist during the observation. In general one can say that, in the microscope, a specimen has little or no contrast by itself, and that the contrast is produced either by defocussing the objective lens (thus producing phase contrast) or by introducing a relatively small objective aperture (producing diffraction contrast).

5-1 Diffraction contrast

Here, the objective aperture is used to select either the zero transmitted beam, or a diffracted beam. The beams not selected impinge on the aperture and hence do not contribute to the image. The contrast obtained is then just a map of electron intensities for the range of scattering angles selected, at the exit face of the specimen. If we consider the zero beam (this is called bright field imaging), light areas correspond to regions which are weakly or not scattering, whilst dark regions are those which are strongly scattering outside the area of diffraction (reciprocal) space defined by the objective aperture. On the other hand, if one of the diffracted beams is selected (dark-field imaging), then the light regions corresponds to regions which are strongly scattering into this beam, i.e. regions which are in the Bragg position, and darker regions are then either not in the Bragg condition (e.g. see Figure 3), or thinner, or consist of material which scatters less. This simplified reasoning is only valid, strictly, in the framework of elastic single scattering. Dynamical effects can lead to effects such as the non-complementarity of bright and dark field images in two-beam excitation conditions, but these are not likely to be major in the study of multilayers. Diffraction contrast is extensively covered in the standard texts [1, 2, 3, 4, 5].

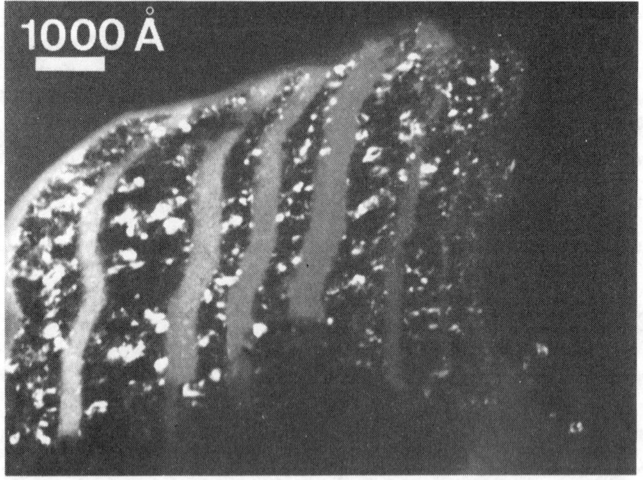

Fig. 3 Dark-field image of an ultramicrotomed Ni/Zr multilayer. Crystallites which satisfy the Bragg condition appear as bright. The thinner layers are Zr and the crystallites are smaller than in the (thicker) Ni layer. Micrograph courtesy of E. Gaffet [32].

This mode of contrast is immediately useful in the examination of multilayers, either in cross-section or in plan-view. In cross-section, for a multilayer consisting of elements with considerably different atomic number (and hence different atomic scattering factors for electrons), the heavier element layers will always appear dark in bright field images, since more electrons will be scattered to higher angles (outside the objective aperture) than for the lighter element. Dark-field imaging using electrons scattered from one or other of the phases in the multilayer (assuming that these can be separated in diffraction space) enables the grain size, for polycrystalline layers, to be determined (e.g. figure 3). This mode of contrast also enables amorphous phases to be identified due to their particular 'speckle' contrast [13, 14]. Diffraction contrast can also be used to image strain fields or dislocations at interfaces, especially in the case of single crystal substrates. This has been particularly studied for semiconductor interfaces (e.g. [11]). The reference work by Hirsch et al [1] gives an exhaustive description of diffraction contrast from different microstructural features such as dislocations, stacking faults, twins, domain boundaries, precipitates, etc...

5-2 Phase Contrast

The mechanism for phase contrast is different. Defocussing the objective lens leads to the introduction of an extra phase shift for the diffracted beams with respect to the zero transmitted beam. This phase shift varies from 0 to $\pi/2$, and then oscillates from $\pi/2$ to $-\pi/2$ as a function of scattering vector **q**, for a given value of defocus. For a thin specimen, it can be shown that the effect of the specimen on the electron wave is to introduce a phase shift ('phase object approximation', see [16, 17]).

The wave function at the exit face of the specimen, $\Psi_s(x, y)$ can then be written as

$$\Psi_s(x, y) = \exp[-i\sigma\Delta(x, y)] \qquad (10)$$

for a unit amplitude incident plane wave, where

$$\sigma = 2\pi me\lambda/h^2 \qquad (11)$$

is an interaction constant, and $\Delta(x, y)$ is the projected potential of the specimen.

If $\sigma\Delta$ is very small (this is known as the 'weak phase object approximation', see [16, 17]), i.e. the specimen is both very thin and weakly scattering, then the exponential term can be expanded to give

$$\Psi_s(x, y) \sim 1 - i\sigma\Delta(x, y) \qquad (12)$$

Here we identify the first term, 1, as being the incident wave, whilst the term $i\sigma\Delta$, corresponds to the scattered wave. This latter term is $\pi/2$ out of phase with respect to the incident wave, and so no interference can occur to first order. The extra phase shift introduced by the objective lens defocus then enables interference to occur, thus leading to strong contrast. The image intensity is then given by

$$|\Psi_s(x, y)|^2 \sim 1 - 2\sigma\Delta(x, y) \tag{13}$$

In fact, as these equations are written, the projected potential $\Delta(x, y)$ corresponds to the projected atomic potential of the specimen, and thus we are in the realm of high resolution electron microscopy (see Spence [16]). It is important to realise that the image is related to the projected potential. The effects of this projection are extremely serious for structures and microstructures which vary parallel to the direction of projection. This is, of course, the case for surface roughness in the study of interface structure and for specimen tilt for the measure of layer thickness.

Recently high resolution techniques have been extensively used to study, for example, Mo/Si [40, 41], W/C and W/Si [27, 30, 40, 42], Ti/Si [20, 43], Mn/Sb [23], and WC/Co [44] multilayers. These studies have sought to evaluate the abruptness and roughness of interfaces within the multilayer, to identify whether the layers are polycrystalline or amorphous, to determine the existence of new phases at the interface and also, simply, to measure the layer thicknesses. For multilayers consisting of light/heavy elements, the contrast on the scale of the layer thickness is still dominated by diffraction contrast effects, despite the use of a large objective aperture which includes at least the first set of Bragg diffracted beams. This is because the scattering by the heavy element is still considerably stronger out to very large angles. Incidentally this strong large angle scattering difference can be used to produce contrast in images made using a large angle annular detector in a dedicated scanning transmission electron microscope [e.g. 45].

Thickness measurements have been made by several authors (e.g. [27, 40]), and once the microscope magnification has been calibrated, the remaining problems concern the orientation of the cross-sectioned multilayer rigorously parallel to the electron beam (i.e. the projection direction), interface abruptness (i.e. where does a layer really end), roughness, and thinning artefacts. Thinning artefacts have been well discussed by Ruterana [27, 30]. In the case of W/C multilayers, the C layer can, in some cases be partially sputtered away, decreasing the C layer thickness, and eventually leading to the coalescence of the W layers. For Mo/Si multilayers, annealing leads to the formation of interfacial phases, denser than the constituents, and this decreases the layer spacing [41]. The major problem, however, for the measurement of layer thicknesses, is that of specimen orientation. In regions of thickness around 500 Å, a tilt of the order of 10^{-2} rad. leads to an increase in the thickness of the heavier layer by about 5 Å. It is therefore essential to carry out these measurements in regions of thickness less than 200Å (e.g. [40]). For multilayers deposited on an oriented single crystal substrate, this can be avoided since the substrate can be very accurately oriented,

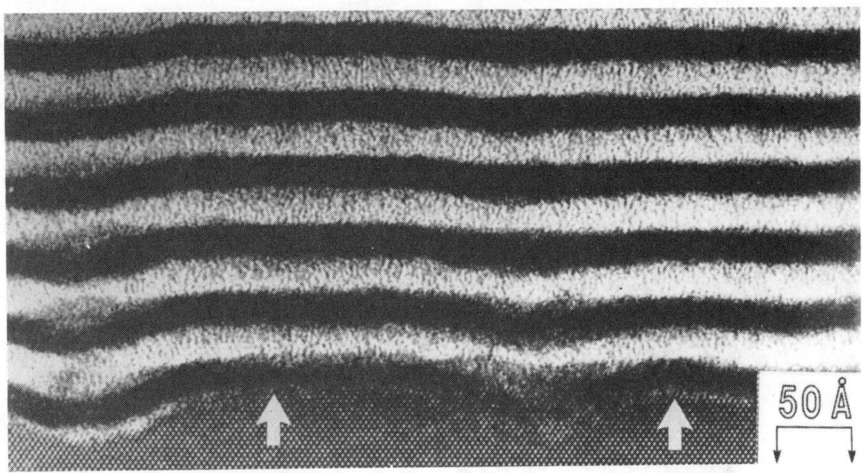

Fig. 4 High Resolution image of part of a W/C multilayer on a (100) GaAs substrate. Layer thickness measurements can only be made in the relatively flat regions (arrowed). It is interesting to note from this micrograph that the residual substrate roughness is rapidly (3-4 bilayers) reduced by the layer deposition. Micrograph courtesy of P. Ruterana.

using Kikuchi lines or seeking the highest symmetry for the diffraction pattern in Laüe conditions. The thicknesses can then be readily measured using the substrate lattice fringes as internal calibration [27, 40]. Petford-Long et al [40] quote an accuracy of between 3 and 5 Å for layers of 60 Å or so in thickness. For non-crystalline or poly-crystalline substrates, this is not so easy, and it is then necessary to tilt the multilayer and seek the minimum thickness for the heavy element layer [46]. Unless there is clear evidence for the formation of a new interfacial phase [41], the diffuse width of the interface is generally less than the accuracy in measurement quoted by Petford-Long et al [40], and thus is not a problem. Roughness, however is more of a problem, and Ruterana et al [27] suggest picking out only the flatter portions for thickness measurements (e.g. see Fig. 4). Finally it should be stated that it is better to measure the average periodicity from the small angle scattering in diffraction (see § 4) and then to use the high resolution image to evaluate mistakes in the layer stacking sequence, as well as thickness fluctuations and roughness.

The projection effect is the major source of error in evaluating roughness at interfaces between layers. It is obviously essential that the layers be rigorously parallel to the electron beam, but even so, the projected roughness will be less in the 2-dimensional image than at the interface in three dimensional space. This has been discussed and modelled by Goodnick et al [47] in the context of roughness measurement at the SiO_2/Si (100) interface. To minimise this projection effect, the simplest approach is to evaluate the roughness in the very thinnest regions of the cross-section. Another approach has been suggested by Lepêtre et al [25]. Using cleaved specimens they show that part of the 3-dimensional information lost through projection can be recovered by examination of the thinnest part of the cleavage wedge tilted through about 15° (see figure 5).

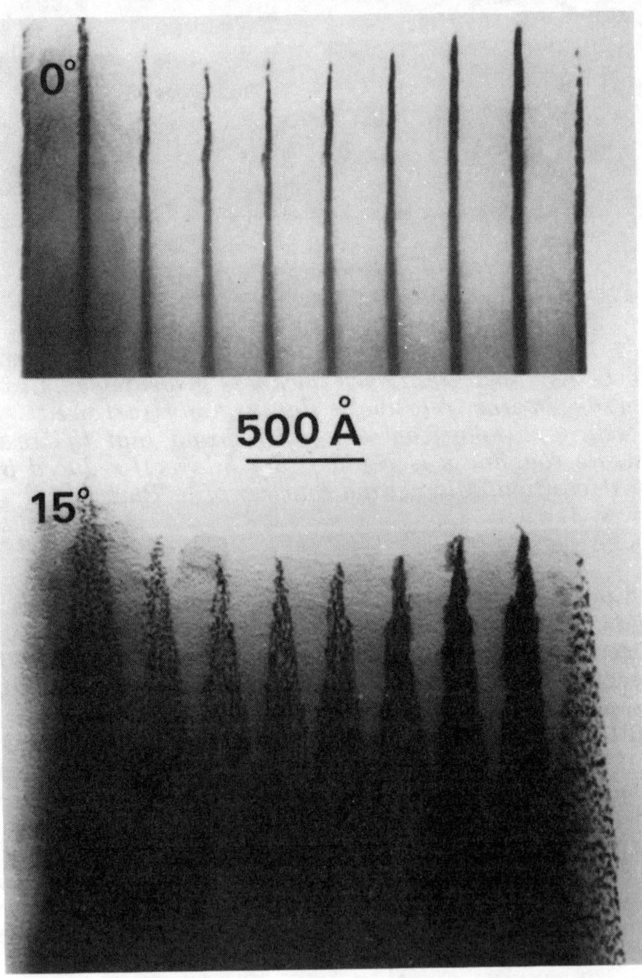

Fig. 5 Set of bright field images for untilted (top) and tilted (bottom) W/C multilayer with variable W thickness. The thinnest layer is at the left of the image, and the last layer on the right is due to pollution from the substrate. Note that layers which appear continuous in projection with 0° tilt turn out to be discontinuous when viewed with a different projection. Micrographs courtesy of Lepêtre [25].

High resolution imaging is well suited to determine the structure within the layers. The first stage is to determine whether the deposited layers are crystalline or amorphous. For thin specimens, and reasonably large lattice spacings with respect to the resolution of the microscope, then the total absence of lattice fringes anywhere within the layer, coupled with the existence of the typical contrast from amorphous specimens (e.g. see [12, 13, 14] and references therein) strongly suggests that the layer is amorphous. Although not usually done, it is undoubtedly wise to confirm this using dark-field imaging and diffraction contrast, where both the contrast from crystallites and the classic 'speckle contrast' from amorphous phases are well characterised [13]. It should also be stated that imaging is the only technique which can yield this information when the the area of interest is too small to diffract significantly. In general all authors report that C or Si layers are always amorphous, whilst the W or Mo layers tend to be crystallised for thicknesses exceeding a value ranging from 10-20 Å [40] to 40 Å [27].

In this context, Ruterana et al [27] have investigated the effect of the ion beam thinning of the cross-section in crystallising amorphous W layers. Using a multilayer with W layers of increasing thickness they were able to establish the critical thickness (around 40 Å) for the appearance of crystallites in the W layer, using cleaved specimens. The same experiment on a cross-section prepared by ion milling yielded a critical thickness some 10 Å less, showing clearly that ion bombardment can induce some crystallisation, and this may explain some of the differences between different authors. It is also interesting to note that imaging makes it possible to establish spatial correlations between different features : as such, Ruterana et al [27] show that layer roughness is less when the layer is amorphous, and that in partially crystallised layers, roughness can be directly attributed to crystallites protruding.

Another advantage of high resolution imaging is that it enables crystallographic relationships between different layers to be established. Petford-Long et al [40] found that Mo crystals tended to align (110) planes close to the layer interfaces, resulting in a facetted interface. In a somewhat special case, Nakayama et al [23] show continuity of the Sb(10.1) lattice fringes through a 1 Å MnSb layer in a Mn(1 Å)/Sb(50 Å) multilayer. Here, high resolution microscopy gives particularly clear results, and enables these authors to propose structural models for the interfacial regions.

Interfacial phase changes, notably in the case of as-deposited layers, where the reacted region is likely to be very thin, can also be detected using high resolution microscopy. For the W/C system, both Petford-Long et al [40] and Ruterana et al [27] have identified WC crystallites at the interface (e.g. see figure 6). A more subtle phase change has been observed by Holloway and Sinclair [20] in Ti/Si multilayers. They report the formation of an amorphous Ti-Si alloy at the Ti/Si interface for rapid thermal annealing (e.g. 30 s at 455°C). This amorphous phase appears very clearly as a layer of intermediate contrast between the (dark) micro-crystalline Ti and the (light) amorphous Si layer. No possible confusion between a true interfacial layer and a projection effect for a rough interface is possible here since the layers obtained have a residual roughness much less than the interfacial layer thickness. Finally it is clear [27, 40] that for W/C and W/Si the interfaces are not symmetric; i.e. the Si growing on W interface is sharper than that for W on Si and the C growing on W is much sharper than that for W on C. Petford-Long et al have proposed a model for this and they suggest that this asymmetry is due to the implantation of heavy W ions into loosely packed amorphous Si and vice versa.

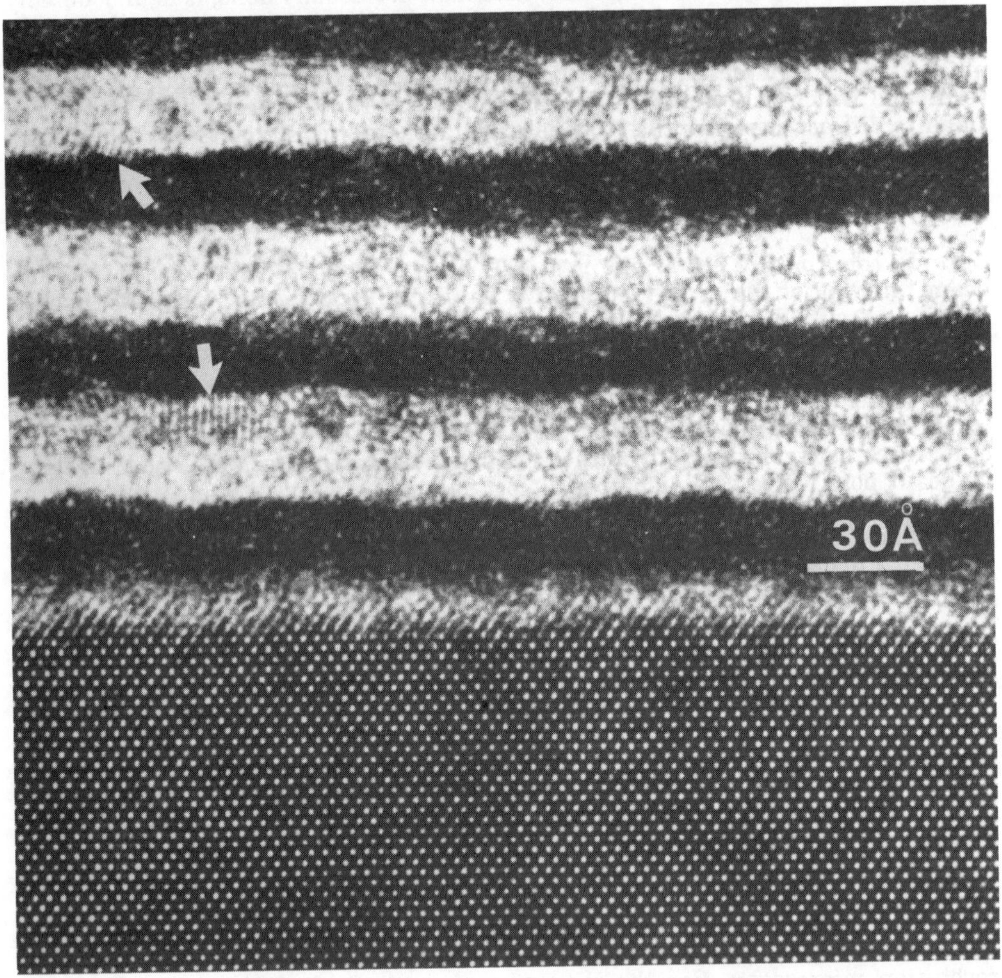

Fig. 6 High resolution image of part of a W/C multilayer on a (100) Si substrate. The presence of crystallites at the W-C interface can be seen (arrowed).Otherwise, note that both the W and the C layers are mainly amorphous. Also note the light contrast between the Si substrate and the first W layer : this is due to residual SiO_2 Micrograph courtesy of P. Ruterana.

Fresnel contrast, which is another phase contrast mechanism, should be considered in the study of metallic multilayers, even though it is not a very high resolution technique. This contrast arises by refraction of electrons [1] at a discontinuity in the projected mean inner potential in a specimen, and by defocussing the objective lens, produces a bright or dark line in the image. Such techniques have recently been considerably used in the microscopy of interfaces (e.g. [48, 49, 50]), notably when looking for interfacial phases. In the context of metallic multilayers, and especially in the case of reacted multilayers, this contrast mechanism can be used to image voids (e.g. Kirkendall voids) especially if these are too small to produce sufficient diffraction contrast. Stobbs has demonstrated this for voidites in ultra fine grain polycrystalline Nb foils [51].

6 CONCLUSIONS

Electron microscopy should now appear to the reader as a powerful tool for the study of the structure and microstructure of metallic multilayers, once the difficult stage of specimen preparation, with the potential dangers of artefacts, has been mastered. Here, I have indeed stressed the structural and microstructural investigation, but the composition of the products from reacted multilayers can also be determined using energy dispersive X-ray microanalysis [19] and the use of electron energy loss spectroscopy [18] makes it possible to investigate the plasmon spectrum and dispersion, which is another field in itself.

ACKNOWLEDGMENTS

It is a pleasure to thank Y. Lepêtre, E. Gaffet and P. Ruterana for discussion and for kindly allowing me to use their micrographs to illustrate this paper. The W/C multilayers were prepared by P. Houdy at the L.E.P. Figures 1, 2 and 4 are essentially the same as those published in the Journal of Applied Physics [27], and figure 5 as that published in Applied Physics Letters [25].

REFERENCES

[1]- "Electron Microscopy of Thin Crystals", P.B. Hirsch, A. Howie, R.B. Nicholson, D.W. Pashley and M.J. Whelan, pub. R.E. Krieger, New York (1977)
[2]- "Electron Beam Analysis of Materials", M.J. Loretto, Chapman and Hall, London (1984).
[3]- "Electron Microscopy in Materials Science" ed. U. Valdrè and A. Zichichi, pub. Academic Press, New York (1971).
[4]- "Electron Microscopy in Materials Science" ed. U. Valdrè and E. Ruedl (4 volumes), pub. Commission of the European Communities, Brussels (1976).
[5]- "Méthodes et techniques nouvelles d'observation en métallurgie physique", ed. B. Jouffrey, pub. Société Française de Microscopie Electronique, Paris (1972).
[6]- "Microscopie électronique en Science des Matériaux", ed. B. Jouffrey, A. Bourret and C. Colliex, pub. Les Editions du CNRS, Paris (1983)
[7]- "Introduction to Analytical Electron Microscopy" ed. J.J. Hren, J.I. Goldstein and D.C. Joy, pub. Plenum Press, New York (1979).
[8]- Hutchison J.L. : Ultramicroscopy, 15, 51 (1984)
[9]- Hutchison J.L. : Ultramicroscopy, 18, 349 (1985).
[10]- Humphreys C.J. : p. 105 of "Electron Microscopy 1986" ed. T. Imura, S. Maruse and T. Suzuki, pub. Jap. Soc. Electron Microscopy, Tokyo (1986).
[11]- Chevalier J-P. : p. 109 of Springer Proceedings in Physics N° 22 "Interfaces in Semiconductors : Formation and Properties", ed. G. LeLay, J. Derrien and N. Boccara, pub. Springer-Verlag, Berlin (1987).
[12]- Howie A. : J. Non-Cryst. Sol., 31, 41 (1978).
[13]- Chevalier J-P. : Ch. XIV of ref. [6]
[14]- Chevalier J-P. : J. Microsc. Spectrosc. Electron., 10, 149 (1985).
[15]- "Principles of Optics" M. Born and E. Wolf, Pub. Pergamon Press, Oxford (1975).
[16]- "Experimental High-Resolution Electron Microscopy", J.C.H. Spence, pub. Clarendon Press, Oxford (1981).
[17]- "Diffraction Physics", J.M. Cowley, pub. North Holland, Amsterdam (1975)
[18]- "Electron Energy Loss Spectroscopy in the Electron Microscope", R.F. Egerton, pub. Plenum Press, New York (1986).
[19]- "Quantitative Electron-Probe Microanalysis", V.D. Scott and G. Love, pub. Ellis Horwood, Chichester (1983).
[20]- Holloway K. and Sinclair R. : J. Appl. Phys. 61, 1359 (1987).
[21]- Meng W.J., Fulz B., Ma E. and Johnson W.L. : Appl. Phys. Lett., 59, 661 (1987).
[22]- Herd S., Tu K.N. and Ahn K.Y. : Appl. Phys. Lett., 42, 597 (1983).
[23]- Nakayama N., Moritani I., Shinjo T., Ishizaki A. and Hajimoto K. : Philos. Mag. A 59, 547 (1989).
[24]- Lepêtre Y. and Rasigni G. : Opt. Lett., 9, 433 (1984).
[25]- Lepêtre Y., Ziegler E. and Schuller I.K. : Appl. Phys. Lett., 50, 1480 (1987).
[26]- Lepêtre Y., Schuller I.K., Rasigni G., Rivoira G., Philip R. and Dhez P. : Optical Engineering, 25, 948 (1986).
[27]- Ruterana P., Chevalier J-P. and Houdy P. : J. Appl. Phys., 65, 3907 (1989).

[28]- Ruterana P., Buffat P.A. and Ganière J-D. : J. Microsc. Spectrosc. Electron., 13, 421 (1988).
[29]- Dupuy M. : J. Microsc. Spectrosc. Electron., 9 , 163 (1984).
[30]- Ruterana P. : Thèse d'Etat, Université de Caen (1987).
[31]- Schröder H., Samwer K. and Köster U. : Phys. Rev. Lett., 54, 197 (1985).
[32]- Gaffet E. : Thèse de l'Université Paris VI (1988).
[33]- Marshall A.F. and Dobbertin D.C. : Ultramicroscopy, 19, 69 (1986).
[34]- Thompson G. E. and Wood G.C. : J. Microsc. Spectrosc. Electron. 12 , 391 (1987).
[35]- Timsit R.S., Waddington W.G., Humphreys C.J. and Hutchison J.L. : Ultramicroscopy, 18 , 387 (1985).
[36]- Stobbs W.M., Kallend J.S. and Williams J.A. : Acta Met. 24 , 1083 (1976)
[37]- Glanvill S.R., Kwietniak M.S., Pain G.N., Rossow C.J., Warminski T., Wielunski L.S. and Wilson I.J. : Phil. Mag. Lett., 59 , 17 (1989).
[38]- Steeds J.W. : Ch. 15 of [7].
[39]- Buxton B.F., Eades J.A., Steeds J.W. and Rackham G.M. : Phil. Trans. Roy. Soc., A281, 171 (1976).
[40]- Petford-Long A.K., Stearns M.B., Chang C.-H., Nutt S.R., Stearns D.G., Ceglio N.M. and Hawryluk A.M. : J. Appl. Phys., 61, 1422 (1987).
[41]- Holloway K., Ba Do K. and Sinclair R. : J. Appl. Phys., 65, 474 (1989).
[42]- Vidal B.A. and Marfaing J.C. : J. Appl. Phys., 65, 3453 (1989).
[43]- Holloway K. and Sinclair R. : J. Less-Common Metals, 140, 139 (1988).
[44]- Moustakas T.D., Koo J.Y., Ozekcin A. and Scanlon J. : J. Appl. Phys., 65, 4256 (1989).
[45]- Pennycook S.J. and Narayan J. : Appl. Phys. Lett., 45, 385 (1984).
[46]- Takeoka H. and Ishii Y. : p. 1467 of "Electron Microscopy 1986", ed. T. Imura, S. Maruse and T. Suzuki, pub. the Japanese Society for Electron Microscopy, Tokyo (1986).
[47]- Goodnick S.M., Ferry D.K., Wilmsen C.W., Lilienthal Z., Fathy D. and Krivanek O.L. : Phys. Rev. B , **32** , 8171 (1985).
[48]- Mellul S., and Chevalier J-P. : p. 241 of 'Electron; Microscopy. and Analysis 1987', ed. L.M. Brown, pub. Inst. of Phys., Bristol (1987).
[49]- Mellul S. : Thèse de l'Université Paris VI (1988).
[50]- Ness J.N., Stobbs W.M. and Page T.F. : Phil. Mag. A54, 679 (1986).
[51]- Stobbs W.M. : p. 253 of 'The Structure of Non-Crystalline Materials', ed. P.H. Gaskell, pub. Taylor and Francis, London (1976).

Materials Science Forum Vols. 59 & 60 (1990) pp. 161-228

X-RAY ABSORPTION, PHOTOEMISSION, AND AUGER ELECTRON SPECTROSCOPY: ELECTRONIC STRUCTURE AND GROWTH OF METAL/METAL INTERFACES

B. Carrière (a) and G. Krill (b)
(a) I.P.C.M.S., Groupe "Surfaces-Interfaces"
U.M. CNRS - ULP - EHICS 46
4, rue Blaise Pascal, F-67070 Strasbourg, France
(b) Laboratoire de Physique des Solides
U.R.A. C.N.R.S. 155
Université de Nancy I, BP 239
F-54506 Vandoeuvre-les-Nancy, France

1 INTRODUCTION

These lectures are devoted to the use of X-ray spectroscopy for the determination of electronic properties of metal interfaces. The first "chapter" will be devoted to those techniques involving the absorption of a photon, like X-ray Absorption (XAS) or X-ray Photoemission (XPS) spectroscopies. The second one is devoted to the presentation of results obtained by electron spectroscopies such as the Auger effect (AES).

The aim of these lectures is to give to the reader the fundamentals necessary for a general understanding of the mechanisms taking place in these spectroscopies, and also to show the power of such techniques in the study of multilayer compounds. We do not pretend to cover the entire literature which exists in this domain but, by using well chosen examples, to illustrate the methods of investigation.

We would first of all like to underline the fact that spectroscopy which involves electron detection, like XPS or AES, is essentially devoted, due to the mean free path of electrons in matter, to the determination of **surface** properties. Thus it is

The authors would like to thank M. Gerl and J.P. Deville for their help in the writing and oral presentation of these lectures.

hopeless to try to obtain bulk information on those systems where a great number of interfaces are present. Such a restriction is obviously no longer valid as soon as we consider photon detection (XAS in transmission or in fluorescence modes).

In the first paragraph, we present the basic principles governing the absorption of a photon by an atom. Some readers may be disappointed by this quite formal approach, however we consider that the understanding of this first paragraph is important and suggest the reading of some general books on this subject. The lectures given at the summer school "Rayonnement Synchrotron dans le domaine des Rayons X", Aussois 1986 seem particularly well suited to an introduction to the subject we discuss here.

Starting from a general formulation, we introduce step by step the main hypothesis used in the one electron picture which is the model generally used for the interpretation of the spectro-scopic results (2.1). In the second paragraph (2.2), we shall comment on the simularities and the differences between Absorption (XAS) and Photoemission (XPS), the aim being to underline what kind of specific information can be extracted from each technique. In paragraph (2.3), we shall discuss briefly the role played by the spin in the spectroscopic experiments. The interest of such a discussion is obviously related to the possible determination of the magnetic properties of surfaces and interfaces by spin-dependent spectroscopy. Paragraph (2.4) is devoted to a brief discussion of many-body effects in XAS and XPS. This subject is really difficult and, to many aspects, is out of the scope of this summer school. We want just here to draw the attention of the reader to the fact that such effects may exist and that in several circumstances they can be used to increase the sensitivity of spectroscopic investi-gations. In the last paragraph of part 2 (2.5), we discuss more precisely what kind of spectroscopic experiments can be performed on interfaces.

The second paragraph is a direct application of all the concepts we have developed in part 2. Starting from recent experi-mental work, we want to show how the spectroscopic experiments yield direct informations on the electronic structures of the interfaces between two metals. In considering electronic structures we must include not only the collective properties of the electrons (band structures, density of states, interface state, localisation/delocalisation of the electrons, existence of magnetic moments at the interface, etc...) but also the properties related to the structural parameters (growth mechanisms or stabi-lity of the interfaces).

2 GENERALITIES ABOUT X-RAY SPECTROSCOPY

2.1 Interaction between matter and radiation

The processes we want to discuss, are, to a first approximation rather simple. They are summarized in Fig 1.

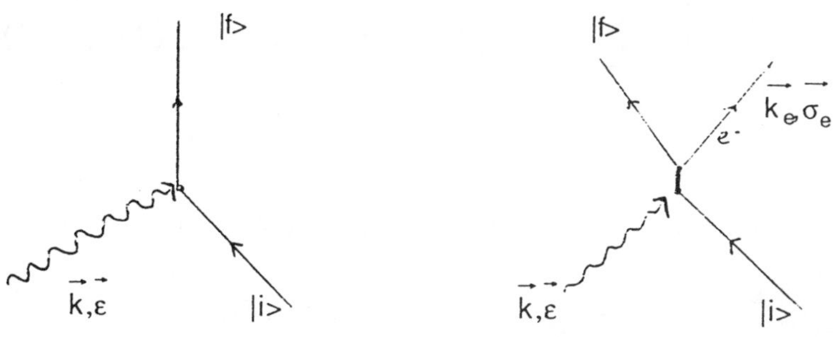

Fig. 1 Schematic representation of the absorption and photoemission processes.

We consider a system (atom **a**) in the initial state |i>, at time **t** a photon (wave vector **k**, polarisation ϵ) is absorbed, the atom reaching an excited final state |f>. If the energy of the photon is high enough a photoelectron can be produced (wave vector $\mathbf{k}_e$, spin σ_e). The correct description of this process, which evidently needs a quantum mechanical treatment, is based on "simple" considerations :

- Energy conservation,
- Momentum conservation,
- Angular momentum conservation.

The goal is simple. We have to describe properly how the sytem undergoes the transition from the initial state |i> to the final state |f>, under the photon (electromagnetic field)/electron interaction. Here the situation becomes more complex. We have first to define the form of both the initial and final states.

2.1.1 Particle Hamiltonian

The initial state |i> of an N electron system is a stationary state, the solution of a Schrödinger equation involving Ho which is time independent

$$H_o|i\rangle = E_o|i\rangle \quad \text{where,}$$

$$H_0 = \sum_{i=1}^{N} \frac{p_i^2}{2m} + \sum_{i=1}^{N} \frac{-Ze^2}{r_i} + \sum_{i>j}^{N} \frac{e^2}{r_{ij}} + \sum_{i=1}^{N} \frac{-e\hbar}{2mc^2} \vec{s}_i(\vec{E} \times \vec{p}_i) \qquad (1)$$

$$H_{ok} \quad + \quad H_{ov} \quad + \quad H_{o\,s.o}$$

The spin-orbit interaction $H_{o\ s.o}$ can be written as

$$H_{o\ s.o} = \sum_{i=1}^{N} \xi_i \vec{l}_i.\vec{s}_i \quad \text{où} \quad -\xi_i = \frac{\hbar^2}{2mc^2}\left(\frac{1}{r} \cdot \frac{\delta\phi}{\delta r}\right)_{r=r_i}$$

Thus |i> represents the initial state of an N electron system. In a 'r' representation, it is associated with a wave function for which the simplest form is a Slater determinant (NxN). We must underline the fact that the use of a single Slater determinant for the description of an N electron system is only a rough approximation of reality. Often, a better approximation is to use a linear combination of such Slater determinants. This is known as the Interconfiguration Mixing formalism [see 1].

The final state of the system |f> is also a stationnary state of an Hamiltonian which has the same form as (1). The transition is described by an Hamiltonian of interaction that we will now discuss.

2.1.2 Hamiltonian of interaction between a photon and an electron

A time-dependent perturbation theory (see G.Krill. Ecole d'Aussois, September 1986, p.254 [1]), yields a simple description of the transition which occurs between two non stationary states |I> and |F> under the photon-electron interaction.

|I> et |F> are solutions of a general Schrödinger equation

$$i\hbar \frac{\delta}{\delta t} |I> = H_0 |I> \quad ; \quad |I> = e^{-i\frac{E_i t}{\hbar}} |i>$$

$$i\hbar \frac{\delta}{\delta t} |F> = H_0 |F> \quad ; \quad |F> = e^{-i\frac{E_f t}{\hbar}} |f>$$

Assuming $h\omega_o$= Ef - Ei, we obtain the Fermi "Golden rule", giving the transition probability per unit time between states |i> and |f>

$$\Gamma_{if} = \frac{2\Pi}{\hbar^2} |<f| H_{int} |i>|^2 \delta(\omega - \omega_o) \qquad (2)$$

If we suppose that both |i> and |f> are known (in fact all the problem is here), we have just to get H_{int} in order to deduce the oscillator strength : $<f|H_{int}|i>$ and then to calculate the transition probability T_{if}. Then the determination of the absorption or photoemission cross-sections is easy.

The interaction Hamiltonian between a photon (electromagnetic field characterized by **A**, **B**, **E**) and charged particles is a well-known problem of general physics (see for instance [2], [3], [4]). In the Coulomb gauge (div **A** = 0)

$$H_{int} = \frac{e^2}{2mc^2} \sum_i^N |\overrightarrow{A(r_i)}|^2 - \frac{e}{mc} \sum_i^N \overrightarrow{A(r_i)}.\overrightarrow{p_i} - \frac{e\hbar}{mc} \sum_i^N \vec{s_i}.\overrightarrow{B(r_i)} - \frac{e\hbar}{2mc^2} \sum_i^N \vec{s_i}\left(\frac{\delta \overrightarrow{A(r_i)}}{\delta t} \times \overrightarrow{A(r_i)}\right) \qquad (3)$$

If we restrict ourselves to the domain of linear optics (i.e. to low photon fluxes) which is always correct for X-ray spectro-scopy, we can neglect the higher order terms in the Hamiltonian (3). The term which involves the spin $\mathbf{s}_i$ can be safely ignored in the spectral range we consider here (and particularly in XPS where λ > 10Å). This does not mean that we are insensitive to the spin of the particles since we must keep in mind the spin-orbit interaction. Finally, the only significant term is

$$H_{int} = - \frac{e}{mc} \sum_i^N \overrightarrow{A(r_i)}.\overrightarrow{p_i} \qquad (4)$$

This time-independent Hamiltonian which does not depend explicitly on the spin can be simplified by using the electric dipolar approximation.

2.1.3 The electric dipolar approximation

In identity (4), we notice that the photon is defined via the potential vector **A** alone. In quantum electrodynamics, the vector potential can be written by using the creation (a^+) and annihilation (a) operators of a photon

$$\overrightarrow{A(r(i))} = \sum_{\vec{k},\vec{\varepsilon}} A_{\vec{k},\vec{\varepsilon}}\, \vec{\varepsilon}\,.\, \frac{e^{i\vec{k}.\vec{r}_i}}{\sqrt{V}} + A^{+}_{\vec{k},\vec{\varepsilon}}\, \vec{\varepsilon}^{\,*}\,.\, \frac{e^{-i\vec{k}.\vec{r}_i}}{\sqrt{V}} \qquad (5)$$

(a) (a^+)

In (5), ϵ is a unit vector describing the photon polarization. By expanding the exponentials around **r** = 0 and keeping only the first term we get (writing **k**= i∇)

$$H_{int} = \frac{i\hbar e}{mc} A_{\vec{k},\vec{\varepsilon}}\, \vec{\varepsilon}\,.\, \vec{\nabla} \qquad (6)$$

which can be written as

$$H_{int} = \frac{iA}{\hbar} (E_f - E_i)\, e\, \vec{\varepsilon}\,.\, \vec{r} \qquad (7)$$

This new approximation supposes in fact that **k.r** << 1. We find here the term ϵ.**r** which is associated to the classical electric dipole, and explains the name of this approximation

2.1.4 The one electron approximation

In the above section, we get a simplified version of the interaction Hamiltonian from which we could expect to calculate directly the matrix elements of the transition probability. Unfor-tunately, the main problem is still not resolved because we do not have the exact form of |i> and |f>. The problem is still too complicated because we have to solve a true many-body problem with N electrons in the initial state and N-1 electrons plus a deep core hole in the final state. Thus a further approximation has to be made which is really a drastic one. We shall suppose that, except for the electron which absorbs the photon, all the other electrons remain passive. Evidently such an approximation, in which we neglect all the correlations between the electrons is never justified. Fortunately, most of the spectroscopic experiments can be explained in this over-simplified picture. However, as we briefly discuss in section 2-4, we shall underline the role played by many-body processes in X-ray spectroscopy, particularly in the case where metallic compounds are investigated.

In the context of this one electron picture, the XAS and XPS experiments are relatively simple to describe. We can build the final state from the initial one by changing only one electronic state in the Slater determinant and then calculate E_f and E_i. Remembering the form of H_o (1) and neglecting the spin-orbit term which shifts the levels in the same way both in the initial and final states, we get

$$E_i = \sum_{i=1}^{N} \varepsilon_i^o + \sum_{i>j}^{N} [J_{ij}-K_{ij}] \tag{8}$$

ϵ_i^o is the kinetic energy of the ith electron.

J_{ij} is the matrix element of the direct Coulomb interaction.

K_{ij} is the matrix element of the exchange Coulomb interaction.

It is convenient, but not compulsory, to think in terms of orbitals. In this case (8) becomes

$$E_i = 2 \sum_{i=1}^{N_{orb}} \varepsilon_i^o + \sum_{i>j}^{N_{orb}} [2J_{ij}-K_{ij}] \tag{9}$$

K_{ij} has not to be multiplied by the factor 2 because the electrons in the same orbital have opposite spins and thus the exchange is zero.

A similar calculation can be made for the final state leading to E_f. If we suppose that the photon has been absorbed by an electron located on the *l* ith orbital

$$E_f = \varepsilon_\ell^o + 2 \sum_{i=1\neq\ell}^{N_{orb}} \varepsilon_i^o + \sum_{i>j}^{N_{orb}} [2J_{ij}-K_{ij}] - \sum_{i=1}^{N_{orb}} [2J_{i\ell}-K_{i\ell}] \tag{10}$$

We obtain finally for $E_f - E_i$

$$E_f - E_i = -(\varepsilon_\ell^o + 2\sum_{i=1}^{N_{orb}} [2J_{i\ell}-K_{i\ell}]) = -\varepsilon_\ell \tag{11}$$

This is the well-known Koopmans' theorem which allows the determination of the kinetic energy of a photoelectron produced by the absorption process. The conservation of energy implies

$$E_K + E_B + \Phi = h\nu \tag{12}$$

where
- E_K is the kinetic energy of the photoelectron.
- E_B is the binding energy which equals $E_f - E_i$
- Φ is the work function for the photoelectron.

Koopmans' theorem, in its simplest form, is at the origin of a quite "trivial" use of X-ray spectroscopy in the determination of the electronic structure of a given atom imbedded in any material. Indeed, the energy position of an absorption edge, or a photoemission peak, can be simply related to the binding energy of the electron which has absorbed the photon. This binding energy is obviously dependent on the other electrons (particularly from the valence electrons) and thus a direct relation between the energy position and the electronic structure can be made [5]. Although correct in the framework of the one electron picture, this approach can yield unpredictable results in many cases and thus have to be used with caution. In fact, we intuitively guess that a correct description of the energy position parameter has to take into account the relaxation of the electronic system in the presence of the core hole. Here we leave the one electron approximation.

2.1.5 The dipolar selection rules. The multiplet effects

In the case of a central force problem, we know that the angular momentum (J) is a constant of motion. It implies that H_o commutes with J and thus that the eigen states of H_o can be built from those of J. J and M_J are good quantum numbers

$|i\rangle \equiv |JM_J\rangle$ (general case) or $|LSM_LM_S\rangle$ (LS coupling)

$|f\rangle \equiv |J'M'_J\rangle$ (general case) or $|L'S'M'_LM'_S\rangle$ (LS coupling)

The LS coupling (Russel-Saunders coupling) is relevant if L and S, the orbital and spin angular momenta, can be built separately, *i.e.* if the spin-orbit term is small compared to the Coulomb interaction.

The oscillator strength can be written as

$$\langle f|\,\varepsilon\,.\,\mathbf{r}\,|i\rangle = \langle J'M'_J|\,\varepsilon\,.\,\mathbf{r}\,|JM_J\rangle$$

If we apply the Wigner-Eckart's theorem (see reference [1]), recognising that $\epsilon \cdot \mathbf{r}$ can be written as an irreductible tensor of rank 1, we get

$$<f| \vec{\varepsilon}.\vec{r} |i> = <J'M_J'| r \ Y_1^{-1,0,+1} |JM_J>$$

or

$$<f| \vec{\varepsilon}.\vec{r} |i> = (-1)^{J-M_J'} \begin{pmatrix} J & 1 & J' \\ -M_J & q & M_J' \end{pmatrix} <J'M_J'|| r \ Y_1^q ||JM_J> \qquad (13)$$

In (13), Y_l is the spherical harmonic associated to $l = 1$, q can take the values 0, ±1. The interesting properties of the 3j symbols (term in parenthesis in 13), identical to those of the Clebsch-Gordon coefficients, are at the origin of the dipolar selection rules which are of considerable interest in X-ray spectroscopy.

For an atom

General case		LS coupling
$\Delta J = 0, \pm 1$		$\Delta L = 0, \pm 1$
		$\Delta S = 0$
$\Delta M = 0$	<-- Linear polar -->	$\Delta M_L = 0$
$\Delta M = -1$	<--Circular polar (R)-->	$\Delta M_L = -1$
$\Delta M = +1$	<--Circular polar (L)-->	$\Delta M_L = +1$

For an electron

General case		LS coupling
$\Delta j = \pm 1$		$\Delta l = \pm 1$
		$\Delta s = 0$
$\Delta m = 0$	<-- Linear polar -->	$\Delta m_l = 0$
$\Delta m = -1$	<--Circular polar (R)-->	$\Delta m_l = -1$
$\Delta m = +1$	<--Circular polar (L)-->	$\Delta m_l = +1$

These selection rules are of considerable interest in the use of high energy spectroscopies, like those we discussed here. They are the direct consequence of the use of the electrical dipolar approximation and thus their validity is restricted to the cases where this approximation holds. They show clearly that the photoelectron, created by the absorption process, reaches, in the final state, an orbital with a well-defined symmetry. This gives to these spectroscopies their high selectivity. The situations of particular interest are those where the outermost valence electrons are well localized and thus contribute directly to the angular momentum (both orbital and spin) of the atom. The coupling between the core hole and these valence electrons induces the existence of multiplet states defined by $J'M_{J'}$ ($L'M_{L'}$) in the final state and thus direct transitions from $JM_J \rightarrow J'M_{J'}$ will be observed in the absorption (photoemission) spectra. These multiplet structures can be considered as real fingerprints of the electronic configuration of the atoms in the initial state. As we shall see later, such multiplet structures may be used to study, in an indirect way, the magnetic properties of surfaces and interfaces simply because they are related to localized states. As an example, we show in Fig. 2 and 3 the multiplet structures observed experimentally on the absorption and photoemission spectra of several rare-earth metals where the localized character of the 4f states is well-known.

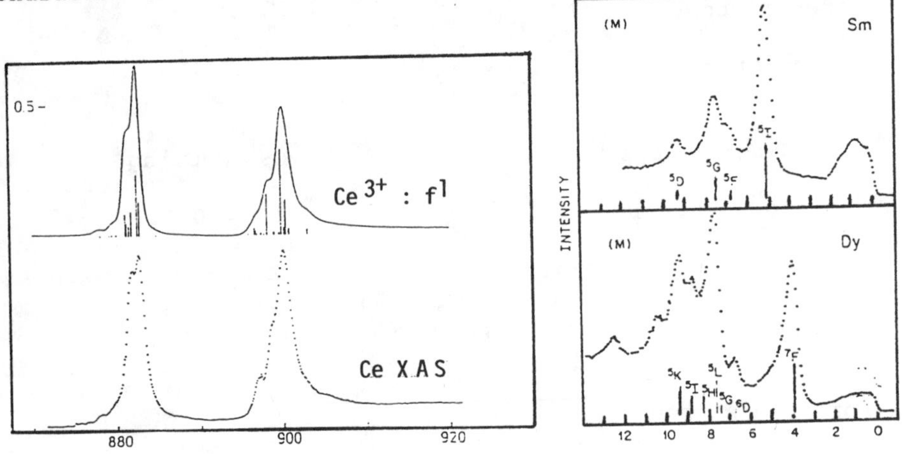

Fig. 2 M_{IV} XAS spectrum of Ce metal.

Fig. 3 Photoemission spectra of Sm and Dy metals.

2.1.6 Absorption and photoemission cross-sections

We shall comment about the differences between absorption and photoemission in sections 2.2 and 2.4. In the one-electron picture, the absorption and photoemission cross-sections are quite identical, except for some aspects which are important for practical use.

These cross-sections are directly proportional to the transition probability

$$\sigma_{XAS}(E) \propto \sum_{i,f} |\langle f| \vec{\varepsilon}.\vec{r} |i\rangle|^2 \, \delta(E_f - E_i - \hbar\omega) \qquad (14)$$

In the case where the photoelectron, in the final state, reaches an unoccupied band of l symmetry, characterized by the density of states $\rho_l(E)$ we get

$$\sigma_{XAS}(E) \propto \sum_{i,f} \int_{E_a}^{E_b} |M_{i,f}|^2 \, \delta(E-\hbar\omega)\, \rho_l(E)\, dE \qquad (15)$$

$$M_{i,f} = \langle f| \vec{\varepsilon}.\vec{r} |i\rangle|$$

In most cases $|M_{i,f}|$ can be considered over constant on the whole range of integration, i.e on the bandwidth, thus (15) can be written as

$$\sigma_{XAS}(E) \propto \sum_{i,f} |M_{i,f}|^2 \int_{E_a}^{E_b} \delta(E-\hbar\omega)\, \rho_l(E)\, dE \qquad (16)$$

We get the important conclusion that the X-ray absorption cross section is a convolution product of the projected density of empty states ρ_l , the symmetry l being fixed by $l = l_i \pm 1$. It is thus possible to obtain directly the partial density of (empty) states from an XAS experiment near the absorption threshold. We illustrate such a possibility on Fig. 4, where the correlation between the absorption structures and the density of unoccupied states is well indicated.

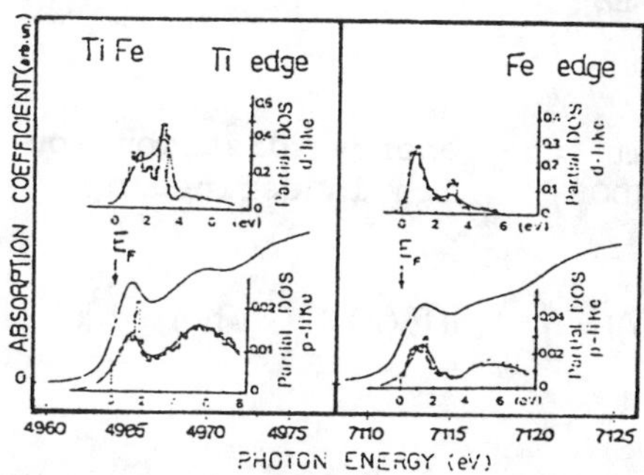

Fig. 4 Calculated and d density of unoccupied for Ti and Fe in TiFe. Comparison with K edge experiments [8].

In the case of photoemission, the situation is quite similar except that

-the final state is made of N - 1 electrons plus one core-hole, whereas it is N electrons plus one core-hole for XAS;
-if the kinetic energy of the photelectron is high enough, it lies in a structureless continuum.

An interesting situation occurs if we consider the excitations of valence or conduction electrons induced by low energy photons (10 < hυ <50 eV), which are typical of Ultraviolet Photoemission Spectroscopy (UPS). If we restrict ourselves to those cases where the momentum (**k**) is conserved (direct transitions, three-step model [9] Fig.5), the photoemission cross-section is

$$\sigma_{UPS}(E,\omega) \propto \sum_{i,f} \int_{BZ} |M_{i,f}|^2 \delta(E_f(\vec{k})-E_i(\vec{k})-\hbar\omega)\ \delta(E_f(\vec{k})-E)\ d^3k \qquad (17)$$

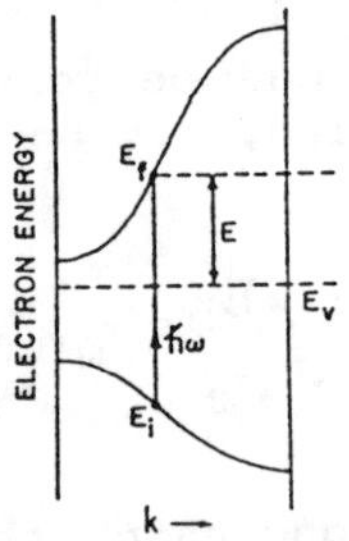

Fig. 5 Direct transitions in UPS. E_V is the vacuum level E is the kinetic energy of the photoelectron.

If $M_{i,f}$ is constant

$$\sigma_{UPS}(E,\omega) \propto \sum_{i,f} |M_{i,f}|^2 \int_{BZ} \delta(E_f(\vec{k})-E_i(\vec{k})-\hbar\omega)\ \delta(E_f(\vec{k})-E)\ d^3k \qquad (18)$$

The integral is the energy distribution of the joint density of states (JDOS) which is defined by

$$J(h\omega) = \int_{BZ} \delta(E_f(\vec{k})-E_i(\vec{k})-\hbar\omega)\ d^3k \qquad (19)$$

This is precisely the quantity which is measured in UPS experiments. It contains both information on the occupied and unoccupied density of states. It is easy to show that (18) can be written in a more explicit form [10]

$$\sigma_{UPS}(E) \propto \overline{|M_{i,f}|^2}\ \rho(E_f)\ \rho(E_i) \qquad (20)$$

We shall see later that such UPS techniques can be used for the determination of the band structures of single crystal materials via the Angular Resolved Photoemission Spectroscopy (ARPES). Let us underline the fact that for kinetic energies of the photoelectron exceding 20 eV, the influence of the unoccupied density of states can be safely ignored. On the other hand, we have to be very cautious with the description of experiments in the regime $E_K < 20$ eV. As an example, we show on Fig 6a,b the UPS and XPS spectra of the valence band of gold and a comparison between the calculated integrated density of states.

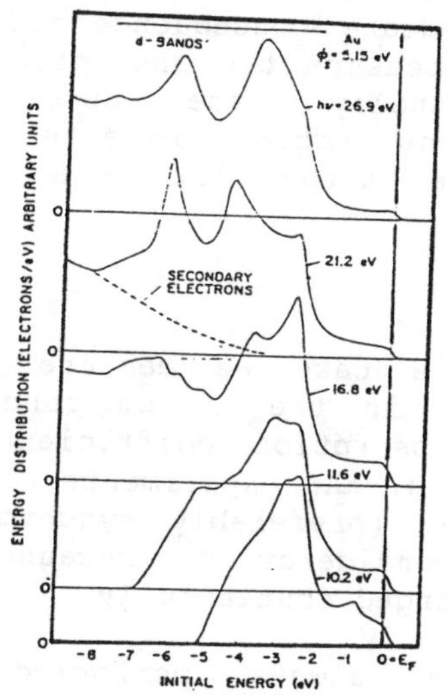

Fig. 6a UPS spectra of gold [11].

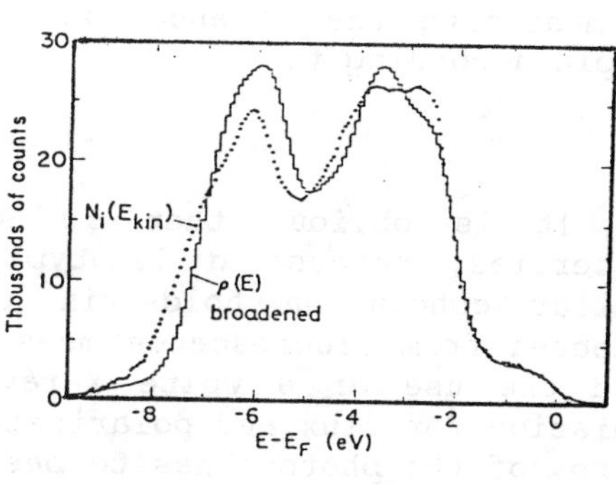

Fig. 6b XPS spectra for the the valence band of gold. Comparison with calculation.

Before closing this discussion on the problem of XAS and XPS (UPS) cross-sections, we want to briefly mention the fact that, before using the identities 16-20 for applications, we have to take into account the finite lifetime of the core-hole created by the absorption process. This lifetime, which decreases rapidly as the binding energy increases, is typically in the 10^{-14} sec (i.e. some fractions of an eV) for low binding energies ($E_B < 500$ eV), but reaches a few eV (≈ 5) for levels whose binding energies are more than 5 keV. In order to take into account this finite life-time, we have to convolute all the expressions we gott previously with a Lorentzian whose FWHM (Full Width at Half Maxima) is preci-sely the core-hole lifetime. In the same way, all other sources of broadening (experimental resolution, etc...) have to be taken into account by finally convoluting with a Gaussian-like function.

2.2 Differences and simularities between X-ray absorption (XAS) and photoemission (XPS)

Up to now, except for the difference in the final state of XAS and XPS, we did not comment on the particularities of each technique. Here we want to discuss this point, excluding for the moment all the problems related with the relaxation mechanisms which will be discussed in section 2.4. An essential point comes from the fact that, in photoemission, the measurement is performed on the photoelectron itself and the major differences between XAS and XPS result from this simple fact. For instance, it has an immediate consequence related with the methods that are used to measure the absorption or photoemission cross section. Usually, the absorption coefficient (i.e. the absorption cross section) is determined by measuring the attenuation of the photon flux in a given sample according to

It is obvious that in such a case we measure bulk properties, because d is typically in the 10 μm range. A similar conclusion holds if the absorption coefficient is detected from fluorescence measurments. XAS experiments always need the use of a white X-ray beam (preferably synchrotron radiation for flux and polarisation considerations) because the energy of the photons has to been changed continuously.

Photoemission experiments are always performed by measuring the kinetic energy of the photoelectrons and thus, ultra high vacuum techniques are always mandatory. In contrast to XAS expe-riments, they do not need absolutely the use of white beam radia-tion but can be performed at a fixed wavelength. Thus experiments in the laboratory are possible. However, and this is the essential difference with XAS, photoemission experiments can only yield information on surface properties, because we are limited by the finite mean free path of photoelectrons in a solid. To a first approximation this mean free path is only dependent from the kinetic energy of the photoelectrons as shown on Fig. 7.

Indeed, only elastic processes can contribute to the photo-emission cross section. As can be noticed in Fig. 7, the mean free path of the photoelectrons never exceeded 50 Å in the usual range E_K < 2 keV. Thus, photoemission experiments will essentially give information on surface properties. This conclusion is obviously important for our purpose and restricts considerably the use of this technique in the study of multilayer compounds.

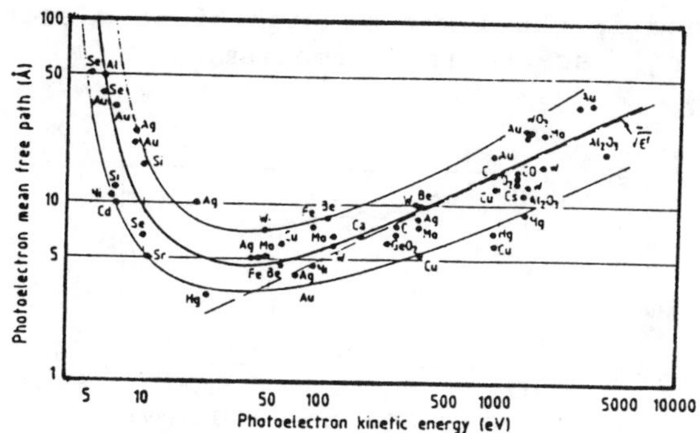

Fig. 7 Mean free path of electrons as a function of E_K.

We must underline the fact that the difference between surface/bulk techniques is rather ambiguous, at least in the case of XAS experiments. Indeed, it is possible to measure the absorption cross section from the Auger electrons which are produced by the absorption process or by the electron total yield. In those cases, we deal with surface absorption spectroscopy (SXAS).

Up to now, we have considered the absorption process in the immediate vicinity of the absorption singularity (E_o). It is clear that the absorption cross section does not vanish if $h\nu > Eo$. This is precisely the photoemission process ! Except that, in this situation, the photoelectron which is produced never goes out of the solid but will be scattered by the neighbouring atoms surrounding the central atom which has absorbed the photon (Fig.8).

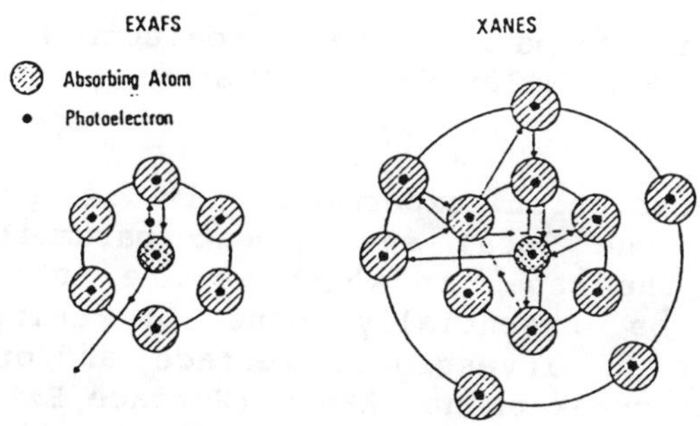

Fig. 8 Scattering mechanisms for the photoelectron.

These scattering mechanisms induce an oscillating behaviour of the absorption coefficient (Fig.9), which can be easily explained in the limit of high kinetic energy ($E - E_o >$ 100eV) by a single scattering formalism.

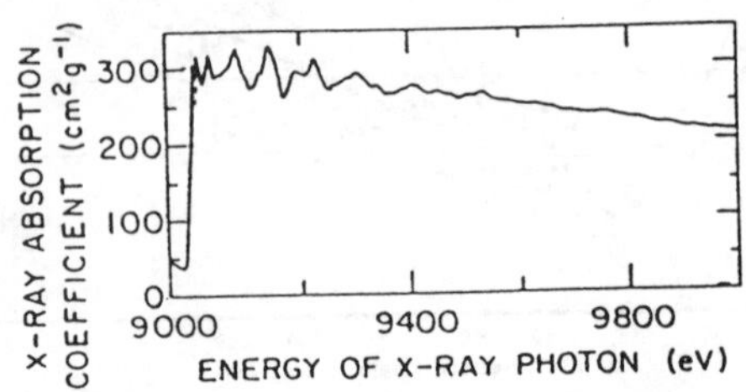

Fig. 9 Behaviour of the absorption coefficient of Cu at the K edge [12].

These are the EXAFS structures (Extended Fine Structure Absorption Spectroscopy) which can be described by the well-known formula [13]

$$\chi(k) = \sum_j \frac{N_j}{kR_j^2} e^{-\frac{2R_j}{\lambda}} e^{-2\sigma_j^2 k^2} |f_j(k,\pi)| \sin(2kR_j + \delta_j(k)) \quad (21)$$

where :

- k is the photoelectron wave vector
- N_j is the number of neighbours atoms in shell j.
- $|f_j(k,\pi)|$ et $\delta j(k)$ are the amplitudes and the phase shifts for the diffusion which can be either calculated or extracted from reference compounds.
- R_j is the average distance between the central atom and the neighbours.
- λ is the mean free path of the photoelectron.
- σ_j is the Debye-Waller factor of atoms j.

From X-ray absorption experiments it is possible, in principle, to get the local environment parameters of the central atom. For the cases in which we are interested, the experiments will be essentially done by using electron detection in order to investigate surface and/or interface properties. We will speak about SEXAFS (Surface Extended X-ray Absorption Spectroscopy). In these lectures, we will say nothing about the intermediate regime of XANES (X-ray Absorption Near Edge Spectroscopy), simply because in that case we deal with multiple scattering processes which are evidently more complicated to handle. Evidently much valuable structural information may be obtained from this XANES regime (in particular the angles), and we think that, in the near future, the development of multiple scattering calculations will be useful for the study of interfaces.

2.3 The role of spin in high energy spectroscopy

When we obtained the Hamiltonian of interaction between a charged particle and a photon (see section 2.1.4), we noticed that this Hamiltonian was spin independent. This implies that the spin of the particle remains **unchanged in the absorption process** and this conclusion presents some advantages and disadvantages for the study of magnetism by high energy spectroscopy. Here we enter a new and interesting field of research that, unfortunately, we cannot discuss in detail here. We refer the reader, who may be interested, to the paper by C.Brouder, D.Malterre and G.Krill [14] given at the Spring school "Rayonnement Synchrotron Polarisé, Electrons polarisés et Magnétisme" held in Mittelwihr (March 1989) in which we try to address this problem.

The main advantage, presented by the spin conservation, lies in the possibility to perform direct spin resolved photoemission experiments on the valence bands. Because of spin conservation, it is only necessary to make an analysis of the spin polarisation of the photocurrent (by using a Mott detector for instance) in order to get directly the spin resolved density of occupied states.

As a consequence of the one electron approximation, it is possible in photoemission experiments to use the exchange interaction between the core hole and the localized valence electrons in order to reveal the existence of a magnetic moment on the atom which has absorbed the photon. This is a simple way to study directly the magnetic properties of surfaces and we will give some examples in part 3.

The main disadvantage of the spin conservation is founded in XAS experiments, where we cannot make direct measurements on the photoelectron itself. In that case, it is possible to show that the spin dependence of the absorption cross section is made indirectly by the spin-orbit interaction of the photoelectron in the final state [14]. Except for highly localised electrons (f for instance), this spin dependence can only be revealed by the use of circularly polarized light. We will show, in part 3, how it is possible to study the magnetic properties of multilayers by using Spin-Resolved Absorption Spectroscopy.

2.4 Many-body effects in high energy spectroscopy

As underlined in the general introduction of these lectures, we do not want to enter into the details of the highly complicated problem of many-body effects in high energy spectroscopy. We refer the reader to chapter IV and V of reference [1] for an intro-duction to these problems. Here we just want to underline some interesting results which may be useful for the general under-standing of the examples described in part 3.

For our applications, we need only to consider the many-body effects on the two following points

1) Energy position of the X-ray Absorption edges and the core level XPS spectra.
2) Modification in the line shape of core level XPS spectra.

Both of these effects have the same physical origin : the relaxation of **all** the electrons to the presence of the core hole and its description is beyond the one electron picture.

2.4.1 Energy position of an absorption edge or of a photoemission core level spectrum

In the one electron approximation, the energy position of an absorption edge (or of a photoemission peak) is simply given by the Koopmans' theorem [11]. It is easy to understand that, under the influence of the core hole, **all** the electrons of the system and particularly the conduction electrons in a metal, will participate to the screening of the core hole. This screening appears to the experimentalist as a decrease in the binding energy which can be larger than 10 eV. This is the reason why the binding energies have to be used with caution when one tries to use them in order to describe the electronic structure of various compounds.

Here, we want to underline the major difference which exists between XAS and XPS. Indeed in XAS, and near the threshold energy, the photoelectron is always bound to the atom and thus can participate efficiently to the screening mechanisms, whereas in XPS (or in XAS in the EXAFS regime) the photoelectron leaves the atom. This does not mean that the many body effects will be less important in XAS as compared to XPS, but simply that the rela-xation in XAS will be quasi adiabatic and surely more difficult to describe than in XPS. From a practical point of view this is the reason why the energy position of an absorption edge will never be used (except in a few particular cases) to describe the electronic structure of an atom in a solid. The occurrence of many body effects in XAS spectra is well illustrated in Fig. 10, where the K edges of several transition metals are compared with one electron calculations. If the agreement is correct in the EXAFS range (i.e. in the photoemission regime) it is less good near the threshold energy except in the case of metallic copper. The better agreement we observe in the case of copper is not fortuitous but results from the fact that the correlations between electrons in copper are small and explain why the one electron model is correct. This is obviously not true for nickel, where it is known that the electrons are highly correlated, and the XAS structures are no longer explained by the one electron approximation.

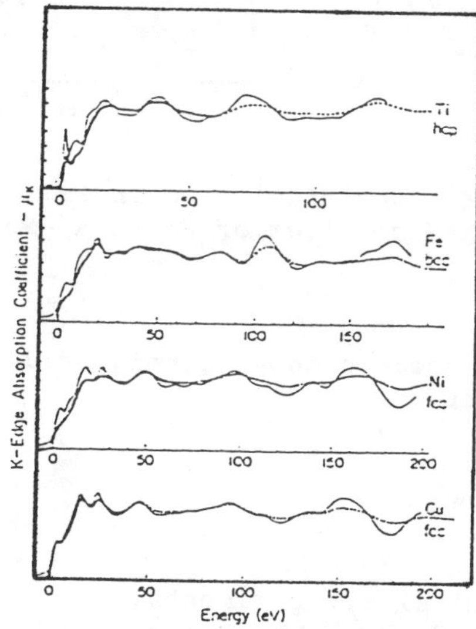

Fig. 10 K absorption edges (experimental and calculated) of the transition metals of the first series [15].

2.4.2 Lineshapes of the core levels photoemission spectra

The existence of strong relaxations of the electrons under the influence of the core-hole yields important modifications of the lineshape of the photoemission spectra, particularly in the case of metals. Departures from a simple Lorentzian line shape (see 2.1.5) are experimentally observed (an asymmetry of the XPS lineshape is easily detected on the high binding energy side) which can be directly related to the collective excitations of the electron gas in the presence of the core-hole. From a correct treatment of the core-hole screening, it is possible [16], to relate the asymmetry parameter of the XPS spectrum to the **density of states at the Fermi level.** This correlation is well demon-strated on Fig. 11a and b, where the

XPS spectra of Pd_{3d} core levels in Pd metal (having a high density of states at E_F) and in $Ag_{0.9}Pd_{0.1}$ alloy (having a low density of states /≈ 0/ at E_F) are reported. We notice the difference in the asymmetry of the two XPS lineshapes.

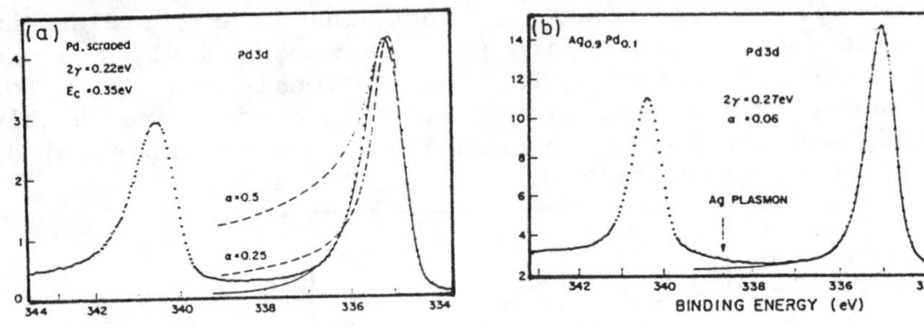

Fig. 11 a) 3d XPS spectrum of Pd in Pd metal [17]
b) 3d XPS spectrum of Pd in Ag-Pd.

2.5 High-energy spectroscopy techniques used for the study of bi(multi)layer compounds

2.5.1 Photon sources

For the photoemission experiments, it is possible to use conventional sources in the 10-1500 eV energy range. For the ultraviolet range (HeI, He_{II}, etc), discharge lamps are used and the soft X-ray range is obtained with usual Mg and Al anodes. These conventionnal sources present the advantages to be used in the laboratory and to deliver relatively high photon fluxes (≈ 10^{12}ph/sec). However, as mentionned earlier, they cannot be used for X-ray absorption studies and synchrotron radiation becomes mandatory. Today, storage rings can deliver synchrotron radiation with a wide range of energy (10-15000 eV) with flux intensities which are comparable to those delivered by classical sources. The obvious advantages presented by the use of synchrotron radiation are related to its polarisation properties which allow more selective experiments.

2.5.2 Determination of electronic structures

a) Occupied states : surface and interface states.

In these studies, we are mainly interested to get information on the occupied density of states from UPS, or eventually from XPS, measurements like those discussed in section I-1-5. Here it is interesting to present the basic principles of the angular resolved photoemission technique (ARPES) which is of first importance in the study of epitaxial systems. In this technique, the photoelectrons are collected in a very small angular range (typically 1°-3°), from the kinetic energy of the photoelectron it is possible to deduce the angular dependence (both azimuthal Φ and polar θ) of the photoemission cross section.

The conservation of the **k** vector, in the range of specular reflexion, implies the conservation of the k component which is parallel to the surface :

$\mathbf{k}_{//} = \mathbf{K}_{//} + \mathbf{G}_{//}$ where **G** is any vector of the reciprocal lattice and **K** is the photoelectron wave vector which can be simply deduced from the kinetic energy itself $E_K = h^2K^2/2m$.

Here we see that we can experimentally determine the dispersion curves, i.e. the band structure of the material itself. In practice we are limited by the fact that the parallel (//) and perpendicular (⊥) components of **K** are indeterminate. There are two situations where such a problem may be solved :

i) when the photoelectrons are detected at the normal : in this situation, the $\mathbf{K}_{//}$ component is zero and thus $\mathbf{K} = \mathbf{K}_{\perp}$. By changing continuously the photon energy (here we need the synchrotron radiation), it is thus possible to deduce $E=f(\mathbf{k})$. Indeed to each value of $|\mathbf{K}^2|$ (and thus of E_K and $h\nu$) we associate a given point of Brillouin zone ,

ii) when bidimensional systems are studied : lamellar compounds or epitaxially grown mono(bi)layers. In this situation it is obvious that we do not have to consider any $\mathbf{k}_{\perp}$ component. The Brillouin zone is of 2D character and is fully characterized by $\mathbf{k}_{//}$. Such experiments can be performed at a fixed energy and thus conventional equipment may be used. The continuous change of $\mathbf{k}_{//}=(2mE_K/h^2)^{1/2}\sin\theta$, is achieved by changing θ. It is thus possible, as in the first case, to determine the dispersion curves in well defined directions of the bidimensional Brillouin zone.

b) Unoccupied states : absorption and inverse photoemission.

We have underlined above the fact that XAS experiments, near the threshold, may yield the unoccupied density of states. In fact such studies are limited because of the core-hole lifetime which becomes shorter and shorter as the binding energy of the core increases. This results in a broadening of the structures of the unoccupied states, thus in most cases the unoccupied density of states are studied by a complementary technique of photoemission where the "bremstrahlung" radiation of the electrons is used (BIS). A monokinetic beam of electrons impinges on the surface of a material and the radiation at a fixed wavelength is detected. By changing continuously the kinetic energy of the electrons, keeping constant the wavelength of the detected photons, we get directly the density of unoccupied states. In that case we deal with a real inverse photoemission process and the sensitivity to surface properties is comparable to that observed in photoemission experiments.

2.5.3 Studies of structural properties

In this paragraph, we can mention all the techniques which may be used to study the oscillating part of the absorption coefficient (EXAFS,SEXAFS). Such investigations can be made in three different modes : i) either in the transmission mode, ii) either in the total reflexion regime or iii) by detection of electrons, the sensitivity to surface properties increasing from i) to iii).

In the transmission mode, it is possible to study thick samples . The structural information is averaged over the whole sample, and the specific properties of the interfaces are thus difficult to obtain. In the next part we will give an example of an EXAFS study performed on a multilayer system in this transmission mode. In case ii), EXAFS is measured by transmission after a total reflexion on the sample under investigation. The probing depth depends strongly on the electronic properties of the material and is typically in the 50 Å range. When EXAFS is detected from photoelectrons (iii), we shall speak about SEXAFS, the sensitivity to surface properties is rather similar to that of photoemission experiments.

The information that we try to get from such experiments is always related to the structural properties of the interfaces themselves, either related to their topology (abrupt or diffuse) or to their electronic properties (particularly those related to phonon properties).

3 EXAMPLES FOR THE USE OF HIGH ENERGY SPECTROSCOPIES FOR THE STUDY OF MULTILAYER SYSTEMS

In this chapter, we will discuss some recent results which will illustrate how X-ray absorption spectroscopy and/or photoemission can be used in order to obtain information on the physical properties of multi(bi)layers compounds. In the first section (3.1), we shall mainly focus our attention on the problem of the determination of the electronic structure of bilayers, principally from UPS and XPS experiments whereas the second section (3.2) will be essentially devoted to the study of multilayers by XAS.

3.1 Electronic structures of metal-metal interfaces

Our aim, here, is to show some direct applications of photoemission for the determination of the electronic structure of metal-metal interfaces. Although it is possible to perform experiments on polycristalline surfaces (except obviously the angular studies), we shall restrict ourselves to the cases of epitaxial systems because the interpretation of the results is easier. As it is not possible to illustrate on a single example all the possibilities of the technique, we shall discuss several systems. Such a procedure although it may be rather confusing for the reader, is necessary if we want to reach the aim in which we are interested.

3.1.1 Study of an abrupt interface: Cu/Pt (111) [18]

A.E.S (Auger) and LEED (Low Energy Electron Diffraction) studies have shown that, when the growth of Cu on Pt(111) is made in the layer-by-layer mode, there is no diffusion in this system and thus the interface is abrupt. This is the simplest case one can imagine. The first interesting property, from which photomission benefits, is the behaviour of the photoemission cross sections of Cu and Pt in the energy range $h\nu$ = 80-180 eV. This variation, reported on Fig. 12, shows the existence of a minimum in the photoemission cross section of Pt near 150 eV (Cooper's minimum), which indicates that for experiments performed in that energy range, we will get essentially information on copper atoms at the surface.

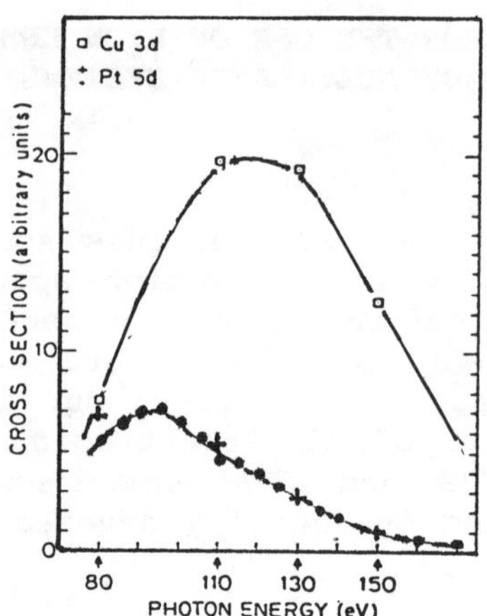

Fig. 12 Behaviour of the photoemission cross sections of Cu and Pt.

a) Study of Cu 3d states

In figures 13a and 13b, we report the valence band photoemission spectra obtained at hν = 150 eV (Cooper's minimum of Pt) for several Cu coverages. These spectra reflect the modifications of the Cu 3d states as the copper thickness increases.

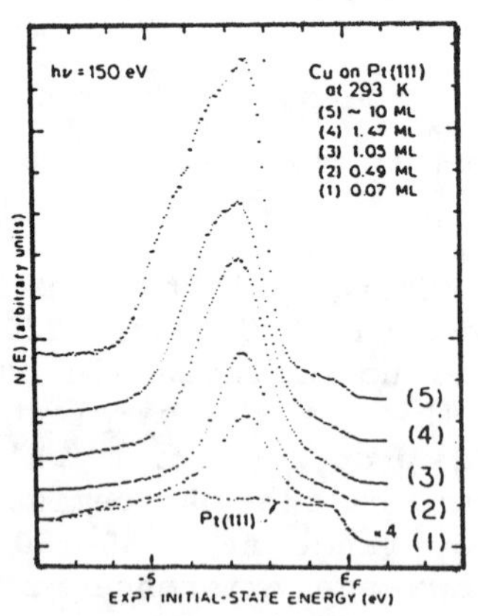

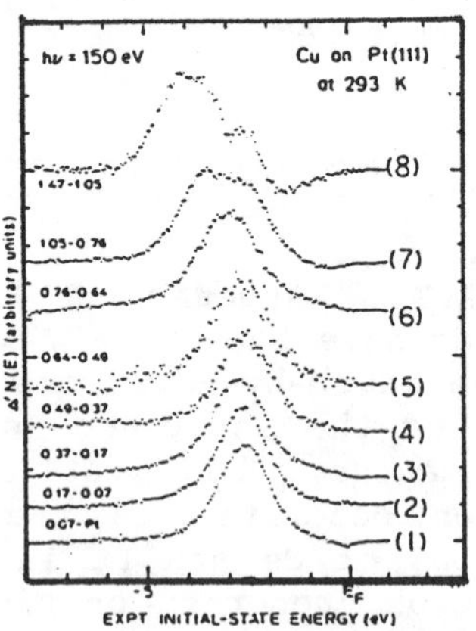

Fig. 13a) Cu 3d states for Cu/Pt (111)   b) Difference spectra.

We notice that for high copper coverages ($\theta \geq$ 2ML), the typical valence band of bulk copper is observed. The interesting features are observed for lower copper coverages, below one monolayer, as shown on Fig. 13b. The narrow Cu 3d structures (FWHM ≈ 1.5eV) observed for 0.1 < θ < 0.7 ML are rather independent from θ and are characteristics of "isolated" copper atoms at the surface of platinum. The Cu-Cu interaction in the layer becomes significant for θ > 0.7ML, where we notice a broadening of these Cu 3d structures and, finally, the formation of a 3d band characterized by a shift of the structures to lower binding energies. After the formation of the first layer, we notice clearly new structures in the spectra located at - 3.5 eV below E_F. These structures are due to copper-copper interaction between **two layers** and their intensities follow roughly the copper coverage.

b) Study of core level spectra.

The Cu $2p_{3/2}$ and Pt $4f_{7/2}$ core level spectra are reported in Fig.14a and 14b. They give a clear illustration about the kind of information which can be obtained from such experiments.

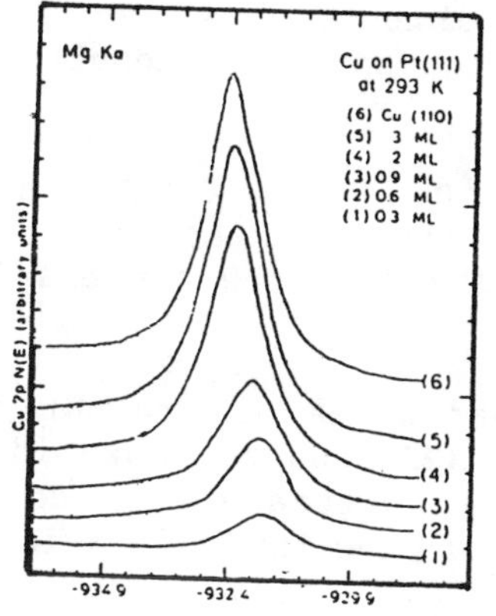

Fig.14a Cu $2p_{3/2}$ XPS spectra.

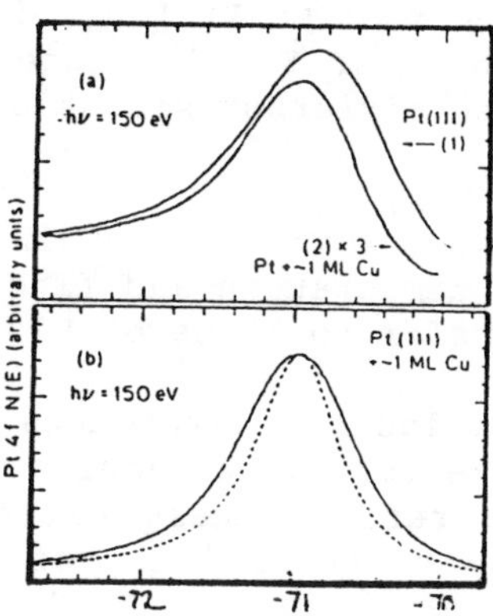

Fig.14b Pt $4f_{7/2}$ XPS spectra.

In Fig. 14a, a clear evolution is observed between the submonolayer range and the thicker copper coverages (> 1 ML). The spectra shift progressively to higher binding energies, but no significant changes occur in the line shape. This result is in agreement with those obtained for the Cu 3d states, however the fact that the XPS peak shifts to higher binding energy is not easy to understand. Indeed, if we suppose that the number of conduction electrons increases with the copper coverage, it is intuitively evident that the screening of the 2p core hole (for instance) should be easier and thus we expect a shift to lower binding energies. The reader should refer to [18] to see how a model which takes into account the relaxation in the final state gives the correct answer to this problem. This illustrates clearly what we have said about the direct use of Koopmans' theorem in the one electron approximation. The fact that the copper XPS lineshape does not change reflects simply the fact that i) there are no multiplet interactions in that case (trivial because the 3d electrons are delocalized) and ii) the density of states at the Fermi level remains small enough (also trivial for Cu) so that no asymmetry can occur. On Fig.14b, we observe a narrowing of the Pt $4f_{7/2}$ XPS peak for $\theta \approx 1$ ML. The explanation of this narrowing is simple. The $Pt4f_{7/2}$ XPS line of pure platinum is made of two contributions [19], a surface component which is located at a lower binding energy (- 0.4 eV) than the bulk component. When 1 Cu ML is deposited at the Pt surface, we observe the disappearance of this surface component which is characteristic of the Pt/vacuum interface.

c) Sensitivity to local environment

What can we say about the sensitivity of photoemission experiments to local environment properties ? From the analysis of the experiments on Cu/Pt [20], it is possible to show that this sensitivity exists. For instance, let us compare the results obtained for these different situations :

Pt(111), Pt(111) + 0.2ML Cu(2), $Pt_{0.98}Cu_{0.02}(110)$ alloy (1), Pt(111)-7.7 % Cu (3) obtained by diffusion of copper in platinum at high temperature and finally Cu(110) (4) (the numbers in parentheses refer to those of Fig.15).

The local environments of copper atoms are rather different for these various situations and the XPS results shown in Fig. 15a and 15b reflect these differences. Unfortunately space does not allow us to enter into the details of the interpretation of such results.

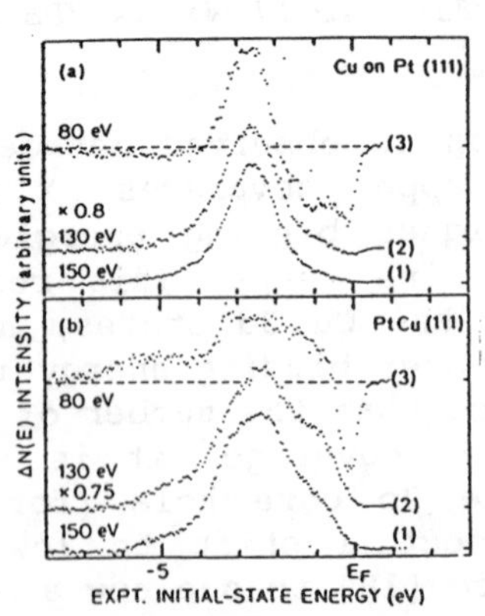

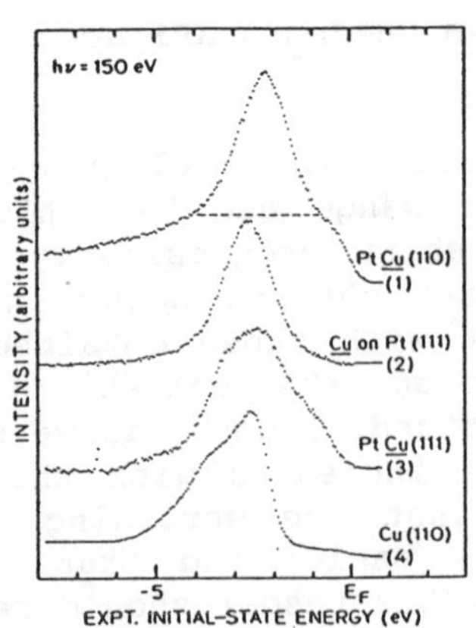

Fig.15 a) Valence band spectra of the Cu/Pt systems b) difference spectra.

In Fig.16, we report the behaviour of the Pd $3d_{5/2}$ core level XPS spectrum during the formation of the Yb-Pd interface [21]. Due to the strong diffusion of palladium in ytterbium, the interface is diffuse. The Pd XPS spectra reveal the formation of Yb_xPd_{1-x} alloys which are characterized by lower binding energies and smaller densities of states at the Fermi level (the Pd 4d band is filled by the conduction electrons of Yb) which explains the progressive "symmetrisation" of the Pd $3d_{5/2}$ core level spectra.

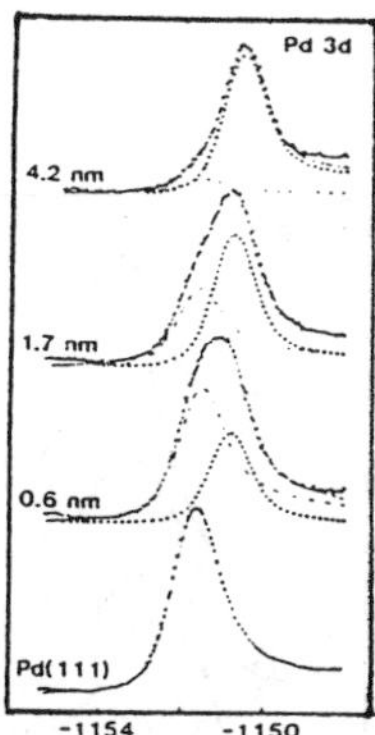

Fig. 16 Pd $3d_{5/2}$ core level as a function of the Yb coverage on Pd (111).

3.1.2 Study of diffuse interfaces. Magnetic properties

We want here to discuss the cases of both Cr/Au(100) [22,23,24,25] and Cr/Ag(100) [26] interfaces. The comparison between these two systems is interesting because a strong diffusion occurs in the Cr/Au system (like in Yb/Pd discussed above), whereas no diffusion is present for Cr/Ag. Moreover, the magnetic properties of chromium (and Cr alloys) are well known and can be easily studied by photoemission. We want to focuss our discussion of these systems on this last point.

a) Core level spectra

The behaviour of Cr2s(3s) and $Cr2p_{3/2}$ (only in the case of Cr/Au(100)) core level spectra with the chromium coverage is reported in Fig. 17 a,b and c. We notice, particularly for the Cr $2p_{3/2}$ XPS spectra, that these behaviours are rather different from those observed in the previous case of Cu/Pt(111). For the Cr 2s (3s) spectra, we notice clearly the existence of two well-separated contributions whose intensities are highly dependent on the chromium coverage. The change in the intensity of these two contributions is drastic in the Cr/Ag system whereas it is quite continuous for the Cr/Au system.

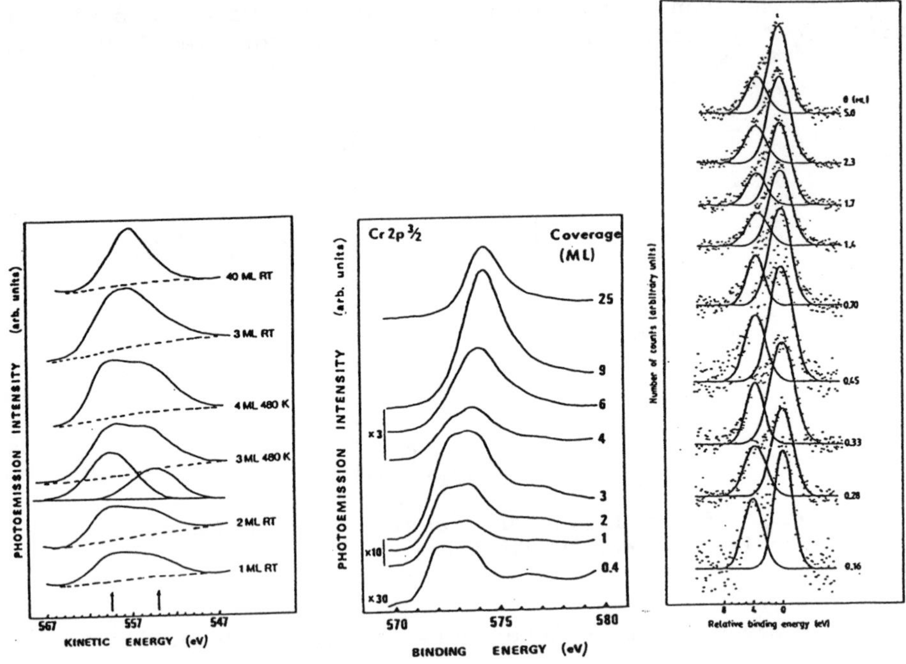

Fig. 17 a) Cr 2s XPS spectra in Cr/Au (100) b) Cr $2p_{3/2}$ XPS spectra in the same system c) Cr 3s XPS spectra in Cr/Ag (100).

Let us first comment on the line shape of Cr 2s (3s) spectra. As the spin-orbit interaction is zero for an s level ($l = 0$), the only explanations for the presence of two well separated contributions in the XPS spectra are either related to different local environments for Cr at the surface or to the existence of multiplet interactions like those discussed in 2.1.5. The first hypothesis can be rejected by an analysis of the Cr $2p_{3/2}$ XPS spectra. The energy difference we observe between the two contributions of the Cr $2p_{3/2}$ is only 3eV whereas it is at least 6eV for the Cr 2s, thus the existence of several contributions on the Cr 2s (3s) XPS lines are due to multiplet splitting.

b) Consequence of the existence of multiplet splitting

As discussed in section 2, a multiplet splitting is due to the coupling between the core hole and the valence electrons. In the case of s levels, this coupling is very simple and we need just to consider two different final states for the Cr atom :

-one state where the spin of the s core hole couples parallel to the spins of the 3d electrons,

-one state where the spin of the s core hole couples antiparallel to the spins of the 3d electrons.

The energy separation between these two different final states, ΔE, is directly proportional to the exchange energy. In the case of an s state we get [27]

$$\Delta E = \frac{2S+1}{2\ell+1} G^{\ell}(s,\ell)$$

In this expression S is the spin of the initial state, *l* the symmetry of the valence electrons (in our case *l* = 2) and G is the exchange Slater's integral. Thus a linear relation exists between the spin (and thus the magnetic moment) and the energy splitting of the two XPS contributions in the Cr 2s (3s) core level spectra. It is thus possible to measure **directly** the magnetic moment of chromium at the Cr/Au or Cr/Ag interfaces (Fig. 18). We notice the difference between the two systems which reflects the difference between the interfaces themselves. In Cr/Ag(100) the interface is "abrupt" (no diffusion), thus the magnetic moment of chromium decreases quickly in order to reach the value of bulk metallic chromium. On the contrary, in Cr/Au(100) the diffusion of Au in Cr leads to the formation of Cr_xAu_{1-x} alloys which are still magnetic and thus explain why the multiplet splitting on the Cr 2s is still observed for high chromium coverages.

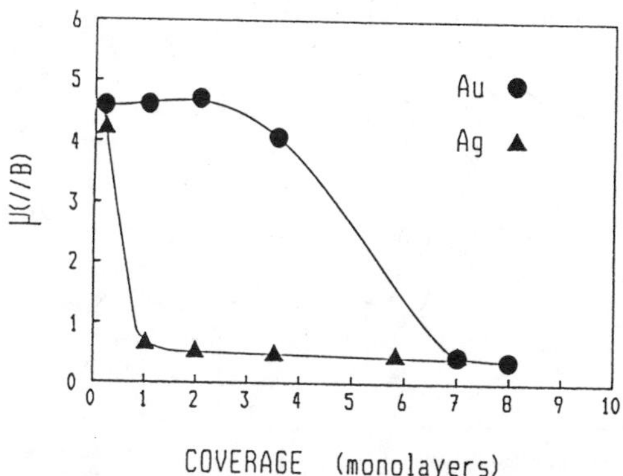

Fig. 18 Magnetic moment of chromium at Cr/Au and Cr/Ag interfaces as deduced from XPS experiments.

3.1.3 Angular resolved photoemission studies : Ni(111)/W(110)

Angular resolved photoemission experiments are now commonly used for the study of the electronic structure of interfaces. We choose to discuss here a recent study of the Ni(111)/W(110) system [28], because it illustrates well all the possibilities of the technique and particularly those related to the use of the polarisation of the synchrotron radiation.

The epitaxial growth of Ni on W(110) is rather complex. LEED experiments yield to the following conclusions : i) for low coverage rates, $\theta < 0.5$ ML, the growth is quasi pseudomorphic, ii) for $\theta > 0.5$ ML, the situation becomes more complex. If the growth is still pseudomorphic in the (110) direction, it is no longer possible in the other directions. A distortion occurs and a superstructure (7x1) takes over, iii) the distortion continues to reach a (111) f.c.c structure of Ni on W(110).

a) Evidence for the existence of polarisation effects

On Fig. 19, we report the photoemission spectra obtained with unpolarized (HeI) and polarized light (Synchrotron radiation) on several Ni/W interfaces. The photoelectrons have been detected at the normal and the polarisation (ϵ) was in the W(110) direction. We notice first that for $\theta > 3$-5 ML the photemission spectra resemble that of pure nickel. The effects of polarisation are clearly seen and are particularly important for the low coverage rates $\theta < 1$ ML. The fact that these polarisation effects become small for high values of θ can be explained by the appearance of an f.c.c (111) structure (see [29]). The important changes observed in the $\Theta \approx 0.5$ ML range are a direct consequence of the (7x1) superstructure mentioned above.

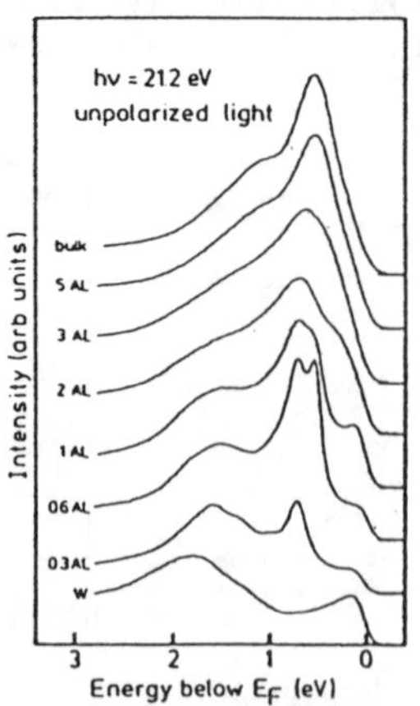

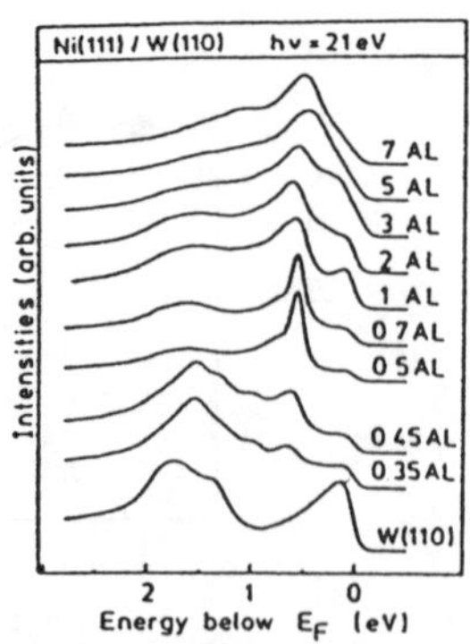

Fig. 19 UPS spectra of Ni/W obtained with unpolarized (left curve) and polarized light (right curve).

b) Surface band structures

For θ = 0.7 ML, the photoemission spectra (for normal detection) are quite independent of the photon energies. This is a clear indication of the 2D character of the nickel layer. It is thus possible to study directly the surface band structures by measuring the dispersion curves as a function of $\mathbf{k}_{//}$. This is shown in Fig. 20. By changing continuously the polar angle **h**, we describe particular directions of the Brillouin zone (here the ΓM direction of the 2D Brillouin zone of an f.c.c lattice). It is important to notice the weak dispersion of the d states near 0.5 eV, this reflects the weak hybridization between the nickel atoms (in other words the bandwidth is small). Near the M point in the Brillouin zone, we notice the existence of a more symmetric band whose dispersion is more important. Its origin is as yet unclear.

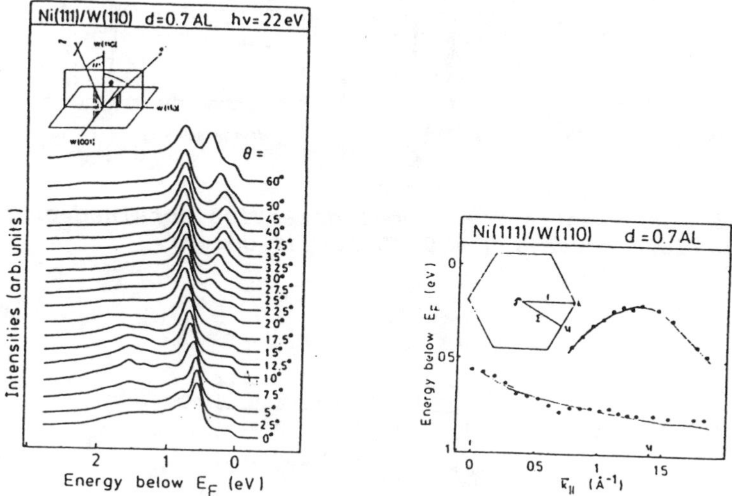

Fig. 20 UPS spectra as a function of θ (left curve), dispersion curves E = f(k) (right curve).

c) Bulk band structures

In Fig. 21, the photoemission spectra (always recorded at the normal) for θ = 2ML are reported. The situation is completely different from that discussed above and a clear variation with the photon energy is observed. This implies that the processes including $\mathbf{k}_{\perp}$ become important and thus, that 3D behaviour takes place. For this special value of θ (2ML), there is an admixture between 2D and 3D behaviour and the interpretation is complicated.

The situation becomes simpler for higher coverage rates, as shown in Fig. 21. As shown in Fig. 22, it is indeed well established that for θ = 5 ML we get the band structure of bulk Ni(111).

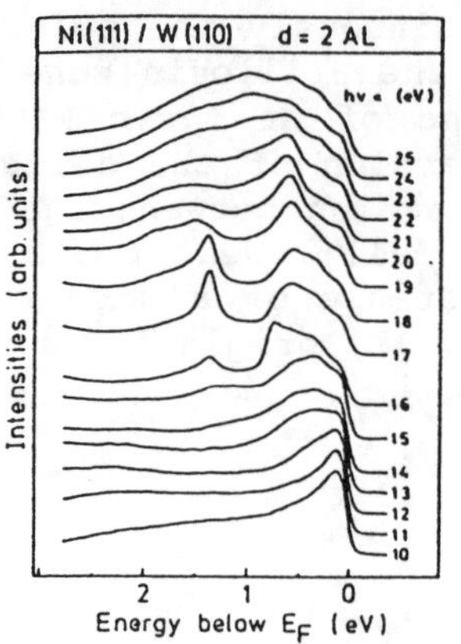

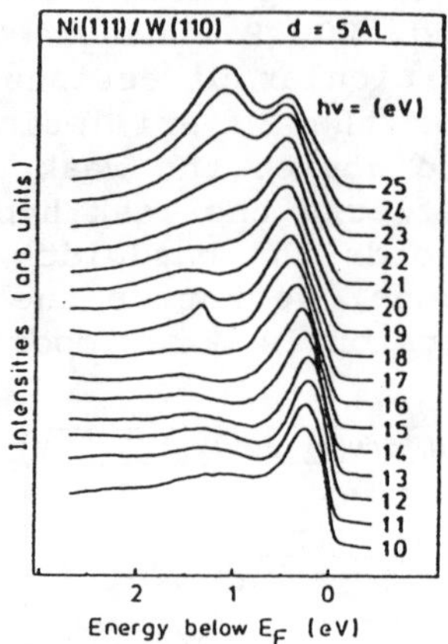

Fig. 21 UPS spectra for normal detection obtained on Ni/W. θ = 2 ML (left curve), θ = 5 ML (right curve).

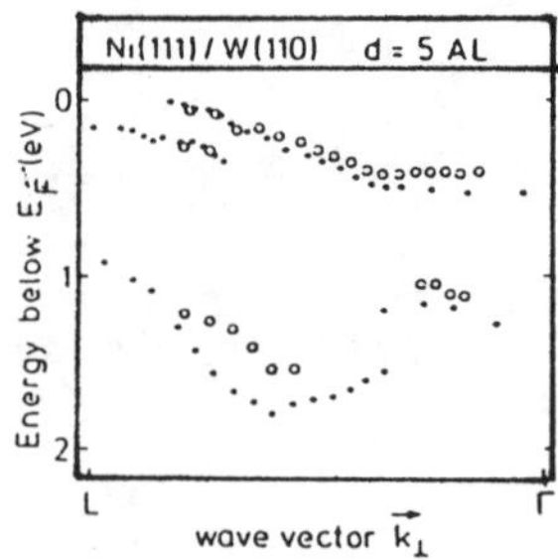

Fig. 22 Dispersion curves E=f(k) obtained from experiments shown in Fig. 21 (o : 5ML Ni/W ; • : Ni (111)).

3.1.4 Spin resolved photoemission experiments

In principle, <u>all</u> the photoemission experiments can be done by resolving the spin of the photoemitted electrons, i.e. by measuring for all energies (E) and all angles (θ,ϕ) the polarisation of the electrons defined by :

$$P(E,\theta,\phi) = \frac{n_\uparrow - n_\downarrow}{n_\uparrow + n_\downarrow}$$

Thus, the spin resolved band structure can be obtained. Up to now, only a few results have been obtained in the case of metal-metal interface by such a technique. We can mention, for instance, the work done on Gd/W [30]. This lack of results can be easily understood by the experimental difficulties encountered in such experiments, particularly those related to the weakness of the photoemitted current after spin resolution. In the future, with the developments of new synchrotron sources which will be able to deliver circularly polarized light, there is no doubt for us that such experiments will be done more systematically. For the moment, the experiments are carried out in the total yield mode (i.e. without any resolution in energy and/or angles) either in Auger [31] or in photoemission spectroscopy [32,33].

On Fig. 23 and 24, we show the results obtained on (f.c.c.) Fe/Cu(001) [32] and (b.c.c.) Fe/Ag(001) [33]. We notice that the measurement of the polarisation P is equivalent to the determination of the surface magnetization of the samples.

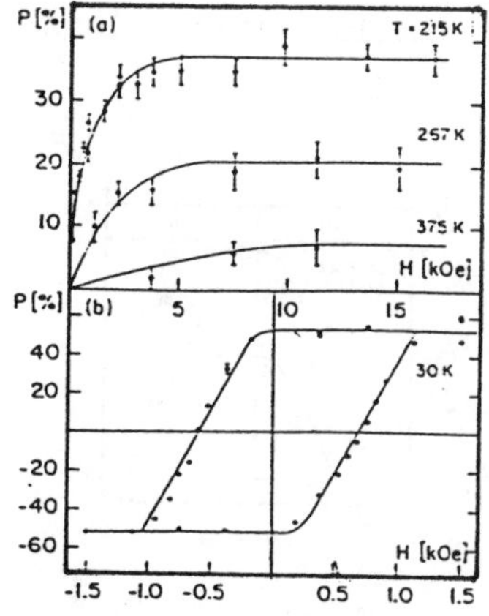

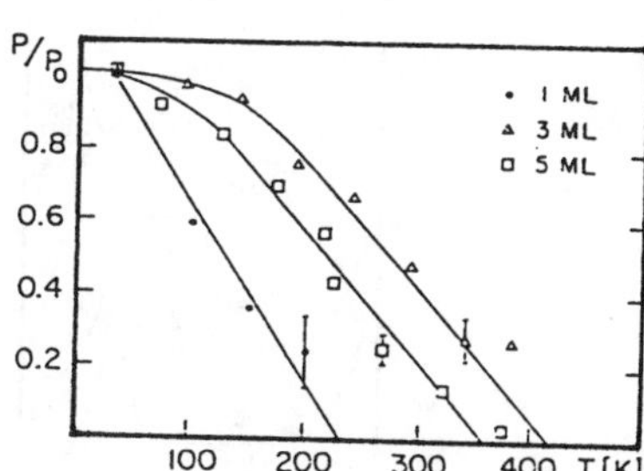

Fig. 23 Behaviour of the polarisation P with applied magnetic field (left) and temperature (right) for fcc Fe/ Cu (100).

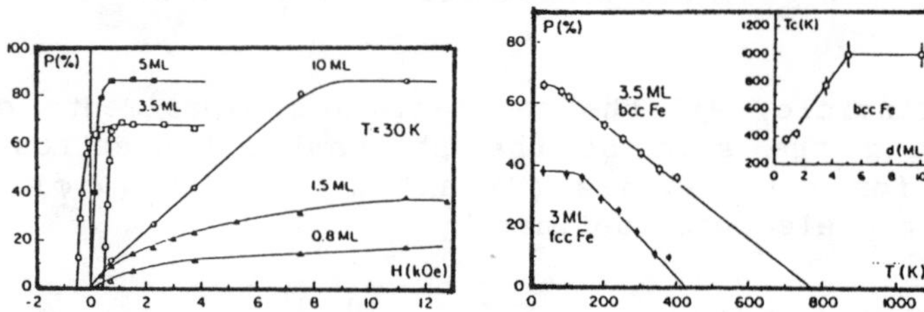

Fig. 24 Behaviour of the polarisation P with applied magnetic field (left) and temperature (right)for (b.c.c.) Fe/Ag(100). The curve reported 3 ML (f.c.c.) Fe is referred to Fe/Cu.

3.1.5 Quantum wells in metallic superlattices

The technological applications developed from the existence of quantum wells in semiconducting superlattices, like GaAlAs for instance [34], are well known. Because the electrons are obviously more delocalized in metallic systems, the existence of such quantum wells in metallic superlattices are less probable. However, recent experiments performed on Ag/Au (111) superlattices [35] have clearly demonstrated the existence of such quantum wells in this system.

On Fig.25, we report the photoemission spectra, recorded at the normal, of Ag(111), Au(111) and Ag(111) + 20 ML Au, Au(111) + 20 ML Ag. On the same curve, the dispersion curves in the ΓL direction, of Au(111) and Ag(111), are shown. Except for the existence of a surface state in the case of the Au/Ag interface, the photoemission spectra of Au/Ag are quite identical to those of bulk Au(111). Clearly we cannot draw similar conclusions for the Ag/Au(111) interface.

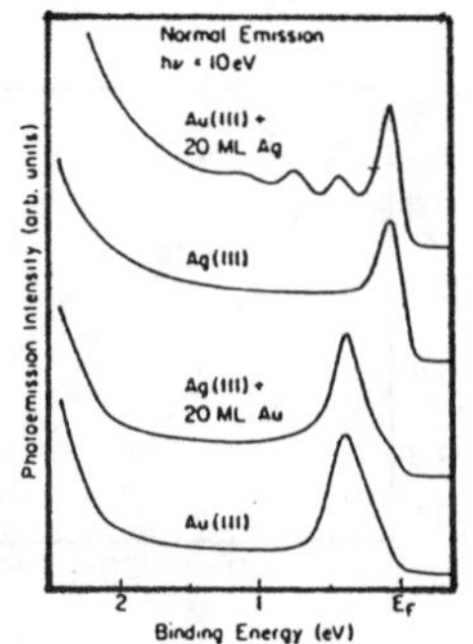

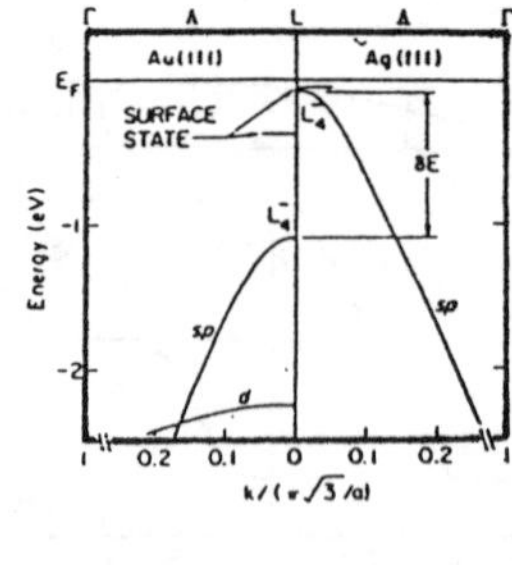

Fig.25 Photoemission spectra for the Au(111), Ag(111) + 20 ML Au, Ag(111) and Au(111) + 20 ML Ag sytems (left) Dispersion curve E=f(k) for Au(111) and Ag(111) (right).

The demonstration that the extra structures, observed for the Ag/Au (111) superlattices, are the result of the existence of quantum wells is rather simple. Indeed, the existence of a stationary state of the electrons implies an obvious relation between the wave vector of the electrons ($|\mathbf{k}|$) and the thickness of the silver film (d) :

$$2kd + \delta = 2n\pi$$

By changing continuously d, the energies where these extra structures are located must follow the dispersion curves E(k). This is well demonstrated in Fig. 26. We notice that stationnary states can only be generated in the energy interval δE. The explanation of this result requires that we take into account the k conservation at the Ag/Au [35] interface and thus explains why such stationary states can never be observed in the case of the Au/Ag interface.

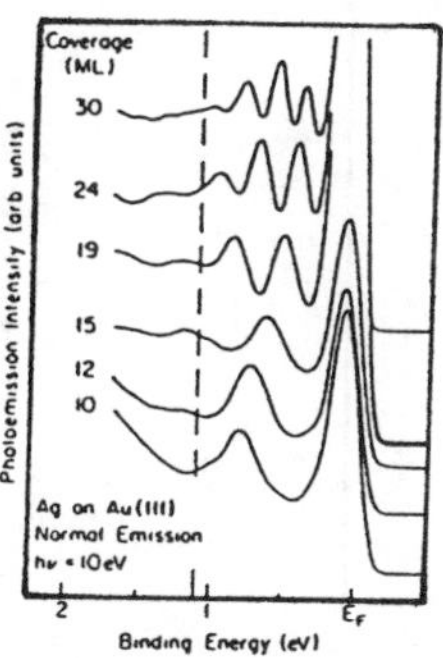

Fig. 26 Behaviour of the normal photoemission spectra at hν=10 eV in Ag/Au (111) with the Ag coverage.

3.1.6 Use of inverse photoemission spectroscopy for the study of epitaxial systems

As discussed in chapter 2, it is possible from IPS to obtain exactly the same informations for unoccupied states as those we obtain for occupied states with XPS. Here we illustrate this point by considering two examples :

a) Existence of interface states in the Pt/Au(001) system [36]

The behaviour of the IPS spectra in the Pt/Au(110) system are reported in Fig. 27. These spectra have been obtained in the case of normal incidence of the electrons. We clearly notice the appearance of two structures, labelled A and B, in the IPS spectra and the evolution of their relative intensities with the Pt coverage is also given in Fig.27. The A structure clearly presents an anomalous behaviour which cannot be ascribed to a Pt state (this conclusion comes from the study of Au/Pt (111)). The authors demonstrate that this A structure is characteristic of the existence of an **unoccupied** interface state in the Pt/Au(100) system. Moreover, from angular studies (they change the incidence angle of the electrons) they can precisely locate this interface state in the Brillouin zone (see ref [36]).

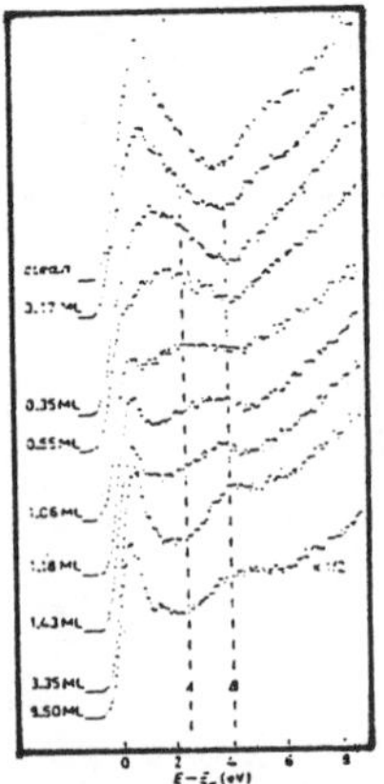

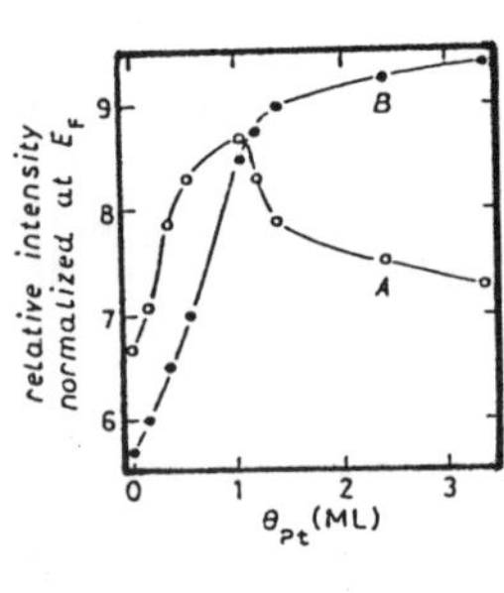

Fig. 27 IPS spectra of Pt/Au (111) as a function of the Pt coverage (left); behaviour of the A and B intensities (right).

b) Study of the unoccupied states of Mn/Ag (111) [37]

The interest of the Mn/Ag(111) system is similar to that discussed previously for the Cr/Ag(100) interface. IPS experiments can yield **directly** the density of unoccupied states with minority spin in the systems where the exchange Coulomb energy is high (≈ 5 eV). This is well demonstrated in Fig. 28. The comparison with the experiment performed on a well-known spin-glass system like Ag(111)-15%Mn, confirms that the unoccupied states are related to minority spins. Such a conclusion is in good agreement with recent theoretical calculations [38]. The splitting of the spin ↓ in two sub-bands is due to crystal field effects which split the 3d states in $e_g(\Gamma_{12})$ and $t_{2g}(\Gamma 25)$ components.

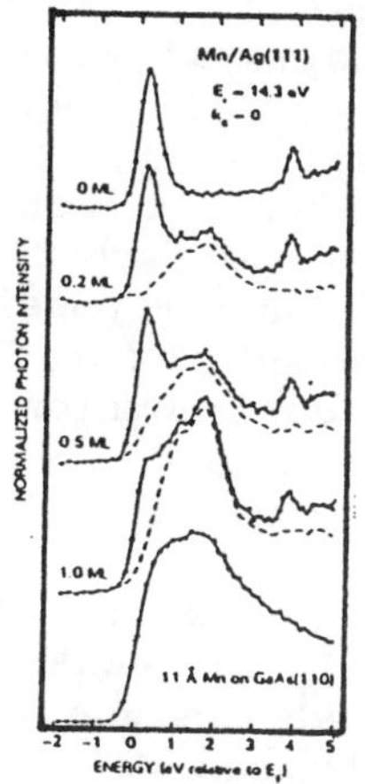

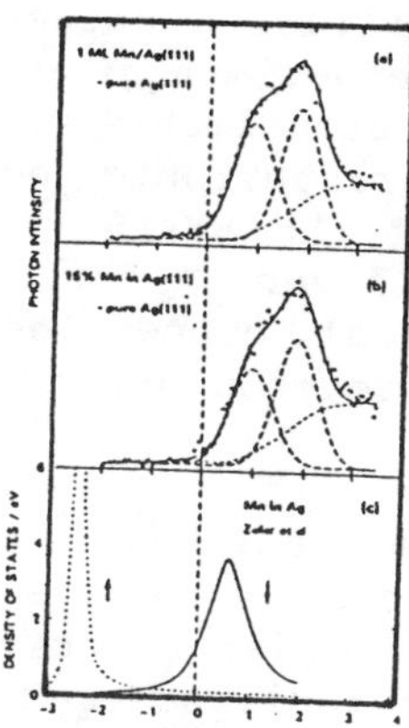

Fig. 28 IPS spectra of Mn/Ag (111) (left); comparison between experiment and theory (right).

3.2 Study of multilayers by X-ray absorption spectroscopy

In the above section, we essentially discuss the properties of a single interface. The X-ray absorption technique can be used to study more complex systems simply because we are no longer limited by the photoelectron escape depth. This last part of chapter 3 will be essentially devoted to the discussion of EXAFS experiments performed on multilayers and thus, we will focus on the local environment properties of such systems. However, in several cases we will also point out the interesting information that we can get on the electronic properties of such systems from XAS experiments.

3.2.1 XAS experiments in the transmission mode

In such experiments, the measurements are carried out directly on a sample made of n doublets like $n.(A_xB_y)$. Here x and y are the thicknesses of the A and B elements and $\Lambda = x+y$ is the modulation length of the multilayer system. The obvious advantage is that XAS experiments can be done independently on the A and B elements; the obvious disadvantage being that we get an average signature of the local environment of A and B in the system. Thus it is easy to understand that the informations one can get will depend strongly on the stability of the interfaces.

In the case of stable interfaces (i.e. for those systems where the diffusion between A and B is small at the temperature where the experiments are performed) it is possible to carry "classical" experiments where the absorption will be measured both on the characteristic edges of elements A and B. From the behaviour of the EXAFS structures with the structural parameters of the multilayer (Λ for instance) one can see possible changes in the structural properties and, in some favourable cases, get direct information on the interface contribution to the EXAFS signals.

In the case of unstable interfaces (i.e. when thermal diffusion becomes important), we have to follow the kinetics of diffusion in order to try to understand its mechanism. Such experiments are really complicated and need the use of a dispersive EXAFS station where the absorption data can be collected in very short times (less than 1 second). We will not discuss such experiments here; we refer interested readers to references [39, 40, 41] which are devoted to this subject. However a special case will be discussed as an example.

EXAMPLE : EXAFS study of a stable system NbZr [42]

Nb/Zr multilayers have been studied by EXAFS at the K absorption edge of both niobium and zirconium and for L values between 4 and 200 Å. The atomic ratio between Nb/Zr is kept constant and equal to 1. The EXAFS parameters, deduced from a conventional analysis using formula I-21, are summarized on Fig.29. The number of Nb neighbours of a Zr atom increases continuously from 0 to 4 as Λ decreases from 200 to 4 Å. The total number of neighbours (i.e. the coordination of Zr) is equal to 12 for the high values of Λ but goes to 8 for the small values. We also notice a rapid decrease of the Zr-Zr distances for $\Lambda < 50$Å and a consequent increase of the Debye-Waller factor in the same range of Λ. All these EXAFS results show clearly the occurrence of a structural phase transition of Zr which undergoes a h.c.p -> b.c.c transition as $\Lambda \to 0$.

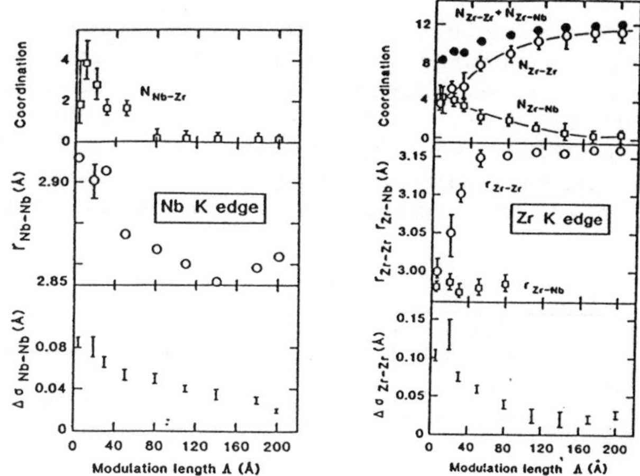

Fig. 29 EXAFS parameters of the Zr/Nb multilayers as a function of the modulation length Λ.

Moreover the EXAFS results reveal the existence of diffuse interfaces. This is demonstrated on Fig.19, where it is shown how it is possible to reproduce the behaviour of the coordination number with Λ by supposing the existence of diffuse interfaces whose width are in the 15 Å range.

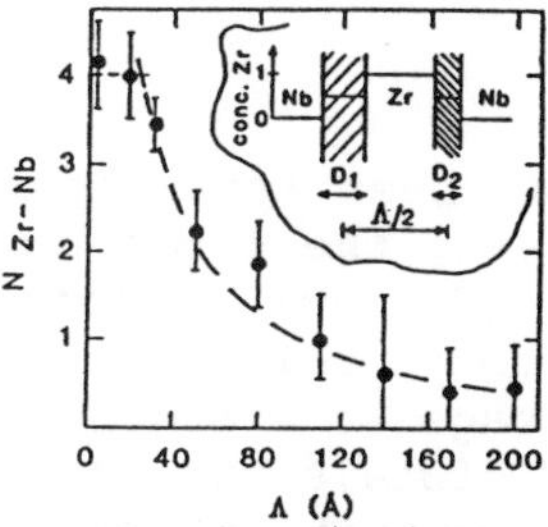

Fig. 30 Adjustement of the number of Nb-Zr neighbours (diffuse interface model) [42].

3.2.2 XAS experiments in the reflexion mode

The main disadvantage of the above method comes from the fact that we get simultaneously the contributions of the interfaces and that of the "bulk" of the multilayers. Experiments performed in the total reflexion mode (the photons impinging on the multilayer with an angle of incidence which is smaller that the critical angle) increase considerably the contribution of the interface.

This extra sensitivity of the XAS experiments, performed in the total reflexion regime, is well demonstrated by the evolution of the EXAFS structures observed at the Cu K edge in Cu/Al multilayers [43] (Fig. 31a). As the angle of incidence increases we notice that the EXAFS spectra resemble more and more to that of pure copper. The analysis of the spectra obtained for θ = 4 mrad (Fig. 31b) shows the existence of a $CuAl_2$ compound at the interface.

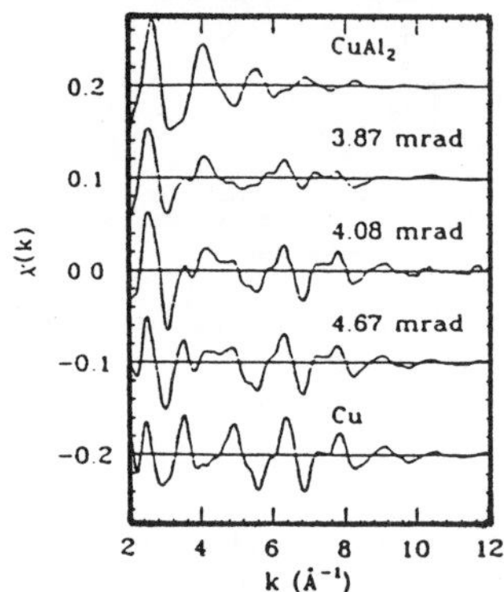

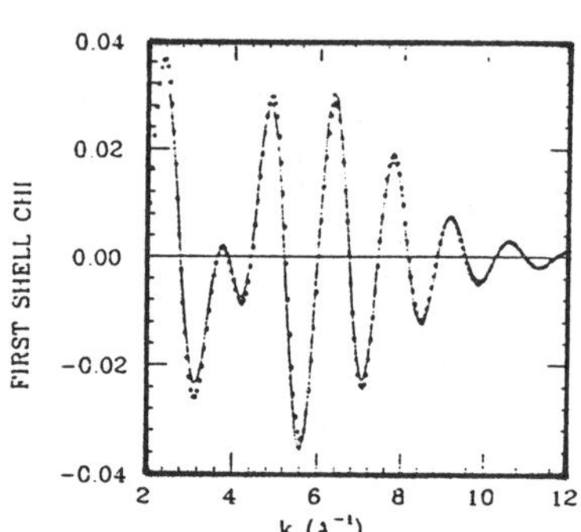

Fig. 31 a) Evolution of the EXAFS spectra of Al/Cu as a functiun of the angle of incidence (Cu K edge); b) Adjustement of the EXAFS spectra for the parameters in $CuAl_2$.

A similar study has been recently performed on an epitaxial system: Al/Ni/Fe(001) [44]; the unique role ofaAluminium (e = 75Å) being to prevent the sample from oxydation. The EXAFS experiments have been made, in the total reflexion regime, on the K edge of Ni and for two nickel coverages : 6 and 37 ML. The Fourier transforms (F.T.) of the two EXAFS signals are shown on Fig. 32. For the high nickel coverage (37 ML) we notice that the F.T resembles that of pure Ni, the diminution of the contribution at large distances (4 - 5Å) reveals however the existence of structural disorder which becomes important for the 3rd and 4th shells. In the case of the low nickel coverage (6 ML), the situation is clearly different from that of pure Ni. The first neighbour distance is significantly shorter and the quantitative analysis shows that this is due to the formation of an Al_3Ni_2 compound at the interface between Al and Ni. Thus an interdiffusion between Al and Ni takes place at the interface.

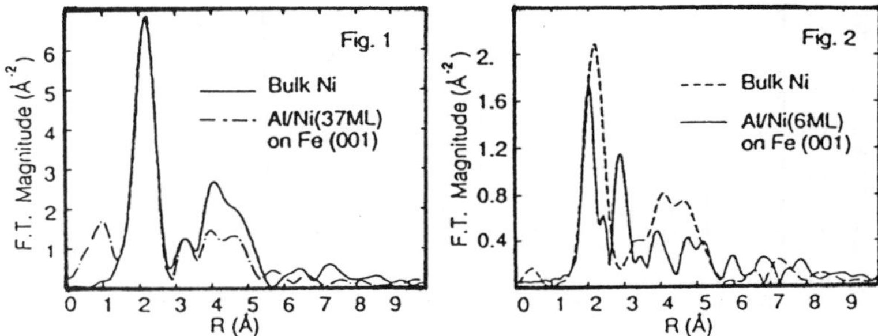

Fig. 32 Fourier transforms of EXAFS data obtained at the K edge of nickel in Al/Ni/Fe(100) systems.

Such XAS experiments in the reflexion mode are obviously interesting but we should take note of the fact that this technique is really efficient for the study of single interface systems. Thus they should be compared with the photoemission experiments described in section 3.1. Here we want to mention that one can study a multilayer system by XAS, without encountering the diffi-culties mentioned in section 3.2.1. We refer to the standing wave technique which has been recently applied to the case of Ti/Ni multilayers [43-45]. The principle of the method is really simple. We know that if we are under Bragg conditions, a standing wave of the electrical field takes place in the material and the modulation of the electrical field depends strongly on the Bragg angle (see Fig. 33). The interesting point is that the regions where the electrical field is maximum (i.e. the potential vector **A**) will contribute more efficiently to the EXAFS signals than those where the electrical field is a minimum. It is thus possible, in principle, to change continuously the "contrast" between the two components of the multilayer. This effect is clearly shown in Fig. 33. We must underline the fact that several technical problems have to be solved before this technique becomes widespread. Indeed, it is really difficult to always keep the Bragg conditions when the photon energy is changed. However there is no doubt that such standing wave experiments will be generalised in the future and will be very useful for the study of multilayers and superlattices.

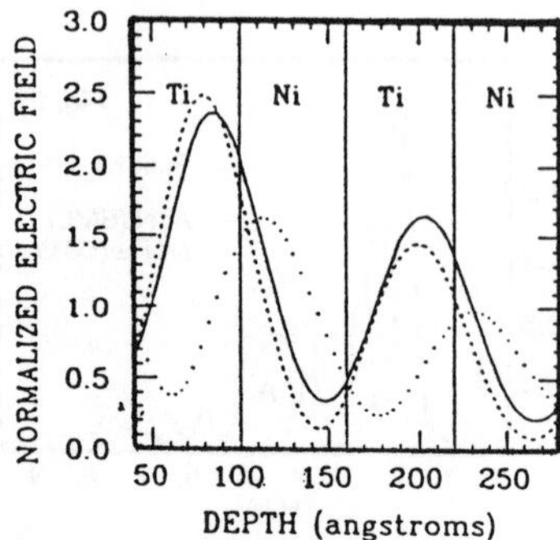

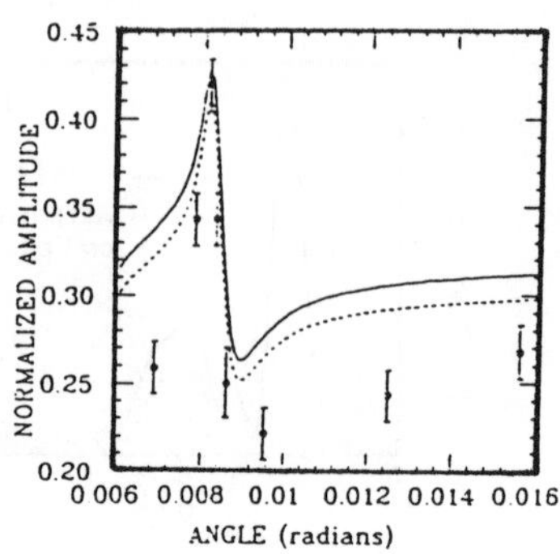

Fig. 33 Modulation of the electric field in a Ti-Ni multilayered system due to stationary wave (left); Variation of the EXAFS signal of nickel obtained when varying the angle of incidence.

3.2.3 Surface EXAFS (SEXAFS) experiments

When the detection of EXAFS is made by using the photoemitted electrons, this technique becomes a very efficient tool for the study of ultrathin metal layers. We want to illustrate this point by discussing several recent results obtained at LURE on the system Co/Cu(111) [47, 48, 49].

At room temperature the cobalt atoms do not diffuse in copper and the growth of cobalt on copper is made in the layer-by-layer mode. The LEED experiments show that the structure remains 1 x 1 during the growth of Co on Cu(111) and is of the type shown in Fig. 34.

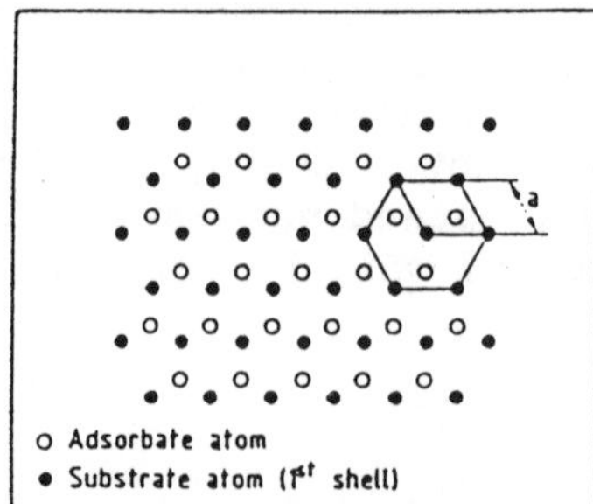

Fig. 34 Schematic structure of the (fcc) Cu (111) surface (•) covered by a cobalt monolayer (°).

When SEXAFS experiments are performed on single crystals, one can take advantage of the linear polarisation of the synchrotron radiation in the orbit plane. Indeed two different situations can be encountered :
i) either the X-ray beam is along the normal of the surface, the electrical field is then parallel to the surface and, in such a case, we will get essentially information on the structure of the adsorbate itself,
ii) or the X-ray beam is perpendicular to the normal of the surface, and then the electrical field is perpendicular to the surface and information on substrate-adsorbate interactions will be obtained.

In the general EXAFS formula, this can be taken into account by replacing the number of neighbouring atoms Nj, by an effective number of neighbours defined by

$$N_j' = 3\sum_{i=1}^{N_i} \cos^2 \alpha_i$$

where α_i is the angle between the electrical field (i.e. the polarisation vector) and the $\mathbf{r}_i$ vector which joins the central atom to the ith scattering atom of the jth shell.

On Fig. 35 we show the fit of the EXAFS curves, obtained by a Fourier filtering of the total EXAFS signal at the K edge of cobalt, and by using the structural parameters given below.

	Spectre a	Spectre b	Spectre c	Spectre d
N_{Co}	0	6	0	3
$R_{Co\text{-}Co}$	---	2.51Å	---	2.51Å
N_{Cu}	3	3	3	3
$R_{Co\text{-}Cu}$	2.47Å	2.47Å	2.51Å	2.51Å
$\Delta\sigma/\sigma$	0	0	10%	0

In the case of one ML of Co on Cu, and for grazing incidence (i.e. the polarisation is perpendicular to the surface), we do not see the contribution from the Co neighbours and the EXAFS spectrum can be reproduced by 3 Cu atoms located at 2.47 Å, in agreement with the structrure shown on Fig. 34. For normal incidence (fig.35b), the contribution of the adsorbate is clearly seen, the distances Co-Co are indentical to those observed in pure cobalt (2.51Å). We can see that this distance is smaller than the Cu-Cu distance of the substrate (2.55Å). In both situations, as expected, the Co-Cu distances are the same (2.47Å).

In the case of 1/3 ML of Co on Cu, we get interesting results (Fig. 35c,d). For instance the reduction in the number of first neighbours of a Cobalt atom (6->3) is well demonstrated. This reduction indicates the presence of small clusters of Co (≈ 7) at the surface of Cu. An interesting result which we will not describe further is the increase in the Debye-Waller factor observed for grazing incidence.

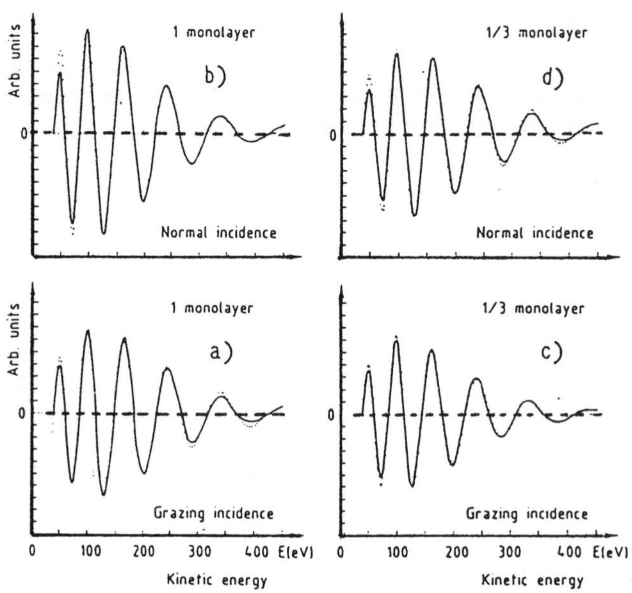

Fig. 35 Adjustement of the SEXAFS spectra of the cobalt first neighbours for 1 ML Co/Cu(111) and 1/3 ML Co/Cu(111).

3.2.4 Magnetic X-ray absorption studies.

In section 2, we briefly discussed the possibility of obtaining some informations about the magnetic properties of a given material by performing absorption experiments. Such a possibility implies the use of circularly polarized light and, in ferromagnetic substances, enables the determination of the magnetic moment carried by the atom which has absorbed the photon. Experimentally, what is measured is the difference between the absorption cross section for two directions of the magnetization (or for left and right circular polarisation). This difference contains the information about the magnetic moment. This method has been tested in the very last years (see Chapter 2 [14]). As a conclusion of this chapter devoted to the use of high energy spectroscopy in the study of multilayers, we want to show very recent results we obtained by magnetic X-ray absorption on Nd/Fe multilayers [50].

At ambient temperature the Nd/Fe multilayers are stable, the diffusion between Nd and Fe occurs only at 500 K. The magnetic properties of such multilayers are interesting and have been published elsewhere [51]. For our purpose it is just necessary to know that, at T = 300K, the iron is in a ferromagnetic state as soon as the Fe thickness is more than 15 Å, whereas the Nd atoms are in a **paramagnetic state.** In Fig. 36, we report the magnetic X-ray absorption results obtained on the L_{II} edge of Nd for several Fe thicknesses. The important point here is to notice **the existence of a magnetic X-ray absorption signal on the neodymium atoms which are expected to be in a paramagnetic state.** This magnetic signal is proportional to the thickness of the iron layer as shown in Fig. 36. Clearly, this is a signature of the properties of the interfaces themselves, either by direct polari-sation of the Nd atoms which are close to the interfaces (and thus to the Fe atoms) or by the existence of diffusion at the interface yielding to highly Fe concentrated Fe-Nd compounds (alloys). Note that, after diffusion at 500 K for 6 hours, the magnetic signal is strongly reduced (see Fig. 36). This point can be explained by the fact that the local environment of Nd atoms is different (there are less Fe atoms in the first neighbours). Recent experiments have been made, in which the Fe thickness was kept constant (35 Å) whereas the Nd thickness was varied from 15 to 60 Å. These experiments seem to confirm that the Nd magnetic ions, we observe in the magnetic X-ray signal, are those which are very close to the interface (≈ 15 Å). Although such conclusions have to been considered as preliminary, we think that magnetic X-ray absorption could be interesting in the near future for the study of such magnetic multilayers.

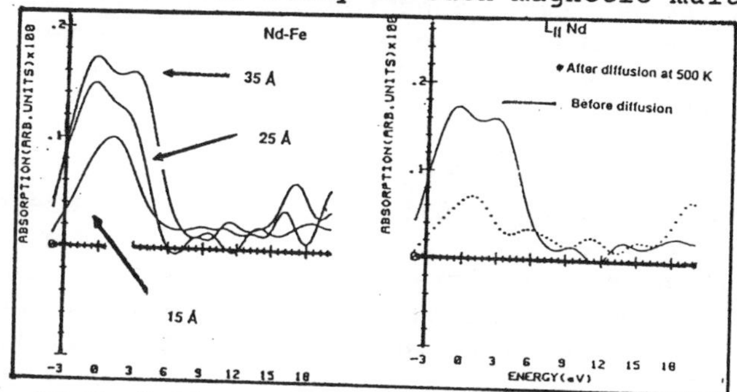

Fig. 36 Evolution of the magnetic absorption at the L_{II} edge of Nd in Nd-Fe systems as a function of thickness (left); and after diffusion at 500 K for a Nd (35 Å)- Fe (35 Å) multilayer (right).

4 THE AUGER PROCESS AND AUGER ELECTRON SPECTROSCOPY

The Auger process is the result of a non-radiative decay of an atom excited through the interaction of electrons, photons or ions with matter. The subsequent ionization of a core subshell of an atom is followed by two competing processes : X-Ray emission (fluorescence) and Auger electron emission [52].

The intensity of Auger emission depends upon the relative probability of non-radiative to radiative decay. Fig. 37 shows the predominance of Auger transitions with respect to fluorescence for transitions to K shell vacancies of low Z elements : Auger emission is then the most important mechanism for energy relaxation.

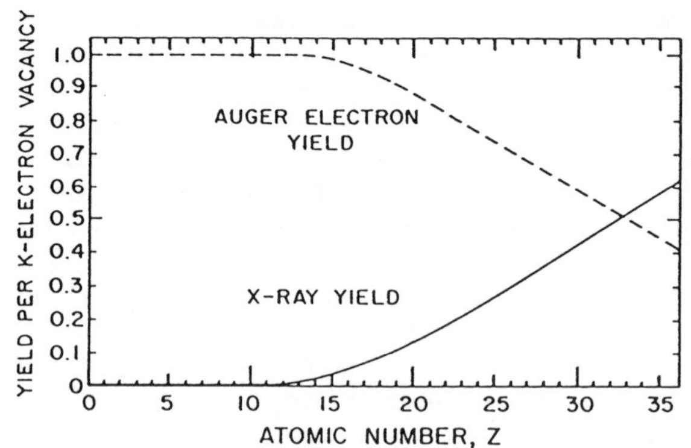

Fig. 37 Auger electron and X-Ray fluorescence yield as a function of atomic number.

In fact, in the way Auger Electron Spectroscopy is carried out experimentally, the energy of the primary beam is always less than 10 keV and the Auger process is the dominant de-excitation mode whatever the initial hole.

The Auger process is the basis of AES and X-AES. The aim of Auger Electron Spectroscopy (AES) is to study the energy distribution of electrons emitted from a surface under a primary electron beam. Auger electron transitions can also be investigated in the course of X-Ray Photoelectron Spectroscopy (XPS) experiments, *i.e* under photon irradiation. In this case, the acronym X-AES is used.

As for the other electron spectroscopies, Auger analysis is carried out under ultra-high vacuum conditions to avoid rapid contamination of the surface under examination.

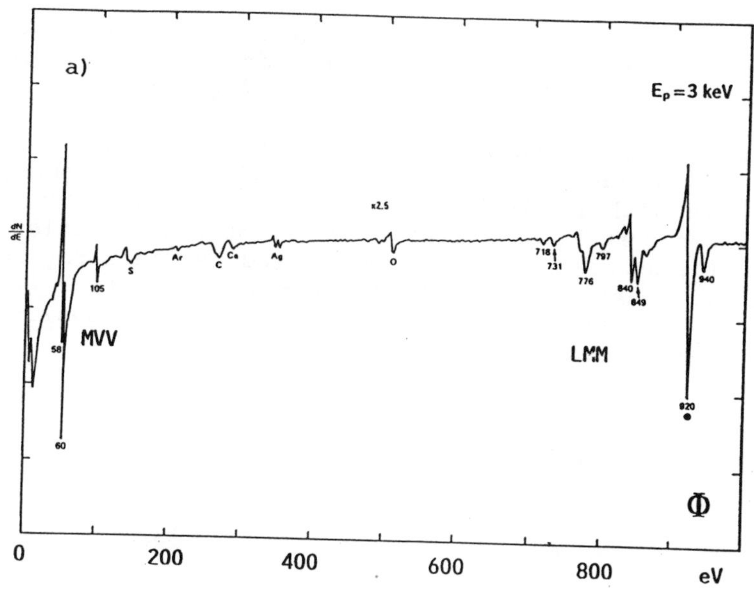

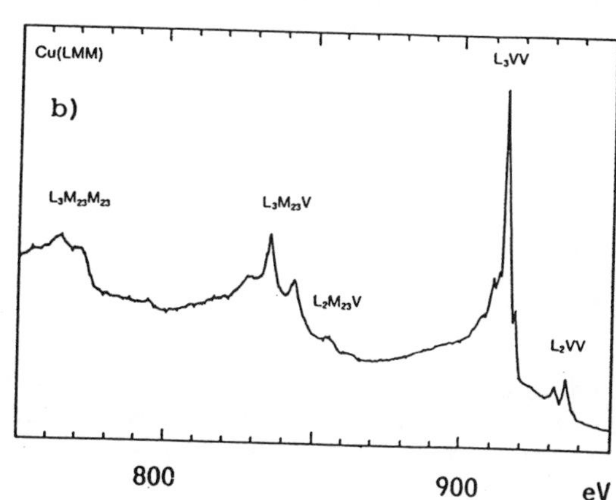

Fig. 38 Auger spectra of copper
a) in the derivative mode dN(E)/dE
b) in the direct mode N(E).

A description of the energy analyzers used in AES and of the experimental procedures which are beyond the scope of this course, can be found in several books [52, 54, 55, 56].

As an example, Auger Electron spectra of copper obtained respectively by AES [in the derivative mode, dN(E)/dE], and by X-AES [in the direct mode N(E)] are given in Fig. 38 [96].

AES is a final state spectroscopy. In the Auger process, an electron is ejected from an atom by the radiationless decay of an initial state with a single core hole into a less bound final state with two core holes. Even if it is not always satisfactory, Auger transitions are still usually denoted by the X-ray notation for the levels involved, *i.e.* KL_1L_1 stands for a [1s] ---> [$2s^2$] transition. A schematic diagram for various Auger transitions and the usual nomenclature are shown in Fig. 39.

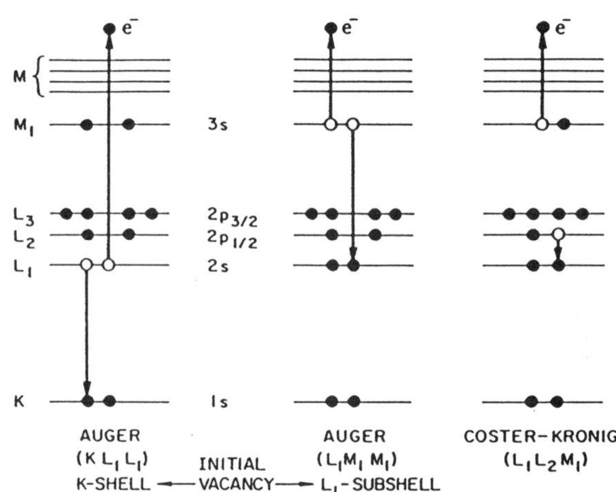

Fig. 39 Schematic diagram of various two-electron deexcitation processes : KL_1L_1 and LM_1M_1 Auger transitions; Coster-Kronig $L_1L_2M_1$ transition.

From energy conservation considerations, the kinetic energy for a KLL Auger electron is found to be in a one electron picture

$$E^A_{KLL} = E_K - E_L - E_L - \varphi_C \qquad (1)$$

where E_i corresponds to the positive level binding energy with respect to the Fermi level and φ_C the collector work function.

This basic equation gives rise to three comments :

(i) E^{A}_{KLL} does not depend upon the primary energy nor upon the mode of excitation,

(ii) E^{A}_{KLL} is characteristic of the excited atom and of its chemical bonding state. We will show in Section 2 how the changes of energy and shape in Auger spectra can be used as a fingerprint of the chemical state.

(iii) Considering the transitions allowed in AES, the escape depth of Auger electrons, monitored by the inelastic mean free path (IMFP), always remains small. From Fig. 40, the IMFP for Cu MVV and Cu LMM Auger electrons are respectively 6 and 12 Å.

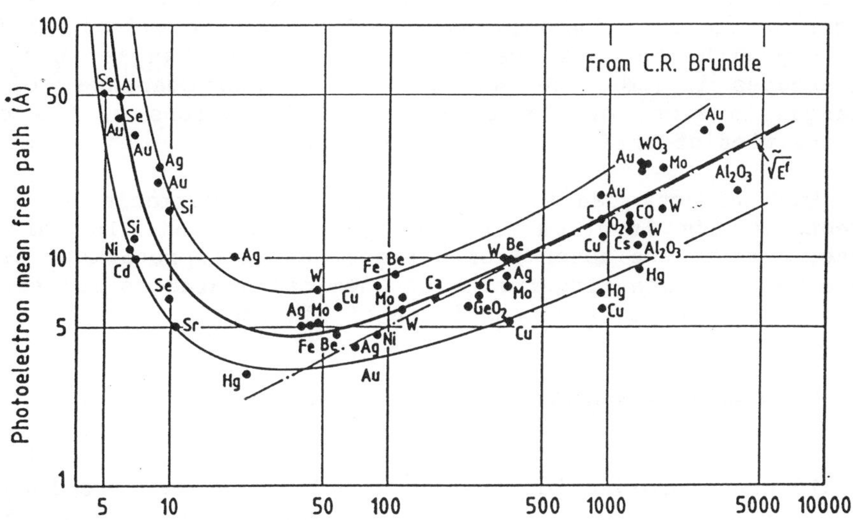

Fig. 40 Electron mean free path as a function of their kinetic energy.

This last comment explains the specificity of AES as a surface analytical tool. By choosing the appropriate transition and detection method, the process can be extremely surface sensitive.

Depth resolution and chemical informations obtained through AES measurements are such that it is particularly well suited to the characterization of metallic thin films. It will also be shown that it is useful for investigating their growth mode.

5 CHEMICAL INFORMATION FROM AES

5.1 General Considerations

From equation (1), it is to be expected that chemical changes generally are accompanied by a shift in the Auger transitions.

To interpret the changes in the Auger spectra strictly, it is necessary to use a more precise equation for the Auger transition and to know each term involved in it. Generally, these conditions are not fulfilled. In practice, the experimental approach consists of considering a reference spectrum as well defined as possible and to study, in the same experimental conditions, the evolution of this spectrum with the changes of chemical environment. More generally, the chemical information available from AES measurements is contained in the energy positions of the spectral features and also in the lineshapes. The energy shifts of the Auger peaks are commonly referred to as "chemical shifts" by analogy with the ones observed in XPS. The changes in the Auger lineshapes include the appearance (or disappearance) of peaks, changes in width of the lines and modifications of the relative intensities of various spectral features.

Such effects can be very strong and remarkable when valence electrons are involved in an Auger transition. In this latter case, even in the absence of a rigourous interpretation, AES leads to a very precise fingerprint of chemical states as illustrated in Fig. 41 for the $L_{2,3}VV$ transition of silicon.

5.2 Auger energies

In the approximation that core-hole lifetimes are long enough for considering their creation and decay as to be considered as independent processes [57,58,59], the kinetic energy of the emitted electrons is the difference in binding energy of the initial and final states, which, in the free atom, is :

$$KE^{atom}(abc:\ ^{2S+1}Lj) = E^{atom}(a) - E^{atom}(bc :\ ^{2S+1}Lj) \qquad (2)$$

where a, b, and c are the three levels involved in the Auger transition and S, L and J the quantum numbers which describe the multiplet structure in the two-hole final state.

The final state multiplet structure is determined by the Coulomb repulsion terms which can be expressed as Slater $F^k(n_bl_b, n_cl_c)$ and $G^k(n_bl_b, n_cl_c)$ integrals [60] and the spin-orbit coupling ξ_{nl} L.S .

For KLL spectra, the multiplet splitting depends on the various coupling schemes according to the Z values as discussed in several review articles [61,62,63,64] on the basis of atomic structure calculations. The L, M and N shell Auger spectra are much more difficult to interpret than the KLL series since the presence of various possible subshells in the initial state gives rise to many overlapping spectral series. Moreover, the involvment of d and f orbitals creates more final states than occur in the KLL transitions. The spectra are further complicated by the effects of fast Coster-Kronig transitions which change the hole distribution amongst the initial state subshells. Haynes [65] has given an element by element survey of work done on L and M shell Auger spectra and recently Aksela [66] has assessed work on the most intense groups of the LMM, MNN and NOO series.

Auger energies can be also interpreted with the aid of empirical expressions which describe the binding energy of the two-hole final state as a combination of one-hole binding energies.

Adopting a two-step description of the Auger process, Shirley *et al.* have developed an expression for the energy of Auger transitions [67] :

$$KE^{metal}(abc{:}^{2S+1}Lj)=E_B(a)-E_B(b)-E_B(c)-F(bc{:}^{2S+1}Lj)+R^a+R^{ea} \quad (3)$$

in which the terms F and R can be considered as artefacts of the process of expressing the Auger final states $[n_bl_b, n_cl_c]$ as a combination of separate $[n_bl_b]$ and $[n_cl_c]$ states. The static relaxation R_s is the difference between the relaxation energy of the two-hole states $[n_bl_b, n_cl_c]$ and the sum of the relaxation energies of the one-hole states $[n_bl_b]$ and $[n_cl_c]$. The F term corresponds to the reduction of the electron-electron interaction in the two-hole state taking into account the multiplet splitting arising from the Coulomb and spin-orbit coupling interactions. In the condensed phase it is necessary to add a term of extraatomic relaxation R_{ea}.

The Auger kinetic energy KE and the binding energies E_B of the single hole states can be determined from the same XPS spectrum. It allows an accurate value for the sum of the three remaining terms to be obtained.

Equation (3) has become the usual tool for evaluating the contribution of chemical environment to the Auger energy [67]. For example, in the case of metallic screening R^{ea} defined as the energy difference observed between the metal and the free atom. Table 3 in ref. [97] shows experimental data of the shifts corresponding to metallic screening in XPS and AES. One can see, as it was pointed out by Wagner and Biloen, that the shift in AES R^{ea} is three times the one measured in XPS. Since the photoelectron emission results in a singly ionized final state whereas for AES it is a doubly ionized one, then it follows that differences in polarization energy are a source of systematic differences between Auger and photoelectron chemical shifts.

It should be noted that Lang and Williams [68] defined a so-called Auger parameter ξ as the difference :

$$E^{metal}(a) - E^{metal}(b) - E^{metal}(c) - KE^{metal}(abc : {}^{2S+1}L_j) \quad (4)$$

They argue that this parameter may be associated with metallic screening.

For Wagner [69] the "Auger parameter" α is the sum of the Auger electron kinetic energy and the kinetic energy of a photoelectron, studied in the same XPS apparatus.

The main interest in this parameter is to provide, without reference problems (due for example to charge effects), the change in extra-atomic relaxation between various environments. Wagner has surveyed existing results and given tables of XPS and Auger lines on most elements [70] and suggested that the chemical information available from the comparison of XPS and Auger energies could be displayed on "chemical state" plots. In these, the XPS positions of an elemental core line in a series of compounds are plotted against the corresponding Auger line positions.

Auger parameters can be also used for the determination of screening energies [71,72] by noting the change in the Auger parameter between the isolated atomic state and the chemical state in question.

5.3 Auger spectra of valence bands

Auger transitions which involve valence states are expected to be the most sensitive to changes in chemical environment through modification both in energy and shape of the corresponding Auger peaks. On the other hand such Auger spectra show either valence band structures, containing information on the occupied states of the valence band, or quasi-atomic structures which cannot be related to the valence electron distribution. These two kinds of spectra, band like and atomic like, can be very useful for explaining changes in the chemical environment [73].

5.3.1 Band-like spectra

Before recalling how some Auger spectra involving valence electrons may nevertheless be atomic-like, let us first mention the factors which determine the profile of band-like CVV Auger peaks.

For such transitions, electron and hole delocalisation in the valence band leads to an Auger profile which, in a first approximation, is proportional to the self-convolution of the occupied valence states (SCDOS). The Coulomb interaction between the final-state holes is negligible due to a nearly complete decoupling of the holes. We could thus think of extracting the density of occupied states from the Auger spectrum by means of a deconvolution technique.

This expectation should hold provided the correlation energy between the two holes may be effectively neglected and if it is assumed that the transition probability remains constant over the whole valence states. But this conjecture is not supported by experiments. So, in the case of sp band materials, the valence Auger spectra have a width and kinetic energy qualitatively consistent with the energy difference between the initial state and the self-convolution of the density of states. In fact, the shapes are quite different from those of the SCDOS and the expectation that the transition probability is constant over the whole valence states has no foundation. Theoretical and experimental studies on sp band materials have shown that initial and final state effects should indeed be taken into account.

Auger transition matrix elements depend on the angular momentum of the valence electrons involved and transitions leading to *pp* final states are favoured relative to *sp* final state [73]. Besides, it appears that the Auger profile is mainly determined by the local charge distribution on the excited atom and only to a lesser extent by the bonding charge distribution. To add further difficulties, calculations and experimental studies on *sp* band metals [74] showed that relaxation of the initial core hole involves mainly an *s*-type screening, modifying in consequence the local charge population.

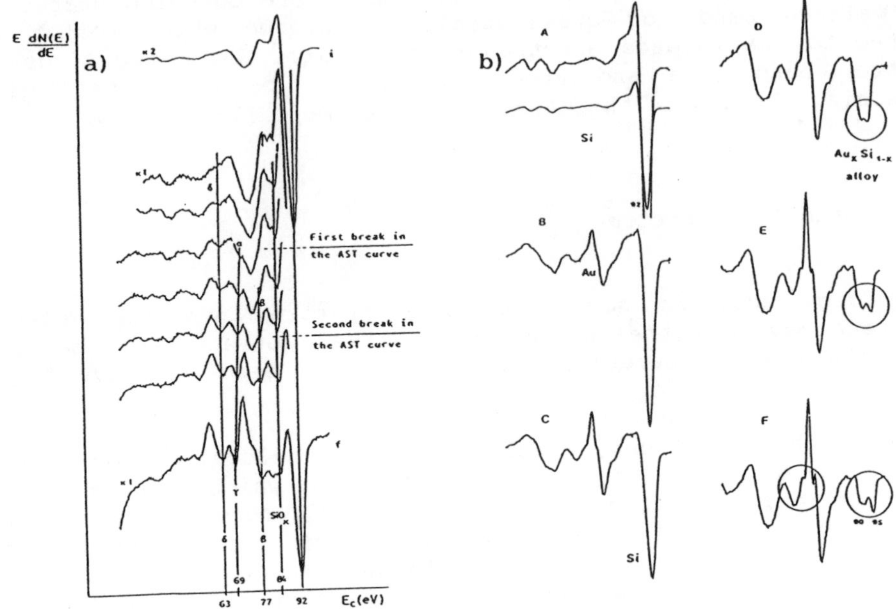

Fig. 41 Si $L_{2,3}VV$ spectra
a) at various stages of oxidation Si(111)
b) during gold deposition on Si(100).

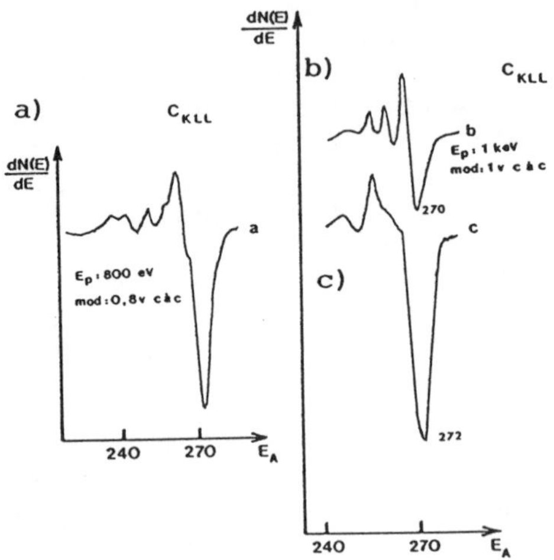

Fig. 42 Carbon KLL spectra
a) contamination overlayer on poly-Si
b) pure carbide
c) pure graphite.

Despite these difficulties of interpretation, band-like Auger spectra provide, in numerous cases, a very sensitive fingerprint of changes in the chemical environment at the site of the initial core hole. Such chemical information can be illustrated by considering a series of Si $L_{2,3}VV$ spectra (Fig. 41). These spectra which present energy shifts and lineshape changes can be used, in a fingerprint mode, to identify silicon bonded to oxygen or alloyed to gold [75,76].

An other useful example is given by the carbon KVV line shape (Fig. 42, [77]).

5.3.2 Atomic like spectra ; d bands materials

For some CVV spectra of metals such as Cu, Zn, Ag or Cd, Auger spectra present a fine structure which cannot be related to any valence band density of states, but which is very similar to the spectral features observed for the corresponding free atom. These spectra are therefore termed as "quasi-atomic" and their overall lineshape is determined by the various final-state multiplets in the appropriate coupling scheme (Fig. 43, [71]). These spectra are shifted to lower kinetic energies than the ones expected from the difference in binding energies of the initial core hole and the self-convolution of the occupied valence states. This indicates that the two hole final states, $[d^2]$, formed in the Auger process are more tightly bound than the sum of the binding energies of two single hole states [d]+[d]. The shift in energy of these "atomic like" spectra is denoted U_{eff} and takes into account the effective interaction between the two final state holes.

Note, as proposed by Antonides [78], that this correlation energy U_{eff} corresponds to the three terms F, R^a and R^{ea} in equation (3) (5.2) :

$$U_{eff} = E_B(a) - E_B(b) - E_B(c) - KE\ (abc{:}\,^{2S+1}L_j) \qquad (5)$$

and can be determined from XPS and Auger measurements.

With this definition, Cini [79] and Sawatzky [80] formulated a simple criterion based on the $U_{eff}/2\Gamma$ ratio, where Γ is the valence band width, to determine when CVV transitions present a quasi-atomic or a band-like behaviour.

A quasi-atomic spectra is expected when $U_{eff} \gg 2, \Gamma$ whereas for $U_{eff} \ll \Gamma$ the solid will behave like an ordinary metal.

Fig. 44 shows a plot of U_{eff} and Γ for the metal having electron configurations d^6 to d^{10} [95]. The studies of Auger spectra in the case of "transitional" metals underlines the success of the Cini and Sawatzky approach.

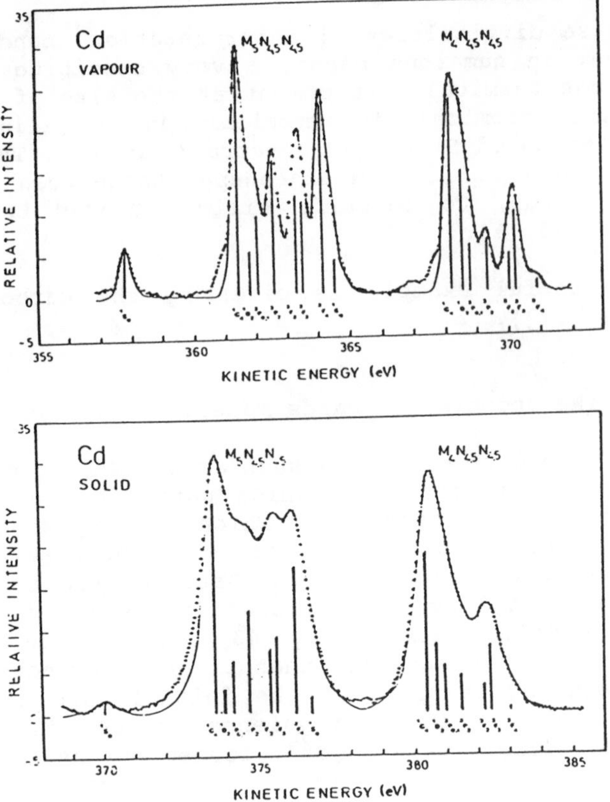

Fig. 43 Vapour and solid phase $M_{4,5}N_{4,5}N_{4,5}$ Auger spectra of cadmium.

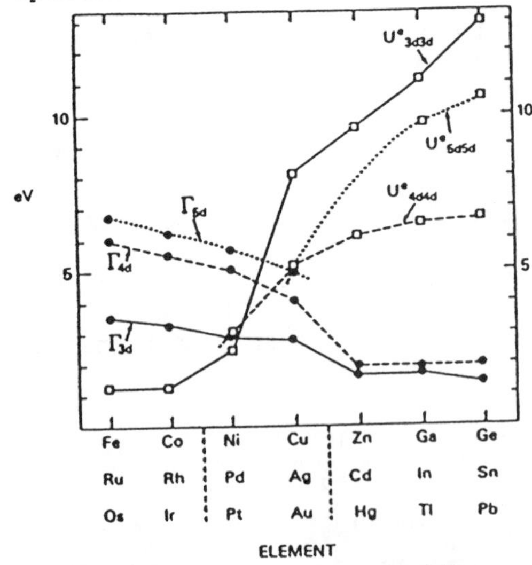

Fig. 44 Comparison of Γ and U_{eff} for the metals with the electronic configuration d^6s^2 to $d^{10}s^2p^2$.

This may be illustrated in Fig. 45 for $M_{4,5}$ $N_{4,5}$ $N_{4,5}$ transitions of Pd in pure Pd, Al_{80} Pd_{20}, Mg_{75} Pd_{25} [81] and at the interface between Yb and Pd (111) [21]. Fig. 45a shows a study of the influence of the Pd [4d] band on the Auger spectrum ($[4d^2]$ two-hole final states).

The $M_{4,5}$ $N_{4,5}$ $N_{4,5}$ transitions of Pd in Mg_{75} Pd_{25} have narrow spectral features and an overall profile characteristic of the $M_{4,5}$ $N_{4,5}$ $N_{4,5}$ spectra of pure Ag. The Auger spectra of Al_{80} Pd_{20} and pure Pd are broader with "band-like" character. The extent of the broadening correlates with an increase in the Pd [4d] bandwidth and with the density of states at the Fermi energy.

In the case of Yb deposition on Pd (111) at room temperature (Fig. 45b), the behaviour of the $M_{4,5}$ $N_{4,5}$ $N_{4,5}$ Pd Auger spectrum can be used as a signature of an interdiffusion process between Yb and Pd. The spectra observed for coverages above 4 nm look quite similar to the one reported for the Mg_{75} Pd_{25} alloy confirming the progressive dilution of Pd in Yb near the surface as the Yb coverage increases.

An other interesting example is given by the behaviour of the Auger $M_{4,5}$ $N_{4,5}$ $N_{4,5}$ Ag spectra in pure Ag and alloys [82] .

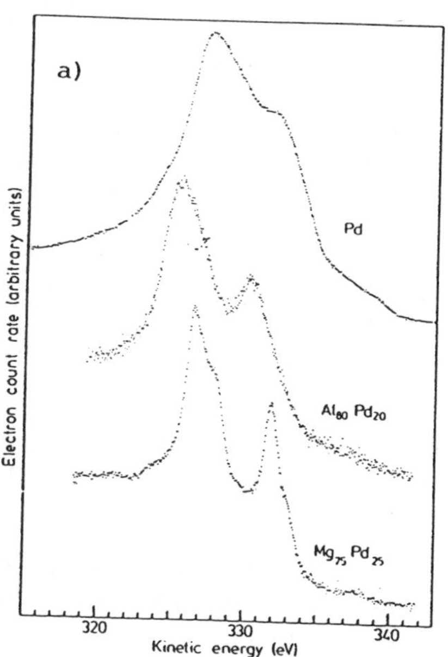

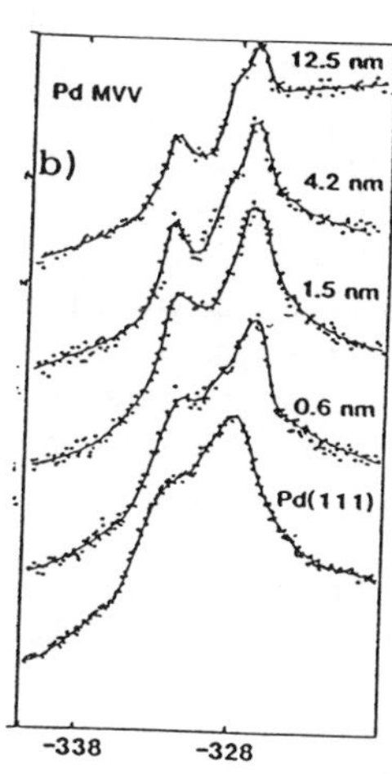

Fig. 45 a) Pd $M_{4,5}N_{4,5}N_{4,5}$ Auger spectra observed from Pd, Al80Pd20 and Mg75Pd25

b) Pd $M_{4,5}N_{4,5}N_{4,5}$ Auger spectra for different coverages of Yb on Pd (111).

On the basis of these studies, it is clear that the Cini-Sawatzky criterion for the origin of quasi-atomic Auger spectra is correct and that the valence band spectra are very useful for studying transition metals and their alloys.

The Cini-Sawatzky approach can be also used in order to explain the Auger spectra of metallic clusters. In this case, in principle, the correlation should be increased because the expected value of the Coulomb interaction U_{eff} of the two localized holes increases as the cluster size decreases [83]. As a result, the Auger lines should be narrowed for the smallest particles because U increases and the bandwidth decreases. In contrast for clusters deposited on non-interacting substrates, the unexpected lineshape broadening is attibuted to the cluster size distribution, as explained by De Crescenzi in the case of clusters deposited on graphite.

6 QUANTITATIVE AUGER ANALYSIS

Assuming that for an element x the intensity of the Auger signal is proportional to the density of x atoms in the analyzed area, the first problem is to measure the real intensity of the peaks. On the other hand, the purpose is to retrieve the total amount of "true" Auger electrons from the background. First of all the distribution N(E) must be generated by integration or directly recorded. The second step is to determine and substract the background in order to obtain the distribution $N_A(E)$ of "true" Auger electrons [84, 85, 86]. For example, a problem arises from the knowledge of how far the generally asymmetric Auger peak extends to the low energy side. Not only must the Auger electrons resulting directly from a transition be taken into account, but also those which have suffered subsequent losses.

In this $N_A(E)$ distribution, the area under each peak is a measure of the Auger current which results from the corresponding transition.

Another important point is to relate the number of Auger electrons to the absolute atomic densities of the given element. The determination of an absolute concentration of an element x in a matrix from the yield Y_A of Auger electrons is complicated by the influence of the matrix on the electron backscattering and escape depth.

A simple expression for the yield for Auger electrons produced from a thin layer of thickness Δt at a depth t in the sample can be given by :

$$Y_A(t)=N_x\Delta t \, . \, \sigma_e(t).[1-\omega_x].e^{-t \cos\theta/\lambda} \, I(t).d\Omega/4\pi \qquad (6)$$

where

N_x= the number of x atoms/unit vol;
$\sigma_e(t)$= the ionization cross section at depth t;
ω_x= the fluorescence yield;
λ= the escape depth;
θ= the analyser angle;
T= the transmission of the analyzer;
$d\Omega$= the solid angle of the analyzer;
I(t)= the electron excitation flux at depth t.

with

$$I(t) = I_p+I_B(t) = I_p(t)\,[1+R_B(t)] \qquad (7)$$

where I_p is the flux of primary electrons at depth t and I_B is the flux due to backscattered primary electrons and R_B is the backscattering factor.

It is clear that such a formula cannot be used directly. In practice, the simplest determination of surface composition is obtained from the measure of Auger peak to peak heights in the derivative mode dN/dE (APPH) and using standards or relative sensitivity factors for all the elements involved. From these Auger data a number of methods for calculation of surface composition have been proposed. These differ mainly in consideration of matrix effects on the yield of Auger electrons (such as the escape depth of Auger electrons and/or backscattering effects) [87].

This approach has been successfully applied to the studies of superficial segregation in metallic alloys. So segregation kinetics on binary or ternary alloys can be established from Auger data according to appropriate models of segregation [88].

Two additional remarks can be done :

(i) More generally when external standards are used with a known concentration of element x close to that of the test sample, the element composition can be determined directly from the ratio of Auger yield provided the measurements are made under the same experimental conditions.

(ii) To determine the "true" surface composition, problems can arise from the ion bombardments and/or heat treatments used for surface cleaning.

Another quantitative approach of the Auger data consists of considering in the same experimental conditions the evolution of the Auger intensities as a function of appropriate treatment without trying to determine the elemental concentrations.

In this case the evolution of intensities is followed in comparison with an initial situation i.e. clean surface before heating, ion sputtering, adsorption, deposition.

This approach, termed here as "semi-quantitative", is widely used in two kinds of studies in connection with the field of this school :

a) characterization of different modes of film growth
b) Auger depth profiles

5.1 Characterization of growth modes

The possibility of determining growth modes from the evolution of Auger intensities is based on the exponential probability of inelastic scattering for an electron emerging from a solid.

After an homogeneous deposit of an overlayer thickness x on a substrate, the Auger intensity I of an Auger electron from the substrate can be written as :

$$I = I_o \, e^{-x/\lambda} \qquad (8)$$

where I_o is the intensity from the clean surface before deposition and λ the mean free path of the substrate Auger electron through the overlayer material.

From this equation and on the basis of simple topologic considerations [89,94], one can deduce the typical sets of curves corresponding to the different modes of film growth (Fig. 46).

(i) For the layer by layer growth (or Frank-Van der Merwe mode), the characteristic curve is a series of straight lines with breaks at coverages corresponding to the completion of the successive monolayers. The envelope of points corresponding to each completed monolayer is an exponential having the $e^{x/\lambda}$ form,

(ii) For the second mode (termed Stranski-Krastanov), the deposition of a single layer or two uniform layers is followed by the formation of islands,

(iii) The third mode (or Volmer-Weber mode) presented here corresponds to the formation of islands without growth of an initial uniform layer.

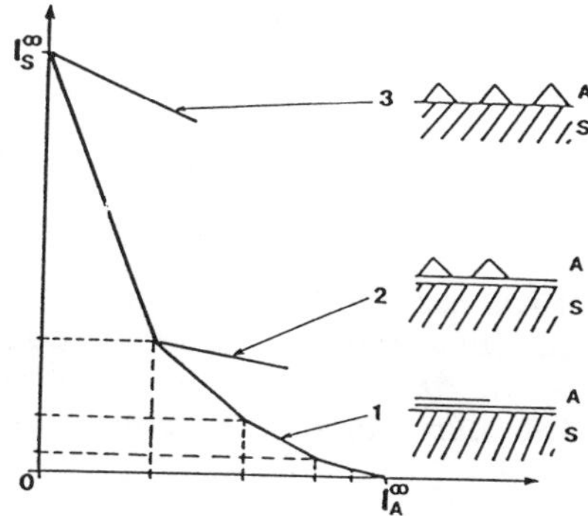

Fig. 46 Variation of Auger intensity of the substrate versus Auger intensity of the adsorbate for various growth mechanisms
1. Frank-van der Merwe
2. Stranski-Krastanov
3. Volmer-Weber.

To explain rigorously the experimental curve obtained by plotting the Auger intensities of absorbate and substrate as a function of coverage (or time of deposition), it is necessary to get an accurate determination of the absolute coverage and a knowledge of the involved mean free path.

In Fig. 47, two typical examples corresponding respectively to the formation of Au/Pt and Pt/Au interfaces are described [90]. We can see clearly that the difference in the growth mode can be revealed through this type of diagram.

This possibility to characterize the growth modes of thin films is one of the most prominent uses of AES. Moreover numerous problems can occur mainly in connection with the measurement of peak intensities, calibration of coverage and more generally of standardization of the experimental conditions.

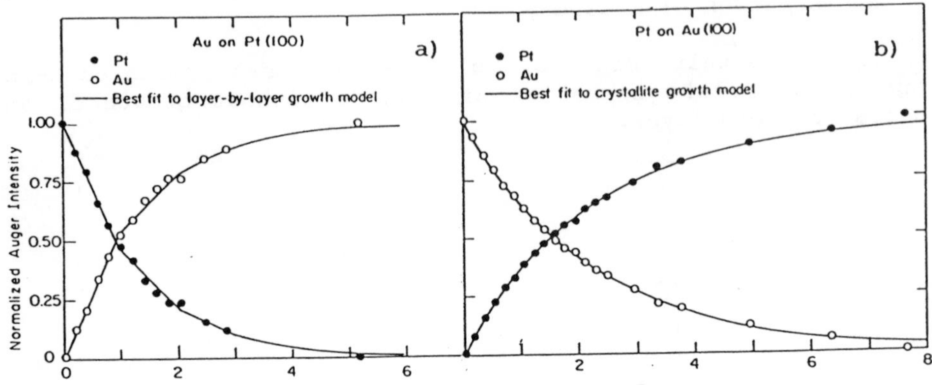

Fig. 47 a) Plot of the Auger intensities of gold and platinum versus Au coverage on Pt(100) b) Plot of the Auger intensities of gold and platinum versus Pt coverage on Au(100).

In order to overcome experimental difficulties, another way of representating the intensities as a function of coverage can be to consider the ratios between the adsorbate and substrate signals [91]. This is illustrated on Fig. 48 for the case of the growth of Cr on W(110) [92].

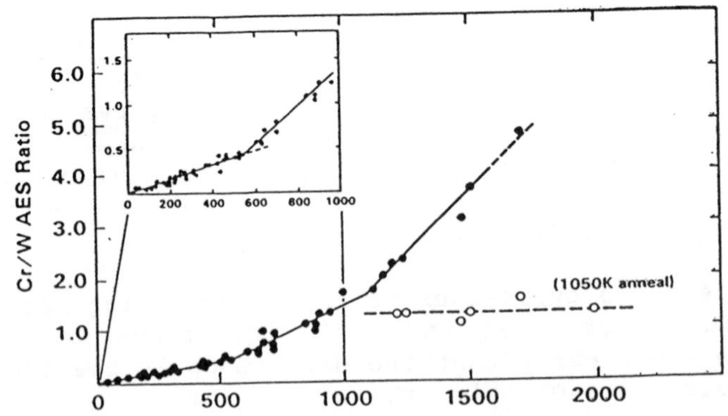

Fig. 48 Plot of the peak-to-peak Cr(529)/W(169) Auger intensity versus Cr coverage on W(110).

Whatever representation is chosen, it is clear that the characterization of thin-film growth modes must be also based on all the Auger data available (changes in shape and intensities), and on the combination of other methods such a LEED, RHEED, Photoemission or SEXAFS.

6.2 Auger Depth Profiles [56]

One of the major uses of AES at present is the determination of the composition as a function of depth in thin films and multilayers. In this approach the Auger spectrometer is coupled with an ion-sputtering device and one records the Auger signal heights against the sputtering time in order to obtain an Auger depth profiling.

As compared to RBS, Auger depth profiling provides better depth resolution. Another advantage of AES is its sensitivity to low mass impurities such as carbon or oxygen.

Fig. 49 shows sputter depth profiles of multilayered Cr/Ni thin film structures deposited on a Si substrate [93]. This result illustrates the ability of AES to profile a multilayer film in a semi-quantitative manner and underlines one of the experimental difficulties. The surface roughness which appears in the case of a stationary sample (upper portion) could be minimized by rotating the sample during sputtering (lower portion).

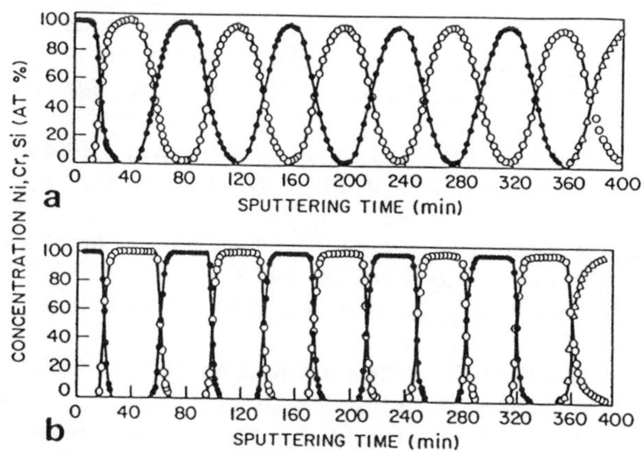

Fig. 49 Auger electron spectroscopy sputter depth profiles of multilayer Cr/Ni thin film structures deposited on a Si substrate.

REFERENCES

1) Lectures of the Summer School : "Rayonnement Synchrotron". Aussois Septembre 1986. Edited by CNRS AD11,Grenoble.
2) Loudon R. : The Quantum Theory of Light. Clarendon Press, Oxford 1973.
3) Cohen-Tannoudji C., Dupont-Roc J. et Grynberg G. : Photons et atomes. Savoir actuel, Editions du CNRS 1988.
4) Jackson J.D. : Classical electrodynamics. New-York, Wiley and Sons, 2nd ed. 1975.
5) Cardona M. and Ley L. : Photoemission in Solids (Tome I) Cardona M. and Ley L. ed. : Topics in Applied Physics Vol 26. Springer-Verlag Berlin 1978 p 60
6) Thole B.T., van der Laan G., Fuggle J.C., Sawatzky G.A., Karnatak R.C. and Esteva J.M. : Phys.Rev.B, 1985, 32, 5107.
7) Campagna M., Wertheim G.K. and Baer Y. : Photoemission in Solids (Tome II) Cardona M. and Ley L. ed. : Topics in Applied Physics, Vol.27. Springer-Verlag Berlin 1978 p 231
8) Balzarotti A. : EXAFS and Near Edge Structure. Bianconi A., Incoccia L. and Stipcich ed. : Chemical Physics, Vol 27. Springer-Verlag Berlin 1983 p. 138
9) Berglund C.D. and Spicer W.E. : Phys.Rev., 1964, 136 , A1030
10) Fadley C.S : Electron Emission Spectroscopy. Dekeyser et al eds. D.Reidel Publishing Company, Dordrecht-Holland p. 181
11) Eastman D.E. : Electron Spectroscopy. Shirley D.A ed. North-Holland Publishing Company, Amsterdam 1972 p. 487
12) Stern E.A. : X-Ray Absorption. Koningsberger D.C. and Prins R. eds. John Wiley and Sons, New-York 1988 p 4
13) Sayers D.E, Stern E.A. and Lytle F. : Phys.Rev.Lett., 1971, 27, 1204
14) Brouder C., Malterre D. and Krill G. : Ecole de Printemps "Rayonnement Synchrotron Polarisé, Electrons Polarisés et Magnétisme ". Carrière. B. ed., Mittelwihr March 1989. To be published.
15) Müller J.E. and Wilkins J.W. : Phys.Rev.B., 1984, 29, 4331
16) Doniach S. and Sunjic M. : J.Phys.C. Solid State, 1970, 3, 285
17) Wertheim G.K. and Citrin P.H. : Photoemission in Solids (Tome I). Cardona M. and Ley L. ed. : Topics in Applied Physics, Vol 26. Springer-Verlag Berlin 1978 p. 232
18) Shek M.L., Stefan P.M., Lindau I. and Spicer W.E. : Phys.Rev.B., 1983, 27, 7277.
19) Spanjaard D., Guillot C., Desjonquères M.C., Tréglia G. and Lecante J. : Surface Science Reports, 1985, 5, 1-85
20) Shek M.L., Stefan P.M., Lindau I. and Spicer W.E. : Phys.Rev.B 1983, 27, 7288
21) Beaurepaire E., Carrière B., Légaré P., Krill G., Brouder C., Chandesris D. and Lecante J.: Surface Science, 1989, 212, 448
22) Hanf M.C., Pirri C., Peruchetti J.C., Bolmont D. and Gewinner G. : Phys.Rev.B 1987, 36, 4487

23) Hanf M.C., Haderbache L., Wetzel P., Pirri C., Peruchetti J.C., Bolmont D. and Gewinner G. : Solid. State. Com., 1988, 68, 113
24) Hanf M.C., Pirri C., Peruchetti J.C., Bolmont D. and Gewinner G. : Phy.Rev.B 1989, 39, 1546
25) Hanf M.C., Pirri C., Peruchetti J.C., Bolmont D. and Gewinner G. : Phys.Rev.B 1989, 39, 3021
26) Newstead D.A., Norris C., Binns C. and Stephenson P.C. : J.Phys.C. Solid State, 1987, 20, 6245
27) Van Vleck J.H. : Phys.Rev., 1934, 45, 405
28) Kämper K.P., Schmitt W., Güntherodt G. and Kuhlenbeck H. : Phys.Rev.B., 1988, 38, 9451
29) Hermanson J. : Solid State. Com., 1977, 22, 9
30) Weller D., Alvarado S.F., Gudat W, Schröder K. and Campagna M. : Phys. Rev. Lett., 1985, 54, 1555
31) Landolt M., Allenspach R. and Taborelli M. : Surface Science, 1986, 178, 311
32) Pescia D., Stampanoni M., Bona G.L., Vaterlaus A., Willis R.F. and Meier F. : Phys. Rev. Lett., 1987, 58, 2126
33) Stampanoni M., Vaterlaus A. and Meier F. : Phys. Rev. Lett., 1987, 59, 2483
34) Kolbas R.M. and Holonyak N. : Am. J. Phys., 1984, 52, 431
35) Miller T., Samsavar A., Franklin G.E. and Chiang T.C. : Phys. Rev. Lett., 1988, 61, 12, 1404
36) Michel E.G. and Ferrer S. : Eur. Phys. Lett., 1987, 4, 603
37) Drube W. and Himpsel F.J. : Phys. Rev. B., 1987, 35, 8, 4131
38) Zeller R., Podloucky R. and Dederichs P.H. : Z. Phys.B ., 1980, 38, 165
39) Brouder C., Krill G., Dartyge E., Fontaine A., Tourillon G., Guilmin P. and Marchal G. : Materials Science and Engineering, 1988, 97, 109
40) Brouder C., Krill G., Dartyge E., Fontaine A., Tourillon G., Guilmin P. and Marchal G. : Phys. Rev. B., 1988, 37, 2433
41) Brouder C., Krill G. and Piecuch M. : J.Less Com. Met., 1988, 140, 67
42) Claeson T., Boyce J.B., Lowe W.P. and Geballe T.H. : Phys. Rev. B, 1984, 29, 4969
43) Heald S.M., Chen H. and Lamble G.M. : Physica B 1989, 158, 309
44) Jiang D.T., Alberding N., Seary A.J., Heinrich B. and Crozier E.D. : Physica B 1989, 158, 662
45) Heal S.M. and Tranquada J.M. : to be published
47) Chandesris D., Roubin P., Rossi G. and Lecante J. : Surface Science, 1986, 169, 57
48) Chandesris D., Roubin P., Rossi G., Lecante J. , Desjonquères M.C. and Tréglia G. : Phys. Rev. Lett., 1986, 56, 1272
49) Roubin P. : Thèse de Doctorat d'Etat ès Sciences. Orsay 1987
50) Dartyge E., Krill G., Baudelet F., Brouder C., Kappler J.P. and Fontaine A. : Rapport d'Activité LURE 1989; to be published in the Proceedings of XSR89 Conference Rome 1989

51) Piecuch M., Baczewski L.T., Durand J., Marchal G., Delcroix P. and Nabli H. : J.Phys., 1988, C8-12, 175
52) Burhop E.H.S. : J.Phys.Radium, 1955, 16,625
53) Briggs D and Seah M.P. : Practical Surface Analysis, J. Wiley, 1983
54) Brundle C.R. and Baker A.D. : Electron Spectroscopy, Academic Press (London), 1981
55) Ertl and Küppers : Low Energy Electrons and Surface Chemistry, Verlag Chemie, 1974
56) Feldman L.C. and Mayer J.W. : Fundamentals of Surface and Thin Films Analysis, North Holland, 1986
57) Ohno M. and Wendin G. : J.Phys.B ,1979, 12, 1305
58) Günnarsson O. and Schönhammer K. : Phys.Rev.B, 1980, 22, 3710
59) Aberg T. and Howat G. : Hand.Phys., 1981, 31
60) Slater J.C. : Quantum Theory of Atomic Structure, N.Y., McGraw-Hill, 1960
61) Burhop E.H.S and Asaad W.N. : Adv. Atom. Molec. Phys. 1972,8, 163
62) Melhorn W. : Lectures on Electron Spectroscopy , 1978
63) Bambinek W.,Crasemann B., Fink R.W., Freund H.U., Mark H. : Rev. Mod. Phys. ,1972, 44, 716
64) Chattarji D. : The Theory of Auger Transitions, N.Y. Academic, 1976
65) Haynes S.K. : Proc. Int. Inner Shell Ionisation Phenomena, R.W.Fink Ed. , 1973, 559
66) Aksela H. : J.Electron Spectrosc. ,1980,19, 371
67) Shirley D.A. : Phys. Rev.A, 1973, 7, 1520
68) Lang N.D. and Williams A.R. : Phys.Rev.B, 1979, 20, 1369
69) Wagner C.D. and Biloen P. : Surf.Sci.,1973, 35, 82
70) Wagner C.D.and Joshi A.: J.Electron.Spectrosc. : 1988, 47, 283
71) Vayrinen J., Aksela S. and Aksela H.: Phys.Scr.: 1977, 16, 452
72) Aksela S., Kumpula R., Aksela H. and Vayrinen J. : Phys.Scr., 1982, 25, 45
73) Carrière B.,Deville J.P. and Humbert P. : Springer Series in Surface Science, 1989, (in press)
74) Jennisson D.R., Madden H.H. and Zehner D.M. : Phys.Rev. B, 1980, 21, 430
75) Carrière B., Deville J.P. and El Maachi A. : Phil.Mag., 1987, B55,721
76) Carrière B., Deville J.P. and El Maachi A. : Surf. Sci., 1986, 168, 149
77) Carrière B., Deville J.P. and Burggraf Ch.:Analusis, 1981, 9, 236
78) Antonides E., Janse E.C.and Sawatzky G.A. : Phys.Rev.B, 1977, 15, 1669
79) Cini M. : Phys.Rev.B, 1978, 17, 2788
80) Sawatzky G.A. : Phys.Rev.Lett., 1977, 39, 504
81) Weightman P. and Andrews P.T.: J.Phys.C, 1980,13, L815
82) Hedegard P. and Johansson B. : Phys.Rev.Lett. 1984, 52, 2168
83) De Crescenzi M., Diociaiuti, Lozzi L,Picozzi P. and Santucci S. : Surf.Sci., 1986, 178, 282

84) Sickafus E.N.: Rev.Sci.Instr., 1971, 42,933
85) Sickafus E.M.: Phys Rev.B, 1977, 16, 1448
86) Tougaard S. : Surf.Sci., 1989, 216, 343
87) Lejcek P. : Surf.Sci., 1988, 202, 493
88) Boudjemaa F. and Mosser A. : J. of Less-Comm. Met., 1988, 145, 55
89) Kern R. : Bull. Minéral., 1978, 101, 202
90) Sachtler J.W.A, Van Hove M.A., Bibérian J.P. and Somorjai G.A. : Surf.Sci. , 1981, 110, 19
91) Ossicini S., Memeo R. and Ciccacci F. : J. Vac. Sci. Techn., 1985, A3, 387
92) Berlowitz P.J. and Shinn N.D : Surf.Sci., 1989, 209, 345
93) Zalar A. : Thin Solids Films, 1985, 124, 223
94) Rhead G.E. : J. Vacuum Sci. Technol., 1972, 32, 703
95) Ramaker D.E. : Applic. of Surf.Sci. , 1985, 21, 243
96) Handbook of Auger Electron Spectroscopy , Perkin-Elmer Corp. , 1978
97) Weightman P. : Reports on Progress in Physics, 1982, 45, 753

Materials Science Forum Vols. 59 & 60 (1990) pp. 229-286

PRACTICAL ASPECTS OF MOLECULAR BEAM EPITAXY

Ch. Chatillon (a) and J. Massies (b)
(a) Laboratoire de Thermodynamique et Physico-Chimie Métallurgiques
Ass. au CNRS - URA 29
E.N.S.E.E.G. - Domaine Universitaire - B.P. 75
F-38402 St-Martin-d'Hères, France
(b) Laboratoire de Physique du Solide et Energie Solaire
CNRS - Sophia Antipolis
F-06560 Valbonne, France

Molecular beam codeposition methods are now used to provide single crystal thin films and multilayers of various materials including semiconductors, metals and ceramics. From the practical point of view all these molecular beam techniques have in common two main aspects: molecular beam generation and the control of the beam interaction on the surface under growth. The present chapter deals with -first the production and physical chemistry of molecular beams, and the matter flow balance at the growing surface (part A) - second, the use of grazing reflection electron diffraction to control the growth process (part B).

A THERMODYNAMICS and PHYSICO-CHEMICAL ASPECTS of MOLECULAR BEAM DEVICES

1 INTRODUCTION

Atomic and molecular beams are usually obtained by high temperature vaporization of pure elements or compounds, and for this reason the vaporization processes may be a priori calculated after careful compilation of the literature concerning high temperature studies that have been performed, mainly by mass spectrometry, an analytical method which is able to determine the vapor composition. Once the vaporization process is known, the experimentalist must be able to control the molecular beam which is the molecular flow exchanged between the source and the substrate used for the deposition. The useful molecular flow will be analysed in the light of the physico-chemical processes that have been studied in the case of the Knudsen-cell effusion method and our conclusions will be directed to the production of molecular beams directed of good quality. Finally, gross molecular flows will be analysed at the growth surface, taking into account condensation and evaporation flow balances to calculate the deposition rate.

2 MOLECULAR SOURCES and THERMODYNAMICS of VAPORIZATION

2.1 Vaporization laws

The vaporization of a substance in equilibrium conditions is rigourously attained for a closed system (fig 1a), in which the condensation and evaporation flows are equal at the surface. The determination of vapor pressures under molecular flow is performed by the way of a small, thin orifice (fig 1b) machined on a wall that allows gaseous species to escape the so-called Knudsen-cell [1,2]: assuming the section of the orifice is small compared to the evaporation area of the sample, the kinetic theory of gases leads to the evaluation of the escaping molecular flow by calculating the collision probability of the gas molecules on an equivalent orifice surface *ds* of the wall (fig 2) according to the following equation

$$dn/dt = n\ v\ (d\Omega/4\ \pi)\ ds \cos\theta \qquad (1)$$

with

$$n = p/RT$$

$$v = (8RT/\pi M)^{1/2}$$

where dn/dt is the molecular flow in the θ direction for an elementary solid angle $d\Omega$, n the density of molecules in the gaseous phase, v their average velocity and $d\Omega/4\pi$ the directional probability. T is the temperature, p the pressure, R the gas constant and M the molar mass of the gaseous species. The integration of this relation, over all the space outside the cell leads to the total molecular flow escaping from a cell

$$dn/dt = p\ s\ /\ (2\ \pi\ MRT)^{1/2} \tag{2}$$

which is the Hertz-Knudsen relation.

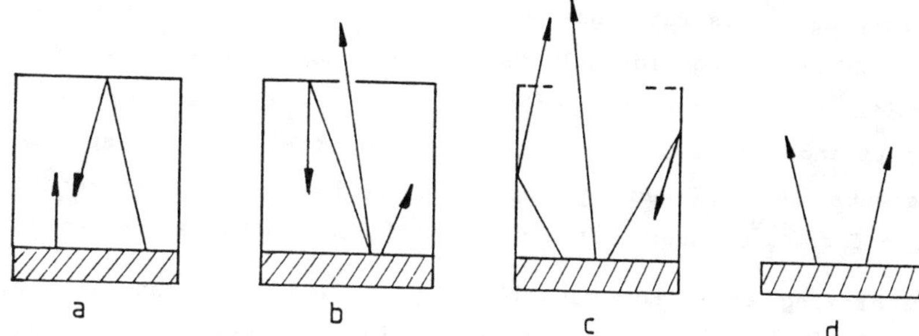

Fig.1 (a) Closed system: vaporization at equilibrium,. (b) Knudsen-cell quasi-equilibrium,. (c) Open crucible as used in Molecular Beam Epitaxy (M.B.E.), (d) Free or Langmuir vaporization.

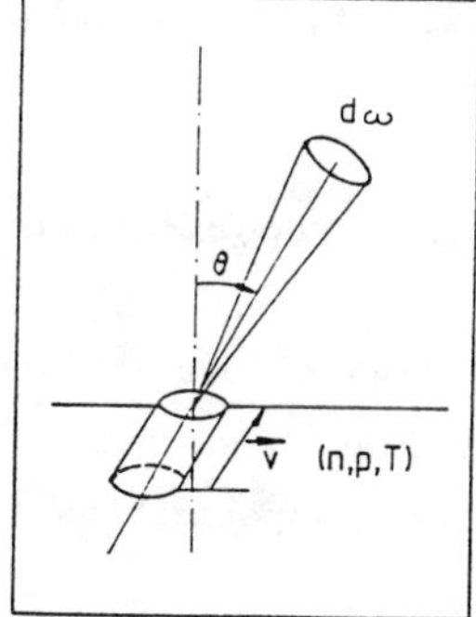

Fig.2 The molecular emission of a surface is calculated according to the flow of molecules coming from the vapor in equilibrium and escaping through an aperture of the same surface.

By enlarging the effusion orifice for a constant evaporating surface, the open crucible (fig 1c) is obtained. This type of crucible is often used in MBE, and called incorrectly Knudsen-cell since the evaporation is not necessarily performed at equilibrium on the extreme. The limiting situation is the free or "Langmuir" [3] vaporization (fig 1d), where there is no molecular flow coming back to the surface, at least in the range of experimental uncertainties. In this case the evaporation flow is related to the equilibrium vaporization flow and the evaporation coefficient α_L^{ev}:

$$(dn/dt)_L = \alpha_L^{ev}\ p\ s\ /\ (2\ \pi\ MRT)^{1/2} \qquad (3)$$

with $0 < \alpha_L^{ev} \leq 1$. For an open crucible, the α^{ev} values lie between α_L^{ev} and 1 as a function of the ratio between molecules that come back to the surface and those escaping ; this ratio has been calculated by Clausing [4] for cylindrical orifices and by analogy for cylindrical crucibles. The evaporation coefficients are specific of each gaseous species but are also correlated to the structure of the vaporizing surface, its defect density, the bulk and surface diffusion coefficients as reviewed by Rosenblatt [5] : useful references are also available for α_L^{ev} values, but a main difficulty is the actual knowledge of the real vaporizing area. For metallic liquids and polycristalline solids, the vapors of which are monoatomic, the evaporation coefficients are equal to unity, but for solids with polymeric vapors these coefficients are generally lower than unity and may be very low [10^{-6} for P_4 (g) over P(red, am)]. Moreover the vaporization flow will depend on the stabilization time for the vaporizing surface as well as on its height in the crucible determining the transmission probability of molecules from the vaporizing area to the crucible aperture.

2.2 Vaporization thermodynamics

Free or equilibrium vaporization flows are calculated using thermodynamic relations, according to the stability of the different phases. As an exemple for gold vaporization at temperature T

$$\text{Au (s or l)} = \text{Au (g)} \qquad (I)$$

$$\Delta G^{\circ}_{vap}(\text{Au, T}) = -RT\ln p(\text{Au, T})$$
$$= \Delta G^{\circ}_f(\text{Au, g, T}) - \Delta G^{\circ}_f(\text{Au, s or l, T})$$

relation in which ΔG is the Gibbs energy, vap and f account respectively for vaporization and formation. Also for every gaseous species, as for Au_2 (g)

$$2\ \text{Au(s or l)} = Au_2\ \text{(g)} \qquad (II)$$

$$- RT \ln p(Au_2, T) = \Delta G^{\circ}_{f}(Au_2, g, T) - 2 \Delta G^{\circ}_{f}(Au, s \text{ or } l, T)$$

For pure solid or liquid elements the Gibbs energy of formation in the standard state is equal to zero. For the case of compounds, the standard state must be the same for all the species involved in the vaporization reaction. When the gaseous phase contains several molecules, partial pressures may be calculated using as many partial vaporization or dissociation reactions as the number of involved species.

Free vaporization flows can be rigourously deduced from equilibrium flows only when evaporation coefficients are known as well as their temperature variations.

2.3 Gaseous Phases Stability

The gaseous phases of pure elements may be either monoatomic or polyatomic. For metals, the monomers are dominant and the relative abondancy of dimers often lower than 10^{-2} or 10^{-3}. For non-metallic elements, at the right side of the C, P, As, Sb, Bi line (including these elements) in the periodic table, the polymers are dominant. The partial pressures of gaseous species in equilibrium with pure elements have been published in thermodynamic data compilations, the Hultgren's one [6] gathering all the pure elements in one book.

For compounds, and by extension for alloys and mixtures, heteroatomic gaseous species are numerous and very often they are stable relatively to the pure elements. As the molecular sources operate at high temperatures, the container material must be taken into account when searching for gaseous binary entities that can be formed, even if the crucible seems compatible with the vaporizing material. A careful literature check must be done every time some new material is to be evaporated : a two step action is to be performed:

- for compounds in compilations [7-9], the gaseous phase composition can be calculated by thermodynamics as done in the next part,
- for compounds that are not compiled, high temperature Knudsen (or effusion) mass spectrometric original papers give useful informations.

About 3000 papers about this method have been published since 1954, the earlier one's being compiled [10,11].

2.4 Compounds and alloys vaporization

The partial vaporization pressure of a gaseous species from a mixture is directly related to its activity in the condensed phase according to the partial vaporization equilibrium

$$A \text{ (solution or compound)} = A(g) \qquad K(T) = p_A/a_A$$

refering to the same equilibrium for the pure phase

$$A(s \text{ or } l) = A(g) \qquad Kp(T) = p_A^{o}/1$$

and the A activity is defined by $a_A = p_A/p_A^{o}$. For heteroatomic gaseous species, their partial pressures are depending simultaneously on the activity of each of their elements as illustrated by

$$A \text{ (sol.)} + B \text{ (sol.)} = A_2B \text{ (g)} \qquad K_p(T) = p_{A2B} / a_A^2 a_B$$

When A_2B exists in the condensed phase, the partial pressures in equilibrium with this compound may be taken as a reference. This solution is useful for thermodynamic calculations in subsystems as for exemple Al_2O_3 - CaO or InAs - GaAs.

The thermodynamics of complex systems is generally formulated using partial Gibbs energies [12] and nowadays assessed from original thermodynamic data in correlation with phase diagram data [13]. Partial pressures are then deduced for any compositions using suitable computer calculations [14,15]. From the point of view of molecular sources, the most interesting case is the vaporization of compounds, either these compounds may be the molecular source itself or the crucible material.

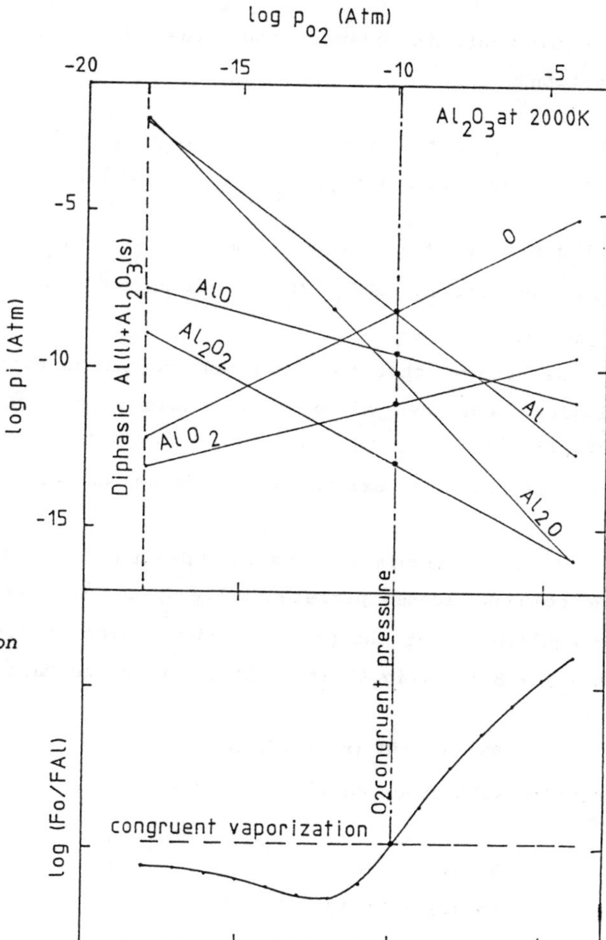

Fig.3 Thermodynamic calculation of the Al_2O_3 vaporization as a fonction of O_2 pressure and corresponding vaporization pressures.

According to the gaseous and condensed phase stability, the vaporization of compounds may be congruent or not. Congruent vaporization is obtained when the evaporating molecular flow has the same composition as the condensed phase. For instance, during vaporization Al_2O_3 (s) is in equilibrium with several gaseous molecules (7) as illustrated by the following reactions

$$Al_2O_3(s) = 2\,Al\,(g) + 3/2\,O_2\,(g) \quad (III)$$

$$Al_2O_3(s) = Al_2O\,(g) + O_2\,(g) \quad (IV)$$

$$Al_2O_3(s) = 2\,AlO\,(g) + 1/2\,O_2\,(g) \quad (V)$$

$$Al_2O_3(s) = Al_2O_2\,(g) + 1/2\,O_2\,(g) \quad (VI)$$

$$Al_2O_3(s) + 1/2\,O_2 = 2\,AlO_2\,(g) \quad (VII)$$

$$O_2(g) = 2\,O\,(g) \quad (VIII)$$

Aluminium atomic evaporation flow is calculated from the Hertz-Knudsen relation

$$F_{Al} = [\ s/(2\ \pi\ RT)^{1/2}\ ]\ [p_{Al}/(M_{Al})^{1/2} + 2p_{Al_2O}/\ (M_{Al_2O})^{1/2} + p_{AlO}/ (M_{AlO})^{1/2} + 2p_{Al_2O_2}/\ (M_{Al_2O_2})^{1/2} + ...]$$

and similarly for oxygen atomic flow. Using the oxygen pressure p_{O_2} as a parameter, the ratio $F_{Al}\ /F_O$ is calculated and compared to alumina ratio 2/3 on figure 3.

We observe that the congruent condition exists in the stability range of alumina, and then alumina is stable under vacuum. It is also possible to analyse the Al_2O_3 vaporization taking into account of the background oxygen pressure [16] that exists in the vacuum chamber.

Similar calculations performed with BN(s) show incongruent vaporization : the compound looses preferentially N_2 and the partial pressures should be those in equilibrium at the phase boundary which corresponds to the diphasic mixture BN (s) + B (s) (fig.4) in agreement with the main vaporization reaction

$$BN\ (s) = B\ (s) + 1/2\ N_2\ (g) \qquad (IX)$$

occuring with other equilibria

$$B\ (s) = B\ (g) \qquad (X)$$

$$BN\ (s) = BN\ (g) (XI) \qquad (XI)$$

Free evaporation studies of BN(s) [17] show that this compound vaporizes with very low evaporation coefficients. Comparing free vaporization pressures and those calculated for the diphasic BN(s) + B(s) (fig 4), the evaporation coefficient may be mainly explained assuming that the N_2(g) flow is congruently limited by the B(g) maximum flow that exists over pure B (s). This situation results from lack of diffusion in this III-V compound. When bulk diffusion is not as important as the evaporation flow of at least one component, the high temperature vaporization produces the apparition of a new phase at the surface which may "catalyse" the vaporization process as observed for other III-V compounds [18,19] and the evaporation coefficients increase. Such vaporization behaviour is usual for ceramics or intermetallic stable compounds.

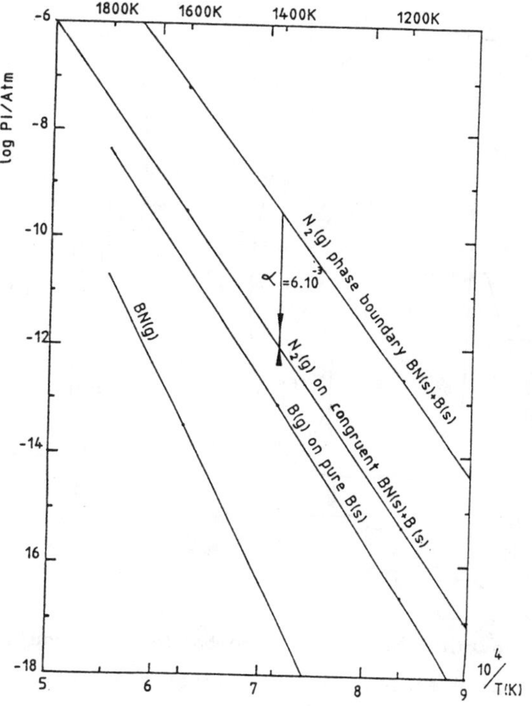

Fig. 4 Thermodynamic calculation of the BN(s) vaporization and free evaporation pressures corresponding to the measured evaporation coefficient $\alpha = 6.10^{-3}$.

2.5 Source technology

The usual furnaces for molecular beam metal co-deposition use electron bombardment guns, the metallic source being located in a water cooled crucible so that chemical interactions between the source and the crucible materials are minimized. This is very important for materials the reactivity of which is very high, correlated with very high temperature for vaporization , like Zr, Ti, Nb, Hf, W, Mo, HfO_2, Al_2O_3.Commercial electron bombardment guns are usually built to obtain high molecular flows but laboratory devices have been built [20] to enlarge the evaporation area and tentatively decrease the evaporation temperature (fig 5) of the material.

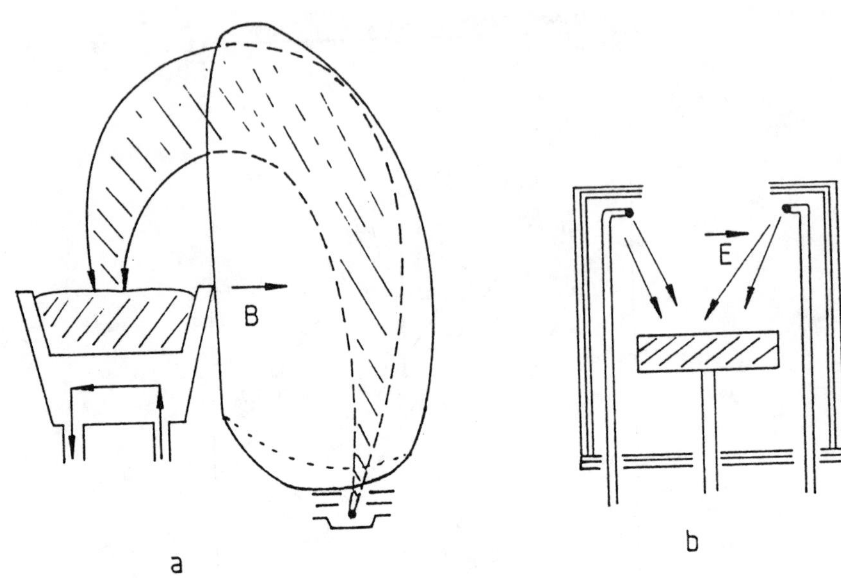

Fig.5 Two kinds of electron bombardment molecular sources (a) with an electron gun, (b) direct electron bombardment.

When possible, Knudsen-cells are more convenient for better control of molecular flows, and their construction must be independent of the heating mode of the surrounding furnace to allow the choice of containers adapted to each source material.

The production of molecular beams from Knudsen-cells is closely related to the ability to find convenient containers. A first step to choose the container material is to compile original mass spectrometric works on vaporization (6,10,11), and then the experimenter must follow two guide-lines.

-The first one is to analyse the chemical compatibility between the source and container materials using binary [21] and ternary [22] phase diagrams that predict a possible reciprocal solubility at working temperatures. When an intermediate compound is to be formed, it is convenient to check its high temperature mechanical behaviour. For instance, liquid silicon in dense graphite forms a compact SiC layer which prevents a continuous chemical attack of graphite and then the vaporization of liquid silicion is possible.

-The second one concerns the dynamical aspect of source working in M.B.E. For instance, the vaporization of Al in BeO or ThO_2 crucibles is possible but the effusion of Al_2O (g) together with Al(g) removes oxygen continously from the crucible, hence Be or Th is continuously solubilized in the liquid Al source. The evaporation flow of Al then decreases since its activity decreases continously [23], except if the metallic component coming from the crucible is volatile, which is the case of Be that induces pollution in the Al beam. A third exemple is the Al vaporization in BN crucibles. The crucible itself loses N_2 according to the reaction (IX). An AlN(s) coating is not formed, and the Aluminium reacts with B (s) to get borides (AlB_n), the structure of which are not compatible with BN (s). Then the crucible becomes brittle. The gaseous phase remains pure aluminium after a first outgasing of residual B_2O_3 at high temperature which is enhanced by Al_2O (g) departure.

Physico-chemical interactions may remain important at high temperature that can limit the use of some crucibles : creeping of liquids occurs at some kind of transition temperature, and intergranular penetration shortens their working duration. Basic studies of these phenomena which deal with interfacial energy determination are not directly useful for the choice of the crucibles since these studies must be performed with very well characterized surfaces to be significant.

Polyatomic or polymeric molecular beams impinging the growth surface very often present low sticking coefficients i.e. the probability for condensation is <1 and a part of the incident flow is immediatly evaporated or reflected. It may be more convenient to increase this sticking coefficient to provide the absorbed layer on its surface with impinging gaseous species, the stoichiometry and structure of which are close to the equilibrium absorbed species on this layer. For this reason As_4(g) and P_4(g) molecular flows are nowadays replaced by As_2(g) and P_2(g) flows obtained by cracking saturated vapors for the growth of III-V compounds. Molecular sources are then built like tandem Knudsen-cells, the lower one providing the saturated vapor, the upper one working at higher temperatures to crack the vapor (fig 6). In such sources, the decomposition is performed according to kinetics or thermodynamics depending on their conception. For the latter, the gas composition can be calculated by thermodynamics and the source working is not time-dependent. Feed gas-lines as Organo-Metallics can replace the lower cell.

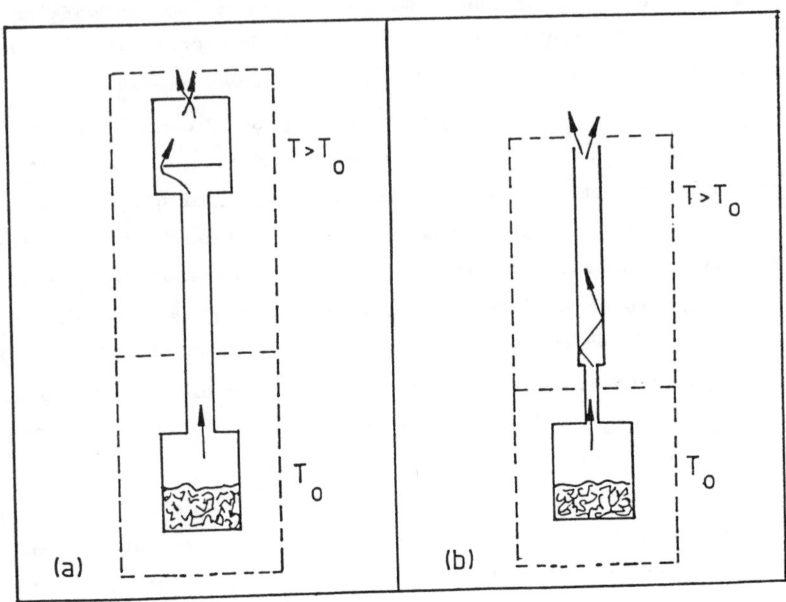

Fig.6 Molecular sources with cracked vapors, (a) cracking in quasi-equilibrium conditions, (b) cracking under kinetic conditions.

3 MOLECULAR FLOWS BETWEEN SOURCES AND SUBSTRATE

The composition of the growth layer may be *a priori* known if the impinging flows are also known. These flows exchanged between sources and substrates can be either calculated or measured. Whatever the way chosen by the experimenter, the difficulties to know accuratly these flows are the same as those encountered for the partial vapor pressure determinations in high temperature chemistry. Thus, we shall present the molecular flow behaviour on the basis of studies that have been performed on the Knudsen-effusion method.

3.1 Emitted molecular flows

The flow of molecules that is emitted from an elementary source surface ds towards an elementary substrate surface ds' is obtained (1) from the relation :

$$dn/dt = (nv/4\pi) . ds \cos \theta \, ds' \cos \theta'/l^2 \quad (4)$$

Integrating this relation over source and substrate surfaces theoretically leads to the real flow of impinging molecules. Talalaeff [24] and Curless [26] have done these calculations in the real context of III-V molecular beam epitaxy with open crucible sources. Figure 7 shows the respective contributions of molecules comming from the source surface and from the crucible walls as a function of the surface level of the source material. Morever, as the source level moves by continuous loss of material, we observe that the total impinging flow of molecules will decrease. Summing these contributions Talalaeff [24] concluded that the crucible apertures must have at least the same size as the

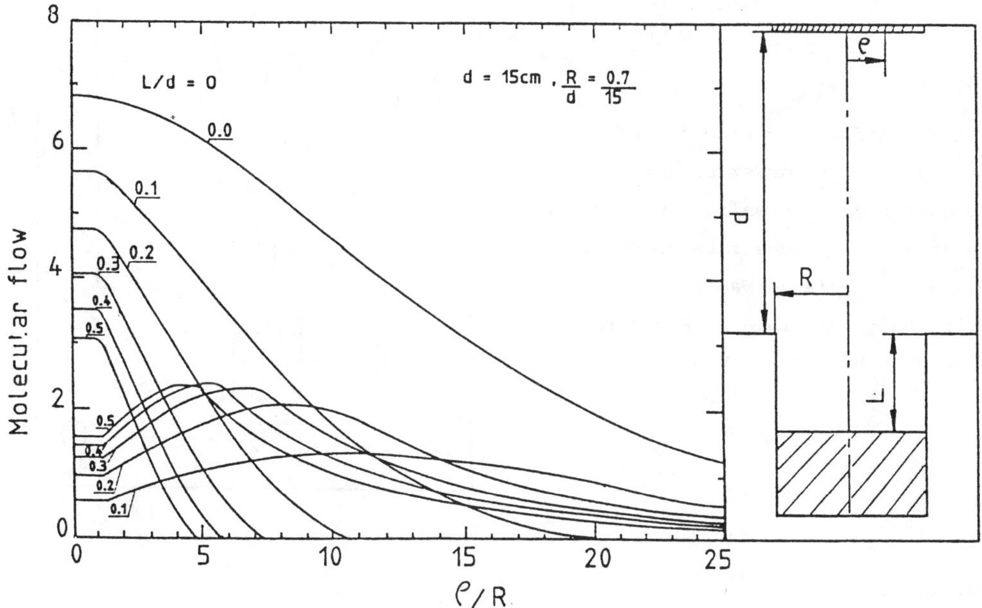

Fig. 7 Respective molecular flow contributions coming from the source surface and from the crucible walls according to ref [25].

substrate to grow homogeneous thickness layers. Curless [26] calculated from relation (4) the impinging molecular density on a 2x2 cm square substrate tilted by 33° from the crucible axis. He found a maximum variation of about 30 %.

Comparing calculated results and experimental observations the main deviations are coming - first from misaligments of the crucible axis and the substrate centre [24] - second from temperature gradients at the source surface. The first cause may be overcome by minimizing or discarding those of the molecules that come from the crucible walls, i.e. usually employing covers that transform open crucibles into Knudsen-effusion cells (fig 8). So, from the point of view of the substrate, the true surface for molecular emission is the lid aperture. The molecular flow is calculated from relation (4), for two coaxial

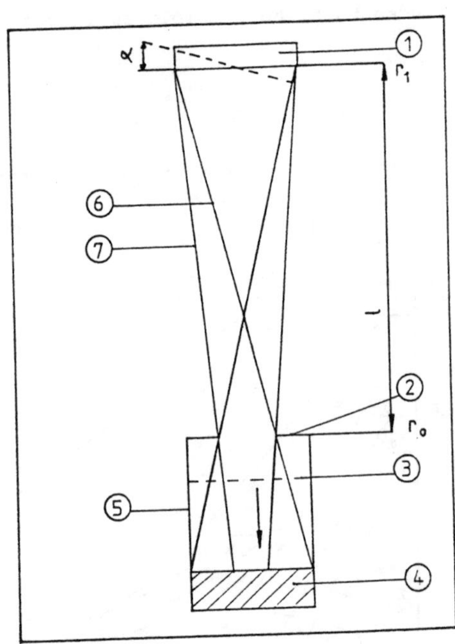

Fig.8 Effusion source fitted with a cover in order to discard molecular wall emission toward the substrate. 1. Substrate 2. Cover 3 and 4. Sample surfaces 5. Effusion crucible 6. Penumbra zone 7. Umbra zone.

disks, to be

$$F(r_0,r_1,l) = (nv\pi/8)[(l^2 + r_0^2 + r_1^2) - ((l^2 + r_0^2 + r_1^2)^2 - 4r_0^2r_1^2)^{1/2}] \quad (5)$$

r_0, r_1, and l are defined on figure 8. When the substrate is tilted by an angle α, the flow may be calculated approximatly as $F(r_0,r_1,l)\bullet\mathrm{Cos}\alpha$, provided that $\alpha<45°$, $l>10r_o$ and $r_1 \approx r_o$. The second cause is more difficult to overcome since as large molecular emission apertures as the substrate area [24] are necessary to obtain homogeneous growth thickness that is contrary to black cavity cell behaviour. In the case of free vaporizing surfaces as used with electron bombardment guns, the molecular emissivity depends strongly on the surface temperature according to the Clausius-Clapeyron relation (see further).

3.2 Physico-Chemistry of the molecular emission

Physico-chemical interactions between the source material and their containers have been studied in the frame of high temperature Knudsen-effusion to improve pressure determinations. Experimental studies of molecular distribution in beams were explained in relation with four phenomena

- the surface diffusion of absorbed molecules on the orifice (or crucible) walls towards the outer surfaces and competing with vaporization
- the specular reflection of razing incident molecules on the walls
- the anisotropy of the beam due to collisions at the source when the pressure increases
- the dissociation of complex gaseous species when colliding the walls.

The two first phenomena have been summarized and quantified by Ward [27] from the point of view of the angular distribution of molecules and the total emitted flow as compared to the cosine law of genuine effusion. The specular reflexion concerns mainly the razing incident molecules on the walls. It is not a well-known phenomenon, and the Ward's estimates show that its importance along the normal direction to the surface may increase by 10 % the molecular flow for as large L/R (defined on fig.9) as 10. This situation occurs only when deep crucibles are used during a very long time. The surface diffusion must be considered as systematically occuring at high temperature since absolutely inert systems are scarce and it is always an additional contribution that can be 20 to 30% of genuine effusion. Figure 9 shows the influence of the mean diffusion path $\bar{X}$ on the surface and of the ratio L/R on the molecular flow n compared to the genuine effusion flow n_{eq} along the normal direction. The real problem for estimating these contributions is the lack of mean diffusion path data.

The anisotropy of molecular flow [28] at emission is to be taken into account when the mean free path of molecules λ is close to the size of the source. The pertinent parameter is the Knudsen number, $K = \lambda/d$, d being the orifice source diameter. The emission law is corrected by an anisotropic factor as soon as K< 8. The proposed factor is reliable for 0.4 < K < 8 [29]. The mean free path is

$$\lambda = 1/\sqrt{2}\pi\delta^2 n \quad (6)$$

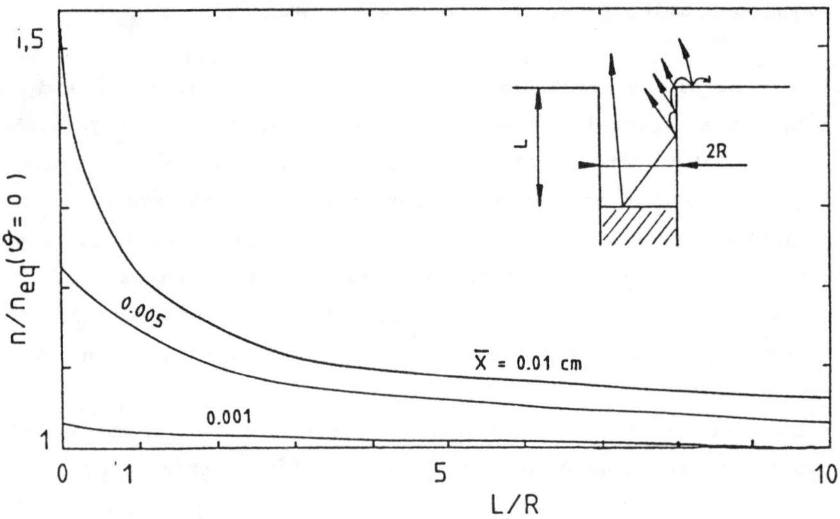

Fig.9 Total molecular flow n including surface diffusion along the walls and outer surfaces compared to genuine effusion flow n_{eq} as a function of the source level L and of the mean surface diffusion distance $\overline{X}$, according to ref [26].

where n is the number of molecules per unit volume and d their diameter. Applying the ideal gas law

$$\lambda = RT/\sqrt{2}\pi\delta^2 N P \quad (7)$$

N being the Avogadro's number. At 1500 K, for δ = 3 Å, the λ values are : λ(m) $\approx$ 0.052/P (Pa). The anisotropic factor does not apply for pressures lower than 1 Pa (10^{-5} Atm) for Knudsen-cell sources devices. For Electron bombardment gun sources the vaporization pressures are in the operation range of this coefficient, and then corrections are necessary to calculate the molecular emission. The differences in these two sources behaviour come mainly from their location relatively to the substrate : electron gun sources are located farther to favour homogeneous thickness, and consequently their vaporization temperature has to be higher.

The dissociation of complex gaseous species when colliding with the walls, has been calculated for dimers [30] and experimentally observed [31]. The resulting angular distribution law is different for each species, and the central part of the beam is slightly richer with the decomposition products than for equilibrium flow. This phenomenon is more pronounced when the vapour is largely unsaturated at the orifice of the source.

3.3 Technical limitations and their impact

Technical set up of sources and their environment (furnaces, shutters, traps...) leads often to perturbations in molecular beams. Thermal equilibrium at the sources is difficult to establish since the attended molecular flows originate from large apertures in Knudsen sources or from open surfaces in electron gun furnaces. At high flow rates, the evaporating surface may be also cooled by providing the necessary vaporization enthalpies. For open crucibles, deposits are often observed at the upper part of the walls, and for electron guns, the apparent homogeneous temperature zone is usually extended to a circle of a few millimeters. Under saturated conditions the emission pressure of a surface varies quickly with temperature according to Clausius-Clapeyron relation

$$d(\ln p)/d(1/T) = -\Delta H_{vap}/R$$

or

$$\delta P/P = (\Delta H_{vap}/R)\ (\delta T/T^2)$$

because the enthalpies of vaporization have often large values. For Ti at 2000 K, $\delta P/P \sim 0.06\ \delta T$. So, any temperature gradient at the source will have an influence on the growth layer's thickness. For open crucibles, it would be preferable to overheat their apertures, since with unsaturated conditions, $\delta P/P = R \cdot \delta T/T$ according to the law of ideal gases, which gives $\delta P/P \sim 0.001\ \delta T$ for any monoatomic gas at 2000 K. The last difficulty in the control of the source is the determination of the temperature of the evaporating surface that would lead to a reliable calculation of the molecular beam flow values. Some geometrical conceptions as true Knudsen-cells or use of two stage crackers allow a better control and knowledge of the temperature of the source.

As the molecules effuse in all directions, the non-useful beam may be reevaporated from hot parts of the devices like thermal shields, shutters, sample holders ... Such reevaporations which cause parasitic flows of molecules have been observed in Knudsen-cell mass spectrometry [32], and they are usually of the order of 10 to 30% of the useful beam, when vaporizing pure metals. In some cases, depending on the stability of gaseous molecules, the parasitic flows become richer with complex molecules, as observed for $Al_2O(g)$ which was coming directly from tantalum shields [32]. Careful sampling of the molecular beam as well as intelligent choice of furnace materials can minimize these contributions.

3.4 Improving molecular sources and beams

As it is not possible for each source - container - furnace group of materials to study and quantify any physico-chemical interactions, we propose to discard any parasitic contributions of molecules to the useful beam by a new architecture of sources and their beams. Up to now the position of the sources, the shutters and the substrates are optimized according to builders experience. We think that a more rationalized way is to improve the molecular beam devices in two directions:

- improve the thermal equilibrium at the sources by using true Knudsen-cells or modulating more widely the electron gun spots,
- use cooled field diaphragms to sample the molecular beams as done in mass spectrometry [33].

These principles are illustrated on figure 10. Application of these principles will set up the apparatus in the exact field of assumptions that are the basis of gas kinetic calculations of molecules exchanged between the source and the substrate.

4 THERMOCHEMICAL MODELLING OF GROWTH

The maximum deposition rate is determined by the equilibrium situation of the deposited material at the substrate, i.e. using thermodynamics of vaporization and condensation phenomena. The flow balance can be written as the net flow rate of condensation [5] according to the Hertz-Knudsen relation :

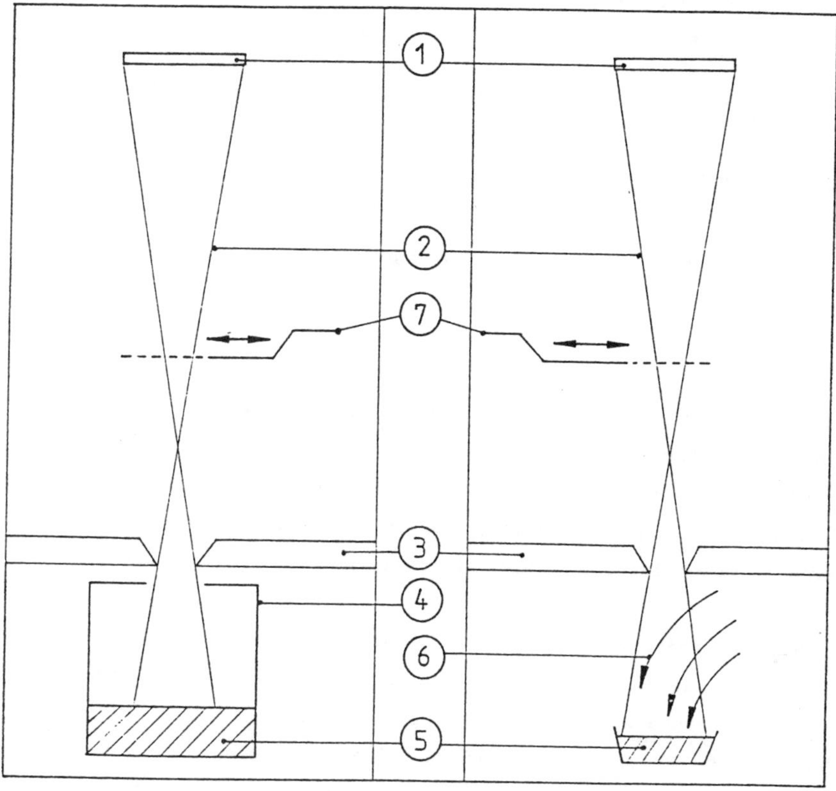

Fig.10 Principle of molecular beam sampling that discards parasitic contributions 1. Substrated 2. Penumbra zone 3. Cooled trap with field diaphragm 4. Knudsen-cell 5. Source samples 6. Electron beam

$$Fg = \alpha \, (p_i - p_{eq}) \; / \; (2 \, \pi \, M \, RT)1/2$$

where p_i is the incident pressure of molecules which are thermally accomodated [$p_i = p_i^\circ \, (T^\circ / T)^{1/2}$, p° and T° being related to the source of molecules], and p_{eq} is the equilibrium pressure of the evaporating molecules from the substrate at the growth surface and T the substrate temperature. The sticking coefficients account for the non-equilibrium situation: $\alpha<1$. As pointed out by Rosenblatt [5], it is preferable to use gross flow rates for each phenomenon, condensation and evaporation, since the condensation α^{cd} and evaporation α^{ev} coefficients are not usually equal, refering to different mechanisms when approaching the equilibrium situation from each side :

$$F_{cd} = \alpha^{cd}\, p_i \;/\; (2\pi\, M\, RT_i)^{1/2}$$

$$F_{ev} = \alpha^{ev}\, p_{eq} \;/\; (2\pi\, MRT)^{1/2}$$

The deposition or growth flow F_g rate results from the flow balance

$$F_g = F_{cd} - F_{ev}$$

The equilibrium pressures of each gaseous species at the growth surface are obtained using thermodynamics of bulk deposited material, and the α values are automatically fixed by the growth process in close relation to the necessary supersaturation, but they cannot be > 1. Thus different situations can be observed in molecular beam codeposition :

- The substrate temperature is low enough to impose the total condensation of all impinging flows and to prevent any appreciable partial vaporization pressure of gaseous species, an impinging one or a new one which may be created at the surface. Thus, the partial vaporization pressures are very low compared to growth rate, and the deposited material composition is entirely determined by the impinging flow composition. This kind of elaboration occurs for non-volatile metals with their impinging pure atomic vapors and also for some ceramics, the vaporization of which is usally congruent.

- The substrate temperature has to be high enough to increase the mobility of the absorbed species at the growth surface or to facilitate the crystal formation. Thus the partial vaporization pressures of some gaseous species, either the impinging one's or decomposition and recombination products, may be comparable to those in the incident beams. The resulting growth composition will be different from the one expected from the lone impinging flows and the experimenter have to balance the evaporation losses. This kind of set-up is typically the case for III-V and II-VI growth, for which thermodynamic calculations were applied [34-36] to control the crystal growth composition. Looking ahead, we can predict such behaviour with Organo-Metallic or Hydride sources since molecules with intermediate valencies, or low oxidizing degrees for cations in the gaseous phase, become stable at low pressures and low activities as imposed at the surface of the growth layer. The departure from equilibrium has been evaluated by *in-situ* determination of the α coefficients [37] during the GaAs growth at a surface transition observed by RHEED. For the future, we think that the thermodynamic analysis of the MBE will become an important tool for together control the growth rate and the composition. The main assumptions are presented and applied to the $Ga_y\ In_{1-y}\ As$ compound epitaxy in Shen's Thesis [38]. This thermodynamic analysis combines complex equilibrium calculations and flow balance to characterize the steady state of growth.

B THE USE OF RHEED FOR REAL TIME CONTROL OF MOLECULAR BEAM EPITAXY

1 INTRODUCTION

From the beginning of molecular beam epitaxy (MBE) of semiconductors, reflection high-energy electron diffraction (RHEED) has been of general use in controlling the growth process [39]. Actually, it has rapidly become the indispensable growth control tool even in MBE machines dedicated to production. Recently, the power of RHEED in controlling MBE has increased further owing to the discovery of the RHEED intensity oscillation phenomenon [40-42]. This is because it has clearly been shown that the RHEED oscillation period corresponds to the growth time for a monolayer of the growth material [41,42], at least for adequate growth and diffraction conditions [43]. This unambiguous correlation has been extensively used in various studies of both thermodynamic and kinetic aspects of the MBE growth mechanism. Belonging to this area of investigation are the studies dealing with group III element desorption [44-46] and diffusion [47,48] at the growth interface of .III-V compounds, or interaction kinetics of group V molecules with the surface [49]. In addition to these studies, RHEED oscillations have also largely been used to improve the growth of single or multi-quantum wells (QW) and superlattice (SL) semiconductor structures [50].

RHEED is now coming into general use, in conjonction with MBE, for the study of epitaxial growth of other classes of materials such as oxides or metals [51]. We review in the following the main information which can be easily extracted from the routine use of RHEED within the framework of MBE.

2 BASIC NOTIONS : A BRIEF SURVEY

2.1 Crystal structure

A crystal structure is constituted by the combination of a lattice and a basis. A lattice is a periodic arrangement of points in space related by

$$\mathbf{R} = m_1\,\mathbf{a}_1 + m_2\,\mathbf{a}_2 + m_3\,\mathbf{a}_3$$

where m_i are integers and $\mathbf{a}_i$ are the fundamental translation vectors of the crystal. The basis is formed by one or a group of atoms identically attached to each lattice point with positions defined by

$$\mathbf{r}_j = x_j\,\mathbf{a}_1 + y_j\,\mathbf{a}_2 + z_j\,\mathbf{a}_3$$

with j = 1,2... and x_i, y_i, z_i fractional numbers.

2.2 Reciprocal lattice

This is defined as the wave vector set **G** which gives plane waves with the lattice peridodicity such as

$$\exp\,(-\,i\,\mathbf{G}(\mathbf{r} + \mathbf{R})) = \exp\,(-\,i\,\mathbf{G}\,\mathbf{r})$$

$$\Longrightarrow \exp\,(-\,i\,\mathbf{G}\,\mathbf{R}) = 1$$

with **R** defined as before and

$$\mathbf{G} = h\,\mathbf{b}_1 + k\,\mathbf{b}_2 + l\,\mathbf{b}_3$$

with h, k, l integers then

$$\mathbf{b}_i = 2\Pi\,\frac{\mathbf{a}_j \wedge \mathbf{a}_k}{\mathbf{a}_l \cdot \mathbf{a}_j \wedge \mathbf{a}_k}$$

with $i \neq j \neq k$
$\mathbf{b}_1$, $\mathbf{b}_2$, $\mathbf{b}_3$ are the reciprocal lattice fundamental vectors.

2.3 Structure factor

The amplitude of the diffracted wave in the direction **G** is proportional to the geometric structure factor defined as

$$S_G = \sum_{j=1}^{p} f_j \exp(-i\,\mathbf{r}_j \,.\, \mathbf{G})$$

$$= \sum_{j=1}^{p} f_j \exp\left[-2\Pi i\,(x_j h + y_j k + z_j l)\right]$$

where p is the number of atoms in the basis, $\mathbf{r}_j = x_j\,\mathbf{a}_1 + y_j\,\mathbf{a}_2 + z_j\,\mathbf{a}_3$ is the position of the jth atom in the basis and f_j is the atomic form factor. (h k l) is the indexation of the plane on which the scattering occurs so

$$\mathbf{G} = h\,\mathbf{b}_1 + k\,\mathbf{b}_2 + l\,\mathbf{b}_3$$

This structure factor depends only on the matter distribution in the unit cell.

2.4 The kinematic model (single-scattering approximation)

The kinematic model, which corresponds to a single-scattering approximation, gives a satisfactory account of the main features of the diffraction patterns, such as the production of discrete scattered electron beam directions, and gives us the possibility to calculate the beam intensities. However it neglects the attenuation of the incident beam as it passes through the successive atomic layers, excitation of secondary beams and also interference effects between primary and other diffracted beams (for example diffraction + diffusion effects). Neglecting absorption effects, this theory implies that the amplitude of the first diffracted beams is very low compared to the incident beam, which is in fact the so-called "single-scattering approximation". This is far from being fully justified in the case of electron diffraction since interactions between electrons and the crystal are strong.

However, even in that case the kinematic model is sufficient to qualitatively understand the diffraction features.

According to this model, the diffracted wave from a plane electron wave interacting with a single atom is given by

$$\psi = \frac{\exp(-i\,\mathbf{k}\,\mathbf{r})}{r} \,.\, f$$

where f is the atomic form factor. For the diffraction by a crystal lattice we need to know the phase difference Φ arising when the electron beam is scattered by different atoms at positions defined by

$$\mathbf{R} = \sum_{i=1}^{3} (m_i + i_i)\,\mathbf{a}_i$$

with $\mathbf{a}_i$ unit vectors of the direct lattice, m_i integers and i_i fractional numbers. At the distance r from the crystal, the diffracted wave will be of the form :

$$\psi = \frac{\exp(-i\,\mathbf{k}\,\mathbf{r})}{r}\, f \exp(i\,\Phi)$$

To obtain Φ it is easier to use the reciprocal lattice. Then the path difference
$\mathbf{G} = \mathbf{k} - \mathbf{k}_i$ (with $|\,\mathbf{k}\,| = |\,\mathbf{k}_i\,|$) is

$$\mathbf{G} = \sum_{i=1}^{3} g_i\,\mathbf{b}_i$$

where $\mathbf{g}_i$ are scalars and $\mathbf{G}_i$ the reciprocal lattice vectors. Then

$$\Phi = -2\Pi\,\mathbf{G}\,\mathbf{R} = -2\Pi \sum_{i=1}^{3} (m_i + i_i)\,g_i$$

The diffracted wave is therefore

$$\psi = \frac{\exp(-\mathrm{i}\,\mathbf{k}\,\mathbf{r})}{\mathrm{r}} \sum_{\mathrm{i}} f_i \exp\left[-2\,\mathrm{i}\,\Pi\,(\mathrm{i}_1\,g_1 + \mathrm{i}_2\,g_2 + \mathrm{i}_3\,g_3)\right]$$

$$\times \sum_{\mathrm{m}} \exp\left[-2\,\mathrm{i}\,\Pi\,(\mathrm{m}_1\,g_1 + \mathrm{m}_2\,g_2 + \mathrm{m}_3\,g_3)\right]$$

where the first sum is the geometric structure factor while the second is the lattice factor. The diffracted intensity is then :

$$I = \psi\,.\,\psi^* = \frac{1}{r^2}\left[\sum_{\mathrm{i}}\sum_{\mathrm{i'}}\right]\left[\sum_{\mathrm{m}}\sum_{\mathrm{m'}} \exp\left[2\,\Pi\,(\mathrm{m}_1 - \mathrm{m'}_1)\,g_1 + \ldots\right]\right]$$

$\psi.\psi^*$ has a sharp maximums when all the g_i are integers i.e. at the reciprocal lattice nodes.

2.5 Ewald reflection sphere construction

The Ewald sphere construction gives the direction **k** and enables the calculation of the intensity $|\Phi|^2$ for all the diffracted beams produced for a given monokinetic incident beam of direction $\mathbf{k}_i$ and such as $\mathbf{G} = (\mathbf{k} - \mathbf{k}_i) / 2\Pi$ where **G** is a reciprocal lattice vector. To construct this sphere a vector of length $\lambda^{-1} = |\mathbf{k}_i| / 2\Pi$ is drawn to the origin O of the reciprocal lattice in

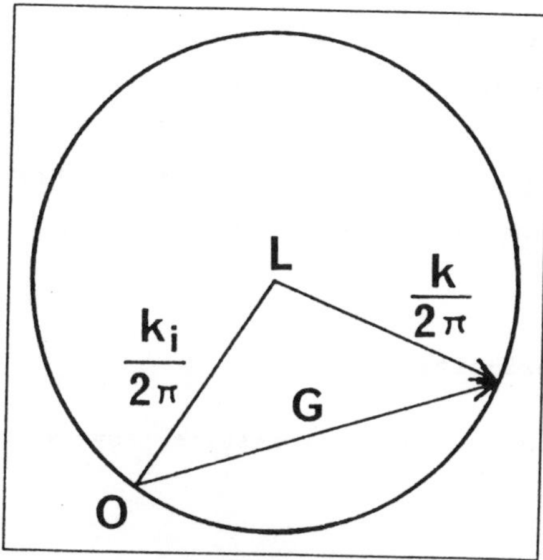

Fig.11 Ewald sphere construction.

the direction $\mathbf{k}_i$ from a point L (λ is the electron wavelength). A sphere of radius λ^{-1} is then drawn with L as center (fig.11).

Then, the diffraction pattern of a crystal in this model is constituted of a group of diffracted spots regularly situated at the reciprocal lattice nodes and restricted in the volume of the Ewald sphere. It should be noted that this construction is based on the kinematic approximation of the diffraction theory and thus the loss of energy of the primary beam in the diffraction process is neglected i.e. $|\mathbf{k}| = |\mathbf{k}_i|$.

2.6 Failures of the kinematic model

Effect other than single scattering events corresponding to the kinematic approach of diffraction (i.e. primary Bragg diffraction) are of importance in RHEED. This is for example shown in figure 12 (taken from [52]) where rocking curves for the 00 rod are presented for the GaAs (001) surface with different reconstructions. It is clear from this figure that the Bragg positions alone cannot explain all the features of the intensity variation measured when changing the primary beam incidence angle. These extra features are due to multiple diffraction processes including multiple excitations between allowed reflections and surface resonances (see the following paragraph).

Another important feature which is not accounted for by the kinematic model is the phenomenon of Kikuchi lines schematically explained in figure 13. The propagation in the crystal transforms the incident beam into divergent and quasimonokinetic beams. These beams can be then diffracted by a set of (h k l) planes at the Bragg angle Θ. They are generators of cones of semi-angle $\pi/2 - \Theta$ whose axes are the normal to the diffracting planes. Strictly, their intersection with the screen gives hyperbolas. In fact because the Bragg angle Θ is small, the hyperbolas appear as lines (D and D') parallel to the trace of the (h k l) plane family (T).

2.7 RHEED pattern characteristics

To understand the usual RHEED pattern characteristics, several particular features should now be discussed.

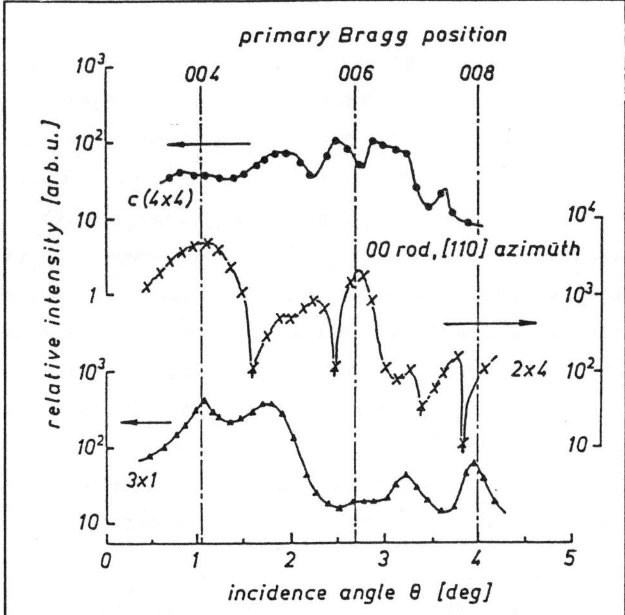

Fig. 12 Rocking curves for the 00 rod corresponding to three different reconstructions of GaAs (001) (from [52]).

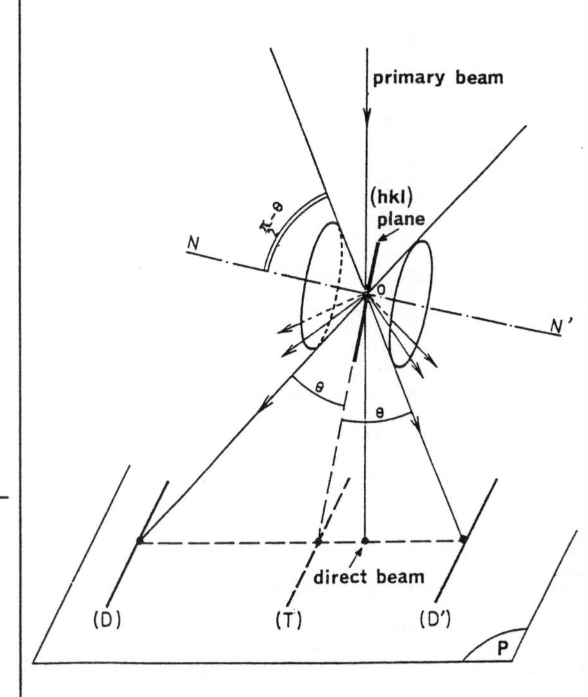

Fig. 13 Schematic interpretation of the formation of Kikuchi lines (Θ is the Bragg angle, (hkl) is the diffracting plane, NN' is normal to the (hkl) plane and P is the recording plane).

2.7.1 origin of the streaks

Due to grazing incidence, the diffraction of electrons occurs within a very thin surface layer of the crystal which can be idealized as a 2D object, at least if the surface sample is sufficiently flat. The reciprocal space is then formed by rods normal to the surface as indicated in figure 14 for the sphalerite lattice (from [53]). However in the real case, since the third dimension is not completely missing, we have to consider the superposition of the 3D reciprocal lattice on the 2D one as shown in figure 14. In fact, for a smooth surface, the contribution of 3D Bragg diffraction to the RHEED pattern is slight and results only in a modulation of the intensity along the streaks, which are the main features of the pattern. To explain the formation of these streaks, it should be noted first that both the reciprocal lattice rods and the Ewald sphere have finite thicknesses due respectively to lattice imperfections and to the electron energy spread and beam divergence. Since, in addition, the

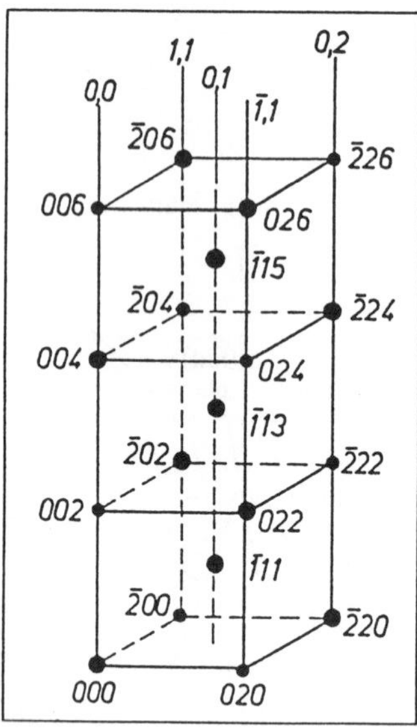

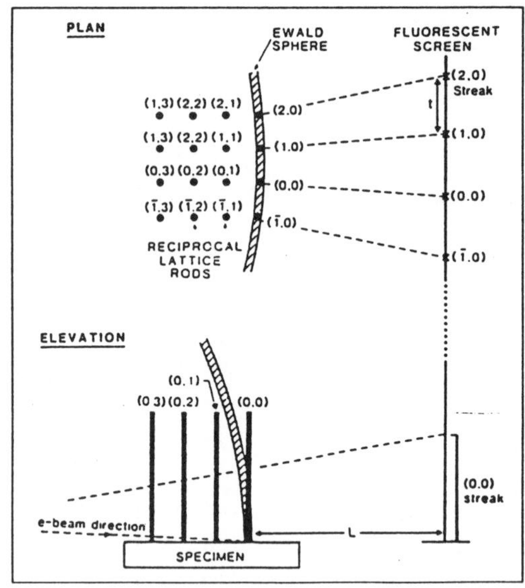

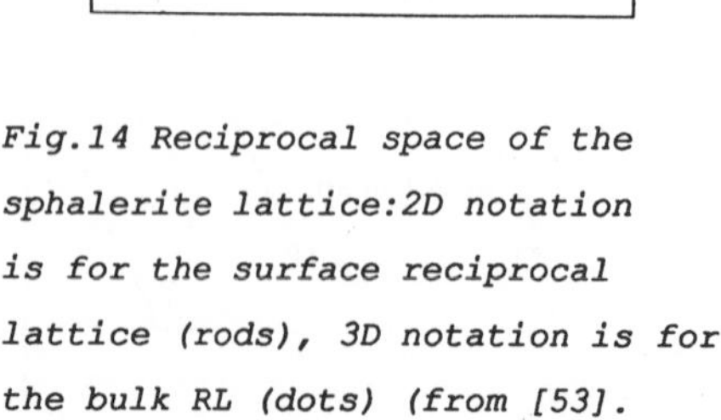
Fig.14 Reciprocal space of the sphalerite lattice:2D notation is for the surface reciprocal lattice (rods), 3D notation is for the bulk RL (dots) (from [53].

Fig.15 Schematic interpretation of the origin of elongated spots (or streaks) on the RHEED pattern (from [54]).

radius of the Ewald sphere is very much larger than the distance between rods, the RHEED pattern of a flat surface is constituted by streaks instead of by spots as indicated in figure 15 (from [54]). However, as it is obvious from the same figure, this is true only for the zero-order Laue zone. But it corresponds to the main visible part of the RHEED diagram.

2.7.2 surface reconstruction effects

In many cases, a rearrangement of the surface atoms occurs leading to a periodicity different but still correlated to the bulk one. This structural change is called "surface reconstruction" (see paragraph 3.2). If the surface periodicity is, for example, twice in one direction in the real space then the associated reciprocal lattice vector is only half the undisturbed one corresponding to the unreconstructed surface. This leads to the appearance of half order rods between each Laue zone (half-order Laue zone). As a consequence, the RHEED pattern will be formed by diffraction features associated with both the unreconstructed and reconstructed lattices, respectively indexed by integer and fractional numbers (in respect to the bulk lattice symmetry).

2.7.3 multiple scattering

As discussed earlier, multiple scattering is inherent to electron diffraction process. Although it does not change the geometric aspect of the RHEED pattern described above, it is important to consider when intensity measurements are performed. Important diffraction possibilities including multiple diffraction and resonance effects are illustrated in figure 16 (from [53]) by the intersection of the reciprocal lattice and the Ewald sphere for the zero-order Laue zone in the case of the GaAs (001) surface.

In this figure, $\mathbf{k}_\perp$ is the wave vector along the surface normal, $\mathbf{G}_\perp$ a reciprocal vector along this direction and $\mathbf{k}_{//}$ and $\mathbf{G}_{//}$ are similarly defined for the direction parallel to the surface. The smallest reciprocal lattice normal to the surface is $\mathbf{g}_\perp = 4\,\Pi / \mathbf{a}$ (a = lattice constant). $\mathbf{G}_{//}$ defines the lattice rod

$$\mathbf{G}_{//}^{pq} = \frac{2\sqrt{2}\,\Pi\,\sqrt{(p^2+q^2)}}{\mathbf{a}}$$

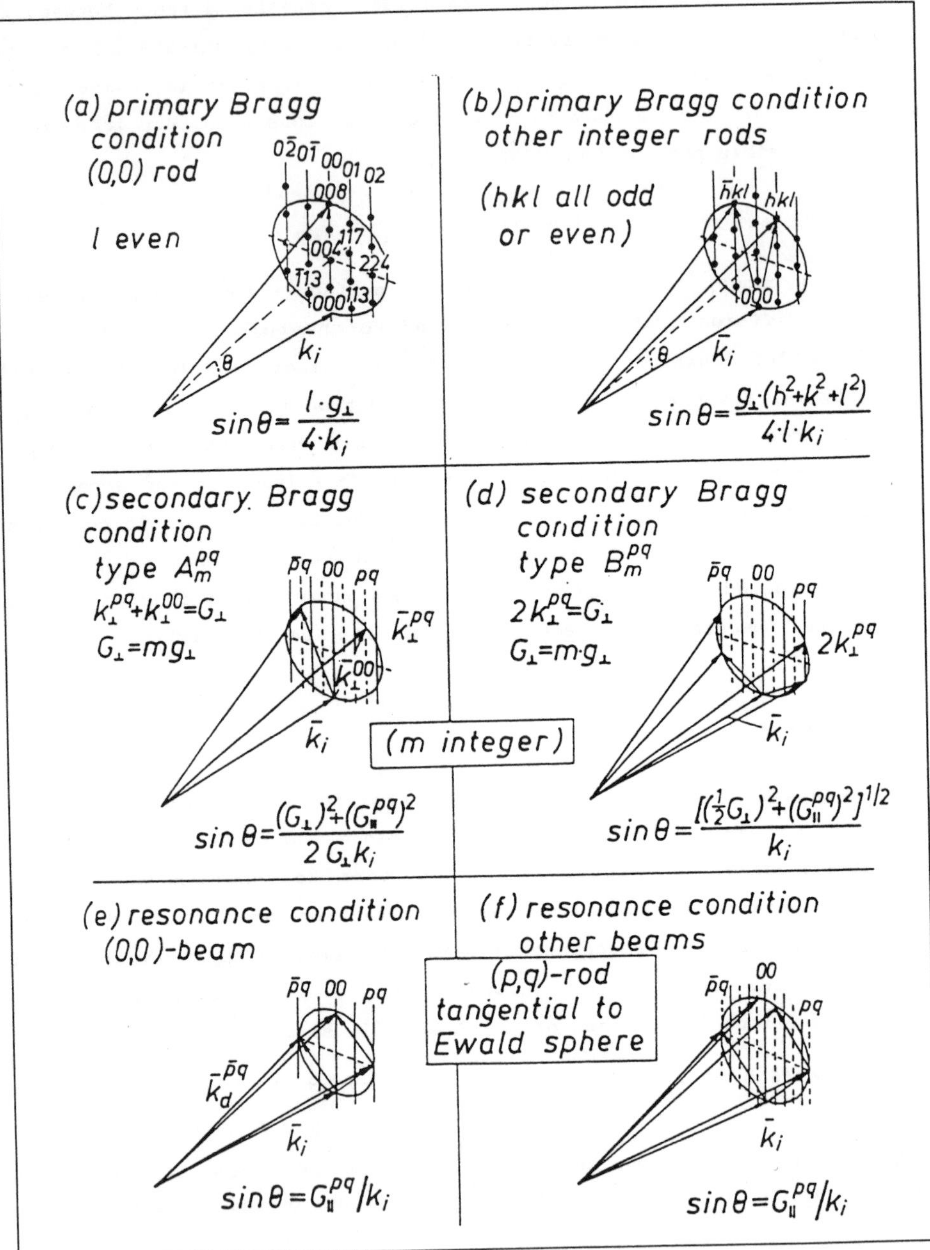

Fig. 16 Ewald sphere construction illustrating different possible diffraction conditions at the GaAs (001) surface : (a,b) primary Bragg reflections, (c,d) secondary Bragg diffraction conditions and (e,f) surface resonance conditions (from [53]).

indexed by p and q, where these are integers for an unreconstructed surface but also fractions for a reconstructed one (fractional order rods are indicated by dashed lines). Each diffraction condition is defined by the equation indicated in the corresponding panel : (a) and (b) are for h k l primary Bragg reflections, (c) and (d) for secondary Bragg reflections (with m referring to the momentum $\mathbf{G}_\perp$ given in units of $g_\perp$, (e) and (f) surface resonance conditions.

2.7.4 inner potential effect

In order to compare the positions of the expected diffraction features and the actual RHEED pattern, it is necessary to take into account the so-called inner potential effect. It is the consequence of the refraction of electrons when they cross the potential barrier of the surface and enter into the crystal. This effect changes the electron energy into the solid in a way related to V_0, the inner potential (≈ 10 - 15 V). The angle between the surface and the incident beam outside the crystal Θ_0, is related to the equivalent angle inside the crystal Θ by $\Theta = \Theta_0 + \varepsilon$ with $\varepsilon = V_0 / 2P \cot \Theta_0$, where P is the accelerating potential of the electrons. As a consequence of this refraction effect Θ is always greater than Θ_0. This affects the angular positions as well as the emergence conditions of the diffracted beams as discussed in detail in [53].

3 INFORMATION DEDUCED FROM THE RHEED PATTERN

3.1 Substrate preparation

Within the framework of molecular beam epitaxy, the first use of RHEED is to get qualitative information on the substrate surface preparation procedure used. Even if, for this particular purpose, RHEED observations cannot replace the use of powerful chemical surface analysis techniques such as Auger electron spectroscopy (AES) or X-ray photoelectron spectroscopy (XPS), simple basic information on the substrate surface can rapidly be deduced from the RHEED pattern. After chemical etching and handling in air before transfer into the UHV environment, the RHEED pattern will be formed by diffuse bulk type spots on a high background intensity if the amorphous or badly crystallized oxide layer resulting from the chemical etching process, or from the air exposure, is thinner than the material thickness involved in standard RHEED experiments (typically <≈ 50 Å). If this oxide layer, or any other extrinsic (intrinsic) poorly crystallized layer is thicker, the situation is different : a ring-like pattern is observed for polycrystalline thick films while only a homogeneous diffuse intensity results when the film is amorphous. Such an oxide layer can be eliminated by heating in UHV at a suitable temperature mainly depending on the substrate and oxide characteristics (from 200°C for GaAs covered by As_2O_3 up to 1800°C for oxide covered refractory metals such as W or Mo) or by the ion-bombardment and annealing procedure. Using the simple annealing method, a clear RHEED pattern with bright spots appears on the screen when the oxide is just leaving the surface. This sudden and spectacular change observed when the oxide evaporation temperature is attained is generally used in III-V semiconductor MBE growth to calibrate the thermocouple or IR-pyrometer measurements of the surface temperature of the sample (example 580-600°C for GaAs : Ga_2O_3 desoxidation).

When the substrate oxide layer is eliminated, the grazing incidence (~ 1-3°) RHEED pattern is generally constituted by spots, more or less elongated, but intermediate between a 2D diffraction pattern (reflection mode) and a 3D one (transmission mode). This is due to the roughness of the substrate surface at a microscopic scale, giving rise to a mixture of reflected and transmitted electron beams. The 3D like features can in general be easily eliminated by depositing a few monolayers of the substrate material with a suitable set of growth parameters in order to ensure a 2D growth mode (see paragraph 4-1). The RHEED pattern then becomes streaked, which is characteristic of a 2D diffraction process. If homoepitaxial growth is not possible, the growth of another material (heteroepitaxial growth), with no or sufficiently low lattice mismatch and capable of wetting the substrate surface in order to obtain a 2D growth mode, would produce the same effect, i.e. the smoothing of the surface substrate. The growth of such a so-called buffer (homo or heteroepitaxial) layer is generally pursued for several hundred (or even thousand) angstroms to obtain a smooth, clean and unstrained surface, far from any remaining effects of the substrate/epitaxial layer interface. At this stage, the intended active structure can be successfully grown. However, it should be emphasized that if the chemical etching before the oxidation of the substrate is insufficient, heating, for instance to evaporate the oxide layer, may produce a facetting of the surface which can be easily detected owing to the RHEED pattern observation. In that case elongated spots or even streaks in form of "seagull wings" should be visible due to diffraction from micro-facets forming an angle with the nominal surface. Such a situation may occur largely below the so-called roughening transition if any remaining adsorbate changes the free energy of the nominal surface in such a way that more stable planes develop. It is for example the case of carbon on a GaAs (001) surface [55,56]. To recover the morphology of the surface it is then necessary to grow a rather thick buffer layer, but with no firm guarantee of success.

3.2 Surface reconstruction

Because of bond breaking at the surface, the surface plane is not identical to a bulk one. Atomic positions at the surface can only be relaxed from their bulk positions without breaking the initial bulk symmetry of the plane, thus preserving the unit cell dimensions and symmetry. This case is generally observed for metals. For material with highly localized bonds a rehybridization occurs due to bond breaking at the surface: atomic displacements are then more important causing changes in the unit cell dimensions and, generally, the surface symmetry (fig.17).

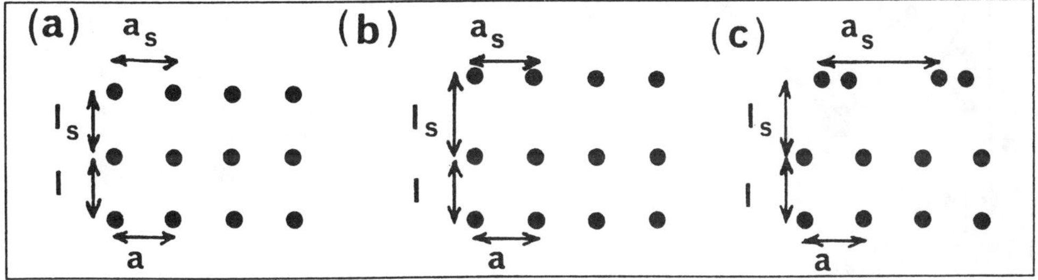

Fig.17 Schematic cross-sections of (a) an ideal surface (a_S = a, l_S = l), (b) a relaxed surface (a_S = a, $l_S \neq$ l), (c) a reconstructed surface ($a_S \neq$ a, $l_S \neq$ l).

This process of energy minimization gives rise to the surface reconstruction and is most often observed for covalent semiconductors such as those corresponding to IV and III-V elements (with the well known exception of the (110) cleavage face of III-V semiconductors). To indicate a surface reconstruction the Wood notation [57] is generally used. It is expressed as :

$$(S_1/B_1 \times S_2/B_2) \ \Theta$$

where S_1, S_2 and B_1, B_2 are respectively the modulus of the vectors defining the surface mesh M_S and the bulk one M_B and Θ is the angle between **S** and **B** vectors. When the chosen surface mesh is centered a prefix C is simply added to this expression. As an example, the schematic representation of the reciprocal lattice (RL) and the direct lattice (DL) corresponding to a C(2x8) reconstruction, normally observed for the GaAs (001) surface, is given in figure 18.

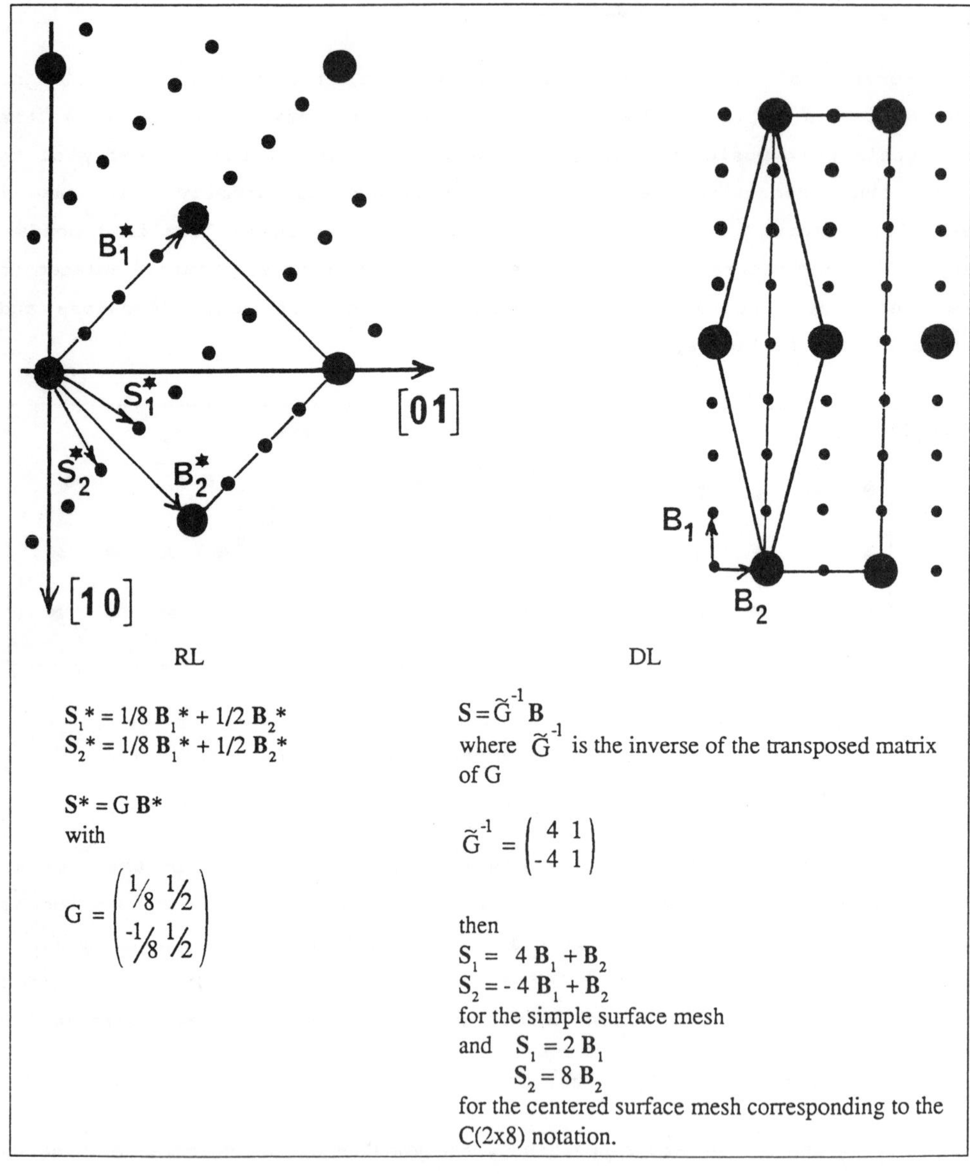

B1*
S1*
S2*
B2*
[01]
[10]
B1
B2
RL
DL
S1* = 1/8 B1* + 1/2 B2*
S2* = 1/8 B1* + 1/2 B2*
S* = G B*
with
G = (1/8 1/2 / -1/8 1/2)
S = G̃^-1 B
where G̃^-1 is the inverse of the transposed matrix of G
G̃^-1 = (4 1 / -4 1)
then
S1 = 4 B1 + B2
S2 = - 4 B1 + B2
for the simple surface mesh
and S1 = 2 B1
S2 = 8 B2
for the centered surface mesh corresponding to the C(2x8) notation.

Fig.18 Schematic representation of the reciprocal (RL) and direct (DL) latticesfor a c(2x8) reconstruction.

When using a surface sensitive diffraction technique such as RHEED, the effect of the surface reconstruction is to give rise to extra diffraction features (i.e. in addition to the expected bulk-type ones). As noted before, these extra spots are always of fractional order with respect to the bulk lattice since the reconstructed surface unit mesh is larger (i.e. smaller in the reciprocal space) than the ideal bulk derived one. The creation of a reconstructed surface is generally taken as a proof of the quality of the surface preparation procedure or to the growth itself. However, great care should be taken because the origin of the observed reconstruction may not be an intrinsic property of the surface. It could simply be a result of an ordered adsorption (segregation) of an impurity coming from the gas (solid) phase. In case of doubt, a surface sensitive chemical analysis technique, such as AES or XPS, should be performed *in situ* in order to distinguish between intrinsic and extrinsic origins of the reconstruction (note however that in some cases impurity traces below the detection level of AES or XPS may induce a surface reconstruction). Also, apparent extra spots can be simply due to the presence of an ordered step array on the surface. This may occur when the surface is slightly misoriented (vicinal surface). In that case, the surface can be considered as the convolution of the single terrace unit with the step lattice, and diffraction features occur when the product of the two corresponding reciprocal lattices is nonzero [58]. As a result, streak or spot splitting (= apparent extra spots) is observed in the RHEED pattern even if there is no surface reconstruction.

The determination of the reconstructed surface unit mesh using RHEED is quite easy provided that observations following different azimuths are possible. If the surface structure determination is now the aim, RHEED alone is generally not sufficient because of complications arising mainly from multiple diffraction effects in the RHEED intensity profiles. It is then necessary to use other diffraction methods such as grazing incidence X-ray diffraction (GIXD) [59] or direct space investigation techniques such as scanning tunneling microscopy [60].

3.3 Surface disorder

With a simple kinematic treatment based on the reciprocal lattice (RL) and Ewald sphere constructions it is possible to extract information on surface disorder from the RHEED streak shape [61,62]. As a first example of surface disorder effect on the RHEED pattern, figure 19 (from [62]) represents RL and Ewald sphere constructions for a polygonized surface corresponding to a mosaic structure which contains a large number of crystallites with a small angular spread. The reciprocal lattice then consists of cones and the diffraction pattern, defined as the intersection of the RL and the Ewald sphere, is in form of the shaded area of figure 19 (fan shapped streaks). Such a situation has been observed for example for an epitaxial Au (111) grown on a mosaic crystal of W (110) [62].

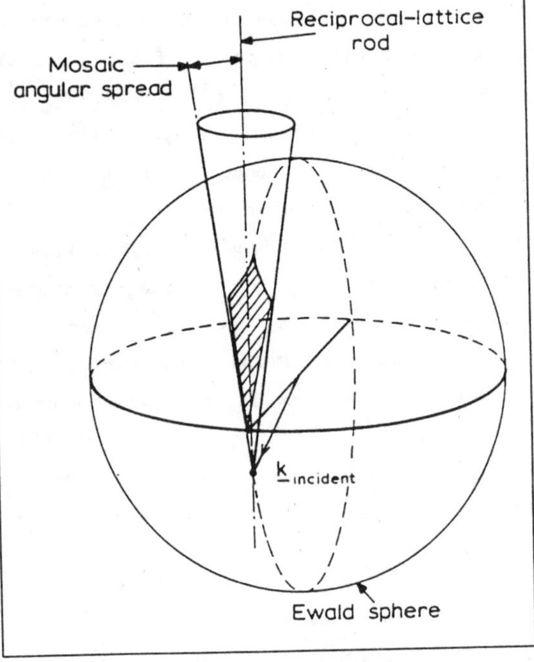

Fig.19 Reciprocal lattice and Ewald construction for a mosaic structure (from [62]).

As also shown in [62], 2D disorder effects can also be deduced from the streak shape. In that case, observations following different azimuths are necessary as indicated in figure 20 (from [62]). Under the grazing incidence diffraction condition used in RHEED, the reciprocal lattice is ideally made by rods normal to the surface due to the relaxation of the third Laue condition. The effect of a lack of perfect ordering in a particular direction will be to restrict the average size of the ordered regions. As a consequence the RL rods become 2D, forming ellipsoidal cylinders as schematically shown in figure 20. This figure corresponds to the case where domains are restricted in one direction while they become extensive in the perpendicular direction. When the electron beam is parallel to the short domain side, the streaks are long and narrow (fig. 20a), while short and broad streaks are seen when the beam is parallel to the long domain side (fig. 20b). This situation occurs for the 2x4 and C(2x8) reconstructions of GaAs (001). Details of the domain structure giving rise to such a disorder effect can be found in [63].

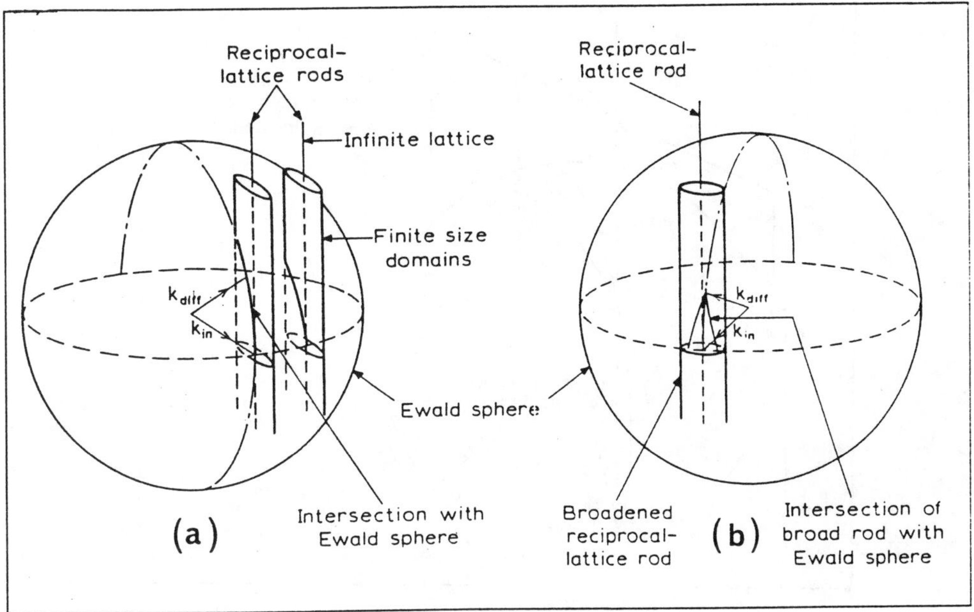

Fig.20 Ewald sphere and reciprocal lattice construction showing disorder effects for GaAs (001) surface : elongated and broad streaks are respectively formed in [110] (a) and [$\bar{1}$10] (b) azimuths (from [62]).

The existence of 2D surface disorder implies the presence of 1D boundaries separating the perfectly ordered domains. These antiphase boundaries are represented by planes in the reciprocal lattice as indicated in figure 21. If such a lattice plane is in the plane of incidence, long streaks normal to the shadow edge (i.e. to the surface) are seen on the RHEED pattern, while when there is an angle between them, curved streaks are produced. This remarkable example of a surface disorder effect has been observed and discussed in detail by Joyce et al [62] in the case of the 2x4 and C(2x8) reconstructions of GaAs (001) and by Delescluse and Masson for S adsorbed on Ni (111) [61].

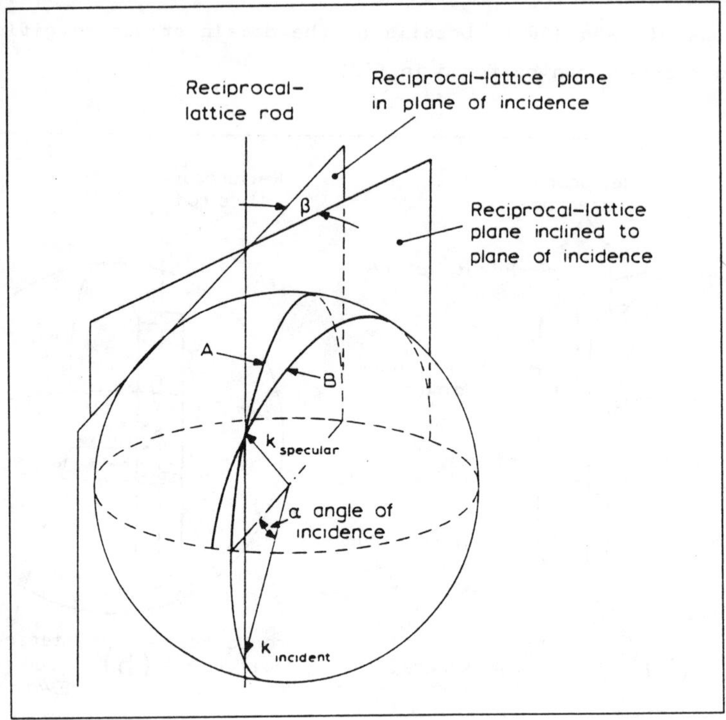

Fig. 21 Formation of curved streaks by surface disorder : A and B define the streaks due to the intersection of the Ewald sphere by a disorder related reciprocal plane in the plane of incidence and inclined at an angle β respectively (from [62]).

3.4 Epitaxial relationships

One of the basic application of RHEED when studying heteroepitaxial growth is the determination of the epitaxial relationships between the substrate and the deposit. When different azimuthal observations are possible, this determination is straightforward by simply comparing the substrate bulk type unit mesh and that for the deposit. In some particular cases where the surface of the deposit is sufficiently rough to observe a transmission diffraction-like pattern, the observation of the RHEED pattern along a single azimuth can be sufficient to determine the epitaxial relationship. However, in the general case corresponding to streaked patterns where 3D Bragg diffractions give rise only to a slight intensity modulation along the streaks, it is necessary to use several azimuths, as for the determination of the surface reconstruction. When the growth is commensurate (i.e. the deposit lattice is strained by the substrate) a perfect superposition of the two lattices is observed. If the lattice parameter difference between the two materials is too large (or if plastic relaxation of the elastic strain occurs beyond a critical thickness) the growth is incommensurate and the deposit takes its expected bulk lattice parameters.

Although epitaxial relationship determination is usually an easy task, care should be taken for the following reasons :

a) the epitaxial relationship is not always the one expected from simple lattice matching considerations. As an example, in the case of growth of Ag on GaAs(001), the observed relationship is [64-66]

Ag (001) [100] // GaAs (001) [100] (Δ a/a ≈ 30 %)

instead of

Ag (001) [100] // GaAs (001) [110] (Δ a/a ≈ 2 %)

expected on the basis of the lattice match criterion (this latter relationship is however observed in the similar case of the growth of Al on GaAs (001) [67]) ;

b) different epitaxial relationships may coexist giving rise to a RHEED diagram which, depending on the domain size, is a result of the superposition of the two orientations or changes when moving the electron beam over the sample surface

c) in some cases the observed orientation of the deposit may correspond to an unstable (or even unknown) bulk phase [68,69]

d) in spite of the similarity of both the structure and the parameters of two materials they can take very different orientations on a given substrate even in the same growth conditions. This is, for example, the case of Fe and Cr grown on GaAs (001) [70] : the observed epitaxial relationships are

Fe (001) [100] // GaAs (001) [100]

Cr (112) [110] // GaAs (001) [110].

4 INFORMATION DEDUCED FROM THE RHEED INTENSITY BEHAVIOUR DURING GROWTH

4.1 Growth mode

In order to obtain nearly perfect single crystal epitaxial layers, the mode of growth of the layer on the substrate should be of the Frank van der Merwe type, i.e. two-dimensional layer-by-layer growth mode. This is only easy for homoepitaxial growth. In the general case of heteroepitaxy, the classical criterion of the occurence of the 2D or 3D growth mode is based on the comparison of the surface free energy of the substrate (γ_S) and of the deposit (γ_D). Taking into account the interface free energy γ_{SD} between the two materials, if

$$\gamma_S > \gamma_D + \gamma_{SD}$$

the deposit material wets the substrate and a 2D growth may occur. Alternatively, a 3D (or Volmer-Weber) growth mode is predicted for the opposite sign of this inequality, i.e. the deposit will not wet the substrate. An intermediate situation referred as the Stranski-Krastanov growth mode is obtained when there is wetting of the substrate but the strain imposes the formation of 3D islands on top of a first 2D elastically strained layer. However, even for these 3D growth modes, when the growth is pursued until the substrate interface is sufficiently distant to obtain coalescence of the islands and a complete relaxation of the strain, a 2D growth mode can be achieved. In other words, we come back to the homoepitaxial case, with however some important limitations exemplified in the well known case of the growth of GaAs on Si [71]. Also in order to overcome the occurence of 3D growth, different growth procedures have been proposed such as the growth of intermediate layers or the use of surfactants on the substrate [72].

Considering now the study of the growth mode, RHEED has the unique capability to be used during the growth, giving real time information. As noted before, the occurence of a 2D or 3D growth mode can simply be detected by the general aspect of the RHEED pattern (streaked or spotty). Moreover, the behaviour of the RHEED intensity during the growth has been demonstrated to be closely related to the growth mode [41,42]. In particular, when the growth proceeds according to a 2D mode, an oscillation of the RHEED intensity is observed, as shown in figure 22 which corresponds to the growth of Ag on GaAs (001) (far from the interface however) [73].

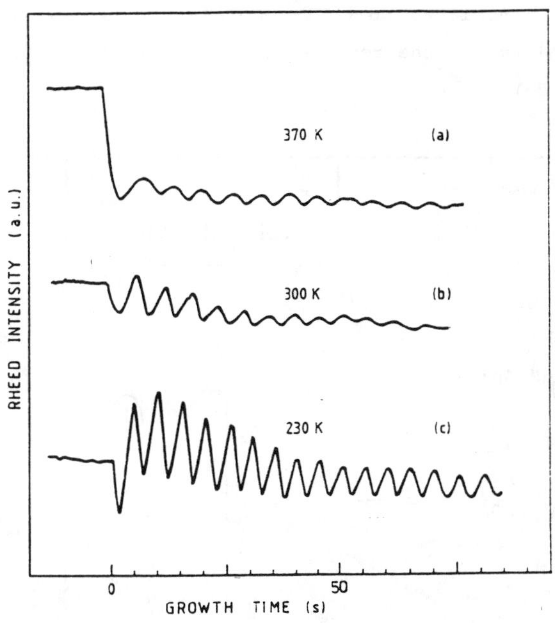

Fig. 22 RHEED intensityoscillations for Ag grown at different temperatures (a, b, c) on GaAs (001) (from [73]).

So far the detailed study of the RHEED intensity oscillation characteristics has been mainly done for the GaAs and Si homoepitaxial growth. However RHEED intensity oscillations are now observed for the growth of various materials including all the III-V semiconductors but also some of the II-VI [74], IV-VI [75], and recently, metals [51,76-81]. The oscillatory nature of the RHEED intensity during the growth is interpreted in terms of surface reflectivity which is obviously maximum for a complete monolayer formation (i.e. minimum of the step density) and minimum for half-monolayer coverage (maximum of the step density). Since most of the coherent scattering is concentrated into the specular beam, its intensity is in turn very sensitive to the variation of the diffuse incoherent or inelastic scattering associated with the sequential change of the step density (or mean terrace size) corresponding to a layer-by-layer growth. For this reason, and also because, as a corollary of the above statement, it is the most intense RHEED pattern feature, intensity oscillations are generally measured on the specular beam spot. If the primary beam intensity is kept constant, the sum of elastic and diffuse scattering intensities is also constant and therefore when the diffuse scattering is maximum, i.e. for the maximum step density associated to half-monolayer coverage, the specular beam intensity will be minimum and the reverse will be observed for integer monolayer deposition (see fig. 23).

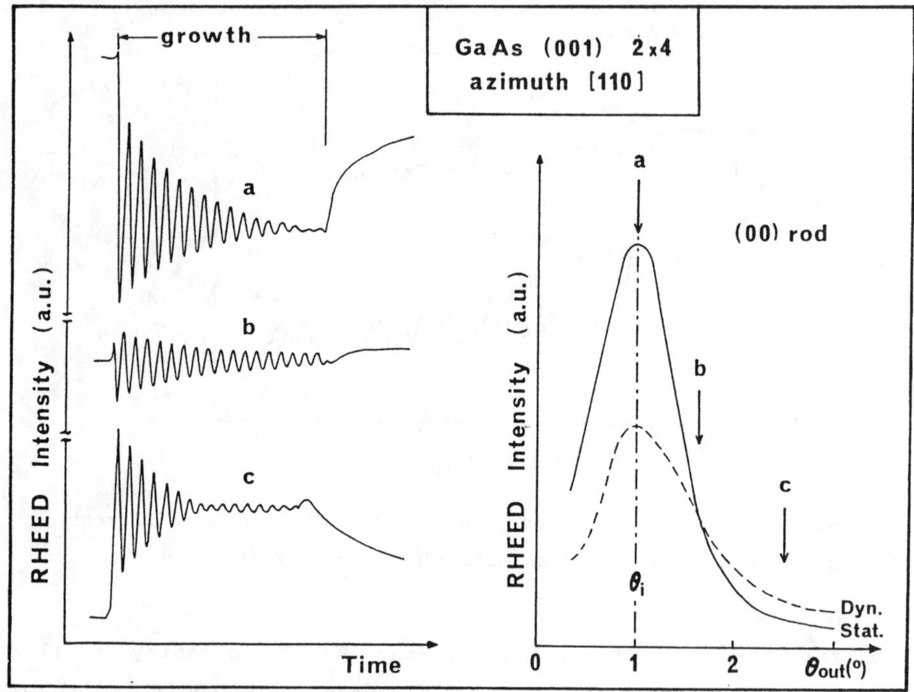

Fig.23 RHEED intensity oscillations corresponding to different recording positions (a,b,c) along the 00 rod for a GaAs (001) 2x4 surface. The specular beam is defined by Θ_i in the right panel. In this panel the full line corresponds to the static intensity variation along the 00 rod (no growth), while the dotted line is for the dynamic one (under growth).

Even for nearly perfect growth situations such as the one occuring for GaAs and $Al_xGa_{1-x}As$ compounds, a damping of the specular beam oscillation is generally observed at the growth beginning (fig. 23).The first interpretation of this phenomenon [41] invokes a progressive transition from 2-D ideal growth mode to a 3-D like one. However, such a 3-D like growth mode would be incompatible with chemical and structural interface sharpness (at most within 2 monolayers) demonstrated in MBE growth of quantum-well and superlattice semiconductor structures. Also, this model should degenerate, i.e yield an ever-increasing growth front roughness with continued growth which would result, in all cases, in a continuously decreasing RHEED average intensity, in contrast to experimental data. It is therefore suggested that the observed damping is mainly a transient effect arising from the fact that the surface evolves from the static pseudo-equilibrium condition (no growth) to the dynamic ones (i.e. under growth). This transient corresponds to the time needed for the mean terrace sizes (or step densities) associated with complete and half-complete monolayer coverages to reach their equilibrium values. At this stage a faint RHEED oscillation is still visible because even if the mean terrace sizes for complete and half-complete monolayers are closer than in the beginning they are still different.

The important point to emphasize is that the RHEED intensity oscillations arise from 2D nucleation on a smooth surface. Since steps are always present even on a smooth surface, this implies that the average terrace width l corresponding to the smooth initial surface is very large compared to the surface diffusion length L of the less mobile atomic species controlling the growth mechanism (for example, Ga in GaAs MBE) (fig. 24a). If now the growth temperature T_S is increased, the diffusion coefficient

$$D = D_O \exp(- E_D / k T_S)$$

(where E_D is the activation energy of the surface diffusion for the atom considered) increases and for a critical temperature T_C, L will be equal to l and therefore the adatoms reach the step edges bounding the terraces (fig. 24b). At this temperature, the RHEED intensity oscillation vanishes because the growth is changing from the 2D nucleation mode to the step propagation (or step "flow") mode where the roughness is no longer a function of the growth development. The same effect, i.e. a transition from 2D nucleation to step flow growth mode, is obtained if, for a given set of growth conditions, the average terrace width is decreased so that $l \leq L$. This is possible by using intentional misoriented surfaces (vicinal surfaces). Such vicinal surfaces can be used to evaluate the diffusion coefficient D through the measure of the temperature for which RHEED oscillations disappear, equated to T_C [82]. Defining N_S/J_I, where N_S is the number of surface sites and J_I the incident flux, as the average time interval between the arrival of atoms at a specific site and a sticking coefficient of unity, and assuming that the terrace width L is equivalent to the mean atomic displacement distance, we can obtain D from a simple model of surface diffusion based on the Einstein relation

$$D = (L^2 /2) (J_I /N_S)$$

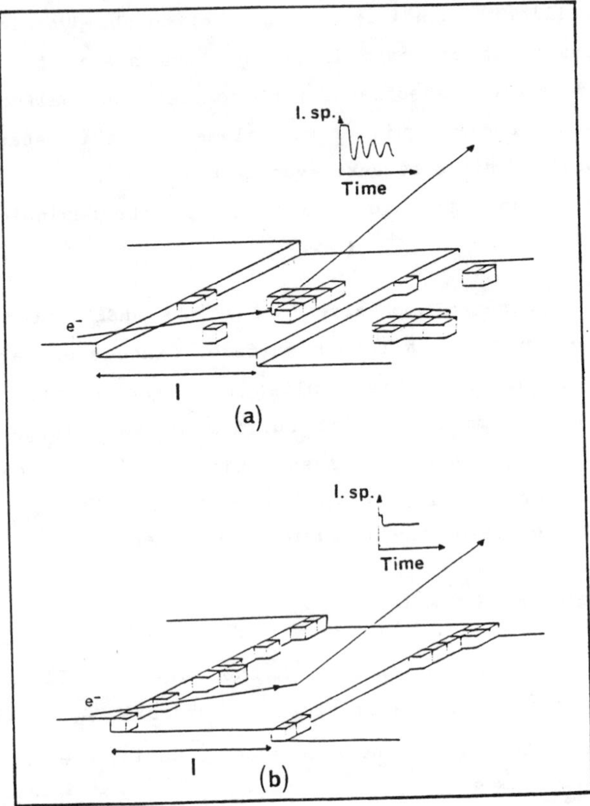

Fig. 24 Schematic illustration of (a) the 2D nucleation growth mode and (b) the step flow mode. RHEED oscillations are expected only in (a).

Measuring T_C for different incident fluxes J_I, an Arrhenius plot of log D versus 1/T gives the activation energy of surface diffusion E_D. Implicitly, this model assumes no anisotropy. This is not always the case, as for example for the GaAs (001) surface where Ga and As dangling bonds are in perpendicular <110> planes. Step edges occupied by As atoms or Ga atoms are then respectively oriented parallel to [110] or to [110] directions. As shown in figure 25, when using misoriented surfaces such as [110] or [$\bar{1}$10] steps are formed, T_C differs significantly, indicating that $D_{[110]}$ is greater than $D_{[110]}$.

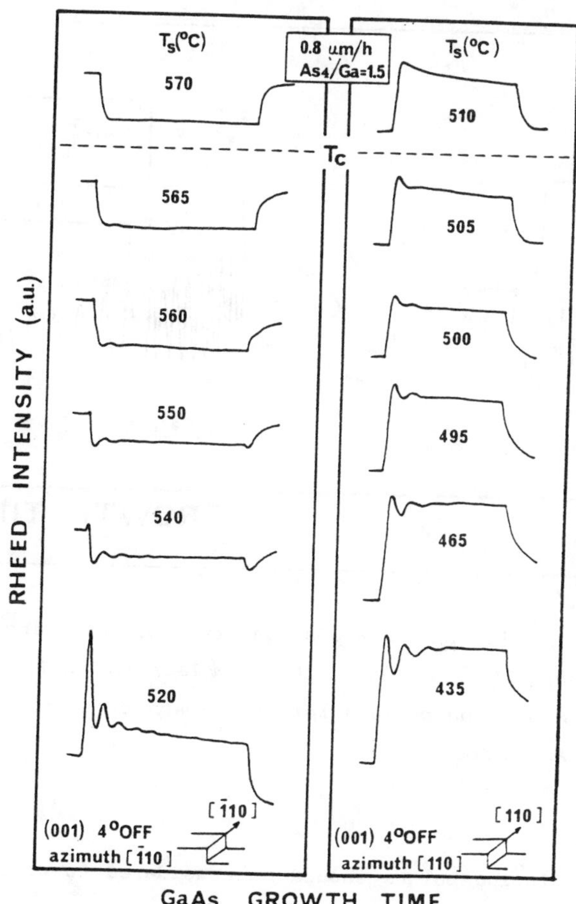

Fig. 25 RHEED intensity behaviour as a function of growth temperature for GaAs grown on misoriented substrates. T_C is the temperature for the transition between 2D and step flow growth modes.

4.2 Growth rate

Since the RHEED intensity oscillation period corresponds to the time needed for the growth of a monolayer, its measurement provides a continuous growth rate monitor. Moreover, alloy composition can also be determined by simply comparing changes in growth periods, as indicated in figure 26 for the (Ga,Al)As system (in this figure X is the Al mole fraction) [83].

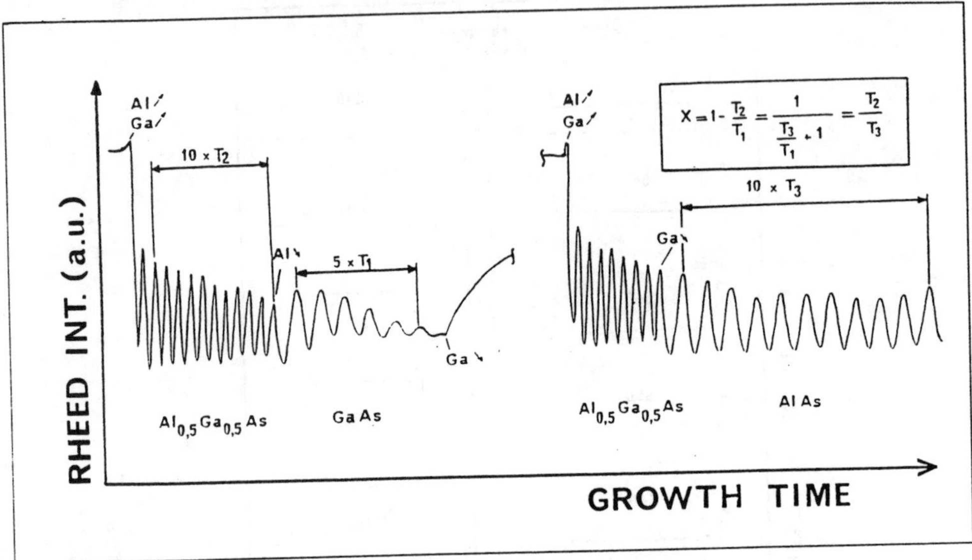

Fig. 26 Determination of $Al_xGa_{1-x}As$ composition using RHEED intensity oscillations (x is the Al mole fraction T_1, T_2, T_3 are the oscillation periods for the growth of GaAs, $Al_xGa_{1-x}As$ and AlAs respectively).

However the following points should be considered to correctly interpret the observed oscillation period in all cases :

a) for the growth of binary compounds such as GaAs the growth period corresponds to a "molecular" monolayer (Ga plane + As plane)

b) depending on the structural characteristics of the material both mono and biatomic layer growth mode can be observed. This is, for example, the case for Si [84]

c) when the lattice of the deposit is strongly strained by the substrate, a change in the RHEED intensity oscillation period can be observed when plastic relaxation (by dislocations) of the strain occurs. If the growth is still 2D, which is however unusual, the new period would reflect the change in the density of surface sites associated with the parameter relaxation. Such a behaviour has been recently observed in the growth of CdTe on ZnTe [85]

d) last but not least, under diffraction conditions for which elastic specular and diffuse scattering are both important (as in b of figure 23), a frequency doubling may appear (fig. 27, from [43]). This is interpreted as a superposition effect of the elastic and diffuse scattering intensities. Since diffuse scattering is increasing with step density (as the disorder increases) while the elastic specular intensity is reduced, their respective intensity oscillations are in opposite phase and their summation can give rise to harmonics as schematically indicated in figure 28 (from [43]).

From the practical point of view, the important thing to realize is that RHEED intensity oscillations are strongly dependent on the diffraction conditions, which hinders a simple straightforward interpretation of all of their features in terms of crystal growth. As a simple recommendation, RHEED intensity oscillations should be recorded for a position of the specular beam (fixed by the incidence angle which should in any case remain very low) far from other diffraction features, in particular those of 3D origin such as Bragg reflections or Kikuchi line crossing.

4.3 Growth mechanisms

So far thermodynamic and kinetic aspects of the MBE growth mechanisms have been mainly studied for III-V and II-VI semiconductors. The study of these mechanisms using RHEED intensity oscillations is based on the growth period variation when changing the growth thermodynamic parameters (e.g. substrate temperature, growth rate, flux ratio of the incoming species,...). As thermodynamics in this sense applied mainly in conditions where desorption is no more negligible (i.e. non unity sticking coefficients) the detailed discussion of such studies is not relevant to the MBE of metals which is generally performed at low temperature [79]. Therefore we will not pursue it further here. The interested reader is referred to references [36], [44-46], [49], [74], and [86].

4.4 Multilayer sequence determination

One of the main interests of RHEED intensity oscillations is to allow an accurate control of the sequence parameters of multilayer structures. In order

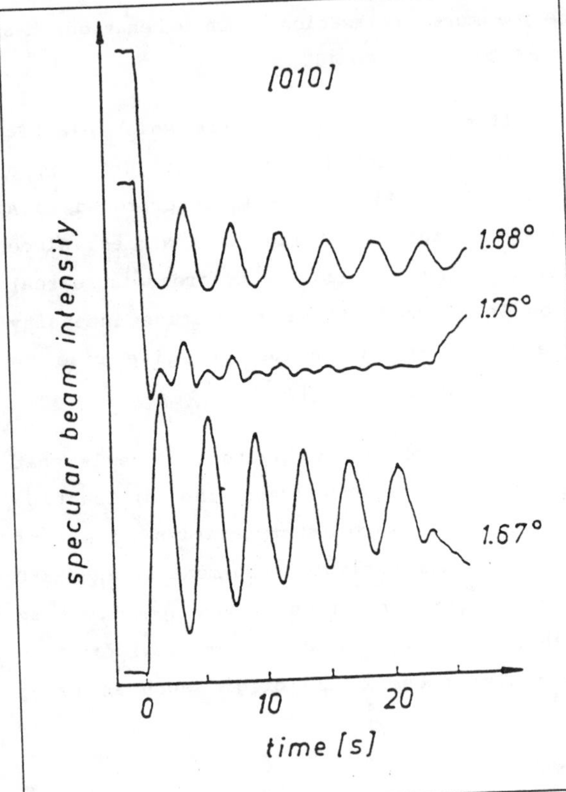

Fig. 27 Observation of harmonics in specular beam intensity oscillations for a particular incidence angle (1.76°) on a GaAs (001) 2x4 surface ([010] azimuth) (from [43]).

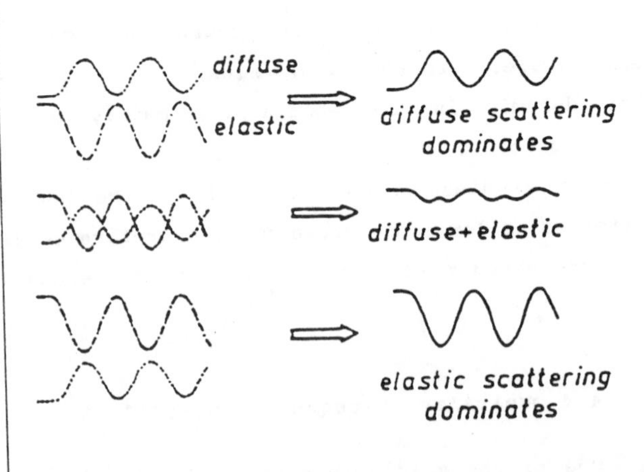

Fig. 28 Schematic interpretation of the origin of harmonics in RHEED intensity oscillations (from [43]).

to grow a superlattice structure, we can first try to precisely adjust the time needed for the growth of the desired number of monolayers for the different materials involved, by simply measuring their growth periods. Then, the continuous recording of RHEED oscillations will provide a real time control of the structure parameters and, in particular, the flux stability. This is illustrated in figure 29 for the growth of $Al_{0.3}Ga_{0.7}As$ / GaAs superlattice of 160 periods (ª 1 mm). Each superlattice period would be ideally formed by 11 monolayers (ML) of $Al_{0.3}Ga_{0.7}As$ (the barriers L_B) and 11 ML of GaAs (the wells L_W). It is worth noting here that growth interruptions, of time given in the figure, have been performed at each interface, as is now usual in the MBE growth

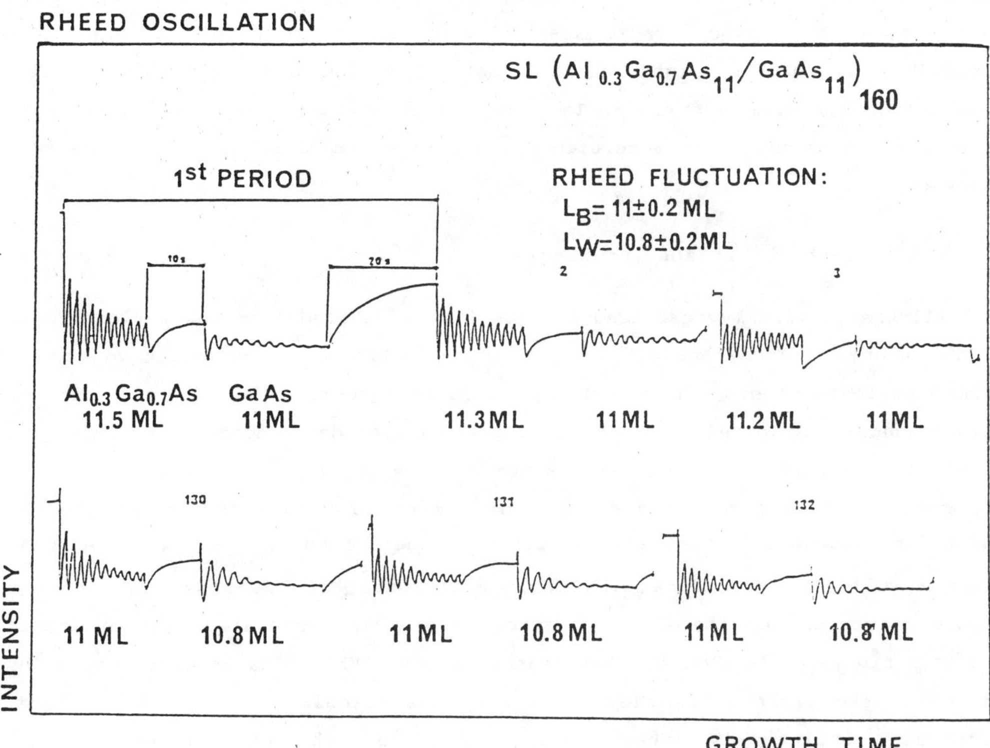

Fig. 29 RHEED intensity oscillations during the growth of a $Al_{0.3}Ga_{0.7}As$/GaAs superlattice. Growth interruptions of 10s and 20s were performed after AlGaAs and GaAs growth respectively. Note that oscillations are stronger for AlGaAs than for GaAs due to the lower surface mobility of Al atoms (i.e. stronger nucleation). For this superlattice, X-ray diffraction gives $L_B + L_W = 21.5$ and $L_B = L_W$.

of such structures [87]. This procedure is based on the recovery of the RHEED intensity when the growth is interrupted, mainly associated with a smoothing of the surface (i.e. an increase of the mean terrace size). The important point to note here is the fluctuation of the number of monolayers deposited for each material, at the beginning of the growth, indicating that the fluxes are not stable. After these very first superlattice periods, the growth rates become stable. However, the growth of GaAs corresponds to a non-integer number of monolayers (it is only fortuitous that it is not the case for AlGaAs). In principle, this can be avoided by coupling the computer controlling the evaporation cell shutter operation to the oscillation phase. Such a procedure has been called "phase-locked" epitaxy [88]. However, a word of caution is needed about this procedure because of the very local nature of the RHEED information and of the slight but noticeable variation of the growth rate which exists over the sample surface. In other words, even if an integer number of monolayers is deposited in a certain place of the sample it may not be the case elsewhere.

4.5 Growth interface roughness

All the previously described information is deduced from the period of the RHEED intensity oscillation. However, since RHEED is very sensitive to the atomic scale roughness, in principle, it could also serve as an evaluation of this roughness. A correlation, at least qualitative, between the RHEED intensity level and the interface roughness is demonstrated by the experiment whose data are reported in figure 30. The upper part of this figure corresponds to the RHEED intensity oscillation recording during the growth of a $Al_{0.3}Ga_{0.7}As/GaAs$ quantum well (QW) structure with three different QWs. The lower part of the figure correspond to the 2K photoluminescence (PL) associated with excitonic recombinations in the QWs. The key feature of the RHEED data is that during the growth of the first thick AlGaAs barrier, the specular beam intensity level decreases continuously after roughly 10 ML growth, this decrease being associated with a temporary disfunction of the liquid-nitrogen cooling of the growth chamber cryo-panels. Although the growth interruption after growing this first layer has practically no effect on the RHEED intensity level, the growth of the first GaAs QW (28 ML) seems sufficient to smooth the growth interface. Then, the usual growth interruption enables the RHEED intensity level to be significantly improved. After this step, the RHEED intensity behaviour is similar to the one usually observed in such a structure. If now the PL spectrum of the grown structure is considered, the crucial point is that the emission from the QW (28 ML), grown just after the perturbed $Al_{0.3}Ga_{0.7}As$ layer, has collapsed. A 10 times increase in the detection sensibility, with respect ro the subsequently grown wells, is necessary to observe its luminescence. Moreover, the PL linewidth (≈ 17 meV) is considerably enlarged compared to the one we usually observe for 28 ML GaAs QW (2.5-3.5 meV). This indicates a fluctuation of the well width of more than ± 5 ML, i.e. one or both interfaces extending over 5 levels separated by one ML. On the other hand, the improvement of the growth interface observed via RHEED intensity behaviour after the growth of this QW is well correlated to the PL characteristics of the subsequent QWs. In particular, the observed PL linewidth of the 14 ML (≈ 5 meV) and 7 ML (≈ 7.9 meV) QW excitonic recombinations are similar to those obtained for normally grown identical structures.

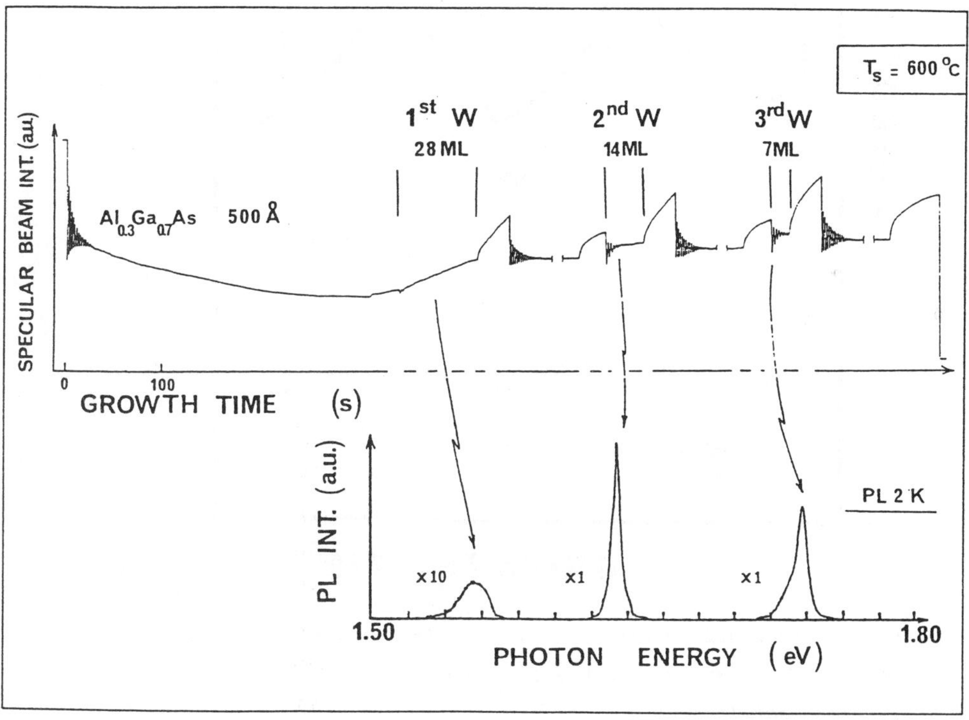

Fig. 30 RHEED oscillations recorded during the growth of a three $Al_{0.3}Ga_{0.7}As$/GaAs QW structure (Upper part); photoluminescence associated with the three QWs (Lower part).

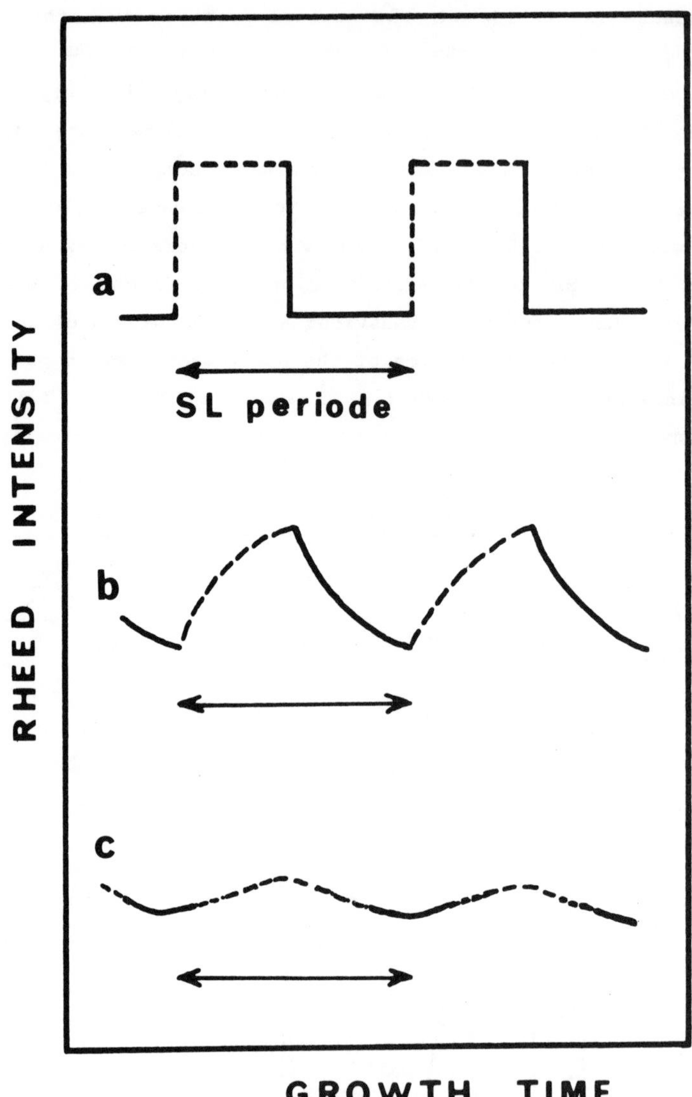

Fig.31 Schematic RHEED intensity variation during the growth of a multilayer of materials A and B in absence of growth induced oscillations :

a) ideal growth with sharp interface formation

b) multilayer growth where B layers are rougher than A layers

c) growth conditions giving a 3 D growth mode or a strong interdiffusion of A and B.

4.6 Control of the multilayer sequence in absence of growth induced RHEED intensity oscillations

Even if no RHEED intensity oscillation is observed during the growth of a multilayer of A and B materials, recording of the RHEED intensity level variation can give information in particular on the interface properties. This is schematically shown in figure 31a, b and c. In the case of a perfect Franck-Van der Merwe system, but for growth in conditions of "step propagation" (i.e. above the critical temperature defined in 4-1), the RHEED intensity variation will be of the type of fig. 31a. Obviously, such an intensity profile is only possible if sharp interfaces are formed between A and B. The intensity variation is due to the difference in the structure factors of the two materials and the period of the observed "oscillation" is then simply the superlattice period. If now the two materials exhibit different growth interfacial roughness, a tooth-shaped RHEED intensity profile would be observed as schematically shown in fig. 31b, where the step-density at the growth interface is supposed to be higher in B than in A. However, the intensity variation can be still correlated to the superlattice period. On the contrary, when interdiffusion between A and B becomes significant or if the growth occurs following a 3D mode, the intensity profile becomes smooth and no superlattice effect is expected for the grown structure (Fig. 31c).

5 REFERENCES

[1] Carlson K.D., "The Characterization of High Temperature Vapors", Ed. J.L. Margrave, John Wiley and sons, N.Y., 1967, chap 5

[2] Cater E.D., "The Characterization of High Temperature Vapors and Gases", NBS special publication 561/1, Washington, 1979, 3-38

[3] Paule R.C. and Margrave J.L., "The Characterization of High Temperature Vapors", Ed. J.L. Margrave, John Wiley and Sons, N.Y., 1967, chap 6

[4] Dushman S., "Scientific Foundations of Vacuum Technique", John Wiley and sons, N.Y., 1958, chap 2

[5] Rosenblatt G.M., "Treatise on Solid-State Chemistry, VI Surfaces", Ed. N.B. Hannay, Plenum Press, N.Y., 1976, chap 3

[6] Hultgren R., Desai P.D., Hawkins D.T., Gleiser M., Kelley K.K., Wagman D.D., "Selected Values of the Thermodynamic Properties of the Elements", Amer. Soc. for Metals : Metals Park, Ohio 44073, 1975

[7] JANAF Thermochemical Tables, Third Ed., J. Phys. Chem. Ref. Data, 1985, 14, suppl. 1, Part I and II

[8] Barin I. , Knacke O. and Kubachewski O., "Thermochemical Properties of Inorganic Substances", Springer, Berlin, 1973 et Suppl. 1977

[9] Gurvich L.P. et al."Thermodynamic Properties of Individual Substances". Nauka : Moscou , 1981

[10] Inghram M.G. and Drowart J., "Proc. of an Int. Symp. on High Temperature Technology", Mc Graw-Hill Book Co, Inc, N.Y., 1959, 219-240

[11] Drowart J., "Condensation and Evaporation of Solids", Gordon and Breach, N.Y., 1964, 255-310

[12] Prigogine I. and Defay R., "Thermodynamique Chimique", Dunod, 1950

[13] Nabot J.P. and Ansara I., Bull. Alloy Phase Diag., 1987, 8, 246

[14] Bernard C. , Deniel Y., Jacquot A, Vay P. and Ducarroir M., J. Less-Commom Met., 1975, 40, 165

[15] Barbier J.N. and Bernard C., CALPHAD XV, 7-11 July 1986, "Melange. A versatile computer program for all kinds of complex equilibria"

[16] Laurent V., Chatain C., Chatillon C. and Eustathopoulos N., Acta Met., 1988, 36, 1797

[17] Hildenbrand D.L. and Hall W.F., "Condensation and Evaporation of Solids", Gordon and Breach, N.Y., 1964, 399-416

[18] Schoomaker R.C., Buhl A. and Lamley J., J. Phys. Chem., 1965, 69, 3455

[19] Lou C.Y. and Somorjai G.A., J. Chem. Phys., 1971, 55, 4554

[20] Heiblum M., Bloch J. and O'Sullivan J.J., J. Vac. Sci. Technol. 1985, A3, 1885

[21] Binary Alloys Phase Diagrams. Ed. Massalski T.B. , ASM, Metals Park, Ohio 44073-9989 USA, 1986

[22] Ageev N.V. and Ivanov O.S., Ed.: "Phase Diagrams of Metallic Systems" Nauka, Moscou, 1971

[23] Chatillon C., Allibert M. and Pattoret A., "Characterization of High Temperature Vapors and Gases", NBS Special Publication 561/1, Washington 1969, 181-210

[24] Talalaeff G., Thin Solid Films, 1987, 150, 369

[25] Talalaeff G., Note Technique NT/LAB/ICM/201, CNET, BP.40, 22301 Lannion Cédex, France, 1986

[26] Curless J.A., J. Vac. Sci. Technol. 1985, B3, 531

[27] Ward J.W , Bivins R.L. and Fraser M.V., J.Vac.Sci.Technol., 1970, 7, 206

[28] Wahlbeck P.G., J. Chem. Phys. 1971, 55, 1709

[29] Wey S.J. and Wahlbeck P.G., J. Chem. Phys. 1972, 57, 2932 and 2937

[30] Voronin G.F., Russ J. Phys. Chem., 1967, 41, 889

[31] Grimley R.T. and Larue J.L., Ber. Bunsenges. Phys. Chem., 1976, 80, 167
[32] Chatillon C., Allibert M. and Pattoret A., High Temp. Science, 1976, 8, 233
[33] Martin-Garin L., Chatillon C. and Allibert M., J. Less Common Met., 1979, 63, 9
[34] Heckingbottom R. and Davies G.J., J. Cryst. growth, 1980, 50, 644
[35] Gaillard J.P., Revue Phys. Appl., 1987, 22, 457
[36] Turco F., Guillaume J.C. and Massies J., J. Cryst. growth, 1988, 88, 282
[37] Harmand J.C., "Possibilités offertes par l'épitaxie par jets moléculaires dans la croissance d'hétérostructures GaAs/GaAlAs pour transistors bipolaires.Thesis at Université de Paris VII, 9 March 1987. Paris, France, chap.3
[38] Shen J.Y., "Thermodynamique des systèmes III-V, As-Ga-In et Al-As et Analyse de leur épitaxie par jets moléculaires". Thesis at Institut National Polytechnique de Grenoble, 6 July 1989, Grenoble, France
[39] Cho A.Y. and Arthur J.R., Progr. Solid State Chem., 1975, 10, 157
[40] Wood C.E.C., Surf. Sci., 1981, 108, L441
[41] Neave J.H., Joyce B.A., Dobson P.J. and Norton N, Appl.Phys.A, 1983, 31,1
[42] Van Hove J.M., Lent C.S., Pukite P.R. and Cohen P.I, J.Vac.Sci.Technol.B, 1983, 1, 741
[43] Dobson P.J., Joyce B.A., Neave J.H. and Zhang J.,J.Cryst.Growth, 1987, 81,1
[44] Van Hove J.M. and Cohen P.I., Appl. Phys. Lett., 1985, 47, 726
[45] Ralston J., Wicks G.W. and Eastman L.F., J.Vac.Sci.Technol. B, 1986, 4, 594
[46] Turco F. and Massies J., Appl. Phys. Lett., 1987, 51, 1989
[47] Neave J.H., Dobson P.J., Joyce B.A. and Zhang J., Appl.Phys.Lett., 1985,47, 100
[48] Van Hove J.M. and Cohen P.I., J.Cryst.Growth, 1987, 81, 13
[49] Garcia J.C., Neri C. and Massies J., J. Cryst.Growth, 1989, 98, 511
[50] See for example "Proceedings of the IVth Int.Conf. on MBE" :J.Cryst.Growth, 1987, 81, 1-569
[51] Purcell S.T., Heinrich B. and Arrott A.S., Phys.Rev.B, 1987, 35, 6458
[52] Joyce B.A., Dobson P.J., Neave J.H., Woodbridge K., Zhang J., Larsen P.K. and Bölger B., Surf.Sci., 1986, 168, 423
[53] Larsen P.K., Dobson P.J., Neave J.H., Joyce B.A., Bolger B. and Zhang J. Surf. Sci., 1986, 169, 176
[54] Cho A.Y., in "The technology and physics of molecular beam epitaxy" edited by Parker E.H.C., Plenum 1985
[55] Massies J., Etienne P. and Linh N.T., Rev.Techn. THOMSON-CSF, 1976, 8, 5

[56] Laurence G., Simondet F. and Saget P., Appl.Phys., 1979, 19, 63

[57] Wood E.A., J. Appl. Phys., 1964, 35, 1306

[58] See for example Lagally M.G. in Methods of Experimental Physics, Vol. 22, "Solid State Physics : Surfaces", edited by Park R.L. and Lagally M.G. Academic, 1985, 237

[59] Sauvage-Simkin M., Pinchaux R., Massies J., Claverie P., Jedrecy N., Bonnet J. and Robinson I.K., Phys.Rev.Lett., 1989, 62, 563

[60] Pashley M.D., Haberern K.W., Friday W., Woodhall J.M. and Kirchner P.K., Phys.Rev.Lett., 1988, 60, 2176

[61] Delescluse P. and Masson A., Surf.Sci., 1980, 100, 423

[62] Joyce B.A., Neave J.H., Dobson P.J. and Larsen P.K., Phys.Rev. B, 1984, 29, 814

[63] Larsen P.K., Van der Veen J.F., Mazur A., Pollmann J., Neave J.H. and B.A. Joyce , Phys. Rev. B, 1982, 26, 3222

[64] Massies J. and Linh N.T., J.Cryst.Growth, 1982, 56, 25

[65 Ludeke R., Chiang T.C. and Eastman D.E., J.Vac.Sci.Technol., 1982, 21, 599

[66] Nguyen-Van-Dau Thesis, Orsay University 1989

[67] Massies J. and Linh N.T., Surf.Sci., 1982, 114, 147

[68] Prinz G.A., Phys.Rev.Lett., 1985, 54, 1051

[69] Maurer M., Ousset J.C., Ravet M.F. and Piecuch M., Europhys.Lett., 1989, 9, 803

[70] Etienne P., Creuzet G., Friederich A., Nguyen-Van-Dau F., Fert A. and Massies J., Appl.Phys.Lett., 1988, 53, 162

[71] See for example Freundlich A., J.Phys. (Paris) C, 1989, 5, 499

[72] Copel M., Reuter M.C., Efthimios Kaxiras and Tromp R.M., Phys.Rev.Lett., 1989, 63, 632

[73] Etienne P., Massies J., Nguyen-Van-Dau F., Barthélémy A. and Fert A., Appl.Phys. Lett. to be published (1989)

[74] Turco F. and Tamargo M.C., J. Appl. Phys., 1989, 66, 1695

[75] Fuchs J., Feit Z. and Preier H., Appl.Phys.Lett., 1988, 53, 894

[76] Steigerwald D.A. and Egelhoff W.F., Surf.Sci., 1987, 192, L887

[77] Koziol C., Lilienkamp G. and Bauer E., Appl.Phys.Lett., 1987, 51, 901

[78] Heinrich B., Urqhart K.B., Dutcher J.R., Purcell S.T., Cochran J.F. and Arrott A.S., J.Appl.Phys. 1988, 63, 3863

[79] Egelhoff W.F. and Jacob I., Phys.Rev.Lett., 1989, 62, 921

[80] Jalochowski M. and Bauer E., Surf.Sci., 1989, 213, 556

[81] Wowchak A.M., Kuznia J.N. and Cohen P.I., J.Vac.Sci.Technol. B, 1989, 7,733

[82] Neave J.H., Dobson P.J., Joyce B.A. and Zhang J., Appl.Phys.Lett., 1985, 47, 100

[83] Turco F., Massies J. and Contour J.P., Revue Phys.Appl., 1987, 22, 827

[84] Sakamoto T., Kawamura T. and Hashiguchi G., Appl.Phys.Lett., 1986, 48, 1612

[85] Cibert J. et al. to be published and private communication

[86] Harmand J.C., Alexandre F. and Beerens J., Revue Phys. Appl., 1987, 22, 821

[87] Sakaki H., Tanaka M. and Yoshino J., Jpn. J. Appl. Phys., 1985, 24, L417

[88] Sakamoto T., Funabashi H., Ohta K., Nakagawa T., Kaway N.J., Kojima T. and Bando Y., Superlattices and Microstructures, 1985, 1, 347

Materials Science Forum Vols. 59 & 60 (1990) pp. 287-360

THERMODYNAMICAL ASPECTS

J.C. Joud (a), J.L. Bocquet (b) and M. Gerl (c) [1]

(a) LTPCM (LA 29) INPG, BP 75, F-38402 St-Martin-d'Hères, France
(b) D. TECH/SRMP, CEN-Saclay, F-91191 Gif-sur-Yvette Cedex, France
(c) Université Nancy I, Lab. de Physique des Solides
BP 239, F-54506 Vandoeuvre Les Nancy Cedex, France

In this chapter the properties of modulated structures are studied from a thermodynamical point of view. These structures can be observed in many cases, as in spinodal decomposed alloys or can be artificially synthetized in the form of epitaxial or non-epitaxial crystalline multilayers, or as multilayers of amorphous metals or semiconductors. The main characteristics of these structures are the following (i) the scale of the concentration modulation is of the order of 10 to 1000 Å ; (ii) they contain an anomalously large number of interfaces between different materials ; (iii) they are often observed in a metallurgical state which is not predicted by the usual equilibrium phase diagram and (iv) the kinetics of their thermal evolution is controlled by interdiffusion coefficients which can be markedly different from the intrinsic diffusivities. The three sections this chapter consists of deal with these problems, more or less specific to modulated structures.

In the first section the concept of a free surface or of an interface between two materials is carefully defined. The excess quantities attached to a surface are recalled and it is shown that the so-called surface tension is a superficial density of the grand canonical potential. The surface tension can be calculated using accurate electronic models, which account fairly well for the trends of the surface tension along a transition metal series, or a simple metal series. For alloys the main phenomenon of interest is the segregation (or diffusion) of a constituent species in the vicinity of a surface. Simple phenomenological models can often be used to predict segregation properties of alloys but refined approaches are necessary to understand most of the segregation profiles and their evolution with temperature.

In epitaxial multilayers stress effects arising from the difference in the atomic distance, or in the atomic size of the two alloy species play an

---

[1] Listed in the order of the three main sections.

important role. Thus the second section of this chapter is concerned with the changes expected in coherent phase diagrams when size effects are included. First the stability of solid solutions against two types of concentration fluctuations is examined. It is shown that the critical temperature for unmixing is more depressed (i.e. the solid solution is more stabilized) when the wavevector of the fluctuation is perpendicular than when it is parallel to the substrate. Secondly the effect of stresses on the thermodynamical behaviour of two-component systems is studied, and applied to the problem of the coherent equilibrium of two phases of different composition. Finally, it is shown how elastic effects can be introduced at the atomic level by incorporating strains in a refined statistical model of alloys called the Cluster Variation Method (CVM).

In the third section the kinetics of atomic diffusion in multilayers are linked with the thermodynamic properties of the system. Matter transport is controlled by the interdiffusion coefficient, which is not a purely kinetic quantity. This coefficient is always positive for stable or metastable systems but can become negative for unstable solid solutions, thus producing up-hill diffusion. According to the properties of the solid solution, up-hill diffusion may favour long wavelength concentration fluctuations (as in clustering systems) or short wavelength modulations (as in ordering systems). Moreover the coupling between stresses and diffusion must be introduced to understand the behaviour of practical systems. The influence of stresses is essentially non local and may lead to unexpected phenomena in transitory regimes. Finally, because of the steep composition variation in multilayers, many relevant quantities strongly vary with distance and a non-linear behaviour of the diffusion equation can be expected, which again may lead to surprising results.

I SURFACE TENSION AND SURFACE SEGREGATION IN METALLIC SYSTEMS

The general plan is as follows: in paragraph I.1 thermodynamic properties of capillary system are described using Gibbs' model. Paragraph I.2 presents some empirical and theoretical calculations of surface energy of pure metals. In paragraph I.3 statistical treatments used in the evaluation of the free energy of binary alloys are discussed. The various contributions to the segregation driving force are also considered. Paragraph I.4 briefly presents the kinetics of surface segregation.

I.1 THERMODYNAMIC PROPERTIES OF CAPILLARY SYSTEMS

I.1.1 Thermodynamic definitions

A simple relationship between surface tension and surface free energy can be deduced using the classical Gibbs model describing a two phase capillary system in equilibrium with a planar interface, as shown in Fig. I.1.

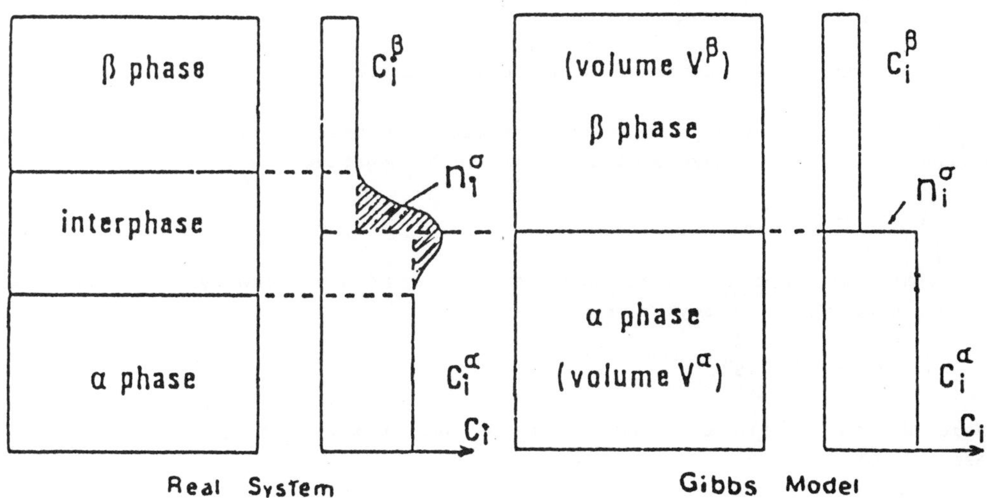

Fig. I.1 Comparison between the real system and Gibbs' model in the case of a planar α-β interface, from [I.14].

The variation of the free energy of the entire system can be expressed in the form

$$dF = - Pdv - Sdt + \sigma\, d\Omega + \sum_i \mu_i\, dn_i \tag{I.1}$$

where S is the entropy, T the absolute temperature, P the pressure (for a planar interface $P_\alpha = P_\beta = P$), V the volume, μ_i the chemical potential of the component i and n_i the number of moles in the ith component. The extra term $\sigma\, d\Omega$, where σ is the surface tension and Ω the surface area, expresses the mechanical work required to create an additional surface area $d\Omega$. From eq. I.1 the surface tension may be defined as

$$\sigma = \left(\frac{\partial F}{\partial \Omega}\right)_{T,V,n_i} \tag{I.2}$$

In the Gibbs treatment [I.1] of capillary systems, bulk phases are maintained homogeneous up to a dividing surface assumed to be of zero thickness. Local concentrations of each extensive property, such as free energy or number of atoms, are constant up to the geometrical surface in each bulk phase. Accordingly the balance of the extensive property is assigned to the dividing surface (Fig. I.1).

I.1.2 Surface excess properties

We denote by C_i^α the concentration of component i in the bulk α phase, far from the interface. In Gibb's model this concentration is considered to be constant up to the dividing surface (corresponding volume V^α) and defines a number of moles $n_i^\alpha = C_i^\alpha\, V^\alpha$. So the surface excess number of moles may be written

$$n_i^\sigma = n_i - n_i^\alpha - n_i^\beta$$

where n_i is the total number of moles of component i.

Dividing by Ω the area of the surface we obtain the Gibbs adsorption of component i, Γ_i which may be either positive or negative

$$\Gamma_i = \frac{n_i^\sigma}{\Omega} \qquad (I.3)$$

Similarly from the classical expression of free energy we define the surface excess free energy F^σ as

$$F^\sigma = \mathcal{U}^\sigma - TS^\sigma$$

where $\mathcal{U}^\sigma$ and S^σ are the surface energy and surface entropy of the system, respectively.

The variation of free energy corresponding to the bulk α phase in Gibbs' model is

$$dF^\alpha = - P\, dv^\alpha - S^\alpha\, dT + \sum_i \mu_i\, dn_i$$

and similarly for the β phase.

The chemical potential μ_i of the ith component has the same value in both the bulk phase and at the interface because the system is in chemical equilibrium.

By subtracting the contribution dF^α and dF^β from eq. I.1 we obtain the excess free energy associated with the dividing surface

$$dF^\sigma = - S^\sigma\, dT + \sigma\, d\Omega + \sum_i \mu_i\, dn_i^\sigma \qquad (I.4)$$

At constant temperature, the excess surface free energy F^σ is a first order homogeneous function of Ω and n_i^σ. Thus, from eq. I.4 and using Euler's theorem, the surface tension may be written

$$\sigma = \frac{F^\sigma}{\Omega} - \sum_i \mu_i\, \Gamma_i \qquad (I.5)$$

This relation shows that in general the surface tension cannot be identified with the surface free energy per unit area. For a pure component, the location of the dividing surface such that adsorption is zero leads to

$$\sigma = \frac{F^\sigma}{\Omega} = f^\sigma \qquad (I.6)$$

the validity of this relation is limited to a pure system (liquid or solid) in which surface and bulk chemical potentials remain equal after an equilibrium displacement. In particular, for solids [I.2], this condition restricts the validity of eq. I.6 to high temperatures.

Differentiating eq. I.5 and substituting dF^σ from eq. I.4 we obtain the well known Gibbs' adsorption equation

$$d\sigma = -\frac{S^\sigma}{\Omega}\, dT - \sum_i \Gamma_i\, d\mu_i \qquad (I.7)$$

I.1.3 Effect of temperature

A general expression for the temperature coefficient $\frac{d\sigma}{dT}$ may be deduced from eq. I.7

$$\frac{d\sigma}{dT} = -\frac{S^\sigma}{\Omega} - \sum_i \Gamma_i \frac{d\mu_i}{dT}$$

In a one-component system and choosing a dividing surface corresponding to zero adsorption we have

$$\frac{d\sigma}{dT} = -\frac{S^\sigma}{\Omega}$$

In this case the temperature coefficient of the surface tension is equivalent in magnitude, but opposite in sign, to the excess surface entropy per unit area. This quantity which is an indication of the surface disorder, can be estimated in the case of a liquid-vapour interface. At the critical temperature the surface tension must be zero, consequently the surface tension of a one liquid component should decrease with increasing temperature. Accordingly, an empirical relation has been proposed to correlate the surface tension variations with temperature [I.3, I.4]. The Eötvös equation [I.4] is the best known

$$\sigma_{LV}(M/\rho_L)^{2/3} = K\ (T_c - T)$$

where M is the molar weight, ρ_L the liquid density and T_c the critical temperature. The constant K is about 2.1 $erg.K^{-1}$ for numerous organic liquids but approaches 0.64 for liquid metals [I.5].

In a multi-component system, Gibbs' adsorption equation may be written

$$\frac{d\sigma}{dT} = -\frac{S^\sigma}{\Omega} - \sum_i \Gamma_i \left(\frac{\partial\mu_i}{\partial T} + \sum_j \frac{\partial\mu_i}{\partial N_j}\frac{dN_j}{dT}\right) \qquad (I.8)$$

The first term of this relation is purely entropic, the second corresponds to the variation of chemical potentials with temperature. The surface tension coefficient may thus be negative or positive, according to the thermodynamic behaviour of the solution [I.6]. Positive values have been observed for several binary or ternary liquid alloys and interpreted on the basis of a strong surface segregation of the solute [I.6, I.7].

I.1.4 Effect of composition

For binary alloys, at constant temperature, Gibbs' adsorption equation (I.7) is reduced to

$$d\sigma = -\Gamma_A\, d\mu_A - \Gamma_B\, d\mu_B$$

Substituting the Gibbs-Duhem relation we obtain

$$d\sigma = -\left(\Gamma_B - \frac{x_B\,\Gamma_A}{x_A}\right) d\mu_B = -\,\Gamma_{BA}\,d\mu_B \qquad (I.9)$$

Where x_A, x_B are the bulk molar fractions. The relative adsorption Γ_{BA}, (defined by the previous equation), may be obtained from the experimental measurements of the variations of the surface tension with the chemical potential of one component.

In the case of a dilute solution and assuming Henry's law for the solute B, we have

$$\frac{1}{RT}\,d\sigma \,/\, d\ln x_B \cong -\,\Gamma_{BA} \cong -\,\Gamma_B$$

In dilute solutions, $x_B \cong 0$, and the relative adsorption is essentially the solute adsorption. Consequently, the more the solute lowers the surface tension, the more it segregates to the surface. These equations have been frequently used to interpret the effects of surface active additions. Some applications are discussed in ref. [I.14].

I.2 SURFACE TENSION OF PURE METALS

I.2.1 Empirical calculations

The most simple method to calculate the surface tension of pure metals involves the summation of the energies associated with "broken bonds" due to the absence of similar atoms on the vapour side of the surface. In the simplest model of this type, it is considered that only the nearest neighbour interactions are responsible for the cohesion energy of the bulk phase. On this basis Skapski [I.8] deduced that, at absolute zero, the molar surface energy of a pure liquid σ^{00}_{LV} is a fraction of the molar heat of vaporisation ΔH_v

$$\sigma^{00}_{LV} = \frac{Z - Z_s}{Z}\,.\,\Delta H_v = m\,.\,\Delta H_v \qquad (I.10)$$

where Z and Z_s are the atomic coordination numbers in the bulk and in the surface respectively.

A plot of σ^{00}_{LV} v.s. ΔH_v is presented in Fig. I.2. σ^{00}_{LV} is calculated from experimental values of the surface tension σ^{0}_{LV} and the temperature coefficient $d\sigma^{0}_{LV}/dT$

$$\sigma^{00}_{LV} = \Omega\left(\sigma^{0}_{LV} - T\,\frac{d\sigma^{0}_{LV}}{dT}\right) \qquad (I.11)$$

where Ω the molar surface area is calculated from the molar volume assuming the most densely-packed plane for arrangement of the surface atoms ({111} for the f.c.c. structure).

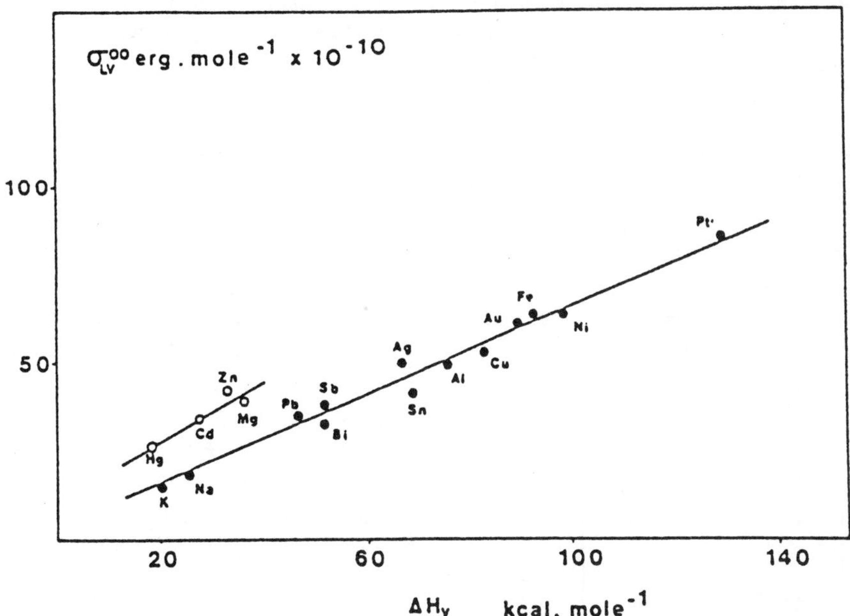

Fig. I.2 Molar surface energy of some liquid metals vs heat of vaporization, from [I.14].

Fig. I.2 illustrates the proportionality between σ_{LV}^{00} and ΔH vap indicated by eq. I.10 but the ratio $\sigma_{LV}^{00}/\Delta H_{vap}$ is found to be equal to 0.16 whereas the value predicted on the basis of the idealized model of the lattice liquid surface is 0.25. This difference has been attributed to the change of the pair bonding energy in the vicinity of the surface resulting from a structural and electronic relaxation effect [I.9, I.10]. A similar correlation exists between the molar surface energy of solid metals and the heat of sublimation ΔH_s ; the ratio $\sigma_{SV}^{00}/\Delta H_s$ is also found to be 0.15. Similar correlations have been proposed by Miedema [I.11] and Overbury et al. [I.12].

Neglecting the surface entropy difference between liquid and solid metals a useful expression can be deduced on the same basis as eq. I.10. This can be used to calculate at the melting point the surface tension of pure solid for the most densely-packed atomic plane, from the generally well known value of the surface tension of the liquid

$$\sigma_{SV}^{0} = (V_L / V_S)^{2/3} \sigma_{LV}^{0} + m L_f / \Omega \qquad (I.12)$$

where V_L and V_S are the molar volumes of the metal in the liquid and solid state and L_f the latent heat of melting. Some improvements of this formula have been proposed [I.13] which lead to a ratio of calculated values $(\sigma_{SV}^{0} - \sigma_{LV}^{0}) / \sigma_{LV}^{0}$ close to 0.20, in agreement with experimental values [I.14].

From a practical point of view, experimental measurements of solid or liquid tensions are rare. This is due to the difficulty of performing reliable measurements at high temperatures without contamination (oxygen). This is particularly true for the "zero creep" technique used in solid surface tension measurements.

Some experimental values, presented in a review by Eustathopoulos and Joud [I.14], are gathered in Table I.1.

Metal	Tm K	σ^0_{LV} erg.cm^{-2}	d σ^0_{LV} / dT erg.cm^{-2}.K^{-1}	T K	σ^0_{SV} erg.cm^{-2}	d σ^0_{SV} / dT erg.cm^{-2}.K^{-1}
Cu	1356	1295	-0.2 ± 0.1	1243	1650 ± 100	-0.5 ± 0.4
Au	1336	1134	-0.1	1273	1400 ± 500	-0.5 ± 0.1
Ag	1234	915	-0.15 ± 0.05	1183	1140 ± 900	-0.4 ± 1.5
Pb	600.5	462		590	610 ± 20	
Ni	1724	1756	-0.30 ± 0.15	1488	2385 ± 100	-1.1 ± 0.8
Co	1768	1884		1678	2282 ± 300	
Fe^{γ}				1648	2150 ± 300	
Fe^{δ}	1809	1806	-0.42 ± 0.15	1723	2220 ± 250	-1.5 ± 2
Ti	1940	1450		1873	1700	
Zn	692	767		653	830	
In	429	558		420	674 ± 74	
Sn	505	554		488	685	
Mo	2890	2200		1700	2200 ± 200	
Nb	2740	1900		2523	2100 ± 100	
Ta	3269	2230		1773	2820 ± 500	
W	3650	2450		2030	2900 ± 300	

Table I.1 Experimental surface tension values for some liquid and solid metals, from [I.14].

I.2.2 Classical pair potential calculations

Because surface atoms are in an asymmetrical environment they may take up equilibrium positions quite different from those in the bulk, so a tangential and normal relaxation can be obtained for the last atomic plane at the surface [I.15].

Generally the last plane remains coherent with the crystalline substrate and the tangential relaxation is analysed in terms of dislocations. The normal relaxation has been most extensively studied from both theoretical and experimental points of view.

The first calculations have been done, considering pair potentials such as the Morse or Johnson [I.16] potential between every pair of atoms in the crystal lattice terminating in a surface plane. The pair potential is generally assumed to be the same for surface atoms and for atoms in the bulk.

Results of such calculations using the Morse parameters proposed by Girifalco and Weizer [I.17] are given in Table I.2 [I.18, I.19].

Element	E_S^0 eV/at	E_S^0 erg/cm^2	Δ_1 %	Δ_2 %	$-\frac{(E_S - E_S^0)}{E_S^0}$ %
Ag_{111}	1.117	2475	2.8	0.3	0.97
Cu_{111}	1.496	4226	4.6	0.7	1.9
Al_{111}	1.293	2910	5.1	0.7	2.1
Fe_{110}	1.792	4917	5.0	0.9	2.1
Cr_{110}	1.475	4015	3.1	0.4	1.1
W_{110}	3.390	7645	3.1	0.2	1.0

Table I.2 Surface energy calculations for a Morse potential, from [I.18, I.19].

Δ_1 (Δ_2), the first (second) interlayer spacing modification, indicates an expansion of the lattice. This result is classic for a pair potential with a single minimum, owing to the necessarily repulsive interaction betwen first neighbours. Surface energy values are clearly overestimated (roughly by a factor of 2) while surface energy relaxations are close to 1-2 %.

LEED experiments on various metals nearly always show a contraction of the first layer distances [I.15, I.20] which is weak on a close-packed surface and larger (5-15 %) on a more opened one.

Better agreement is obtained when using this type of pair potential for the calculation of surface tension anisotropy. The variations of E_{hkl}/E_{110} calculated with Mie's potential by Dreschler and Muller for W are well correlated with experimental values [I.21]. Similar results are deduced from both the calculations of Nicholas [I.22] using the same kind of potential, and the experimental values of surface tension anisotropy measured on α Fe [I.23].

I.2.3 Electronic structure

The electronic structure of a surface is linked to its atomic structure and vice-versa. So both structures should be self-consistently determined in a complete theory. Except for particular cases, this complete theory is still lacking and electronic structure calculations are generally performed assuming that the atomic structure is known. Because it breaks the periodicity of the crystal in one direction, the surface modifies the electronic structure in its neighbourhood and gives rise to wave functions of two types : 1) extended wave functions corresponding to electrons which move towards the surface and are reflected into the bulk and 2) localized wave functions which decrease exponentially on both sides (bulk and vacuum) of the surface and propagate along the surface (surface state). Some intermediate situations corresponding to resonant states are obtained for electrons which are fairly localized in the surface region but can come back into the bulk [I.24] (Fig. I.3).

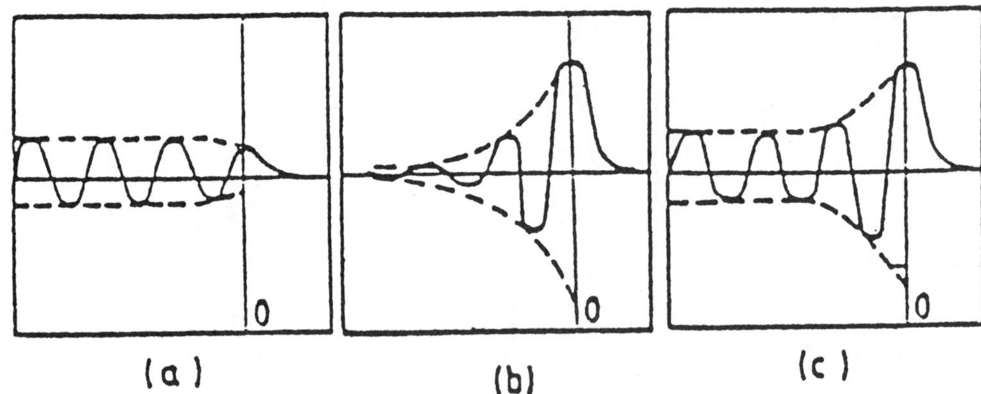

Fig. I.3 Schematic representation of the wave functions near a surface : a) Extended states, b) Surface state, c) Resonant state, from [I.24].

Several theoretical methods have been developed to calculate this surface electronic structure with a physical background deduced in each case from classical bulk calculations.

i) The "muffin tin" approximation :
In this method the potential around each atom has a spherical symmetry and is limited to a radius R_{MT} such that the potentials of two neighbouring atoms do not overlap. The potential between spheres is assumed to be a constant. This approximation is well suited to metals (normal or transition-metals) [I.25, I.26] but it is much less appropriate to semiconductors or insulators which have fairly anisotropic potentials.

ii) The jellium model :
The valence electrons are assumed to be free, in interaction with their own average charge and with an ionic charge uniformly spread in half the space, equilibrating the whole electronic density. This essentially applies to normal metals.

iii) The pseudo potential method :

In this method the potential of each ionic site is replaced by an effective potential which produces the correct free-atom energies for the valence electrons and gives the right behaviour of the wave functions far from the ionic cores.

iv) The tight-binding approximation :
In this approximation the wave functions are written as a linear combination of atomic orbitals centered at each lattice site. This approximation applies to fairly localized electrons such as d electrons of transition metals.

These various approximations have already been reviewed in detail [I.24, I.27], so we shall only emphasize the main results about surface energy calculations using the jellium model and the tight-binding approximation.

I.2.3.1 Jellium model

We shall first consider a simplified model in which the valence electrons are independent free electrons. For an infinite bulk, the wave functions can be written as plane waves ; using Born Von Karmann conditions on a cube of side L

$$\psi_{\mathbf{K}}(\mathbf{r}) = \frac{1}{(L^3)^{1/2}} \exp i\, \mathbf{K}\, \mathbf{r} \tag{I.13}$$

The existence of the surface is expressed by particular boundary conditions in the direction Z normal to the surface ; the electron wave functions are assumed to vanish at the surface, i.e. for Z = 0 and Z = L (infinite potential barrier). The wave functions have now the form

$$\psi_{\mathbf{K}}(\mathbf{r}) = \left(\frac{2}{L^3}\right)^{1/2} \exp i\, \mathbf{K}_{//}\, \mathbf{r}_{//} \sin K_Z Z \tag{I.14}$$

The electronic density n(Z) is calculated (I.28)

$$n(Z) = n_0 - 3\, n_0 \frac{\sin 2 K_F Z - 2 K_F Z \cos 2 K_F Z}{(2 K_F Z)^3} \tag{I.15}$$

with $n_0 = K_F^3 / 3 \pi^2$ being the electron number per unit volume.

This density n(Z) is presented in Fig. I.4a. We see that n(Z) differs significantly from n_0 only in the neighbourhood of the surface (2, 3 atomic planes). At larger distances the asymptotic form of n(Z) is

$$n(Z) = n_0 + 3\, n_0 \frac{\cos 2 K_F Z}{(2 K_F Z)^2} \tag{I.16}$$

It is clear from an electrostatic point of view that the exit of electrons to form a dipole layer is advantageous for the system.

This point is illustrated by the high surface energy values which are calculated. For example we obtain for Cu and Ag 4050 and 2520 erg cm^{-2}, respectively.

$$E_S^0 = \frac{1}{4\pi} \int_0^{K_F} K\, dK\, (E_F - E(E)) = \frac{1}{16\pi} K_F^2\, E_F \tag{I.17}$$

If the potential barrier is now taken as finite, a similar behaviour is obtained with the same exponentially decaying function beyond the surface which gives rise to a dipole layer (Fig. I.4b). The asymptotic form n(Z) becomes

$$n(Z) = n_0 + 3\, n_0 \frac{\cos\,(2\, K_F\, Z + \phi)}{(2\, K_F\, Z)^2} \qquad (I.18)$$

where ϕ is a phase shift determined by the boundary conditions. In this case the calculated surface energy values are considerably lower as indicated by Huntington [I.29] who obtained a value of $E_S^0 = 860\ erg.cm^{-2}$ for Cu.

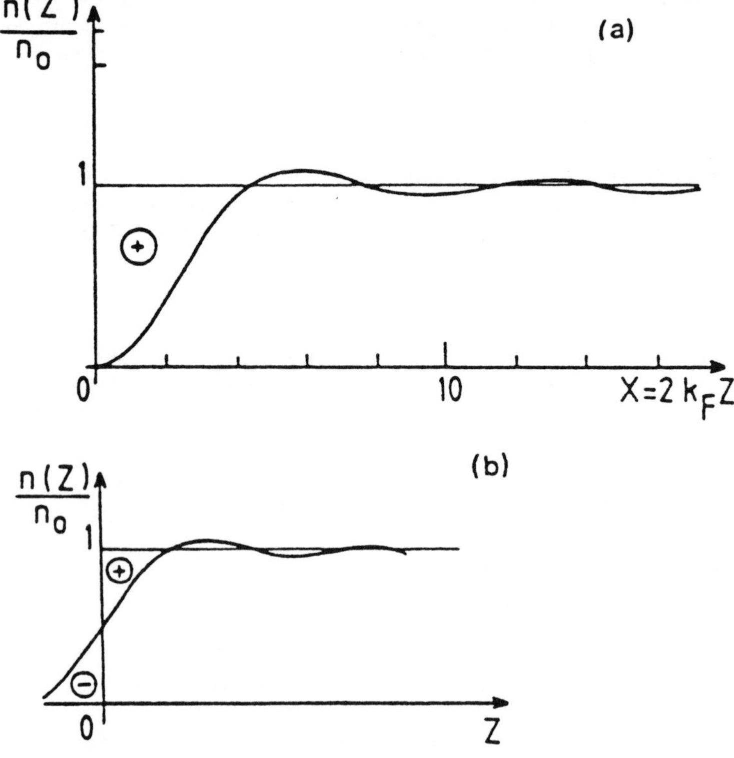

Fig. I.4 Electronic charge densities as a function of the distance to the surface. a) Assuming an infinite potential barrier at the surface, b) Assuming a finite barrier, from [I.24].

Lang and Kohn [I.30] have considerably improved this model using the local density formalism [I.31].

The total energy of the crystal is written as a sum of three contributions, the ionic charges being uniformly spread with a density n_{ion}.

$$\mathcal{U}_{tot} = T + \mathcal{U}_{es} + \mathcal{U}_{xc} \qquad (I.19)$$

where T is the kinetic energy of electrons ; $\mathcal{U}_{es}$ the electrostatic energy of the system, with

$$\mathcal{U}_{es} = \frac{|e|}{2} \int \phi\ (\mathbf{r})\ \{n_{ion}\ (\mathbf{r}) - n(\mathbf{r})\}\ d3\mathbf{r}$$

$\phi(r)$ being the potential resulting from the charge distribution.

$$\phi(\mathbf{r}) = |e| \int \frac{n_{ion}\ (\mathbf{r}') - n(\mathbf{r}')}{|\mathbf{r} - \mathbf{r}'|}\ .\ d3\mathbf{r}'$$

$\mathcal{U}_{xc}$ is the exchange and correlation energy which is calculated with the local approximation of Kohn and Sham [I.32]. $n(\mathbf{r})$ is self consistently calculated, giving charge oscillations near the surface (more pronounced for a low electronic metal with low electronic density).

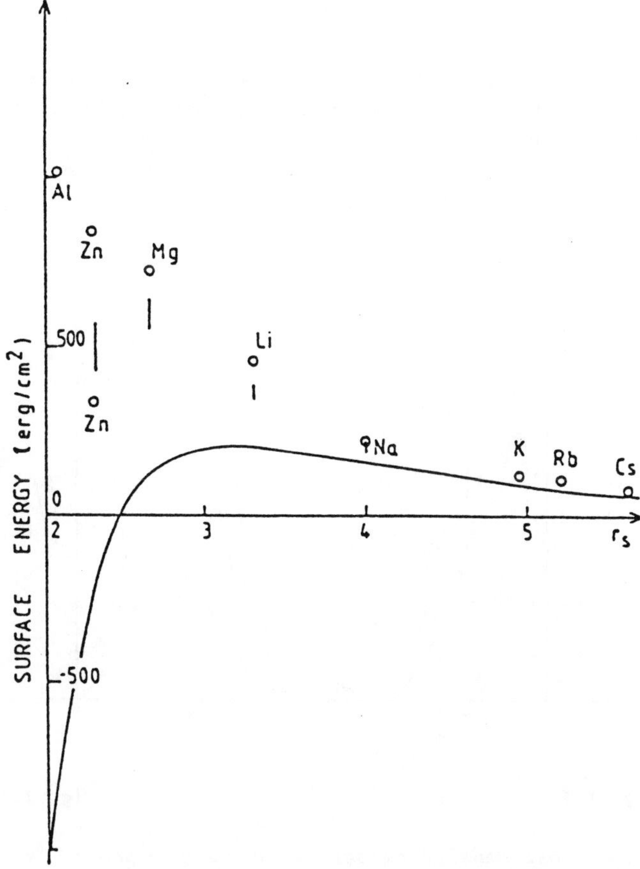

Fig. I.5 Surface energy of simple metals : (0) experimental results : continuous line : jellium model, vertical lines : Lang and Kohn theory with the ionic pseudo potential corrections, from [I.30].

The corresponding surface energy values are presented on Fig. I.5. We observe a qualitative agreement with experimental data for large values of r_s (alkaline metals) and a negative contribution for low values of r_s (large electronic density metal). This negative contribution is due to the kinetic term.

To include the discrete distribution of ionic charges, corrections are introduced by means of an ionic pseudo-potential treated within the perturbation theory. In this case Kohn and Lang obtain a fair agreement with experiment. However the correction is large for small r_s.

I.2.3.2 The tight binding approximation

This model has been largely reviewed [I.33, I.34] and we shall only recall some results concerning surface electronic structures and surface energy values of transition metals.

The local density of states (LDOS) may be calculated using the continued fraction technique [I.35]. Some results are presented on Fig. I.6 and I.7 for the (110) and (100) surfaces of Mo.

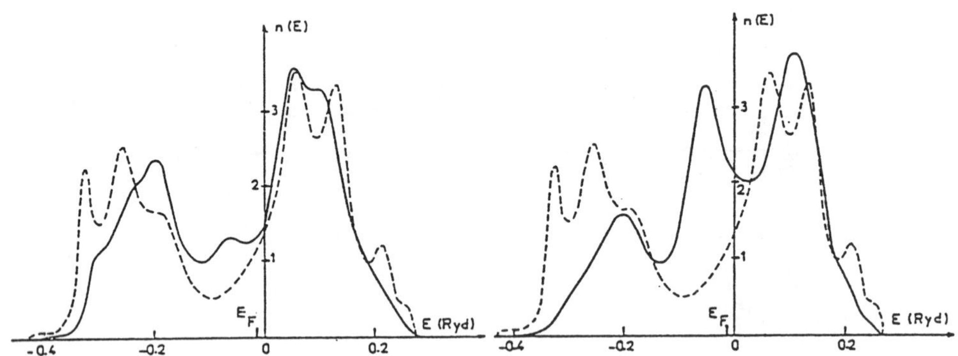

Fig. I.6 *Fig. I.7*

Fig. I.6 Local density of states on the surface of Mo (110) (full line) compared with the bulk density of states (broken line), from [I.24].
Fig. I.7 Local density of states on the surface of Mo (100) (full line) compared with the bulk density of states (broken line), from [I.24].

Calculations have shown that the only noticeable changes in the LDOS are limited to the surface plane. On close packed surface planes ((110) in BCC, (111) in FCC...) the LDOS reproduces roughly the bulk one (Fig. I.6) [I.36] with a small reduction of the effective width proportional to $\sqrt{Z_s}$, Z_s being the surface coordination number. On non close packed surface planes ((100) and (111) in BCC, (110) and (100) in FCC...) a well defined central peak is obtained in the middle of the d band which contains an important fraction of the total number of d states and disappears on the second planes. This peak corresponds to d states of surface atoms which are weakly broadened [I.36].

The specific surface energy (which is an excess energy) can be calculated in a similar way to the cohesive energy.

The first contribution which is positive is due to the variation of the delocalization of d electrons

$$E_s = \sum_{P} \{\int^{E_F} E(n_p(E)) - n(E)\, dE - Nd\, \delta V_p\} \tag{I.20}$$

$n_p(E)$ is the LDOS on plane P, n(E) is the bulk density of states, Nd is the number of d electrons per atom, δV_p is the potential variation of surface atoms. The change of the LDOS at the surface (reduced mean width of the LDOS at the surface correlated with the reduced coordination number) gives rise to a charge rearrangement which modifies the potential of surface atoms [I.35]. This problem has been consistently solved [I.37] and the main results are :

- Very small charge transfers are obtained which can be limited in a first approximation to the first plane.

- The resulting dipolar layer is characterized by a potential barrier δV_1 which can be calculated using a strict charge neutrality condition. The variation of δV_1 along the transition series has been calculated for various surface orientations [I.38]. δV_1, which increases with the number of broken bounds is positive at the end of the transition series and negative at the beginning.

At the beginning of a series, a negative atomic level correction is necessary for surface atoms to accomodate the number of d electrons with a density of states narrower at the surface than in the bulk and inversely for the end of the series. This term δV_1 can be experimentally measured by core level spectroscopy of surface atoms [I.39].

The first contribution to the surface energy given by eq. I.20, can be calculated using rectangular density of states. The width W_s (W_b) of the surface (bulk) DOS which is determined from the second moment, appears directly in the expression of surface (bulk) cohesive energy. So we obtain for the surface energy

$$E_s = \frac{W_b - W_s}{20} Nd\ (10 - Nd) \tag{I.21}$$

The parabolic evolution of E_s with the filling of the d band is in fair agreement with experimental data [I.40].

A correction to take into account electronic correlation effects has been proposed [I.40]. This negative correction is maximum for a half-filled band and explains the local minimum of E_s observed in the middle of 3d or 4d series.

Assuming that hopping integrals between two sites have the same value at the surface as in the bulk, the band widths W_b and W_s are respectively proportional to $(Z)^{1/2}$ and $(Z_s)^{1/2}$, so we can write eq. I.21 in a condensed form.

$$E_s = \left(1 - \left(\frac{Z_s}{Z}\right)^{1/2}\right) \cdot E_{coh} \qquad (I.22)$$

Eq. I.22 gives a ratio E_s/E_{coh} which, for close packed surface planes, is close to the mean value deduced from the experimental correlation (σ_s vs ΔH_{sub}).

The last contribution to surface energy is provided by the short range repulsive interatomic potential assumed to be pairwise. This negative contribution is proportional to the number of broken bounds at the surface [I.41].

Surface relaxation corresponding to a minimization of surface energy with the last interlayer spacing has been calculated [I.42] giving rise to a contraction in agreement with experiments.

I.3 EQUILIBRIUM SURFACE SEGREGATION IN BINARY ALLOYS

Surface properties of binary alloys depend on a subtle balance of energies. For instance, the heat of segregation for a dilute binary alloy is the difference between the energy for the solvent with a solute atom in the surface and the energy for the solvent with a solute atom in the bulk. The total energy of a solvent with an impurity is large compared with the mentioned difference and it is difficult to achieve the accuracy required to obtain the segregation energy. In addition to the description of the energy, we have to use a statistical description of the system for the evaluation of the partition function. The most widely used statistical description is based on the Bragg and Williams approximation (B.W.). Some calculations have been performed using the quasichemical approximation and more recently surface segregations have been calculated using Monte-Carlo simulations which are, in principle, free from statistical approximations.

I.3.1 Bragg and Williams approximation

In the B.W. approximation the atoms A (molar fraction X) and B (molar fraction 1 - X) are randomly distributed on the lattice sites ; this distribution corresponds to the high temperature limit. We assume that the crystal consists of P atomic layers each of them containing N atoms. It is further assumed that the energy E can be written as a function of all the layer concentrations X_k with k = 1, 2, ...P.

Considering only the configurational entropy we can write the free energy of the system as

$$F = \mathcal{U} - TS = NE\ (X_1, ..X_k..) + N\ KT \sum_{k=1}^{P} (X_k \ln X_k + (1-X_k)\ln(1-X_k)) \quad (I.23)$$

where K is the Boltzmann constant, T the absolute temperature, the number of atoms per plane N is sufficiently large to justify the Sterling approximation.

The equilibrium concentration profile can be obtained by minimizing the free energy with respect to the layer concentrations X_k with the constraint that the total number of atoms of each type is constant. This constraint is expressed by a Lagrange multiplier μ, so we have to obtain the stationary points of the function

$$F - \mu N \sum_{k=1}^{P} X_k$$

this gives

$$\frac{\partial E}{\partial X_k} + KT \ln \frac{X_k}{1 - X_k} = \mu \qquad \text{for } k = 1, 2 \ldots \quad (I.24)$$

The Lagrange multiplier μ (chemical potential differences) is the same for all layers, so eq. I.24 can be written for the bulk concentration

$$\frac{\partial E}{\partial X_b} + KT \ln \frac{X_b}{1 - X_b} = \mu \quad (I.25)$$

and we get the system of equations

$$\frac{X_k}{1 - X_k} = \frac{X_b}{1 - X_b} \exp - \frac{1}{KT} \left(\frac{\partial E}{\partial X_k} - \frac{\partial E}{\partial X_b}\right) \quad (I.26)$$

with k = 1, 2, ...P

These equations give the concentration profile for the alloys, the main problem being to calculate the energy with sufficient accuracy.

I.3.1.1 Monolayer approximation

This approximation with a simple description of the energy (ideal solution or regular solution) has been widely used [I.43, I.44]. In the case of a regular solution model, using a nearest neighbor pairwise approximation, the calculation of the energy of the system may be performed easily. In such a case the term $\partial E/\partial X_1 - \partial E/\partial X_b$ which represents the exchange energy of a bulk A atom with a surface B atom is written

$$\frac{\partial E}{\partial X_1} - \frac{\partial E}{\partial X_b} = \frac{Zm}{2} (\varepsilon_{BB} - \varepsilon_{AA}) + 2\Omega\ Zl\ (X_b - X_1) + Zm\ \Omega\ (2 X_b - 1) \quad (I.27)$$

where ε_{AA}, ε_{BB}, ε_{AB} represent the energy of pair of atoms. Ω is the regular solution parameter defined by $\Omega = \varepsilon_{AB} - (\varepsilon_{AA} + \varepsilon_{BB})/2$. Z is the coordination number, Zl representing the number of nearest neighbors of a given atom in the plane parallel to the surface and Zm the number of nearest neighbors between two adjacent planes. The surface tension σ_A^0 of pure component A can be expressed (at absolute zero) in terms of bond energies as

$$\sigma_A^0 = -\frac{Zm}{2}\frac{\varepsilon_{AA}}{\omega} \qquad (I.28)$$

where ω is a mean atomic area.

So taking into account eq. I.27 and I.28 we get

$$\frac{X_1}{1-X_1} = \frac{X_b}{1-X_b} \exp - \frac{1}{KT}\left((\sigma_A^0-\sigma_B^0)\,\omega - Zm\,\Omega + 2\Omega\,(Z(1+m)\,X_b - Zl\,X_1)\right) \qquad (I.29)$$

In the case of a highly dilute solution i.e. $(X_B)_b \cong 0$ the enrichment factor $(1 - X_1)/(1 - X_b)$ tends to a non vanishing finite limit

$$\frac{(X_B)_s}{(X_B)_b} = \frac{1 - X_1}{1 - X_b} = \exp + \frac{1}{KT}\left((\sigma_A^0 - \sigma_B^0)\,\omega + Zm\,\Omega\right) \qquad (I.30)$$

This formula indicates that a strong segregation of solute B is obtained in the case of a low surface tension value of the solute with respect to the solvant and a positive value of the regular solution parameter (i.e. the system has a tendancy to unmix in the bulk). Such results have been proposed for liquid solutions (organic or metallic solutions) described by a quasi-lattice assumption [I.44, I.18].

A second contribution to the enthalpy of segregation may be due to the strain energy associated with a solute atom in a solid solution, arising from the difference in atomic volumes between the solute and the solvent. MacLean [I.45], in order to calculate the segregation to grain boundaries in dilute alloys, postulates that the heat of segregation must be equal to the gain of elastic strain energy in the matrix when a solute atom in the bulk is exchanged with a solvent atom in the grain boundary. More recently Wynblatt and Ku [I.46] have proposed a unified segregation model in which contributions arising from surface energies, chemical interactions and solute strain energy are taken into account. These authors have assumed a simple additivity of the three contributions neglecting the correction of the enthalpy of mixing which itself contains a contribution due to the latice strain energy. At almost the same time Kumar [I.47] and Molinari and Joud [I.48] proposed a correction to this point. The elastic strain energy associated with a solute atom B in matrix A is evaluated from the linear continuum elasticity theory. We retain the following expression which has been proposed by Friedel [I.49]

$$W_0 = \frac{24\,\pi\,K_B\,G_A\,r_A\,r_B\,(r_B - r_A)^2}{3\,K_B\,r_B + 4\,G_A\,r_A} \qquad (I.31)$$

where K_B is the bulk modulus of the solute, G_A is the shear modulus of the solvent, r_A and r_B are the radii associated respectively with the solvent and the solute. This expression is strictly valid only in the limit of infinite dilution since interactions between solute atoms are not taken into account. Chemical interactions are described using a standard regular solution model as previously noted. In this approximation, the enthalpy of mixing of the alloy may be written

$$\frac{\Delta H_m}{(X_A)_b\,(X_B)_b} = Z\,\Omega + W_0 = Z\,\Omega' \qquad (I.32)$$

Assuming high dilution of the solute B i.e. $(X_b)_b \cong 0$ we obtain a new expression for the enrichment factor defined in eq. I.30

$$\frac{(X_B)_s}{(X_B)_b} = \exp + \frac{1}{KT}\left((\sigma_A^0 - \sigma_B^0)\ \omega + m\ (Z\Omega' - W_0) + W_0\right) \qquad (I.33)$$

This relation emphasizes the three contributions which arise in the segregation driving force. The first term corresponds to the variation of surface tension between pure components and, in all cases, induces a segregation of the tensioactive component. The second term indicates the effect of chemical interaction taking into account the number of broken bonds. The third term is due to the contribution arising from elastic strain energy due to size effect. As indicated by Lambin [I.50] and Tsaï et al. [I.51] this last contribution must be taken into account only in the case of dilute alloys with an oversized solute atom with respect to the solvent. The validity of this assumption has been checked by the analysis of experimental results performed by Seah [I.52]. This author, using a least square fit of experimental results, proposes the following expression of the solute surface enrichment factor β_B^s

$$\beta_B^s = \frac{(X_B)_s\ (X_A)_b}{(X_B)_b\ (X_A)_s}$$

$$\ln\beta_B^s = \left(0.64\ (\sigma_A - \sigma_B)\omega + 1.86\ \Omega' + 4.64\ 10^7\ M.a_A(a_B - a_A)^2\right)/RT \qquad (I.34)$$

a_A^3 and a_B^3 are the atomic volume of the element A and B, M = 1 for $a_B > a_A$ and M = 0 for $a_B < a_A$ (the coefficient of the strain term is in $J.nm^{-3}$).

The correlation of the experimental results with the prediction is shown in Fig. I.8. The scatter of the experimental results about the estimations deduced from eq. I.34 corresponds to a standard deviation of 8.7 KJ/mole over a range of free energy of segregation from - 80 to + 20 KJ/mole.

The uncertainty arising in the calculation of the strain energy term given by eq. I.31, makes a direct verification of eq. I.33 difficult. Nevertheless the relatively good fit of experimental results to eq. I.34 confirms the validity of the previous theoretical analysis including three contributions (surface energy term, chemical energy term and strain energy term) in the segregation driving force.

I.3.1.2 Multilayer treatment

The first attempt to solve eq. I.26 has been carried out by Defaye and Prigogine [I.44] using a classical regular solution theory. With the same model Williams and Nason [I.53] have clearly correlated the shape of the concentration profile, whose extension is limited to 3 or 4 layers, with the mixing enthalpy of the alloy ; monotonous (oscillating) profile when the alloy tends to unmix (to order) in the bulk. For example in the case of an ordering alloy a strong surface enrichment in the B element (induced by a low surface tension value of B element or a strong size effect) involves a weak enrichment of the first underlayer in the A element, maximizing the number of AB pairs which are energetically favorable.

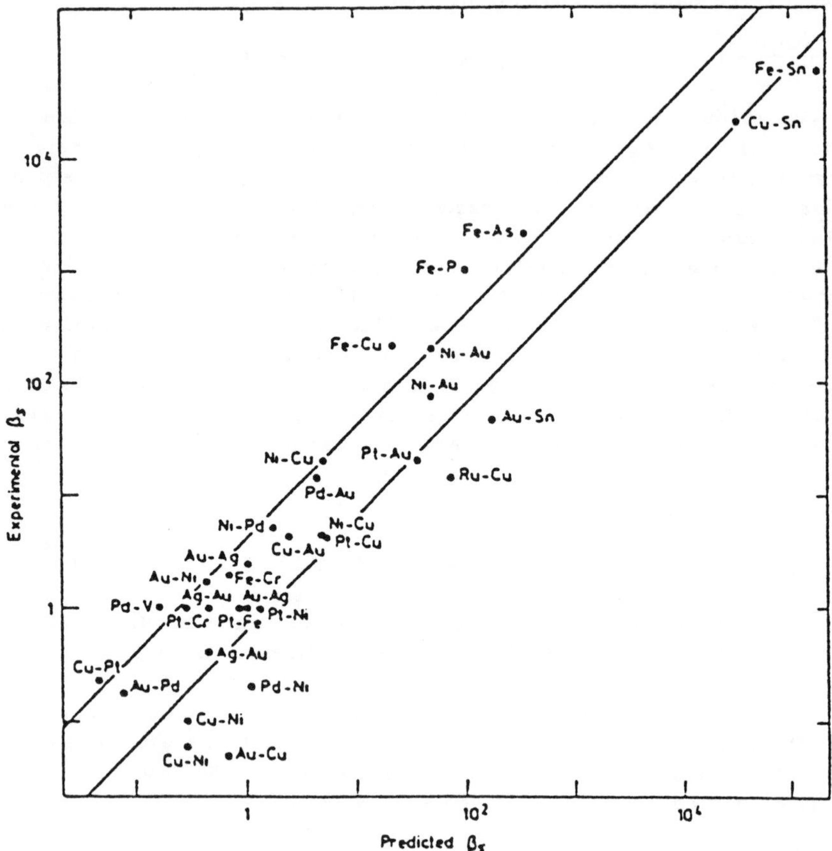

Fig. I.8 The relation between the predicted surface segregation enrichment ratio and that observed experimentally, from Seah [I.52].

More recently Tréglia and Legrand [I.54] have proposed a new treatment based on a simple mean field approach (including a surface tension term and a chemical effect) and a strain energy term treated atomistically. In order to overcome the limitation of the classical elastic theory, this last contribution is calculated using, simultaneously, an atomic model and a relaxation process to minimize the strain energy of one impurity successively in the bulk and at the surface of the solvent matrix. The total energy is written as the sum of an attractive d band term and a repulsive Born Mayer term. The strain energy contribution is then calculated in the dilute approximation for an impurity in various situations (surface, first underlayer...). Between the two conjugate dilute solutions a simple interpolation (sinusoidal shape) is used to estimate the strain energy contribution for intermediate concentrations.

The corresponding results are presented on Fig. I.9 for the Pt-Ni system. The curves corresponding to various surface planes are compared to the classical elastic contribution. We see that similar results are obtained when the size of impurity is larger than that of a solvent ($C_{Pt} \rightarrow 0$). On the other hand, when the impurity is the smallest, the sign of the elastic contribution ΔH_{se} can be reversed.

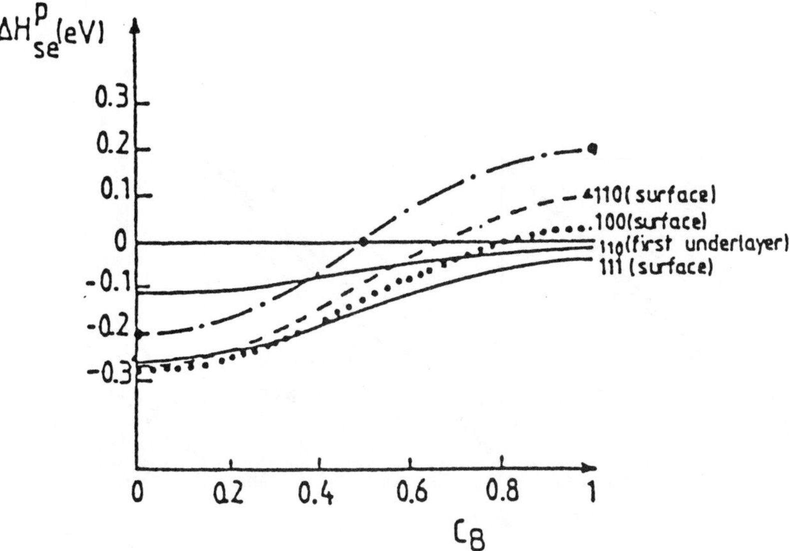

Fig. I.9 Interpolation of the strain energy contribution between its extremal values as a function of Pt bulk concentration for a Pt-Ni alloy. For comparison the symetric prediction of the elasticity theory is indicated, from [I.54].

The concentration profile of the A element is given for an alloy A_xB_{1-x}, by the following equation, similar to eq. I.26

$$\frac{X_k}{1 - X_k} = \frac{X_b}{1 - X_b} \exp - \frac{1}{KT}\left(\frac{\partial E}{\partial X_k} - \frac{\partial E}{\partial X_b} + \Delta H^K_{s.e}\right) \quad (I.35)$$

with k = 1, 2... P : number of planes. In this equation E is the energy of the system calculated using a simple regular solution model.

Application of this equation to the Pt-Ni system is schematized on Fig. I.10. Calculations have been made neglecting the difference in surface tension between Pt and Ni.

We observe in the whole concentration range a Pt enrichment of the surface, a depletion on the first underlayer and again a slight enrichment of the second underlayer. This oscillation is much more damped for the 111 face than for the 100 one. These results are in qualitative agreement with the experiments [I.55].

The interest of this kind of approach has been recently enhanced by the theoretical work of Tréglia, Legrand and Ducastelle [I.56]. They have shown that in the case of surface segregation processes in transition metal alloys, the segregation energy can be written in formal agreement with those derived from a classical Ising model. Such an Ising hamiltonian is recovered by expanding the total energy relative to that of a reference medium (the disordered alloy) as in the case of bulk ordering processes [I.57].

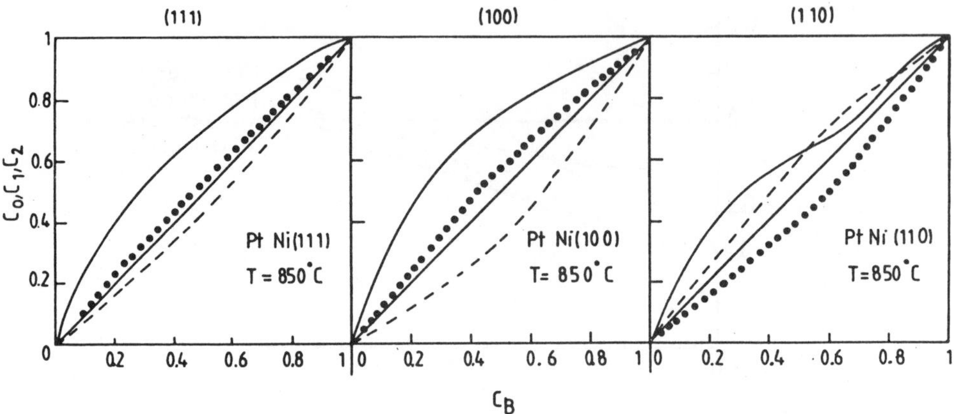

Fig. I.10 Variations of the Pt concentration at the surface (——) on the first (---) and second (···) underlayers as a function of the Pt bulk concentration, for the (111), (100) and (110) surface planes of the Pt-Ni alloy at T = 850°C, from Tréglia et al. [I.54].

The first term of this effective hamiltonian, which is generally the leading one, is quasi concentration independent and related to the difference in surface tensions between the pure elements. The second term is a quadratic one involving effective pair interactions larger at the surface than in the bulk. More precisely, they calculate the variation of the effective pair interactions $V_{nm} = \frac{1}{2} (V_{nm}^{AA} + V_{nm}^{BB} - 2 V_{nm}^{AB})$ (which is similar to $- \Omega$, Ω being the regular solution parameter) for the three low index faces of f.c.c. alloys. Their trends, as a function of the average d band filling and of the concentration are similar to those encountered in the bulk but their amplitudes are much larger. More quantitatively they find

$$V_{ss} \cong V_{sb} \cong 2 V_{bb} \text{ for the (110) plane}$$

$$V_{ss} \cong V_{sb} \cong 1.5 V_{bb} \text{ for the (100) and (111) plane}$$

This variation may be important for surface segregation phenomena. In particular, it explains [I.58] the segregation reversal observed for Pt-Ni between (111) and (110) planes [I.59].

Another energetic model deduced from the embedded-atom method [I.60] has been recently applied to surface segregation of a binary alloy [I.61, I.62]. In the EAM the total electron energy is written as a superposition of contributions from individual atoms. The electron density is approximated by a uniform electron density calculated from a superposition of the atomic electron densities. This crude approximation is improved by adding the core-core interactions.

The total energy is written

$$E_{tot} = \sum_i F\ (\rho_h(r_i),\ Z_i) + \frac{1}{2} \sum_{i,j} \phi\ (R_{ij},\ Z_i,\ Z_j) \tag{I.36}$$

The ith atom is supposed to sit at position r_i and to have the atomic number Z_i, ρ_h (r_i) is the electron density of the host atoms surrounding the ith atom, the function $F(\rho_h\ (r_i),\ Z_i)$ gives the energy to embed the ith atom in the host. $\phi(R_{ij},\ Z_i,\ Z_j)$ is the repulsive core-core interaction between two atoms with atomic number Z_i and Z_j at a distance $R_{ij} = (r_i - r_j)$.

The core-core terms ϕ and the embedding functions F are parametrized and fitted to experimental values of bulk properties for the pure metals and for dilute binary alloys.

Eq. I.36 can be used to calculate the total energy for the surface region of an alloy A_xB_{1-x}. The surface is considered as a slab of P atomic layers k = 1, 2, ...P having particular concentration X_k. The composition profile can be found as a solution to eq. I.26. The advantage with this method is that the energies depend explicitely on the geometry so, the relaxation of the interlayer distances near the surface can be obtained.

The application of this method to the Pt-Ni system has been performed by Lundberg [I.62] and reproduces the observed plane related segregation with Pt enrichment on the (111) plane and the Ni enrichment on the (110) plane [I.59] with a qualitative agreement of the calculated interlayer distances with the experimental ones. In addition this treatment shows interesting results about the existence of a metastable state of segregation on the (110) plane of $Pt_{0.5}Ni_{0.5}$. This point corresponds to the fact that eq. I.26 does not always give a unique solution X_k of the concentration profile.

I.3.2 The quasichemical approximation and Monte Carlo simulations

The quasichemical treatment has been applied to the segregation problem by Joud et al. [I.63] using the monolayer approximation and by Kumar [I.64] in the more general case of a multilayer segregation. In this treatment the total configurational energy of the system is written as the sum of the energies of various pairs considering only the nearest neighbor interactions. The configurational entropy can be expressed using the point probability P_i and the pair probability P_{ij} for a crystal A_xB_{1-x} consisting in various layers (k = 1,... ∞)

$$S = N K \sum_{k=1}^{\infty} \{ (Z - 1) \sum_{i=A,B} p_i^k \ln p_i^k - \frac{1}{2} Zl \sum_{ij} p_{ij}^{kk} \ln p_{ij}^{kk}$$

$$- Zm \sum_{ij} p_{ij}^{k\ k+1} \ln p_{ij}^{k\ k+1} \} \qquad (I.37)$$

Minimization of the free energy is performed with respect to X_k the molar fraction in layer k and with respect to the short-range order parameters α_k and β_k defined by

$$X_k (1 - X_k) \alpha_k = p_{AB}^{kk}$$

$$X_k (1 - X_{k+1}) \alpha_k = p_{AB}^{k\ k+1}$$

The first minimization gives the following expression

$$\frac{X_k}{1 - X_k} = \frac{X_b}{1 - X_b} \exp - \frac{\Delta F_k}{KT} \qquad (I.38)$$

with ΔF_k is the free energy of segregation in the layer k.

Minimization with respect to the S.R.O. parameters gives the classical pair equilibrium equation

$$\frac{\left(p_{AB}^{11}\right)^2}{p_{AA}^{11}\ p_{BB}^{11}} = \frac{p_{AB}^{12}\ p_{BA}^{12}}{p_{AA}^{12}\ p_{BB}^{12}} = \exp - \frac{2\ \Omega_s}{KT}$$

$$\frac{\left(p_{AB}^{KK}\right)^2}{p_{AA}^{KK}\ p_{BB}^{KK}} = \exp - \frac{2\ \Omega}{KT} \qquad (I.39)$$

with $\Omega = \varepsilon_{AB} - \frac{\varepsilon_{AA} + \varepsilon_{BB}}{2}$ and similarly for the surface.

Application of this treatment to various systems (Au-Ag, Cu-Ni) predicts surface compositions which are close to that calculated using the B.W. approximation. The calculated S.R.O. parameters cannot be actually compared with experimental results. Nevertheless we have obtained a good agreement between the calculated S.R.O. values and those deduced from Monte Carlo simulations for three low-index surface planes of F.C.C. alloy with various interchange energies Ω [I.65].

The Monte Carlo simulation has been applied to surface segregation phenomena, the interest being the calculation of concentration profiles without any assumption about configurational entropy. Donnelly and King [I.66], Lambin [I.50] and more recently Arroyo and Joud [I.67] have widely studied the various applications of this method.

The application to the Cu-Ni system is presented in Fig. I.11 for various surface orientations [I.68]. According to the bond breaking theory, the component with the lowest bond strength will segregate i.e. Cu. Due to positive interchange energy (Ω = 0.01 eV) the concentration profiles are monotonously decreasing on 3 or 4 layers in qualitative agreement with experimental results.

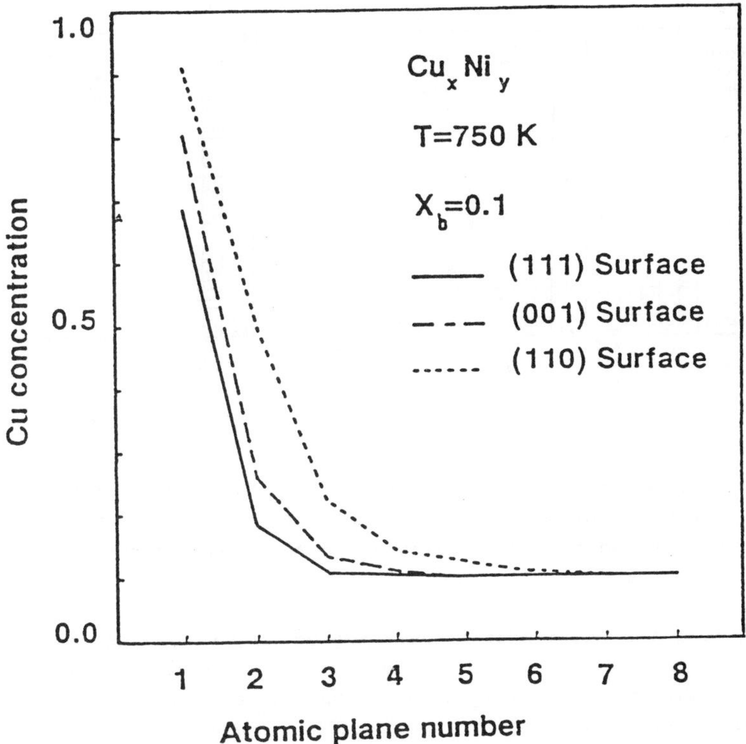

Fig. I.11 Cu composition depth profile for (111), (001) and (110) surface of $Cu_{0.5}Ni_{0.9}$ alloy at T = 750 K, from [I.68].

The cluster distribution of the various sites on the surface defined in a similar way as Donnelly and King is shown on Fig. I.12 for the three surface orientations of Cu-Ni system. The distribution is given by the parameter f_i

$$f_i = \frac{i\, n_x^i}{N_x^1} \qquad (I.40)$$

where n_x^i is the number of clusters of size i ; N_x^1 is the total number of X atoms on the surface.

These distributions are strongly influenced by crystallographic orientation, surface layer composition and bulk composition. We also note the decrease of the distribution with respect to B.W. calculations for clusters of size 1 and 2, due to the positive sign of the interchange energy.

The Cu-Ni alloy is representative of systems with weak size effects. The leading term of the segregation energy is the surface tension contribution and explains the observed Cu segregation. The sign of the interchange energy Ω explains the monotonous concentration profile.

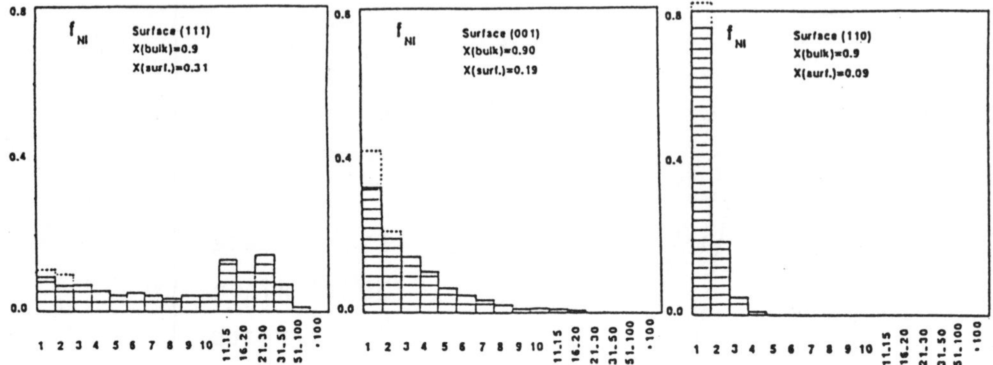

Fig. I.12 Ni cluster size distribution for (111), (001), (110) surfaces of a $Cu_{0.1}Ni_{0.9}$ alloy at T = 750 K. Dotted lines correspond to a B.W. calculation for clusters of size 1 and 2, from [I.68].

Conversely, for the Pt-Ni system whose segregating properties are dominated by a competition between chemical and elastic energy contributions, it is necessary to introduce a strain energy term in the Markov chain of the simulation. This has been done for the results presented on Fig. I.13. The energetic model which has been used is the nearest neighbor model proposed by Donnelly and King where the pair energy is modulated by the coordination number of atoms [I.66]. In addition a partial elastic energy has been attributed to the Pt atoms located in the surface layers as proposed by Tréglia and Legrand [I.54].

The concentration profiles for the (111) and (110) surface orientations exhibit oscillations in agreement with the thermodynamical prediction based on the sign of interchange energy (Ω = - 0.038 eV). We also note the inversion of the segregating element for the two surface orientations (Pt for (111) orientation, Ni for (110) orientation) in agreement with the experimental observations [I.55, I.59].

I.4 THE KINETICS OF SURFACE SEGREGATION

In the practical situation where segregation is important, the segregating atoms need enough time to reach their equilibrium concentration as defined by the segregation theories. Most models of kinetics are derived from McLean's approach [I.45] applied to the grain boundary. Application to the segregation at the free surface has been done by Lea and Seah [I.69]. Diffusion in the crystal

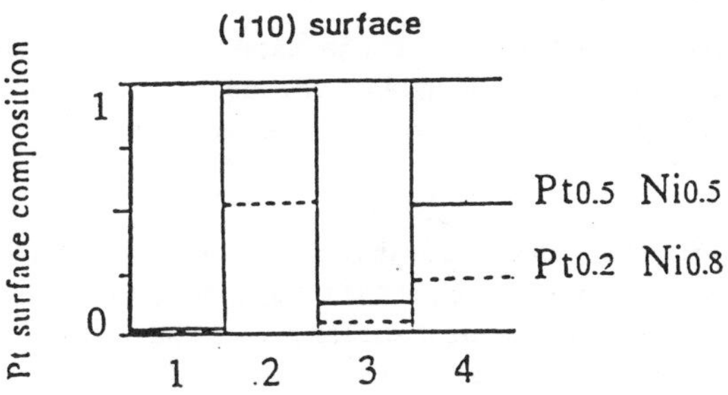

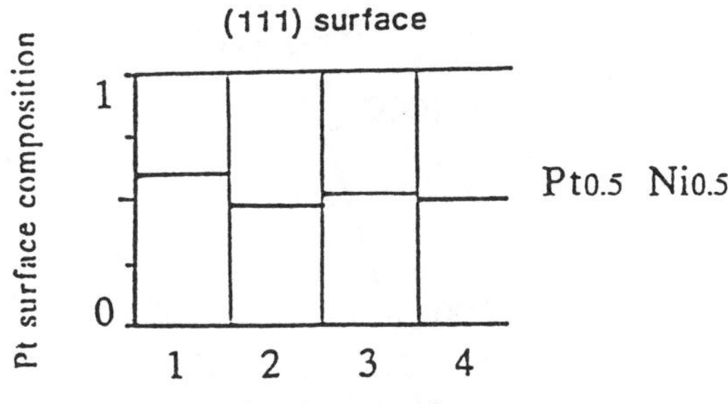

Fig. I.13 Depth concentration profiles calculated by Monte-Carlo simulation for (110) and (111) surfaces of Pt-Ni alloys at 1200 K, from [I.68].

is described by Fick's law and the ratio of the solute at the surface to that in the adjacent atomic layer of the bulk is expressed by a constant enrichment ratio β. The kinetics of the segregation is thus described by

$$\frac{C_s(t) - C_s(0)}{C_s(\infty) - C_s(0)} = 1 - \exp \frac{Dt}{\beta^2 d^2} \operatorname{erfc} \left(\frac{Dt}{\beta^2 d^2}\right)^{1/2} \qquad (I.41)$$

This equation can be approximated for $\frac{Dt}{\beta^2 d^2} < 1$ by

$$\frac{C_s(t) - C_s(0)}{C_s(\infty) - C_s(0)} = \frac{2}{\beta d} \left(\frac{Dt}{\pi}\right)^{1/2} \qquad (I.42)$$

D is the solute diffusion coefficient and d is the solute monolayer thickness.

The main problem of this equation is the enrichment ratio β which is constant only for dilute systems with low segregation levels.

In general the enrichment ratio β decreases as the segregation proceeds. If the initial values are high, the reduced equation (I.42) may be used practically up to saturation, giving rise to a $t^{1/2}$ dependence.

This comment is supported by the experimental work of Lea and Seah [I.69] on the Fe-Sn 1% alloy presented on Fig. I.14.

From measurements of the enrichment ratios with surface concentrations Lea and Seah are able to calculate a family of curves corresponding to eq. I.41, giving rise to a graphical solution for the time dependence of the segregation. This solution obtained for time dependent values of $\beta(t)$ is in close agreement with experimental results and differs from McLeans's curve for constant β^∞ = 175.

A detailed analysis for the saturation occurring in the McLean theory has been presented by Rowlands and Woodruff [I.70]. The filling of the surface by solute atoms induces a decreasing probability for the solute atoms on the first underlayer to jump into the surface.

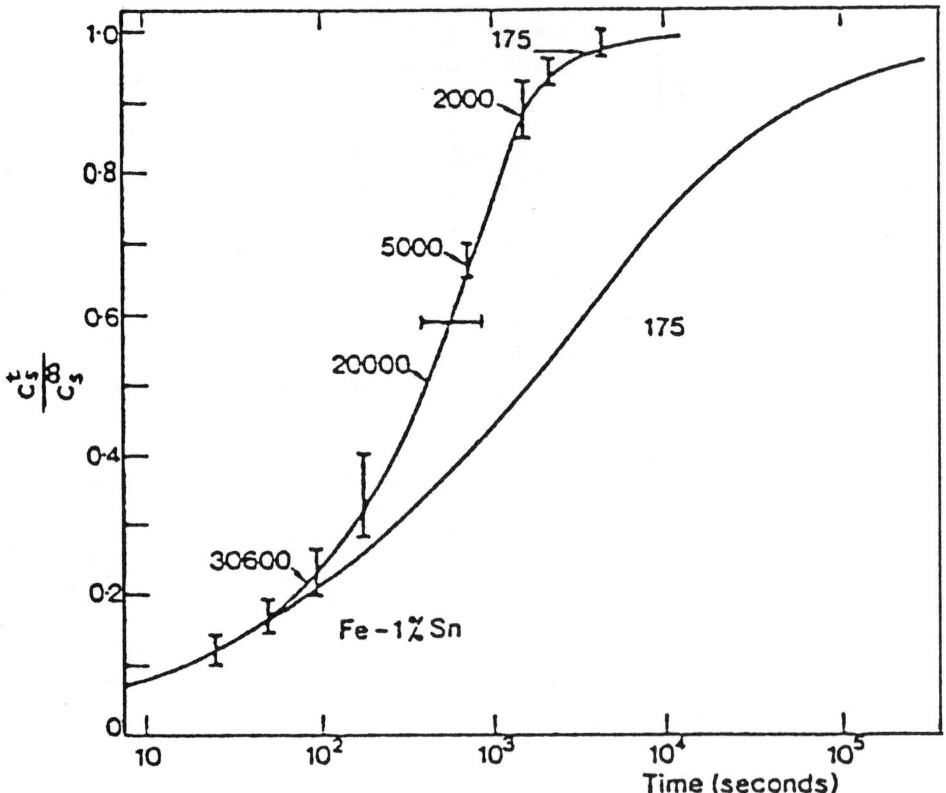

Fig. I.14 Kinetics of segregation in the Fe-Sn 1% system at 550°C. Experimental results and theoretical curve corresponding to the constant β^∞ case and the $\beta(t)$ case varying as indicated by numerical values, from [I.69].

For surface segregation experiments in a vacuum system, evaporation of the segregant is possible. Lea and Seah [I.69] evaluated the effect of surface evaporation concomitant with surface segregation. Initially no divergence from the McLean equation I.41 is obtained. However, as the segregation builds up, the evaporation rate increases and the bulk alloy begins to be depleted of solute.

The segregation passes through a maximum and eventually falls down to a low value.

I.5 CONCLUSION

As a conclusion we would like to emphasize the recent progress in the understanding of the energetic aspects in the surface energy and surface segregation calculations. This situation has to stimulate careful experimental measurements of the surface tension of pure solid metals which appears as a critical parameter in theoretical calculations. In addition, experimental determinations involving surface concentration, geometrical surface relaxation and chemical order effect measurements are actually necessary for some reference systems.

II EQUILIBRIUM OF ELASTICALLY CONSTRAINED PHASES

We explain below how elastic constraints influence the stability of solid solutions and modify the compositions of two phases which coexist at equilibrium. The first part recalls basic notions on spinodal decomposition and nucleation when a size mismatch is taken into account. The second part deals with the stability of epitaxial solid solutions against several types of perturbations. The third part focusses on coherent phase diagrams where two phases of different composition are in mutual equilibrium, and possibly with an underlying substrate which imposes a given value for the lattice parameter. The fourth part mentions a recent model which tries to incorporate the strain effects at the atomic level.

II.1 NUCLEATION-GROWTH : SPINODAL DECOMPOSITION

Let us consider a solid solution (A,B) which exhibits a miscibility gap at low temperatures. Representing the free enthalpy of mixing g_o per mole as a function of the concentration of component B (denoted hereafter by c) yields the curve drawn in Fig. II.1.a. Using the chemical potentials μ_A, μ_B, the free enthalpy becomes

$$g_o = (1 - c)\ \mu_A + c\ \mu_B \qquad \text{(II.1)}$$

The graphical interpretation of this equation is straightforward (remember that, in this representation, the slope is $dg_o/dc = \mu_B - \mu_A$).
The equilibrium of two phases α, β (concentrations C_o^α, C_o^β with $C_o^\alpha \le C_o^\beta$) is characterized by the uniformity of each chemical potential throughout the system

$$\mu_A^\alpha = \mu_A^\beta \quad \text{and} \quad \mu_B^\alpha = \mu_B^\beta \qquad \text{(II.2)}$$

The graphical interpretation shows that C_o^α, C_o^β are the common tangency points to the free enthalpy curve. Drawing the locus of C_o^α, C_o^β as a function of temperature defines the miscibility gap shown in figure II.1.b : the maximum defines the critical temperature T_c for unmixing and occurs for a particular composition Co^c somewhere inbetween (Co^{α}, Co^{β}). A solid solution Co is stable, i.e. remains single-phased, whenever $Co < Co^{\alpha}$ or $Co > Co^{\beta}$, due to the convexity of g_o.

Conversely, when $Co^{\alpha} < Co < Co^{\beta}$, the stable state of the alloy is no longer single-phased, and consists of a mixing of two phases (α, β). The proportion z of the β-phase obeys the mass conservation (lever-rule) and is given by

$$(1 - z)\, Co^{\alpha} + z\, Co^{\beta} = Co \qquad \text{(II.3)}$$

Let us denote by (S^{α}, S^{β}) the inflection points of g_o and by ($Co^{S\alpha}$, $Co^{S\beta}$) the corresponding concentrations with $Co^{S\alpha} < Co^{S\beta}$. The locus of S^{α} and S^{β} when temperature is varied is the spinodal curve ; it is tangent to the miscibility gap curve at point (Co^c, T_c) and lies entirely inside the latter (Fig. II.1.b). When an homogeneous solid solution is quenched from a high temperature down to a low one, it is experimentally observed that its decomposition follows two different kinetics according to the closeness to the miscibility gap boundary. Near the boundary, the system is metastable and the precipitation is traditionnally described by the nucleation and subsequent growth of small particles of the β-phase in equilibrium with the remaining matrix. This approach accounts for the existence of an incubation time, its decrease when Co moves further away from the boundary and its divergence at $Co = Co^{\alpha}$. On the other hand, far inside the miscibility gap, the system is unstable and the starting manifestation of a second phase is the development of a concentration modulation with a characteristic wave length : this spinodal decomposition is accounted for by cooperative models which take into account the change in free energy attached to such a modulation which is, unlike nuclei of β-phase, small in amplitude but large in (spatial) extent.

The two above mentioned models have been traditionally considered as mutually exclusive, the spinodal line being considered as a sharp transition between both types of decomposition.

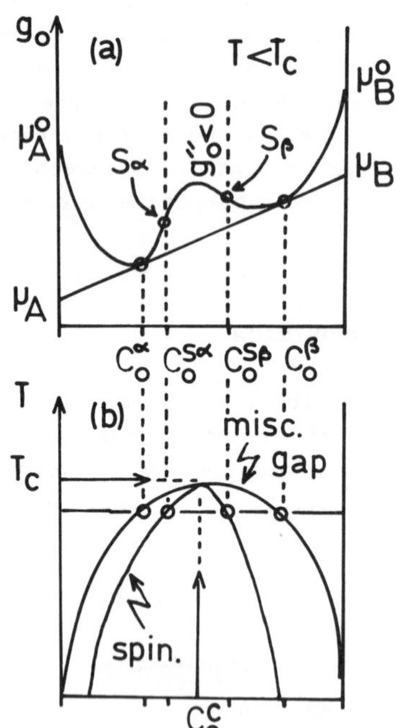

Fig. II.1 Unmixing : a) free enthalpy curve ; b) incoherent miscibility gap and spinodal line.

However, the reader should remember the two following alterations in this too simple picture :

- Theoretically, cooperative models have been extended to metastable regimes and a cluster approach has been used to describe the spinodal decomposition [II.1].
- Experimentally, only a few systems with long-range forces undergo a true spinodal decomposition. In most cases, the unmixing gradually evolves from a nucleation-growth mechanism to a non-linear spinodal decomposition, the pertinent kinetic parameter being the depth of the quench inside the miscibility gap [II.2].

For sake of simplicity we will however keep the usual framework of spinodal decomposition and will make use of the underlying linear analysis of stability which holds at least for the very early stages of the decomposition process.

II.1.1 Nucleation regime without stress [II.3]

The occurrence of the β-phase is controlled by a nucleation-growth-coarsening process whenever $Co^{\alpha} < Co < Co^{S\alpha}$. Any concentration fluctuation occurs in a solid solution which acts as a reservoir and which fixes the chemical potentials μ_A, μ_B at their value for Co : the energy associated with such a fluctuation is represented by a point on the tangent to the free enthalpy curve at Co (Fig. II.2).

It is easily seen that the solid solution is stable against concentration fluctuations of small amplitude (smaller than $|C_b - Co|$) but that for sufficiently large ones, it will decrease its free energy if it chooses instead to convert a significant fraction of its volume to the β-phase in which the chemical potentials have more favourable values. The corresponding decrease in free enthalpy ("the chemical driving force for nucleation") is represented by the vertical arrow of length $\Delta g^{c}_{\alpha\beta}$ pointing downwards ($\Delta g^{c}_{\alpha\beta} < 0$) : this driving force is zero when $Co = Co^{\alpha}$ and is a maximum when Co reaches $Co^{s\alpha}$. It plays a leading role in the kinetics of nucleation.

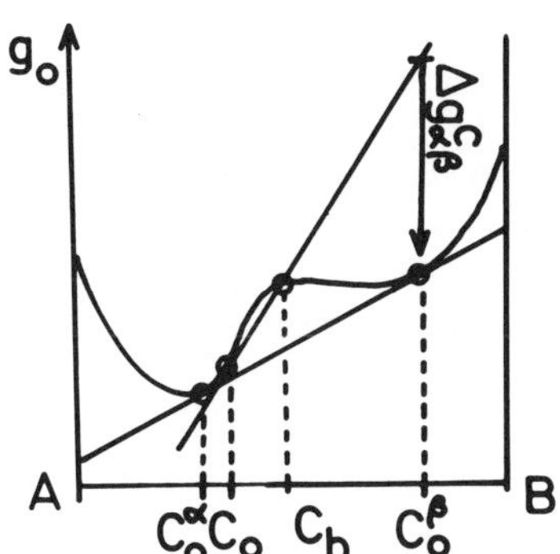

Fig. II.2 Schematic representation of the chemical driving force for the nucleation of a β-phase embryo in an α-matrix of composition Co.

Denoting by $\Delta g(n)$ the free enthalpy for a β-nucleus containing n moles, we have

$$\Delta g(n) = n\, \Delta g^{c}_{\alpha\beta} + A\, n^{2/3}\, \gamma_{\alpha\beta} \quad \text{(II.4)}$$

where A depends on the geometrical shape of the nucleus [II.1] and $\gamma_{\alpha\beta}$ is the energetic contribution of the newly created interface α/β. $\Delta g(n)$ passes through a maximum Δg^* obtained for a critical size n^*

$$n^* = \left(\frac{2\ A\ \gamma_{\alpha\beta}}{-\ 3\ \Delta g^{c}_{\alpha\beta}}\right)^3 \qquad \Delta g^* = -\frac{1}{2}\, n^*\, \Delta g^{c}_{\alpha\beta} \qquad \text{(II.5)}$$

The steady state nucleation current is the net number of nuclei crossing the critical size per unit time. Among the N nuclei per unit volume, only $N \exp -\frac{\Delta g^*}{kT}$ have the required energy ; if w* is the frequency at which they cross the critical size, then the nucleation current J_s is given by

$$J_s = w^* N \exp - \frac{\Delta g^*}{kT}$$

This is an overestimation however ; thermal fluctuations allow clusters which are slightly beyond the critical size n* to regress to a subcritical one and subsequently dissolve. A corrective multiplication factor Z (Zeldovich's correction) must be introduced in J_s to account for this effect

$$J_s = Z N w^* \exp - \frac{\Delta g^*}{kT} \qquad \text{(II.6)}$$

Z is related to the curvature of the potential barrier $\Delta g(n)$

$$Z = \left(- \frac{1}{2\pi\, kT} \frac{\partial^2 G}{\partial n^2} \Big|_{n^*} \right)^{1/2}$$

and is typically of the order of 10^{-2}. It can be shown that 1/Z is approximately the width of the potential barrier $\Delta g(n)$ at a distance kT below the maximum Δg^* (Fig. II.3). In the same spirit the incubation time τ_s which is necessary to establish the steady state value of J_s is reached when the clusters have grown sufficiently to escape a possible dissolution, i.e. when their size n is at least n* + 1/2Z ; since the potential barrier is nearly flat in the neighbourhood of n* the nuclei propagate by random walk in this region with a frequency approximately equal to w*. Hence

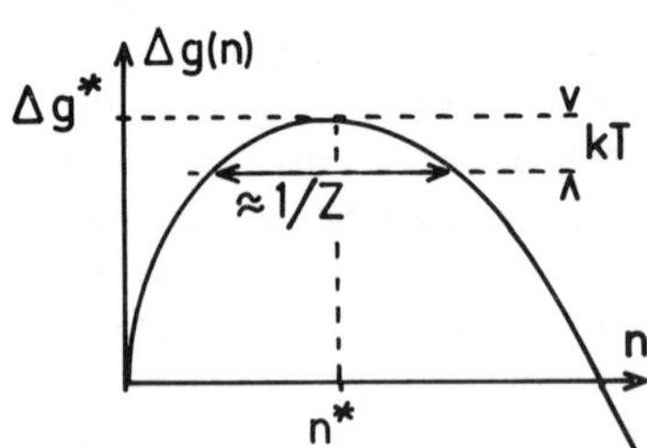

Fig. II.3 : Free enthalpy curve for a nucleus containing n atoms.

$$\tau_s = \frac{1}{2\, Z^2\, w^*}$$

It can be shown that τ_s varies as $n^{*4/3}$; it decreases when moving further from the miscibility gap boundary but gets infinite at the boundary for $Co = Co^{\alpha}$.

The most controversial point of such a model is the attribution of macroscopic β-phase properties μ_i, $\gamma_{\alpha\beta}$ to small sized nuclei and the implicit assumption of no interaction between nuclei.

II.1.2 Spinodal regime without stress [II.4]

The solid solution is unstable with respect to any concentration fluctuation occurring in the medium whenever $Co > Co^{S\alpha}$. Assuming for simplicity that a plane wave fluctuation a small amplitude takes place along Oz in the whole system, we get

$$c = Co + A(q) \exp [R(q)\ t] \exp iqz \qquad (II.7)$$

where q is the wave vector.

The time evolution of the system is calculated by solving the diffusion equation. Neglecting the cross-terms in the phenomenological expressions of the fluxes, it can be shown that the flux of species B with respect to the Matano plane, and denoted hereafter by J, is given by

$$J = - M \nabla (\mu_B - \mu_A) \qquad (II.8)$$

where M > 0 is a function of the mobilities of both species. Hence

$$J = - M g_0'' \nabla c \qquad \text{with } g_0' = \frac{dg_0}{dc} \ ; \quad g_0'' = \frac{d^2g_0}{dc^2}$$

Focussing on the very early stage of decomposition, we neglect the spatial dependence of M and g_0''. The diffusion equation for the B component becomes

$$\frac{\partial c}{\partial t} = - \operatorname{div}(J) = - \operatorname{div} (- M g_0'' \nabla c) \approx M g_0 \frac{\partial c^2}{\partial z^2}$$

Using the above expression for c yields

$$\frac{\partial c}{\partial t} = R(q)\ (c - Co) = - M g_0''\ q^2\ (c - Co)$$

The amplification factor R(q) is thus given by

$$R(q) = - M g_0''\ q^2 \qquad (II.9)$$

It is positive whenever $Co^{S\alpha} < Co < Co^{S\beta}$. i.e. when g_0'' is negative, and whichever the value of the wave vector q. In other words the stability limit is described by the equation $g_0'' = 0$ (Fig. II.4). The expression for R(q) is however incorrect : the modulations which are expected to be actually observed in the solid solution are those associated with the maximum R(q), that are the very small wavelengths (in a discrete medium, the lower bound is the lattice spacing). It is not observed in the experiments which show much larger wavelengths typically of the order of 100 Å or more. This failure stems from an incorrect evaluation of the free enthalpy g_0 for a system in which steep gradients develop : indeed, whenever the concentration is likely to change significantly within the range of atomic interactions, the energy of the system at r cannot be a function of the local concentration only. Introducing the derivatives of c(r) gives

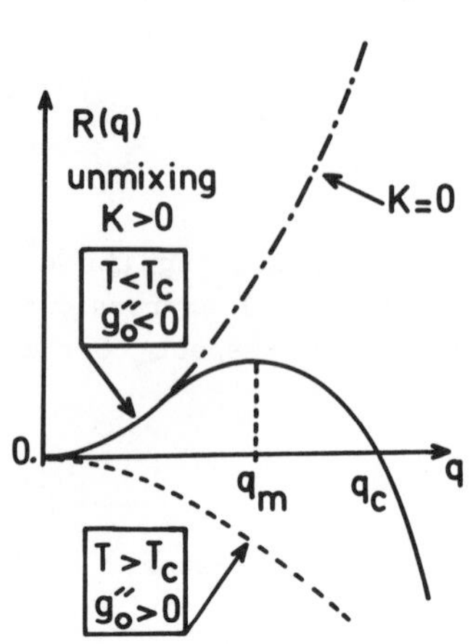

Fig. II.4 Time-amplification rate R(q) as a function of the wave vector. The dash-dotted curve (K = 0) corresponds to the case where the gradient energy term is neglected.

$$g(r) = g_0\ (c(r)) + g_1(\nabla c)^2 + g_2\nabla c^2 + ..$$

Here g_0 $(c(r))$ is the molar free enthalpy for an homogeneous solid solution of uniform composition c(r) which has been implicitely used up to now ; the gradients are envaluated at point r. The first order term is dropped according to parity arguments. It can be further shown that the second order tems can be recombined into a sngle one according to

$$g(r) = g_0\ (c(r)) + K\ (\nabla c)^2 \qquad \text{(II.10)}$$

where K is the gradient energy coefficient defined by

$$K = g_1 - \frac{dg_2}{dc}$$

The occurrence of this new term in the molar free enthalpy g_0 opposes the preferential growth of concentration modulations of too small wavelengths.

It is then shown that the new diffusion potential which drives the matter fluxes is no longer g_0' but $g_0' - 2K\ \nabla^2 c$. Introducing this new expression into the diffusion equation finally yields

$$R(q) = -\ M\ (g_0'' + 2K\ q^2)\ q^2 \qquad \text{(II.11)}$$

The gradient energy term is preponderant at large q (small wavelengths) and imposes a negative sign to R(q) which implies a regression of the very small wavelengths. The solution is easily shown to be unstable for all wavelengths larger than a critical one λ_c

$$\lambda_c = \left(\frac{8\ \pi^2\ K}{-\ g_0''}\right)^{1/2}$$

It will preferentially develop concentration fluctuations which correspond to the maximum of R(q) occurring for

$$q = q_m = \frac{q_c}{\sqrt{2}} \quad \text{and} \quad \lambda_m = \lambda_c\ \sqrt{2} \quad \text{(Fig. II.4)}$$

In the frame work of such a description, the interface is always coherent and its energy is attached to the gradient of the corresponding concentration profile. In a system where the concentration depends only on coordinate z, the energy of a planar interface perpendicular to Oz is given by [II.4]

$$\gamma = N \int_{-\infty}^{+\infty} (g_0 + K\ (\nabla c)^2)\ dz - Co\ \mu_B^{\alpha} - (1 - Co)\ \mu_A^{\alpha}$$

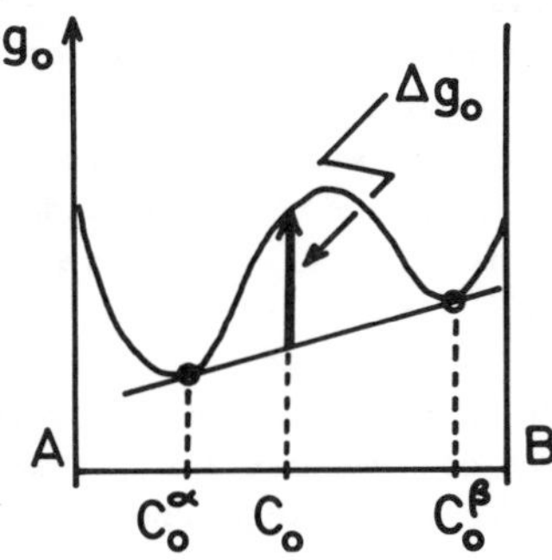

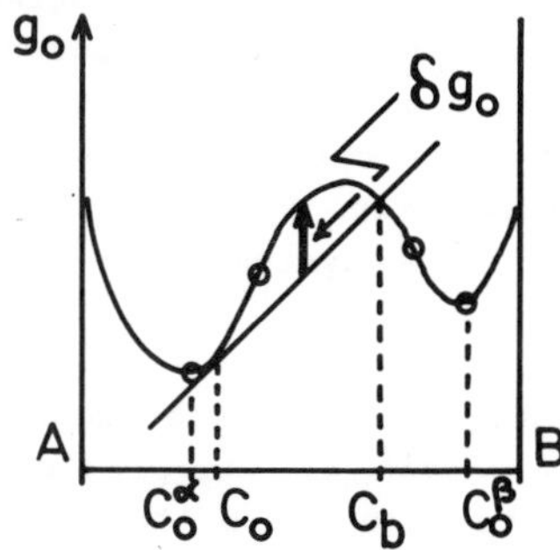

Fig. II.5 Representation of Δg_0 a) and δg_0 b) in spinodal decomposition.

$$= N \int_{-\infty}^{+\infty} (\Delta g_0 + K\,(\nabla c)^2)\,dz \qquad \text{(II.12)}$$

where N is the number of sites per unit volume and $\Delta g_0 = g_0 - (1 - Co)\,\mu_A^\alpha - Co\,\mu_B^\alpha$ is depicted on Fig. II.5.

The profile which minimizes γ is found to obey

$$\Delta g_0 = K\,(\nabla c)^2$$

Hence

$$\gamma = 2N \int_{-\infty}^{+\infty} K\,(\nabla c)^2\,dz = 2N \int_{Co^\alpha}^{Co^\beta} (K\,\Delta g_0)^{1/2} dc$$

The thickness l of the interface is defined by

$$l = \frac{Co^\beta - Co^\alpha}{\nabla c|_{max}} = (Co^\beta - Co^\alpha)\left(\frac{K}{\Delta g_0|_{max}}\right)^{1/2}$$

It is to be noted that l gets infinite at the coexistence curve when $Co = Co^\alpha$ or Co^β.

In the same spirit, the free enthalpy Δg associated to a concentration fluctuation along Oz in a solid solution Co is given by

$$\Delta g = \int_{-\infty}^{+\infty} (g_o(c) + K\,(\nabla c)^2 - g_o\,(Co))\,dz = \int_{-\infty}^{+\infty} (\delta g_o + K\,(\nabla c)^2)\,dz$$

where $\delta g_o(c)$ is represented in Fig. II.5.b.

The critical nucleus is defined by that concentration profile which minimizes Δg under the constraint of mass conservation. The main difference from the classical nucleation theory is the fact that the nucleus has no longer a uniform concentration. It can be shown further [II.4] that its center is necessarily stable with respect to the surrounding matrix and has a composition larger than C_b (Fig. II.5.b).

The most disputable aspect of this theory is the very existence of the hump exhibited by the free enthalpy diagram between Co^{α} and Co^{β} which plays a crucial role. The portion of the g_o curve with upwards curvature corresponds to metastable states which have finite lifetimes and are, as such, observable ; the portion of the free enthalpy between $Co^{s\alpha}$ and $Co^{s\beta}$ which corresponds to unstable states is, conversely, a hypothetical guess ; it can only be understood as the free enthalpy of a system which would be artificially constrained to remain homogeneous.

In the above models, the growth of the β-phase could be achieved without implying any strain of the precipitate or of the matrix : this corresponds actually to the case where the lattice mismatch between the α and β phases is accomodated by dislocations at the interface α/β, the latter being incoherent. This is the reason why the curves defined above (miscibility gap, spinodal line) represent the incoherent phase diagram of the binary A-B.

In all that follows, we focus on the case where the lattice mismatch is not compensated by plastic accomodation and where the maintenance of coherency across the interface α/β involves a strain energy.

II.1.3 Strain contribution

Among many difficulties, the contribution to the energy of the strain induced by a composition heterogeneity depends on the shape of this heterogeneity : for a non-spherical one, it depends on its orientation with respect to the axes of the crystal. At last the finite dimension of the solids implies the existence of image forces, the contribution of which depends on the size and the shape of the whole solid itself in which the heterogeneity is embedded. As a consequence, the contribution of strains to the driving force for unmixing can only be estimated for simple models. In the following, we assume that replacing an A atom (atomic volume Ω_A) by a B atom (Ω_B) produces an isotropic and purely dilatational stress-free strain of the solid solution : B is considered as a sphere of radius r_B which is forced into a smaller hole of radius r_A. And the corresponding linear stress-free strain per unit concentration change η is defined by $\eta = \dfrac{\Omega_B - \Omega_A}{3\ \Omega_A}$.

II.1.3.1 Nucleation regime with stress

It is usually assumed that the average atomic volumes Ω^{α} and Ω^{β} of the α and β phases are different but concentration independent. The calculation of the elastic energy E^{el} which is stored in the matrix (α) and the precipitate (β) containing n moles yields a classical result [II.5]

$$E^{el} = n\ \Delta g^{e}_{\alpha\beta} \text{ with } \Delta g^{e}_{\alpha\beta} = \frac{9\ \Omega^{\alpha}\ (1 - 2\nu)\ B_{\alpha}\ B_{\beta}}{2\ (1 - 2\nu)\ B_{\alpha}\ r_{\alpha}/r_{\beta} + (1 + \nu)\ B_{\beta}}\ \eta^2 \quad \text{(II.13)}$$

B_{α} and B_{β} are the bulk moduli of α and β phases ; ν, the Poisson ratio (~ 0.33) ; r_{α} and r_{β}, the atomic radii ($\Omega_{\alpha} = 4/3\ \pi\ r_{\alpha}^3$).

It is seen that $\Delta g^{e}_{\alpha\beta}$ is always positive since it depends on the square of the strain ; as a consequence it always reduces the chemical driving force for nucleation (equation II.4) and the total free enthalpy of the nucleus of β-phase becomes

$$\Delta g(n) = n\ (\Delta g^{c}_{\alpha\beta} + \Delta g^{e}_{\alpha\beta}) + A\ n^{2/3}\ \gamma^{coh}_{\alpha\beta}$$

where $\gamma^{coh}_{\alpha\beta}$ is now the energetic contribution of the coherent interface α/β.

The critical size n* is still given by equation II.5, after replacing $\Delta g^{c}_{\alpha\beta}$ by $\Delta g^{c}_{\alpha\beta} + \Delta g^{e}_{\alpha\beta}$. In usual system $\gamma^{coh}_{\alpha\beta}$ is much smaller than $\gamma^{inc}_{\alpha\beta}$ and its decrease may compensate the decrease of the denominator of n*. This explains why, in some cases, coherent metastable phases nucleate first in a solid solution, despite the fact that the chemical driving force is smaller than for the incoherent precipitation. It is also understandable that beyond a critical volume of the precipitate, the interface cannot sustain any longer the strains imposed by the maintenance of the coherency ; the size misfit is then partially cancelled by dislocations and the incoherent β-phase will precipitate instead.

II.1.3.2 Spinodal regime with stress

As above a concentration fluctuation around Co takes place along Oz according

$$c = Co + A \cos qz$$

where A includes the time-amplification factor R(q).
This concentration fluctuation induces a corresponding variation of the lattice parameter along Ox, Oy, Oz. Let us consider two neighbouring planes at z and z + a : their lattice parameters along Ox (and Oy) are not equal. However a solid of infinite extension along Ox (and Oy) cannot tolerate any lattice parameter misfit as small as it might be, which would imply an infinite energy. This is the reason why internal stresses σ_{xx} and σ_{yy} must establish in order to keep the lattice parameter along Ox (and Oy) to a constant value equal to a_o the equilibrium value of the lattice parameter of an alloy of uniform concentration Co. The lattice planes are however free to relax along Oz.

For an isotropic solid with concentration independent elastic constants, it is easy to show that the planar stresses are given by

$$\sigma_{xx} = \sigma_{yy} = -A \frac{E}{1-\nu} \eta \cos qz \qquad \sigma_{zz} = 0 \qquad \text{(II.14)}$$

and that the elastic strains along Ox and Oy must cancel exactly the stress-free strains

$$\varepsilon_{xx} = \varepsilon_{yy} = -A\eta \cos qz \qquad \varepsilon_{zz} = \frac{2\nu}{1-\nu} A\eta \cos qz$$

where E is Young's modulus.

The elastic energy density per unit volume is then

$$E = \frac{1}{2}\sigma_{ij}\,\varepsilon_{ij} = A^2 \frac{E}{1-\nu}\eta^2 \cos^2 qz = \eta^2 \frac{E}{1-\nu}(c - Co)^2 \qquad \text{(II.15)}$$

(with the convention that a repeated subscript means a summation) and its average over one wavelength is equal to $\frac{1}{2}\frac{E}{1-\nu}A^2\eta^2 = \frac{1}{2}\frac{E}{1-\nu}\varepsilon^2$, that is, independent of the wavelength (notice that η is the strain per unit concentration change whereas ε is the actual strain). The new expression of the total free enthalpy becomes

$$g = g_o(c) + K(\nabla c)^2 + \frac{E}{1-\nu}\eta^2 (c - Co)^2 \Omega_0 \qquad \text{(II.16)}$$

Since g has been defined per mole, the last term due to the strain energy has been multiplied by the average atomic volume ; it is always positive, cancels only at points where c = Co, and shifts the free enthalpy from an amount proportional to $(c - Co)^2$.

In cubic systems, the elastic energy density depends on the <u>direction</u> of the wave vector q and $E/(1-\nu)$ must be replaced by an effective modulus Y(q) which has been calculated as a function of the elastic constants C_{11}, C_{12}, C_{44} and the direction cosines of q [II.6].

The instability limit (or <u>coherent</u> spinodal) is no longer dictated by the condition $g_o'' = 0$ because of the presence of the strain energy term but instead by

$$g_o'' + 2\eta^2 Y(q)\Omega_0 = 0$$

It is clear that the coherent spinodal curve lies now entirely <u>inside</u> the incoherent one and it is easy to calculate the depression of the critical temperature T_c for unmixing with a simple model of regular solution. Assuming the $g_o(c) = c(1-c)\omega + kT(c \operatorname{Log} c + (1-c)\operatorname{Log}(1-c))$ it is found that

$$T_c^{inc} = \frac{\omega}{2k} \quad \text{and} \quad T_c^{coh} = (\omega - \eta^2 Y(q)\Omega_0)/2k \qquad \text{(II.17)}$$

In some cases where η is significant, T_c^{coh} can even become negative [II.7].

<u>The first conclusion is that strain energy stabilizes the solid solution and depresses the critical temperature for unmixing.</u>

In addition the solid solution in cubic crystals will first decompose along the directions q which correspond to the higher critical temperatures, that is, to the lower values Y(q) : namely Y(111) < Y(110) < Y(100) if $2C_{44} < C_{11} - C_{12}$ (the order is reversed in the other case).

II.1.3.3 Order-disorder transition with stress

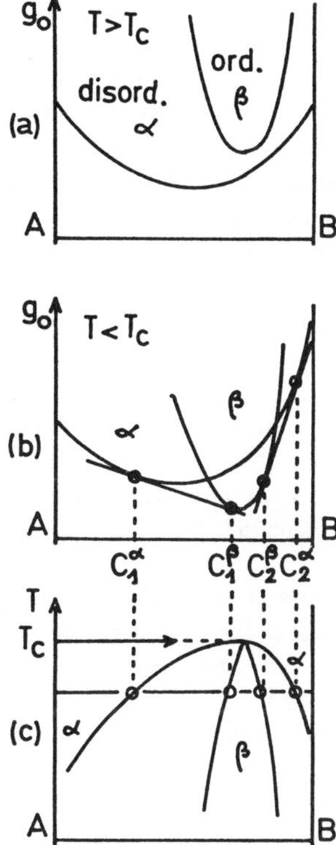

Fig. II.6 Ordering : free enthalpy curves above (a) and below (b) the critical temperature for ordering ; (c) corresponding phase diagram (β is the ordered phase).

All that has been done for unmixing can be modified to describe an order-disorder transition [II.8, II.9]. But the precipitation of an ordered structure, which is of lower symmetry than the solid solution, cannot be described by the continuous deformation of a unique free enthalpy curve which can represent at the same time both of the coexisting phases. The free enthalpy curve of the ordered phase includes a further parameter, the degree of long-range order, and is distinct from that of the solid solution ; it exhibits a minimum in the neighbourhood of the stoichiometric composition ; the lower the temperature, the sharper this minimum. For $T > T_c$, the critical temperature for ordering, this curve lies entirely above the free enthalpy of the solid solution ; at T_c they are tangent and below T_c, the usual common tangent construction determines the composition of the phases in equilibrium (Fig. II.6 a, b, c). But in all cases the free enthalpy curve of the solid solution retains its positive (upward) curvature ($g_o'' > 0$). In the same way as above, the ordered structure can be represented by a concentration wave with a wavelength of the order of the lattice parameter, and the tendency of the solid solution to precipitate an ordered phase is accounted for by a negative value of the energy gradient term ($K < 0$). Since small wavelengths are essential, a discrete approach is necessary, unlike spinodal decomposition. It is found that the time amplification factor R(q) can now be written

$$R(q) = - M\,(g_o'' + 2K\,B^2)\,B^2 \qquad \text{(II.18)}$$

where

$$B^2 = \frac{1}{a^2}\sum\,(1 - \cos qr)$$

the summation running over the nearest neighbours, and a being the lattice spacing in direction q. It is seen that B^2 reduces to q^2 for small q : the discrete approach becomes identical with the continuous one for large wavelengths ("large" when compared to a).

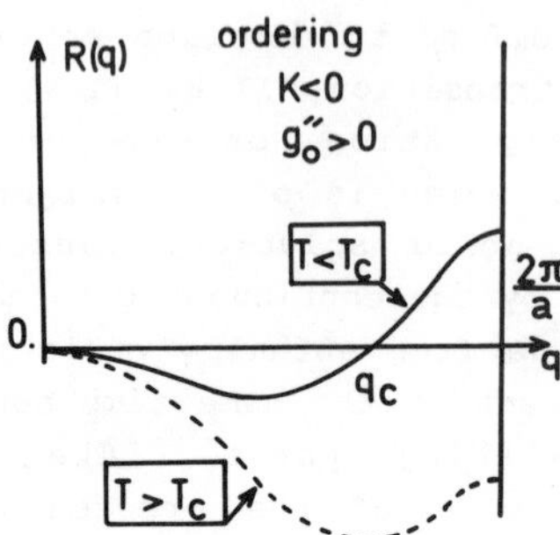

Fig. II.7 Time-amplification rate R(q) for a system exhibiting an order-disorder transition.

It is easy to check that the higher the temperature, the larger the second derivative g_o". Thus at high temperature, g_o" imposes its sign and the amplification factor R(q) is negative for all q. Decreasing the temperature lets the system cross the critical temperature for ordering T_c where the negative gradient energy coefficient becomes preponderant and accounts for a positive amplification factor in the neighbourhood of $2\pi/a$ (Fig. II.7).

The elastic energy, when calculated with this discrete model, is shown to depend not only on the direction of the wave vector q but also on its magnitude and the same holds for the effective elastic modulus Y(q), which converges towards Cahn's result at small q. This wave vector magnitude dependence of Y(q) induces a gradient energy coefficient for the elastic energy, analogous to the "chemical" gradient energy coefficient introduced above, but its contribution remains negligible as long as the atomic size difference remains reasonable.

It is shown at last that the effective modulus Y(q) may change significantly with q ; for the ordered β-brass structure, Y(q) even goes to zero at the boundary of the first Brillouin zone in <100> and <111> directions. This implies that the elastic energy involved in the formation of the solid solution is entirely relaxed on forming the ordered structure, the latter being therefore considerably stabilized [II.9].

II.2 EFFECT OF EPITAXIAL STRAINS ON THE STABILITY OF SOLID SOLUTIONS

This second part uses the results of the preceding section and works in the same spirit. It is assumed now that a solid solution is deposited on a substrate surface which imposes its lattice parameter a_s along the Ox and Oy directions. The thin film is free to relax perpendicularly to the surface along Oz. The substrate is supposed to be infinitely rigid. The instability limit is checked with respect to two different types of perturbations.

II.2.1 Concentration fluctuations along Oz

The concentration in the film is supposed to be a function of z only and to be uniform in xOy planes

$$c(z) = Co + A \cos qz$$

* If the lattice parameter a_o = a (Co) corresponding to the uniform concentration Co is equal to a_s (Co is equal to C_s, the "lattice-matching composition"), then the energy term to be added to the free enthalpy is $Y(q)\ \eta^2\ (c - Co)^2\ \Omega_0$. The elastic energy due to the coherency with the substrate stabilizes the solid solution against this type of perturbation. This might explain why an alloy, which is two-phased in a bulk configuration, can remain single-phased when deposited on a substrate because of the above mentioned decrease of the critical temperature for unmixing.

* If a_o is not equal to a_s, it is however assumed, for a coherent deposit to be possible, that there is some concentration C_s such that a $(C_s) = a_s$; in this case the lattice-matching composition C_s is different from the average composition Co of the thin film, and the energy term to be added to the free enthalpy is still given by $\Omega\ Y(q)\ \eta^2\ (c - C_s)^2$ but gives additional terms in the second derivation [II.10]. If the film were homogeneous over its whole thickness z_m, then the elastic energy per atomic area (a_s^2) would be

$$E_{hom} = \int_0^{z_m} \Omega_0\ Y(q)\ \eta^2\ (Co - C_s)^2\ dz = \Omega_0\ Y(q)\ \eta^2\ (Co - C_s)^2\ z_m = \Omega_0\ Y(q)\ \varepsilon^2\ z_m$$

In the presence of a sine fluctuation around Co defined by δc = c Co = A cos qz, the energy becomes

$$E_{inh} = E_{hom} + \int_0^{z_m} \frac{\partial^2 E_{hom}}{\partial c^2} (\delta c)^2\ dz \qquad \text{(II.19)}$$

The stability limit is crossed as soon as the summation is negative, that is, as soon as $\frac{\partial^2 E}{\partial c^2} = 0$, which writes

$$g_o'' + 2Y(q)\ (\varepsilon')^2\ \Omega_0 + \varepsilon\Omega_0\ [Y''\ \varepsilon + 2Y\ \varepsilon'' + 4Y'\ \varepsilon'] = 0 \qquad \text{(II.20)}$$

where prime and double-prime denote concentration derivatives.

* The term $2Y(q)\ \varepsilon'^2$ is the term already introduced in the spinodal decomposition (since $\varepsilon' = \eta$) and is always positive.

* The remaining terms come from the fact that Co is not the lattice matching composition C_s :

$2Y(q)\ \varepsilon\ \varepsilon''\ \Omega_0$ is non-zero whenever Vegard's rule is not obeyed ($\varepsilon'' = \eta' \neq 0$) and changes its sign when C_s is crossed.

$Y''\ \varepsilon^2\ \Omega_0$ is a modulus effect ; it is negative when the elastic energy is reduced by a phase separation.

$4\ \varepsilon\ \varepsilon'\ Y'\ \Omega_0$ is a combined modulus and coherency strain effect which changes its sign when the lattice matching composition is crossed.

The fact that their sign can change with Co shows that their effect on the critical temperature T_c can be either a further decrease or an increase with respect to the depression induced by the first term only : it can also change significantly with the orientation of the substrate through Y(q). Hence the expression retaining only the first term which has been widely used by numerous authors to explain their experimental results is actually valid in the only case of the lattice matching composition Co = C_s where $a_o = a_s$ [II.11, II.12]. The models used up to now are often crude ones, namely regular solid solutions without or with [II.13] short-range order.

II.2.2 Concentration fluctuations along Ox

Such fluctuations have been indeed observed in some systems. It is assumed for simplicity that the film and the substrate are isotropic, have the same elastic constants and are lattice-matched ($C_o = C_s$; $a_o = a_s$).

Let us assume that a concentration fluctuation takes place in the film parallel to the substrate (say along Ox). The stress-free modulation of the lattice parameter is then

$$a(x) = a_s\,(1 - \varepsilon \cos qx) \qquad 0 \le z \le z_m$$
$$a(x) = a_s \qquad z < 0 \quad \text{(in the substrate)}$$

The strain and stress-field are calculated in the film and the substrate ; the density of the elastic energy stored in the systems is shown to depend on z_m/λ where $\lambda = 2\pi/q$ is the wavelength of the concentration modulation

$$E = \frac{1}{2}\frac{E}{1-\nu}\varepsilon^2\left(1 - \frac{(1+\nu)}{q\,z_m}\left(1 - e^{-q\,z_m}\right)^2\right) \qquad \text{(II.21)}$$

The first factor is nothing but the energy density in a bulk material for the same modulation of the lattice parameter (see § II.1.3.2). The bracket is a function of the total thickness z_m of the film ; it is always smaller than unity and reaches a minimum of 0.455 for $q\,z_m \sim 1.27$. Thus, in this case, the elastic energy stored in the system can be approximately twice smaller for a concentration fluctuation parallel to the substrate than for the same fluctuation perpendicular to it. As a consequence, the decrease of the critical temperature T_c is only one half of that predicted by the theory of spinodal decomposition. The most probable evolution of the film is therefore a spinodal-like decomposition with a wave vector parallel to the substrate and a wavelength of the order of the film thickness [II.14].

II.3 MACROSCOPIC MODELS FOR TWO-PHASE COHERENT EQUILIBRIA

The most general problem is complicated since the solution depends on the morphology of the two phases (shape, distribution, volume) as well as on the mechanical boundary conditions. The phases in mutual equilibrium need no longer to be chemically as well as elastically homogeneous : chemical and stress gradients can be present. But a new generalized diffusion potential can be built, which will be uniform throughout the system at equilibrium : it is a function of the whole stress tensor σ_{ij} in the general case. It will depend only on the trace σ_{ii} whenever the solids are elastically isotropic or when the stress-free strains induced by replacing an A by a B are pure dilatations. The latter case will be the only one to be considered in the following, since it is the only one in which the elastic perturbation can be introduced through a simple scalar (the concentration). The solution of the equilibrium equation stating that the potential is uniform in the system involves usually both elastic and chemical effects in the same step : in some instances however, the problem can be simplified in such a way that the chemical problem of determining a concentration profile can be decoupled from the elastic problem, which can be solved with new effective elastic constants. At last in simple cases the coexisting phases at equilibrium can be homogeneous and the elastic energies depend only on their volumes ; but it is not a general result. The last point to be mentioned is the fact that Gibbs' phase rule generally cannot be used for a coherent two-phase equilibrium.

II.3.1 Derivation of the equilibrium conditions for stressed solids

The conditions for thermomechanical equilibrium in each phase and at the interface are derived by considering the total change in energy which is associated with a mass-transfer across the interface : in a closed and isolated system, the variation must be an extremum and its first derivatives must vanish, subject to constraints of total entropy conservation and mass conservation for A and B. The corresponding Lagrange multipliers appearing in the minimization procedure are respectively the temperature T and the chemical potentials denoted by λ_A, λ_B (the notation μ_A and μ_B will be restricted to the case without any stress but an hydrostatic pressure p). These Lagrange multipliers must be constant and uniform throughout the system to ensure equilibrium.

The new relevant thermodynamical function to be used is the grand canonical potential ω defined usually per unit volume by

$$\omega = f - g = f - \rho_A\lambda_A - \rho_B\lambda_B \tag{II.22}$$

where ρ_A and ρ_B are densities (number of moles/unit volume). It reduces to $-$ p for a fluid (by convention $p > 0$). Unlike fluids, a solid can sustain shear stresses and the expression of ω must take a full account of the whole stress tensor σ_{ij} and not only of its hydrostatic part (by convention $\sigma_{ij} > 0$ for a tension).

A further constraint is imposed by the existence of a crystalline lattice : in each region of a solid where the number of sites is conservative (no free surface, no incoherent interface, no dislocations able to climb...), it implies that only the <u>difference</u> $\lambda_B - \lambda_A$ can be calculated in the volume. Indeed each site which is not occupied by an A atom is occupied by a B one ; and the elastic deformation due to a size effect is calculated after an A has been replaced by a B atom. The <u>difference</u> $M_{BA} = \lambda_B - \lambda_A$ <u>is called the diffusion potential</u> and this is the quantity which must be uniform throughout the system to ensure that no matter flux takes place.

The general expression for M_{BA} has been derived [II.15]

$$M_{BA}\ (c, \sigma_{ij}) = \mu_B - \mu_A - \Omega_0\, \eta_{ij}\, \sigma_{ij} - \frac{\Omega_0}{2}\, \frac{d\, S_{ijkl}}{d_c}\, \sigma_{ij}\sigma_{kl} \tag{II.23}$$

where :

* μ_A and μ_B are the chemical potentials with no stress

* η_{ij} is the linear dilatation for a unit concentration change. In the assumption of a pure dilatation, the stress-free strain is given by

$$\varepsilon^c_{ij} = \varepsilon\delta_{ij} \text{ and } \eta_{ij} = \frac{d\varepsilon}{dc}\,\delta_{ij} = \eta\,\delta_{ij} \text{ with } \eta = \frac{\Omega_B - \Omega_A}{3\Omega_A}$$

* S_{ijkl} are the compliances defined by

$$\varepsilon_{ij} = \varepsilon^c_{ij} + S_{ijkl}\,\sigma_{kl}$$

conversely the elastic constants C_{ijkl} are such that

$$\sigma_{ij} = C_{ijkl}\,(\varepsilon_{kl} - \varepsilon^c_{kl})$$

The new expression of the free enthalpy per unit volume is

$$g_o\,(c,\,\sigma_{ij}) = \rho_o\,(c\,\mu_B + (1-c)\,\mu_A) + \frac{1}{2}\,S_{ijkl}\,\sigma_{ij}\,\sigma_{kl} \qquad \text{(II.24)}$$

where ρ_o is the average density and $\Omega_0 = 1/\rho_o$ the average atomic volume ; the last term represents the elastic energy density. For the particular case of an isotropic medium, the above expressions simplify and read

$$M_{BA}\,(c,\,\sigma_{ij}) = \mu_B - \mu_A - \Omega_0\,\eta\,\sigma_{ii} + \frac{\Omega_0}{2}\left(\frac{d}{dc}\left(\frac{\nu}{E}\right)(\sigma_{ii})^2 - \frac{d}{dc}\left(\frac{1+\nu}{E}\right)\sigma_{ij}\sigma_{ij}\right)$$

$$g_o\,(c,\,\sigma_{ij}) = \rho_o\,(c\,\mu_B + (1-c)\,\mu_A) - \frac{1}{2}\left(\frac{\nu}{E}\right)(\sigma_{ii})^2 + \frac{(1+\nu)}{2E}\,\sigma_{ij}\sigma_{ij} \qquad \text{(II.25)}$$

The equilibrium equations separate into volume equations to be obeyed everywhere in the system (interface α/β included) and into interface equations which are only valid along the interface α/β and which depend on the type of the interface.

Volume conditions

* Mechanical equilibrium with no body forces

$$\sum_j \frac{\partial\,\sigma^\alpha_{ij}}{\partial x_j} = \sigma^\alpha_{ij,j} = 0 = \sigma^\beta_{ij,j} \qquad \text{(II.26)}$$

* Chemical equilibrium

$$M^\alpha_{BA}\,(c^\alpha,\,\sigma^\alpha_{ij}) = M^\beta_{BA}\,(c^\beta,\,\sigma^\beta_{ij}) = \text{cte} \qquad \text{(II.27)}$$

This is a local relationship between stress and composition. Once the stress distribution is known, the composition profile is determined by the above equality. In the general case both are intimately interconnected and they must be solved consistently at the same step of the calculation.

Interface conditions

We restrict ourselves to planar interfaces which remain planar during the transformation. The mechanical equilibrium condition depends on the type of the interface to be considered.

* For a solid (α)-liquid (β) interface where the solid is embedded in a liquid at pressure p we have

$$\omega^{\alpha} - \omega^{\beta} = 0 \; ; \; \omega^{\beta} = - p \qquad \text{(II.28)}$$

If the solid is immersed in the liquid, it is submitted to the same hydrostatic pressure, its state of stress is uniform and the condition of constant diffusion potential implies a uniform chemical composition. At the interface $\lambda_A^{\alpha} = \mu_A^{\beta}$; since the solid is uniform this condition is also valid in the bulk. In this particular case of hydrostatic stress, the solid behaves like a fluid and the chemical potentials for each species are known everywhere in the system.

* For an incoherent solid (α)-solid (β) interface, each phase is free to slip with respect to the other along the interface : only the normal strains must be continuous across the interface. The normal to the interface must be a principal axis of the stress tensors and the corresponding principal values equilibrate each other.

$$\omega^{\alpha} - \omega^{\beta} = 0 \; ; \; \omega^{\alpha} n_i^{\alpha} = \sigma_{ij}^{\alpha} n_j^{\alpha} \; ; \; \omega^{\beta} n_i^{\beta} = \sigma_{ij}^{\beta} n_j^{\beta} \qquad \text{(II.29)}$$

where n_i^{α} are the components of the normal to the interface oriented from α to β. Since $n_i^{\alpha} + n_i^{\beta} = 0$ it is easy to deduce that

$$\sigma_{ij}^{\alpha} n_j^{\alpha} = - \sigma_{ij}^{\beta} n_j^{\beta} \qquad \text{(II.30)}$$

This equality expresses that the normal forces are equal along the interface and that no tangential force is allowed. It can be shown that in this case too, λ_A can be determined at the interface by the equality

$$\lambda_A = \Omega_0^{\alpha} \left(g_o^{\alpha} - M_{BA}^{\alpha} \rho_o^{\alpha} - \sigma_{ij}^{\alpha} n_j^{\alpha} \right) = \Omega_0^{\beta} \left(g_o^{\beta} - M_{BA}^{\beta} \rho_o^{\beta} - \sigma_{ij}^{\beta} n_j^{\beta} \right) \qquad \text{(II.31)}$$

But unlike the preceding case, the interface is the only place in the system where both λ_A <u>and</u> λ_B can be determined.

* For a coherent solid (α)-solid (β) interface, the very existence of a unique lattice common to both phases implies that the normal <u>and</u> transverse strains are continuous across the interface. The equality of the forces at the interface is still required

$$\sigma_{ij}^{\alpha} n_j^{\alpha} = - \sigma_{ij}^{\beta} n_j^{\beta}$$

but in the present case they are in general not normal to the interface. Transverse forces are allowed and their work during matter transfer across the interface adds an extra term to the difference of the grand potentials

$$\omega^{\alpha} - \omega^{\beta} - \left(\varepsilon_{ij}^{\alpha} - \varepsilon_{ij}^{\beta} \right) \sigma_{ij}^{\alpha} = 0 \qquad \text{(II.32)}$$

where the strains are measured from the unconstrained state of each phase.

In the case where some excess surface energy γ is attached to the interface (for very thin films or curved interfaces) an extra term $2\gamma/R$ must be added on the left-hand side of equation II.32 to take account of the surface curvature ; γ is the excess grand-canonical potential and R the mean radius of curvature [II.16).

We give below two examples to illustrate these general results.

II.3.2 Liquid phase epitaxy : lattice pulling for III-V compounds [II.17]

It is observed that thin films (α) of $A_cC_{1-c}B$ which are grown from a ternary melt (β) of composition (C_A, C_B) by liquid phase epitaxy on a given substrate, exhibit a nearly constant composition whichever the composition of the liquid ternary. Ternary III-V compounds can be considered for most purposes as quasi-binary $(AB)_c(CB)_{1-c}$ compounds : only one composition c^α is needed to describe their composition. We assume for simplicity that the film is elastically isotropic, that is, elastic constants are concentration independent and that the substrate is infinitely rigid. The stress tensor in this case reduces to

$$\sigma_{xx} = \sigma_{yy} = -\eta\ (c^\alpha - C_s)\ Y \qquad \text{(II.33)}$$

where C_s is the lattice matching composition and Y some effective elastic modulus depending on the orientation of the substrate.

The only variation of c^α is to be expected along the growth direction Oz. The equality of the diffusion potentials in both phases according equation II.23 gives

$$M^\alpha_{AC} = \mu^\alpha_{AB} - \mu^\alpha_{CB} - \Omega^\alpha\ \eta\ \sigma^\alpha_{ii} = M^\beta_{AC} = \mu^\beta_A - \mu^\beta_C \qquad \text{(II.34)}$$

At the interface located at $z = z_m$, $\lambda^\alpha_{AB} = \mu^\beta_A + \mu^\beta_B$ and $\lambda^\alpha_{CB} = \mu^\beta_C + \mu^\beta_B$

$$\omega^\alpha = \omega^\beta = -\ p \cong 0$$

Starting from equation II.25 and remembering that $Y = \frac{E}{1 - \nu}$ yields

$$\Omega^\alpha\ g_o{}^\alpha = c^\alpha\ \mu^\alpha_{AB} + (1 - c^\alpha)\ \mu^\alpha_{CB} - \frac{\Omega^\alpha}{2}\frac{\nu}{E}\ (\sigma_{ii})^2 + \frac{\Omega^\alpha}{2}\frac{1+\nu}{E}\ \sigma_{ij}\sigma_{ij}$$

$$= c^\alpha\ \mu^\alpha_{AB} + (1 - c^\alpha)\ \mu^\alpha_{CB} + \Omega^\alpha\ \eta^2\ (c^\alpha - C_s)^2\ Y$$

subtracting $c^\alpha\ \lambda^\alpha_{AB} + (1 - c^\alpha)\ \lambda^\alpha_{CB} = c^\alpha\ M^\alpha_{AC} + \lambda^\alpha_{CB}$ and expressing λ^α_{CB} as a function of the chemical potentials in the liquid β yields finally

$$\omega^\alpha = \mu^\alpha_{CB} - \eta^2\ Y\ \Omega^\alpha\ \left((c^\alpha)^2 - C_s{}^2\right) - \left(\mu^\beta_C + \mu^\beta_B\right) \cong 0 \qquad \text{(II.35)}$$

The first equation II.34 can be solved for $c^{\alpha}(z)$; since z does not appear explicitely, c^{α} must be a constant ; therefore the film is of uniform concentration and so is its state of stress. Notice that this simplification is a direct consequence of the stresses described by equation II.33 ; if the substrate would not have been supposed infinitely rigid, the stress would have included a dependence on z as well as on the effective modulus of the substrate and the equilibrium would not have yielded a uniform concentration.

When the departure of c^{α} from C_s is large, the layer is expected to lose its coherency with the substrate and release its elastic energy by generating dislocations. The physically accessible solutions are in the neighbourhood of C_s for c^{α} and C_A*, C_C* for the liquid. A simplified system can be obtained from equations II.34 and II.35 by linearizing around C_s for the film and C_A*, C_C* for the melt.

A simple differentiation yields after some algebra

$$\frac{dc^{\alpha}}{dC_A} = Q.N \text{ where } Q = \frac{\dfrac{\partial\mu^{\alpha}_{AB}}{\partial c^{\alpha}}}{\left(\dfrac{\partial\mu^{\alpha}_{AB}}{\partial c^{\alpha}} + (1 - c^{\alpha})\ 2\eta^2\ Y\ \Omega^{\alpha}\right)} \qquad (II.36)$$

and N is a function of chemical potential derivatives in the solid and the liquid, including no elastic term. Particular models can be used to calculate them. As a result, the composition c^{α} in the film is given by

$$c^{\alpha} - C_s = \frac{dc^{\alpha}}{dC_A}(C_A - C_A*) \qquad (II.37)$$

and the state of stress is given by : $\sigma_{xx} = \sigma_{yy} = - \eta\ QN\ (C_A - C_A*)\ Y$.

In the absence of any elastic effect (incoherent equilibrium), Q is unity ; on the contrary, if Q is small, c^{α} will remain close to C_s whichever the departure of C_A from C_A* : this is the effect of lattice pulling and Q is a measure of its magnitude. It is seen that Q is independent of the properties of the liquid ; and depends on the orientation of the substrate through Y.

II.3.3 Equilibrium of a coherent two-phase film on a coherent substrate [II.18]

Let us assume that a bi-layer $\alpha + \beta$ is deposited on the substrate ; the interfaces substrate/α and α/β are planar and parallel. The substrate imposes its lattice parameter a_s to α and β along Ox and Oy but the film is free to relax along Oz. The film is thin and of infinite extension along Ox and Oy. Vegard's rule is assumed to be followed by α and β ; the elastic effective moduli Y^{α} and Y^{β} are concentration independent, but different from each other. The substrate is infinitely rigid. Co^{α} and Co^{β} denote the incoherent equilibrium compositions of the phases α and β : c^{α} and c^{β} their compositions under stress. The change in the lattice parameter a^{α}(or a^{β}) is due partly to the stress-free strain which stems from the difference $(c^\alpha - Co^\alpha)$ and to the elastic strain which matches a^α to a_s

$$\varepsilon^\nu = \frac{a^\nu - a_s}{a_s} = \varepsilon_o^\nu + \eta^\nu\ (c^\nu - Co^\nu) \text{ with } \eta^\nu = \frac{\Omega_B^\nu - \Omega_A^\nu}{3\Omega_s} \qquad (\nu = \alpha \text{ or } \beta)$$

$$\varepsilon_o^\nu = \frac{a_o^\nu - a_s}{a_s}$$

where Ω_s is the average atomic volume corresponding to a_s.

The only non vanishing stresses are $\sigma_{xx}^\nu = \sigma_{yy}^\nu = - Y^\nu\ \varepsilon^\nu$.

The equality between the two diffusion potentials is easily established

$$M_{BA}^\alpha = \mu_B^\alpha - \mu_A^\alpha - \Omega_s\ \eta^\alpha\ \sigma_{ii}^\alpha = \mu_B^\alpha - \mu_A^\alpha + 2\Omega_s\ Y^\alpha\ \eta^\alpha\ \varepsilon^\alpha$$

$$= \mu_B^\beta - \mu_A^\beta + 2\Omega_s\ Y^\beta\ \eta^\beta\ \varepsilon^\beta \qquad \text{(II.38)}$$

In the same way the jump in the thermodynamical potential if expressed by

$$\mu_A^\alpha - \mu_A^\beta + Y^\alpha\ (\varepsilon^\alpha)^2\ \Omega_s - Y^\beta\ (\varepsilon^\beta)^2\ \Omega_s - 2c^\alpha\ \eta^\alpha\ \varepsilon^\alpha\ Y^\alpha + 2c^\beta\ \eta^\beta\ \varepsilon^\beta\ Y^\beta = 0 \qquad \text{(II.39)}$$

The solution of equations II.38 and II.39 gives the values of c^α and c^β. It can be seen for this particular geometry that the state of stress does not depend explicitely on z ; the diffusion potential M_{BA}^α is uniform if the composition is uniform. Linearizing the two equations as before yields the solution

$$c^\alpha - Co^\alpha = \frac{Y^\beta\ (\varepsilon_o^\beta)^2 - Y^\alpha\ (\varepsilon_o^\alpha)^2 + 2\eta^\alpha\ \varepsilon_o^\alpha\ Y^\alpha\ (Co^\alpha - Co^\beta)}{(Co^\beta - Co^\alpha)\ \chi^\alpha}$$

$$c^\beta - Co^\beta = \frac{Y^\beta\ (\varepsilon_o^\beta)^2 - Y^\alpha\ (\varepsilon_o^\alpha)^2 + 2\eta^\beta\ \varepsilon_o^\beta\ Y^\beta\ (Co^\alpha - Co^\beta)}{(Co^\beta - Co^\alpha)\ \chi^\beta} \qquad \text{(II.40)}$$

with $$\chi^\nu = \frac{kT}{\Omega_s\ Co^\nu\ (1 - Co^\nu)} \left(1 + \frac{\partial \text{ Log } \gamma B}{\partial \text{ Log } c}\right)$$

and γB the activity coefficient.

The first obvious conclusion is that the result of combined modulus and size-mismatch effects is difficult to predict in the general case ; only particular cases can be easily commented.

If the partial molar volumes are equal $\eta^\alpha = \eta^\beta = 0$, then the change in equilibrium composition is determined by the relative changes in the elastic energy of each phase : if a_o^α and a_o^β are similar the elastically hard phase is destabilized to the benefit of the elastically soft one ; similarly if a_o^α is close to a_s, the α phase is stabilized to the expense of the β phase.

It can be easily checked that in the present case the equilibrium compositions under stress c^{α}, c^{β} can also be determined by a common tangent construction to the free enthalpy curves including the elastic contributions. And, for this case, although an elastic contribution to the energy is added for each phase, the equilibrium is not determined by the minimization of the total elastic energy which depends on the volume fraction z of the β-phase ; and Gibbs' phase rule can be used.

II.3.4 Equilibrium of a coherent two-phase film on an incoherent substrate [II.19]

It is supposed now that the interface substrate/α is incoherent and that α can slip on the substrate. The role of the latter consists simply in preventing the bilayer α + β from bending. The equilibrium value of the common lattice parameter $a^{\alpha} = a^{\beta}$ is no longer dictated by a coherency constraint with a substrate, but rather by the relative amounts of elastic energy which are stored in α and β phase.

For simplicity both phases have the same elastic constants, are assumed to be homogeneous and are described by a regular solution model. Co is the overall composition of the bilayer ; Co^{α} and Co^{β} the equilibrium compositions of α and β without stress, c^{α} and c^{β} their compositions under stress. z being the proportion of β in α + β, the problem is to minimize the total free enthalpy g

$$g = (1 - z)\, g_o^{\alpha} + z\, g_o^{\beta} + g_{el}$$

where

* g_o^{v} is the free enthalpy of α and β phase without stress which is given by

$$g_o^{v} = (1 - c^{v})\, \mu_A^{o} + c^{v} \mu_B^{o} + \gamma\, (c^{v} - Co^{v})^2$$

μ_A^{o} and μ_B^{o} are the chemical potential with no stress at compositions Co^{v} and γ is the second derivative with respect to concentration (γ^{α} is taken equal to γ^{β} for sake of simplicity).

* g_{el} is the elastic contribution to the total energy which is due to the coherent mixing of α and β and which is assumed to depend only on their relative amounts z

$$g_{el} = z\, (1 - z)\, \frac{E}{1 - v}\, \varepsilon^2\, \Omega$$

Defining reduced compositions x^{α}, x^{β}, x_o by

$$x^{v} = 1 - \frac{2\,(c^{v} - Co^{v})}{Co^{\beta} - Co^{\alpha}} \quad (\text{with } v = \alpha, \beta) \text{ and } \quad x_o = 1 - \frac{2\,(Co - Co^{\alpha})}{Co^{\beta} - Co^{\alpha}}$$

a dimension-less free enthalpy is defined for convenience

$$\phi = \frac{4\,(g - Co\, \mu_B^{o} - (1 - Co)\, \mu_A^{o})}{\gamma\, (Co^{\beta} - Co^{\alpha})^2}$$

$$= z\ (1 - x^{\alpha})^2 + (1 - z)\ (1 + x^{\beta})^2 + A\ z\ (1 - z)$$

The problem is to minimize ϕ with respect to x^{α}, x^{β} and z with the further constraint of mass conservation

$$x_o = z\ x^{\beta} + (1 - z)\ x^{\alpha}$$

The solution when the elastic term vanishes (A = 0) is composed of three parts : when the reduced composition x_o is smaller than - 1 or larger than + 1, the equilibrium state is single-phased (β and α respectively) and the free enthalpy is a portion of parabola centered around - 1 and + 1 respectively. In-between, the equilibrium state is two-phased ; the free enthalpy is a portion of straight line (Fig. II.8 a).

The solution x^{α}, x^{β} are given as functions of the reduced composition x_o in Fig. II.8 b. Of course $x^{\alpha} = x_o$ when $x_o > + 1$ and $x^{\beta} = x_o$ when $x_o < - 1$; in-between x^{α} and x^{β} are constant and the equilibrium state of the system consists of a mixing of two-phases of constant composition, the proportion z of which is dictated by the lever-rule.

When A is not zero, the two-phase domain shrinks ; the β phase is stable up to the reduced concentration $- 1 + \frac{A}{4}$ and the α phase is stable down to the reduced composition $1 - \frac{A}{4}$. In-between the system is two-phased and its free energy if a portion of parabola centered around 0 with negative curvature and which is tangent to the free enthalpy curves of β and α phases at $-1 + \frac{A}{4}$ and $1 - \frac{A}{4}$ (Fig. II.8 c).

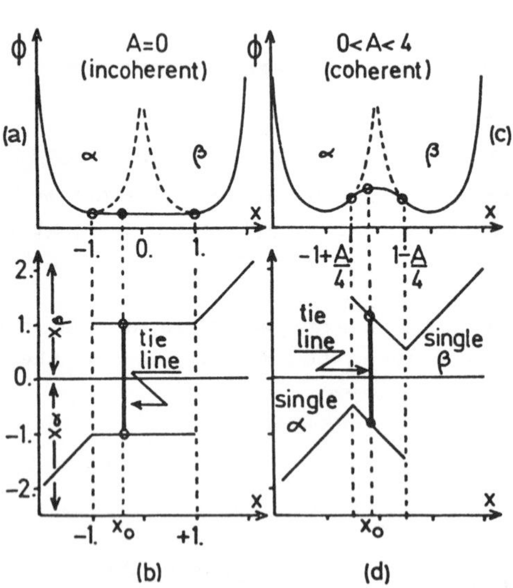

Fig. II.8 Comparison between incoherent (a,b) and coherent (c,d) equilibrium ; ϕ is a reduced free enthalpy ; x_{α}, x_{β}, x_o are reduced compositions. Notice the phase rule violation for the coherent case (d).

When plotting x^{α} and x^{β} as functions of x_o, both concentrations exhibit a linear variation even in the two-phase domain. The equilibrium two-phase state of the system is a mixing of two phases, the composition of which vary linearly with x_o (Fig. II.8 d).

The conclusion is that Gibbs' phase rule is no longer valid for coherent constrained systems in general but particular cases : in some instances the number of degrees of freedom which are left for the system is even independent on the number of coexisting phases [II.19].

II.4 MICROSCOPIC MODELS

Only macroscopic models have been considered up to now. We consider below a recent attempt to incorporate elastic effects at a microscopic level.

During the last decade the problem of binary alloys has been mapped onto an Ising problem in a rigid lattice : A and B atoms are modelled by + and - spins, the coupling of which was given as input parameters. First neighbour interactions were usually considered and the problem reduced to the calculation of a free energy with this hamiltonian. Most of the effort has been devoted to improve the statistical description of the alloy and evaluate more accurately its configurational entropy. The cluster variation method (CVM) has been proved to be a powerful tool and it has been widely used to propose temptative phase diagrams [II.22]. Its essence is to model the alloy by a population of clusters of small size, the energies of which are input parameters : they are easily expressed by bond counting in phenomenological models, but they can be evaluated through first principle calculations. The numbers of clusters with a given occupancy of their vertices are functions of their energies, of the concentration and temperature ; they are submitted to constraints due to their spatial overlapping : indeed a given lattice site, or a given bond belong to several distinct clusters. The CVM is a mean field approximation and, provided the basic cluster is carefully chosen, the larger the cluster the better the evaluation of the partition function [II.23].

We focus in the following on the case of a FCC lattice containing N sites ; the minimum cluster to be chosen in order to describe the ordered structures L_{10} and L_{12} is a tetrahedron made up of four first-neighbour sites : the lattice is partitioned into four interpenetrating simple cubic lattices α, β, γ, δ. The lattice contains 2N tetrahedra and 6N pairs.

Let z_{ijkl} be the proportion of tetrahedra having atoms i, j, k, l on sites α, β, γ, δ respectively ; $y_{ij}^{\nu\nu'}$ the proportion of bonds having i, j atoms on sites ν, ν' ($\nu \neq \nu'$) and x_i^{ν} the proportion of sites ν having an A atom. These variables are not independent and are linked by the relationships

$$\sum_{i,j,k,l} z_{ijkl} = 1 \qquad \sum_{i,j} y_{ij}^{\nu\nu'} = 1 \qquad \sum_{i} x_i^{\nu} = 1$$

$$\sum_{k,l} z_{ijkl} = y_{ij}^{\alpha\beta} \qquad \sum_{j,l} z_{ijkl} = y_{ik}^{\alpha\gamma} \; .... \text{ etc}$$

$$\sum_{j} y_{ij}^{\nu\nu'} = x_i^{\nu} \qquad \sum_{i} y_{ij}^{\nu\nu'} = x_j^{\nu'} \tag{II.41}$$

The energy E_{ijkl} attached to each tetrahedron is given by

$$E_{ijkl} = \frac{1}{2}\,[E_{ij} + E_{ik} + E_{il} + E_{jk} + E_{jl} + E_{kl}] \tag{II.42}$$

where the E_{ij}'s are nearest neighbour pair energies and the factor $\frac{1}{2}$ accounts for the fact that each pair belongs to two distinct tetrahedra. The energy of the alloy is then

$$E = 2N \sum_{i,j,k,l} z_{ijkl} E_{ijkl} \qquad \text{(II.43)}$$

The configurational entropy S is given by

$$S = kN \left(\sum_{\nu\nu'} \sum_{ij} \mathcal{L}(y_{ij}^{\nu\nu'}) - 2 \sum_{ijkl} \mathcal{L}(z_{ijkl}) - \frac{5}{4} \sum_{i} \sum_{\nu} \mathcal{L}(x_i^{\nu}) \right) \qquad \text{(II.44)}$$

where $\mathcal{L}(x)$ stands for x Log x.

The phase diagram which is obtained with constant pair energies E_{ij} shows a significant improvement with respect to cruder approximations. The symmetry of the phase diagram around 0.5 which is not met in practice can be suppressed by introducing multiatom interactions E_{ijkl} which do not reduce to a sum of pair energies as above, or equivalently, by keeping additive pair energies which can vary with the nominal concentration of the alloy. The resulting phase diagrams are much improved [II.24]. At last interactions beyond first neighbour distance can be introduced by using larger clusters ; it is encouraging that a small number of pair energies is sufficient to account for most of metallic ordered structures ever encountered so far [II.25].

The parallel drawn with a magnetic system however restrains the problem to perform A ⇔ B exchanges only on a rigid lattice. It is well known that in practice the atomic volumes of the constituents are unequal and that some strain of the lattice is present. In the frame of the CVM, taking into account only the "chemical" energies E_{ijkl} but neglecting the elastic contribution which is always positive leads to some drawback :
* A positive mixing enthalpy ΔH_m leads always to clustering at lower temperatures, whereas a negative one leads to ordering. Both tendencies are mutually exclusive : as a consequence, such models cannot account for those alloys which exhibit a positive mixing enthalpy at high temperature and an order-disorder reaction at lower temperatures. In other words it is impossible to choose the E_{ijkl} in such a way that both T_c and ΔH_m agree with the experimental results. This failure can be cured by the so-called ε-G approach in which an elastic contribution to the free energy is included [II.26, II.27].

II.4.1 Inclusion of elastic effects at a microscopic level

The alloy is considered as a population of tetrahedra B_nA_{4-n}, the proportion of the latter being z_n. The energy required to build such tetrahedra from the pure constituents is calculated in two formal steps :

i) A pure tetrahedron A_nA_{4-n} is expanded or compressed to adjust its atomic volume Ω_A to Ω : the same is done for B_nB_{4-n}. The elastic energy required for this step is :

$$E_{el}(n,\Omega) = \frac{4-n}{4} \left(E_{el}(A,\Omega) - E_{el}(A,\Omega_A)\right) + \frac{n}{4} \left(E_{el}(B,\Omega) - E_{el}(B,\Omega_b)\right)$$

$$= (1-C_n)\,\alpha(\Omega) + C_n\,\beta(\Omega) \qquad \text{(II.45)}$$

ii) Then n A atoms are converted into B in the tetrahedra A_nA_{4-n} to match the composition $C_n = \frac{n}{4}$; an equivalent interchange is operated on B_nB_{4-n} ; this step involves a "chemical" energy E_n of exchange at constant volume. Although it is defined at the equilibrium value Ω_n of the ordered structure n, it is implicitely assumed not to depend on Ω. It is also assumed that no further relaxation takes place inside each tetrahedron during the interchange, which means that the whole elastic contribution comes from the first step only. The enthalpy required to build the ordered phase n at atomic volume Ω is then

$$\Delta H^{o}_{m} (n,\Omega) = E_n + E^{o}_{el} (n,\Omega) \tag{II.46}$$

with

$$E^{o}_{el} (n,\Omega) = (1 - C_n)\, \alpha(\Omega) + C_n\beta(\Omega)$$

The equilibrium volume Ω_n of the ordered phase n is defined by

$$\frac{\partial E^{o}_{el}}{\partial\Omega} = (1 - C_n)\, \alpha' + C_n\beta' = 0 \tag{II.47}$$

where the prime denotes the volume derivative and the formation enthalpy of the ordered phase n is

$$\Delta H_n = \Delta H^{o}_{m} (n,\Omega_n) \tag{II.48}$$

The enthalpy of the disordered alloy at composition c is taken as a superposition of the contributions of each kind of tetrahedron n which is representative of the ordered phase n

$$\Delta H^{D}_{m} (c,\Omega) = \sum_n z_n\, \Delta H^{o}_{m} (n,\Omega) = \sum_n z_n\, E_n + E^{D}_{el} (c,\Omega) \tag{II.49}$$

with $\quad E^{D}_{el} (c,\Omega) = (1 - c)\, \alpha(\Omega) + c\beta(\Omega)$

since $\quad \sum_n z_n = 1 \quad$ and $\quad \sum_n z_n\, C_n = c$

The equilibrium atomic volume $\Omega(c)$ of the disordered alloy is defined by

$$\frac{\partial E^{D}_{el} (c,\Omega)}{\partial\Omega} = (1 - c)\, \alpha' + c\beta' = 0 \tag{II.50}$$

and the mixing enthalpy of the disordered alloy

$$\Delta H^{D}_{m} (c) = \Delta H^{D}_{m} (c,\, \Omega(c)) \tag{II.51}$$

It is seen that equations (II.47) and (II.50) yield the same value of the atomic volume Ω if $c = C_n$; in other words, this approach neglects any change of the atomic volume with respect to the change of the state of order for a given concentration. It is reasonable since in most cases the volume change on an order-disorder transition is smaller than the volume change due to a varying concentration.

Equation (II.50) defines $\Omega(c)$ or $c(\Omega)$ as well provided in the latter Ω is the <u>equilibrium</u> atomic volume at concentration c. The bulk modulus B is defined by

$$\frac{\partial^2 E_{el}}{\partial \Omega^2}\Big|_{\Omega(c)} = \frac{B}{\Omega(c)} \qquad \text{(II.52)}$$

General expressions can be found for α and β. At equilibrium when $\frac{\partial E_{el}^D}{\partial \Omega} = 0$, we get

$$\frac{dE_{el}}{dc} = \frac{\partial E_{el}}{\partial c} + \frac{\partial E_{el}}{\partial \Omega}\frac{d\Omega}{dt} = \frac{\partial E_{el}}{\partial c} = \beta - \alpha$$

$$\frac{d^2E_{el}}{dc^2} = \frac{d(\beta - \alpha)}{dc} = (\beta' - \alpha)\frac{d\Omega}{dc}$$

$$= \frac{\partial^2 E_{el}}{\partial \Omega^2}\left(\frac{d\Omega}{dc}\right)^2 + \frac{\partial E_{el}}{\partial \Omega}\frac{d\Omega}{dc} + \frac{\partial^2 E_{el}}{\partial c^2} = \frac{\partial^2 E_{el}^D}{\partial c^2}\left(\frac{d\Omega}{dc}\right)^2$$

Hence $\beta' - \alpha' = \frac{B}{\Omega}\frac{d\Omega}{dc} = \frac{B}{\Omega}\left(\frac{dc}{d\Omega}\right)^{-1}$

Integrating on Ω or on c yields a summation together with an integration constant μ

$$\beta - \alpha = \int_{\Omega_A}^{\Omega} \frac{B}{\Omega'}\left(\frac{dc'}{d\Omega'}\right)^{-1} d\Omega' + \mu = \int_{o}^{c} \frac{B}{\Omega'}\left(\frac{d\Omega'}{dc'}\right)^{2} dc' + \mu$$

Defining for short $Z(\Omega) = -\frac{B}{\Omega}\left(\frac{dc}{d\Omega}\right)^{-1}$ and $Z(c) = -\frac{B}{\Omega}\left(\frac{d\Omega}{dc}\right)^{2}$ gives

$$\beta - \alpha = -\int_{\Omega_A}^{\Omega} Z(\Omega')\, d\Omega' + \mu = -\int_{o}^{c} Z(c')\, dc' + \mu \qquad \text{(II.53)}$$

Integrating equation II.50 to get α yields

$$\alpha = \int_{\Omega_A}^{\Omega} c(\Omega')\, Z(\Omega')\, d\Omega' + \nu = \int_{o}^{c} c'\, Z(c')\, dc' + \nu$$

The integration constants μ and ν are defined in such a way that the elastic energy of mixing at both ends is zero $E_{el}^D(o,\Omega) = E_{el}^D(1,\Omega) = 0$.

After some rearrangements, the final expression of $E_{el}^D(c)$ (which is denoted by G(x) in [II.27]) becomes

$$E_{el}^D(c) = (1-c)\int_{o}^{c} c'\, Z(c')\, dc' + c\int_{c}^{1} (1 - c')\, Z(c')\, dc' \qquad \text{(II.54)}$$

It is clear that B and Ω can be determined for each concentration from first principle calculations : but the phase diagrams computed in this way are still of poor quality [II.26], a failure which is due to very small departures of B or Ω from the experimental values. Equation II.54 is used instead as a functional form for the elastic contribution ; polynomial interpolations of B(c) and Ω(c) through the experimental values determined on the ordered structures (A_3B, A_2B_2, AB_3) and on both ends (pure A, pure B) allow to calculate the elastic part of $\Delta H_m^D(c)$. The difference of the latter with the experimental value of ΔH^D_m allows to calculate concentration independent multi-atom interaction E_n : and it is ascertained that the critical temperatures obtained with these E_n's are in good agreement with the experimental ones. The single phase domains are found also to shrink at lower temperatures in agreement with basic physical arguments and the final diagrams are much improved (see Fig. II.9 a, b).

Although it has been (somewhat arbitrarily) rejected by the authors of the ε-G approach [II.28], the choice of additive pairwise interactions which depend on the concentration c, in place of multi-atom interactions E_n which are concentration independent, could have been performed as well. But the separation of "chemical" and elastic effects would not have been so easy.

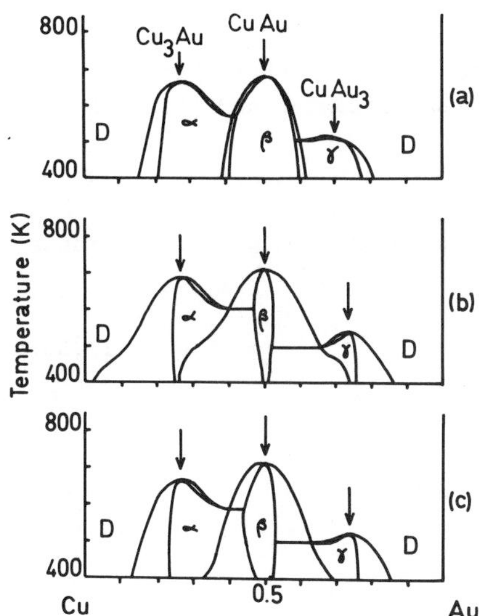

Fig. II.9 Phase diagram for (Au,Cu) system calculated by various approximations : a) CVM with multi-atom interactions only. b) C.V.M. with multi-atom interactions and elastic effects. c) Epitaxial deposit on Cu-Au I.

II.4.2 Epitaxial cubic alloys [II.29, II.30]

The coherency with a substrate strains the initially cubic cell (a_n,a_n,a_n) of the ordered phase n into a tetragonal one (a_s,a_s,b_n). It can be shown that the difference between the mixing enthalpies of the epitaxial alloy ΔH^{ep}_m and the bulk alloy ΔH^m_{bk} can be expressed by

$$\Delta H^m_{ep}(a_s,c) = \Delta H^m_{bk}(c) + \frac{9}{8} B' a (a_s - a)^2$$

where a is the equilibrium lattice parameter and B an effective modulus depending on the epitaxially constrained population z_n' of tetrahedra.

For Au_cCu_{1-c} alloys deposited on the ordered CuAu I phase, the change in the phase diagram is small but detectable (see Fig. II.9 c) ; the single-phased domains are broadened and the critical temperatures are slightly decreased.

For $GaAs_cSb_{1-c}$ alloys deposited on Ga_2AsSb substrate, the changes are, conversely, drastic : the decrease of the critical temperature for unmixing is large (≅ 1000 K) ; ordered phases which were calculated as metastable become stable (see Fig. II.10 a, b) and the solid solution is stable down to low temperatures.

In the same frame, by expanding the epitaxial c_{ep} and bulk compositions c_{bk} around the lattice matching composition c_s, the lattice pulling factor Q can be related to spinodal temperature τ_{ep} and τ_{bk} for the epitaxial and the bulk alloys

$$Q = \frac{c_{ep} - C_s}{c_{bk} - C_s} \sim \frac{T - \tau_{bk}(C_s)}{T - \tau_{ep}(C_s)} \qquad (II.55)$$

where $\tau_{bk}(C_s)$ is the critical temperature for unmixing at the lattice matching composition and $\tau_{ep}(C_s)$ is given by

$$\tau_{ep}(C_s) = \tau_{bk}(C_s) - \frac{9}{4}\frac{B'}{k}\, a \left(\frac{da}{dc}\right)^2 \qquad (II.56)$$

where the derivative da/dc is evaluated for $c = C_s$.

This equation makes clear the connection between composition pinning (Q small) and epitaxial stabilization ($\tau_{ep} < \tau_{bk}$).

II.5 COHERENT-INCOHERENT TRANSITION

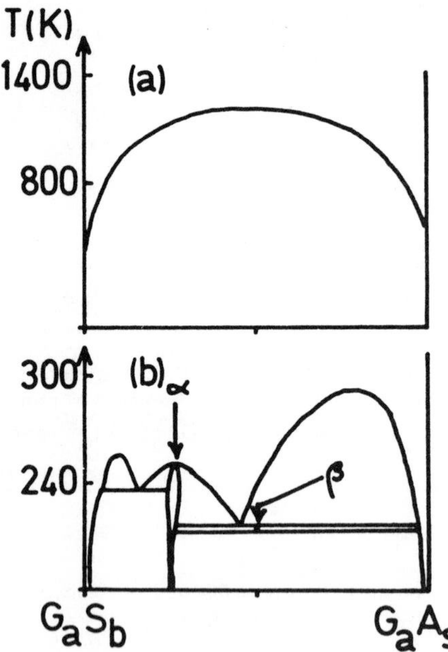

Fig. II.10 : Phase diagram for (Ga, As, Sb). a) CVM with elastic effects included. b) Epitaxial deposit on Ga_2AsSb (α and β stand for the ordered structures Ga_4AsSb_3 and Ga_2AsSb).

It has been implicitely assumed up to now that the layer deposited coherently on the substrate was of infinite extension along Ox and Oy and of finite (but macroscopic!) thickness along Oz. This is possible only under restrictive conditions :

- The infinite extension parallel to the substrate is allowed only for a low lattice mismatch η : if the latter reaches a critical value η_c, a commensurate-incommensurate transition occurs.
- The finite (and macroscopic) extension z_m along Oz is possible below a critical thickness z_c ; otherwise misfit dislocations are generated (coherent-incoherent transition).

We refer the reader to recent review papers [II.31, II.32].

III DIFFUSION IN COMPOSITIONALLY MODULATED MATERIALS

III.1 INTRODUCTION

Compositionally modulated materials are unstable systems with steep concentration gradients and strong chemical potential variations. Thus atomic diffusion processes are expected to operate, even at relatively low temperatures. It is therefore of primary importance to investigate the diffusion behaviour of a given modulated structure in order to determine its 'stability' conditions against various external perturbations such as temperature, irradiation, etc. It is also important to know the variation with temperature of the interdiffusion rate to determine the preparation conditions of the specimens. For epitaxial multilayers for example the temperature must be high enough to allow for a sufficient amount of surface mobility, but must be kept low enough to insure steep interfaces.

Modulated structures are also very efficient to accurately determine very low diffusion coefficients. The wavelength and amplitude of the modulation can be determined from observation of X-ray or neutron diffraction satellites. As the modulation wavelength can be made very small (≤ 50 Å), a small number of atomic jumps is sufficient to alter considerably the composition profile. Thus the intensity of the satellites varies very quickly with time, which makes the technique very sensitive. Interdiffusion constants as low as 10^{-20} $cm^2.s^{-1}$ can easily be measured. This technique was introduced by Dumond and Youtz in 1940 [III.1], and employed by Cook and Hilliard in Au-Ag modulated films [III.2]. Because of the presence of large concentration gradients the gradient energy term of Cahn and Hilliard [III.3] must be included in the free energy and shows up in the diffusion constant. Thus atomic transport measurements allow us to determine thermodynamic quantities such as the gradient energy parameter K, or g_o'', the second derivative of the free energy with respect to concentration. This technique is particularly useful in amorphous alloys because (i) diffusion constants must be measured at low temperature to avoid structural relaxation or crystallization ; (ii) there is no conventional means to determine K or g_o'' in such metastable or unstable alloys.

In the following we shall recall the essential elements of diffusion theory, that can be found in excellent textbooks or treatises [III.4, III.5, III.6] and specialize to diffusion in multilayers, including stress effects. Most of the material necessary to understand diffusion in modulated structures is also involved in spinodal decomposition theory, and we shall rely heavily on fundamental papers on this subject [III.3, III.7, III.8]. Let us also mention previous review articles on diffusion in modulated structures [III.9, III.10].

III.2 DIFFUSION IN SYSTEMS WITH SMALL CONCENTRATION GRADIENTS

Quite generally the diffusion flux J_B of the solute B in a matrix A can be written

$$J_B = n_B \, v_B \tag{III.1}$$

where by definition J_B is the number of B atoms crossing a plane of unit area per second, n_B is the number of B atoms per unit volume and v_B is their drift velocity. This velocity may arise from external forces (as is the case for the ionic conductivity in a uniform system) or from thermodynamic forces arising from chemical potential gradients. In this last case the drift velocity is proportional to the mobility M_B of B atoms and to the thermodynamic force $F_B = -\nabla \mu_B$

$$v_B = M_B F_B = - M_B \nabla \mu_B \qquad (III.2)$$

where μ_B is the chemical potential of B. This potential can be calculated in the framework of the regular solution theory as follows. The free energy of a mixture of n_A and n_B atoms interacting via nearest neighbour interactions ε_{AA}, ε_{AB}, ε_{BB} can be written

$$G = \frac{1}{2} z \, (n_A \varepsilon_{AA} + n_B \varepsilon_{BB}) + n_A z \varepsilon C_B^2 + n_B z \varepsilon C_A^2 + kT \, (n_A \ln C_A + n_B \ln C_B) \qquad (III.3)$$

where z is the coordinence, $C_i = n_i/N$, $N = n_A + n_B$ and $\varepsilon = \varepsilon_{AB} - \frac{1}{2} (\varepsilon_{AA} + \varepsilon_{BB})$ is the ordering energy. It is straightforward to generalize this expression of G to more complicated interactions. By definition the chemical potentials μ_A and μ_B are such that

$$g = n_A \mu_A + n_B \mu_B \qquad (III.4)$$

so that the chemical potential of B is simply

$$\mu_B = \frac{1}{2} z \varepsilon_{BB} + z \varepsilon C_A^2 + kT \ln C_B \qquad (III.5)$$

It is customary to write this expression in the following way

$$\mu_B = \mu_B{}^o + kT \ln \gamma_B C_B \qquad (III.6)$$

where $\mu_B{}^o$ is the reference potential and γ_B the activity coefficient

$$\gamma_B = \exp \left(\frac{z \varepsilon C_A^2}{kT} \right) \qquad (III.7)$$

Using this expression of the chemical potential it is possible to determine the flux of species B.

a) If the concentration C_B of B is very low throughout the system the activity coefficient can be considered as constant (Henry law) so that

$$J_B = - n_B M_B \nabla \mu_B = - kT M_B \frac{dn_B}{dx} = - D_B^* \frac{dn_B}{dx} \qquad (III.8)$$

This equation defines the diffusion constant $D_B^* = kT M_B$ in a system where the concentration of B is very small ; D_B^* can be measured using radioactive isotopes.

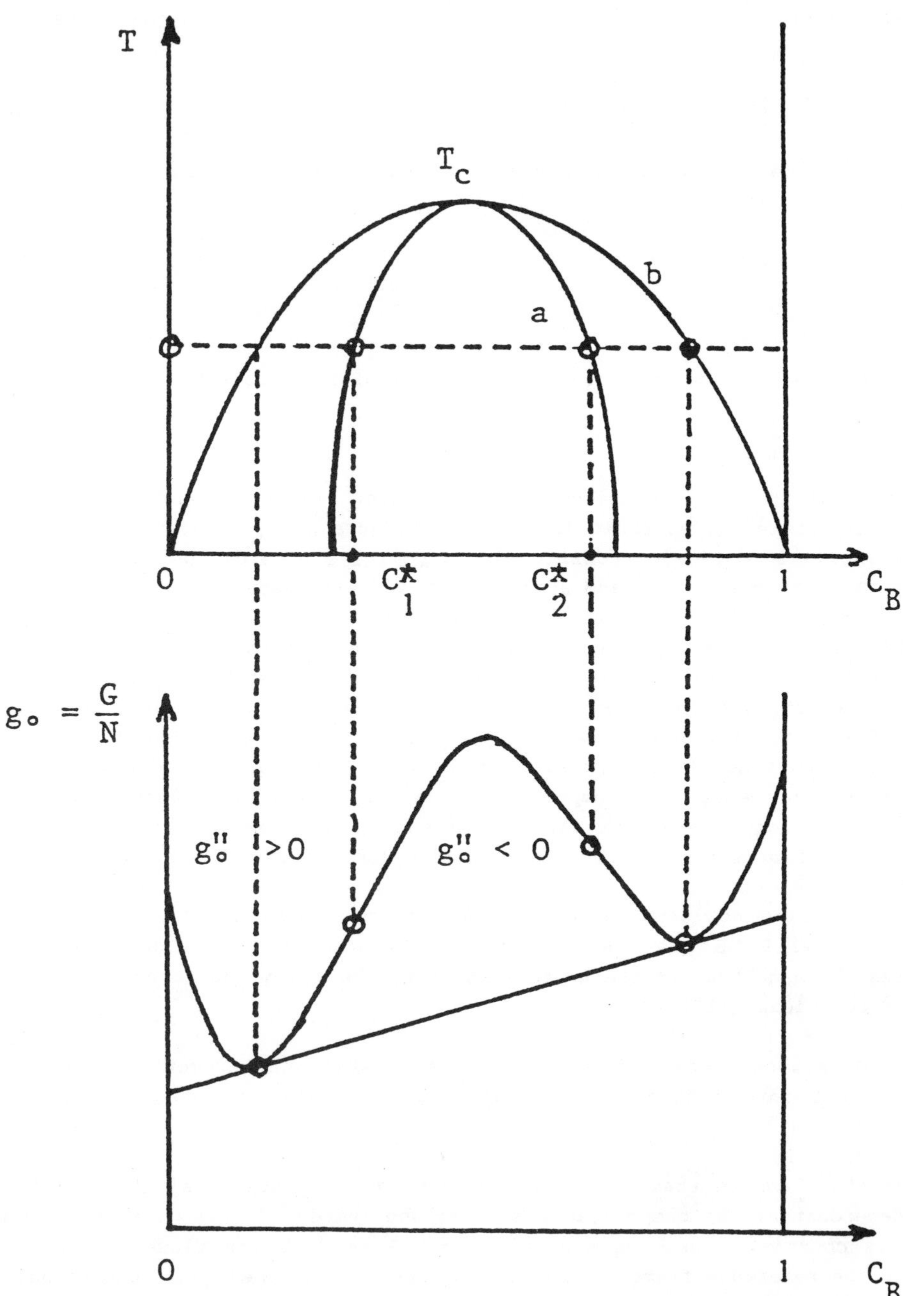

Fig. III.1 Average free energy g_o per atom and corresponding phase diagram for a binary alloy. a) Spinodal line. b) Coexistence curve.

b) When the concentration of B increases, it is necessary to use the full expression (III.6) of μ_B and it is easy to show that the diffusion coefficient can now be written

$$D_B = D_B^* \left(1 + \frac{d\ ln\ \gamma_B}{d\ ln\ C_B}\right) \qquad (III.9)$$

D_B is the intrinsic diffusion coefficient of B in a relatively concentrated A B alloy.

The correction factor

$$1 + \frac{d\ ln\ \gamma_B}{d\ ln\ C_B}$$

is very important. Using the expression (III.7) of γ_B we obtain

$$1 + \frac{d\ ln\ \gamma_B}{d\ ln\ C_B} = 1 - \frac{2\ z\ \varepsilon\ C_B\ (1 - C_B)}{kT} = \frac{C_B\ (1 - C_B)}{kT}\ g_o'' \qquad (III.10)$$

where $g_o = G/N$ is the free energy per atom and g_o'' is the second derivative of g_o with respect to C_B. As shown in Fig. III.1 for a phase separating alloy ($\varepsilon > 0$), the intrinsic diffusion coefficient is <u>positive</u> outside the spinodal line, i.e. for $C_B < C_1^*$ and $C_B > C_2^*$ and <u>negative</u> below the spinodal.

This change of sign of D_B is a consequence of the change of the character of the solid solution. For $C_B < C_1^*$ the solution is stable (or metastable) against composition fluctuations. If the concentration of B increases in a given region of an initially homogeneous alloy, B atoms diffuse out of this region and the fluctuation tends to disappear. For $C_1^* < C_B < C_2^*$, D_B is negative and the alloys have a tendency to decompose spontaneously. In this concentration range composition fluctuations (of large wavelength) increase, B atoms drifting towards regions of already large concentration (uphill diffusion).

Let us also notice that in an alloy of composition C_1^* a concentration fluctuation will be quasi-stationary as $D_B = 0$ along the spinodal. This expresses the fact that at the concentration C_1^* the alloy hesitates between two forms of behaviour [III.11].

c) We now consider a diffusion couple in which the elements A and B are initially put into contact and are allowed to diffuse in one another (Fig. III.2).

We shall assume that the atomic volumes of each are equal ($\Omega_A = \Omega_B = \Omega$) and independent of the composition, so that the number of atomic sites per unit volume is constant : $n_A + n_B = N$. We denote J_A^o and J_B^o the fluxes of A and B atoms in the reference frame of the end surfaces of the specimen, that we assume maintained at fixed positions. Because of the law of conservation of A and B species it is possible to write

$$\frac{\partial}{\partial t}\ (n_A + n_B) = - \frac{\partial}{\partial x}\left(J_A^o + J_B^o\right) = 0$$

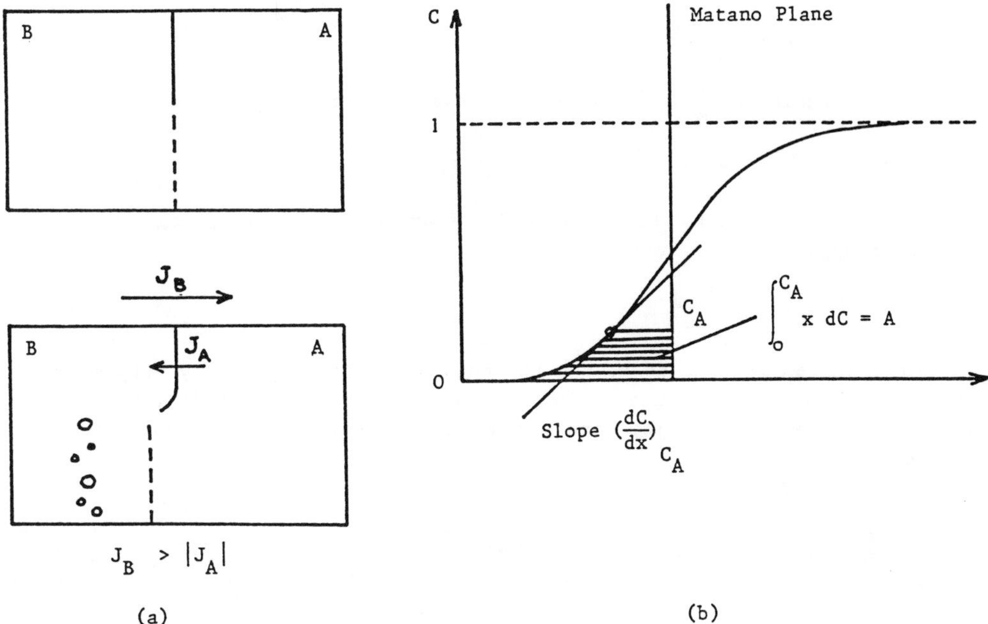

Fig. III.2 (a) When $J_B > |J_A|$, matter is exchanged from the left to the right of the plane containing the markers. (b) Determination of the interdiffusion coefficient by the Matano-method [III.4, III.5] : $\tilde{D} = + \frac{1}{2t} \frac{|A|}{(dc/dx)_{C_A}}$

because $n_A + n_B$ = constant. From this equation it follows that

$$J_A^o + J_B^o = \text{constant} = 0$$

because the ends of the sample are fixed.

Within the sample however, and because the fluxes of A and B are not equal (in absolute value), the atom lattice experiences a local velocity v that varies along the specimen. In the reference frame moving at velocity v, we define the <u>intrinsic</u> diffusion fluxes of A and B

$$J_A = - D_A \frac{\partial n_A}{\partial x}$$
$$J_B = - D_B \frac{\partial n_B}{\partial x} \qquad \text{(III.11)}$$

where D_A and D_B are the intrinsic diffusion coefficients given by equation III.9.

The fluxes of A and B in the reference frame are thus

$$J_A^o = J_A + n_A\, v$$

$$J_B^o = J_B + n_B\, v$$

and from the condition $J_A^o + J_B^o = 0$, it follows immediately that

$$Nv = -\,(J_A + J_B)$$

This shows that the local velocity v arises from the <u>imbalance of the diffusion fluxes</u> J_A and J_B and

$$v = (D_B - D_A)\,\frac{\partial\, C_B}{\partial x} \qquad \text{(III.12)}$$

The detailed meaning of v, and the methods to experimentally determine it, will be explained at the end of this paragraph. In practice, it is the fluxes J_A^o and J_B^o that are experimentally determined and from the preceding equations

$$J_A^o = C_B\, J_A - C_A\, J_B$$

$$J_B^o = C_A\, J_B - C_B\, J_A$$

so that

$$J_A^o = -\,\mathbf{D}\,\frac{\partial\, n_A}{\partial x}$$

$$J_B^o = -\,\mathbf{D}\,\frac{\partial\, n_B}{\partial x} \qquad \text{(III.13)}$$

where D is called the <u>interdiffusion</u> coefficient

$$\mathbf{D} = C_A\, D_B + C_B\, D_A$$

$$\mathbf{D} = (C_A\, D_B^{*} + C_B\, D_A^{*})\left(1 + \frac{\partial\, ln\ \gamma_B}{\partial\, ln\ C_B}\right) \qquad \text{(III.14)}$$

The first factor insures that $\mathbf{D} = D_B$ in almost pure A ($C_A \sim 1$; $C_B \sim 0$) and $\mathbf{D} = D_A$ in pure B. Here again the second factor governs the sign of **D** when the spinodal line is crossed.

In order to interpret the meaning of the velocity v, let us consider a diffusion couple as shown in Fig. III.2 and assume that $D_B > D_A$ so that $|J_B| > |J_A|$. Inert markers (tungsten wires or oxide particles for example) have been placed at the initial interface. As more B atoms tend to drift to the right than A atoms to the left, there is a tendency for the system to accumulate matter to the right of the initial interface and to loose matter in the left region. But because we assumed a continuous and incompressible solid, the right part of the sample will expand to the expense of the left part. This process shows up as a shift of the markers, called the <u>Kirkendall effect</u>. It may also happen that the relaxation process, by which the right part of the sample develops and the left part shrinks, is not complete so that holes may form in the left region. These holes are called <u>Kirkendall voids</u>. From this qualitative analysis we can conclude that the inert markers drift towards the fast diffuser and that Kirkendall voids may appear within the bulk of the fast diffuser.

This Kirkendall process is independent of the microscopic mechanism by which diffusion operates. However the origin of the Kirkendall voids is particularly clear in the case of the vacancy diffusion mechanism. Because $|J_B| > |J_A|$ (Fig. III.2) vacancies must be created in the right part of the sample, and eliminated in the left part, after crossing the plane containing the markers. If we imagine that a full atomic plane of vacancies is created in the right region, and eliminated to the left, it is clear that the overall process is equivalent to extracting an atomic plane at the left and forcing it into the right region. This process induces a displacement of the markers, to the left, of an interatomic distance. Thus the velocity v of the markers is directly related to the vacancy flux J_l^o in the laboratory reference frame

$$J_l^o = n_l \, v_l^o = N \, v$$

Finally, to conclude this paragraph it is useful to mention the method currently used to determine the interdiffusion coefficient D in a diffusion couple of two pure elements A and B. From the conservation of species A, equation (III.13) can be written

$$\frac{\partial n_A}{\partial t} = - \frac{\partial J_A^o}{\partial x} = \frac{\partial}{\partial x} \left(D \, \frac{\partial n_A}{\partial x} \right)$$

Using a suitable transformation [III.4, III.5], it is easy to show that, at a point where the concentration of A is C_A, the interdiffusion coefficient is given by (Fig. III.2)

$$\mathbf{D} \, (C_A) = - \frac{1}{2t} \, \frac{\int_o^{C_A} x \, dC_A}{(dC_A/dx)_{C_A}}$$

The distance x must of course be determined from the Matano plane, defined by

$$\int_0^1 x \, dC_A = 0$$

which determines two surfaces of equal area on the concentration profile.

III.3 DIFFUSION IN SYSTEMS WITH LARGE CONCENTRATION GRADIENTS

The expression III.3 of the free energy was obtained for a homogeneous system. In the preceding paragraph we used this form of the free energy (through the chemical potentials deduced from it) to calculate the diffusion fluxes. Thus we implicitely assumed that the system could be decomposed into small volume elements $d_3 \, r$, and that each element could be considered as a piece of a homogeneous alloy of local concentration C $(\mathbf{r})$. The free energy of the whole system was therefore expressed as

$$G = \int g_o \, (C \, (\mathbf{r})) \, \frac{d_3 \, r}{\Omega} \qquad \text{(III.15)}$$

where g_o is the free energy per atom in a homogeneous alloy of concentration $C_B = C(\mathbf{r})$ and Ω is the average atomic volume.

In a system with strong concentration gradients this approximation is no longer valid. This is because the energy of the alloy involves interaction pairs : starting from atom A at the origin one has to sum up all pair interactions starting from A and leading to more or less distant A or B atoms. It is clear that in this process one probes regions where the concentration is different from the local concentration at the starting point.

This analysis can be carried out in detail [III.3, III.7] and results in the following form for the total free energy of a system

$$G = \int \frac{d_3 r}{\Omega} \left(g_o (C) + K \, |\nabla C|^2 \right) \tag{III.16}$$

where the free energy per atom at point $\mathbf{r}$ depends on the concentration and on the concentration gradient at this point.

The gradient energy coefficient K can actually be calculated as a function of the interatomic interactions. It turns out that, in a model with nearest neighbour interactions, K is directly proportional to the ordering energy ε [III.7]

$$K = \frac{1}{2} \lambda^2 z \varepsilon \tag{III.17}$$

where λ is of the order of the interatomic distance.

Let us recall that, in equation III.16, g_o is the free energy per atom of an alloy constrained to remain homogeneous at the concentration C, and that

$$g_o'' = \frac{d^2 g_o}{dC^2} = \frac{kT}{C(1-C)} - 2 z \varepsilon = 4 kT - 2 z \varepsilon \tag{III.18}$$

for an equiatomic binary mixture.

The expression III.16 for the free energy allows us to determine the concentration profile of an alloy at equilibrium, using appropriate boundary conditions. In a one-dimensional system, the minimization of G, under the constraint of mass conservation, leads to the following equation, to be obeyed by the atomic concentration C (x)

$$- 2 K \frac{d^2 C}{d x^2} + \frac{\partial g_o}{\partial C} = \alpha \tag{III.19}$$

where the constant α can be interpreted as follows. If the gradient energy term vanishes, $\alpha = \partial g_o / \partial C = \mu_B - \mu_A$ and the equation III.19 expresses the fact that, at equilibrium, $\mu_B - \mu_A$ is constant throughout the specimen. In other words, there is no interdiffusion flux in the alloy. In the presence of gradient energy contributions the equilibrium is realized when

$$\frac{\partial g_o}{\partial C} - 2 K \frac{d^2 C}{dx^2}$$

remains constant in the system ; this expression can therefore be interpreted as a generalized chemical potential α. The gradient of α is now the driving force for interdiffusion. The flux of constituent B can thus be written

$$J_B = - n_B M_B \frac{d\alpha}{dx} = - C_B M_B g_o'' \left(\frac{\partial n_B}{\partial x} - \frac{2K}{g_o''} \frac{\partial^3 n_B}{\partial x^3} \right) \qquad (III.20)$$

where M_B is the mobility of B atoms and $g_o'' = \partial^2 g_o / \partial C_B^2$. To a first approximation we may assume that $C_B M_B g_o''$ is nearly independent of the concentration and write

$$J_B = - \mathbf{D}_B \left(\frac{\partial n_B}{\partial x} - \frac{2K}{g_o''} \frac{\partial^3 n_B}{\partial x^3} \right) \qquad (III.21)$$

with

$$\mathbf{D}_B = C_B M_B g_o'' \qquad (III.22)$$

With this approximation the second Fick's equation, which expresses mass conservation, is linear

$$\frac{\partial n_B}{\partial t} = - \frac{\partial J_B}{\partial x} = \mathbf{D}_B \left(\frac{\partial^2 n_B}{\partial x^2} - \frac{2K}{g_o''} \frac{\partial^4 n_B}{\partial x^4} \right) \qquad (III.23)$$

It is usually convenient to expand the concentration $n_B(x,t)$ in Fourier components [III.12] around the average concentration $\mathbf{n}$ of the alloy :

$$n_B(x,t) = \mathbf{n} + \sum_q \mathbf{n}(q,t)\, e^{iqx} \qquad (III.24)$$

where $\mathbf{n}(q,0)$ and the possible values of q are determined by the shape of $n_B(x,0)$, the composition profile at t = 0. Because of the linearity of equation III.23, the Fourier amplitudes $\mathbf{n}(q,t)$ change in time independently of one another, according to the law :

$$\mathbf{n}(q,t) = \mathbf{n}(q,0) \exp(- q^2 \mathbf{D}_q t) \qquad (III.25)$$

where

$$\mathbf{D}_q = \mathbf{D}_B \left(1 + \frac{2K}{g_o''} q^2 \right) \qquad (III.26)$$

is the wavelength dependent interdiffusion coefficient.

The growth or decay of the amplitude $\mathbf{n}(q,t)$ is determined by the sign of $\mathbf{D}_q$. In the framework of the regular solution model

$$g_o'' = \frac{kT}{C_B (1 - C_B)} - 2 z \varepsilon$$

$$K = \frac{1}{2} \lambda^2 z \varepsilon \qquad (III.27)$$

so that, according to the sign of g_o'' and K, three different cases can be distinguished (Fig. III.3) [III.14, III.15].

a) $\varepsilon > 0$ (phase separating system) : the gradient energy term is positive, but g_o'' can be positive or negative (Fig. III.1).

a1) Below the spinodal line, g_o'' is negative so that $\mathbf{D}_B = C_B M_B g_o''$, which represents the diffusion constant for long wavelength modulations, is

<u>negative</u> (uphill diffusion). More precisely, modulations of wave-vector larger than

$$q_c = \left(\frac{|g_o''|}{2K}\right)^{1/2}$$

decay, because of the large number of regions where the concentration quickly varies (high density of "interfaces"). Modulations with $q < q_c(\lambda > \lambda_c)$ <u>grow in time</u> (spinodal decomposition), and the amplification factor $-q^2\mathbf{D}_q$ is maximal for $q_o = q_c/\sqrt{2}$ (Fig. III.3).

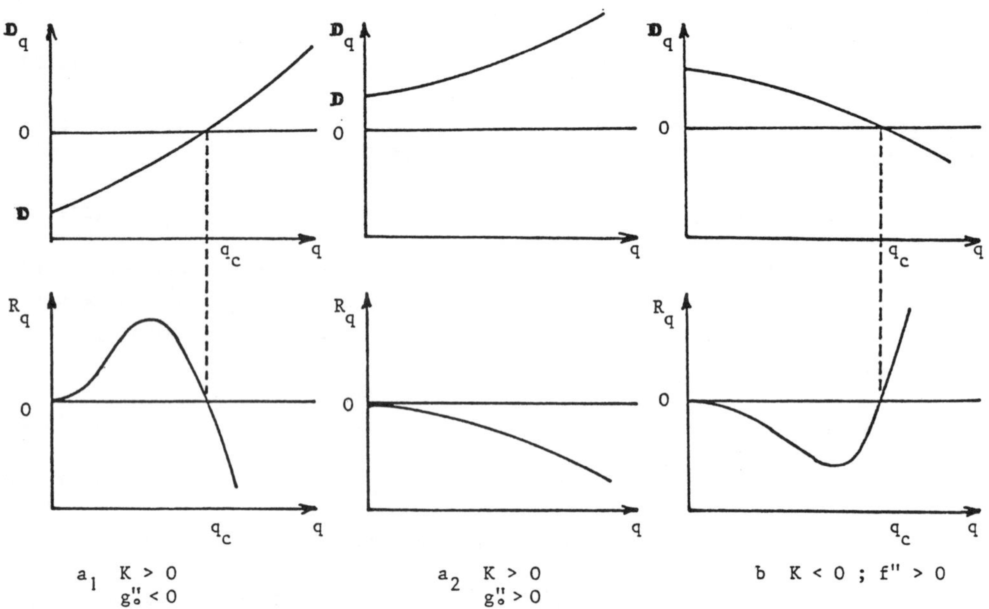

Fig. III.3 Wavelength dependent diffusion coefficient in a phase separating system (a_1, a_2) and in an ordering system (b). $R_q = -q^2\mathbf{D}_q$ is the amplification factor (equation III.25) of the modulation Fourier components.

a2) Outside the spinodal region, in a phase separating system, K > 0 and g_o" is positive. The q-dependent diffusion coefficients are positive for all wavelengths and the modulation amplitudes decay whatever their wavelength. This result is the expression of the stability of the solid solution against composition fluctuations outside the spinodal region.

b) $\varepsilon > 0$ (ordering system) : In this case, g_o" is always positive, **D** is positive and modulations of long wavelength ($q < q_c$) decay whereas composition fluctuations of short wavelength ($q > q_c$) grow. As the critical wavelength $\lambda_c = 2\pi/q_c$ is usually of the order of the interatomic distance, this growth of short wavelength fluctuations can be interpreted as the ordering of the solution.

Let us notice that the preceding results have been obtained in the framework of the continuum approximation, which must break down for large wave-vectors, especially in crystalline systems. This problem has been studied in detail [III.2, III.13, III.16]. Broadly speaking the main change in the theory amounts to replacing q^2 in equation III.26 by the dispersion relation

$$B^2\,(\mathbf{q}) = \frac{1}{a^2} \sum_{n.n} (1 - \cos \mathbf{q}.\mathbf{r}) \qquad \text{(III.28)}$$

where a is the lattice parameter and the summation is over the nearest neighbours of a given atom.

In artificial multilayers it is possible to choose the wave-vectors of the modulation. The initial composition of the alloy can be expanded in a Fourier series (equation III.24). The allowed values of q are $q_n = n\, q_o$ where $q_o = 2\pi/l$ in a specimen where l is the periodicity of the multilayer. Each of these modulation components gives rise to diffraction satellites around the (000) peak, or to satellites around the Bragg diffraction peaks in a crystalline specimen.

In an isoconfigurational specimen and if the linear approximation leading to equation III.23 is satisfactory, the variation with time of the intensity $I_n(t)$ of these satellites gives a direct measurement of $\mathbf{D}_{q_n}$ as

$$ln \frac{I_n(t)}{I_o(t)} = - 2\, q_n^2\, \mathbf{D}_{q_n}\, t \qquad \text{(III.29)}$$

In principle a given sample should provide several values of $\mathbf{D}_{q_n}$ corresponding to the decay (or growth) of the first Fourier components of the composition. Actually, because of non-linearities, it is usually found that only the first Fourier components varies according to equation III.29, after a transient period due to structural relaxation or grain boundary motion [III.9]. In order to get experimental information about M_B, g_o'' and K, it is customary to prepare different multilayer specimens with different modulation wavelengths, and use the decay of the first satellite only [III.13, III.14, III.15, III.17].

III.4 DIFFUSION IN SYSTEMS IN THE PRESENCE OF STRESSES

It is well known that stress gradients may induce atomic diffusion ; let us mention the Gorski effect, i.e. the diffusion of light interstitials in a specimen subjected to external non-uniform stresses, the Cottrell atmosphere around dislocations, ... Diffusion problems are difficult ones however because atomic motions induce a change in the elastic state of the solid so that the (generalized) chemical potentials of the constituent species change. This interplay between diffusion and elastic stresses is difficult to describe in all generality because elasticity is essentially non local. If the composition of the specimen is changed in a given region, long range stresses are induced that change the chemical potential at large distances. Moreover this change in principle depends on the shape and size of the solid, on the boundary conditions, etc. This problem has been studied in detail by Cahn and Larché [III.18].

Let us define $\eta = da/adc$ which describes how the lattice parameter a depends on composition. It can be shown that in some conditions (infinite and isotropic system, ..) the free energy of the system can be written [III.19]

$$G = \int \frac{d_3r}{\Omega} [g_o (C) + K |\nabla C|^2 + \Omega \eta^2 Y (\Delta C)^2] \qquad (III.30)$$

where Ω is the average atomic volume, $Y = E/(1-\nu)$ where E is the Young modulus and ν is the Poisson ratio ; $\Delta C = C - C_o$ is the local deviation of the concentration with respect to the average composition of the solid.

A variational calculation similar to that we performed to obtain equation III.19 leads to the new definition of the generalized chemical potential

$$\alpha = -2K \frac{d^2C}{dx^2} + \frac{\partial g_o}{\partial C} + 2 \Omega \eta^2 Y \Delta C \qquad (III.31)$$

Using this expression it is again possible to calculate the diffusion flux

$$J_B = - n_B M_B \nabla\alpha = - C_B M_B g_o'' \left(\frac{\partial n_B}{\partial x} - \frac{2K}{g_o''} \frac{\partial^3 n_B}{\partial x^3} + \frac{2 \Omega \eta^2 Y}{g_o''} \frac{\partial n_B}{\partial x} \right) \qquad (III.32)$$

The wave-vector-dependent diffusion coefficients can thus be written

$$D_q = D_B \left(1 + \frac{2K}{g_o''} q^2 + \frac{2 \Omega \eta^2 Y}{g_o''} \right) \qquad (III.33)$$

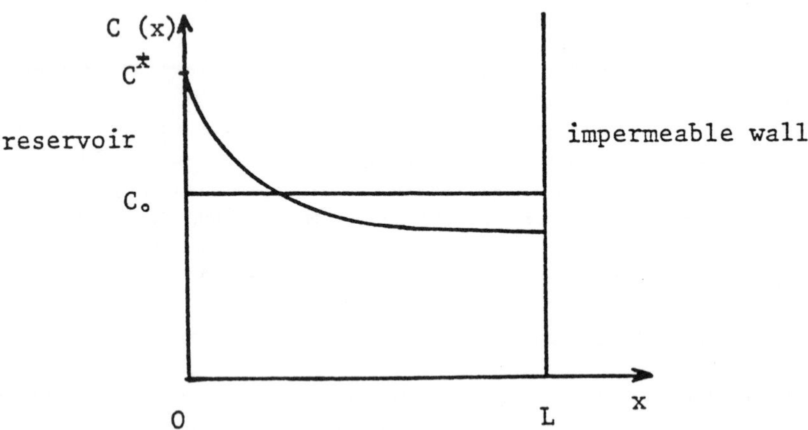

Fig. III.4 Concentration profile in a slab of initial concentration C_o in the first stages of the diffusion process [III.18]

This definition of an apparent diffusion constant can be used to interpret diffusion data in macroscopic samples. It must be emphasized that stresses make diffusion equations non-local. This may lead to surprising phenomena. Let us consider for example [III.18] a thin plane located at $x = 0$ (Fig. III.4) and with an impermeable wall at $x = L$. The initial concentration in the slab is C_o and the chemical potential of the solute is larger in the reservoir than in the specimen. Solving the elasticity problem corresponding to this situation, Larché showed that the diffusion flux can be written

$$J = - \mathbf{D}\ (1 - y)\ \frac{\partial n}{\partial x} + \frac{12\ \mathbf{D}\ y}{L^3} \int_o^L (x - \frac{L}{2})\ \Delta n(x)\ dx \qquad \text{(III.34)}$$

where $y = 2\ \Omega\ n^2 Y/g_o''$ and $\Delta n = n(x) - n_o$ where $n(x)$ denotes the number of solute atoms per unit volume. Thus the diffusion flux does not obey a Fick equation anymore, J depending on the concentration everywhere in the sample and on the shape of the specimen.

In the first stages of the diffusion process (Fig. III.4) we may assume that the concentration has only changed in a narrow layer of thickness $\sim (D_{app}\ t)^{1/2}$ around $x = 0$. If the concentration in the plane $x = 0$ is C^*, the integral I involved in equation III.34 can be estimated

$$I = \int_o^L (x - \frac{L}{2})\ \Delta n(x)\ dx \sim - \frac{L}{2}\ \Delta n^*\ (D_{app}\ t)^{1/2} \qquad \text{(III.35)}$$

Thus the flux induced by stresses in the sample is <u>negative</u> and, contrary to intuition, leads to the <u>depletion</u> of the layer in regions that have not yet been reached by the incoming flux of solute atoms.

As the integral I is negative it is also easy to check that $\left(\frac{\partial n}{\partial x}\right)_{x=L}$ is positive, as shown in Fig. III.4.

III.5 NON LINEAR DIFFUSION

In the theory of spinodal decomposition we assumed that the mass conservation equation III.23 was linear, $\mathbf{D}_B$ and $2K/g_o''$ being independent of the composition. In specimens with steep concentration gradients this approximation may not be sufficient to understand the time variation of the Fourier components of the composition modulation.

Some authors [III.12, III.20, III.21, III.22] explicitely took account of the composition dependence of M_B, g_o'', η and Y in equation III.33 and solved the one-dimensional non-linear diffusion equation using a perturbation analysis. They were then able to interpret the peculiar variation of $ln\ ((It)/I_o)$ with time for the first two satellites in a CuNi foil with wavelength 45.6 Å and 52 at % average Cu concentration (Fig. III.5).

Similar results have been found in amorphous multilayers [III.20, III.21] or in compositionally modulated metallic glasses.

Let us also notice that the simple diffusion equations derived previously are not valid any longer if a new phase is formed at the interface between individual layers. This is for example the case when amorphization by diffusion or by mechanical alloying takes place at the interfaces [III.24, III.25].

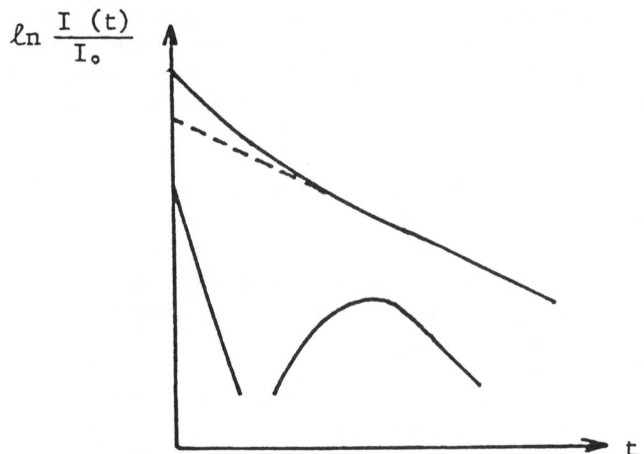

Fig. III.5 Variation with time of the intensity of the two first satellites, in the presence of strong non linearities [III.12].

REFERENCES

I.1 Gibbs J.W., The collected Works Vol. I, Thermodynamics (Yale University Press, New Haven Connecticut) (1928)

I.2 Murr L.E., Interfacial Phenomena in Metals and Alloys, Addison-Wesley Pub. Comp. (1975)

I.3 Guggenheim E.A., J. Chem. Phys. 13, 253 (1945)

I.4 Eötvös R.V., Wied Ann 27, 456 (1986)

I.5 Grosse A.V., J. Inorg. Nucl. Chem. 24, 147 (1962)

I.6 Joud E.C., Eustathopoulos N., Bricard A. and Desré P., J. Chim. Phys. 9, 1290 (1973) ; 7-8, 113 (1974)

I.7 Joud J.C., Desré P., Acta Astronautica 8-5, 407 (1981)

I.8 Skapski, Acta Met. 4, 576 (1956)

I.9 Oriani R.A., J. Chem. Phys. 18, 575 (1950)

I.10 Eustathopoulos N., Joud J.C., Desré P., J. Chim. Phys. 1, 42 (1973)

I.11 Miedema A.R., Niessen A.K., Cohesion in Metals, North Holland Publishing Company (1988)

I.12 Overbury S.H., Bertrand P.A., Somorjai G.A., Chem. Rev. 75, 547 (1975)

I.13 Eustathopoulos N., Joud J.C., Desré P., J. Chim Phys. 1, 49 (1973)

I.14 Eustathopoulos N., Joud J.C., Interfacial tension and adsorption of metallic systems from Current Topics in Materials Science 4, 281, E. Kaldis Editor, North Holland Publishing Company (1980)

I.15 Friedel J., Desjonqueres M.C., Catalyse par les métaux, p. 9, Imelik Editor, CNRS, Lyon (1984)

I.16 Johnson R.A., Phys. Rev. 134 A, 1329 (1964)

I.17 Girifalco L.A., Weizer V.G., Phys. Rev. 114, 687 (1959)

I.18 Joud J.C., Thèse d'Etat, Université de Grenoble (1973)

I.19 Wynblatt P., Gjostein N.A., Surf. Sci. 22, 125 (1970)
I.20 Adsorption on metals surface, Ed. J. Benard, p. 1, Elsevier Scientific Pub. Comp. (1983)
I.21 Drechsler M., Muller A., J. Cryst. Growth 3-4, 518 (1968)
I.22 Nicholas J.F., Aust. J. Phys. 21, 21 (1968)
I.23 Grenga H.E., Kumar R., Surf. Sci. 61, 283 (1976)
I.24 Desjonqueres M.C., Spanjaard D., Introduction à la physique des surfaces, Ecole d'Hiver sur la physico-chimie des surfaces, Les Menuires (Janvier 1987)
I.25 Spanjaard D., Jepsen D.W., Marcus P.M., Phys. Rev. B 19, 642 (1979)
I.26 Noguera C., Spanjaard D., Jepsen D.W., Phys. Rev. B 17, 607 (1978)
I.27 Gerl, Interfaces et Surfaces en Métallurgie, Ecole d'Eté de Métallurgie Physique, Gassin (1973), Trans. Tech. Publications
I.28 Bardeen J., Phys. Rev. 49, 653 (1936)
I.29 Huntington H.B., Phys. Rev. 81, 1035 (1951)
I.30 Lang N.D., Kohn W., Phys. Rev. B 1, 4555 (1970)
I.31 Hohenberg P., Kohn W., Phys. Rev. B 136, 864 (1964)
I.32 Kohn W., Sham L.J., Phys. Rev. 140, 1133 (1965)
I.33 Friedel J., Transition metals, J.M. Ziman Editor, Cambridge University Press, 340 (1969)
I.34 Gautier F., Métaux et alliages de transition, Masson Ed., 255, Paris (1973)
I.35 Desjonqueres M.C., Cyrot Lackmann F., J. Phys. F Metal Physics 5, 1368 (1975)
I.36 Desjonqueres M.C., Cyrot Lackmann F., J. Phys. F Metal Physics 6, 567 (1976)
I.37 Allan G., Electronic structure of transition metal surfaces in Handbook of Surfaces and Interfaces, L. Dobrzynski Ed., STPM Press, 299 (1978)
I.38 Desjonqueres M.C., Spanjaard D., Lassailly Y., Guillot C., Solid State Com. 34, 807 (1980)
I.39 Spanjaard D., Guillot C., Desjonqueres M.C., Tréglia G., Lecante J., Surf. Sci. Report 5, 1 (1985)
I.40 Friedel J., Sayers C.M., J. Physique 38, 697 (1977)
I.41 Ducastelle F., J. Physique 31, 1055 (1970)
I.42 Tréglia G., Desjonqueres M.C., Spanjaard D., J. Phys. C Solid State Physics. 16, 2407 (1983)
I.43 Guggenheim E.A., Mixtures, Clrendon Press, Oxford (1952)
I.44 Defay R., Prigogine I., Surface tension and adsorption, Longmans London (1966)
I.45 Mac Lean D., Grain boundaries in metals, Oxford University Press, London (1957)
I.46 Wynblatt P., Ku R.C., Surface Sci. 65, 511 (1977)
I.47 Kumar V., Surface Sci. 84, L231 (1979)
I.48 Molinari C., Joud J.C., Desré P., Surface Sci. 84, 141 (1979)
I.49 Friedel J., Advanc. Physics 3, 446 (1954)
I.50 Lambin P., Thèse d'Etat, Liège (1981)
I.51 Tsaï N.H., Pound G.M., Abraham F.F., J. Catal. 50, 200 (1977)
I.52 Seah M.P., J. Catal. 57, 450 (1979)
I.53 Williams F.L., Nason D., Surf. Sci. 45, 377 (1974)
I.54 Tréglia G., Legrand B., Phys. Rev. B 35-9, 4338 (1987)

I.55 Baudoing R., Gauthier Y., Lundberg M., Rundgren J., J. Phys. C 19, 2825 (1986)
I.56 Tréglia G., Legrand B., Ducastelle F., Europhys. Letters 7 (7), 575 (1988)
I.57 Ducastelle F., Gautier F., J. Phys. F 6, 2039 (1976)
I.58 Legrand B., Tréglia G., The structure of surfaces II, J.F. Van der Veen, M.A. Van Hove Editors, Springer Verlag (1987)
I.59 Gauthier Y., Baudoing R., Lundberg M., Rundgren R., Phys. Rev. B 35, 7867 (1987)
I.60 Daw M.S., Baskes M.I., Phys. Rev. B 29, 6443 (1984)
I.61 Foiles S.M., Phys. Rev. B 32, 7685 (1985)
I.62 Lundberg M., Phys. Rev. B 36, 1 (1987)
I.63 Joud J.C., Mathieu J.C., Desré P., Bonnier E., J. Chimie Physique 1, 131 (1972)
I.64 Kumar V., Phys. Rev. B 23-8, 3756 (1981)
I.65 Arroyo P.Y., Joud J.C., J. Physique 48, 1721 (1987)
I.66 King T.S., Donnelly R.G., Surf. Sci. 141, 417 (1984)
I.67 Arroyo P.Y., Joud J.C., J. Physique 48, 1733 (1987)
I.68 Eymery J., Joud J.C., to be published
I.69 Lea C., Seah M.P., Phil. Mag. 35, 213 (1977)
I.70 Rowlands G., Woodruff D.P., Phil. Mag. 40, 459 (1979)

❁❁❁❁❁❁❁

II.1 Martin G., in "Solid state phase transformations in metals and alloys" Ed. de Physique, Orsay, France (1980) p. 337
II.2 Simon J.P., Guyot P., Lyon O., Solid State Phenomena 324(1988) p. 179
II.3 Russell K.C., in "Nucleation III", A.C. Zettlemoyer Ed., Marcel Dekker, N.Y. (1975)
II.4 Cahn J.W., Hilliard J.E., J. Chem. Phys. 28 (1958) p. 258 ; 30 (1959) p. 1121 and 31 (1959) p. 688
II.5 Friedel J., Advances in Physics 3 (1954) p. 446
II.6 Hilliard J.E., in "Phase transformations", ASM, USA (1970) p. 497
II.7 Cahn W.E., Acta Met. 9 (1961) p. 795
II.8 Cook W.E., de Fontaine D., Hilliard J.E., Acta Met. 17 (1969) p.765
II.9 Cook W.E., de Fontaine D., Acta Mt. 17 (1969) p. 915 ; 19 (1971) p. 607
II.10 Larché F.C., Johnson W.C., Chiang C.S., Martin G., J. Appl. Physics 64 (1988) p. 5251
II.11 Stringfellow G.B., J. of Electr. Mat. 11 (1982) p. 903
II.12 de Crémoux B., J. de Physique 43 (1982), Colloque C5, Supp. au n°12, p. 19
II.13 Marboeuf A., Guillaume J.C., J. de Physique 43 (1982), Colloque C5, Supp. au n°12, p. 47
II.14 Glas F., J. Appl. Phys. 62 (1987) p. 3201
II.15 Larché F.C., Cahn J.W., Acta Met. 33 (1985) p. 331
II.16 Johnson W.C., Alexander J.I.D., J. Appl. Phys. 59 (1986) p. 2735
II.17 Larché F.C., Cahn J.W., J. Appl. Phys. 62 (1987) p. 1232
II.18 Johnson W.C., Chaing C.S., J. Appl. Phys. 64 (1988) p. 1155
II.19 Cahn J.W., Larché F.C., Acta Met. 32 (1984) p. 1935
II.20 Johnson W.C., Metall. Trans. 18A (1987) p. 1093
II.21 Ariosa D., Fischer O., Karkut M.G., Triscone J.M., Phys. Rev. B 37 (1988) p. 2415

II.22 Kikuchi R., J. de Physique 38 (1977), Colloque C7, Supp. au n°12, p. 307
II.23 Finel A., Thèse à l'Université P. et M. Curie (1987)
II.24 Kikuchi R., Sanchez J.M., de Fontaine D., Yamauchi H., Acta Met. 28 (1980) p. 651
II.25 Kanamori J., Kakehashi J., J. de Physique 38 (1977), Colloque C7, Supp. au n°12, p. 274
II.26 Wei S.H., Mbaye A.A., Ferreira L.G., Zunger A., Phys. Rev. B 36 (1987) p. 4163
II.27 Ferreira L.G., Mbaye A.A., Zunger A., Phys. Rev. B 37 (1988) p. 10547
II.28 Ferreira L.G., Mbaye A.A., Zunger A., Phys. Rev. B 35 (1987) p. 6475
II.29 Wood D.A., Zunger A., Phys. Rev. Letters 61 (1988) p. 1501
II.30 Wood D.M., Zunger A., Phys. Rev. B 38 (1988) p. 12756
II.31 Van der Merwe J.H., Jesser W.A., J. Appl. Phys. 64 (1988) p. 4968
II.32 Bruinsma R., Zangwill A., J. de Physique 47 (1986) p. 2055

❁❁❁❁❁❁❁

III.1 DuMond J., Youtz J.P., J. Appl. Phys. 11, 357 (1940)
III.2 Cook H.E., Hilliard J.E., J. Appl. Phys. 40 (5), 2191 (1969)
III.3 Cahn J.W., Hilliard J.E., J. Chem. Phys. 28, 258 (1958) ; J. Chem. Phys. 30, 1121 (1959)
III.4 Adda Y., Philibert J., "La Diffusion dans les Solides", Bibliothèque des Sciences et Techniques Nucléaires, Presses Universitaires de France (1966)
III.5 Philibert J., "Diffusion et Transport de Matière dans les Solides", Les Editions de Physique (1985)
III.6 Shewmon P.G., "Diffusion in Solids", Mc Graw-Hill (1963)
III.7 Martin G., in "Solid State Transformations in Metals and Alloys", Les Editions de Physique (1978)
III.8 de Fontaine D., J. Appl. Cryst. 8, 81 (1975) ; J. Phys. Chem. Sol. 33, 297 (1972) ; 34, 1285 (1973)
III.9 Spaepen F., Mat. Res. Soc. Symp. Proceedings 37, 207 (1985)
III.10 Piecuch M., Rev. Phys. Appliquée 23, 1727 (1988)
III.11 Bocquet J.L., Brébec G., Limoge Y., in "Diffusion in Metals and Alloys", Chapt 3, 440 (1983)
III.12 Tsakalakos T., Scripta Met. 20, 471 (1986)
III.13 Tsakalakos T., Hilliard J.E., J. Appl. Phys. 55 (8), 2885 (1984)
III.14 Greer A.L., Scripta Met. 20, 457 (1986)
III.15 Gammarata R.C., Greer A.L., J. Non-Crystalline Solids 61-62, 889 (1984)
III.16 Khachaturyan A.G., "Theory of Structural Transformations in Solids", John Wiley, New York (1983)
III.17 Bruson A., Marchal G., Piecuch M., J. Appl. Phys. 58, 1229 (1985)
III.18 Larché F.C., Cahn J.W., Acta Metal. 30, 1835 (1982)
Larché F.C., in "Non Linear Phenomena in Materials Science", Trans. Tech. Publications, 205 (1987)
III.19 Bitter F., Phys. Rev. 37, 1527 (1931)
Crum M.M., Nabarro F., Proc. Roy. Soc. A 175, 519 (1940)
III.20 Hilliard J.E., in "Phase Transformations", American Soc. Metals, Metals Park, Ohio, 497-560 (1970)
III.21 Langer J.S., Ann. Physics 65, 53 (1971)

III.22 Fleming R.M., McWhan D.B., Gossard A.C., Wiegmann W., Logan R.A., J. Appl. Phys. 51, 357 (1980)

III.23 Janot Chr., Roth M., Bruson A., Marchal G., Piecuch M., Journal of Non-Crystalline Solids 81, 41 (1986)

III.24 Gerl M., Guilmin P., in "Non Linear Phenomena in Materials Science", Trans. Tech. Publications, 215 (1987)

III.25 Johnson W.L., Prog. Mat. Science 30, 81 (1986)

Materials Science Forum Vols. 59 & 60 (1990) pp. 361-438

SURFACES-INTERFACES AND ULTRATHIN FILMS OF TRANSITION METALS
GROWTH, ELECTRONIC STRUCTURE AND MAGNETISM

F. Gautier
Institut de Physique & Chimie des Matériaux de Strasbourg
(IPCMS - UMR 46)
4, rue Blaise Pascal
F-67070 Strasbourg Cédex, France

1 ELECTRONIC STRUCTURE OF TRANSITION METALS

1.1 Band structure and cohesion of transition metals: a qualitative description

The band structure of the transition metals can be qualitatively described [1, 2] as follows:

(i) The sp bands issued from the (4 sp) atomic states are broad.

(ii) The high localization of the "d" states ($r_d \sim 0.5$ Å for the end of the first series) requires that the corresponding bands are narrow ($W \sim 5 - 10$ eV).

(iii) In ferromagnets, the "d" bands are splitted by an exchange splitting Δ roughly proportional to the magnetization ($\Delta = Im$).

(iv) The "sp" and "d" bands are degenerate and hybridized in localized regions of the reciprocal space (Fig. 1). The "d" bands become increasingly stable and narrow when the atomic number increases along a series (from Ti to Ni for example). The noble metals are in an intermediate situation for which the "d" bands are filled but in the vicinity of the Fermi level so that they still contribute to some physical properties (cohesion, optical properties...). The peculiar properties of the transition metals -large cohesive energies, high temperature of fusion...- are explained by the existence of the narrow "d" bands [1] (Fig. 2). In the simplest model (neglect of electron electron interaction, and of the "sp" electrons) the center of the "d" bands is assumed to be the same as the atomic "d" energy ϵ_d, the number of the "d" electrons N_d is the same as in both the metallic and atomic states: the cohesive energy E_c i.e. the energy gain $E - E_{at}$ going from the atomic to the metallic state is then equal to the band structure binding energy E^d_{bs} and is given by (for a paramagnetic state)

$$E^d_{bs} = 2\int^{\epsilon_F} n_d(\epsilon)\epsilon \, d\epsilon - N_d\epsilon_d = \int^{\epsilon_F} N_d(\epsilon)(\epsilon - \epsilon_d)d\epsilon \qquad (1.1)$$

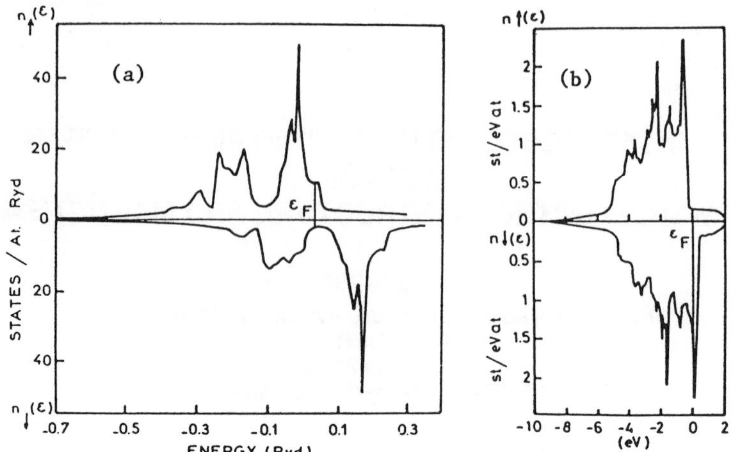

Fig. 1a Densities of states of bcc iron from [5]. Iron is a weak ferromagnet characterized by unfilled d↑ and d↓ band ($N_{d\downarrow} \simeq 2.5$ $N_{d\uparrow} \simeq 4.7$).

Fig. 1b Densities of states of fcc nickel from [5]. Nickel is a strong ferromagnet characterized by filled d↑ bands ($N_{d\uparrow} \simeq 5$) and unfilled minority bands ($N_{d\downarrow} \simeq 4.4$).

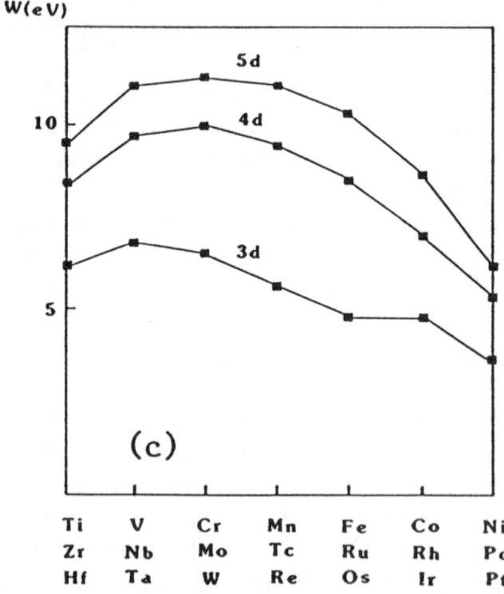

Fig. 1c d band width W_d for the 3d, 4d and 5d transition metals (schematic from [7b]).

The number of d electrons N_d par atoms and ϵ_d are related to the density of states (per spin direction, atom and unit energy $n_d(\epsilon)$

$$N_d = 2\int^{\epsilon_F} n_d(\epsilon)d\epsilon \qquad \epsilon_d = \int \epsilon\, n_d(\epsilon)d\epsilon \tag{1.2}$$

E^d_{bs} is always negative; it's zero for $N_d = 0$ and $N_d = 10$ (both bonding and antibonding states are filled) and is maximum for $N_d = 5$ (all the bonding states are filled and the antibonding states are empty). An order of magnitude of E^d_{bs} is obtained assuming that the density of states $n_d(\epsilon)$ is constant and equal to $n_d(\epsilon) = \frac{5}{W}$ where W is the "d" band width

$$\left|E^d_{bs}\right| = \frac{W}{20}(N_d)\ (10 - N_d) \tag{1.3}$$

It varies parabolically with N_d (in qualitative agreement with experiment) and is maximum for $N_d = 5$. The order of magnitude of $\left|E^d_{bs}\right|$ we obtain in this way ($W \simeq 5$ eV, $N_d = 5$, $|E_c| \sim 6$ eV/at) is also correct. The previous conjecture -E^d_{bs} is the main contribution to the cohesive energies- has been supported later in the LSD framework by a detailed analysis of all the contributions to the cohesive energies [3 - 8] (see 1.2, 1.3). The deviations from the parabolic behaviour which are observed experimentally -mainly for the elements of the first series- is explained by the effects of magnetic order and "d" correlations in the atomic and condensed phases [9 - 12].

The role of the Coulomb exchange intraatomic interactions can be qualitatively understood (Fig. 3) assuming that

(i) the Coulomb (exchange) intraatomic integrals are equal to an average value U (J) taken over all possible pairs of atomic orbitals,

(ii) the electron-electron interaction effects can be treated by a perturbation theory $\left(\frac{U}{W}, \frac{J}{W} << 1\right)$ from the band states for non interacting electrons. A simple calculation shows that [9]

$$E^d_c = E^d_{bs} + E^{d(1)}_c + E^{d(2)}_c + O(U^3, UJ, J^2) \tag{1.4a}$$

$$E^{d(1)}_c = \frac{U}{20} N_d(10 - N_d) + 2S(2S - 1)\frac{J}{4} \tag{1.4b}$$

$$E^{d(2)}_c = -\frac{45}{W}\left[\frac{N_d}{10}\left(1 - \frac{N_d}{10}\right)U\right]^2 \tag{1.4c}$$

$$2S = N_d \text{ for } N_d \leq 5\ ;\ 2S = 10 - N_d \text{ for } N_d \geq 5 \tag{1.4d}$$

The first order term $E^{d(1)}_c$ is the Hartree-Fock correction: it's positive, the charge and spin fluctuations being larger in the metallic state. More precisely, in the metallic state, the electrons are randomly distributed on the atomic orbitals; each electron interacts on each site with $\frac{N_d}{10} \times 9$ electrons occupying the 9 other orbitals.

(i) The Coulomb interaction in the metallic state, $E^{(1)U}_{met} = \frac{U}{2}\left(\frac{9}{10}N_d\right)N_d$, is then larger than in the atomic state, $E^{(1)U}_{at} = \frac{U}{2}N_d\ (N_d - 1)$, and contributes to the cohesion energy $E^{(1)U}_{met} - E^{(1)U}_{at} = \frac{U}{20}N_d\ (10 - N_d)$ (1st term of (1.4b)).

(ii) The exchange energy appears also in (1.4b), the atom being magnetic and the metal being assumed to be non magnetic; it produces a cusp in $E^{d(1)}_c$ which is associated to the special stability of the half filled "d" atomic shell (Hund's rule).

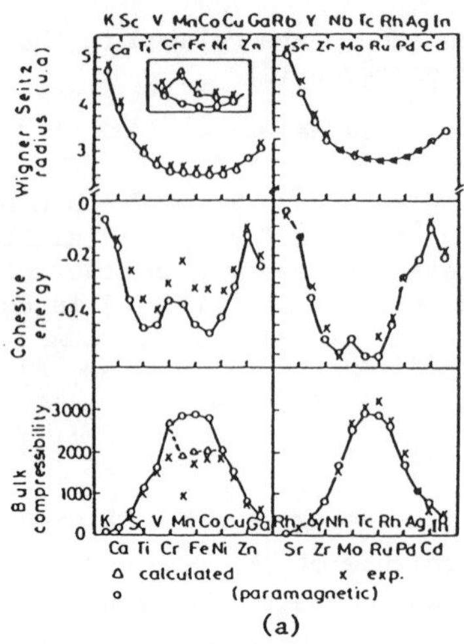

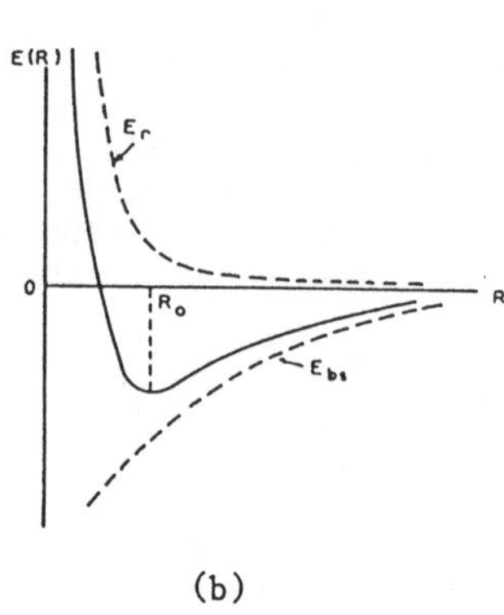

Fig. 2a Experimental (×) and theoretical variations of the cohesion energies E_c (Ryd), lattice parameters and bulk compressibilities of transition metals (from [5]). The agreement between theory and experiment is satisfactory for elements of the second series. (o: local density calculation with m = 0; Δ: local spin density calculation). The one electron treatment of ferromagnetism is not sufficient to explain the discrepancy between theory and experiment for the "3d" metals.

Fig. 2b The equilibrium position R_0 results from the competition between the band structure (E_{bs}) and repulsive (E_r) energies (schematic).

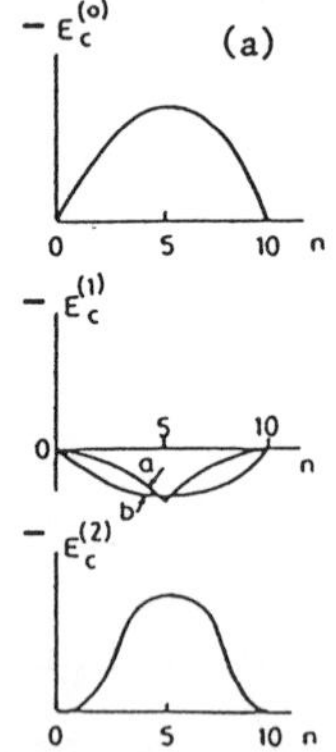

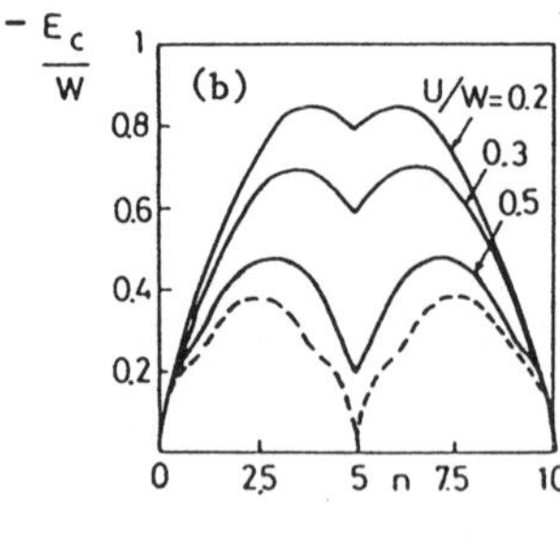

Fig. 3a Qualitative variations of the contributions to the cohesion energies in paramagnetic transition metals; the curves (a) and (b) represent respectively the first order exchange and Coulomb corrections to the cohesion energy (from [9]).

Fig. 3b Cohesion energy as a function of band filling N_d for different values of the Coulomb energy U/W . Dotted lines show the HFA result for U/W = 0.5. The results are obtained for J/W = .2 from [12] using a variational method.

The last term $E_c^{(2)}$ comes only from the metallic contribution: the correlations reduce the charge fluctuations so that $E_c^{(2)}$ is negative. This contribution can be estimated by a second order perturbation theory

$$E_c^{d(2)} = \sum_{n\neq 0} \frac{|\langle 0|V|n\rangle|^2}{E_0 - E_n} \qquad (1.4e)$$

where the states $|n\rangle$ are the eigenstates of the non interacting electrons (with energies E_n) and V is the electron-electron contribution to the hamiltonian. Qualitatively $E_0 - E_n$ can be replaced by an average energy which is equal to the band width W; each of the 45 Coulomb interactions between two different orbitals induce transitions from the occupied to two unoccupied orbitals; each orbital having the probability $\frac{N_d}{10}$ to be occupied and $1 - \frac{N_d}{10}$ to be empty, $E_c^{(d)(2)}$ is given by (1.4c).

Using reasonable values of $\frac{U}{W}$ (going roughly from 1/2 to 1/4 when on goes from the first to the third series), the corrections due to correlations reduce the cohesion $|E_c^*|$ by 10 % to 20 %, the exchange interactions producing the central cusp in the cohesion energies for half filled bands.

The parabolic variation of the equilibrium atomic volumes Ω_a with N_d cannot be explained only by the corresponding variation of E_{bs}^d. When the atomic volume Ω decreases, the hopping integrals t and the band width increase (roughly as $-\,d\,\mathrm{Log}\,W/d\,\mathrm{Log}\,\Omega \sim 1$ to $\frac{5}{3}$) so that E_{bs}^d decrases: a non zero equilibrium atomic volume Ω_a is only obtained if we add to this energy a repulsive energy E_r which represents qualitatively all the effects we did not take into account previously [13]

$$E = E_r + E_c^b \qquad (1.5)$$

The simplest form for E_r is a Born-Mayer term

$$E_r = Ae^{-pR} \qquad A > 0 \qquad (1.6)$$

between nearest neighbours separated by R; the phenomenological parameters A and p have to be determined separately from experimental data. Assuming that these parameters are independent of N_d and that W is also varying exponentially with R ($W(R) = W_0 \exp - q R$), the equilibrium volume and interatomic distance R_0 are obtained by the condition $\frac{dE}{dR}(R_0) = 0$ with

$$E(R) = A \exp - pR - B \exp - qR \qquad B = \frac{W}{20} N_d(10 - N_d) \qquad (1.7)$$

The experimental trends (parabolic variations of E_c and Ω_a with N_d) are then reproduced by choosing $\frac{p}{q} = 3$ and $qR_0 = 3$; most of the cohesive energy ($\simeq$ 75 %) comes thus from the band structure energy in agreement with the previous discussion $\left(E_{bs}^d(R_0) = -\frac{p}{q} E_r(R_0)\right)$; the study of the bulk modulus (i.e. of $\frac{\partial^2 E}{\partial R^2}(R_0)$) leads to the same conclusion.

The previous discussion shows that calculation of Ω_a must take account of the terms we neglected previously: shift of the center of the "d" band ϵ_d relative to the corresponding atomic energies ϵ_d^{at}, "sp" electrons and "sp-d" electron transfers, electron electron interactions...

For example, the metallic "d" functions $\varphi_d(r)$ are different from the corresponding atomic functions $\varphi_d^{at}(r)$. Several effects must be considered [2]:

(i) There is a change of the number of s electrons going from atom to metal (configuration effects).

(ii) The wave function is more or less uniformly enhanced in most of the atomic volume (renormalization effect $\varphi_d(r) \sim \alpha\varphi_d^{at}$; $\alpha > 1$) each electron being compressed in the Wigner-Seitz cell.

(iii) The wave functions are modified midway between atoms and ϵ_d is shifted towards lower energies by the potentials of the surrounding atoms (environment effects). Detailed quantitative calculations have determined the relative importance of these effects. They show that the equilibrium "d" levels are shifted upwards by about 3 eV from the atomic energy [4]

$$\epsilon_d(R) = \epsilon_d^{at} + \Delta\epsilon_d + Q \exp - q_\epsilon R \tag{1.9}$$

this comes both from the configuration effect ($\Delta\epsilon_d$) and from the other two terms. It's important for a quantitative study of the cohesive energy (see above) and of the work function.

The wave functions are energy dependent and are more extended for the bottom of the bands (bonding states) than for the top of the bands (antibonding states). The electronic density n(r) in each atomic cell is thus the sum of renormalized and bonding densities.

A detailed discussion of the equilibrium volume using the virial theorem in the LDA [4] shows that both "s" and "d" electrons must be considered to explain the observed values of the atomic volumes: the pressure p is the sum of "sp" and "d" contributions ($p = p_s + p_d$) and the equilibrium is obtained by a compensation between these contributions. For the equilibrium volume ($p = 0$), the "sp" pressure ($p_{sp} > 0$) comes both from the increase of the bottom of the "sp" band when Ω decreases and from the corresponding increase of the kinetic energy. The "d" pressure is negative at equilibrium distance R_0; it results from the competition between the structure energy E_{bs}^d ($p_{d\,bs} < 0$) and from a repulsive energy ($p_{d\,r} > 0$) which comes itself from the increase of ϵ_d when Ω decreases (see (1.9)). The neglect of p_{sr} would give much too small atomic volumes: the sp contribution is a small (but essential) part of p which must be considered in order to obtain good orders of magnitude for Ω_a.

1.2 The local spin density approximation

The electronic structure is always determined via one electron states $\psi_\alpha(r)$ which are the eigenstates of a Schrödinger equation (eigenenergies ϵ_a); each of these states represents "a priori" the motion of each electron in the average potential v(r) of the others. Two separate problems must then be solved:

(i) determine the corresponding one electron potential v(r) and relate the physical properties to the ψ_α and ϵ_α,

(ii) solve the Schrödinger equation for an electron moving in this potential. The last problem can be solved in a perfect crystal using a number of accurate methods which are currently used now: KKR, APW and linearized methods (ASA, LMTO, LAPW...)... Usually such methods assume that the potential

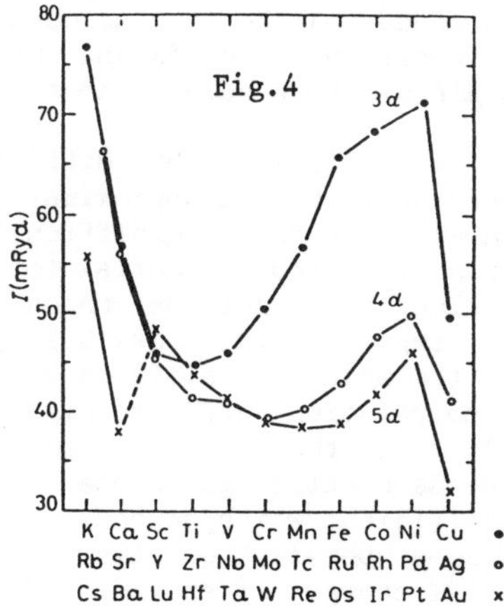

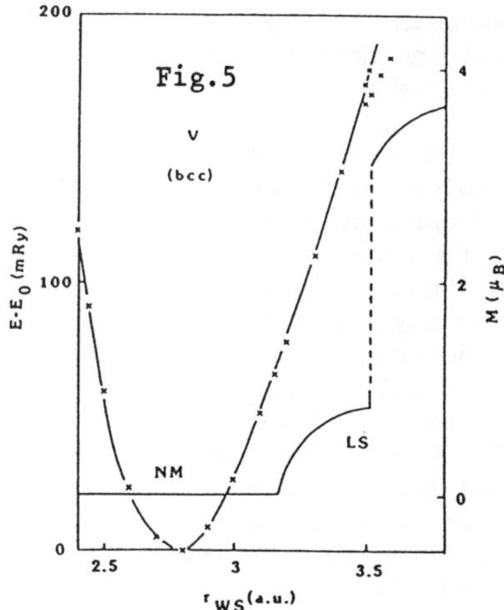

Fig. 4 The Stoner parameter I of transition metals calculated from LMTO wave functions (from [8]).

Fig. 5 Magnetic moment M and total energy E as a function of the Wigner-Seitz radius r_{ws} for bcc vanadium: the LDA results show a type III transition from a non magnetic (NM) to a low spin (LS) and than to a high spin (HS) state (from [23]).

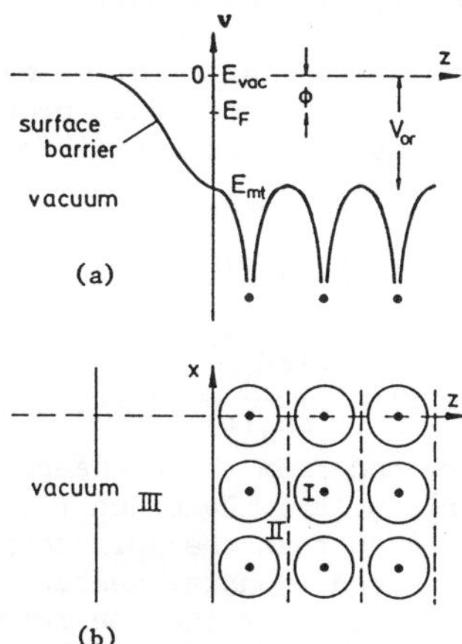

Fig. 6a One electron potential $v(r)$. E_{vac}, E_F and φ denote the vacuum, Fermi level and the work function respectively.

Fig. 6b The real space is separated into three regions denoted I (spheres around the nuclei), II (interstitial region in the bulk) and III (vacuum region). In the FLAPW method no shape approximation is made to $n(r)$ and $v(r)$ but the wave functions are expanded in terms of simple basis functions for each of these regions i.e. products of radial functions and spherical harmonics in region I, plane waves in region II and products of functions which depend only on z and two-dimensional plane waves in region III.

can be treated in the muffin-tin approximation when the local environment of each atom is nearly isotropic. They can be extended to take into account the potential anisotropy for covalent materials, insertion compounds and surface problems (see section 2) for example; we do not discuss here such methods.

The first problem -i.e. the determination of v(r)- is much more difficult. The simplest self-consistent methods -Hartree and Hartree-Fock approximations- cannot represent the physical properties of an electron gas, the difficulties coming from the long-ranged Coulomb interactions. It's thus essential to take into account the correlations between the electronic motions which are qualitatively important; as an example they suppress the possibility of ferromagnetism in the free electron model for the actual metallic densities. Slater first suggested that the exchange correlation effects might be approximately $v(r) = \alpha\, v_s(r)$: $v_s(r) \sim n(r)^{1/3}$ is the average exchange potential over the occupied states of an homogeneous electron gas whose density is equal to the actual density n(r); α is a coefficient which depends only on the atomic number and is determined by fitting a physical quantity to the experiment or by the requirement that the total energy of an isolated atom calculated with v(r) is equal to the HF energy (Xα approximation). We do not discuss this semi-empirical approach since the very successful local density (LDA) and local spin density (LSDA) approximations were developed by Kohn, Sham, Hedin, Lundqvist, Gunnarsson...: such methods allow to get accurate properties for the ground state properties (cohesive energy, electronic density, spin density, atomic volumes, bulk compressibility, etc...).

This spin density functional theory is based on two theorems [14, 15]:

(i) The Hohenberg-Kohn theorem states that the ground state energy E_0 can be expressed in terms of an universal function of the density n(r) and of the magnetization density m(r) if the system is ferro- or ferrimagnetic ($m(r) = n_\uparrow(r) - n_\downarrow(r)$) where $n_\sigma(r)$ is the electronic density of the electrons of spin σ ($\sigma = \pm 1$).

(ii) The Kohn-Sham theorem introduces a set of one electron states $\psi_{\alpha\sigma}(r)$ which allow to calculate exactly the ground state properties. These states are solutions of Schrödinger equations whose potential for each spin σ ($\sigma = \uparrow, \downarrow$), $v_\sigma(r)$, is given by

$$v_\sigma(r) = v^{ext}(r) + v^{H}(r) + v_{xc\sigma}(r) \tag{1.8}$$

where $v^{ext}(r)$ is the potential due to the nuclei (or the ions), $v^{H}(r)$ is the Coulomb potential and $v_{xc\sigma}(r) = \dfrac{\delta E_{xc}}{\delta n_\sigma(r)}(n_\uparrow(r), n_\downarrow(r))$ is the exchange correlation contribution determined from the exchange correlation contribution to the total energy, $E_{xc}(n_\uparrow(r), n_\downarrow(r))$. Equation (1.8) assumes that the spin orbit interaction is negligible so that the spin component σ in the direction of the magnetization $\underline{m}(r)$ is a good quantum number. If $v_{xc\sigma}(r)$ is known, the self-consistent one electron equations for the non interacting system determine the exact ground state properties but the one electron energies $\epsilon_{\alpha\sigma}$ and wave functions $\psi_{\alpha\sigma}(r)$ have "a priori" no direct physical meaning: the energies $\epsilon_{\alpha\sigma}$ can be considered as "excitation energies" $\bar{\epsilon}_{\alpha\sigma}$ but there is a priori no justification for such an interpretation [2].

For practical purposes it's necessary to specify $v_{xc\sigma}(r)$: the most popular approximation is the local density approximation (LDA); for a <u>non magnetic</u> system E_{xc} is assumed to be equal to $E_{xc} = \int n(r)\epsilon_{xc}(n(r))d\underline{r}$ where $\epsilon_{xc}(n)$ is the exchange correlation energy per electron for an homogeneous gas of density n. The exchange correlation potential $v_{xc}(r) = \dfrac{d}{dn}(\epsilon_{xc}(n)n)\Big|_{n=n(r)}$ is thus obtained from the exchange correlation part of the potential of an homogeneous gas, a quantity which is approximately known for the actual densities. The LDA can be extended to a magnetic ground state (local spin density approximation: LSDA); each exchange correlation potential $v_{xc\sigma}(r)$ is deduced, as previously, from the exchange correlation energy of an homogeneous ferromagnetic gas, $\epsilon_{xc}(n_\uparrow, n_\downarrow)$ and is dependent on both the local electronic $n(r) = n_\uparrow(r) + n_\downarrow(r)$ and magnetization $m(r) = n_\uparrow(r) - n_\downarrow(r)$ densities [15].

1.3 Application of the LSDA to ferromagnetic transition metals and alloys

In a ferromagnet, the spin splitting $\Delta_\alpha = \epsilon_{\alpha\downarrow} - \epsilon_{\alpha\uparrow}$ is determined by the difference between the corresponding exchange potentials $\Delta v_{xc} = v_{xc\downarrow} - v_{xc\uparrow}$. If Δv_{xc} is small, it can be expanded in successive powers of $m(r)$, $\Delta v_{xc} \sim \frac{\delta \Delta v_{xc}}{\delta m} m(\underline{r})$, and the corresponding spin splitting Δ_α can be obtained by a first oder perturbation theory: $\Delta_\alpha = \langle \psi_\alpha^0 | \Delta v_{xc} | \psi_\alpha^0 \rangle$ where $|\psi_\alpha^0\rangle$ is the corresponding eigenstate for $m = 0$. The magnetization density $m(r)$ comes mostly from the states near the Fermi level ϵ_F so that $m(\underline{r}) = m\, \varphi^2(r)$ where $\varphi(r)$ is the "d" radial wave function at the Fermi level. The splitting Δ_α is then given by

$$\Delta_\alpha = I_\alpha m \qquad I_\alpha = \int \varphi_\alpha^2(r) \quad \varphi^2(r) \frac{\delta \Delta v_{xc}}{\delta m} d\underline{r} \tag{1.10}$$

It's α independent ($I_\alpha = I$) if the radial wave function $\varphi_\alpha(r)$ is itself α independent: the "d" states are rigidly shifted by the exchange correlation term. This approximation was shown numerically to be valid [2].

Let us recall that the unstability of the paramagnetic state ($m = 0$) with respect to ferromagnetism comes from the competition between:

(i) the exchange correlation decrease: $\Delta E_{xc} = E_{xc}(m) - E_{xc}(0) \simeq - \frac{Im^2}{4}$,

(ii) the kinetic (band) energy increase: $\Delta E_k = \frac{m^2}{4n(\epsilon_F)}$.

$$\Delta E = E(m) - E(0) = \frac{m^2}{4\, n(\epsilon_F)} - \frac{Im^2}{4} \tag{1.11}$$

so that the paramagnetic state becomes unstable when $\Delta E < 0$ i.e. when the Stoner criterion

$$I\, n(\epsilon_F) > 1 \tag{1.12}$$

is satisfied. I increases with the localization of the "d" functions, i.e. from Pt to Pd to Ni for example, and ferromagnetism is found theoretically to occur only for iron, cobalt and nickel [2 - 8] (Fig. 4).

The LSD approximation provides an accurate description of the properties of the magnetic ground state [2, 16]. The spontaneous magnetization, the cohesive energy and even the hyperfine fields are in suprisingly good agreement with experiment if we note that these calculations do not involve any empirical adjustment: they depend only on the choice of the exchange correlation potential and on the accuracy of the numerical determination of the one electron eigenstates.

Systematic errors induced by the classical LSDA are observed; they become important for transition metals with half-filled "d" bands and especially for 3d metals (Fig. 2a):

(i) the cohesive energy is too large (up to few eV),

(ii) the atomic volume is underestimated, the error becoming large for 3d metals and typically 3-4 %. The corresponding anomalies of the bulk modulus for the same metals are not reproduced. Finally, the crystalline phase stability is often difficult to reproduce when the corresponding energy difference is small (some 10^{-2} eV/at).

The anomalous properties of the transition metals have been interpreted qualitatively in terms of correlation effects either by a perturbation theory in successive powers of I/W (see [9 - 11] and section 1.1) or by an extension of the Gutzwiller approximation [12].

Quantitative calculations need an improvement of the LSDA. Several attempts have been done recently: systematic improvements have been obtained [17] but they remove only partially the errors of the LSDA. Note finally that the one electron energies $\epsilon_{\alpha\sigma}$ can be considered qualitatively as the one electron excitation energies deduced by angular resolved photoemission experiments. Quantitative deviations from the LSDA energies $\epsilon_{\alpha\sigma}$ have been observed for the elements of the end of the 3d series: shifts and broadenings of the excitation energies increase with the importance of the correlations (i.e. with I) from chromium to nickel and introduce a narrowing of the "d" band excitations (≃ 20 % for nickel) [2].

1.4 Magnetic transitions and metastable phases

It's expected that all transition metals become magnetically ordered when the atomic volume Ω increases; the non magnetic transition metals must undergo magnetic transitions and, in the limit of large volumes, the magnetic moments must increase up to values consistent with the Hund's rule and the stable free atom configuration. The interest for phases with large Ω and m values has been recently stimulated by the obtention of several transition metals with metastable crystalline structures by epitaxy on convenient substrates (see Table 1 for some examples).

	a (Å)	c (Å)	n	Ref.
COBALT (hcp)	2.51	4.07		
bcc Co/GaAs (110)	2.82		200	[24]
fcc Co/Ni (100)	3.29		30	[25]
IRON (bcc)	2.86			
fcc Fe/Cu (001)	3.61	3.54	12	[26]
bcc Fe/GaAs (110)	2.82		100	[27a]
hcp Fe/Ru (1000)	2.63	4.15	6	[27b]
MANGANESE (fcc) (T = 1095°C)	3.86			
bct Mn/Ag (100)	2.89	3.31	10	[28a]
fcc Mn/Pd (100)	3.80		20	[28b]

Table 1 Some metastable phases obtained by epitaxy on convenient substrates; the typical number (n) of layers for which such phases have been obtained is also indicated.

Systematic LSDA calculations have been reported to study the magnetic transitions induced by the increase of the atomic volume. The phase stability is obtained by a calculation of the total energy as a function of both volume and atomic moment. The magnetic moment is determined for each atomic volume Ω or Wigner-Seitz radius r_{ws} $\left(\Omega = \frac{4\pi}{3} r_{ws}^3\right)$ by the energy minimization; stable and metastable magnetic states can be obtained depending upon the location of the Fermi level in the up and down spin density of states.

Different magnetic states are favoured in different volume ranges so that we expect continuous or discontinuous magnetic transitions; such transitions can be classified (type I, II, III...) according to the number of singularities or critical r_{ws} values: for example bcc Fe, fcc Co, bcc V and fcc Fe are found to undergo type I, II, III and IV transitions respectively (see Table 2).

	r_{ws} (a.u.)	ground state	transition type	critical r_{ws} values			
bcc V	2.79	NM	III	3.17	3.17	3.53	3.47
bcc Cr	2.59	NM	II	3.46			3.12
bcc Mn	2.59	LS	III	2.53	2.53	2.92	2.88
bcc Fe	2.62	FM	I	2.28			2.28
fcc Fe	2.54	NM	IV	2.685	2.67	2.685	2.66
				NM	LS		HS

Table 2 Equilibrium Wigner-Seitz radii (r_{ws} in a.u.), ground states, transition types and critical r_{ws} values. The non magnetic (NM) critical r_{ws} value refers to the termination of the NM state and the high spin (HS) critical value refers to the begenning of HS or ferromagnetic (FM) behaviour. Systems which have type III transitions exhibit LS behaviour in the region founded by the indicated LS critical r_{ws} values (from [23]).

The type III transition metals undergo three transitions from a non magnetic to a high spin state via a low spin (LS) ferromagnetic state (Fig. 5). The existence of such magnetic states has been related to the peaky structure of the density of states characteristic of each of the crystalline structures (fcc, hcp, bcc) and to the band filling.

These calculations suggest a number of experiments on epitaxial thin films with selected lattice parameters determined by a convenient substrate: for example, fcc cobalt can become non magnetic by a 4 % lattice contraction. They also show that the various magnetic phases and crystalline structures are often nearly degenerate at the relevant experimental volumes: for example, the non magnetic bcc nickel phase is only 5×10^{-2} eV less stable than the usual ferromagnetic fcc phase. Moreover, for a given crystalline structure (fcc Fe for example), the various spin polarized band calculations are not in agreement and determine significantly different critical Wigner-Seitz radii for the magnetic transitions: this is related to the non muffin-tin effects, to small errors induced by the numerical integration of the one electron Schrödinger equation [8] and to the choice of the exchange correlation potential [17]. The determination of the equilibrium crystalline structure of iron is very difficult for these reasons: the equilibrium NM fcc and FM bcc phases are nearly degenerate so that the authors find either the fcc or the bcc phase to be stable according to the assumptions of the LSD calculations. The most recent calculation finds that the fcc Fe NM ground state has a marginally higher energy than ferromagnetic bcc Fe, but the non magnetic ground state of hcp Fe has a lower total energy than FM bcc Fe by 5×10^{-2} eV/atom [21b].

2 ELECTRONIC STRUCTURE OF TRANSITION METAL SURFACES, ADSORBATES AND OVERLAYERS

2.1 Electronic structure of perfect surfaces - Applications to transition metals

2.1.1 Introduction [29 - 31]

In this section, we summarize briefly some general results on the electronic structure of surfaces; such results will be useful when we will study overlayers, superlattices and the magnetism of such systems. It's first essential to note that:

(1) the translational symmetry has been lost perpendicular to the surface so that the eigenstates of a semi-infinite crystal are no more Bloch functions: they can be labelled by the quasi-momentum $\underline{k}_{\|}$ parallel to the surface;

(2) moreover, the local symmetry of the surface atoms is so anisotropic that the muffin-tin approximation which is widely used for compact systems is no more valid. This is why LDA calculations were developed using different approximations near the considered surface [32]:

(i) Below the surface z = 0, the muffin-tin approximation is assumed to be valid.

(ii) Above this plane, the surface barrier v(z) and the corresponding charge densities are self-consistently determined within the local density functional theory: most of these calculations were done in the all electron full potential linearized augmented plane wave (FLAPW) method (Fig. 6). In order to save computational time, the calculations are usually done for films of ν layers ($\nu \leq 9$), hoping that the central layer is really representative of the bulk. The convergence of this process can be verified increasing ν up to 7 or 9 and by a comparison with the bulk density of states. It's generally satisfactory -except in special situations such as magnetic Cr(100) where the perturbation induced by the surface is long ranged (see section 2.1.5).

The range of the screening in transition metals and, more generally, the physical properties of transition metal surfaces are much more easily discussed using semiempirical tight binding models: the corresponding features are briefly summarized in this section considering only -for simplicity- the "d" band structure.

2.1.2 Surface densities of states - Surface charge transfers

The local density of states $n_i(\epsilon)$ for each atom of the i-th layer (i = 1, 2,... ∞ where i = 1 labels the surface layer) is roughly characterized by the average energy $\epsilon_i = \int \epsilon\, n_i(\epsilon) d\epsilon$ and the width $W_i = \left[\int n_i(\epsilon)\ (\epsilon - \epsilon_i)^2 d\epsilon\right]^{1/2}$. The difference $\epsilon_i - \epsilon_d = U_i$ between ϵ_i and its bulk value comes from the change of local environment near the surface and from the corresponding electronic redistribution. The band width W_i is related qualitatively in a tight binding approximation to the number Z_i of nearest

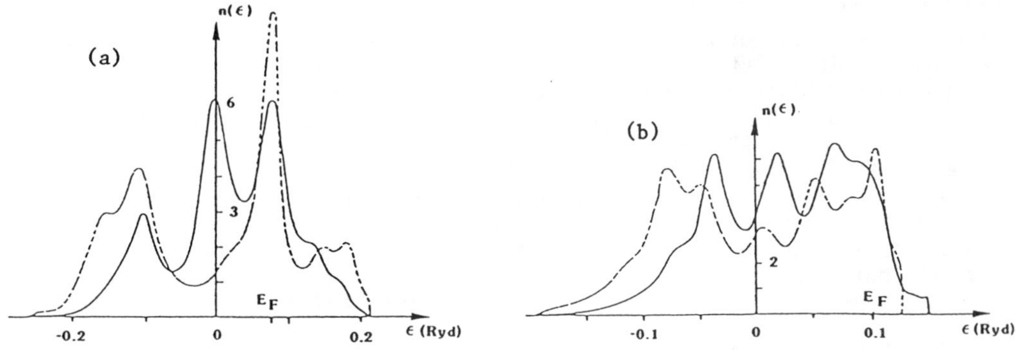

Fig. 7 Surface "d" densities of states $n_1(\epsilon)$ of paramagnetic bcc(100) iron (a) and fcc(100) nickel (b): (—). The corresponding bulk densities of states $n(\epsilon)$ (---) are also given.

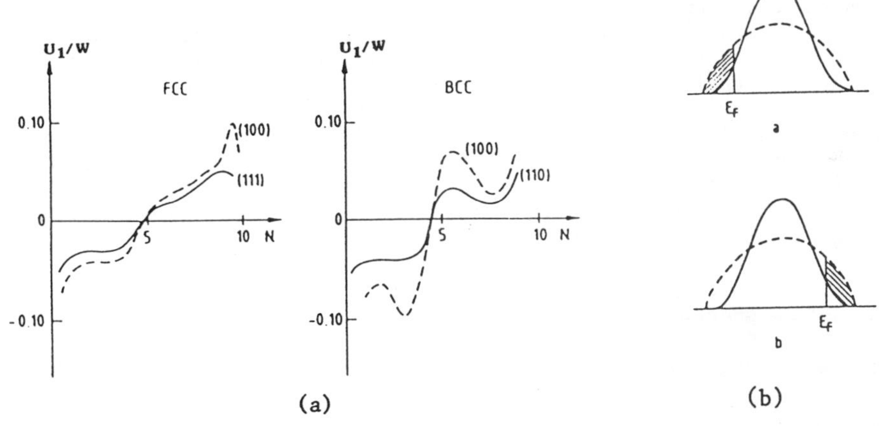

Fig 8a Variation of U_1/W as a function of the bulk number of d electrons N for fcc and bcc surfaces (from [33]).

Fig. 8b Assuming $U_1 = 0$ there is a lack or an excess of electrons at the surface ($N_1 < N$ or $N_1 > N$) when $N < N_0$ or $N > N_0$ (resp.) (— $n_1(\epsilon)$, --- $n(\epsilon)$, $N_0 \sim 5$) (from [56]).

neighbours and to the host band width W, $W_i = \left(\frac{Z_i}{Z}\right)^{1/2}$ W, where Z_i and Z are (resp.) the number of nearest neighbours for an atom of the i-th layer and of the bulk; this relation is valid when only the transfer integrals between nearest neighbours are non zero and when these integrals are not modified by the surface [33 - 35]. Surface broken bonds induce the narrowing of the surface "d" bands $n_1(\epsilon)$: if the bulk and surface transfer integrals are identical, the surface band width W_1 is related to the bulk one, W, by $W_1 \sim W(1 - Z_B/Z)^{1/2} \sim W - Z_B W/2Z$ where Z_B is the number of surface broken bonds (Fig. 7). The surface band width decreases when the atomic density of the considered surface decreases. The local environment of each atom is different from the bulk <u>only</u> for the surface layer (and perhaps the second one) so that the local densities of states are <u>significantly</u> modified <u>only</u> on this layer.

The Fermi level of a semi-infinite crystal is equal to the bulk Fermi level ϵ_F; the surface number of "d" states

$$N_1 = \int_{-\infty}^{\epsilon_F} n_1(\epsilon)d\epsilon \qquad (2.1)$$

is then smaller than N when the "d" bands are less than half-filled and is larger than N otherwise (Fig. 8). The charge transfer on the surface ($\delta N_1 = N_1 - N$), on the other layers δN_i ($i > 1$) and the energy shifts U_i must be determined self-consistently. These calculations show that the screening is so effective in transition metals that, as a first approximation, the self-consistent energy shifts are such that the charge transfers are nearly zero on all the planes (at most 10^{-1} e) on the surface layers [30]. They can be calculated assuming as a first step that all the atomic cells are neutral; Fig. 8 shows the self-consistent energy shift U_1 of the surface layer with the band filling N and its change of sign to keep the local neutrality ($U_1 > 0$ for $N < N_0$, $U_1 < 0$ for $N > N_0$ where $N_0 \simeq 5$) [33]. Fig. 7 shows the self consistent surface density of states for bcc and fcc (100) surfaces.

Surface states (or resonant states) localized near the surface can modify strongly the surface density of states and induce specific narrow peaks whose weights decrease rapidly from their surface value for increasing i values. Such a situation is shown for the bcc(100) surface (Fig. 7a), the supplementary peak corresponding to non bonding states.

2.1.3 Surface energy

The surface energy is defined by $N \gamma_A = E - N_A E_A^0$ where N is the number of surface sites, E the energy of a semi-infinite crystal of N_A atoms A and E_A^0 the bulk energy per atom. A simplified one electron calculation using assumptions similar to those of Friedel for cohesive energies E_c (rectangular densities of states which are identic to the bulk ones, except for the surface layer) lead to the following expression for the "d" band contribution to the surface energy

$$\gamma_{bs}^d = \frac{W - W_1}{20} N(10 - N) = |E_c|(1 - \frac{W_1}{W}) \qquad (2.2a)$$

$$\gamma_{bs}^d = |E_c|(1 - \sqrt{1 - \frac{Z_B}{Z}}) \sim |E_c| \frac{Z_B}{2Z} \qquad (2.2b)$$

The anisotropy of the surface energy is coming from the number of broken bonds. Such a simple formula does not reproduce the details of the surface tensions (Fig. 9) which present (as well as cohesive energies) anomalies for half-filled

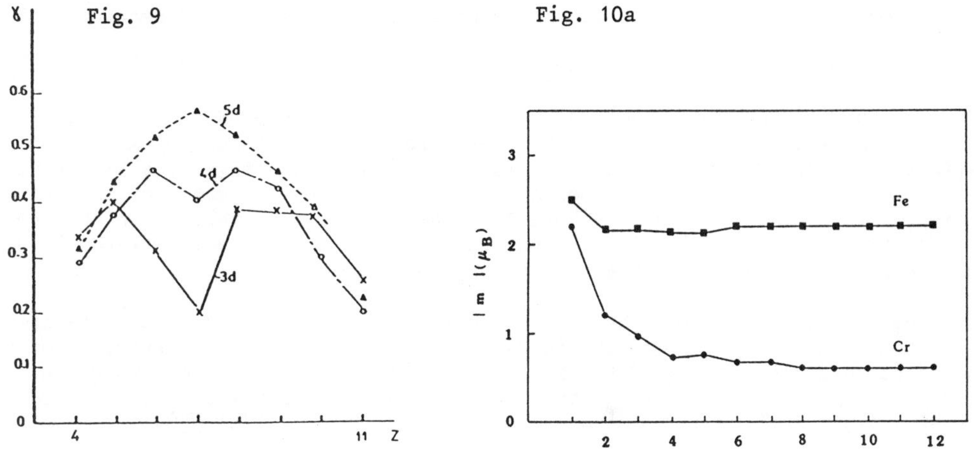

Fig. 9 Surface energy of liquid transition metals in eV per surface atom.

Fig. 10a The magnetization profiles m_i for (100) chromium and (100) iron surfaces (from [45]).

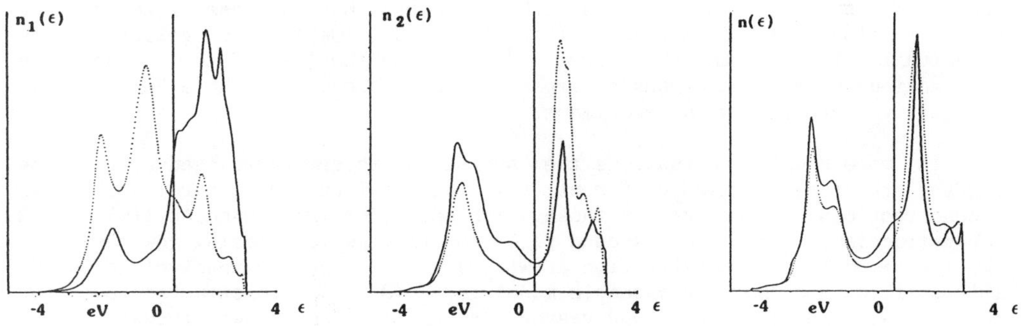

Fig. 10b Local densities of states $n_{i\sigma}(\epsilon)$ for a Cr(100) surface (i = 1, 2) and for the bulk chromium (from [99]).

"d" bands. A more complete calculation must take into account the repulsive contribution (γ_r), the correlation effects $\gamma^{(d)(2)}$ and the band contribution calculated from more realistic densities of states (see (1.4) and section 1.1)

$$\gamma = \gamma^d_{bs} + \gamma^{(d)(2)} + \gamma_r \tag{2.2c}$$

The surface energies, when calculated in this way, reproduce the variation of γ with the band filling but they are smaller than the experimental estimations whereas the band contributions are too large, the difference between theory and experiment being the smallest for elements of the third series with half-filled bands. Note that the experimental values are obtained by an extrapolation of high temperature data to low temperatures [38]; therefore they are known with a rather large uncertainty.

The surface energies have also been calculated within local density functional theory employing the FLAPW method: an idea of the difficulty of a quantitative determination of the surface energy is obtained by a comparison of FLAPW, tight binding (TB) and "experimental" results for W(100): γ(FLAPW) = 5.1 J/m^2 [36], γ(TB) = 3 J/m^2 [37], γ_{exp} = 3.7 J/m^2 [38].

2.1.4 Surface relaxation and reconstruction

In the simplest situation, the layers relax from their ideal positions to the equilibrium state characterized by local contractions ($\Delta d_n < 0$) or dilatations ($\Delta d_n > 0$) where Δd_n is the variation of the distance between the nth and the (n + 1)th layer. Such relaxations are usually oscillating with n. They are small (some % at least) and decrease rapidly with distance [30].

They can be determined using either an ab initio calculation or a simple tight binding model by a generalization of the work done for bulk crystals. Such a multilayer relaxation has been done first for bcc transition metals fitting the A and p parameters (see Chapter 1) to the experimental equilibrium lattice constants and bulk moduli, the transfer integrals varying as R^{-3} [39]. The top layer relaxation increases with the increasing roughness of the surface and for close packed surfaces (bcc(110) and (100)), the relaxation is principally limited to the first plane. For the (111) face, a sequence contraction-contraction expansion-expansion is obtained. All these facts are in qualitative agreement with experiment.

In more complex situations the surface atoms are rearranged so that the surface structure becomes different from the ideal bulk structure. Well-known reconstructions are for example those which occur for Ir, Pt and Au(110) (1 × 2 reconstruction) where every second [1$\bar{1}$0] row is missing (missing row model)... The low temperature reconstruction of W(100) has also been extensively studied; it's still debated but a reasonable model is a c(2 × 2) reconstruction both for the surface and the first underlayer. As an illustration we discuss briefly this last example.

Several models have been proposed to explain the experiments (LEED, surface X ray diffraction):
(i) alternate shifts normal to the surface,
(ii) alternate lateral shifts in the (11) direction,
(iii) a packered model involving both displacements,
(iv) single layer and multilayer reconstructions.
From a theoretical point of view as usual both LDA calculations for films and semi-empirical tight binding models were used. Both calculations lead to multilayer reconstructions, the model (ii) being the most stable.

	N	$d^1_{\parallel}$ (Å)	$d^2_{\parallel}$ (Å)	Δ_1 (%)	ΔE (meV/at)
[43]	4	0.25	0.05	- 1.5 (- 4)	50
[42]	2	0.22	0.05	- 4 (- 6)	60
[41]		0.22	0.45	- 6 ± 10	-

Table 3 Atomic displacements and reconstruction energy of W(100). N: number of mobile layers; $d^i_{\parallel}$: lateral displacement of the i-th layer (i = 1: surface); Δ_1 = inward relaxation of the reconstructed (unreconstructed) surface; ΔE: reconstruction energy per surface atom.

The reconstruction energy defined as the difference in energy between the reconstructed and unreconstructed surfaces is very small as compared to the energies we discussed previously: a reconstruction is thus resulting from a subtle balance between different energies (band, repulsive...) and is "a priori" sensitive to the assumptions which are done to represent the electronic structure.

2.1.5 Surface magnetism of transition metals

The theoretical studies of surface magnetism show that the surface magnetic moment m_1 is larger than the bulk moment m (Table 4)[32]. Both Co, Fe,

m [32]	V cc	Cr cc	Fe cc	Co cc	Ni cfc
Surface (100)	0	2.49	2.98	1.95	0.68
(110)	0		2.65	1.82	0.63
Bulk	0	0.59	2.15	1.76	0.56

Table 4 Surface magnetic moments of transition metals.

Ni and Cr(100) surfaces are ferromagnetic; the V(100) surface is non magnetic but it's nearer from the ferromagnetic instability than bulk V, the V monolayer being itself found to be magnetic (3.09 μ_B). The surface moment enhancement can be qualitatively explained by the narrowing of the surface density of states and the existence of surface states which increase the density of states at the Fermi level $n_1(\epsilon_F)$ and lead to a local instability of the paramagnetic state with respect to ferromagnetism ($I\, n_1(\epsilon_F) > 1$). This "criterion" must be considered with caution for several reasons:

(i) the Stoner criterion is not valid to describe the surface magnetic instability [44],

(ii) it is not related directly to the amplitude of the moment m_1.

The density of states $n_1(\epsilon)$ presents peaks and valleys and the magnetic moment m_1 is determined by the exact position of the Fermi level ϵ_F and the variation of $n_1(\epsilon)$ near ϵ_F (see section 1.4). The surface enhanced magnetism has been first predicted using a simple tight binding model [44, 45] and confirmed later by FLAPW calculations.

The magnetization profile m_i ($i = 1, \infty$) can be theoretically studied either in a tight binding model for a semi-infinite crystal or in a FLAPW

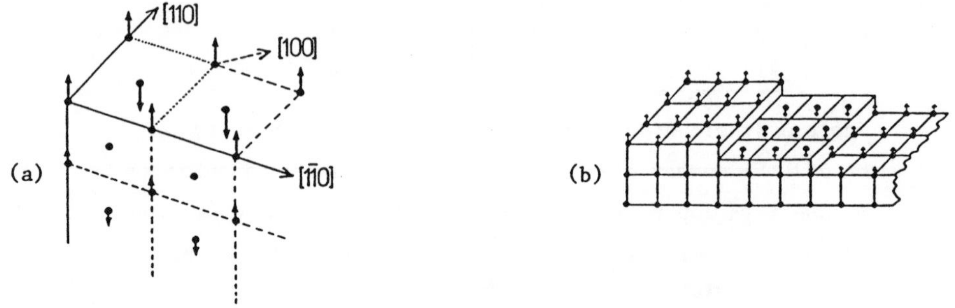

Fig. 11a Surface spins ordered in a C(2×2) antiferromagnetic structure and frustrated spins (•) ... (from [52]).

Fig. 11b Cr(100) terraces separated by monoatomic surface steps are magnetized in opposite directions (from [52]).

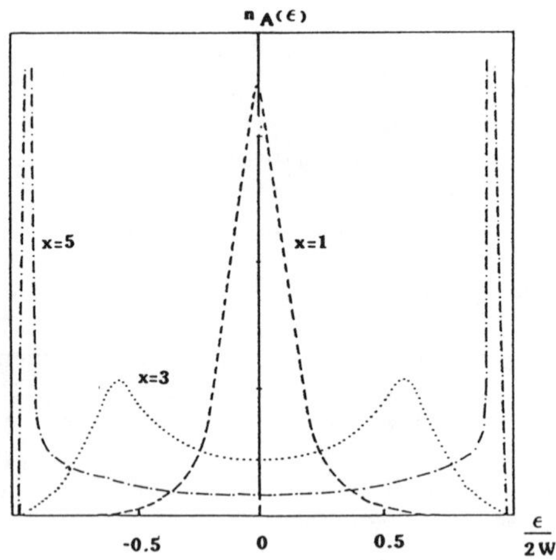

Fig. 12 Local density of states $n_A(\epsilon)$ on an adsorbate A as a function of $x = t_{AB}/t_{BB}$ (schematic) from [56]: strong ($x = 5$) and weak ($x = 1$) coupling regimes.

calculation for a film of ν layers ($\nu \leq 9$). For example, tight binding calculations show that the iron bulk magnetization is recovered near an Fe(100) surface for $i \gtrsim 3$ [46]. The magnetic perturbation introduced by the ferromagnetic Cr(100) surface on the bulk antiferromagnetic chromium is much more extended... It becomes negligible only for $i \gtrsim 7$ [46] (Fig. 10a). In such a case the magnetization of the central layer of a film with $\nu = 7$ layer is not really representative of the bulk.

The temperature dependence of the surface magnetization m_i has also been studied in a tight binding approximation [46] and in the single site spin fluctuation theory. For iron and nickel, the magnetization $m_1(T)$ decreases more rapidly than the bulk one m(T) as obtained in the Ising model with a surface interaction J_s slightly larger than the bulk J_B ($J_S \sim 1.3\ J_B$), the Curie temperature being the same for surface and bulk layers. For Cr(100), the bulk Néel temperature T_N is found to be much smaller than the surface Curie temperature T_{cs} ($T_{cs}/T_N \sim 3$) and the magnetization $m_1(T)$ decreases linaerly with temperature as for Fe and Ni.

The experimental situation is still controversial. The surface magnetizations of nickel and iron have been experimentally studied by SPLEED [47 - 50]. The results are consistent with surface magnetizations larger than the bulk ones but the kinetic energy dependence of the exchange asymmetry for specular reflection of spin polarized electrons is complex and relatively unsensitive to the magnetization profile [47 - 49]. Detailed angular resolved photoemission experiments have also been performed and are found to be consistent with experiment for Fe(100) [51].

Both spin polarized photoemission [51] and SPLEED experiments [50] failed to show the existence of a ferromagnetic Cr(100) surface layer as predicted by theory. This observation is consistent with the antiferromagnetic surface arrangement. The theoretical calculations [52] show that such an arrangement with a strongly perturbed i = 1 underlayer (Fig. 11a) is less stable than the ferromagnetic surface state. Blügel at al suggested recently (Fig. 11b) that surface ferromagnetism is consistent with such experiments if we note that the spin polarized signal comes from several (100) terrasses with opposite magnetizations. Moreover, a detailed study of the temperature dependent angular resolved photoemission [53, 54] shows temperature dependent structures which are consistent with a ferromagnetic splitting, a high Curie temperature (T_{cs} = 780 K, T_N = 312 K) and the theoretical results [46 - 52].

Note finally that electron spin polarization measurements using electron capture spectroscopy suggest that the V(100) surface is ferromagnetic in disagreement with the FLAPW calculations [55].

2.2 Electronic theory of chemisorption

2.2.1. Introduction

The "ab initio" calculation of the electronic structure of single adsorbates is relatively difficult if we note that the translational symmetry is lost and that the local symmetry is far from being simple. In the present section we briefly summarize a simple tight binding model which is sufficient to get the general trends necessary for a study of transition metals on transition metals. We first consider two extreme limits (weak or strong adsorbate substrate couplings) (2.2.2); we deduce from the energy model we introduced in section 1.1 the binding energy, the adsorbate substrate equilibrium distance, the activation energies for surface diffusion. The reader is referred to references [56 - 58] for a more detailed study. Then, we discuss the interaction energies between two adsorbates, the corresponding (band, elastic, coulombic) contributions and the ordering structures which are observed (or predicted) on perfect surfaces.

2.2.2 Adsorbate substrate interactions - Weak and strong couplings

In a tight binding model, the hamiltonian for an adsorbate is given by

$$h = h_B + \epsilon_A |A\rangle\langle A| + \sum_R (t_{AB}(R)|0\rangle\langle R| + hc) + \Delta h_B \qquad (2.3)$$

h_B is the one electron hamiltonian of the B semi-infinite crystal, ϵ_A is the self-consistent energy level of the adsorbate and the second term is the adsorbate one electron hamiltonian; we assumed, for simplicity, that the adsorbate energy state is non degenerate. The third term in (2.3) is the adsorbate-substrate coupling, t_{AB} being the transfer integral for each bond between A and a neigbouring substrate orbital centered at $\underline{R}$, Δh_B is the modification of the substrate hamiltonian induced by the adsorbate; we neglect it as a first approximation.

Two limiting cases can then be defined according to the order of magnitude of $t_{AB}(R)$ with respect to the substrate transfer integrals t_{BB}:

(1) Weak coupling limits $|t_{AB}| \lesssim |t_{BB}|$, $|\epsilon_A - \epsilon_B| \lesssim W_B$): in such case the atomic energy level ϵ_A^0 is shifted ($\epsilon_A^0 \rightarrow \epsilon_A$) and broadened by the adsorbate-substrate interaction (Fig. 12) but, the substrate densities of states $n_R(\epsilon)$ on the neighbouring sites R are only slightly modified: the Fermi energy ϵ_F^B remains unmodified, the B sites are neutral and the energy shift $\epsilon_A - \epsilon_A^0$ is determined by the neutrality condition on the adsorbate site.

(2) Strong coupling limit ($|t_{AB}| > |t_{BB}|$): the covalent bonding is much more efficient between A and neighbouring B sites than between substrate sites (especially when $\epsilon_A \sim \epsilon_B$ i.e. when $\left|\frac{\epsilon_A - \epsilon_B}{W_B}\right| \ll 1$). The local electronic structure is then characterized by bonding and antibonding states below and above the B band continuum (Fig. 12), these states being localized on the "surface molecule" made from A and its B neighbours, shifted and broadened by the interaction with the rest of the B crystal. The energy level shift $\epsilon_A - \epsilon_A^0$ is determined by the neutrality on the surface molecule and the charge transfer from A to B (or the reverse) is determined by the relative position of the A and B energy levels ($\epsilon_A < \epsilon_B$ or $\epsilon_B < \epsilon_A$) and the corresponding weight of A and B states in the bonding states.

The weak limit coupling was shown to be relevant for transition metals whereas the other limit is qualitatively valid for sp adsorbates (M, N, O, S) on transition metals. For the weak coupling situations we are interested in these lectures the charge transfer is thus very small from the highly effective screening of transition metals so that, as a first step, we will consider that this transfer is zero. This assumption is not generally valid (see alkali-atoms on transition metals and on semiconductors for example) when the energy levels ϵ_A are much higher or lower than ε_B (see section 2.2.4).

2.2.3 Binding energies, bonding length and activation energies for surface diffusion

The simplified model we present here has been studied extensively in ref. [56 - 58]. The bonding energy results from band, repulsive and correlation contributions (see section 1.1). We neglect here for simplicity the correlation effects and assume that the L_A adsorbate orbitals which are initially degenerate interact with Z_a substrate atoms via Z_a identical bonds (transfer integral $t_{AB}(R) = t_{AB}$). The energy level ϵ_A is determined by the condition of neutrality

$$L_A \int^{\epsilon_F^B} n_A(\epsilon)\, d\epsilon = N_A \tag{2.4}$$

where N_A is the number of electrons in the valence shell of the adsorbate. The band contribution to the binding energy E_b is:

$$E_b = L_A \int^{\epsilon_F^B} \epsilon\, n_A(\epsilon)\, d\epsilon - N_A\, \epsilon_A^0 - N_A(\epsilon_A - \epsilon_A^0) \tag{2.5}$$

The first term is the band energy, the second term is the corresponding one electron atomic energy and the last one is an Hartree correction taking into account the fact that the Coulomb energy was counted twice in the band term. As in the case of bulk transition metals, we can, as a first approximation, replace the local density of states $n_A(\epsilon)$ by a rectangular density whose width W_A is chosen so that the second moments of this density and of the exact one are equal

$$\mu_2^A = Z_A t_{AB}^2 = \frac{W_A^2}{12} \tag{2.6}$$

The last term in (2.6) is the second moment of the rectangular density of states whereas the second one is the exact moment t_{AB} being an average transfer integral easily expressed in terms of the Slater-Koster integrals ($t_{AB}^2 = \frac{1}{5}(dd\sigma^2 + 2dd\pi^2 + 2dd\delta^2)$ for "d" A orbitals, $\frac{1}{2}pd\sigma^2 + 2pd\pi^2$ for "p" A orbitals, $sd\sigma^2$ for "s" A orbitals). A calculation similar to the one we developed for bulk cohesion energies allows to get

$$E_b = -\sqrt{Z_A}\, B(N_A) e^{-qR} \tag{2.7a}$$

$$\text{with :} \quad B(N_A) = \sqrt{3}\, \frac{|t_{AB}^0|}{L_A} N_A\, (L_A - N_A) \tag{2.7b}$$

The transfer integral $|t_{AB}(R)| = |t_{AB}^0| e^{-qR}$ has been assumed to decrease exponentially. The $\sqrt{Z_A}$ dependence comes from the scale of the binding energy $W_A = \sqrt{Z_A}\, |t_{AB}^0| e^{-qR}$. The binding energy $E_{ads}^{A/B}$ is given by:

$$E_{ads}^{A/B} = Z_A\, A\, e^{-pR} - \sqrt{Z_A}\, B(N_A) e^{-qR} \tag{2.8}$$

and is minimum when $p > q$ for an equilibrium band length R_0 equal to $\left(\frac{dE_{ads}^{A/B}}{dR}(R_0) = 0\right)$

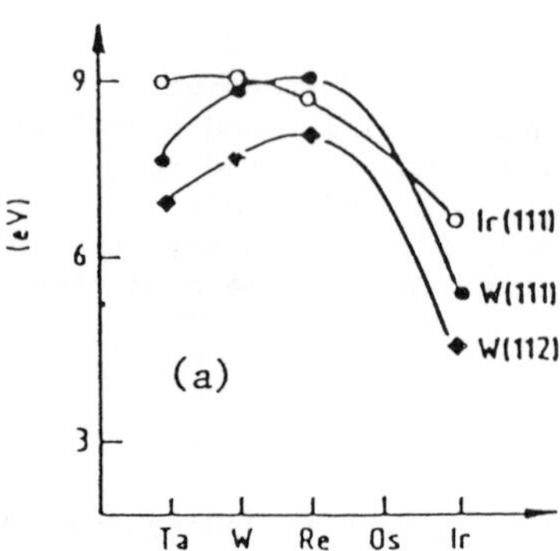

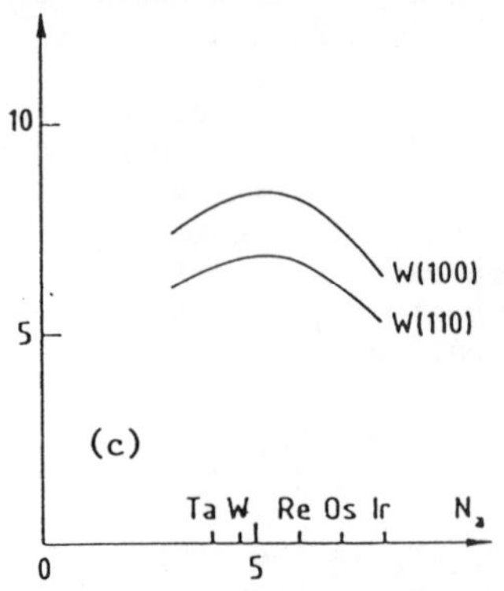

Fig. 13a Adsorption energy $|E_{ads}|$ *of 5d metals on W(111), W(112) and Ir(111) from [74].*

Fig. 13c Theoretical adsorption energy $|E_{ads}|$ *from [56].*

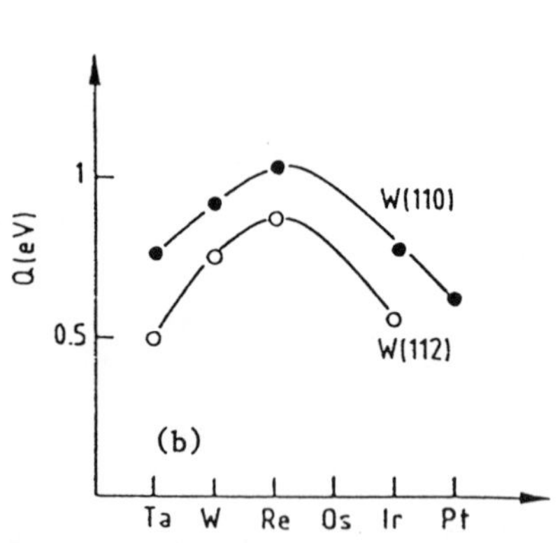

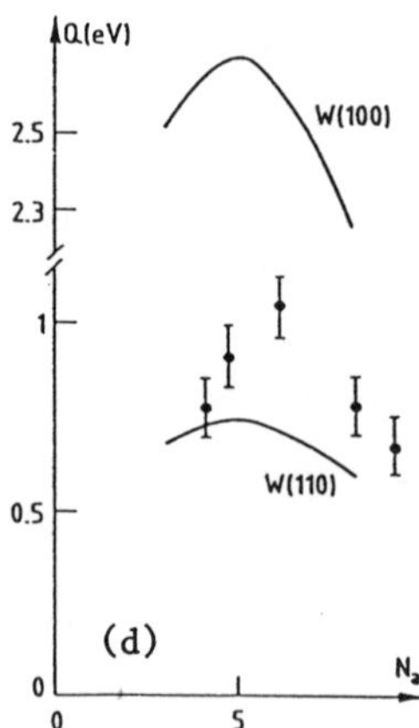

Fig. 13b Experimental activation energy for diffusion of 5d metals on W(110) and W(112) from [76].

Fig. 13d Theoretical activation energy Q from [56].

$$R_0 = \frac{1}{2(p-q)} \mathrm{Log}\, Z_A + \frac{1}{p-q} \mathrm{Log} \frac{pA}{qB} \qquad (2.9)$$

If we increase Z_A keeping R_0 fixed the repulsive term overcomes the band one so that as shown by (2.9) the bond length increases when Z_A increases. The binding energy at equilibrium is given by

$$E^{A/B}_{ads} = \left(\frac{q}{p} - 1\right) B \left(\frac{pA}{qP}\right)^{-q/p-q} Z_A^{(p-2q)/2(p-q)} \qquad (2.10)$$

For $p > 2q$, the binding energy increases with Z_A and the stable adsorption site is the site with the largest possible coordination number. The binding energy is maximum for the half-filled valence shell $N_A = \frac{L_A}{2}$ and it's independent of the substrate band filling in the present approximation.

As far as surface diffusion is concerned, the diffusion path and the saddle points are easily obtained for symmetry points using a similar approximation. If Z_S is the coordination of the saddle point, the activation energy Q is given by

$$Q = E_{ads}(Z_S) - E_{ads}(Z_A) = \left[1 - \left(\frac{Z_S}{Z_A}\right)^{\alpha}\right] \left|E^{A/B}_{ads}\right| \qquad (2.11)$$

where $\alpha = (p - 2q)/2(p - q)$ (cf (2.10)). The activation energy is then varying as the adsorption energy.

The previous calculation was done in the weak coupling limit. A similar calculation can be done in the strong coupling limit assuming that $n_A(\epsilon) = \alpha\, \delta(\epsilon - \epsilon_L) + \beta\, \delta(\epsilon - \epsilon_{AL})$; it results from bonding and antibonding states whose energies ϵ_L, ϵ_{AL} are determined by the neutrality and the exact second moment. In such a case -as physically evident- $\left|E^{A/B}_{ads}\right|$ is maximum when the bonding state is filled and the antibonding state empty. This explains the observed decrease of $E^{A/B}_{ads}$ for N, O, F and H on substrates going from the beginning (V, Nb, Ta column) to the end of the transition series.

In practice, the adsorption energy can be calculated using the recursion method [59] for the adsorbate and the substrate. Some representative results are reproduced in Fig. 13. They take into account the self-consistent change of energy levels and charge transfers on the bulk atoms R (i.e. Δh_B in (2.3)). Fig. 14a shows that the local densities of states $n_A(\epsilon)$ are much narrower than the surface one $n_1(\epsilon)$ ($Z_A < Z$) and that they can present bonding and antibonding states resulting from an interaction with surface states (Fig. 14b).

2.2.4 Dipolar layer and work function [29, 30, 60 - 64]

The work function φ is the minimum energy required to extract an electron from the bulk

$$\varphi = v(\infty) + E_{N-1} - E_N = v(\infty) - \mu \qquad (2.12a)$$

where $v(\infty)$ is the rest energy of the electron infinitely far from the surface, E_{N-1} and E_N are the ground states for N and N - 1 electrons respectively, and μ is the chemical potential. If we introduce the electrostatic potential energy far inside the bulk $v(-\infty)$ and the Fermi level ϵ_F measured from this energy, we get

$$\varphi = v(\infty) - v(-\infty) + v(-\infty) - \mu = D - \epsilon_F \qquad (2.12b)$$

where D is the dipolar layer energy. Usually, the electron cloud expands slightly outside of the positive background so that the surface dipolar moment is directed inwards and D is positive; in a similar way, steps on a perfect surface induce dipolar pointing outwards so that D and φ decrease. For transition metals, the charge transfers and the corresponding dipole layer potentials D are small ($\sim |U_1| \lesssim 0.2$ eV) so that the work function φ is essentially determined by ϵ_F. The Fermi energy is a bulk quantity and is determined by the band filling and by the absolute value of the center of the "d" bands ϵ_d (cf Chapter 1 and (1.9)). It's roughly constant and equal to about 3-4 eV [30].

The work function φ is often used to follow the adsorption and determine structural changes [61 - 63]. The interpretation of this function in the dilute limit is rather simple using the Helmoltz relation [60] for its change $\Delta\varphi$ adsorbing N_A atoms on the B surface

$$\Delta\varphi = - 4 \pi n_A e p \qquad (2.13)$$

where p is the dipolar moment of the adsorbed species and $n_A = \frac{N_A}{A}$ is the density of adsorbates per unit area. In this formula $\Delta\varphi < 0$ if p is directed outwards. Fig. 15 shows as an example the work function changes versus coverage for Cl/Cu(111) and Li/W(111) [64, 65]: p is directed inwards in the first case ($\Delta\varphi > 0$), the charge on chlorine being negative whereas it's the reverse in the last case. The dipolar moment has been determined experimentally in a number of cases:

(1) It's relatively small for transition metals adsorbates; it decreases from 0.5 D for Ta/W(110) to 0.2 D for Pd/W(110) or nearly zero for Pt/W(110) (1 D = 10^{-18} CGS = 0.394 a.u.) and corresponds to a very small charge transfer Δn from the adsorbate to the substrate ($|\Delta n| < 0.1$ e).

(2) It's much larger for alkali earth and also for rare earth adsorbates ($p \sim 2$ - 4D on W(110), W(211), Mo(211), Re(10$\bar{1}$0)... [88] and it's related to large charge transfers $|\Delta n|$.

The change of work function induced by adsorbates has been studied using self-consistent FLAPW calculations and assuming ordered arrays (C(2 × 2)...) of adsorbates. For example, the electron redistribution has been studied for Cs in a C(2 × 2) coverage on W(100) [87c]: it's characterized by (1) an accumulation of charge between the adsorbate and the susbtrate layer occuring from the W 5d and Cs 6s covalent bonding; (2) a counter polarization semicore Cs 5p shell which opposes the reduction of the work function. The alkali atoms form covalent bonds with the substrate and the dipole moment resides primarily between the adsorbate and substrate sites: strictly speaking there is no donation of charge to the tungsten substrate (ionic model). This picture has been recently supported by a photoemission study of the 4f surface core levels: the very small shifts of these levels (0.02 eV) are in agreement with a very small change of electronic density in the W surface atomic cell when alkali atoms are adsorbed.

2.2.5 Adsorbate adsorbate interactions

In the dilute limit, the energy of the N_A adsorbates in interaction with B is obtained from a cluster expansion:

$$E(N_A) = N(E_B^0 + \gamma_B) + N_A E_{ads}^{A/B} + N \sum v_n p_n + \dots \qquad (2.14)$$

where v_n is the interaction energy between two n-th nearest neighbours A atoms on the surface and p_n is the number of such pairs per adsorption site (N is the number of adsorption sites). Such an expansion using even triplets was invoked to explain the ordered structure of sp adsorbates on transition metals.

In this section, we limit the discussion to transition metal adsorbates; we consider first the band and repulsive contributions and then the elastic and coulombic contributions.

2.2.5.1. Pair interactions: band structure contribution [66 - 69]

As usual in these lectures, we first consider the simplest possible situation: two W (or Re) atoms on a W(110) surface (Fig. 16). The adsorbate and bulk "d" bands are half-filled. The nearest neighbouring W atoms on the surface interact directly via a transfer integral $t_{AA}(1)$ (1st nn in the bcc lattice). The adsorbate local density of states is then broadened by the direct binding; the corresponding bonding and antibonding AA states are (resp.) filled and empty so that a strong attractive interaction is predicted:

$$v_1 = E_{ads}^{AA/B} - 2E_{ads}^{A/B} = -2\,[W_{AA} - W_A]\,\frac{N_A(10 - N_A)}{20} \qquad (2.15)$$

$E_{ads}^{AA/B}$ is the adsorption energy for the AA pair; W_A and W_{AA} are the width of the corresponding local density of states for a single adsorbate and for a pair (resp.):

$$\begin{cases} W_A = 2\sqrt{3\,\mu_2^A} = 2\sqrt{3}\,\left(Z_A\, t_{AB}^2\right)^{1/2} \\ W_{AA} = 2\sqrt{3\,\mu_2^{AA}} = 2\sqrt{3}\,\left(Z_A\, t_{AB}^2 + t_{AA}^2\right)^{1/2} \end{cases} \qquad (2.16)$$

In such a model the attractive interaction is maximum for half-filled "d" A orbitals.For two 1st nn W adsorbates on W(110) it's large ($|v_1| \sim 2$ eV for $W_{AA} \simeq 5.7$ eV, $W_A \simeq 4.9$ eV, $N_5 \sim 5$). It must be reduced in practice by the repulsive energy ($E_r \sim 0.6$ eV for two 1st nn W adsorbates) and by the correlation effects (see Chapter 1) ($E_{corr} \sim -U^2 \times 0.4$ with $U \sim 1.5$ eV) [69] so that the resulting interaction is of the order of 0.5 - 1 eV.

This qualitative model can be supported by a more detailed calculation using the recursion method and a general formula for the band contribution to interaction energies [68, 72]

$$v_{nb} = -\frac{\mathrm{Im}}{\pi}\int^{\epsilon_F^B} \mathrm{Tr}\; t_A\, G^0\, t_A^n\, G^0\, d\epsilon \qquad (2.17)$$

where $G^0 = \dfrac{1}{\epsilon - h^0}$ is the host Green function ($h^0 = h_B + h_A$) for non interacting B and A systems, t_A and t_A^n are (resp.) the t matrices of the two adsorbates in

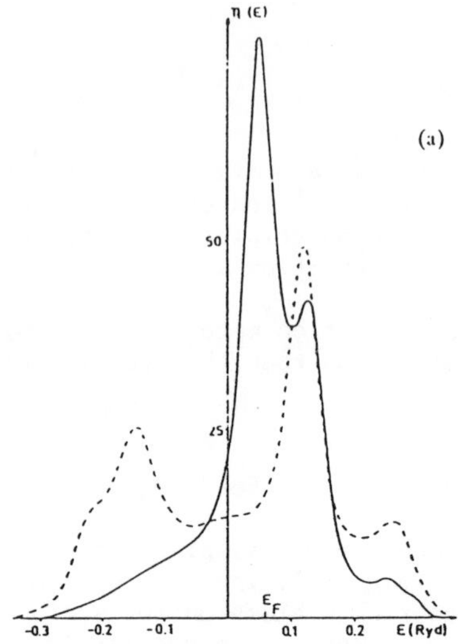

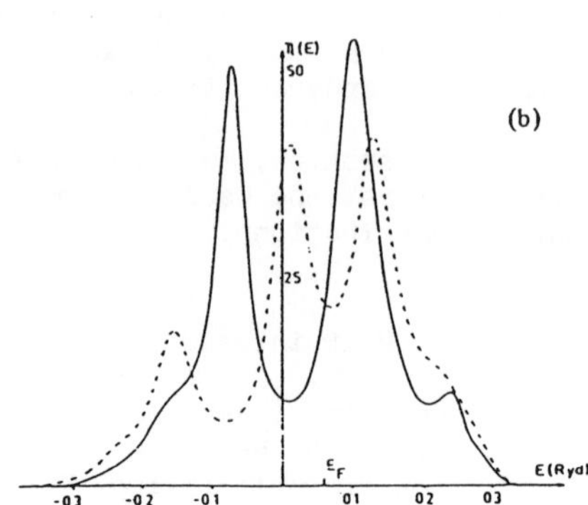

Fig. 14a Surface density of states of a perfect Mo(110) surface (---) and local density of states on a molybdnenum adsorbate Mo/Mo(110) (—) from [56].

Fig. 14b Surface density of states on a perfect Mo(100) surface (---) and local density of states on a molybdenum adsorbate Mo(100) from [56].

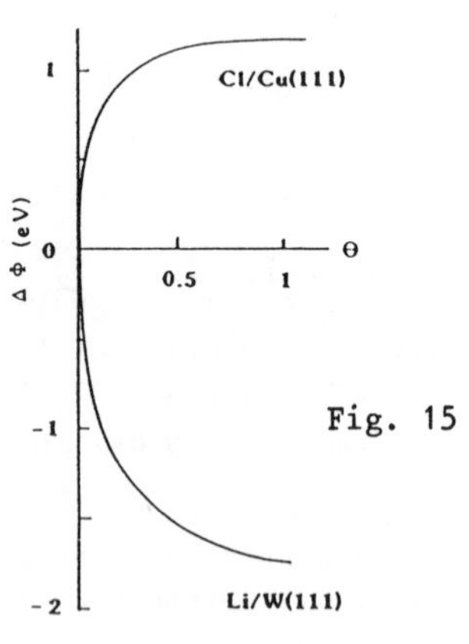

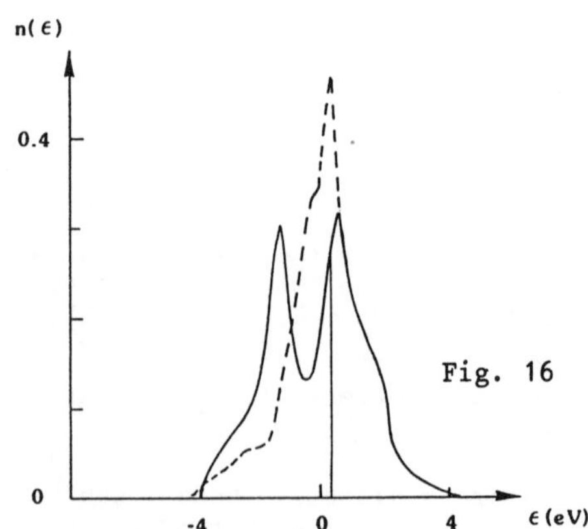

Fig. 15 Work function changes $\Delta\varphi$ versus coverage θ for Cl/Cu(111) and Li/W(111) (schematic) from [64, 65].

Fig. 16 Local density of states of non magnetic monomers (---) and dimers (—) of Re/W(110) from [70].

0 and R_n and express the coupling between both systems. When the adsorbates are near from each other (1st nn in fcc and hcp, 1st and 2nd nn in bcc lattices), they interact directly via the transfer integrals $t_{AA}(R)$ as previously shown (2.16) and indirectly via the substrate

$$v_{nb} = v_{nb}^{dir} + v_{nb}^{ind} \tag{2.18a}$$

$$v_{nb}^{indir} = v_{nb}(t_{AA}) - v_{nb}\,(t_{AA} = 0) \tag{2.18b}$$

When the adsorbates are far apart the indirect interaction is dominant.

The interaction energies were calculated for A = B as a function of the band filling (Fig. 17):

(1) They are strongly attractive for half-filled "d" bands and for 1st and 2nd nn as suggested previously for the (110) surface.

(2) The indirect interactions have the same order of magnitude as the direction interactions for 1st nn and introduce repulsive contributions to v_1 and v_2 on the (100) surfaces and for a narrow range of band filling.

(3) The interactions present oscillations versus band filling which can be predicted by a moment analysis [73] of the v_n vs ϵ_F^B: for example, the indirect interaction between 1st nn presents at least two zeros.

(4) They decrease rapidly with distance and are much smaller than $|v_1|$ for n > 2, $|v_1|$ being itself at most of $\simeq \frac{W_B}{5}$ (compare v_{5b} with v_{1b} in Fig. 17).

Note finally that the pair interactions present long ranged Friedel oscillations ($v(R) \sim \cos D_F R/R^3$ where D_F is a Fermi surface diameter) as in the bulk.

eV/at	A	Ta	W	Re	Os	Ir	Pt
B	$\lvert E_c \rvert$	8.1	8.7	8.1	8.1	6.9	5.9
W(100)	$\lvert E_{ads}^{A/B} \rvert$	7	8	9.5	8	7	5
W(111)		7.5	8.5	9			
W(112)		7	7.3	7.7		4.5	
Ir(111)		9	9	8.5		7	
W(110)	$\lvert v_1 \rvert$ (exp)	0.67	0.37	-	0.07	0.14	0.17
W(112)	$\lvert v_1 \rvert$ (calc)	0.9	0.95	0.80	0.85	0	-

Table 5: Cohesive $|E_c|$ binding and interaction energies $|v_1|$ for adsorbed atoms A on several faces of W and Ir.

The existence of pairs, triplets... can be directly observed by field ion microscopy. For example, the observation of Pt cluster nuclei on Pt(100) shows that the most stable nuclei structure oscillates between <110> chains and island nuclei when the number of adatoms is increased from three to six [78b]. Quantitative informations on diffusion parameters, interaction energies between adsorbed atoms can be obtained from an analysis of the relative frequencies of occurence of dimers, trimers... at a given temperature (see Fig. 18 and Table 5). The orders of magnitude which are experimentally deduced are

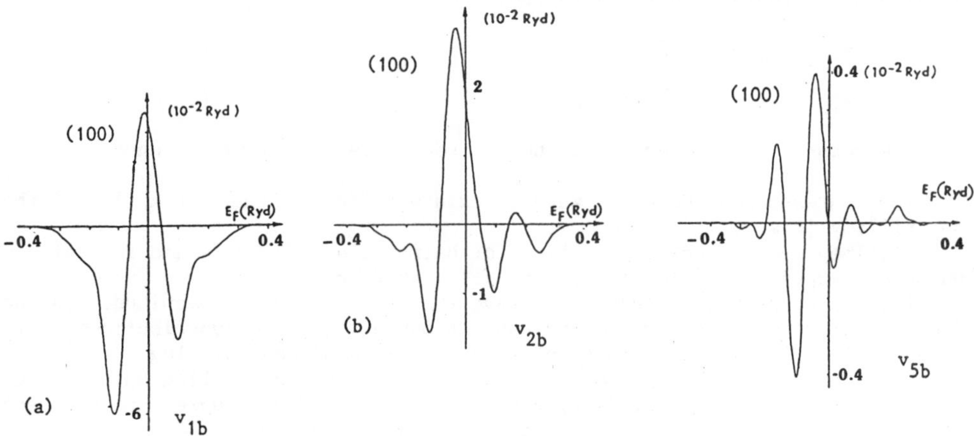

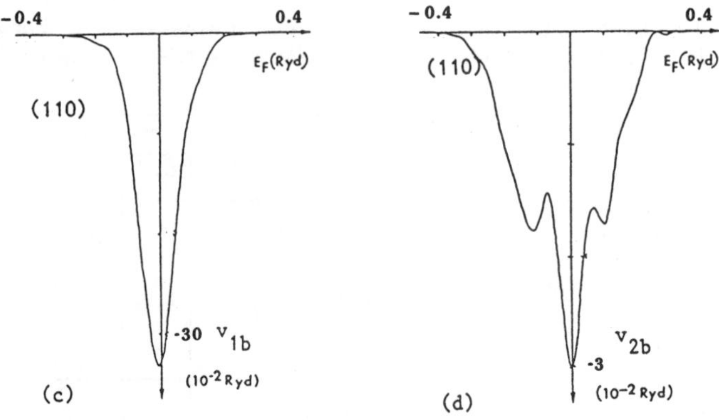

Fig. 17 Plots of the band contribution to the pair interactions versus ϵ_F for first (v_{1b}(a) (c)), second (v_{2b} (b) (d)) neighbouring adsorbates B/bccB(100) and B/bccB(110) (from [68]).

satisfactory when compared to the theoretical results. Note however that for W(110) the interaction energy $|v_1|$ presents a minimum for Re whereas the previous calculations predict a maximum (Table 5): the Re trimers are stable on W(110), the dimers are unstable on W(100), W(211)... but they are stable on W(110). It's thought that the magnetism of monomers is the main reason for the instability of Re on W(110) and that both monomer and dimer are magnetic [69- 71]. The explanation of this effect is still controversed. The corresponding discussion is out of the scope of these lectures.

2.2.5.2. Electrostatic interactions

In the previous section, we considered only the band contribution; this is approximately valid when the local neutrality is satisfied i.e. for weak coupling situations and when the adsorbate energy level ϵ_A is not too far from the Fermi level. When there is a charge transfer between the adsorbate and the substrate, the electron-electron interaction which is counted twice in the band energy must be substracted [72]. Up to now no detailed calculation has been done to take into account both band and electrostatic effects self-consistently. However, the importance of charge transfer effects has been pointed out long time ago in a limiting case corresponding roughly to alkali adsorbates. Let us assume that the adsorbate energy level ϵ_A is much higher than the Fermi level and that the covalent binding (i.e. (t_{AB})) can be neglected. The adsorbate looses valence electrons which are transferred in the substrate surface layer practically at the smallest possible distance a_A between the metallic surface and the adsorbate ($a_A >> d$, the thickness of the surface region). The corresponding dipole moment is then $p_A = z_A a_A$ where z_A is the adsorbate charge.

The electrostatic interaction energy $v^{es}(r_B - r_A)$ between the adsorbate A in $\underline{r}_A$ and another adsorbate B (charge z_B, dipolar moment $p_B = z_B a_B$) in $\underline{r}_B$ is equal to the work done by the charge z_B in moving B from infinity to its position r_B i.e. to $v^{es}(\underline{r}_B - \underline{r}_A) = z_B\,\varphi(r_B)$ where $\varphi(r_B)$ is the electrostatic potential produced by A at $\underline{r}_B$. According to the classical image theory and for large distances $R = |\underline{r}_A - \underline{r}_B|$ $(R >> a_A, a_B)$

$$v^{es}(R) \sim 2\,\frac{p_A\;p_B}{R^3} \sim 0.625\,\frac{p_A\;p_B}{R^3}\;\text{eV} \qquad (2.19)$$

when the distance is expressed in Å and the dipolar moments in debyes. The factor 2 in this dipole-dipole interaction accounts for the fact that the screening charge is distributed on the surface whereas it acts outside of the metal as it was at the image point. The previous expression allows to get an order of magnitude of the electrostatic interactions. It's negligible for transition metals adsorbates ($v^{es}(R) \sim 2$ meV for $p = 0.3$ D, $R = 3$ Å, see section 2.2.4) as compared to experimental (and band structure contributions) but it's significant for alkali or alkaline earth adatoms: for example, on W(112), $v^{es}(R)$ is about 5×10^{-2} eV or 0.3 eV (resp.) for atoms separated by $R = 4.47$ Å, the distance accross the (112) surface channels. This ionic model has been thought to be relevant for alkali atoms but recent band structure calculations and experimental studies (see the discussion of section 2.2.4) show that the dipolar moment results from a complex charge redistribution induced by the adsorbate substrate "covalent" bonding. The previous considerations are thus only qualitative.

2.2.5.3 Elastic interactions [82 - 84]

Each adsorbate induces in the substrate a local deformation. If the substrate is considered as an elastic medium, this deformation is the linear response of the substrate to the forces $\underline{F}(\underline{\lambda})$ acting on each atom $\underline{\lambda}$. The sum of all the forces acting on the adsorbate being zero at equilibrium ($\sum_R \vec{F}(\lambda) = 0$), this distribution is characterized by its dipolar moment:

$$P_{\alpha\beta} = \sum_{\lambda} F_{\alpha}(\lambda)\ \lambda_{\beta} \qquad (2.20a)$$

In the macroscopic limit, the distributions of displacements and strains are determined only by P and by the elastic constants of the substrate and are varying asymptotically (when $R \rightarrow \infty$) as R^{-2} and R^{-3} (resp.) [82]. For two adsorbates (1 and 2) located in $\underline{0}$ and $\underline{R}$ the deformation $e_{\alpha\beta} = e_{\alpha\beta}^{(1)} + e_{\alpha\beta}^{(2)}$ reduces to the sum of the deformations induced separately by each adsorbate ($e_{\alpha\beta}^{(1)}$, $e_{\alpha\beta}^{(2)}$). The elastic energy stored in the substrate is then a quadratic function of the $e_{\alpha\beta}$; the cross terms proportional to $e_{\alpha\beta}^{(1)}\ e_{\gamma\delta}^{(2)}$ are those which contribute to the elastic interaction, and are shown to decrease asymptotically as R^{-3}. For identic adsorbates acting on an isotropic medium and assuming that $P_{\alpha\beta} = P\delta_{\alpha\beta}$ the elastic interaction is always repulsive:

$$v^{el}(R)_{R\sim\infty}\ \frac{1 - \sigma^2}{2\pi E}\ \frac{P^2}{R^3} \qquad (2.20b)$$

where E is the Young modulus and σ the Poisson ratio. It becomes angular dependent and it can be attractive in certain directions of the pair if the susbtrate is elastically anisotropic [84]. The physical origin of the repulsive interaction for isotropic elastic susbtrate is easily understood. The first adatom 1 exerts an attractive force on the atoms in its vicinity; this causes a local contraction of the lattice and a corresponding expansion far from 1. The interaction of an atom 2 (at a large distance R from 1) with the lattice will be less attractive, the substrate being locally expanded: this is equivalent to a repulsive interaction between both adsorbates. The importance of these elastic interactions for transition metals on transition metals is not clearly established: a detailed calculation of the forces, displacements of the discrete anisotropy lattice points... is necessary to estimate the elastic interactions between AA first, second... neighbours and to compare them with the band contributions.

2.2.5.4. Cluster interactions and ordered overlayer structures

The interactions between adsorbates lead either to:

(1) various ordered structures vs coverage θ and temperature T for repulsive interactions or to

(2) a condensation of adsorbates into islands and to a coexistence of the condensate and of the two dimensional adsorbate gas for $T < T_c$ for attractive interactions. In practice the interactions v(R) are oscillating with R and complex phase diagrams are obtained for a system described by (2.14) with dominant repulsive interactions between nearest neighbours; Monte-Carlo calculations were systematically performed in relation with the observation of ordered overlayers and of the corresponding phase diagrams: H/Pd(100), H/Fe(110)..., O/W(110) etc... Short-range interactions -say interactions up to the third nearest neighbours and (possibly) three body interactions between 1st

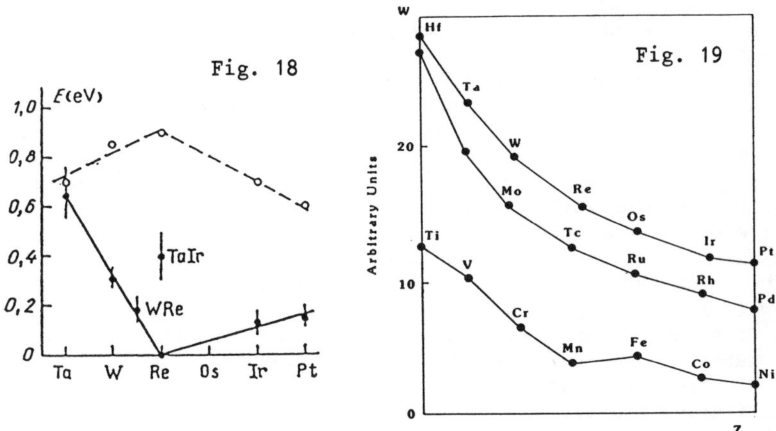

Fig. 18 Experimental interaction energies $|v_1|$ (—) between neighbouring 5d adatoms of W(110) and activation energies Q of single adatoms of the same surface (---) (from [76]).

Fig. 19 "d" band widths of 3d, 4d and 5d transition metals at constant atomic volume (schematic).

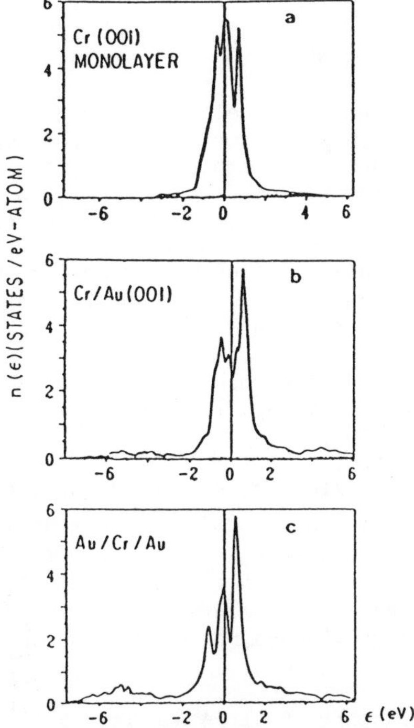

Fig. 20 Local density of states $n_{Cr}(\epsilon)$ (in states/eV at) in the paramagnetic state of Cr(100) monolayer, Cr ML on Au(100) and sandwich Au/Cr/Au(001) (from [91]).

nn- are sufficient to reproduce the large variety of commensurate and incommensurate structures experimentally found [89]. It's important to point out here that

(1) Numerous ordered overlayers have been observed for H, O, CO... [89] and for alkali alkaline earth and some rare earth adsorbates [89, 88].

(2) No ordered structure has been observed for transition metal adsorbates. This suggests that repulsive interactions are dominant in the first case whereas the interactions between transition adsorbates are mostly attractive in agreement with the FIM experiments. The detailed understanding of such a difference from the electronic theory is not still achieved. However, several extreme situations can be qualitatively considered:

(1) The adsorbate substrate covalent bonding is small and the adsorbate energy level is much higher than the Fermi level (alkali metals for example); in such a case the adsorbate is "strongly ionized" and the associated dipolar moment is large (2 - 4 D for alkali on W for example). The interaction energy is mostly electrostatic and repulsive (see 2.2.5.2 and 2.2.4 for a discussion of this qualitative model).

(2) The adsorbate and substrate energy levels are not too much different and the adsorbate substrate coupling is weak: in such a case (transition metals on transition metals), the charge transfer is small and the attractive band interactions are dominant. Finally the sp elements (H, O, N...) are in an intermediate situation for which the substrate adsorbate coupling and charge transfer are larger than for transition metal adsorbates.

2.3 Ultrathin films of transition metals A on transition and noble metals B

2.3.1 Introduction

Here, we are interested in the electronic structure of ultrathin films (1 or 2 monolayers) deposited on a perfect surface. The "epitaxied" overlayers can present an atomic structure which is either commensurate, partially commensurate or incommensurate with the substrate surface according to the importance of misfits, adsorbate-adsorbate interaction v, relative to the adsorbate-substrate interactions u. Apart from perfect epitaxies of identic equilibrium α and β crystalline structures (fcc/fcc, bcc/bcc), let us mention as examples two special situations which allow to realize stable or metastable structures for the A overlayers when $\beta \neq \alpha$:

(1) fcc(100) substrate:

The bcc stable structure of the bulk A can grow in perfect epitaxy with the (100) fcc surface where $a_A \simeq \sqrt{2}\, a_B$: in such a way, it's possible to grow α Fe on Ag(100) or bcc Cr on Au(100);

(2) GaAs(110) substrate:

If the misfit between the (110) fcc surface of GaAs and the (110) bcc surface of A is small it's possible "a priori" to grow from a $C(\frac{1}{2} \times 1)$ surstructure a stable or metastable bcc A phase: this is the case for Fe/GaAs (see table 1).

Let us now restrict our discussion to the case of a perfect epitaxy assuming a layer by layer growth and no interdiffusion.

2.3.2 Electronic structure of ultrathin layers

The physical properties of the transition metal overlayers being mostly determined by the "d" bands we consider only the features of bands or localized states near the interface. The electronic structure is qualitatively

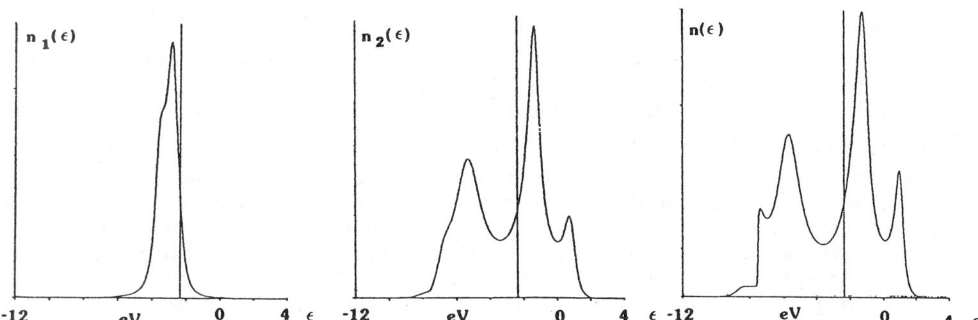

Fig. 21 Local "d" densities of states of a nickel monolayer deposited on bcc Mo(110) (a), of the first molybdenum layer beneath (b) and of bulk molybdenum (schematic from [99]).

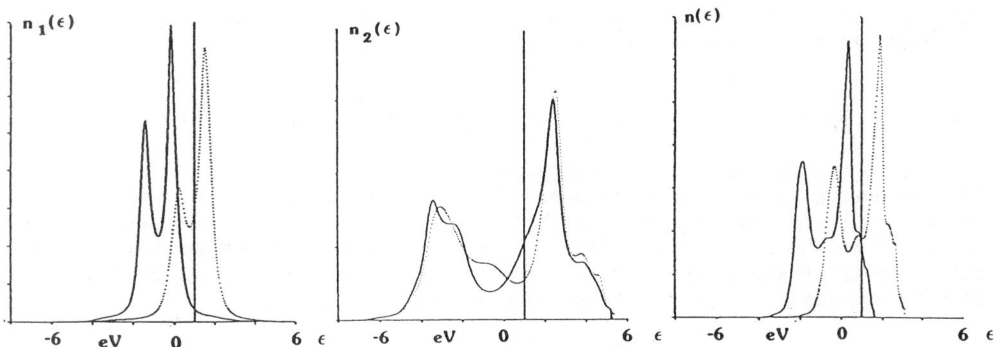

Fig. 22 Local "d" densities of states of a ferromagnetic iron monolayer deposited on W(110) (a), of the first W(110) underlayer (b) and of the iron(100) surface layer (schematic from [99]).

characterized by a small number of relevant parameters:

(1) A and B band centers and widths: the importance of the covalent mixing between A and B near the interface depends on the relative position of the corresponding energy levels i.e. of the relative variation of the band centers ($\delta_d = \frac{\epsilon_A - \epsilon_B}{W_B}$) and band widths ($\delta_{nd} = \frac{W_A - W_B}{W_B}$). Note however, that the epitaxial overlayer is compressed (or expanded) to the volume of the substrate, so that both ϵ_A and W_A must be determined for this peculiar situation (Fig. 19). The energy levels become less stable when the atomic volume decreases (see (1.9)) and the band width vary as Ω^{-n} where $n = \frac{5}{3}$ for most of the transition metals (n = 1 for the first elements of the series). This change is important for our qualitative discussion, the atomic volume varying themselves parabolically with the atomic number: the bandwidths present a maximum in the middle of the series for the equilibrium atomic volume but they decrease with the atomic number at constant volume for each series.

(2) A and B coupling: transfer integrals accross the interface: the coupling between both systems is induced by the transfer integrals $t_{AB}(R)$ across the interface. The linearized methods (LMTO) show that such integrals are given by the Shiba prescription

$$t_{AB}(R) = [t_{AA}(R)\ t_{BB}(R)]^{1/2} \qquad (2.21)$$

We can define strong and weak coupling regimes as for isolated adsorbates and according to the importance of the changes in the local densities of states $n_A(\epsilon)$, $n_B(\epsilon)$ induced by t_{AB}.

The electronic structure of the overlayers can be calculated either using ab initio calculations (FLAPW) on finite films [90 - 97] or semi-empirical tight binding models [98, 99, 100]. Both methods allow to get similar results; in the next section we will present densities of states for the most representative situations (Fig. 20 - 23).

2.3.3 Electronic structure of overlayers: some limiting cases

(1) Quasi 2D electronic structure
If the energy levels of A and B are far from each other ($\delta_d > 1$, $\epsilon_A > \epsilon_B$) and if $W_A \lesssim W_B$, the two dimensional band structure of the A monolayers is weakly perturbed by the substrate: Fe/Ag(100) and Cr/Au are such systems [91, 94, 95]. The density of states of a Cr(100) monolayer is compared in Fig. 20 with those of a Cr monolayer on Au and of the sandwich Au/Cr/Au(100).

(2) Periodic array of quasi-isolated adsorbates
If the overlayer is stronlgy expanded to be in registry with the susbtrate ($\Omega_A < \Omega_B$) and if the corresponding bandwidth W_A is very small as compared with W_B, the adsorbate-adsorbate interaction is very small and the overlayer density of states is about the same as the one of diluted adsorbate. As an example Fig. 21 shows the band density of states of Ni/Mo(110). If a second Ni monolayer is deposited on the first one the surface Ni 1 density of states is still narrowed.

(3) Interface covalent bonding
When the A and B bandwidths and energy levels are nearly identic the covalent bonding between interface layers becomes dominant. An example of the variation of the overlayer electronic structure with the energy levels and bandwidths is illustrated by the recent ab initio calculation of the electronic structure of a monolayer of the transition metals of the first series on Pd(100) [96, 97]: this study exhibits clearly a transition from an ordered array of

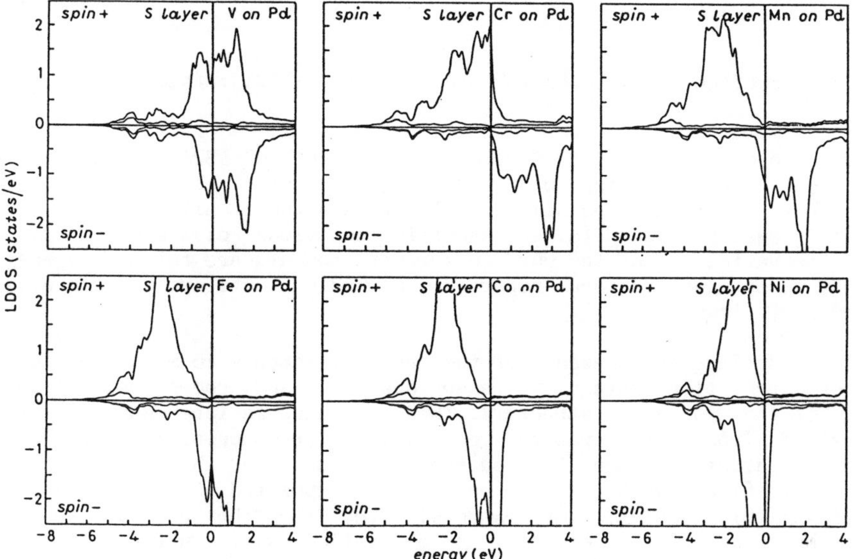

Fig. 23 Local densities of states of the 3d overlayers for the ferromagnetic (1×1) configuration on Pd(001) from [97].

quasi-isolated Ni adsorbates to a covalent bonding between interface atoms for V. For Ni/Pd $W_{Ni} << W_{Pd}$ and $\epsilon_{Ni} \lesssim \epsilon_{Pd}$ whereas for V/Pd $W_{Pd} \sim W_V$ at constant volume and $\epsilon_V \gtrsim \epsilon_{Pd}$ (Fig. 19 and 23).

2.3.4 Band magnetism of ultrathin films: transition metals on noble and transition metals

In this section, we present a survey of the most representative results concerning the itinerant magnetism of ultrathin films. We first discuss the theoretical results for surface magnetism, magnetization profiles, magnetic couplings and surface magnetic order. Then, we point out the difficulty of an experimental study of such systems in relation with growth conditions and interdiffusion.

2.3.4.1 Surface and interfacial magnetism [90 - 101]: theory

From a theoretical point of view both "ab initio" LSDA calculations and semi-empirical tight binding models have been used: they are usually in agreement but, up to now, no simple qualitative argument allows to understand and predict easily the detailed magnetic behaviour of a peculiar system. However, the variations of the surface magnetic moments are relatively easy to understand; some results are reported in Tables 6 to 9 and are briefly summarized as follows:

(1) The surface enhancement of the magnetic moments we mentioned for the surface of transition metals is also observed and comes mostly from the narrowing of the "d" band structure. This increase is large when one or two monolayers of iron are deposited on Ag(100) for example; it's smaller for interface Fe atoms between the surface Fe layer and the silver substrate (2Fe/Ag, Table 6) or when the iron atoms are deposited (1 ML) on transition metals (Fe/W(110), Table 7 for example) i.e. when the substrate adsorbate hybridization is more effective. In a similar way, surface relaxation increasing (compression) or decreasing (expansion) this hybridization can reduce or enhance the surface magnetic moment as shown in Table 7. The surface enhancement of the magnetic moment is the largest for chromium (Cr/Au(100) for example) and is much smaller for nickel probably by spd hybridization.

(2) Surface magnetic order: a theoretical study of the magnetism of transition metals (from V to Ni) deposited on Pd(100) has been done recently [96, 97]. This unique systematic study in the ab initio FLAPW framework can be understood qualitatively from the increasing importance of the surface covalent hybridization going from Ni to V and from the corresponding transition metal band filling: (i) the narrowing of the "d" bands coming from the expansion of the transition metal film and surface effects induces large magnetic moments relative to the bulk; (ii) according to the filling of the local "d" densities of states, the magnetic coupling between two transition atoms is either ferromagnetic ($N > N_0$) or antiferromagnetic ($N < N_0$): this general trend is observed both for the bulks and for impurities in normal metals; it has been explained long time ago for magnetic interactions [103] and for chemical interactions in transition metals [104]. This is also observed in the present case: (1 × 1) ML of V, Cr and Mn are antiferromagnetic (C(2 × 2) magnetic situation) whereas Fe, Co and Ni are ferromagnetic. The energy differences between ferro- and antiferromagnetic configurations are small ($\Delta E \leq 0.26$ eV) and consistent with relatively low Néel temperatures.

(3) Tight binding and "ab initio" calculations vs experiment: the magnetic

SURFACE AND INTERFACE MAGNETIC MOMENTS

Fe/Ag	M (μ_B)	Fe/Cu [90]	M (μ_B)	ΔE (eV)
1Fe/Ag(001) 1Fe/Fe(001)	2.96 2.98	1Fe/Cu(001)	2.85	
2Fe/Ag(001) : S S-1	2.94 2.63	2Fe/Cu(001) 1 FERRO S-1	2.85 2.60	$E_F - E_{AF}$ = -0.2 eV
1Fe/Fe(001) Ag/Fe/Ag(001)	2.98 2.80	2Fe/Cu(001) S ANTIFERRO S-1	2.38 -2.22	

Table 6

Fe/W(110)	FLAPW [101]		TBA [99]	
	M (μ_B)	$\Delta d_1/d_1$ (%)	M (μ_B)	$\Delta d_1/d_1$ (%)
UNRELAXED	2.56	0	2.69	0
RELAXED	2.18	- 15 %	2.20	- 11 %
Ag/Fe/W(110)	2.17	- 4 %	-	-

Table 7

ML Cr and Cr/Au(001) [91]				
Cr ML		M (μ_B)	ΔE = E - E(M = 0)	
F AF		4.19 3.84	- 1.69 - 2.00	
1Cr/Au(100) F AF		M (Cr) 3.70 3.48	M (Au) 0.14	ΔE - 0.78 - 1.13
2Cr/Au(001)	S S-1	2.90 - 2.30	- 0.08	- 0.60
2Au/5Cr/2Au	 I I-1 I-2	 + 1.75 - 1.07	2Au/5Cr/2Au	M (μ_B) 1.55 - 0.67 + 0.67
Au/Cr/Au	 I S S-2	M (Cr) 3.10	M (Au) 0.14 0.13	ΔE - 0.38

Table 8

V/Ag(100)	[90] M (μ_B)
V/Ag(100)	2
Ag/V/Ag	0
2V/Ag S S-1	1.15 0

Table 9

Theoretical layer by layer magnetic moments for specified cases (with estimated uncertainties of ± 0.03 μ_B). S, S-n indicate surface and subsurface layers, I and I-n the interface and subinterface layers. ΔE shows the spin polarization energy (in eV) (from [90], [91], [99]).

moment enhancement and the surface magnetic order are not sensitive to the details of the electronic structure so that the magnetic properties of overlayers determined either by FLAPW calculations or much more easily by tight binding models are in good agreement; the comparison between both calculations for a monolayer of Fe on W(110) is illustrative of such an agreement [99, 101] (Table 7). The theoretical results are usually in qualitative agreement with experiment, except in special situations for which (i) the magnetic moment is in the vicinity of a magnetic transition (see section 1.4), (ii) two different magnetic structures are nearly degenerate from competing ferro- and antiferromagnetic interactions. For example (1) the V monolayer is theoretically predicted to be ferromagnetic when epitaxied on Ag(100) (Table 9) whereas spin polarized photoemission experiments failed to detect magnetism [102]; (2) SPLEED experiments on ν monolayers of iron deposited on Cr(100) ($1 \leq \nu \leq 10$) failed to show ferromagnetism for the first two layers whereas the theoretical calculations suggest a stable ferromagnetic monolayer with a strong magnetic moment [105]; such a discrepancy can be also related to poorly ordered overlayers as discussed in the next section.

(4) Magnetization profiles and interface magnetic couplings: Fe/Cr vs Co/Cr: The magnetic perturbation decreases rapidly when the bulk is ferromagnetic (Fe) whereas it's long-ranged for Fe and Co on Cr(100) (see also section 2.1.5) (Fig. 10). The most stable magnetic arrangement is obtained when the iron overlayer is coupled antiferromagnetically to the last chromium layer and the magnetic moment is large. For cobalt on chromium, a solution with a large moment occurs also reasonably for large I values but it's coupled ferromagneticaly with the last chromium layer; moreover it's nearly degenerate with a solution with a very small antiferromagnetically oriented Co moment [99]. Such ferro- and antiferroamgnetic couplings between interfacial CoCr and FeCr (resp.) layers has also been found theoretically for superlattices [114, 115] (see section 5).

2.3.4.2 Structural perfection, chemical homogeneity and magnetic properties of ultrathin films

Perfectly ordered overlayers are difficult to obtain experimentally from the existence of surface defects on terraces, steps, kinks...; moreover, the deposition temperature T_s must be relatively large to insure surface diffusion but it can be also sufficient to allow interdiffusion, adsorbate agglomeration, formation of superficial ordered compounds... even if the conditions for a perfect wetting of the susbtrate (see section 3) are satisfied. The magnetic properties are highly sensitive to the local environment as shown by the early studies on magnetic alloys (Jaccarino-Walker model): it's necessary to prepare ultrathin films of reproducible structural perfection and chemical homogeneity to get reproducible magnetic results. The experimental difficulty to control such parameters explains probably the controversial results obtained recently for surface magnetism. As examples, let us consider here the properties of two model systems Co/Cu(100) and Fe/Cu(100).

A detailed study of the growth processes has been performed for thin cobalt films on Cu(100) by LEED, RHEED, MEED, TAES, Auger spectroscopy... [106, 107] for typical evaporation rates (2 ML/mn). Several regimes have been identified (using also Monte-Carlo simulations:

(1) At low temperature (T_s < 250 K), the diffusion length is very small ($\ell \sim 1$ Å) and the growth occurs via random 3D nuclei.

(2) At higher temperatures (T_s < 300 K), the surface diffusion becomes effective ($\ell \sim 10$ Å) and the growth occurs via 2D islands which progressively coalesce.

(3) For larger substrate temperatures (T_s < 400 K), the large mobility of

adatoms on the surface allows them to diffuse to the surface steps where they remain bound (the binding energy is larger near the steps than on the terraces, the number of substrate-adsorbate bonds being larger): the growth takes place via step propagation.

(4) Finally, at high temperatures ($T_s \geq 575$ K) interdiffusion takes place.

In order to illustrate the difficult characterization of such ultrathin films, the in plane magnetization at room temperature vs field has been measured for two cobalt films with the same coverage (2 ML), deposited at the same temperature (450 K) but annealed respectively to 525 K and 575 K prior to measurements [106]. The standard techniques of characterization (AES and LEED) can hardly distinguish both overlayers but the magnetic properties are completely different. The first film is ferromagnetic whereas the second one presents no magnetization at room temperature. It's thought that in this case cobalt is partially dissolved in the first few Cu layers; these atoms are non magnetic -cobalt single atoms surrounded by copper are non magnetic in the bulk at T = 300 K- and the surface overlayer has a stronlgy lowered Curie temperature. The high sensitivity of magnetism to atomic arrangement, surface perfection and to interdiffusion is probably one of the reasons of the disagreement between different experiments. Another experimental difficulty is the calibration of the deposited coverage which can be incorrrect if it's obtained only from breaks in the Auger intensity vs deposition time. Several other techniques (MEED, TAES) can be used simultaneously to insure a correct calibration. For example, a cobalt monolayer was found to be ferromagnetic at room temperature (T_s = 300 K) in reference [108] whereas it was reported to be non magnetic in reference [106]. For these last authors the Curie temperature increases linearly with the number ν of overlayers going from 150 K for $\nu = 1.5$ to 600 K to $\nu = 2.5$ and the discrepancy with the results of reference [108] has been suggested to be due to an incorrect calibration, the film being actually thicker than initially thought.

The epitaxial growth of fcc iron on Cu(100) has also been extensively studied. LEED and breaks in the Auger intensity vs evaporation time suggest a layer by layer growth; however, a detailed XPS study -and especially the analysis of the angular distribution of $2p_{3/2}$ core level spectrum of the deposited element [109, 110]- suggests that the growth is not via a Frank-Van der Merwe mechanism but exhibits iron agglomeration at and slightly above room temperature. When $\theta < 2$ ML the film is poorly ordered for $T_s < 200$ K, iron agglomerates for 200 K $< T_s <$ 300 K and copper segregation accompanies iron deposition (intermixing) for higher substrate temperatures. This intermixing can explain the conflicting results reported for the magnetic properties of these systems; note that bulk fcc (γ)Fe is antiferromagnetic whereas Fe_xCu_{1-x} alloys are either non magnetic, antiferromagnetic (spin glasses) or ferromagnetic according to the composition x. "In situ" polar and longitudinal Kerr effect measurements were systematically performed on Fe/Cu(100) vs growth temperature T_s (100 K $< T_s <$ 400 K) and iron thickness (0.5 ML $\leq \nu \leq$ 8 ML). [111]. The magnetization can be either perpendicular or parallel to the surface according to the θ and T_s values:

(1) Square hysteresis loops and perpendicular anisotropy was obtained for films 1.5 ML to 5.7 ML thick for 100 K growth.

(2) The region of stability ($n_1 < n < n_2$) of perpendicular anisotropy decreases increasing T_s and $T_c < T_s$ for $T_s > 350$ K.

(3) At a given growth temperature (T < 350 K), the easy axis reorients in the plane of the film for $\nu < \nu_1$ and $\nu > \nu_2$.

Note that all authors [112, 113] studied iron films obtained by evaporation at room temperature; they claim that the layer by layer growth mode is realized and find perpendicular magnetization from 2 ML to 10 ML at 300 K by spin polarized photoemission spectroscopy.

In conclusion, a detailed study of the growth processes must be done to get well characterized overlayers and relevant magnetic properties. This difficulty occurs mostly for ultrathin films and interfacial magnetism ($\nu \lesssim 3$), the transition metals recovering often their "bulk" properties far from the interface.

3. GROWTH OF EPITAXIAL FILMS, INTERFACE ENERGIES AND ELECTRONIC STRUCTURE

3.1 Introduction

The growth of epitaxial overlayers during the initial stages occurs by one of three well-known mechanisms:

(1) The Frank Van der Merwe (FM) growth mode which is characterized by a layer by layer deposition.

(2) The Volmer-Weber (VM) mode which corresponds to the formation of 3D crystals onto the substrate from the vapor phase.

(3) The Stranski-Krastanov (SK) mode which corresponds to the nucleation of 3D crystals after a layer by layer deposition of few monolayers on the susbtrate [116 - 118].

The growth of such overlayers is usually studied using a macroscopic theory of wetting phenomena [117, 118]. Such a theory assumes that the equilibrium between condensed phases and their vapors is experimentally attained and predicts either nucleation of bulk phase clusters (VW) or wetting of the substrate B by the adsorbate A (FM) in terms of the spreading coefficient S

$$S_{AB} = \gamma_A + \gamma_{AB} - \gamma_B \qquad (3.1)$$

γ_B, γ_A and γ_{AB} are the surface free energies of the substrate, film and substrate-film interface respectively. If $S > 0$, VW mode occurs and if $S \leq 0$, SK or FM growth takes place. Several assumptions must be discussed at this stage:

(1) The previous theory is valid in a macroscopic limit: it cannot be applied to the study of overlayers in the monolayer range. This is why in the microscopic limit several authors used either lattice gas models for wetting and multilayer adsorption [119 - 122] or phenomenological pair potentials (Van der Waals...) [123] to study both equilibrium states and kinetics [124] using simulation methods.

(2) This model was developed for fluid wetting and assumes usually that the film does not support any strain. For solid films, the misfit between the equilibrium macroscopic lattice constant a and the corresponding substrate lattice constant b, $f = \frac{a - b}{a}$, leads at equilibrium either to an homogeneous ($h < h_c$) or to an inhomogeneous ($h > h_c$) film strain to match the substrate lattice: h_c is the critical film height beyond which misfit dislocations are stable. This is why some authors introduced phenomenological elastic terms and inhomogeneous deformations in the macroscopic theory to study the growth modes [125].

(3) The assumption of equilibrium between the condensed phases and their own vapors is not usually realized in practice. In the classical methods for deposition of metals or semiconductors (molecular beam epitaxy, chemical vapor deposition...) the high incident flux Φ leads to a growth mode far from equilibrium. Kinetic theories and simulations must then be developed to describe experiment.

(4) The phenomenological models which have been developed previously in

the microscopic limit cannot be used here: the pair potentials (and the Ising models) cannot describe the cohesion of transition metals and semiconductors from the existence of narrow bands ([1] and section 1). In these lectures we are interested by metallic overlayers in the microscopic limit so that the wetting theories developed previously are not satisfactory and must be modified. Moreover, general trends cannot be understood so that we will consider here directly the electronic structure and the binding energy of a microscopic film deposited on a substrate.

(5) The low vapor pressures of metals and the rapid surface diffusion lead us to follow some recent studies [124, 125] and to assume that, for low enough growth rates, the film morphology is determined by assuming the equilibrium is reached at constant particule number. It's important to mention here the two previous works which use this approach:

(i) in the macroscopic limit R.Bruinsma and A.Zangwill [125] discussed the conditions for VW or SK growths including elastic strains,

(ii) in the limiting cases of few monolayers and using phenomenological potentials Grabow and Gilmer [124] reported recently the results of molecular dynamics computer simulations and studied both homogeneous and inhomogeneous strained films.

(6) The substrate temperature during deposition, T_s, is often sufficient to allow a significant interdiffusion. Up to now, the wetting of the substrate by ultrathin ordered compounds, the selection of these compounds, the dissolution of A in B, the chemical composition and the morphology of the diffusion fronts... are phenomena which have not been considered from a microscopic point of view.

In this chapter, we use the electronic theory of metals and alloys [99, 100] to determine the interface energies, the growth modes and the possible occurrence of ordered compounds. We are interested here only by general trends for transition metals so that we use the simplest possible electronic models i.e. the tight binding scheme summarized in the previous section. However, the general method we use here can be extended to LSDA all electrons calculations.

First, we summarize the qualitative results which we previously obtained from phenomenological models: we point out the essential role of the range of adsorbate-substrate interactions in a lattice gas model (section 3.2) and the interplay of "chemical" and "elastic" interactions to determine growth modes (sections 3.3 and 3.4); we recall briefly the phenomenological criteria used to predict epitaxy or superlattice growth and the relevant parameters which determine in a pair model, bulk and surface segregation, adhesion, surface energies and wetting of the substrate (sections 3.5 and 3.6).

Then, we determine the conditions which must be satisfied to get a peculiar growth mode (SK, VW, FM) when the equilibrium states of a constant number of adsorbates N_A is assumed to be reached (section 3.7). We discuss qualitatively the conditions to get two dimensional islands of segregated or ordered adsorbates, incomplete layers... and we first consider as an example the case of low coverages (section 3.8).

In the section 3.9, we split the interfacial and spreading energies into their "chemical", magnetic, structural, elastic and relaxation contributions. We present a simple model for the electronic structure of overlayers and for the calculation of the wetting conditions, we discuss qualitatively the order of magnitude of the various terms, the trends we get for VW and SK growth modes and the reliability of the phenomenological criteria introduced for example by Van der Merwe and Bauer [133]. We also study the stability of an epitaxial overlayer with respect to the formation of ultrathin ordered compounds and show that such compounds can be stable and wet the substrate even if a VW growth is predicted without interdiffusion.

Finally, we discuss the order of magnitude of the interfacial energies between two semi-infinite crystals in relation with their electronic structure.

3.2 Multilayer adsorption on attractive substrates: lattice gas models

As a first step, let us recall the results of the most simple lattice gas model [119]. In such a model the lattice is assumed to be rigid, each layer parallel to the semi-infinite substrate B is partially occupied by the adsorbate A and the energy of each configuration is determined by adsorbate-adsorbate and adsorbate-substrate interactions. For simplicity, the intralayer interactions are attractive ($v_0 < 0$) so that no intralayer ordering is allowed; the interaction between an A atom and one of its nth neighbouring A plane is labelled v_n whereas the interaction between an adsorbate of the pth layer and the substrate, u_p, is also assumed to be attractive to insure wetting of the substrate. Beyond the first few monolayers, the adsorbate-susbtrate interaction decreases rapidly (as Z^{-3} for Van der Waals forces in physisorbed systems, where Z is the distance from the substrate and even more rapidly in metallic systems from the highly effective screening effects). The excess surface density induced by the adsorbate-substrate interaction n_s and the corresponding phase diagram are then determined at equilibrium with the gas vs the thermal energy k_BT, the chemical potential μ, the adsorbate interaction strength $|u|$ and $|v|$ and their relative ranges. There are roughly three situations: strong substrate systems ($|u| >> |v|$), intermediate substrate systems ($|u| \sim |v|$) and weak substrate systems ($|u| << |v|$).

<u>When the attractive substrate potential is strong</u> ($|u| >> |v|$) <u>and long-ranged,</u> an infinite sequence of transitions occurs: it corresponds to the condensation of successive monolayers. The critical temperatures of these transitions T_{cn} approach when $n \to \infty$ the roughening temperature T_R and at temperatures $T < T_R$ the isotherms for the excess surface density $n_s(\mu)$ show an infinite sequence of sharp steps as $\mu \to \mu_0^-$, μ_0^- is the chemical potential at coexistence between gas g and condensed ℓ liquid phases. $n_s(\mu) \to \infty$ when $\mu \to \mu_0^-$: this situation corresponds to the complete wetting of B by A at coexistence and to the so-called Franck Van der Merwe growth (Fig. 24a).

As an example, such a behaviour is easily exhibited at T = 0 K for a long-range attractive monotonic adsorbate-susbtrate potential $u_1 < u_2 < u_3 \ldots < u_p \ldots < 0$ and a short-ranged attractive adsorbate-adsorbate potential ($v_1 < 0$, $v_n = 0$ for $n \geq 2$). In such a case there are three different sequences:

(1) A complete wetting with the sequence $[0, 1, 2, \ldots \infty, \ell]$ requires $u_1 - u_2 < v_1$.

(2) The sequence $[0, n, n+1, \ldots \infty, \ell]$ occurs when:

$$\sum_{m=1}^{n} u_m - n\, u_{n+1} < v_1 < \sum_{m=1}^{n-1} u_m - (n-1)\, u_n$$ and the sequence $[0, \ell]$ requires

$$v_1 < \sum_{m=1}^{\infty} u_m .$$ A strong substrate (large $|u|$ values) insures the complete wetting of B by A when the interaction u_p is long ranged.

<u>When the interaction $|u|$ decreases,</u> the layer transitions no longer extend to T = 0 K and occur above a wetting temperature $T_W > 0$ which increases when $|u|$ decreases: (1) for $T < T_W < T_R$ the surface is <u>incompletely wet</u> at

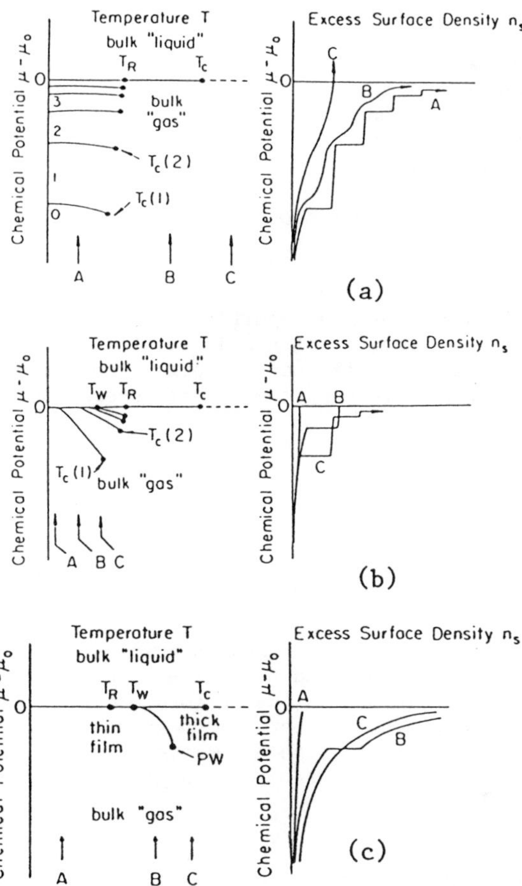

Fig. 24a Strong substrate system ($u < u_W < 0$). Surface phases are labelled 0, 1, 2... according to the number of completed layers at T = 0 K. Isotherm A has an infinite number of sharp steps. Isotherms A and B lead to a complete wetting of the surface when $\mu \rightarrow \mu_0^-$ from [119].

Fig. 24b Intermediate substrate system ($u_W < u < u_R < 0$). At low temperatures there is little adsorption (isotherm A). When T increases the surface phases become stable and the number of steps in the isotherm increases (isotherm B) becoming infinite for $T > T_W$. There is a complete wetting for $T_W < T < T_c$ when $\mu \rightarrow \mu_0^-$ (from [119]).

Fig. 24c In this intermediate situation ($u_R < u < u_{CW}$, $T_W > T_R$) there is a first order transition along the prewetting line joining T_W to PW between thick and thin films behaviour (from [119]).

coexistence: there are a finite number of layer transitions when $\mu \to \bar{\mu}_0$,(2) for $T > T_W$ the surface is completely wet as previously (Fig. 24b).

When |u| is weakened further, $T_W > T_R$, the infinite sequence of layer transitions is replaced by a thin-thick prewetting transition going from $T < T_W$ to $T > T_W$. The wetting becomes complete and the isotherm $n_s(\mu)$ are smooth except when they cross the prewetting phase boundary in which case they undergo a finite jump (Fig. 24c).

Finally, for $u = u_c$, T_W reaches the **critical** temperature T_c and in the regime of weak substrate systems ($u_c < u < 0$) isotherms are smooth and do not exhibit singularities.

This general picture must be modified to take into account the elastic energy. The film is generally strained relative to the bulk so that the equilibrium film thickness does not diverge when $\mu \to \bar{\mu}_0$ even for strong substrates [122, 123]: the role of the elastic energy is briefly discussed in the next section.

3.3. Strained epitaxial films

3.3.1 Introduction

The understanding of epitaxial phenomena must take into account the atomic displacements near the interface and the corresponding energy. Up to now, only phenomenological macroscopic models using the theory of linear elasticity and simulations using pair potentials have been used to study these effects. In this section we summarize briefly some extreme situations according to the film height h; this will be useful later to discuss the deformation part of the interfacial energies.

3.3.2. Interfaces and misfit dislocations ($h > h_c$) [127 - 129]

When the film height h is larger than the critical height h_c, the equilibrium state is characterized by a periodic array of misfit dislocations. In such a case the simplest model was introduced by Van der Merwe et al. for the interface between two epitaxial semi-infinite crystals with the same equilibrium cubic structure with lattice parameters a and b respectively. The deformation results from the atomic interaction across the interface which tends to align A and B atoms across the interface and from the elastic forces of each semi-infinite crystal. The former interaction can be modelled as a shear stress τ having the periodicity in the relative displacement u of interfacial atoms that were in registry in a reference lattice whose lattice parameter is $c = \frac{1}{2}(a^{-1} + b^{-1})^{-1}$. For example, in the parabolic potential model $\tau = \frac{\mu u}{c}$ for $|u| \le c/2$. where μ is the interfacial (unknown) shear modulus. The interfacial energy γ^{el} (per interface atom) can then be represented by an universal relation [128, 129]

$$\gamma^{el}_{AB} = W_t \, F(f\nu_0) \tag{3.2a}$$

$W_t = \frac{\mu c}{2\pi^2}$ determines the order of magnitude of the interfacial energy.

F(x) is an universal function (Fig. 25a); $f = \frac{a-b}{b}$ is the misfit parameter and $\nu_0 = \frac{\lambda c b^2}{\pi W_t}$ is a measure of the ratio between the elastic and interfacial energies. For isotropic A and B media the elastic modulus λ is given in terms of the Poisson ratios (σ_A, σ_B) and shear moduli (μ_A, μ_B) of the two half crystals

$$\frac{1}{\lambda} = \frac{1-\sigma_A}{\mu_A} + \frac{1-\sigma_B}{\mu_B} \qquad (3.2b)$$

The essential feature we must point out is that the interfacial energy is the lowest for the smallest misfits f and elastic constants λ. This model is qualitatively correct, but it does not allow a quantitative determination of the interfacial energy.

3.3.3 Elastic and relaxation energies: epitaxial strained overlayers $h_0 < h < h_c$ [130].

Let us now consider a film A whose height $h = \nu d$ is smaller than h_c so that it's homogeneously strained (without misfit dislocations) but so large ($h > h_0$) that the macroscopic treatment of the deformation energy is reasonable. <u>If the substrate overlayer interaction is short-ranged,</u> the strain is homogeneous both parallel ($\epsilon_{xx} = \epsilon = f$) and perpendicular ($\epsilon_{zz} = \epsilon'$) to the film [123]. The elastic energy stored in the strained epitaxial film $\gamma^{el}_{A/B}(h) = \gamma^{el0}_{A/B}(h) + \gamma^{rel}_{A/B}(h)$ can be splitted into two terms:

(i) $\gamma^{el0}_{A/B}(h)$ is obtained by an hydrostatic compression (or expansion) to the substrate atomic volume.

$$\gamma^{el0}_{A/B} = \frac{B_A}{2}\delta^2 \nu, \qquad \delta = \frac{\Delta\Omega}{\Omega_B}, \qquad \Delta\Omega = \Omega_A - \Omega_B \qquad (3.3)$$

B_A is the bulk modulus, $\frac{h}{d} = \nu$ is the number of A planes parallel to the interface and $\delta = 3f$ for a cubic crystal.

(ii) <u>The relaxation energy</u> $\gamma^{rel}_{A/B}$ is obtained from the previous state leaving the surface atoms relax to their equilibrium position. This gain in energy ($\gamma^{rel}_{A/B} < 0$) is determined from the condition that the normal stresses must be zero. For example, when the crystalline structure is cubic

$$\begin{cases} \gamma^{rel}_{A/B} = -J\,\frac{B_A}{2}\,\delta^2\nu \\ J(100) = \frac{B_A}{C^A_{11}},\; J(111) = \frac{B_A}{B_A + 4C^A_{44}/3},\; J(110) = \frac{B_A}{C^A_{44} + (C^A_{11} + C^A_{12})} \end{cases} \qquad (3.4)$$

This separation is arbitrary: a direct determination of the deformation energy in the elastic approximation is obtained in the Grilhé's lectures. It's nevertheless useful to point out that, in this macroscopic limit, the "relaxation energy" is of the same order of magnitude as the other contribution (for example, J(100) = .58 for tungsten and 0.83 for palladium) so that the relaxation must be taken into account to evaluate $\gamma^{el}_{A/B}(h)$. The interfacial energy $\gamma_{A/B}(h)$ is then given by

$$\gamma_{A/B}(h) = \gamma_{A/B}^{chem} + \gamma_{A/B}^{el0}(h) + \gamma_{A/B}^{rel}(h) \tag{3.5}$$

where $\gamma_{A/B}^{chem}$ is the part of the interfacial energy coming from chemical bonds across the interface (perfect epitaxy).

3.3.4 Ultrathin films [128 - 133]

In this limiting case, a microscopic theory is required using either phenomenological pair potentials or electronic theory. Few calculations were done except for physisorbed systems and for semiconductors. It's interesting, from a physical point of view, to mention here the work of Van der Merwe et al. even if it considers the overlayer as an elastic medium -an approximation which is not really valid in the present case: the epitaxied structure results from a periodic overlayer-substrate potential energy which is in competition with the elastic energy stored in the film. The amplitude of the overlayer-substrate interaction W and the ratio of adsorbate-adsorbate to the adsorbate-substrate interaction ℓ are thus the relevant parameters; in the elastic approximation ℓ is defined by

$$\ell = \frac{\Omega_A\ S_A}{Wr^2} \qquad S_A = \left(C_{11}^{A\,2} - C_{12}^{A\,2}\right) \Big/ 2C_{11}^{A} \qquad r = \frac{b}{a} \tag{3.6a}$$

The dependence on epitaxial parameter W, ℓ, f of the interfacial deformation energy can then be discussed as follows:

When both A and B systems are rigid γ^{el} = W. When the overlayer is allowed to relax to its equilibrium position either the misfit is accommodated by misfit dislocations only, or it's shared between misfit dislocations and homogeneous misfit strains or, below a critical value of the misfit f_c, the overlayer is coherently epitaxied on the substrate. The interfacial energy can be represented as an universal function of $f\ell$

$$\gamma_{A/B}^{el} = + W\ F(f\ell) \tag{3.6b}$$

as in the case 1 we previously mentioned (Fig. 25). Note however that in the present case we do not discuss the selection of ideal epitaxial configurations, the factors which determine misorientations... Moreover, we assume that the substrate is rigid and that the crystalline structures of B and A are the same.

When A and B have different equilibrium structures [131 - 133] and according to the importance of the misfit parameter $r = \frac{b}{a}$ and of the ratio ℓ, phase diagrams can be built which determine the domains in the r,ℓ plane for which perfect epitaxy, partial epitaxy of rows or incommensurate structures are stable. Such a study has been done only in the case of the growth of (111)fcc (or hcp) planes on perfect and rigid (110)bcc surfaces (see in this volume, the lectures by Bermont).

From a microscopic point of view, we start from an ideal surface which is characterized by relaxed planes and even in some cases by a reconstruction (missing rows for Pt(110) surfaces for example) which results from a delicate balance between repulsive and electronic contributions. The epitaxy of A on the surface of B is accompanied by an atomic rearrangement and by the disappearance of the reconstructions which are characteristic of the perfect surface. Increasing the coverage θ, a change of crystalline structure occurs going from epitaxied planes of the B structure to an atomic arrangement characteristic of the equilibrium A structure. It's therefore important to take into account

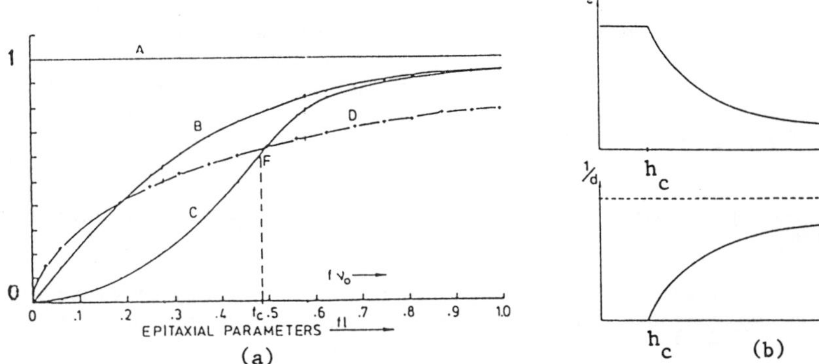

Fig. 25a Interfacial energy γ^{el}/γ_0 *induced by the substrate and overlayer elastic deformations as a function of* $f\nu_0$ *for a thick overlayer* ($\gamma_0 = W_t$) *and of* $f\ell$ *for a monolayer* ($\gamma_0 = W$). *When the substrate and overlayer are both rigid* $\gamma^{el} = \gamma_0$. *Curve C represents the energy of an elastic monolayer on a rigid substrate when the misfit f is accommodated both by dislocations and by homogeneous misfit strain* ϵ *(stable lowest energy configuration); curve B represents the energy of the unstable configuration for which the misfit is accommodated only by dislocations. Beyond a critical misfit* f'_c *the contribution of the homogeneous strain becomes negligible. Curve D represents the deformation energy of a thick overlayer on an elastic medium in the stable misfit dislocations mode* ($h >> h_c$) *(from [131]).*

Fig. 25b Homogeneous strain ϵ ***and misfit dislocation density 1/d as a function of the film thickness h (schematic).***

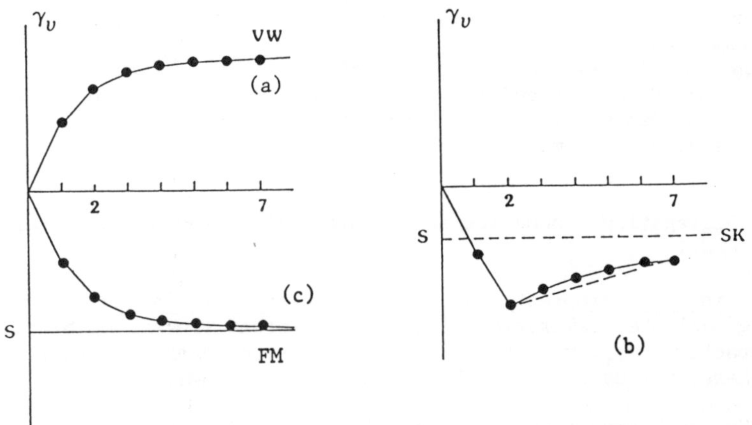

Fig. 26 Variation of the spreading energy versus coverage $\gamma(\theta)$ *for Volmer-Weber (a), Stranski-Krastanov (b) and Frank Van der Merwe (c) growth modes.*

these possibilities if we want to determine the equilibrium structure (and energy) of the film. The energies associated to reconstruction of structural changes are generally very small (10^{-2} eV) as compared with the adsorbate substrate energy W which leads "a priori" to wetting and epitaxy. The orders of magnitude of W and of the energy of desorption are the same and in the eV range (see section 2). This will be verified by the more detailed calculations we present later so that reconstruction and structural changes can be neglected as a first step for a general study of wetting phenomena.

3.4 Macroscopic model for wetting of solids: growth of strained films [125]

In this section we present a generalization of the theory of fluid wetting including the macroscopic treatment of elastic deformations we discussed previously. The thermodynamic potential G of the film is equal to

$$G = \int dx \left[(\gamma_A + \gamma_{A/B}(h) - \gamma_B) + \frac{1}{2}\gamma_A \left(\frac{dh}{dx}\right)^2 + \lambda h\right] \qquad (3.7)$$

In this expression, we assumed that the system is one dimensional (commensurability in the y direction). The first term is the interface energy change which depends on the film height h(x) and is calculated for a flat film. The second term is the energy cost associated with an extra area of a curved surface as compared to a flat film, γ_A being assumed to be independent on the surface orientation. The last term fixes the number of adsorbed particles (λ is a Lagrange multiplier). The interfacial energy $\gamma_{A/B}(h)$ results from (i) the adsorbate-substrate interaction (chemical term) $\gamma_{A/B}^{chem}$ which is assumed to be short-ranged and h independent, (ii) the elastic part of the interfacial energy $\gamma_{A/B}^{el}(h)$ we introduced previously and whose expression is given for example in ref. [126] for all h values

$$\gamma_{A/B}(h) = \gamma_{A/B}^{chem} + \gamma_{A/B}^{el}(h) \qquad (3.8)$$

A study of the Euler-Lagrange equations which express the minimum value of thermodynamic potential with respect to variations in h(x) shows that, as a function of the spreading pressure $S = \gamma_A + \gamma_{AB} - \gamma_B$ there is a first order transition at S = 0 from a phase with droplets on a dry surface (VW) to a phase coexistence of droplets and a thin film of A (SK). The absence of a FM regime comes from the concavity of $\gamma_{A/B}(h)$ i.e. from (i) the assumption of short-ranged AB interactions and (ii) the existence of elastic interactions. The assumption of long-ranged interactions u(r) in physisorbed systems leads to a convex dependence of $\gamma_{A/B}(h)$ and to a FM growth at least when elastic interactions do not become dominant.

3.5 Segregation, adhesion and growth: the relevant parameters in a pair model

Even in the simplest Ising model with only 1st nn interactions, and neglecting all the relaxation effects, different linear combinations of the pair interactions v_{AA}, v_{BB} and v_{AB} must be related to the physical phenomena we consider here. Let us consider the following situations:

(1) Volume segregation and order: it's determined by the sign of

$$w = v_{AA} + v_{BB} - 2v_{AB} \qquad (3.9)$$

($w > 0$ order, $w < 0$ segregation).

(2) Surface segregation of A diluted in B: if Z_B is the number of broken bands at the considered surface, the dilute element is segregating to the surface if

$$\tau_A = Z_B\,(v_{AA} - v_{BB}) > 0 \qquad (3.10)$$

(3) Surface energies: the surface energies being given by $\gamma_i = -\frac{Z_B v_{ii}}{2}$, $\tau_A = 2(\gamma_B - \gamma_A)$ and the A element goes to the surface if the surface energy of A, γ_A, is lower than γ_B.

(4) Adhesion energies and interface energy: for a perfect interface the adhesion energy is given by $\gamma_{adh} = -Z_B\,v_{AB}$ and the interface energy $\gamma_{A/B} = -\frac{Z_B w}{2}$ is related to w... but this relation is no more valid when we take into account interactions between 2nd nn.

(5) Spreading energy: it's given from its definition (3.1) by $S = -Z_B\,(v_{AA} - v_{AB})$.

In conclusion, no simple unique parameter can be used to explain these different physical phenomena: for example, there is no direct relation between the bulk tendency to order or segregate in A_cB_{-c} alloys, or the sign of $\gamma_A - \gamma_B$ and the growth mode (FM, VW, SK...).

3.6 Phenomenological criteria for the prediction of a good epitaxy and of the superlattice feasability [133]

Up to now, neither detailed experimental data for the interface energy γ_{AB} nor theoretical microscopic models were available to describe superlattice growth or even epitaxy of ultrathin films. Bauer and Van der Merwe [133] introduced simple and relevant phenomenological criteria. Using a macroscopic model they assume that the interfacial energies are much smaller than the surface energies: if we want to epitaxy A on B and B on A it's therefore necessary that γ_A and γ_B have the same order of magnitude ($S_{A/B}$ and $S_{B/A} < 0$). From the experimental data on some classical systems they suggest that superlattice can grow if

$$\Gamma_{AB} = 2\,\frac{|\gamma_A - \gamma_B|}{\gamma_A + \gamma_B} < \Gamma_c \sim \frac{1}{2} \qquad (3.11)$$

The surface energy mismatch is reported in Table 10 for some typical metal pairs we will discuss later.

The epitaxy of ultrathin films has been examined by the same authors taking into account both the misfit parameter $r = \frac{b}{a}$ and ℓ i.e. the ratio of the adsorbate-adsorbate to the adsorbate-substrate interactions (see 3.5). Both criteria must be discussed from the electronic theory: this is the aim of the sections 3.8 to 3.10.

	A/B	r_{AB}	Γ_{AB}
bcc/bcc (100)	Nb/Ta Fe/Cr	0.998 0.994	0.01 0.16
fcc/fcc (100)	Ni/Co	0.994	0.17
fcc:bcc (110)	Rh/V Ni/Mo Co/Cr Ru/Fe	1.027 0.914 1.004 1.089	0.002 0.20 0.30 0.15

Table 10 Geometry parameter r = b/a and surface energy mismatch for some metal pairs.

3.7 Epitaxial overlayers: growth, interfacial and wetting energies

In this section, we consider epitaxial clusters of N_A A atoms made of ν overlayers (ν =, 1, 2...). These clusters are in epitaxy with a perfect surface of a semi-infinite susbtrate B. In the microscopic limit it's no more possible to define separately and without ambiguity surface and interface energies. For a single cluster deposited on B, the energy is

$$E = N_B E_B^0 + N_A E_A^0 + N(\gamma_B + \gamma_\nu) \tag{3.12}$$

$N_B E_B^0$ is the bulk energy of N_B atoms of the substrate, $N_A E_A^0$ is the corresponding energy for the deposited material, $N\gamma_B$ represents the surface energy of the N surface atoms and γ_ν is the interfacial spreading energy per surface atom. It will be useful later to introduce an interfacial energy $\gamma_{1\nu}$

$$\gamma_\nu = \gamma_A + \gamma_{1\nu} - \gamma_B \tag{3.13}$$

which is equal to the interface energy γ_{AB} in the macroscopic limit ($\gamma_{1\nu} \to \gamma_{AB}$ when $\nu \to \infty$); note that in the same limit $\gamma_\nu \to S$ where S is the macroscopic spreading coefficient. With such definitions, it's then useful to discuss the stability of deposited clusters and the morphology of the overlayers from the variation of the γ_ν function. We consider here only the thermodynamic limit (N, $N_A \to \infty$ whereas N_A/N remains finite) so that we do not discuss the equilibrium shape of the two or three dimensional clusters.

The Volmer-Weber process is thus realized if $\gamma_\nu > 0$ for all ν values: it's energetically favorable to deposit a macroscopic three dimensional crystal of A on B (Fig. 26a).

When $\gamma_\nu < 0$, the considered 2D cluster is sticked to the B surface by the adsorbate-susbtrate interactions and it's more stable than the corresponding 3D crystals. However, it can be unstable with respect to a situation where the surface is covered by a mixing of clusters of ν, and ν-ν_1 overlayers. As in the classical theory of 3D phase diagrams we introduce the energy $\gamma(\theta,\mathfrak{C})$ per interface atom for each θ value and each atomic configuration $\mathfrak{C}$ of the N_A atoms on the B surface ($\theta = \frac{N_A}{N}$). For each θ value, the most stable atomic configuration will be obtained from the envelope $\gamma(\theta)$ of the set $\gamma(\theta,\mathfrak{C})$ built as usual by using the common tangent construction. The layer by layer growth (FM or complete wetting) is thus obtained if $\gamma(\theta) = \gamma_n + (\gamma_{n+1} - \gamma_n)(\theta - n)$ for $n \le \theta \le n+1$ $\forall$ $n \in N_+$ i.e. when $0 > \gamma_1 > \gamma_2 \ldots > \gamma_p \ldots > S$ and if all the other atomic configurations are less stable. The Stranski-Krastanov growth is obtained if (i) $0 > \gamma_1 > \ldots > \gamma_n$ (ii) $\gamma_p > \gamma_n$ for $p > n$ (ii) the other atomic configurations have an interfacial energy γ larger than $\gamma(\theta)$ (Fig. 26). In another words, the FM growth occurs when the most stable configurations are obtained for complete overlayers and when $\gamma(\theta)$ is decreasing when $\frac{N_A}{N} \to \infty$; the Stranski-Krastanov is obtained when $\gamma(\theta)$ decreases only up to $\frac{N_A}{N} = n$. It's important to note here that (1) the macroscopic condition $S < 0$ is necessary but not sufficient to get a FM growth, (2) the FM growth is obtained if we gain more and more interfacial energy when we deposit more and more A atoms on B.

As an example, let us consider the simplest pair model (see section 3.2) where:

(i) u_p is the interaction between the substrate and an A atom of the pth overlayer and

(ii) v_1 is the adsorbate-adsorbate interaction which we assume to be restricted to neighbouring planes. In such a case we obtain easily

$$\gamma_1 = u_1 - v_1,\ \gamma_2 = u_1 + u_2 - v_1 \ldots \gamma_n = \sum_{p=1}^{n} u_p - v_1 \qquad (3.14)$$

The growth mode is thus determined as in section 3.2 by the relative order of magnitude of AA and AB interactions and by the range of the AB interactions. A FM growth is obtained when $0 > \gamma_1 > \gamma_2 > \ldots > \gamma_p \ldots$ i.e. for large $|u_1|$ values $(u_1 - v_1 < u_2)$ and for long-ranged concave adsorbate-susbtrate attractive interactions $(u_1 < u_2 < u_3 \ldots < u_n \ldots)$.

3.8 Spreading energy, pair interactions and overlayer ordering ($\theta < 1$)

Each atomic configuration $\{p_\lambda\}$ is characterized by the set of the occupation numbers ($p_\lambda = 1$ if the adsorption site λ is occupied by an A atom, 0 otherwise). In the dilute limit the interfacial energy $\gamma(\{p_\lambda\})$ can be written as a rapidly convergent cluster expansion, the number of clusters of n atoms varying as θ^n. If we keep only pair interactions we get easily from (3.12) the relation

$$\gamma(\{p_\lambda\}) = \theta\left[E_A^{coh} + E_{ads}^{A/B}\right] + N \sum_{n>1} v_n\, p_n \qquad (3.14)$$

$E_A^{coh} = E_A^{at} - E_A^0 > 0$ is the cohesive energy of the bulk A, $E_{ads}^{A/B} = E^{A/B} - E_A^{at}$ is the binding energy of an atom A on B, v_n is the pair interaction between two nth neighbouring A atoms and p_n is the corresponding number of AA pairs in the considered configuration (per adsorption site). The quantities which determine γ E^{coh}, E_{ads}, v_n) have been discussed in previous sections (2.2.3 and 2.2.5). Note finally that the expression (3.14) can be also shown to be convergent for <u>all</u> θ values $0 < \theta < 1$ if we use a two dimensional version of the <u>generalized perturbation method</u> [104, 135], the cluster interactions being θ dependent.

If we are interested by the order of the 2D overlayer, it's useful to split the spreading energy into two terms

$$\begin{cases} \gamma(\{p_n\}) = \gamma_{des}(\theta) + \gamma_{ord}(\{p_n\}) \\ \\ \gamma_{ord}(\{p_n\}) = \sum_{n>1} \left(p_n - \frac{Z_n\theta^2}{2}\right) v_n \end{cases} \quad (3.15)$$

$\gamma_{des}(\theta)$ is the value of γ for a completely disordered state whereas γ_{ord} is the contribution to γ of local ordering of the adsorbates. Let us now consider two typical situations which occur for a 2D growth in the submonolayer range:

(1) 2D A islands on B: such islands occur for attractive AA interactions ($v_n < 0$) when $\gamma_1 < E_A^{coh} + E_{A/B}^{ads} < 0$: the first inequality expresses the stability of A islands as compared to an homogenous coverage of the B surface by A atoms and vacant adsorption sites, the last one insures that the growth is bidimensional.

(2) Ordered and disordered states: an homogeneous coverage of the surface by a disordered arrangement of the atoms A occurs when $E_A^{coh} + E_{A/B}^{ads} < \gamma_1$ and when $\gamma_{ord} > 0$ for all possible surface structures (2×2, c(2×2)...). For large repulsive interactions between first neighbours it's possible to stabilize ordered arrays so that at equilibrium the surface is covered by OS and by uncovered or completely covered domains. The nature of the most stable ordered structures for a given θ is known for short-ranged interactions and depends on the relative values of v_1, v_2... (see section 2.2.5.4) (Fig. 27).

3.9 Structural, chemical, elastic and relaxation contributions to the interfacial energies

In this section, we consider only complete overlayers and we split the spreading energies γ_ν, into its separate contributions. The bulk A and B crystals have different equilibrium crystallographic structures (α, β resp.) and different equilibrium volumes (Ω_A, Ω_B resp.). In order to get the energy difference $\Delta E_\nu = E - N_B E_B^0 - N_A E_A^0$ between the substrate covered by ν A overlayers and the separate bulk crystals (i) we first transform the crystal A into the structure β at the volume Ω_A, (ii) we compress (or expand) the crystal A to the volume Ω_B, (iii) then we arrange the atoms on the substrate to get ν A overlayers without any relaxation, finally (iv) the adsorbate and substrate atoms are allowed to relax to their equilibrium position. ΔE_ν, γ_ν and $\gamma_{1\nu}$ (cf (3.12) (3.13)) are thus splitted into four different contributions.

For example

$$\gamma_\nu = \gamma_\nu^{struc} + \gamma_\nu^{el} + \gamma_\nu^{chem} + \gamma_\nu^{rel} \quad (3.16)$$

The structural contribution γ_ν^{struc} comes from the change of structure of A from its equilibrium structure α to the equilibrium structure β of B

$$\gamma_\nu^{struc} = \nu\left[E_A^\beta(\Omega_A) - E_A^0\right] \quad (3.17)$$

$E_A^\beta(\Omega_A)$ is the energy (par atom) of the bulk A with the structure β and the volume Ω_A. The elastic contribution comes from the hydrostatic compression (or dilatation) of A to the equilibirum volume of B, Ω_B

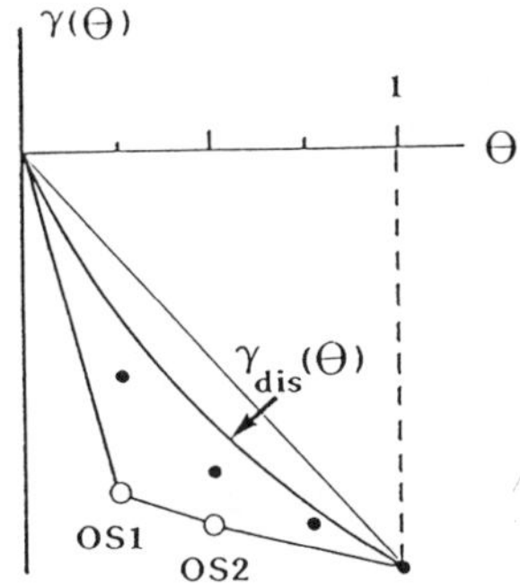

Fig. 27 Spreading energy $\gamma(\theta)$ for disordered and ordered two-dimensional arrays of adsorbates A (schematic). OS1, OS2 are stable ordered structures.

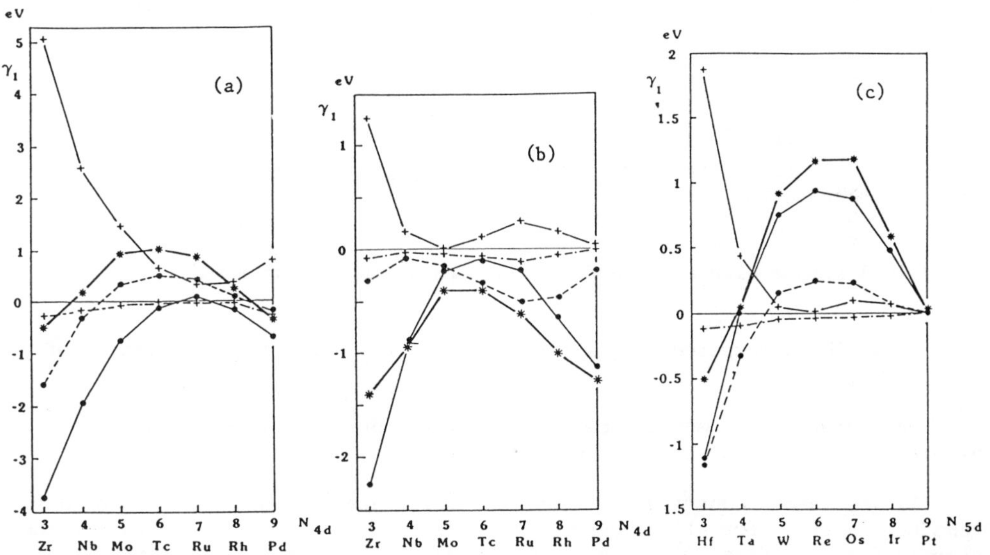

Fig. 28 The contributions to the spreading energies γ_1 of the 4d metals on Fe(100) (a), on W(100) (b) and of the 5d elements on Pt(100) (c) $\gamma_A(\Omega_B^\beta) - \gamma_B$ (•—•), $\gamma_{1,1}^{chem}$(•----•), γ_1^{el} (+—+) and γ_1^{rel} (+—•—•—+), γ_1 (—*—*) (from [99]).*

$$\gamma_\nu^{el} = \nu\left[E_A^\beta(\Omega_B) - E_A^\beta(\Omega_A)\right] \tag{3.18}$$

The chemical contribution γ_ν^{chem} comes from the adsorbate-substrate interactions at constant lattice structure and γ_ν^{rel} is the corresponding relaxation energy. γ_ν^{chem} is related to the interface energy $\gamma_{1\nu}^{chem}$ by

$$\gamma_\nu^{chem} = \gamma_A^\beta(\Omega_B) - \gamma_B + \gamma_{1\nu}^{chem} \tag{3.19a}$$

so that the spreading energy γ_ν is given by

$$\gamma_\nu = \gamma_\nu^{struc} + \gamma_\nu^{el} + \gamma_\nu^{rel} + \gamma_A^\beta(\Omega_B) - \gamma_B + \gamma_{1\nu}^{chem} \tag{3.19b}$$

The chemical contribution can be directly related to the energy changes of each layer induced by deposition. For example, when $\nu = 1$

$$\gamma_1^{chem} = E_A - E_A^\beta(\Omega_B) + \sum_{n\geq 1}\left(E_B^n - E_{Bs}^n\right) \tag{3.19c}$$

where E_A is the energy (per atom) of the overlayer, E_B^n the energy of the n-th underlayer, E_{Bs}^n the energy of the n-th substrate layer before deposition. Its sign results from the competition between (i) the energy loss of an atom A (in the bulk) when it's deposited on the susbtrate, (ii) the energy gain induced in the substrate by the deposit. The interface energy $\gamma_{1,1}^{chem}$ can also be written in a similar way (see (3.19a), (3.19c)); for example, when the surface is assumed to come only from the surface layer and the interface energy from the first two layers

$$\gamma_{1,1}^{chem} = E_A - E_{As}^1 + E_B^1 - E_B^0 \tag{3.19d}$$

Each of the contributions to γ_ν can be splitted as the total energy itself (see (1.5), (2.2)) into its band structure ($\gamma_{\nu bs}^{chem}$...), repulsive ($\gamma_{\nu r}^{chem}$...), exchange, correlation... parts.

3.10 A simple electronic model to determine the growth mode

3.10.1 The assumptions of a tight binding model for the calculation of γ_ν and $\gamma_{1\nu}$

The interfacial energy can be easily obtained from a tight binding calculation using several approximations:

(1) The band contribution can be calculated assuming that the charge transfer is zero for all layers (see section 2.2); the band widths are then determined either from LDA calculations (model 1) or from a fit to the cohesive energies (model 2). The other contributions are neglected in this last model 2 which must be understood as taking into account phenomenologically the repulsive and correlation energies.

(2) The repulsive and relaxation energies and their contributions to the interfacial energies are determined in the model 1 from the minimization of the total energy as done in section 2 for the study of the relaxation and reconstruction of perfect surfaces. The AA and BB repulsive energies (A_i and p_i parameters (see section 1 and formula (1.6)) are obtained using the previous band energies so that the calculated equilibrium lattice constants and bulk

moduli reproduce the experimental results. The transfer integrals t(R) are assumed to vary with distance as R^{-q} where $3 \leq q \leq 5$ according to the considered transition metal. In this work we consider multilayer relaxation up to the third layer and we assume that the repulsive energy between A and B atoms is given by geometrical mean of the corresponding A and B energies in the same way as used for transition integrals. The results we obtain for this model allow to study the importance of atomic rearrangements. We do not take into account the structural contribution γ_ν^{str} we assume to be negligible in agreement with the structural energy differences obtained from the study of the binary phase diagram ($\gamma_1^{str} < 0.1$ eV).

As previously mentioned (section 2.1.3), the surface energies we obtain in this way are smaller than the values "estimated" at 0 K from experiment (up to 30 % for half-filled bands and 50 % for the early and late transition metals). However, their variations along a series and from a series to another one agree with experiment. This crude approxiamtion for the repulsive energy is not important for the determination of the growth mode. It has been verified that the band structure contribution determines the sign of γ_1 except in some exceptional cases for which the γ^{el} is very large. It has been introduced mostly to get relevant orders of magnitude for all the contributions to γ_ν using a single approximation.

3.10.2 Discussion of the results

We calculated band contribution in some special cases (Cr/Fe, Fe/Cr, Co/Cr, Fe/W, Nb/Ta, V/Rh...) using the recursion method to (1) discuss the features of the densities of states in typical situations, (2) study the overlayer magnetism and (3) test the validity of the replacement of $n_i(\epsilon)$ by a rectangular density of states. The comparison between both calculations (Table 11) shows this approximation is valid for a study of wetting phenomena so that the calculations we report are done using rectangular band shapes for the model 1. The results can be summarized as follows (see Tables 12):

$\gamma_{1\ bs}^{chem}$ (eV)	(a)	(b)
Cr/Fe(100)	1.13	1.57
Cr/Fe(110)	0.62	0.91
Fe/Cr(001)	- 0.83	- 1.42
Fe/Cr(110)	- 0.46	- 0.68
Fe/W(001)	- 2.26	- 2.12
Fe/W(110)	- 1.29	- 1.00
Co/Cr(001)	- 1.09	- 1.50
Co/Pt(001)	- 0.36	- 0.11
Co/Pt(111)	- 0.30	- 0.07

Table 11 Band contribution to the spreading energy γ_1 (non magnetic solution): (a) model 1 with rectangular densities of states, (b) recursion method.

(1) Validity of the criterion $\gamma_A < \gamma_B$

The phenomenological criterion (3.11) for the wetting of B by a monolayer of A is roughly in agreement with the results of the electronic theory -even if there is a lot of exceptions to this rule when $\gamma_A - \gamma_B$ is relatively small (section 3.11 below). Such a criterion explains the general trends observed in

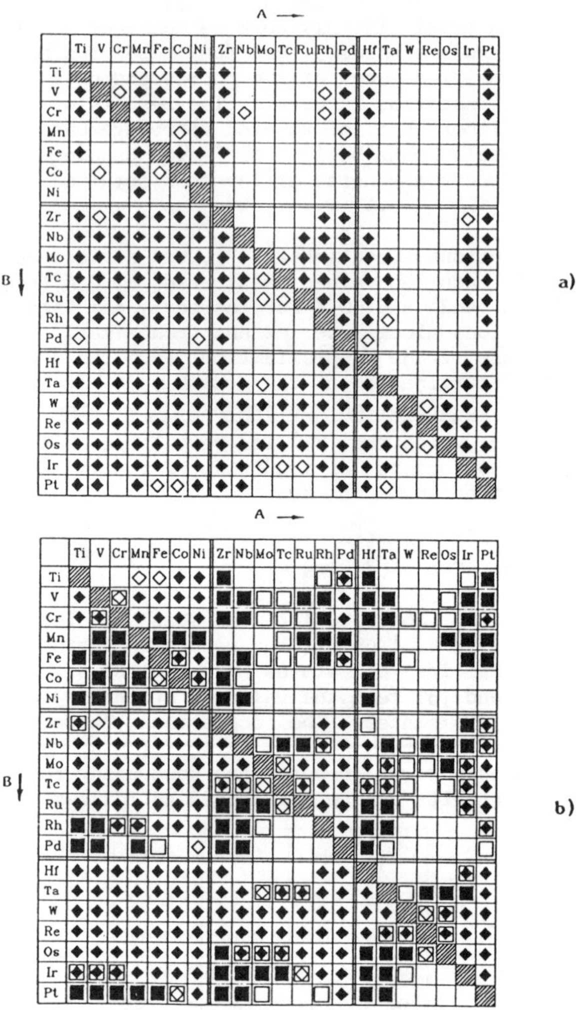

Table 12a Stability of a monolayer of A/B(100) with respect to a cluster of A deposited on B (from [99]):
◆ $\gamma_1 < -0.1$ eV/at, ◇ $-0.1 < \gamma_1 < 0.1$ eV.
The clusters of A are the most stable (Volmer-Weber growth) for other systems ($\gamma_1 > 0.1$ eV).

Table 12b Ranges of stability of a monolayer of A deposited on B (◆ $\gamma_1 < -0.1$ eV, $\gamma_{OS} - \gamma_1 > 0.1$ eV) and of an ordered two dimensional compound
(■ $\gamma_1(OS) < -0.1$ eV, $\gamma_1 - \gamma_1(OS) > 0.1$ eV) (from [99]).
The other systems correspond either to stable clusters of A deposited on B (Volmer-Weber) (white squares) or to systems for which two nearly degenerate configurations are obtained
(◆■ $|\gamma_1 - \gamma_1(OS)| \leq 0.1$ eV, $\gamma_1 < -0.1$ eV or $\gamma_1(OS) < -0.1$ eV;
◇ -0.1 eV $< \gamma_1 < 0.1$ eV and $\gamma_1(OS) - \gamma_1 > 0.1$ eV;
□ -0.1 eV $\leq \gamma_1(OS) \leq 0.1$ eV and $\gamma_1 - \gamma_1(OS) > 0.1$ eV;
◇□ -0.1 eV $< \gamma_1 < 0.1$ eV and -0.1 eV $< \gamma_1(OS) < 0.1$ eV).

Table 12: for example, the wetting condition $\gamma_1 < 0$ is satisfied for the "4d" transition metals on most of the "5d" metals, for "3d" elements on most of the 4d and 5d elements, for all the transition metals on W(100), etc... The same trends are observed using the model 2. The phenomenological Bauer - Van der Merwe criterion (see section 3.6) is thus "justified" from the electronic structure. This is quite surprising: γ_1 is not directly related to $\gamma_A - \gamma_B$ even in the simplest Ising model (see section 3.5 (5)) and results from a subtle balance between its various contributions as shown below.

(2) <u>The contributions to the spreading coefficient</u> γ_1 <u>and the order of magnitude of the "interface energy"</u> γ_{11}

If the signs of γ_1, of $\gamma_A - \gamma_B$ and of $\gamma_A(\Omega_B) - \gamma_B$ are often identic, their values can be very different. The interface γ_{11} and elastic terms are of the same order of magnitude as the surface energies and are thus essential to determine the spreading energy γ_1 -even if they are competing. The relaxation and structural contributions are much smaller in general and do not contribute significantly to γ_1. For illustration, we report in Fig. 28 three typical situations corresponding to (1) 3d transition metals on paramagnetic Fe(100), (2) 4d transition metals on W(100) and (3) "5d" transition metals on Pt(100). The variation of γ_1 is determined mostly by the interface energy γ_{11} in the first case and by $\gamma_A(\Omega_B) - \gamma_B$ in the second one. Both γ_{11} and $\gamma_A(\Omega_B) - \gamma_B$ contribute to γ_1 in the third case. Therefore, <u>the interface energy</u> can be essential to determine γ_1.

(3) <u>The simultaneous wetting of A by B and of B by A</u> is possible in a number of situations for which γ_A and γ_B are not too much different. Let us mention for example the pairs NbRu, PdZr, PtNb, NbRh etc...

(4) <u>The magnetic contribution to</u> γ_1

We investigated the importance of magnetism for wetting in some special situations (FeCr, CoCr, FeW...). We report our results for the band contribution γ_1 in Table 13: the values of γ_1 are slightly modified by magnetism but the sign of γ_1 remains generally unmodified.

γ_1^{chem} (eV)	(a)	(b)
Fe/Cr(001)	- 1.42	- 0.91
Fe/Cr(110)	- 0.68	- 0.31
Cr/Fe(001)	1.57	1.36
Cr/Fe(110)	0.91	0.68
Co/Cr(001)	- 1.50	- 1.88
Co/Pt(001)	- 0.11	- 0.68
Co/Pt(111)	- 0.07	- 0.66
Fe/W(110)	- 0.99	- 0.84

Table 13 Band contribution to the spreading energy γ_1 *(a) paramagnetic, (b) magnetically ordered (recursion method).*

(5) <u>Stranski-Krastanov growth</u>

The elastic interactions and the short range of the interfacial covalent bonding lead -when $\gamma_1 < 0$- to a Stranski-Krastanov mode with n = 1 or 2 (see 3.1). We report some results we obtained for the band contribution γ_1^{chem} by the recursion method. When the energy gain (or loss) $\gamma_{p+1} - \gamma_p$ becomes small, the present calculation is no more significant and "ab initio" calculations including "sp" electrons, exchange correlation effects... must be done.

A	B	γ^{bs}_{AB}	$\gamma^{chem}_{1\ bs}$	$\gamma^{chem}_{2\ bs}$
V	Rh	- 0.6	- 1.07	- 0.8
Ta	Nb	- 0.06	- 0.36	- 0.33
Mo	Ni	-	- 0.23	+ 0.45

Table 14 Interfacial energies γ^{bs}_{AB} (eV) and band contributions to the spreading energies γ_i (eV) for fcc VRh and bcc Ta/Rh and Mo/Nb interfaces from [100].

(6) Stability of an overlayer with respect to the formation of two or three dimensional ordered compounds

Up to now, we assumed that only surface diffusion was possible to realize the most stable atomic configuration of the deposited atoms. However, this configuration can be unstable with respect to (i) the formation of two dimensional ordered structures by an exchange of A and B atoms in the first surface layers, (ii) the formation of three dimensional compounds AB_n deposited on B, (iii) the dissolution of the A atoms in the bulk. For example, even if the substrate is, in principle, wetted by an overlayer of A atoms ($\gamma_1 < 0$), a three dimensional compound AB_n can be observed at the surface of the substrate if $\Delta E_f(AB_n) < \gamma_1$ where $\Delta E_f(AB_n) = E(AB_n) - E^0_A - n\ E^0_B$ is the formation energy of the compound with respect to its bulk constituents. The dissolution of A in the substrate can occur when $\Delta E^{sol}_{A\ in\ B} < \gamma_1$ where $\Delta E^{sol}_{A\ in\ B}$ is the energy of dissolution (per atom) and a two dimensional ordered compound (OS) is stable when $\gamma_1(OS) < \gamma_1$.

Such effects are only possible when the substrate temperature T_s is sufficient to allow interdiffusion either at short distances (surface compounds), or at large distances. Apart from kinetic conditions which depend on the mechanisms for interdiffusion and on the experimental conditions (T_s, Φ, annealing processes...), the selection of the possible ordered compounds is determined by a calculation of the relative stability of all these compounds. Although it's possible to determine in principle the most stable two dimensional ordered structures in the GPM framework [135] as previously done for bulk compounds, we will consider here for illustration only the stability of a monolayer with respect to the most simple 2D structures.

(i) For fcc (001) substrates, we allow an exchange of A and B atoms on the first two (001) layers only and initiate the $L1_0$ structure by occupying the sites in the (100) planes of these layers successively by A and B atoms.

(ii) For bcc(100) substrates we allow an exchange of all the deposited A atoms ($\nu = 1$) with all the B atoms of the first underlayer ($\nu = 2$) to initiate a surface B_2 ordered structure. Table 12b shows the most stable structures we calculated as previously for all transition metal pairs. The general trends we discussed previously for the occurrence of stable monolayers remain qualitatively valid (compare Tables 12a and 12b): (i) surface ordered compounds can be stable and "wet" the substrate when a Volmer-Weber growth is predicted ($\gamma_1 > 0$) for quenched interdiffusion (Ir/V, Nb, Re/Nb, Ta, Nb, V/Pd, Cr/Fe...), (ii) the two dimensional ordered structures occur mostly when the bulk ordered compounds are very stable ($|\Delta E_f|$ large) and sometimes when the bulk contributions are miscible; they are not stable when a miscibility gap is found for the bulk alloys [135c].

These first calculations must be extended to determine the most stable surface ordered compounds, their stability with respect to bulk compounds, etc... Note finally that -to our knowledge- no microscopic study of the mechanism for interdiffusion near surfaces has been achieved up to now although this would be very useful to understand the growth of ultrathin films (see section 3.11 below).

3.11 Growth of ultrathin films: an experimental overview

A detailed knowledge of the atomic structure of ultrathin films, of their growth and morphology is difficult to get experimentally. It's "a priori" necessary to determine both: (1) the atomic concentrations at the surface and in the substrate (Auger experiments), (2) the atomic arrangement of the interface (long- and short-range orders by LEED, SEXAFS, photodiffraction...), (3) the interface relaxation with -for example- the possible existence of misfit dislocations and the observation of the critical height h_c (ion scattering experiments, electron microscopy), (4) the interface relaxation of the strained overlayers (going from bcc to bct for example) and the possibility of structural phase transitions when the film thickness increases, (5) <u>the stability</u> of the observed ordered structure and film morphology by modifications of the deposition temperature incident flux and by thermal annealings, (6) the respective roles of interdiffusion and of the surface defects (see section 2.3.4.2).

Combined LEED and AES studies for given deposition temperature T_s and flux Φ are not sufficient to conclude without ambiguity concerning the growth mode; this is why controversial results are often found in the literature. In this section we try to summarize the experimental results in relation with the theoretical work presented in section 3.10. We select in Table 15 some recent studies we think important for the understanding of growth of ultrathin films. We comment below the experimental results.

(1) <u>Pseudomorphic (1×1) monolayers (ML) A/B</u>

Such ML have been obtained for systems such that $\gamma_A < \gamma_B$. Two exceptions must be mentioned:

(i) transition metals of the first series (Cr, Fe, Co, Ni) on noble metals (Cu, Ag, Au) which, apart from Co/Au, diffuse in the substrate at relatively low temperatures (see section 2.3.4.2),

(ii) some transition metals on palladium (A = Mo, Co, Fe, Pt) and on platinum (A = Re, W).

The experimental results are consistent with theory except for W/Pt(100), Re/Pt(100). The "sp" electrons and the sp-d hybridization has not been included in the rough calculations we summarized in section 3.10. They contribute significantly to the cohesive and surface energies of the noble metals and to a lesser extent of Ni, Pd and Pt: they must be taken into account. Note finally that a comparison between theory and experiment is <u>significant</u> only when the <u>stability</u> of the pseudomorphic monolayer has been verified and when interdiffusion does not occur.

(2) <u>Stranski-Krastanov growth and stability of the pseudomorphic ML</u>

For a number of A/B systems it has been possible to get overlayers for $\nu \leq n$ with large n values but, when a study of the stability has been completed (thermal annealing), it has been shown that these films are metastable: the first (and perhaps the second) layer only is stable up to high temperatures. Such studies have been performed for thin films on W(100) and W(110): as an example we summarize briefly some of these results.

(i) **Cr/W(110) and Cr/W(100) [138]**

A FM growth is observed on both substrates at low temperatures (120 K <

A	W	
Zr	(100)	[137]
Cr	(100)(110)	[138]
Fe	(110)	[139]
Co	(100)(110)	[140]
Ni	(100)(110)	[141]
	(111)	[142]
Pd	(110)	[143]
Cu	(100)	[144]
Ag	(100)	[145]
Au	(100)	[146]

A	Pd	
Mo	(110)	[154]
Mn	(001)	[155]
Fe	(111)	[156]
Co	(001)	[157]
Ni	(111)	[158]
Pt	(100)	[159]
Ag	(100)	[160]
Au	(111)	[161]

A	Nb	
Pd	(110)	[151]
Pt	(110)	[152]
Ag	(110)	[153]
Au	(110)	[153]

A	Mo	
Cu	(110 (100)	[148, 147]
Ag	(100)	[148]
Au	(110)	[149]

A	Ru	
Mn	(0001)	[170]
Fe	(0001)	[171]
	$(10\bar{1}0)$	[172]
Ni	(1000)	[173]
Pd	(1000)	[174]
Ag	(1000)	[174,175]

A	Pt	
Re	(111)(100)	[162]
W	(100)	[163]
Au	(100)	[164]
	(111)	[165]
Ag	(111)(100)	[165]
Cu	(111)	[166]

A	Ta	
Pd	(110)	[150]

A	Ni	
Fe	(100)	[167]
Co	(100)	[168]
Cu	(100)	[169]

Table 15 Ultrathin films on transition metals (Stranski-Krastanov growth).

$T < 400$ K). On W(110) a pseudomorphic (1×1) ML forms at 100 K and is stable up to desorption at 1290 K; a metastable (2×2) structure is observed for $\theta \sim 2$ (500 K < T_s < 800 K) but for $\theta \gtrsim$ 2ML the film is unstable above 400 K and forms 3D clusters which coexist with the (1×1) ML and (2×2) bilayer. On W(100), the pseudomorphic Cr ML remains also stable up to desorption. Increasing θ, a 2ML Cr/W(100) is stable up to 1100 K but is not well ordered (even at high temperatures) and thicker films are unstable above 500 K and form 3D clusters on top of the pseudomorphic Cr ML.

(ii) **Co/W(110) and Co/W(100) [140]**
The structure and stability of cobalt overlayers are qualitatively the same as those we reported for Cr/W. The first pseudomorphic ML is stable up to desorption; the second and third ML grow layer by layer at low temperatures ($T_s \sim 100$ K) and form 3D clusters above 400 K - 500 K.

(iii) **Fe/W(110) [139]**
A detailed study of the growth and stability of Fe/W(100) has been done in relation with magnetic properties of quasi 2D Fe. For iron films prepared at room temperature, the first two monolayers are pseudomorphic on W(110). The stability of the deposit is shown by the fact that the Curie temperature T_{co} = 282 K is independent of annealing up to 800 K. This value of T_c remains also constant in the submonolayer range ($0.5 < \theta < 1$) and equal to T_{co} if the film is prepared at $T_s \geq 200$°C: this suggests that the attractive interactions between iron adsorbate stabilize 2D Fe islands which are so large (when the surface diffusion i.e. T_s is sufficient) that $T_c \simeq T_{co}$.

(3) <u>Growth mode and deposition temperature</u>

The film morphology can be controlled by the substrate temperature T_s during deposition: such a study has been performed for Fe/Pd(111) [156] to grow either flat monolayer platelets, 2D bilayers or islanded multilayer films. Multilayer films are obtained for a deposition on a cold substrate ($T < 300$ K) whereas ML platelets are obtained either by evaporating on a hot substrate ($T_s \simeq 450$ K) or by an annealing of islanded multilayer films for low coverages ($\theta < 0.5$). Iron overlayers are thus growing in a complex way, the growth mode changing significantly with the substrate temperature.

(4) <u>Epitaxial growth, surface defects and atomic steps</u>

The effect of the surface defects on epitaxial growth has been pointed out in section 2.3.4.2 in relation with magnetic properties. A remarkable study of such effects has been recently done using low energy electron microscopy [147], a technique whose resolution is atomic in the direction normal to the surface. The roles of the substrate temperature and of the atomic steps are shown by the micrographs concerning the submonolayer growth of Cu on Mo(110):

(i) At high temperatures (T_s = 700 K) Cu nucleates at the steps and grows anisotropically onto the terraces so that the first layer is pseudomorphic; the second layer is complex, the growth being enhanced in the Mo(100) direction and the islands becoming needle-like and upon completion of the second layer 3D islands grow.

(ii) On the contrary, at low temperatures (T = 500 K), islands grow on terraces and further growth is layer by layer within each terrace.

(5) <u>Metastable phases</u>

Metastable phases can be grown epitaxially for large thickness (see Table 1 of section 1) when the misfit f is relatively small, but the several growth regimes can be observed even if the crystalline structure is relatively perfect, sufficiently far from the interface. For example, a study of the epitaxial growth of bct Mn on Pd(001) at room temperature [155] shows a progressive change of the LEED pattern: starting from the (1×1) pattern corresponding to a pseudomorphic ML, the pattern becomes progressively c(2×2), θ increasing from .5ML to 2ML with relatively high background. The symmetry of the pattern is again (1×1) for $\theta > 7$ up to $\theta = 21$.

(6) Atomic structure and film thickness

Stable pseudomorphic monolayers can be observed for large misfits(f = 9.4 % for Fe/W(110) or 4.8 % for Au/Pd(111)...) despite the large strain energies. However, a sharp transition from the pseudomorphic state to a state with misfit dislocations must be observed in such cases for very small critical heights for h_c. This transition has been observed by ion channeling for Au/Pd(111); the growth mode is layer by layer at room temperature but the shadowing of the scattering from the substrate by the overlayer shows that the critical thickness is about 1ML ($h_c \sim 2$ Å). When the lattice misfit is much smaller (f ~ 2.6 % for Cu/Ni(100) for example), the pseudomorphic regime extends up to larger film heights: a study of the Auger electron diffraction vs film thickness has shown for example that the critical thickness is about $h_c = 14$ Å for Cu/Ni(100). Structural transitions can also occur for ultrathin films when bulk A on B crystals present different equilibrium crystalline structures; the most popular example is the deposition of (111) fcc overlayers on (110) bcc susbtrates as extensively studied and classified by Bauer and Van der Merwe [133]. First order phase transitions from a pseudomorphic ML to incommensurate structures have been observed for Pd/Nb(110) [151] ($\theta > 1.2$) and have been shown to introduce significant modifications of the electronic structure (photoemission experiments and TB calculations [151b]). Similar transitions have been observed for Pt/Nb(110) [152]. At T_s = 300 K the Pt atoms form commensurate 2D islands which become incommensurate. Let us now summarize for illustration the atomic structures observed for Ni(111) films deposited on W(110) at 300°C (for the 1st ML) and 170°C for the other layers [142]. For $\theta < 0.5$ ML the nickel grows epitaxially. For higher coverages, the growth remains pseudomorphic in the W[$1\bar{1}0$] direction (f([$1\bar{1}0$] $\simeq$ 3.6 %) but such a growth is no more possible in the W[001] direction due to the high mistfit in this direction (f([001] $\simeq$ 21.5 % !). By a slight compression ($\simeq$ 12 %) of the nickel lattice along W[001], each ninth Ni atom coincides with each seventh W atom so that a (7×1) superstructure is observed by LEED up to $\theta \simeq$ 3ML. For thicker films the nickel structure changes progressively to a slightly disorted fcc structure.

3.11 Interface energies and electronic structure

In this section, we study the band contribution to the interface energy γ^{bs}_{AB} using the simplest possible band model. We assume that the epitaxy is perfect between two semi-infinite perfect crystals A and B with bulk Fermi energies ϵ^A_F and ϵ^B_F respectively. Once put into contact a dipolar layer results from the flow of electrons from the elements with the highest energy level to the other one and the Fermi levels become aligned and equal to ϵ_F. The potential energy difference between the two sides of the junction has equal magnitude and opposite sign to the difference between the Fermi levels. If V_A and V_B are the electrostatic potentials, thus

$$V_A = -V_B \text{ and } e\,V_A - e\,V_B = U_A - U_B = \epsilon^B_F - \epsilon^A_F \qquad (3.20)$$

$\epsilon_F = \dfrac{\epsilon^A_F + \epsilon^B_F}{2}$ being the new Fermi level. In order to determine the charge transfer and according to the simplest possible Allan-Lannoo-Dobrzynski model (ALD) [136] we assume that the charge transfers $\delta N_A = -\delta N_B$ are limited to the A and B interface layers and that these interface form a planar capacitor. Then,

$$\delta N_A = b(U_A - U_B) \text{ with } b = \frac{1}{4\pi}\left(\frac{S}{Nd}\right) \qquad (3.21)$$

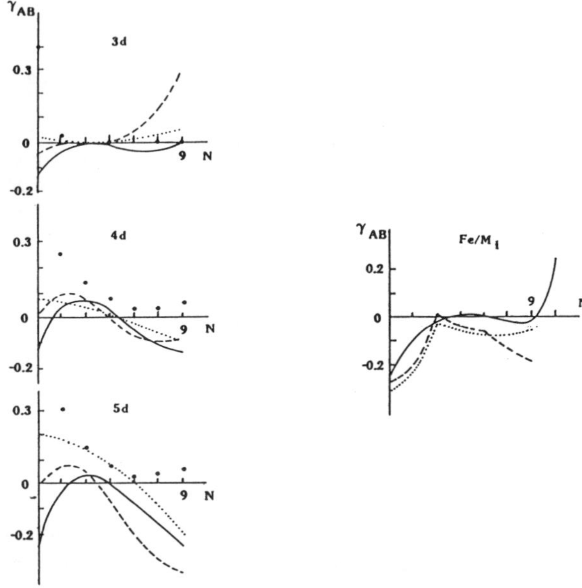

Fig. 29a Interface energy γ_{AB}^{bs} *(B = Cr(110) for constant "d" band width, (...) taking into account the volume dependence of the band widths only (---) and both the volume dependence of the band widths and of the positions of the "d" band centers (—).*

Fig. 29b Interface energies γ_{AB} *calculated by Gerkema and Miedema for the first (full line), second (dotted line) and third transition series (dashed line) on iron [170b].*

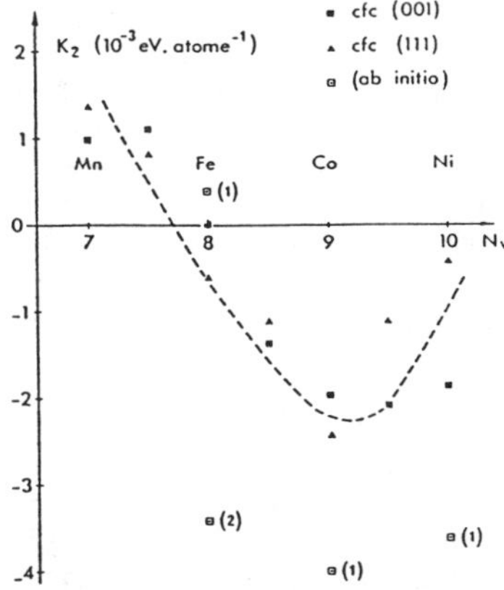

Fig. 30 Anisotropy constant K_S *calculated for monoatomic fcc(100) and (111) films as a function of the number of electrons N [185]. The results obtained in ref. [186] (□(1)) and [187] (□(2)) are also indicated (from [170]).*

where $\frac{S}{N}$ is the surface area per atom. The equation (3.21) and the local density of states $n_A(\epsilon)$ determine self-consistently

$$\delta N_A = \int_{-\infty}^{\epsilon_F} n_A(\epsilon)\, d\epsilon - N_A \tag{3.22}$$

As assumed previously for ultrathin films, we use, as a first step, rectangular band shapes whose width is fitted to the exact moment of the interface layers. Taking into account the band contribution to γ_{AB} and the Coulomb energies we have taken into account twice in the band contribution we get

$$\begin{cases} \gamma_{AB}^{bs} = \gamma_{AB}^{(n)} + \gamma_{AB}^{(ct)} \\ \gamma_{AB}^{(n)} = -\dfrac{\delta W_A^I}{W_A} E_{CA} - \dfrac{\delta W_B^I}{W_B} E_{CB} \\ \gamma_{AB}^{(ct)} = \delta N_A \left[\dfrac{U_A - U_B + \delta_A - \delta_B}{2} + \dfrac{W_A^I}{10}(5-N_A) + \dfrac{W_B^I}{10}(5-N_B) + \dfrac{W_A^I + W_B^I}{20}\delta N_A\right] \end{cases} \tag{3.23}$$

In this expression the first terms $\gamma^{(n)}$ come from the change of binding energies of neutral interface layers. The last one $\gamma_{AB}^{(ct)}$ comes from the charge transfer accross the interface. W_A^I and W_B^I are the interface bandwidths and $\delta W_A^I = W_A^I - W_A$, $\delta W_B^I = W_B^I - W_B$ where W_A and W_B are the bandwidths at the common interface volume, E_{CA} and E_{CB} are the contribution to the cohesive energies in the Friedel's model (section 1); δ_A and δ_B are the self-consistent intraatomic potentials we introduce to achieve charge neutrality within the interface layers. For surfaces such that the A and B bandwidths are nearly the same, the only term which is going to contribute to γ_{AB} is γ_{AB}^{ct}. However, the change in bandwidth (at constant volume) $W_A - W_B$ which occurs not only from one series to another one but also within the same series introduces large values of $\gamma_{AB}^{(n)}$. This term is dominant when the Fermi levels are nearly aligned so that $\delta N_A \simeq 0$.

The band contribution γ_{AB}^{bs} to interface energies γ_{AB} were calculated within the ALD model with rectangular densities of states and taking into account the variations of the bandwidths within each series for the atomic volume of a reference lattice, the volume dependence of the "d" band centers (section 1 (1.9)) and the correlation effects. The first two factors were shown to be essential to determine reliable orders of magnitude of γ_{AB} (Fig. 29). Up to now, only Germeka and Miedema [170] estimated the electronic contribution to the interface energies relating it to the heat of solution of A in B (and vice versa): $\gamma_{AB}^{chem} \sim \dfrac{\Delta H_{A\ in\ B}^{sol}}{V_A^{2/3}}$ the heat of solution without any mismatch, where $\Delta H_{A\ in\ B}^{sol}$ is derived using a semi-empirical approach and V_A is the molar volume of A at T = 0 K. Comparing the values of γ_{AB}^{chem} given by this semi-empirical approach (Fig. 29b) with our results (Fig. 29a), the trends are similar along the transition series and the orders of magnitude are also comparable: they are an order of magnitude smaller than the surface energies γ_i.

In the previous calculations we considered only the band contribution for a perfect epitaxial interface between two semi-empirical infinite crystals with the same atomic volume (reference medium). In practice, the equilibrium state of the interface is characterized by a periodic array of misfit dislocations (section 3.3.2): the interface energy results from the previous bonding contribution and from a relaxation term γ_{AB}^{rel}. Up to now, only phenomenological models were considered to determine γ_{AB}^{rel}: (i) Van der Merwe et al [131] used linear elasticity and phenomenological interaction energy between the two interfacial planes and obtained γ_{AB} in terms of unknown interfacial moduli, (ii) Germeka and Miedema suggested that this term is roughly given by $\gamma^{rel} = 0.15\ (\gamma_A + \gamma_B)$ [170b].

4 MAGNETIC ANISOTROPY

4.1 Introduction

The magnetic anisotropy comes from magnetostatic terms which depend on the direction of the magnetization relative to the macroscopic shape (shape anisotropy) and to the crystalline axis (magnetocrystalline anisotropy). The surface of the magnetic sample being in general different from a sphere and the local crystalline symmetry being always anisotropic (at most cubic), the spontaneous magnetization of a ferromagnet is along "easy axis directions" to minimize the anisotropic part of the free energy F_m. Two microscopic interactions determine mostly the anisotropy energy: (i) the dipolar interactions contribute to the shape anisotropy, (ii) the spin orbit interactions contribute to the magnetocrystalline anisotropy and define the easy axis directions in bulk crystals. For small particles and thin films both anisotropic interactions are important [171].

The anisotropy energy is represented in terms of "anisotropy constants", K_n, by an expansion in successive powers of the direction cosines α_i (i = 1, 2, 3) of the magnetization with respect to the crystalline axis. This expansion is usually rapidly convergent. It must be consistent with the symmetry of the considered sample and invariant by reversal of the magnetization density $\underline{M}(\underline{r})$. For example, thin films are characterized by a planar (or axial) symmetry so that it's often sufficient to consider only the second order terms

$$F_m = K_2 \sin^2 \theta + K_2' \sin^2 \theta \cos 2\varphi + \dots \qquad (4.1)$$

In this expression, θ and φ are the polar and azimuthal angles of the magnetization $\underline{M}$ with respect to the direction z perpendicular to the film; the magnetization will be either in the plane xy or perpendicular to this plane according to the sign of the anisotropy constants. For an uniaxial symmetry $K_2' = 0$, $\underline{M}$ is in the plane for $K_2 < 0$ and in the z direction for $K_2 > 0$ (equilibrium state). An anisotropy field H_A is then defined from $F_m \simeq K_2\theta^2 \simeq 2K_2\ (1 - \cos\theta) = 2K_2 - MH_A \cos\theta$ by $H_A = \frac{2K_2}{M}$; this field expresses the tendency for the magnetization to go back to the equilibirum direction(s).

For cubic systems, the previous expansion is not sufficient (the second order anisotropy constants are zero by symmetry) and the bulk anisotroopy energy can be written as

$$F_m = K_1 (\alpha_1^2\alpha_2^2 + \alpha_2^2\alpha_3^2 + \alpha_3^2\alpha_1^2) + K_2\alpha_1^2\alpha_2^2\alpha_3^2 + \dots \qquad (4.2)$$

In this section, we consider only the simplest situation i.e. uniaxial thin films. From a phenomenological point of view, the symmetry is broken at the surface of the sample: the low symmetry of the surface leads to much larger anisotropies for the surface than for the bulk as first suggested by Néel [172] and experimentally verified (see below). The anisotropy energy of a thin film results from volume and surface contributions:

$$F_m = [K_V V + 2Sh\, K_S] \sin^2 \theta \qquad (4.3)$$

where K_V and K_S are (resp.) the volume and surface anisotropy constants. V and h are the thin film volume and thickness (V = Sh). The experimental data are usually expressed in terms of an "effective" anisotropy constant K_{eff}

$$F_m = V\, K_{eff} \sin^2\theta, \; K_{eff} = K_V + \frac{2K_S}{h} \qquad (4.4)$$

whose dependence with the film thickness allows to determine K_S and K_V. For large h values, the volume contribution (shape anisotropy) is dominant and the magnetization is parallel to film planes ($K_V < 0$). When h decreases, the surface and volume terms become comparable for $\tilde{h}_c = \left|\frac{2K_S}{K_V}\right|$. It's then possible to get a transition from a parallel $h > \tilde{h}_c$ to a perpendicular magnetization ($h < \tilde{h}_c$) if the surface anisotropy K_S is negative. For illustration, Table 16 summarizes the experimental values of the surface anisotropy for some representative situations. The surface anisotropy is 10^2 to 10^3 larger than the volume anisotropy energy but it's restricted to the surface atoms.

	T	K_S (erg/cm^2)
Ni(111) [173]	300 K	- 0.48
Ni/Re(100) [173]	300 K	- 0.19
αFe/Ag(100) [174]	300 K	+ 0.6
αFe(100) [177]	300 K	+ 1
Fe/W(110) [175]	300 K	- 0.5
Au(111)/Co/Au [176]	5 K	+ 0.7
	291 K	+ 0.5

Table 16

Perpendicular magnetization has been observed for surfaces, epitaxial ultrathin films and for sandwiches (Au/Co/Au(111), $\tilde{h}_c \sim 12$ Å). The anisotropy energy results mostly from the shape anisotropy (K^D) and from he magnetocrystalline anisotropy (K^{MC})

$$K = K^D + K^{MC} \qquad (4.5)$$

we discuss in sections 4.2 and 4.3. However, the magnetoelastic coupling and the surface roughness can introduce also significant contributions to the magnetic anisotropy of thin films (section 4.4).

4.2 Shape anisotropy

The anisotropic character of the dipole dipole interactions leads to a very small contribution to the magnetocrystalline anisotropy, but the long range of these interactions ($\sim r^{-3}$) leads to a shape anisotropy which is essential for finite samples and especially for thin films. In a continuum model, the dipolar contribution to the energy $E_d = -\frac{\underline{M}\,\underline{H}'}{2}$ is expressed as a demagnetizing field energy between the magnetization distribution $\underline{M}$ and the field $\underline{H}'$ created by the volume (ρ = div $\underline{M}$) and surface ($\rho' S = \underline{M}.\hat{\underline{n}}$ where $\hat{\underline{n}}$ is the surface normal unit vector) poles equivalent to this distribution. For a thin film with an uniform magnetization density, the shape anisotropy is given by

$$F_D = VK^D \sin^2\theta \qquad K^D = -2\pi M^2 \qquad (4.6)$$

It's minimum when the magnetization is parallel to the film surface ($\theta = \pi/2$). The continuum approximation was shown to be valid for ultrathin films [178].

4.3 Surface and volume magnetocrystalline anisotropy

The spin orbit coupling ($H_{so} = \xi_d \vec{\ell}.\vec{s}$) is responsible for the magnetic crystalline anisotropy, the orbital moment and the magnetization anisotropy. It's very small for 3d magnetic metals (ξ < 0.1 eV) but it introduces new qualitative features: (1) the one electron states are no more labelled by the spin variables σ ($\sigma = \pm 1$) and energy degeneracies are lifted, (2) the one electron energy levels ϵ_α and the total energy are dependent upon the direction of the magnetization $\hat{m}$ with respect to the crystal axes. The magnetocrystalline anisotropy energy is then otbained from band structure calculations including spin orbit coupling [179 - 185]. It comes from the energy shifts and from the electronic redistribution of the occupied and empty states near the Fermi level induced by ξ. Both contributions are very small and have opposite signs, so that it's difficult to get the anisotropy energy with a high accuracy and to relate its value to the peculiar features of the band structure.

The physical origin of these effects can be easily understood in a tight binding model. The basis orthogonal atomic like functions are noted $|i\underline{\lambda}\sigma\rangle$ where $\underline{\lambda}$ labels the crystalline site, i = 1, 2, 3...5 the orbital type (xy, yz, zx, x^2-y^2, $3z^2-r^2$). x,y,z refere to the crystalline axis Oz being choosen perpendicular to the film plane. The spin quantization is along $\hat{m}$ characterized by the standard angular coordinates (θ, φ). The eigenstates for $\xi_d = 0$ are linear combinations of these orbitals.

$$|\psi^0_{\alpha\sigma}\rangle = \frac{1}{\sqrt{N_A}} \sum_{i\underline{\lambda}} a_{i\underline{\lambda}\sigma} \; |i\underline{\lambda}\sigma\rangle \qquad (4.7)$$

The tight binding hamiltonian H_0 can be written as

$$H_0 = \sum_{\substack{m\sigma \\ \underline{\lambda}}} \epsilon_{m\lambda\sigma} |m\underline{\lambda}\sigma\rangle\langle m\underline{\lambda}\sigma| + \sum_{\substack{m,m' \\ \lambda\neq\mu \\ \sigma,\sigma'}} t^{\lambda\mu}_{mm'} |m\lambda\sigma\rangle\langle m'\mu\sigma'| \qquad (4.8)$$

The diagonal elements (atomic energy levels) are σ dependent for a magnetically ordered system (see section 1.3) whose magnetization is along $\hat{m}$

$$\epsilon_{m\underline{\lambda}\sigma} = \epsilon_{m\underline{\lambda}} - \frac{I}{2} m_{\underline{\lambda}}\sigma \qquad (4.9)$$

They are m dependent from crystalline field effects. For a cubic crystal two energy levels only must be considered

$$\begin{cases} \epsilon_m = \epsilon(\Gamma'_{25}) & m = 1, 2, 3 \\ \epsilon_m = \epsilon(\Gamma_{12}) & m = 4, 5 \end{cases} \qquad (4.10)$$

but these degeneracies are partially lifted for surface atoms. For example, near (100) surfaces $\epsilon_2 = \epsilon_3 \neq \epsilon_1$. The second term in (4.8) depends on the d-d hopping terms t which are expressed in terms of free Slater-Koster integrals $dd\sigma(R)$, $dd\pi(R)$, $dd\delta(R)$ for each $\underline{R} = \underline{\lambda}-\underline{\mu}$ vector. The eigenstates (4.7) are obtained for each spin σ by the diagonalization of the (5×5) matrix (see (4.5)). They are not dependent on the direction $\hat{m}$.

The spin orbit interaction is assumed for simplicity to exist only within the same site

$$H_{so} = \xi_d \sum_{\substack{\lambda m \\ \sigma}} \langle \underline{\lambda} m\sigma | \underline{\ell}.\underline{s} | \underline{\lambda} m'_{\sigma}\sigma' \rangle | \underline{\lambda} m\sigma \rangle \langle \lambda m'\sigma' | \qquad (4.11)$$

The spin orbit matrix elements are independent on the site $\underline{\lambda}$ (for identic atoms) but they are dependent on σ, σ' and on the magnetization direction $\hat{m}$. For example

$$\begin{cases} \langle 1\uparrow | \underline{\ell}.\underline{s} | 2\downarrow \rangle = \dfrac{i}{2} \sin\theta \\ \langle 1\uparrow | \underline{\ell}.\underline{s} | 4\downarrow \rangle = \dfrac{1}{2} (\sin\varphi + i \cos\theta \cos\varphi) \\ \langle 1\uparrow | \underline{\ell}.\underline{s} | 2\downarrow \rangle = - \ i \cos\dfrac{\theta}{2} \end{cases} \qquad (4.12)$$

The eigenstate $|\psi_\alpha\rangle$ and energies ϵ_α are obtained by diagonalization of the (10×10) matrix representing $H = H_0 + H_{so}$: they depend on the magnetization direction $\hat{m}$ (cf (4.12)). The total energy E is then obtained by a summation of the one electron energies substracting the electron-electron terms which were counted twice in the band structure energy. However, the spin orbit coupling ξ_d being very small for "3-d" metals, a perturbation theory with respect to H_{so} will be rapidly convergent ($E = E^{(0)} + E^{(2)} + \ldots$). Moreover, for a study of the surface magnetocrystalline anisotropy K_s^{MC}, the second order term $E^{(2)}$ is non zero -even for a 3D cubic structure- and is the most important contribution to the magnetocrystalline anisotropy. $E^{(2)}$ can be written in terms of the eigenstate $|\psi^0_{\alpha\sigma}\rangle$

$$E^{(2)} = \sum_{\alpha\sigma}{}' \sum_{\beta\sigma'} \frac{|\langle \psi^0_{\alpha\sigma} | H_{so} | \psi^0_{\beta\sigma'} \rangle|^2}{\epsilon^0_{\alpha\sigma} - \epsilon^0_{\beta\sigma'}} \qquad (4.13)$$

where the first sum is taken only over the occupied states $(\alpha\sigma)$. It can be determined directly from (4.13). It can be also expressed in the basis of the atomic functions $|i\lambda\sigma\rangle$ replacing $|\psi^0_{\alpha\sigma}\rangle$ by its expression (4.7)

$$E^{(2)} = \frac{1}{2} \sum_{\underline{\lambda},\underline{\mu}} G_{\underline{\lambda\mu}} \qquad (4.14)$$

$G_{\underline{\lambda\mu}}$ is the interaction energy between two atoms (resp. in $\underline{\lambda}$ and $\underline{\mu}$) which results from the spin orbit interaction. It depends on the location of the pair relative to the surface and on its orientation relative to the magnetization $\hat{m}$. For a perfect bulk cubic ferromagnet these interaction energies are only dependent on the distance $R = |\underline{\lambda} - \underline{\mu}|$ and on the orientation of the pair relative to direction of the magnetization $\hat{m}$ so that they can be expanded in series as first suggested by Néel [172]

$$G_{\lambda\mu} = g_1(R)\ P_2\ (\cos \Phi) + g_2(R)\ P_4\ (\cos \Phi) + \dots \qquad (4.15)$$

The P_n $(\cos \Phi)$ are the Legendre polynomials and Φ is the angle between $\underline{R}$ and $\hat{m}$.

The contribution of each atom $\underline{\lambda}$ to the magnetocrystalline anisotropy is then obtained by adding the interactions of the considered atom with all the others. The first term $g_1(R)\ P_2\ (\cos \Phi_i)$ do not contribute to the <u>bulk</u> magnetocrystalline anisotropy, the sum of the P_2 $(\cos \Phi_i)$ over all the pairs i of length R being zero by symmetry for <u>cubic</u> crystals. This is no more true for atoms on (or near) the surface, the cubic symmetry being locally broken. These terms were assumed to determine mostly the surface anisotropy K_s and the magnetostriction. They were estimated from the experimental values of the magnetostriction keeping only 1st nn interactions and values of K_s between 0.1 and 1.3 erg/cm^2 were then deduced in suprisingly good agreement with the recent experimental data (see Table 16). However, the electronic structure -i.e. the $|\psi^0_{\alpha\sigma}\rangle$- are modified by the surface so that the pair interactions G are also dependent <u>-in principle-</u> on the orientation of the pairs with respect to the surface.

The relation between the interactions G and the electronic structure, their range, their dependence with respect to the surface orientations... have not been investigated up to now.

Some direct calculations of the surface anisotropy K_s^{MC} of monolayers or thin films have been reported recently from a direct summation in the reciprocal space and using either LSDA [186, 187] or tight binding calculations [185]. The results obtained for bcc (100)Fe and for fcc (100) and (111) monolayers are reported in Fig. 30 as a function of the number of "d" electrons.

The tight binding results obtained for fcc (100) and (111) monolayers as a function of the total number of electrons N per atom show that the anisotropy constant is probably weakly dependent on the details of the electronic structure and depends mostly on the band filling. The easy axis of magnetization is found to be in the plane of the monolayer for nearly filled bands ($N > 7.8$) and perpendicular to this plane for smaller N values. A comparison of these results with the "ab initio" LSDA calculations [186, 187] shows that the order of magnitude of the anisotropy constant is the same for all calculations and agrees with experiment. The theoretical calculations predict either a perpendicular or a parallel magnetization for iron according to the authors, this disagreement coming probably from the fact that the change of sign of the anisotropy constant occurs in the vicinity if the iron band filling. An "ab initio" determination of the surface anisotropy is difficult: (1) K_s is strongly sensitive to the crystal field effects which determine the anisotropy of the electronic density: consistent values of this anisotropy are hardly reproduced by LDA calculations so that uncontrolled errors are introduced via the crystal field effects, (2) the spin orbit energy is slowly convergent in the number of $\underline{k}$ vectors used to approximate the summation over the occupied states so that an ab initio calculation for thin films is very time consuming and is not very precise.

4.4. Surface anisotropy: magnetoelastic and surface roughness contributions

The strains due to the lattice misfit f introduce a magnetic surface energy whose origin is now briefly discussed [181 - 189]. The magnetoelastic energy of a film which is homogeneously deformed is given by

$$E_{ME} = V\, B\, \epsilon \sin^2 \theta \qquad (4.16)$$

where B is the magnetoelastic coupling constant, ϵ is the strain parallel to the plane of the film we assume to be isotropic ($\epsilon_{xx} = \epsilon_{yy} = \epsilon$) and V = Sh is the volume. When the film height h is larger than the critical height h_c, the lattice mismatch is relaxed by interfacial dislocations and by the homogeneous strain $\epsilon = f \frac{h_c}{h}$. The misfit dislocations do not contribute to E_{ME}, the average value of the strains introduced by these dislocations being zero. The residual strain ϵ decrasing as h^{-1} (Fig. 25b), its contribution to E_{ME} is a term proportional to the surface of the film. This magnetoelastic surface anisotropy constant $K_{ME} = - B\, f\, h_c$ has been evaluated for (1000) Au/Co/Au films using the experimental values of the elastic and of the magnetostriction constants. This term has the same order of magnitude as the magnetocrystalline term ($K_{ME} \sim 1$ erg/cm^2) and must be considered for a study of the surface anisotropy. In the peculiar case of Au/Co/Au magnetocrystalline (Fig. 30) and magnetoelastic surface energies are competing ($K_{ME} > 0$, $K_{MC} < 0$) so that the observed perpendicular anisotropy was suggested to come from K_{ME}.

The effect of the surface roughness on the dipolar and magnetocrystalline anisotropies has been also studied using a simple qualitative model [190]. The surface roughness is characterized by the mean square deviation from the ideally flat surface, σ, and by the average size ξ of flat areas on the surface. When there is some roughness, magnetic charges appear on the surface for in plane magnetization and give rise to a demagnetizing field and to a corresponding anisotropy term

$$E_{SD} = \frac{3\sigma}{4}\, 2\pi S M^2 \left[1 - f(2\pi \frac{\sigma}{\xi})\right] \sin^2\theta \qquad (4.17)$$

The function f(x) decreases from 1 for x = 0 to 0.1 for x = 2π ($\sigma \sim \xi$) so that K_{SD} increases with the roughness σ and with the aspect ratio $\frac{\sigma}{\xi}$. This dipolar term may be important for rough films.

The effect of roughness on the magnetocrystalline anisotropy has also been studied within the Néel's model. For step atoms the asymmetric character of atoms located on perfect terraces is reduced so that the magnetocrystalline surface anisotropy is reduced by surface roughness $\frac{\Delta K_{MC}}{K_{MC}} \sim - 2\sigma/\xi$.

4.5. Ferromagnetic domains and walls in ultrathin films

The competition between surface anisotropy, exchange and dipolar energies determines the domain and wall formation, the nature of these walls and the spatial variation of the magnetization. Up to now, very few studies of these properties of ferromagnetic monolayers and of films of some monolayers have been reported. For illustration, we mention here only two recent results.

(1) From a theoretical point of view, the range of values of the uniaxial surface anisotropy that leads to ferromagnetic strip domains [191] has been

investigated: if $K_s > \tilde{f}_c E_0$ (where $\tilde{f}_c \lesssim 1$ and $E_0 = 2\pi M^2 a^2$ is the dipolar energy) and if the easy axis is normal to the layer a domain pattern is the most stable, the domain width being very sensitive to the value of $\tilde{f} = |K_s / E_0|$.

(2) From an experimental point of view, the magnetic domain structure can be investigated using a scanning electron microscope with spin polarization analysis of the secondary electrons. This technique using a LEED detector for spin polarization analysis has been recently used with a good resolution (≃ 400 Å) to show that the domain walls in the (100) surface of Fe are Néel-like [192] and to study the domain structure in ultrathin fcc cobalt films epitaxially grown on Cu (100) (see section 2.3.4.2). For these films [193] ($3 < \nu < 20$) the easy axis is found in the plane of the film along the <110> directions and the domain boundaries were shown to exhibit an irregular shape which was related to a very small energy of Néel walls and a reduction of the free energy by configuration entropy effects.

Much remains to be done in this field to explain the magnetization processes of ultrathin films (hysteresis, coercivity, annealings and their thermal dependence...) [171].

ACKNOWLEDGEMENTS

I should like to thank D.Stoeffler and A.M.Llois for their contribution to the study of the growth mode from the electronic structure (section 3), V.Pierron-Bohnes for a preliminary version of these lecture notes and all my colleagues of Strasbourg, Orsay and elsewhere for helpful discussions and comments. Finally I should acknowledge the patience of the other authors and of the editor during the preparation of this text.

REFERENCES

[1] J.Friedel, "Transition Metals: Electronic Structure and the "d" Bands", in "The Physics of Metals", ed. by J.Ziman (1969), Cambridge University Press, Cambridge.

[2] F.Gautier, in "Magnetism of Metals and Alloys", ed. by M.Cyrot, p. 1-244 (North-Holland, 1982).

[3] C.D.Gelatt, H.Ehrenreich and R.E.Watson, Phys. Rev. B 15, 1613 (1977).

[4] D.G.Pettifor, J. Phys. F 7, 613 (1977); ibid 8, 219 (1978).

[5] V.L.Moruzzi, A.R.Williams and J.F.Janak, Phys. Rev. B 15, 1854 (1977); see also "Calculated Electronic Properties of Metals" (Pergamon Press, New-York, 1978).

[6] J.F.Janak and A.R.Williams, Phys. Rev. B 14, 4199 (1976).

[7] For reviews on band structure calculations, see for example:
(a) A.R.Mackintosh and O.K.Andersen, the electronic properties of transition metals, in "Electrons at the Fermi Surface", Chapter IV, ed. by M.Springford (Cambridge University Press, Cambridge, 1980);
(b) O.K.Andersen, O.Jepsen and D.Gloetzel, in "Highlights of Condensed Matter Theory", ed. by F.Bassani, F.Fumi and M.P.Tosi, p. 59 (North-Holland, Amsterdam, 1985).

[8] N.E.Christensen, O.Gunnarsson, O.Jepsen and O.K.Andersen, J. de Physique 49, C8-17 (1988).

[9] J.Friedel and C.M.Sayers, J. de Physique 38, 697 (1977).

[10] F.Kajzar and J.Friedel, J. de Physique 39, 397 (1978).

[11] G.Tréglia, F.Ducastelle and D.Spanjaard, J. de Physique 41, 281 (1980).
[12] A.M.Oles, Phys. Rev. B 23, 271 (1981);
A.M.Oles and G.Stollhoff, Europhys. Lett. 5, 175 (1986); A.M.Oles, Acta Phys. Pol. A 69, 197 (1986) and for a recent review, J. de Physique 49, C8-43 (1988).
[13] F.Ducastelle, J. de Physique 31, 1055 (1970).
[14] W.Kohn and L.J.Sham, Phys. Rev. A 40, 1133 (1965).
[15] U. Von Barth and L.Hedin, J. Phys. C 5, 1629 (1972).
[16] O.K.Andersen, J.Madsen, U.K.Poulsen, O.Jepsen and J.Kollar, Physica 86-88, B 249 (1977).
[17] P.Bagna, O.Jepsen and O.Gunnarsson, Phys. Rev. B 40, 1997 (1989).
[18] V.L.Moruzzi, Phys. Rev. Lett. 57, 2211 (1986).
[19] V.L.Moruzzi, P.M.Marcus, K.Scharz and P.Mohn, Phys. Rev. B 34, 1784 (1986).
[20] P.M.Marcus and V.L.Moruzzi, Phys. Rev. B 38, 6949 (1988).
[21] V.L.Moruzzi, P.M.Marcus and J.Kübler, Phys. Rev. B 39, 6957 (1989).
[22] V.L.Moruzzi, P.M.Marcus and P.C.Pattnaik, Phys. Rev. B 37, 8003 (1988).
[23] V.L.Moruzzi and P.M.Marcus, Phys. Rev. B 38, 1613 (1988) and J. Appl. Phys. 64, 5598 (1988).
[24] G.A.Prinz, Phys. Rev. Lett. 54, 1051 (1985).
Y.U.Idzerda, W.T.Elam, B.T.Konker and G.A.Prinz, Phys. Rev. Lett. 62, 2480 (1989).
[25] S.Chambers, S.B.Anderson, H.W.Chen and J.H.Weaver, Phys. Rev. B 35, 2592 (1987).
[26] S.H.Lu, J.Quinn, D.Tian, F.Jona and P.M.Marcus, Surf. Sci. 209, 364 (1989).
[27] (a) G.A.Prinz and J.J.Krebs, Appl. Phys. Lett. 39, 397 (1987);
G.A.Prinz, G.T.Rado and J.J.Krebs, J. Appl. Phys. 53, 2987 (1982);
J.A.Chambers, F.Xu, H.W.Chen, I.M.Vitomirov, S.R.Anderson and J.H.Weaver, Phys. Rev. B 34, 6605 (1987);
(b) M.Maurer, J.C.Ousset, M.F.Ravet and M.Piechuch, Europhys. Lett. 9, 803 (1989).
[28] (a) B.T.Jonker, J.J.Krebs, G.A.Prinz, Phys. Rev. B 39, 1399 (1989); n = 5 ML, a = 2.89 Å, c = 3.31 Å;
(b) D.Tian, S.C.Wu, F.Jona, P.M.Marcus, Solid State Comm. 70, 199 (1989); n = 20, a = 3.80 Å.
[29] A.Zangwill, "Physics at Surfaces" (Cambridge University Press, New-York (1899).
[30] G.Allan, "The Handbook of Surface and Interfaces", ed. by L.Dobrzynski, p. 299 (Garland STPM Press, New-York, 1978).
[31] J.E.Ingelsfield, Rep. Progr. Phys. 45, 223 (1982).
[32] A.J.Freeman, C.L.Fu, S.Ohnishi, M.Weinert, in "Polarized Electrons in Surface Physics", ed. by R.Feder, p. 3 (World Scientific, Singapore, 1985).
[33] M.C.Desjonquières, F.Cyrot-Lackmann, J. Phys. F 5, 1368 (1975) and J. de Physique 36, L45 (1975).
[34] F.Ducastelle, F.Cyrot-Lackmann, J. Phys. Chem. Sol. 31, 1295 (1970); see for an introduction, F.Gautier, "Propriétés Electroniques des Métaux et Alliages", p. 296 (Masson ed., Paris, 1973).
[35] M.C.Desjonquières, F.Cyrot-Lackmann, Suf. Sci. 50, 257 (1975).
[36] C.L.Fu, S.Ohnishi, H.J.F.Jansen, A.J.Freeman, Phys. Rev. B 31, 1168 (1985).
[37] G.Tréglia, M.C.Desjonquières, D.Spanjaard, J. Phys. C 16, 2407 (1983) and Solid State Comm. 55, 961 (1983).
[38] A.R.Miedema, Zeit. für Metallkunde 69, 287 (1978).
[39] J.S.Luo, B.Legrand, Phys. Rev. B 38, 1728 (1988).
[40] C.L.Fu, S.Ohnishi, E.Wimmer, A.J.Freeman, Phys. Rev. Lett. 53, 675 (1984); C.T.Chan and S.G.Louie, Phys. Rev. B 33, 2861 (1986).
[41] I.K.Robinson, M.S.Altman, P.J.Estrup, in "The Structure of Surfaces II"

p. 137, ed. by J.F. Van der Veen, M.V. Van Hove (Springer Verlag, 1987).
[42] C.L.Fu, A.J.Freeman, Phys. Rev. B 37, 2685 (1988).
[43] B.Legrand, G.Tréglia, M.C.Desjonquières, D.Spanjaard, J. Phys. C 19, 4463 (1986).
[44] G.Allan, Phys. Rev. B 19, 4774 (1979) and Surf. Sci. Rep. 1, 121 (1981); see also in "Magnetic Properties of Low dimensional Systems", ed. by L.M.Falicov and J.L.Moran-Lopez, p. 2 (Springer Verlag, Berlin, 1986).
[45] R.H.Victora, ibid, p. 25.
[46] H.Hasegawa, J. Phys. F 16, 347 (1986), 16, 1555 (1986), 17, 165 (1987), 17, 679 (1987).
[47] U.Gradman, G.Waller, R.Feder, E.Tamura, J. Mag. Mag. Mat. 31-34, 883 (1983).
[48] R.Feder, S.F.Alvarado, E.Tamura, E.Kisker, Surf. Sci. 127, 83 (1983).
[49] E.Tamura, R.Feder, Surf. Sci. 139, 191 (1984).
For a review of SPLEED experiments see, for example, U.Gradman and S.F.Alvarado, Chapter 7, p. 320, in "Polarized Electrons in Surface Physics", ed. by R.Feder (World Scientific, Singapore, 1985).
[50] C.Carbone, S.F.Alvarado, Phys. Rev. B36, 2433 (1987).
[51] F.Meier, D.Pescia, T.Schreiber, Phys. Rev. Lett. 48, 645 (1982).
[52] S.Blügel, D.Pescia, P.H.Dederichs, Phys. Rev. B 39, 1392 (1989).
[53] L.E.Klebanoff, S.W.Robey, G.Liu, D.A.Shirley, Phys. Rev. B30, 1048 (1984).
[54] L.E.Klebanoff, R.M.Victora, L.M.Falicov, D.A.Shirley, Phys. Rev. B 32, 1997 (1985).
[55] L.Rau, J. Mag. Mag. Mat. 30, 141 (1982).
[56] (a) D.Spanjaard, M.C.Desjonquières, Electronic Theory of Chemisorption, in "Interaction of Atoms and Molecules with Surfaces", Plenum Press, New-York (to be published, 1990);
(b) M.C.Desjonquières, J.P.Jardin, D.Spanjaard, J. de Physique 85, 827 (1988).
[57] M.C.Desjonquières, D.Spanjaard, J. Phys. C 15, 4007 (1982); ibid 16, 3389 (1983).
[58] C.Thuault-Cytermann, M.C.Desjonquières, D.Spanjaard, J. Phys. C 16, 5689 (1983).
[59] R.Haydock, V.Heine, M.J.Kelly, J. Phys. C 5, 2845 (1972).
[60] L.D.Landau, E.M.Lifchitz, in "Electrodynamics of Continuous Media", p. 99 (Pergamon Press, New-York, 1960).
[61] J.Kolaczkiewicz, E.Bauer, Phys. Rev. Lett. 53, 485 (1984); Suf. Sci. 151, 333 (1985).
[62] J.Hölzl, F.K.Schultz, in "Work Functions of Metals", Vol. 85 of Springer Trends in Modern Physics, ed. by Höhler (Springer Verlag, Berlin, 1979).
[63] E.V.Albano, H.O.Martin, Phys. Rev. B 35, 7820 (1987).
[64] P.J.Goddart, R.M.Lambert, Surf. Sci. 67, 180 (1977).
[65] V.M.Gavrilynk and V.K.Medvelev, Sov. Phys. Sol. State 8, 1439 (1966).
[66] T.L.Einstein, J.R.Schrieffer, Phys. Rev. B 7, 3629 (1973).
[67] T.L.Einstein (a) Phys. Rev. B 12, 1362 (1975) (b) Surf. Sci. 45, 713 (1974) (c) Phys. Rev. B 16, 3411 (1977) (d) Surf. Sci. 84, L497 (1979).
[68] N.R.Burke, Surf. Sci. 58, 349 (1976).
[69] J.P.Bourdin, M.C.Desjonquières, D.Spanjaard, J.Friedel, Surf. Sci. 157, L345 (1985).
[70] J.P.Bourdin, M.C.Desjonquières, D.Spanjaard, G.Tréglia, Surf. Sci. 181, L183 (1987) and 202, 354 (1988).
[71] A.M.Oles, M.C.Desjonquières, D.Spanjaard, G.Tréglia, J. de Physique 49, C8, 43 (1988).
[72] G.Moraitis, F.Gautier, J.C.Parlebas, J. Phys. F 6, 381 (1975).
[73] F.Ducastelle, F.Cyrot-Lackmann, J. Phys. Chem. Sol. 32, 285 (1971).
[74] A.Menand, J.Gallot, Rev. Phys. Appl. 9, 323 (1974).
[75] I.Toyoshima, G.A.Somorjai, Catal. Rev. Sci. Eng. 19, 105 (1979).
[76] D.W.Bassett, Surf. Sci. 53, 74 (1975), J. Phys. C 9, 2491 (1976).

See also in "Surface Mobilities on Solid Materials: Fundamental Concepts and Applications", ed. by Vu Thien Bink, p. 89 (Plenum Press, New-York, 1983).

[77] M.C.Desjonquières, F.Cyrot-Lackmann, Solid State Comm. 26, 271 (1978).

[78] (a) For a review of the experimental studies on single adsorbed atoms by field ion microsocpy, see T.T.Tsong, Rep. Progr. Phys. 51, 759 (1988); (b) P.R.Schwoebel, S.M.Foiles, C.L.Bisson, G.L.Kellog, Phys. Rev. B 40, 10639 (1989).

[79] T.T.Tsong, Phys. Rev. B 6, 417 (1972).

[80] D.W.Bassett, Surf. Sci. 53, 74 (1975).

[81] T.T.Tsong, R.Casanova, Phys. Rev. B 24, 3063 (1981); H.W.Fink, G.Erlich, Phys. Rev. Lett. 52, 1532 (1984) and J. Chem. Phys. 81, 4657 (1984).

[82] K.H.Lau, W.Kohn, Surf. Sci. 65, 607 (1977).

[83] A.M.Stoneham, Solid State Comm. 24, 425 (1977).

[84] W.Kappus, Z. Phys. B 29, 239 (1978) and 45, 113 (1981).

[85] G.L.Kellog, T.T.Tsong, Surf. Sci. 62, 343 (1977).

[86] J.Kolaczkiewicz, E.Bauer, Surf. Sci. 160, 1 (1985).

[87] (a) W.Kohn, K.H.Lau, Solid State Comm. 18, 553 (1976);Comm. 18, 553 (1976). (b) For the adsorption of alkali metals, see also B.N.J.Persson, L.H.Dubois, Phys. Rev. B 39, 8220 (1989); (c) E.Wimmer, A.J.Freeman, M.Weinert, H.Krakauer, J.R.Hiskes, A.M.Karo, Phys. Rev. Lett. 48, 1128 (1982) and E.Wimmer, A.J.Freeman, J.R.Hiskes, A.M.Karo, Phys. Rev. B 28, 3074 (1983).

[88] O.M.Braun, V.K.Medvelev, Sov. Phys. Usp. 32, 326 (1989).

[89] K.Binder, D.P.Landau, Adv. Chem. Phys. 76, 91 (1989).

[90] A.J.Freeman, C.L.Fu, J. Appl. Phys. 61, 3356 (1987).

[91] C.L.Fu, A.J.Freeman, T.Oguchi, Phys. Rev. Lett. 54, 2700 (1985).

[92] C.L.Fu, A.J.Freeman, J. Mag. Mag. Mat. 69, L1 (1987).

[93] S.Onishi, M.Weinert, A.J.Freeman, Phys. Rev. B 30, 36 (1984).

[94] R.Richter, J.G.Gay, J.R.Smith, Phys. Rev. Lett. 54, 2704 (1985).

[95] S.C.Hong and C.L.Fu, J. Appl. Phys. 63, 3655 (1988).

[96] S.Blügel, M.Weinert, P.H.Dederichs, Phys. Rev. Lett. 60, 1077 (1988).

[97] S.Blügel, Europhys. Lett. 7, 743 (1988).

[98] J.Tersoff, L.M.Falicov, Phys. Rev. B 26, 6186 (1982); R.M.Victora, L.M.Falicov, Phys. Rev. B 31, 7335 (1985).

[99] D.Stoeffler, F.Gautier, to be published (1990).

[100] F.Gautier, A.M.Llois, Surf. Sci., to be published (1990).

[101] S.C.Hong, A.J.Freeman, C.L.Fu, J. de Physique 49, C8 1683 (1988).

[102] M.Stampanoni, A.Vaterlaus, D.Pescia, M.Aeschlimann, F.Meier, W.Durr, S.Blügel, Phys. Rev. B 37, 10380 (1988).

[103] T.Moriya, in "Proc. of the Int. School of Physics Enrico Fermi Course XXXVII: Theory of Magnetism in Transition Metals", ed. by W.Marshall, p. 242 (Academic Press, New-York, 1967).

[104] A.Bieber, F.Gautier, G.Tréglia, F.Ducastelle, Solid State Comm. 39, 149 (1981).

[105] S.C.Carbone, S.F.Alvarado, Phys. Rev. B 36, 2433 (1987).

[106] (a) J.J. de Miguel, A.Cebollada, J.M.Gallego, S.Ferrer, R.Miranda, C.M.Schneider, P.Bressler, J.Garbe, K.Bethke, J.Kirschner, Surf. Sci. 211-212, 732 (1989); (b) C.M.Schneider, P.Bressler, P.Schuster, J.Kirschner, J.J. de Miguel, R.Miranda, Phys. Rev. Lett. (to be published, 1990); (c) C.M.Schneider, J.J. de Miguel, P.Bressler, S.Ferrer, P.Schuster, R.Miranda, J.Kirschner, to appear in Vacuum (Proc. of the 11th Vacuum Congress and the 7th Int. Conf. on Solid Surfaces, Köln, 1989).

[107] J.Ferron, J.M.Gallego, A.Cebollada, J.J. de Miguel, S.Ferrer, Surf. Sci. 211-212, 797 (1989).

[108] T.Beier, H.Jahrreis, D.Pescia, Th.Worke, W.Gudat, Phys. Rev. Lett. 61, 1875 (1988); T.Beier, D.Pescia, M.Stampanoni, A.Vaterlaus, F.Meier, Appl. Phys. A 47, 73 (1988).

[109] D.A.Steigerwald, I.Jacob, W.F.Egelhoff Jr, Surf. Sci. 202, 472 (1988).
[110] S.A.Chambers, T.J.Wagener, J.H.Weaver, Phys. Rev. B 36, 8992 (1987).
[111] C.Liu, E.R.Moog, S.D.Bader, Phys. Rev. Lett. 60, 2422 (1988) and J. Appl. Phys. 64, 5325 (1988).
[112] D.Pescia, M.Stampanoni, G.L.Bona, A.Vaterlaus, R.F.Willis, F.Meier, Phys. Rev. Lett. 58, 2126 (1987); ibi, Phys. Rev. Lett. 60, 2559 (1988).
[113] M.Stampanoni, A.Vaterlaus, M.Aeschlimann, F.Meier, J. Appl. Phys. 54, 5321 (1988).
[114] K.Ounadjela, C.B.Sommers, A.Fert, P.M.Lévy, to be published (1990).
[115] H.Hasegawa, F.Herman, Physica B 149, 175 (1988); Phys. Rev. B 38, 4863 (1988).
[116] E.Bauer, Z. Krist. 110, 372 (1958); E.Bauer, H.Poppa, Thin Sol. Films 12, 167 (1971); R.Kern, Bull. Mineral 101, 202 (1978).
[117] J.A.Venables, G.D.T.Spiller, M.Hanbacken, Rep. Progr. Phys. 47, 399 (1984).
[118] For a review of wetting phenomena see, for example, S.Dietrich in "Phase Transitions and Critical Phenomena", Vol. 12, ed. by C.Domb, J.L.Lebowitz, p. 2 (Academic Press, 1988).
[119] R.Pandit, M.Schick, M.Wortis, Phys. Rev. B 26, 5112 (1982).
[120] R.J.Muirhead, J.G.Dash, J.Krim, Phys. Rev. B 29, 5074 (1984).
[121] M.Bienfait, J.L.Sequin, J.Suzanne, E.Lerner, J.Krim, J.G.Dash, Phys. Rev. B 29, 983 (1984).
[122] D.A.Huse, Phys. Rev. B 29, 6985 (1984).
[123] F.T.Gittes, M.Schick, Phys. Rev. B 30, 209 (1984).
[124] M.H.Grabow, G.H.Gilmer, Surf. Sci. 194, 333 (1988).
[125] R.Bruinsma, A.Zangwill, Europhys. Lett. 4, 729 (1987).
[126] J.W.Matthews, in "Dislocations in Solids", p. 461 ed. by F.R.N.Nabarro (North Holland, Amsterdam, 1979).
[127] J.H.Van der Merwe, Proc. Phys. Soc. London A 63, 616 (1950).
[128] J.H.Van der Merwe, Phil. Mag. 45, 127 (1982).
[129] J.H.Van der Merwe, in "Chemistry and Physics of Solid Surfaces", p. 129, ed. by R.Vanse-Low (Boca Raton, CRC Press, 1979).
[130] C.P.Flynn, Phys. Rev. Lett. 57, 599 (1986) and J. de Physique 49, C5-719 (1988).
[131] J.H.Van der Merwe, Phil. Mag. A 45, 145 and 159 (1982).
[132] E.Bauer, Appl. of Surf. Sci. 11-12, 479 (1982).
[133] E.Bauer, J.H.Van der Merwe, Phys. Rev. B 33, 3657 (1986).
[134] C.M.Falco, I.K.Shüller, in "Synthetic Modulated Structures/VLSI", ed. by L.L.Chang and B.C.Giessen (Academic Press, Orlando, 1985).
[135] (a) F.Gautier, F.Ducastelle, J.Giner, Phil. Mag. 31, 1373 (1975);
(b) F.Ducastelle, F.Gautier, J. Phys. F 6, 2039 (1976);
(c) A.Bieber, F.Gautier, Acta Met. 34, 2291 (1986), ibid 35, 1839 (1987).
[136] G.Allan, M.Lannoo, L.Dobrzynski, Phil. Mag. 30, 33 (1974).
[137] C.E.Hill, I.Marklund, J.Martinson, B.J.Hopkins, Surf. Sci. 24, 435 (1971).
[138] P.J.Berlowitz, N.D.Shinn, Surf. Sci. 209, 345 (1989).
[139] (a) M.Przybylski, U.Gradman, J. Appl. Phys. 63, 3652 (1988);
(b) M.Przybylski, U.Gradman, Phys. Rev. Lett. 59, 1152 (1987);
(c) H.J.Elmess, G.Liu, U.Gradman, Phys. Rev. Let. 63, 566 (1989).
[140] B.G.Johnson, P.J.Berlowitz, D.W.Goodman, C.H.Bartholomew, Surf. Sci. 217, 13 (1989).
[141] P.J.Berlowitz, D.W.Goodman, Surf. Sci. 187, 463 (1987).
[142] K.P.Kampers, W.Schmitt, G.Güntherodt, H.Kuhlenbeck, Phys. Rev. B 38, 9451 (1988).
[143] W.Schlenk, E.Bauer, Surf. Sci. 93, 9 (1980).
[144] (a) G.A.Attard, D.A.King, Surf. Sci. 188, 589 (1987);
(b) E.Bauer, H.Poppa, G.Todd, F.Bonezeck, J. Appl. Pys. 45, 5164 (1974);
(c) D.Singh, H.Krakauer, Surf. Sci. 216, 303 (1989).
[145] E.Bauer, H.Poppa, G.Todd, P.R.Davis, J. Appl. Phys. 48, 5164 (1974).

[146] G.W.Wertheim, D.H.Buchanan, V.Lee, Phys. Rev. B 34, 6869 (1986).
[147] (a) M.Mundschau, E.Bauer, W.Zwieck, J. Appl. Phys. 65, 581 (1989); (b)J.Kolaczkiewicz, Surf. Sci. 200, 335 (1988).
[148] F.Soria, H.Poppa, J. Vac. Sci. Technol. 17, 449 (1980).
[149] A.Pavlovska, E.Bauer, Surf. Sci. 177, 473 (1986).
[150] M.W.Ruckman, P.D.Johnson, M.Strongin, Phys. Rev. B 31, 3405 (1985); M.Ruckman, M.Strongin, ibid 29, 7105 (1984).
[151] (a) M.Sugarton, M.Strongin, F.Jona, J.Colbert, Phys. Rev. B 28, 4075 (1983);
(b) V.Kumar, K.H.Bennemann, Phys. Rev. B 28, 3138 (1983);
(c) X.Pan, P.D.Johnson, M.Weinert, R.E.Watson, J.W.Davenport, G.W.Fernando, S.L.Hulbert, Phys. Rev. B 38, 7850 (1988).
[152] Xiao-he Pan, M.W.Ruckman, M.Strongin, Phys. Rev. B 35, 3734 (1987).
[153] M.W.Ruckman, Li Qiang Jiang, Phys. Rev. B 38, 2959 (1988).
[154] Ch.Park, E.Bauer, H.Poppa, Surf. Sci. 154, 371 (1985).
[155] D.Tian, S.C.Wu, F.Jona, P.M.Marcus, Solid State Comm. 70, 199 (1989).
[156] C.Binns, C.Norris, G.P.Williams, M.G.Barthes, H.A.Padmore, Phys. Rev. B 34, 8221 (1986).
[157] F.J.A. den Broeder, D.Kuiper, H.C.Donkersloot, J. de Physique 49, C8-1663 (1989).
[158] Ch.Park, E.Bauer, H.Poppa, J. Appl. Phys. 61, 437 (1987); ibid, J. Mag. Mag. Mat. 66, 351 (1987).
[159] S.L.Beauvais, R.J.Behm, S.L.Chang, T.S.King, C.E.Olson, P.R.Rape, P.A.Thiel, Surf. Sci. 189-190, 1069 (1987).
[160] M.Pessa, M.Vulli, J. Phys. C 16, L629 (1983).
[161] (a) D.L.Weissman-Wenocur, P.M.Stefan, B.B.Pati, M.L. Shek, I.Lindau, W.E.Spicer, Phys. Rev. B 27, 3308 (1983);
(b) Y.Kusk, L.C.Felman, P.J.Silverman, Phys. Rev. Lett. 50, 511 (1983) and J. Vac. Sci. Technol. 11, 1060 (1983).
[162] D.J.Godbey, G.A.Somorjai, Surf. Sci. 202, 204 (1988).
[163] R.W. Judd, M.A.Reichelt, E.G.Scott, M.Lambert, Surf. Sci. 185, 515 (1987); ibid, 529 (1987).
[164] J.W.Sachtler, M.A. Van Hove, J.P.Biberian, G.A.Somorjai, Surf. Sci. 110, 19 (1981).
[165] (a) M.Salmeron, S.Ferrer, M.Jazzar, G.A. Somorjai, Phys. Rev. B 28, 6758 (1983);
(b) D.S.Wang, A.J.Freeman, H.Krakauer, Phys. Rev. B 29, 1665 (1984).
[166] M.L.Shek P.M.Stefan, I.Lindau, W.E.Spicer, Phys. Rev. B 27, 7277 (1983).
[167] S.H.Liu, Z.Q.Wang, D.Tian, Y.S.Li, F.Jona, P.M.Marcus, Surf. Sci. 221, 35 (1989).
[168] S.A.Chambers, S.B.Anderson, H.W.Chen, J.H.Weaver, Phys. Rev. B 35, 2592 (1987).
[169] S.A.Chambers, H.W.Chen, I.M.Vitomitov, S.B.Anderson, J.H.Weaver, Phys. Rev. B 33, 8810 (1986).
[170] (a) See "Cohesion in Metals, Transition Metal Alloys", by F.R. de Boer, R.Boom, W.C.Mattens, A.R.Miedema, A.K.Niessen (North-Holland, Amsterdam, 1988);
(b) J.Germeka, A.R.Miedema, Surf. Sci. 124, 359 (1983).
[171] For a discussion of the magnetic anisotropy of thin films, see P.Bruno, Thesis, University Paris-Sud (1989).
[172] (a) L.Néel, C.R.Ac.Sc. 237, 1468 (1953);
(b) L.Néel, J. de Phys. Rad. 15, 225 (1954).
[173] U.Gradman, J. Mag. Mag. Mat. 54-57, 733 (1986).
[174] K.B.Urquhart, B.Heinrich, J.F.Cochran, A.S.Arrott, K.Myrtle, J. Appl. Phys. 64, 5334 (1988).
[175] U.Gradman, J.Korecki, G.Waller, Appl. Phys. A39, 101 (1986).
[176] C.Chappert, K. Le Dang, P.Beauvillain, H.Hurdequint, D.Renard, Phys. Rev. B 34, 3192 (1986).

[177] S.T.Purcell, B.Heinrich, A.S.Arrott, J. Appl. Phys. 64, 5337 (1988).
[178] H.J.G.Draaisma, W.J.M. de Jongh, J. Appl. Phys. 64, 3610 (1988).
[179] M.Brooks, Phys. Rev. 58, 909 (1940).
[180] G.C.Fletcher, Proc. Phys. Soc. (London) A 267, 505 (1954).
[181] N.Mori, J. Phys. Soc. Japan 27, 307 (1969).
[182] E.I.Kondorskii, E.Stranski, Sov. Phys. JETP 36, 188 (1973).
[183] R.Gesdorf, Phys. Rev. Lett. 40, 344 (1978).
[184] H.Takayama, K.P.Bohnen, P.Fulde, Phys. Rev. B 14, 2287 (1976).
[185] P.Bruno, Phys. Rev. B 39, 865 (1989).
[186] J.G.Gay, R.Richter, Phys. Rev. Lett. 56, 2728 (1986).
[187] W.Karas, J.Hoffke, L.Fritsche, J. de Physique, to be published (1989).
[188] C.Chappert, P.Bruno, J. Appl. Phys. 64, 5736 (1988).
[189] P.Bruno, J.Seiden, J. de Physique 49, C8, 1645 (1988).
[190] P.Bruno, J. Phys. F **18**, 1291 (1988) and J. Appl. Phys. 64, 3153 (1988).
[191] Y.Yafet, E.M.Gyorgy, Phys. Rev. B 38, 9145 (1988).
[192] H.P.Oepen, J.Kirschner, J. de Physique 49, C8 1853 (1988).
[193] H.P.Oepen, M.Benning, H.Ibach, C.M.Schneider, J.Kirschner, to be published in J. Mag. Mag. Mat. (1990).

Materials Science Forum Vols. 59 & 60 (1990) pp. 439-480

MAGNETIC AND TRANSPORT PROPERTIES OF METALLIC MULTILAYERS

A. Fert
Laboratoire de Physique des Solides
Université Paris-Sud, F-91405 Orsay, France

1 INTRODUCTION

The preceding paper by F. Gautier has been devoted to the electronic structure of the metallic multilayers and to their magnetic properties, with special emphasis on the theoretical aspects. This paper follows. In section 2, we present an experimental review of the magnetic properties. Section 3 deals with the transport properties.

2 MAGNETIC PROPERTIES

Before considering the properties and phenomena related to **magnetic coupling** between layers in multilayered structures,we begin with the properties of "**isolated** magnetic layers", which includes multilayers with thick enough non-magnetic layers separating the magnetic layers as well as single films deposited on a substrate or sandwiched between non-magnetic metals.
we discuss the following questions successively:

- magnetization at zero temperature: enhancement or reduction of the magnetic moments due to interface effects, appearance or collapse of ferromagnetism (2-1)
- interface anisotropy (2-2)
- properties of new cristalline phases stabilized by epitaxy (2-3)
- low dimensional thermodynamics (2-4)

We then proceed to the question of the interlayer coupling and the resulting magnetic superstructures (2-5).

We limit our review to ferromagnetic metals (except some data on Dy and Mn). In the future studies of other types of magnetic materials (antiferromagnets, spin-glasses, etc.) will probably be developed but,up to now, most of the research has been done on ferromagnetic layers.

2-1 Magnetic moments (at zero temperature)

The problem is different for rare-earth and transition metal systems. For rare-earths the magnetic moment is carried by the f-shell electrons which, except for some anomalous Ce and Yb compounds, are well localized. Consequently, the magnetic moment of a rare-earth ion is not very sensitive to surface and interface effects. For example, a gadolinium atom at a surface or interface will generally retain its moment of about 7 μ_B. In contrast, for transition metals and transition metal alloys, the magnetic moment is carried by the delocalized electrons of the d-band and is sensitive to environment and proximity effects. In the simple Stoner picture the condition for a ferromagnetic splitting of the d-band (at zero Kelvin) is the Stoner criterion

$$I\, n(E_F) > 1 \tag{1}$$

where $n(E_F)$ is the density of states at the Fermi level and I is a characteristic parameter of the exchange interactions. For atoms on the surface, in the most general case, the coordination number is smaller, the d states are less hybridized with neighbour states, the d-band is narrower, its density of states is higher, which will enhance the magnetism. This is generally observed. At interfaces, the perturbation of the d-band depends on the adjacent metal and only accurate electronic structure calculations can predict the change of the magnetic moment [1,2]. In addition to these environment and proximity effects, the lattice strain of thin layers can also contribute to the change of the magnetic moment, an increase of the interatomic distances generally inducing an enhancement of the magnetism. Finally the stabilization of new crystalline phases by epitaxy can also give rise to completely different magnetic moments. We now present experimental examples of magnetic moment change in thin films and multilayers.

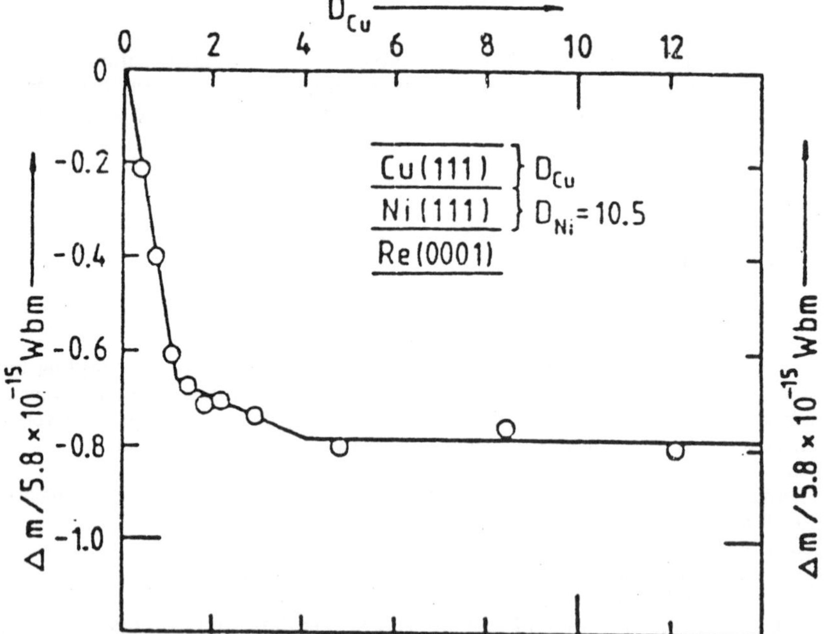

Fig.1 Change of the magnetic moment of a 10.5 ML thick fcc(111) Ni layer induced by a Cu(111) overlayer as a function of the number of Cu monolayers. The magnetic moment unit is the moment of 1 Ni monolayer, m_{ML}=5.8x10^{-15}Wbm. The Ni layer is deposited on a Re(0001) surface. From Gradmann [3].

Fig.1 shows the reduction of the magnetic moment of a 10.5 monolayer (ML) thick Ni(111) layer induced by an overlayer of Cu(111) as a function of the thickness of the overlayer [3]. The reduction reaches 80% of the moment of a Ni ML when two ML of Cu have been deposited and is not changed by further coating. This sort of reduction of the magnetization at an interface has often been described as a dead-layer effect which is, of course, an over-simplified picture.

For Ni, the proximity of a noble metal generally depresses the magnetization. Other examples are shown in fig.2 where the reduced Curie temperature of Ni ultrathin films is plotted versus the number n of atomic layers [4]. For sandwiches of Ni between Au (squares) or Cu (triangles), the ferromagnetism disappears (i.e.$T_c \rightarrow 0$) for about two atomic layers of Ni. In contrast, for Ni films between Rh, T_c is less reduced.

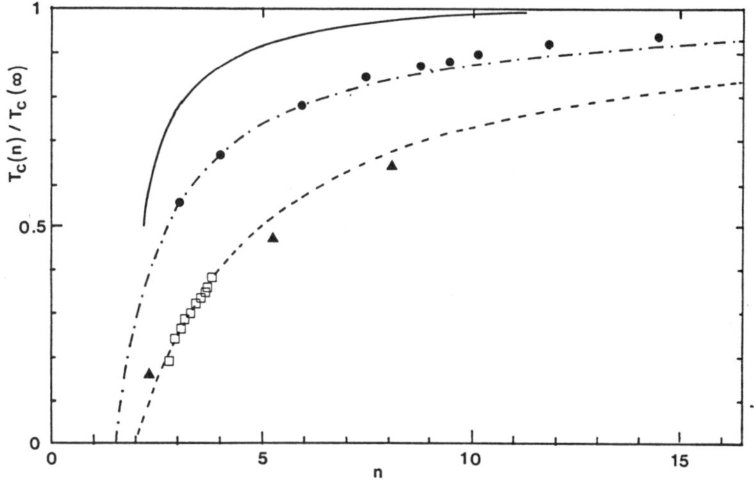

Figure 2 Reduced Curie temperature of ultra-thin Ni films, $T_c(n)/T_c(\infty)$ versus the number of Ni atomic layers. Au/Ni/Au: open squares; Cu/Ni/Cu: triangles; Rh/Ni/Rh: circles. From Renard [4].

The reduction of the magnetic moments at the interface with a non-magnetic metal observed for Ni/Cu or Ni/Au interfaces is not at all a general rule. An opposite example is the enhancement of the magnetic moment of bcc Fe at the interface with Ag. According to the magnetic measurements of Elmers and Gradmann [5], the Fe atoms of a Fe(110) monolayer grown on a W(110) surface and coated with Ag, carry a magnetic moment of $2.60\mu_B$ (extrapolated value to zero Kelvin). This is above the bulk value by 17% and in good agreement with the theorical predictions of Hong et al. [6]. The calculations of Freeman and coworkers [2] also predict an enhancement of the magnetic moment for Fe at the interface with Cu, Ag and Au. Unfortunately, apart from the extensive work of the group of Gradmann, there are not enough **absolute** measurements of the magnetization of ultrathin films to compare with the theoretical predictions.

Two other interesting cases are Cr and V. The bulk metals are not ferromagnetic. V exhibits a conventional Pauli paramagnetism and Cr develops spin density waves with an amplitude of about $0.6\mu_B$. For ultrathin films, Freeman and coworkers [2] have predicted large magnetic moments but the recent calculations of Freeman and Fu [7], and Blügel and Dederich [8] indicate that these moments should order antiferromagnetically. On the experimental side Brodsky et al. [9] have measured the magnetization of Au(001)/Cr(001) multilayers grown on NaCl and, for Cr layers of 1.5Å thickness (about 1 ML), find a ferromagnetic-like behavior with a magnetic moment of $0.4\mu_B$/Cr at 2K. More recently Johnson et al. [10] reported polarized neutron reflection measurements for Ag/Cr/Ag(001) sandwiches. For a Cr layer equivalent to a coverage of 0.33ML, the data suggest long-range ferromagnetic order with a magnetic moment per atom that is significantly greater than the bulk value. For a 3.3 ML thick layer, no ferromagnetism is detected. Recent extensive photoemission studies by Hanf et al. [11] have shown that the question is more complex. For Cr on Au(100), it turns out that up to 3 ML of Cr the Cr diffuses into the Au lattice to form a random substitutional fcc alloy in which the Cr atoms carry a large magnetic moment (~ 4 μ_B as in bulk AuCr alloys). Upon further deposition of Cr a disordered Cr-rich phase is first formed and then, at 7-8 ML, there is epitaxial growth of bcc Cr(100) presenting the properties of bulk Cr. For Cr on Ag(100) the photoemission studies of Hanf et al show that isolated Cr atoms on Ag surface carry a high magnetic moment (~ $4\mu_B$). Then, before completion of the first Cr ML, there is formation of 3-dimensional islands and rapid crossover to bulk-like behavior.The ultrathin films of Cr provide us with a good example of systems in which it is essential to discuss the magnetic properties as a function of the microstructure of the deposited layers.

For ultrathin films of V the experimental results are not clear, with contradictory results obtained by different groups. Rau et al [12] have studied ultrathin (1-7 ML) films of V(100) deposited on Ag(100) by electron capture spectroscopy and find that these films are ferromagnetic. They are able to determine the variation of the surface Curie temperature as a function of the film thickness. The existence of ferromagnetism is also supported by non-spin polarized inverse photoemission experiments of Drube et Himpsel [13]. In contrast the spin polarized photoemission experiments of Stampanoni et al. [14] did not result in the observation of any ferromagnetism.

2-2 Interface anisotropy

2-2-1 General

The magnetic moments at surfaces and interfaces are subjected to strong anisotropy fields. For a ferromagnetic film or multilayer, the anisotropy energy per volume unit of the ferromagnetic element is usually written as

$$E_{an} = K_{eff} \sin^2\Theta \qquad (2)$$

where Θ is the angle between the magnetization and the normal to the surface or interface. K_{eff} includes contributions from surface or interface anisotropy, volume anisotropy and dipolar energy (the so-called demagnetizing field anisotropy)

$$K_{eff} = \frac{2\,K_s}{t_M} + K_v - 2\pi M^2 \qquad (3)$$

K_v arises from the volume magnetocristalline anisotropy and exists only if the film is a uniaxial single crystal with its axis perpendicular to the film (lattice strains of the film can also contribute to K_v). K_s is the surface or interface anisotropy energy per surface unit, so that its contribution per volume unit to K_{eff} is proportional to the density of surfaces or interfaces t_M^{-1} (t_M is the thickness of the ferromagnetic element).The term $-2\pi M^2$, where M is the magnetization, is the contribution from the dipolar interactions calculated for a uniform distribution of the magnetization and perfectly plane surfaces (this dipolar term can be reduced by interface roughness). For relatively thick films or layers (say a few nm for transition metals)the dipolar energy is generally predominant and the magnetization lies in the plane of the film. For ultrathin films (say below 1nm for transition metals) the surface or interface contribution, which is proportional to t_M^{-1}, can overcome the dipolar contribution and the spontaneous magnetization becomes perpendicular to the film. The possible applications of the "perpendicular anisotropy" to vertical recording partly accounts for the present interest in the surface/interface anisotropy.

The main origin of the interface (surface) anisotropy is the lower symmetry at the interface (surface) sites and the resulting enhancement of the magnetocrystalline anisotropy. Additional contributions can arise from anisotropic pair interactions and from magnetoelastic effects induced by the substrate (these magnetoelastic effects contribute to K_v if the strain is uniform and to K_s if the strain is inversely proportional to the thickness, as this approximately occurs in some systems). For transition metals, as described in the lecture by Pr Gautier, the anisotropy of thin films (interface and volume) can be obtained from sophisticated calculations introducing the spin-orbit coupling in band structure models (or by fully relativistic band structure calculations). It is generally hard to derive a physical picture of the calculated effects. For rare-earths with well localized moments, the magnetocrystalline anisotropy is directly related to the crystal field [15] acting on the moments. The surface or interface anisotropy results from an enhancement of the crystal field at the surfaces or interfaces and can be described in relatively simple models. We begin with the case of rare-earths to illustrate the interface anisotropy problem.

2-2-2 Rare-earths

We consider the example of a hcp rare-earth metal with a (0001) surface (or interface). The crystal field interaction can be calculated approximately in the very simple point charge model: the crystal field is described as the interaction of the aspherical charge distribution of the f-shell with the electric field created by point charges on neighbour sites. This is an oversimplified model, especially in metallic compounds but it gives the right order of magnitude of the crystal field effects in many rare-earth systems [15]. For a site of the hcp lattice far from the surface, the lower order crystal field term is written in the usual notation:

$$H_{CEF} \sim B_{20}\left[3J_z^2 - J(J+1)\right] \tag{4}$$

where J_z is the component of the moment **J** along the c axis. The point charge model gives for B_{20} [15]:

$$B_{20} = \frac{2}{21} Ze^2 \frac{\langle r^2 \rangle}{a^3} \alpha_J (1.633 - \frac{c}{a}) \tag{5}$$

where Ze is the value of the point charges, $\langle r^2 \rangle$ and α_J are characteristic parameters of the spatial extension and asphericity of the f-shell, and a and c are the lattice parameters. The crystal field of Eq.4 is uniaxial but, for c/a=1.633, the environment has a higher symmetry, B_{20} cancels out and it remains in the crystal field only higher order terms (B_{40} etc.) which are small (as in cubic crystals). For the hcp structure of the rare-earths, c/a is different from 1.633 but the factor (1.633 - c/a) is always much smaller than 1.

For a site on the (0001) surface, the same type of calculation of the electric field created by point charges gives [16]:

$$B_{20}^s = -\frac{3}{42} \frac{Ze^2 \langle r^2 \rangle}{a^3} \alpha_J \tag{6}$$

Because the point charges on one side of the (0001) plane have been removed, there is no longer the partial balance which, for more symmetrically distributed point charges, gives rise to the factor $(1.633 - c/a)$ in Eq.5, and B^s_{20} is much larger than B_{20}:

$$\frac{B^s_{20}}{B_{20}} = -\frac{3}{4(1.633 - \frac{c}{a})} \tag{7}$$

For example, for Praseodynium, c/a =1.614, which gives:

$$\frac{B^s_{20}}{B_{20}} = -39.5 \tag{8}$$

In the same way, one finds -39.5 for Nd; -14.4 for Tb; -12.5 for Dy; etc.This illustrates the enhancement of the crystal field due to the lower symmetry of surface sites. In hcp structures, a typical value for the volume magnetocrystalline energy associated with B_{20} is 5×10^{-3}eV/atom, which leads to around 10^{-1}eV/atom for the surface anisotropy energy, to be compared to $2\pi M^2 \approx 10^{-3}$eV/atom for the rare-earths with the highest moments (Tb, Dy).

For interfaces, if the point charges introduced on the sites of the second metal are different, they cannot balance the electric fields created by the point charges of the first metal, and a strong enhancement of the crystal field at the interface is expected again.

The contribution to the anisotropy from magnetoelastic effects can also be estimated from the same point charge model. In rare-earth systems the magnetoelastic interaction is generally attributed to the modulation of the crystal field by the lattice strains. According to Eq.7, a variation of c/a by x% will induce a change of B_{20} by about x% of B^s_{20}. Thus only relatively weak effects can be expected from the lattice strains in rare-earth thin films and multilayers.

We proceed to the question of the anisotropic pair interactions between moments on surfaces or interfaces. The exchange interaction between two localized moments is generally written:

$$H_{ech} = -J\,\mathbf{S_1}.\mathbf{S_2} \tag{9}$$

In the most general case, anisotropic interaction terms should also exist, first bilinear antisymmetric terms of the Dzyaloshinsky-Moriya (DM) form [16]:

$$H_{DM} = \mathbf{D_{12}}.(\mathbf{S_1} x \mathbf{S_2}) \tag{10}$$

and higher order terms. The DM term is ruled out when there is a center of inversion symmetry in the system, so that it generally does not exist in ordered crystal lattices, except for some complex structures [16]. However, in disordered systems, in spin glasses for example, there is no longer a center of inversion symmetry, and the DM interactions are known to be significant (~10% of the isotropic exchange, Eq.9) [17]. The calculation of the DM interaction for the standard geometry of Fig.3 -triangle of Mn impurities in Cu for example- gives a DM interaction, Eq.10, with $\mathbf{D_{12}}$ perpendicular to the plane of the triangle [17].

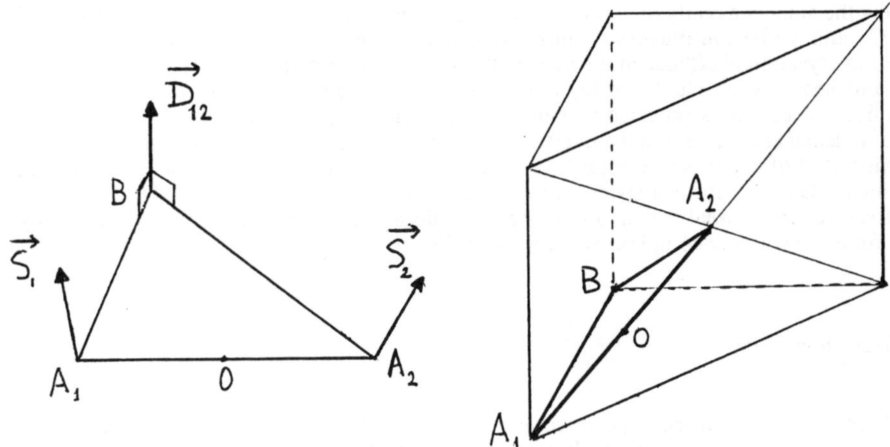

Fig.3 (left) low symmetry geometry giving rise to DM interactions in disordered magnetic alloys. For a triangle of magnetic impurities A_1A_2B, the spins S_1 and S_2 in A_1 and A_2 are coupled by $H_{DM} = \mathbf{D}_{12}.(\mathbf{S}_1 \times \mathbf{S}_2)$ where $\mathbf{D}_{12} \sim \mathbf{BA}_1 \times \mathbf{BA}_2$.

Fig.4 (right) low symmetry for a pair of magnetic atoms on a bcc(110) srface or interface.

In the same way the symmetry breaking at surfaces and interfaces should give rise to DM interactions. For example, in the case of a (110) surface or interface of a bcc lattice , see Fig.4, there is no symmetric atom of B with respect to O (as in the absence of surface) and a DM interaction with $\mathbf{D}_{12}$ perpendicular to the plane of the triangle A_1A_2B is expected. Such DM interactions are also expected at surfaces or interfaces of fcc and hcp lattices. For both rare-earth and transition metals, one estimates $|\mathbf{D}_{12}|/J \approx 10^{-1}$ [18]. The effect of such anisotropic interactions on the macroscopic anisotropy properties ha never been worked out. For non-S rare-earth ions it is likely that the effect of the DM interactions is not important with respect to the very strong surface crystal fields. For gadolinium S-ions the crystal field is much smaller and the DM interactions should play a role in the anisotropy properties. In ferromagnetic transition metals, because the exchange is strong, the DM interactions should be significant and likely should contribute to the anisotropy properties. Of course the anisotropic pair interaction effects we have described suppose a localized moment picture of the ferromagnetism and do not turn out in the band structure models usually applied for transition metal ferromagnets.

While the problem of anisotropy is certainly much simpler with rare-earths than with transition metals, unfortunately there is very little experimental data concerning the interface anisotropy in rare-earth thin films or multilayers. To our knowledge , the most specific information on the interface anisotropy of rare-earth systems is in the work of Piecuch et al. [20] on Nd/Fe and Tm/Fe multilayers.In Nd/Fe the magnetic anisotropy is strong enough to align the spontaneous magnetization in the perpendicular direction. In Tm/Fe the magnetization is in the multilayer plane, which is in agreement with the change of sign of the Stevens coefficient α_J between Nd and Tm, and indicates that the major effect comes from the rare-earth. These results have been interpreted by Piecuch et al. [20] in terms of enhanced crystal fields at the interfaces. However the uncertainty on the interface microstructure does not allow the authors a really quantitative account of the data. Shan et al. [21] have also found in Dy/Co and Tb/Fe multilayers a thickness dependent uniaxial anisotropy which can be ascribed to crystal field effects on the rare-earth ions (this is demonstrated by the cancellation of the anisotropy when the Tb or Dy non-S-ions are replaced by Gd S-ions for which the crystal field interaction is expected to vanish). Shan et al. [21] have interpreted their results in terms of uniaxial crystal field in the interfacial regions. Other interesting data on the anisotropy properties of Dy/Fe and Nd/Fe multilayers can also be found in the work of Hosoito et al. [22]. Finally Gadolinium is a special case. Due to the S-ion character of Gd^{3+}, the crystal field effects are much weaker than for other rare-earths and the interpretation of the anisotropy of thin films is relatively complex. See, for example, the work of Farle [23].

2-2-3 Transition metals

There are much more experimental results on the surface/interface anisotropy for transition metals than for rare-earths. In Fig.3-4 we present typical examples from the work of Draaisma et al. [24] on Co/Pd multilayers (the Co and Pd are fcc and polycrystalline with a [111] texture).

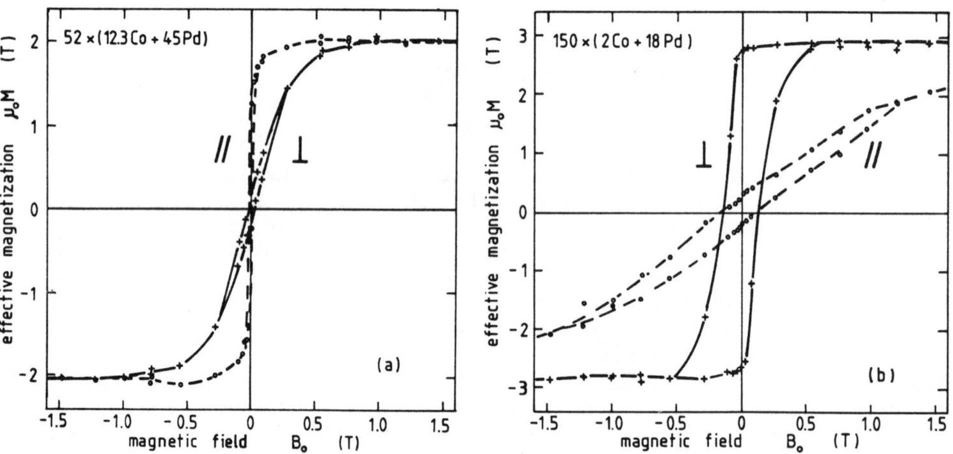

Fig.5 Magnetization curves at room temperature in a magnetic field parallel to the film (dashed curve) and perpendicular to it (solid curve) for two multilayers. (a): 52 x (Co 12.3 Å/Pd 45 Å); (b): 150 x (Co 2 Å/Pd 18 Å). The curves in (b) are typical of films with perpendicular anisotropy. From Draaisma et al. [19].

In these multilayers the interface contribution $2K_s/t_{Co}$ to the anisotropy constant K_{eff} (see Eq.3) is positive and becomes predominant for ultrathin Co layers. Qualitatively this appears in Fig.5: for t_{Co}=12.3Å (Fig.5a) it is easier to saturate the magnetization with in-plane fields; in contrast, for $t_{Co}\approx$2Å (Fig.5b), the saturation is very hard to reach with in-plane fields, while, for transverse field, one observes the square hysteresis loops which are typical of films with perpendicular anisotropy.

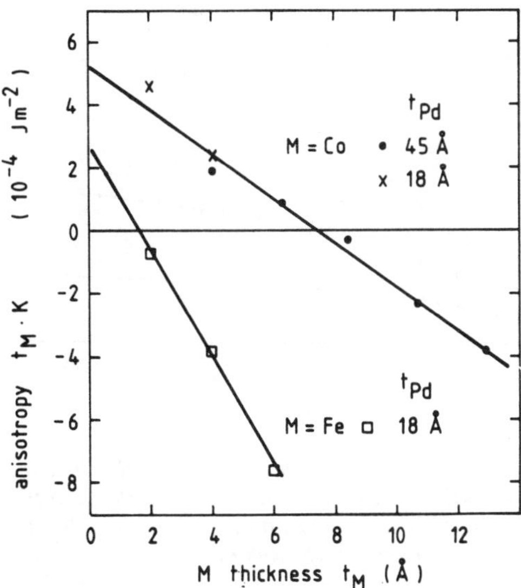

Fig.6 The effective anisotropy constant K_{eff} times the layer thickness of the ferromagnetic element t_M (t_{Co} or t_F) as a function of t_M for Pd/Co and Pd/Fe multilayers with a fixed thickness of Pd. K_{eff} is derived from magnetization curves similar to those of Fig.5. From Draaisma et al. [19].

To derive the interface anisotropy Ks one remarks that Eq.3 can be written as:

$$t_{Co}\, K_{eff} = 2K_s + (K_v - 2\pi M^2)t_{Co} \qquad (11)$$

So that in the plot of Fig.4, $2K_s$ is given by the intercept of the straight lines with the vertical coordinate axis. We see in Fig.4 that the magnetization of Pd/Co multilayers becomes perpendicular for $t_{Co} < 8$Å (intercept of the straight line with the horizontal coordinate axis. Data from measurements on Pd/Fe multilayers are also plotted in Fig.4 and it can be seen that the critical thickness for perpendicular magnetization in Pd/Fe is much smaller than in Pd/Co. The larger thickness for which perpendicular magnetization has been observed for transition metals is 14 ML(~25Å) for the fcc(001)Fe films studied by Stampanoni et al. [25], see section (2-3).

The interface anisotropy of Pd/Co multilayers depends significantly on the interface sharpness of the sample [24]. K_s is different for different methods of preparation and decreases when one anneals the samples to make the atoms interdiffuse. On the contrary, for Co/Au multilayers [24], K_s is enhanced by annealing treatments which, according to X-ray diffraction data, sharpen the interfaces.

The surface or interface magnetic anisotropy of thin films and multilayers has been studied in many systems by various experimental techniques: magnetization and torque measurements, ferromagnetic resonance, spin-polarized photoemission, Mössbauer effect, magneto-optical effects, etc...

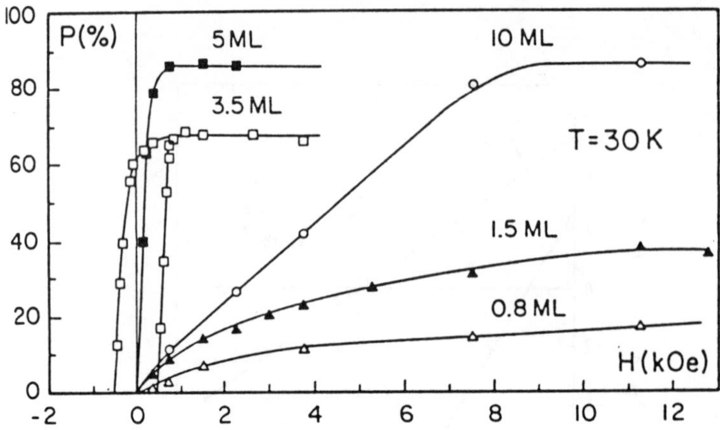

Fig.7 Magnetization curves at T=30K for epitaxial films of bcc Fe on Ag(001) derived from polarization analysis in spin polarized photoemission experiments. The thickness is indicated on the curves. The field H is applied perpendiculary to the film. From Stampanoni et al. [26].

In Fig.7 we show an example of results obtained by spin-polarized photoemission on bcc(001) Fe films grown on Ag(001) [26]. For 10ML, the remanent magnetization is in the plane of the film and an applied field of about 1 Tesla is necessary to align the magnetization along the perpendicular direction. For 5 and 3.5 ML the magnetization curves in transverse fields are typical of perpendicular anisotropy. This perpendicular anisotropy of bcc Fe films on Ag or in Fe/Ag superlattices has been observed by several groups [26-30]. Surprisingly, for 1.5 and 0.8 ML, the magnetization turns back into the plane of the film. This could be ascribed to the defects of the thinnest films. However, the collapse of the perpendicular anisotropy just around 1 ML has also been observed by Spin-polarized photoemission [25] and by Kerr effect [30] for (001) films of fcc Fe, so that , the existence of a specific behavior for a monolayer cannot be ruled out. Data from Kerr effect on fcc Fe films are shown in Fig.8.

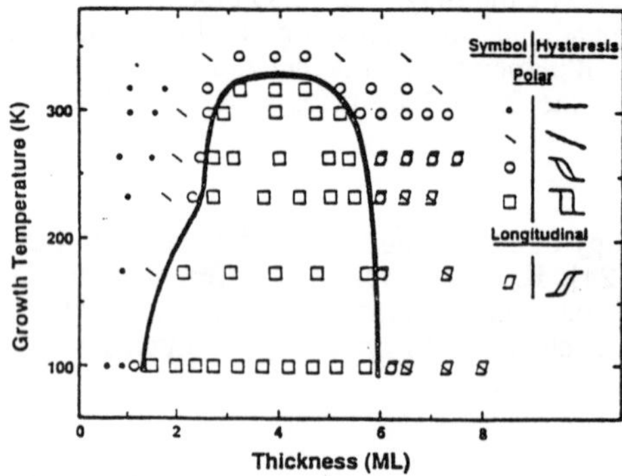

Fig.8 The region of stability of perpendicular anisotropy for fcc Fe/Cu(100) is outlined on a plot of film thickness vs growth temperature. From Kerr-effect experiments by Liu et al. [30].

System	Temperature	K_S (erg/cm^2)	Reference
UHV/Ni(111)	300	- 0.48	[3]
Cu, Pd/Ni (111)	300	- 0.22	[3]
Re/Ni (111)	300	- 0.19	[3]
O_2/Ni (111)	300	0	[32]
UHV/Fe(100)	293	+ 1	[29]
Ag/Fe (100)	293	+ 0.6	[29]
UHV/Fe(100) / Au (100)	293	0	[33]
UHV/Fe(110)/W(110)	293	- 0.5	[34]
Ag/Fe(110)/Ag(111)	77-293	0	[35]
Au/Fe(110)/Au(111)		> 0	[35]
UHV /Fe/Cu(100)	100	+ 1	[30]
Cu/Fe/Cu (100)		+ 0.5	[36]
Au/Co/Au(111)	293	+ 0.1,+ 0.45	[24]
Au/Co/Au(111)	293	+ 0.5	[37]
	5	+ 0.7	[37]
Pd/Co/Pd (111)	293	+ 0.26	[24]
UHV:Co/Cu (100)	293	0	[38]

Table I Experimental data on the surface/interface anisotropy for transition metals. Ks is defined in the text. From Bruno and Renard [31].

We finally present in Table I, a list of experimental values for the surface/interface anisotropy constant K_s for transition metal systems (from Bruno and Renard [31]). Typical values correspond to an energy of 10^{-3}eV/atom , to be compared to 10^{-5}-10^{-6}eV/atom for the volume anisotropy (however the surface/interface anisotropy is even larger for rare earths, 10^{-1}eV/atom, as shown in 2-2-2) . This order of magnitude of K_s is correctly predicted by calculations of the magnetocrystalline anisotropy based on band structure models [31,39]. However, at the present stage, there is really no general quantitative agreement between theory and experiments. In some cases, even the signs of the calculated and observed values are different. A part of the discrepancy is probably due to the additional contributions that we have mentionned at the beginning of 2-2: reduction of the demagnetizing field by the interface roughness, magnetoelastic effects induced by the substrate (they are significant only if the lattice is distorted by 10% or more), reduction of the interface magnetocrystalline anisotropy by the defects of the interface. But the real problem is the difficulty of the calculations: the magnetocrystalline anisotropy energy is a tiny term resulting from the balance of several contributions. Fully relativistic band structure calculations, now in progress, could yield more accurate results.

2-3 New crystal structures stabilized by epitaxy

Epitaxial growth is of interest to stabilize crystal structures which are different from the structure of the bulk material. For example, while bulk Fe is bcc below 1180K, fcc films can be grown on Cu and hcp (0001) Fe/Ru multilayers have also been made recently.Similarly, bulk Co is hcp below 694K and can be grown bcc on GaAs and fcc on Au. Mn can also be obtained in several phases. We now review the magnetic properties of these new phases.

The case of fcc (001) films of Fe grown on the (001) surface of fcc Cu is of particular interest. The calculations of Wang et al. [40] predict that bulk fcc iron should be antiferromagnetic if the lattice constant is less than about 3.7Å and ferromagnetic for higher values. For (001) monolayers and bilayers of fcc Fe on Cu (001) and with the assumption of the lattice constant of Cu (3.61Å), Fu and Freeman [41] find a ferromagnetic ground state with an enhanced magnetic moment ($2.8\mu_B$ for a monolayer).On the experimental side, Stampanoni et al. [25] have found that fcc (001) films of Fe can be grown on Cu (001) up to 14 ML and have the lattice constant of Cu approximatively. Spin polarized photoemission measurements show that these fcc (001) films are ferromagnetic. The hysteresis loops obtained by Stampanoni et al. [25] in these experiments are typical of ferromagnets, see Fig.9. These loops are for magnetic fields perpendicular to the layers, and from the form of the loop, it turns out that, at least for 5, 9 and 14 ML, the surface and interface anisotropy is strong enough to bring the magnetization perpendicular to the film (the 1 ML film does not show this perpendicular anisotropy. This longitudinal magnetization for 1 ML thick films has been discussed in 2-2-3). The Curie temperature of the fcc (001) Fe films (Fig.10) is relatively low and the sharp increase of Tc for films thicker than 15 ML indicates the crossover to bcc structure. Ferromagnetism and perpendicular anisotropy for fcc (001) films of Fe grown on Cu (001) are also found in the Brillouin scattering and resonance experiments of Dutcher et al. [36].

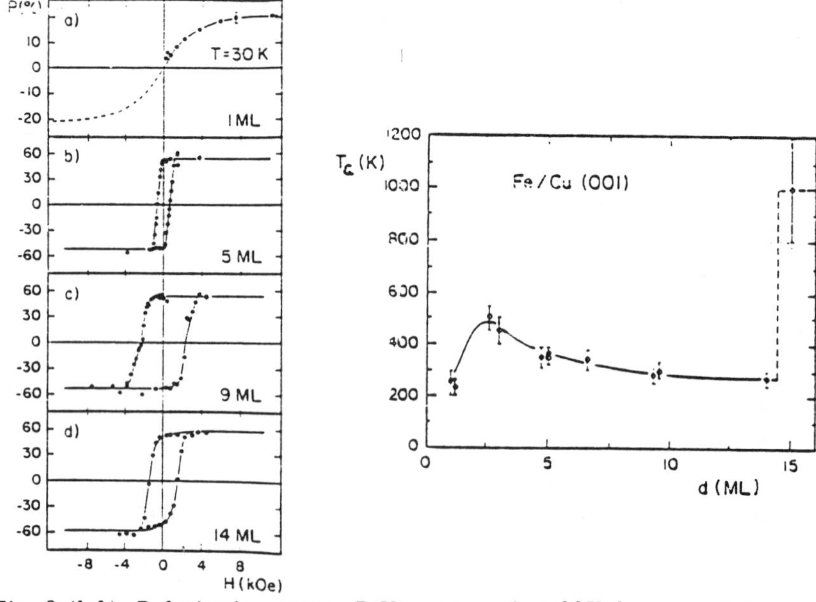

Fig. 9 (left) Polarization curves P(H) measured at 30K for several fcc (001)Fe films on Cu(001). The magnetic field H is perpendicular to the plane of the film. From Stampanoni et al. [25].

Fig. 10 (rigth) Thickness dependence of the Curie temperature T_c for fcc Fe films. Note the maximum of Tc at about 3 ML. The jump to higher T_c above 15 ML corresponds to the collapse of the film to a bcc phase. From Stampanoni et al. [25].

The results of the Mössbauer experiments of Keune and coworkers [42] on fcc (001) films of Fe do not agree with those described above. While the fcc (001) Fe films on Cu_3Au (001) exhibit a ferromagnetic behavior, the fcc (001) Fe films grown on Cu (001) turn out to be either antiferromagnetic (with $T_N \approx 70K$) or ferromagnetic depending on the preparation temperature. These results can be interpreted in terms of a crossover between antiferromagnetism and ferromagnetism as a function of the lattice constant of fcc Fe. Further experiments are needed to understand this interplay between lattice strain and ferro-antiferromagnetic transition.

hcp Fe has also been stabilized recently in Fe/Ru hcp superlattices grown on Ru (0001) substrates [43]. Only preliminary magnetic measurements have been published until now. It turns out that ultrathin hcp Fe layers (4 ML) are only very weakly magnetic. A ferromagnetic behavior appears for thicker films.

We now summarize some experimental data on metastable films of Co. Bulk Co is hcp below 698K and ferromagnetic ($1.74\mu_B$/Co). bcc Co (110) films have been grown on GaAs (110) by Prinz [44] up to very large thicknesses (~ 400Å). These bcc fims are ferromagnetic and the magnetic moment per Co atom is about $1.53\mu_B$. Cu/Co/Cu sandwiches [45,56] and Co/Cu superlattices [46] with the fcc structure have also been prepared and studied.

Bulk Mn has a complex crystal structure and is antiferrromagnetic. Several efforts have been made to prepare thin films with expanded lattices to produce magnetic moments approaching the large value found in some dilute alloys ($5\mu_B$) with possibly ferromagnetic ordering. The most recent achievement is the growth of Mn films with a tetragonal structure on Ag (001) [47]. These films are not ferromagnetic.

2-4 Low - dimensional effects

In comparison with the behaviour of the corresponding bulk metal, the temperature dependence of the magnetic properties in ultrathin magnetic layers is changed by the reduction of the number of neighbours and the crossover from 3 to 2 dimensions. In 2 dimensions (2d) there cannot be long-range order for Heisenberg systems with short-range isotropic interactions. [48]. However, in the presence of anisotropy, a 2d Ising-like behaviour, with a transition at finite temperature, is restored. Bander and Mills [49] have recently discussed the interplay between low dimensionality and uniaxial anisotropy in ultrathin layers and expressed the transition temperature of 2d-systems as a function of exchange and anisotropy parameters.

System	Ref.	$T_C(1)$ (K)	$T_C(1)/T_C(\infty)$	$T_C(1)/T_C(\infty)$Ising
Ag/Fe/W(110)	[52]	296	0.285	0.356
Fe/Au (001)	[33]	315	0.302	0.356
Co/Cu (111)	[53]		0.3	0.372
Co/Cu (001)	[54]	> 400	> 0.29	0.234
" (1.5ML)	[55]	150	0.11	0.234
Au/Co/Au(111) (1.6ML)	[56]	> 300	> 0.22	0.372

Table II Curie temperature of iron and cobalt monolayers $T_c(1)$ compared to the Curie temperature of the bulk metals $T_c(\infty)$. From Renard [51].

The reduction of the Curie temperature T_C in thin films has been studied in many systems. For relatively thick Ni films the variation of T_C with the number of atomic layers n is well described by $[T_C(\infty) - T_C(n)] \sim n^{-\lambda}$ with $\lambda = 1.33 \pm 0.13$ or $\lambda = 1.27 \pm 0.2$ [50], in agreement with the prediction of the Ising model, $\lambda = 1.27$. However, when the nickel is sandwiched between noble metals, its ferromagnetism collapses for $n \leq 2$, as we have seen in Section 2.2 .This clearly shows the limits of localized moment models which assume a fixed value of the moment and cannot describe the change of the band structure with the thickness. In table II, from Renard [51], we present a list of T_C reductions between the bulk metal and the monolayer. For most systems the thickness dependence of T_C is in poor agreement with the prediction of the Ising model. This discrepancy is due to the above mentioned change of the electronic structure with the thickness, and probably also to the effects described by Bander and Mills and related to the thickness dependence of the anisotropy [49]. An example of complete discrepancy is the thickness dependence of T_C for fcc Fe layers grown on Cu(001) [25].As can be seen in Fig. 10, T_C increases as the thickness t_{Fe} decreases and is maximum for 3 ML. This could be due to an increase of the anisotropy as t_{Fe} decreases.

At $T \leq T_c$, while the spontaneous magnetization of 3d ferromagnets is expected from the spin wave theory to vary as $T^{3/2}$, $M(T) = M(O)[1 - bT^{3/2}]$, a variation as T instead of $T^{3/2}$ is predicted for 2d-Ising ferromagnets. Earlier experiments seemed to confirm this linear dependence on T in ultrathin films. For example, Wong et al [57] measured M(T) for Fe/V multilayers and reported the observation of a progressive crossover from $T^{3/2}$ for thick Fe layers to T for a few ML. Fig. 11 shows the results of Wong et al [57]. Similar results have also been found for ultrathin films of Fe on Au(001) ,see Fig. 12.

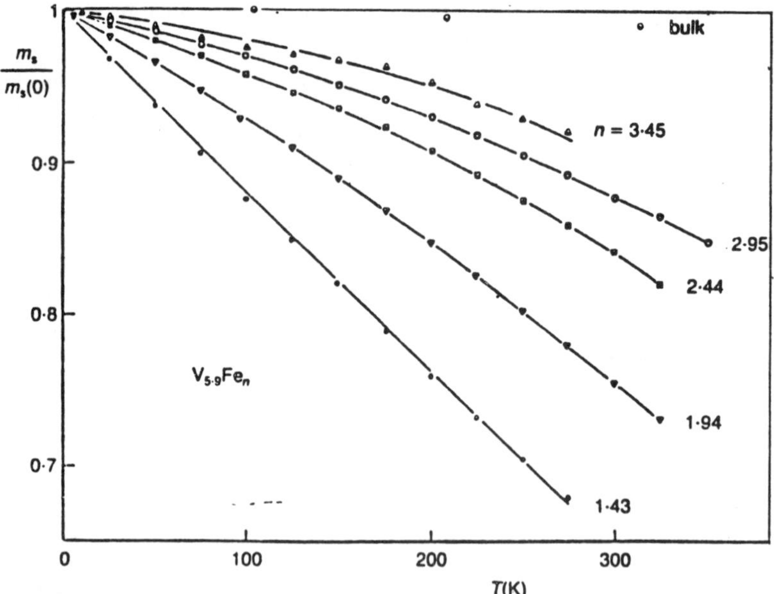

Fig. 11 Temperature dependence of the reduced saturation magnetization for V_{59}/Fe_n. From Wong et al. [57].

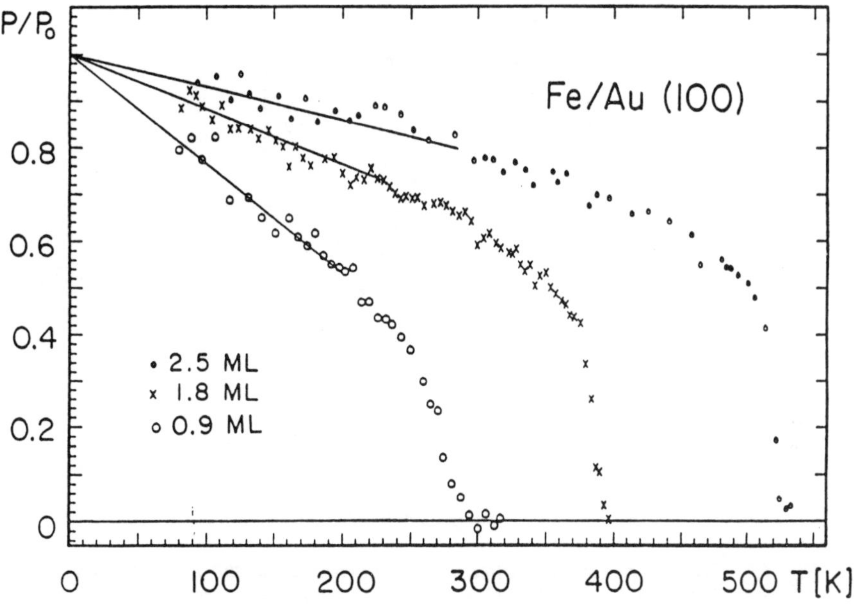

Fig. 12 Temperature dependence of the saturation magnetization for bcc(110) Fe films grown on Au(100) from spin polarized secondary electron emission spectroscopy. From Taborelli et al [58]

However the existence of a linear dependence of M(T) at low temperatures in ultrathin films is questioned in some recent publications. We refer to the work of Mauri et al [59] on the temperature dependence of the spontaneous magnetization M of ultrathin (4.5 - 6 Å) permalloy films . These authors report that the temperature dependence of M in isolated films is much weaker than expected from spin wave theory . Their fit with an exponential variation over a wide range of temperature suggests a gap in the excitation spectrum. They have also reanalysed series of magnetization and Mössbauer data and claim that the conventional spin wave theory always overestimates the variation and that a better fit is obtained with an exponential law again. This leaves the question of the temperature dependence of the spontaneous magnetization at $T \ll T_C$ still open.

The critical behaviour of ultrathin films around the Curie temperature T_C has also not been much studied . Farle and Baberschke [60] have derived the critical behaviour of the susceptibility of Gd films on W(110) from Electron Spin Resonance (ESR) measurements, see Fig. 13. From the fit with $\chi \sim (T - T_C)^{-\gamma}$, they find $\gamma \approx 1.25$ for thick fims (80 Å) and $\gamma \approx 1.8$ for films in the ML range, in good agreement with the predictions of the Ising model, $\gamma = 1.24$ and $\gamma = 1.75$ for 3d and 2d systems respectively.The Ising-like behaviour is supposed to be that of a Heisenberg system with anisotropy. Gadolinium, with well localized moments and weak anisotropy should be an appropriate system to study the crossover from Heisenberg to Ising behaviour.

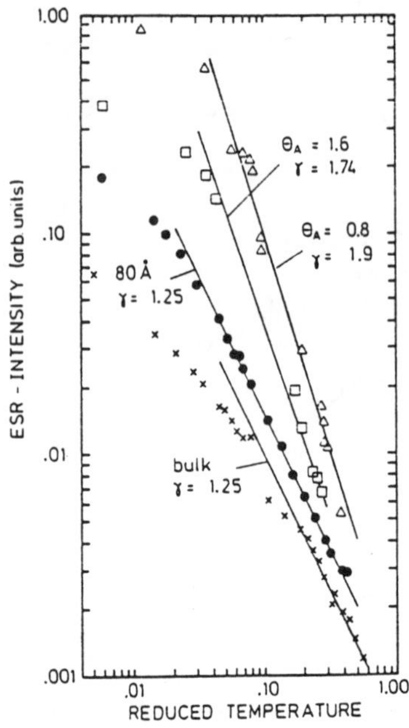

Fig. 13 Log-log plot of ESR intensity (proportional to the susceptibility) of Gd on W(110) for $T > T_C$. The slope gives the exponent γ : $\gamma = 1.9$ for a 0.8 ML coverage, $\gamma = 1.74$ for 1.6 ML, $\gamma = 1.25$ for 80 Å film and bulk Gd. From Farle and Baberschke [60].

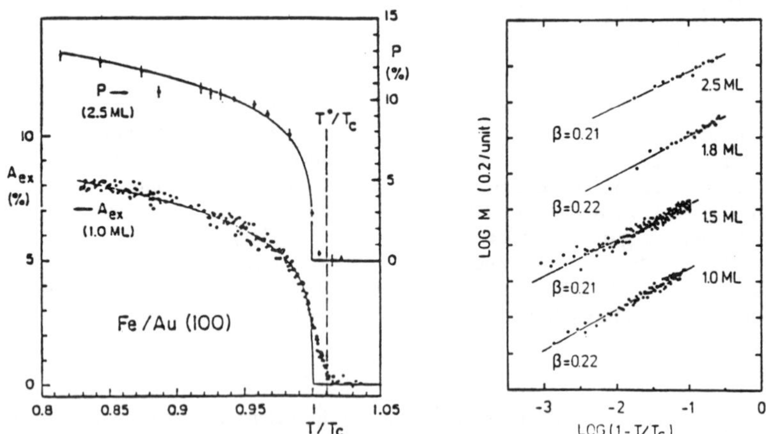

Fig. 14 (left) Magnetization versus T/T_C for bcc (110) Fe films on Au(100). The data are from the exchange asymmetry (A_{ex}) in Spin Polarized Low Energy Electron Diffraction (SPLEED) experiments in the remanent state and from the polarization (P) in Spin Polarized Secondary Electron Emission Spectroscopy (SPSEE). From Taborelli et al. [51].

Fig. 15 (right) Log-log plot of the magnetization vs ($1 - T/T_c$) from the SPLEED and SPSEE experiments of Fig. 13. From Taborelli et al [51].

Dürr et al [61] have studied the critical behaviour below T_C for bcc (110) Fe films on Au(100) by analysing data from Spin Polarized Low Energy Diffraction (SPLEED) and Spin Polarized Electron Emission Spectroscopy (SPEES). By fitting their data to $M \sim (T_C - T)^\beta$, they find $\beta \approx 0.22$ for films in the ML range. The departure from the value expected in 2d-short range systems, $\beta = 0.125$, is ascribed to the long range character of the interactions in Fe. We point out however that the temperature dependences are recorded in a remanent magnetization state, which cannot be the better way to study the temperature dependence of the saturation magnetization. Direct and accurate measurements of the saturation magnetization as a function of T may be hard to carry out.

In conclusion, the low dimensional properties of ultrathin films have not been extensively studied yet. This research field will probably be developped much more in the near future ; this implies accurate measurements on films in which the anisotropy (uniaxial, planar) is well known, preferably on localized moment systems, not only on ferromagnets but also on antiferromagnets, spin glasses, etc.

2-5 Interlayer couplings

Interesting properties of the magnetic multilayers are associated with the coupling between the magnetic moment of different layers. These couplings can give rise to a more or less complex ordering of the magnetic moments of the layers, what is called a magnetic superstructure. For example, Hinchey and Mills [62] have investigated the various superstructure possibilities for a superlattice formed from alternating layers of ferromagnetic and antiferromagnetic materials. We consider the example of Fig. 16 representing a superlattice with an even number of atomic layers in the antiferromagnetic layers. The ground state is the antiferromagnetic (AFM) superstructure shown at the top of the figure. The scheme below displays the successive field induced phase transitions found by Hinchey and Mills [62]. The possibility of creating such artificial superstructures with phase transitions at arbitrary fields is very promissing.

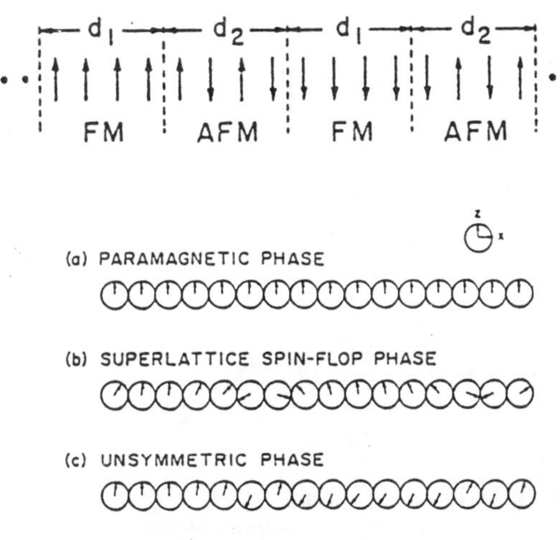

Fig. 16 ***Top*** *: Zero field ground state magnetic superstructure for a ferromagnetic/antiferromagnetic superlattice with an even number (here 4) of atomic layers in the antiferromagnetic layer and with ferromagnetic couplings at the interfaces.*

Below *: Change of the super-structure as a function of the applied field (d) : zero field, (c),(b), (a) : increasing fields. From Hinchey and Mills [62].*

2-5-1 Interlayer coupling accross a non-magnetic layer

The interaction between magnetic moments in metallic systems is an indirect exchange interaction mediated by the conduction electrons. The simplest model, for a free electron conduction band, has been worked out by Rudermann, Kittel, Kasaya and Yosida (RKKY model). In a metallic multilayer, the same type of indirect exchange between magnetic moments in different layers can be mediated by the conduction electrons of the non-magnetic spacing layer. A typical example of such indirect RKKY-like interaction accross a non-magnetic layer is provided by the Gd/Y superlattices (Gd is ferromagnetic and Y is a non-magnetic metal).

The magnetic properties of hcp(c)Gd/Y superlattices grown on Nb(110) have been extensively studied by Kwo and coworkers [63]. They have found that, for certain thicknesses of yttrium, the magnetic moments of the gadolinium ferromagnetic layers are coupled antiferromagnetically through the Y layers.

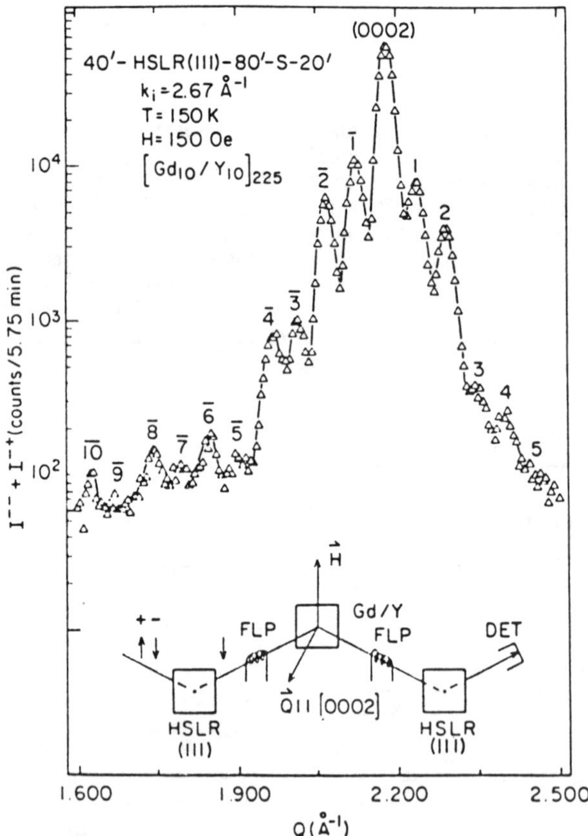

Fig.17 Neutron scattering data for a $(Gd_{10}/Y_{10})_{225}$ superlattice. The odd-numbered satellites begin to appear below T_c at integer multiples of the wave vector for a doubled bilayer thickness with intensities corresponding to spin-flip scattering only. These satellites are associated with the antiferromagnetic superstructure. From Majkrzak et al. [63].

This artificial antiferromagnetic arrangement is clearly shown by the neutron diffraction results of Fig.17. Around the (0002) Bragg peak of the hcp lattice, we can see not only the satellites 2, 4, etc, expected for a superlattice at the wave vector corresponding to the bilayer thickness, but also satellites with odd numbers associated with the doubled period of the antiferromagnetic superstructure (magnetic period = 2 bilayers). The antiferromagnetic satellites collapse when a strong enough magnetic field is applied.

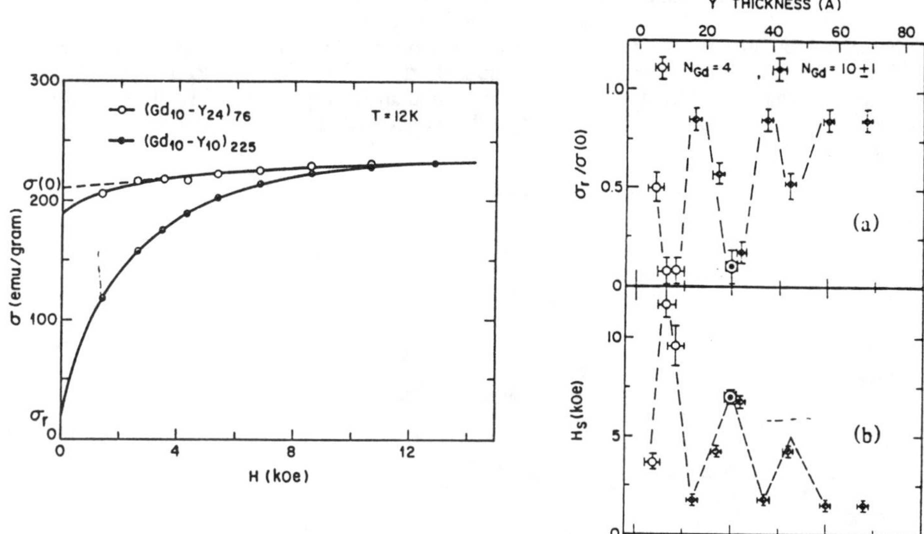

Fig.18 (left) Magnetization curves for $(Gd_{10}\text{-}Y_{24})_{76}$ and $(Gd_{10}\text{-}Y_{10})_{225}$ superlattices. From Kwo et al. [63].

Fig.19 (right) Oscillations of the remanent magnetization μ_r and the saturation field H_S as a function of the yttrium thickness in Y/Gd superlattices. From Kwo et al. [63].

The corresponding magnetic behavior is shown in Fig.18. For the Gd_{10}/Y_{10} superlattice with an antiferromagnetic superstructure a strong magnetic field has to be applied to overcome the interlayer coupling and saturate the magnetization. In contrast, for other Y thicknesses, the magnetization is saturated easily and the remanent magnetization is high, as is expected for a ferromagnetic interlayer coupling. When the saturation field and the remanent magnetization are plotted as a function of the Y thickness, oscillations appear, which means that the interlayer coupling constant is an oscillating function of the Y thickness.

The oscillations of the interlayer coupling between Gd across Y can be accounted for in the model of indirect exchange interaction worked out by Yafet [64]. The model is based on the wave vector dependence of the Fourier transform of the Y susceptibility $\chi(q)$. Due to nesting properties of the Fermi surface of Y, $\chi(q)$ is maximum at a non-zero value q_{max}. Consequently the polarization induced by exchange with a Gd spin oscillates with a period related to q_{max} and the resulting indirect interaction between two Gd layers is an oscillating function of the thickness.

The hcp(c axis) Dy/Y and Er/Y superlattices have been also extensively studied [65,66]. For Dy/Y [65], inside the Dy layers, the spins are in the basal plane of the hcp structure (layer plane) and the turn angle between successive atomic planes changes from 45° at T_N to 31° at 10K. The outstanding result of the neutron diffraction experiments is the phase coherence between the helical structure of successive Dy layers. The turn angle per atomic layer in yttrium is around 52°. The coupling between the phases of the helical structures of successive Dy layers again results from the property of Y to develop spin density waves with a wave vector around q_{max}. In a similar way, the Er/Y superlattices exhibit sinusoidal c-axis modulated magnetic structures with again long range order extending cross the Y layers [66]. Finally the magnetic order of Gd/Dy superlattices corresponds to a super-spiral composed of ferromagnetic Gd layers coherent with helical Dy layers [63]. Fig 20 compares the magnetic structures of the Gd/Y,Dy/Y and Gd/Dy superlattices._

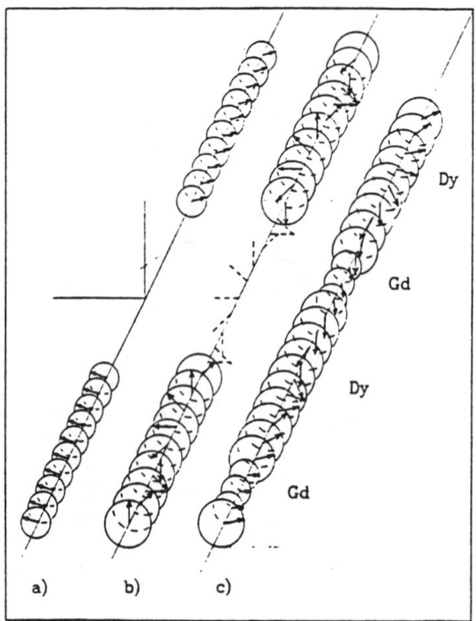

Fig 20 Scheme representating the basal plane components of the magnetization in three RE superlattices : Gd/Y,Dy/Y and Gd/Dy After Kwo et al.[63].

The Fe/Cr multilayered structures too exhibit interesting properties arising from antiferromagnetic interlayer couplings. The existence of antiferromagnetic couplings has been observed first in Fe/Cr/Fe sandwiches by the Brillouin scattering measurements of Grünberg et al. [67] and the SPLEED experiments of Carbone and Alvarado [68]. We show in Fig.20 the SPLEED results: the electron polarization is recorded during the growth of the bcc(001)Fe/Cr/Fe sandwich on Cr(001). The polarization of the second Fe layer turns out to be opposite to that of the first layer, which demonstrates the existence of an antiferromagnetic coupling. When the second Fe layer is thicker that the first one, its magnetization turns to the applied field direction, whereas the magnetization of the first layer is supposed to turn to the opposite direction (see scheme below the figure).

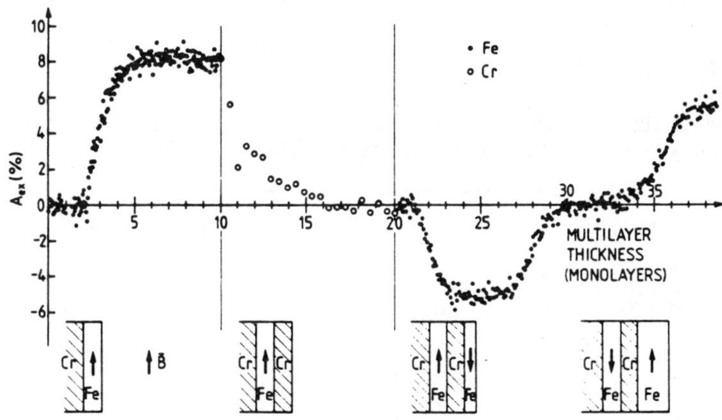

Fig.21 Exchange asymmetry in SPLEED measurements recorded during the growth of a Fe/Cr/Fe sandwich on Cr(001) at 330K. The sandwich is composed of 10 ML of Fe, 10 ML of Cr and 20 ML of Fe. Below: film magnetization deduced from the data. From Carbone and Alvarado [68].

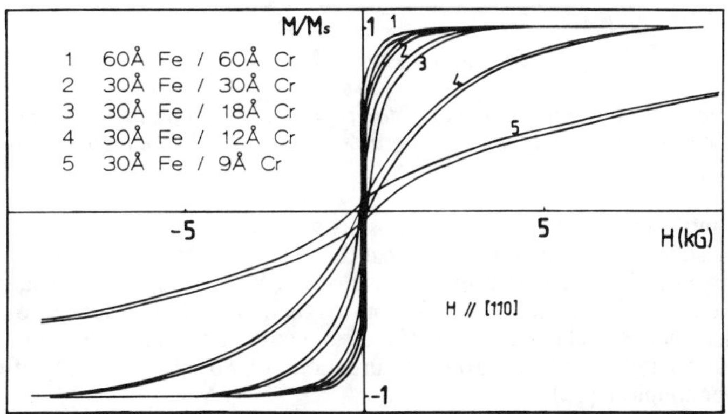

Fig.22 Magnetization curves for a series of bcc(001)Fe/Cr superlattices. As the Cr thickness decreases, higher fields are needed to overcome the antiferromagnetic couplings and saturate the magnetization. From Nguyen Van Dau et al. [69,70].

More recently the interlayer coupling has been studied in bcc(001) Fe/Cr superlattices grown on Ga As(001) [69-71]. The existence of an antiferromagnetic superstructure is demonstrated by polarized neutron diffraction experiments [71] : one observes a small angle Bragg peak corresponding to the doubling of the chemical bilayer periodicity ; this peak appears in the spin-flip scattering and collapses in high magnetic fields, as expected for an antiferromagnetic superstructure. Typical magnetization curves are shown in Fig. 22.As the Cr thickness t_{Cr} decreases the magnetization becomes harder and harder to saturate, which expresses the progressive increase of the interlayer coupling, and the remanent magnetization becomes very small. In the simplest model, which neglects the anisotropy and assumes that the magnetic moment of a Fe layer is field independent, the magnetization along the field directions is expressed as [69]

$$M/M_S = M_S\, t_{Fe} H / 4J \qquad (12)$$

where J is the interlayer coupling constant per surface unit. Analysis of the initial slopes of Fig. 22 yields very large values of J and the thickness dependence shown in Fig. 23.

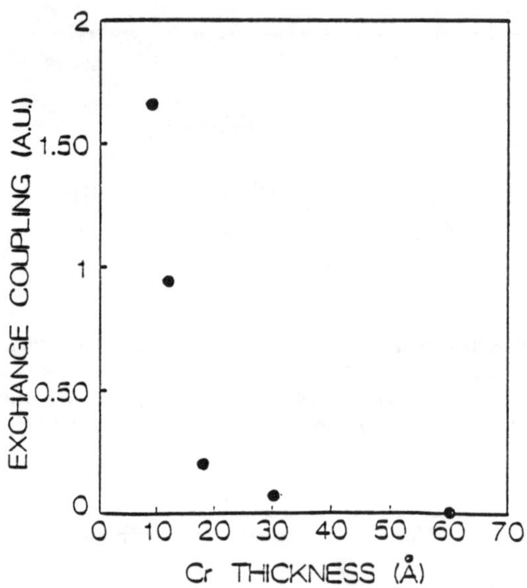

Fig. 23 Interlayer coupling constant J for t_{Fe}= 30Å as a function of the Cr thickness. Data from the magnetization curves of Fig. 22. From Nguyen Van Dau et al [69].

The antiferromagnetic coupling subsists well above the Néel temperature of the spin density wave state of Cr, which indicates that this interlayer coupling is not related to the antiferromagnetism of Cr. In contrast to the case of Gd/Y the coupling constant does not oscillate as a function of the thickness of the non-magnetic metal, which rules out standard RKKY-like models. Numerical calculations of the band structure of the Fe/Cr superlattice are probably necessary to understand the strong perturbation of Cr by the proximity of Fe and the resulting interlayer coupling [72].

2-5-2 Exchange anisotropy

We adopt the definition of exchange anisotropy by Malozemoff [75]. "Exchange anisotropy refers to a complex of unusual properties of exchange - coupled ferromagnetic-antiferromagnetic (F-AF) sandwiches. Typically these properties appear after the sandwich is cooled through the AF Néel temperature T_N with an applied field which single-domains the ferromagnetic layer. At the operating temperature, which is usually the room temperature in systems like Co/CoO and FeNi/FeMn, the applied field can then be removed, but the ferromagnetic now behaves as if it was subject to some sort of "internal exchange field" acting in the same direction as the previously applied magnetic field. Alternatively the effect may be described as a unidirectional anisotropy which breaks time reversal symmetry". A typical shifted hysteresis loop is shown in Fig. 25. The offset or "exchange anisotropy field" H_E is the main relevant parameter. Generally H_E is smaller than 0.1 Tesla. The exchange anisotropy effect is used in magnetic film devices to replace an applied magnetic field.

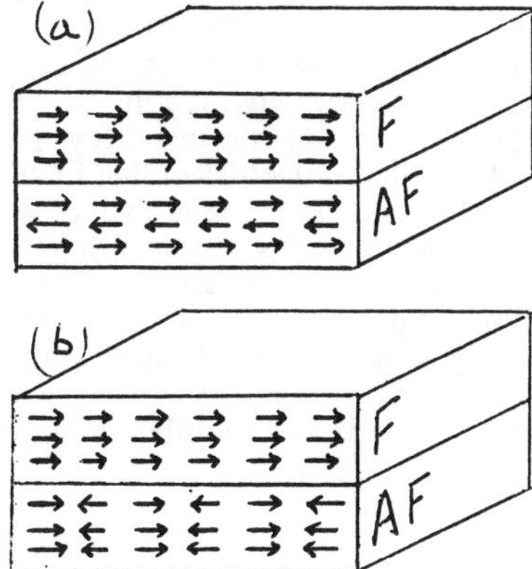

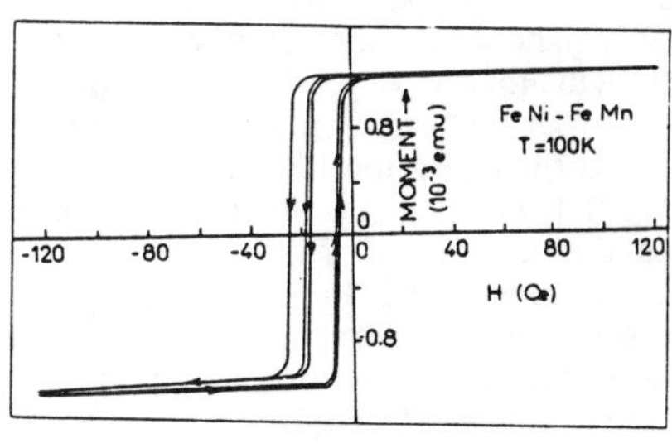

Fig. 24 (left) (a) : Spin configuration at the F/AF interface after field cooling(when one of the two AF sublattices can be placed along the interface); (b) : when the two AF sublattices alternate at the interface, the interfacial exchange energy is zero.

Fig. 25(right) Example of shifted hysteresis loops for Fe Ni/Fe Mn bilayers. FeNi is F, Fe Mn is AF. Note the "contraction" of the shift after several loops. After Schlenker et al [76].

Fig. 24 (a) pictures the mechanism of the exchange anisotropy in the early model of Meiklejohn and Bean [77]. If, for example, the interactions between spins across the interface are ferromagnetic, the spins, during the field cooling , adopt the configuration represented in Fig. 24(a), which minimizes the exchange energy. Below T_N, because the spin configuration is frozen in the antiferromagnet (there is no interaction with the external field to overcome the anisotropy forces), the magnetization of the ferromagnet can be reversed only if the field is strong enough to overcome the interfacial exchange. If σ is "the interfacial exchange" energy per surface unit, it is easy to derive

$$H_E = \sigma / 2M_F t_F \qquad (13)$$

where M_F and t_F are the magnetization and the thickness of the ferromagnetic layer.

This model, when one estimates σ from standard values for the exchange coupling between neighbouring spins across the interface, overestimates H_E by two or three orders of magnitude. Several improvements have been recently introduced in the model to obtain a better quantitative account of the experimental results. H_E can be reduced by the following effects :

- the mixing of the configuration of Fig 24 (a), where only one of the two AF sublattices is along the interface, with the configuration of Fig. 24(b), which does not contribute to the exchange anisotropy.
- the relaxation of the exchange anisotropy energy by introduction of interfacial domains walls which spread the fr tion over several atomic planes.
- the randomness of the interfacial exchange arising from interface roughness [78].

It is not beyond the scope of this lecture to describe these recent theoretical developments and their application to the interpretation of the experimental results. For more details we refer to a review paper by Malozemoff [75].

3 - TRANSPORT PROPERTIES

The first development of the research on the electron transport properties of thin metallic films took place thirty or forty years ago : the Fuchs-Sondheimer theory was representing a fine achievement of the transport models based on the Boltzmann equation and many experiments were performed to show the interplay between the film thickness and the electron mean free path predicted by the theory. A good report on this early stage of the transport properties of thin films is the review paper by Maisel in 1970 [79]. During recent years the progress in the growth of well characterized films and multilayers has revived the subject in new directions : quantum effects, interplay with the magnetic properties, localization phenomena, etc... . We first present the classical Fuchs-Sondheimer theory of the electron transport in thin films in paragraph 3.1 and we review related experiments in 3.2. In 3.3 we describe the recently developed quantum mechanical models. Then we proceed to the transport in multilayers (3.4) and magnetic systems (3.5), and to the localization phenomena (3.6). A review of the superconducting properties of multilayers can be found in the lecture by G. Creuzet.

3-1 Fuchs-Sondheimer classical theory of the resistivity of thin films

The Fuchs-Sondheimer model [80,81] is based on the Boltzmann equation for the electrons [82]:

$$\frac{e}{m}\mathbf{E}.\mathbf{grad}_{\mathbf{v}}\left[f^{0}(\mathbf{v})\right] = \frac{1}{\tau}g(\mathbf{v},\mathbf{r}) + \mathbf{v}.\mathrm{grad}_{\mathbf{r}}[g(\mathbf{v},\mathbf{r})] \qquad (14)$$

where g is the departure of the distribution function of the electron gas $f(\mathbf{v},\mathbf{r})$ from the Fermi-Dirac equilibrium distribution, i.e

$$g(\mathbf{v},\mathbf{r}) = f(\mathbf{v},\mathbf{r}) - f^{(0)}(\mathbf{v}) \qquad (15)$$

with

$$f^{(0)}(\mathbf{v}) = \frac{1}{\exp\left[\frac{(\varepsilon_v - \varepsilon_F)}{k_B T}\right] + 1} \qquad (16)$$

$$\varepsilon_v = \frac{1}{2}mv^2$$

$\mathbf{E}$ is the electric field and τ is the relaxation time supposed to be a function of v only. Eq.11 means that the change of $g(\mathbf{v},\mathbf{r})$ due to the acceleration of the electrons by the electric field $\mathbf{E}$ is balanced by the change due to scattering expressed by the term g/τ and the change due to the diffusion which tends to make the distribution uniform. For a bulk metal , g is uniform, $\mathbf{grad}_{\mathbf{r}}[g] = 0$, and the standard solution is

$$g_B(\mathbf{v}) = \frac{\tau e\,\mathbf{E}}{m}.\mathbf{grad}_{\mathbf{v}}\left[f^{(0)}(\mathbf{v})\right] = e\tau\mathbf{E}.\mathbf{v}\,\frac{\partial f^{(0)}}{\partial \varepsilon_v} \qquad (17)$$

Eq.17 means that the Fermi surface is shifted in k-space by $eE\tau(v_F)/\hbar$ in the opposite direction to E. The current density is proportional to this shift,

$$j = -\frac{ne^2\tau(v_F)E}{m}$$

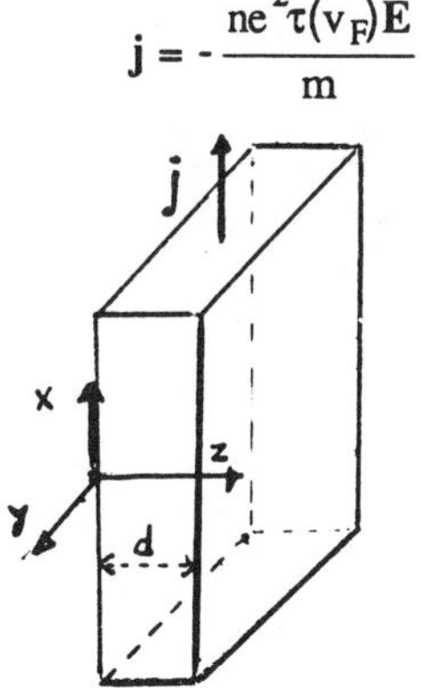

Fig. 26 Thin film geometry.

We now consider the case of a metallic film of thickness d. The z-axis is perpendicular to the plane of the film and the current is supposed to be in the x direction. Due to the diffuse part of the scattering by the surfaces the current density will generally be smaller near the surfaces. This means that g depends on z and that the diffusion term in Eq.14 does not vanish. Eq.14 is written as:

$$\frac{eE}{mv_z} \cdot \frac{\partial f^{(0)}(v)}{\partial v_z} = \frac{g(v,z)}{\tau v_z} + \frac{\partial g}{\partial z} \tag{18}$$

The general solution of Eq.18 is of the form

$$g(\mathbf{v},z) = e\tau \mathbf{E}.\mathbf{v}\,\frac{\partial f^{(0)}}{\partial \varepsilon_v}\left[1 + F(\mathbf{v})\exp(\frac{-z}{\tau v_z})\right] \tag{19}$$

where $F(\mathbf{v})$ is an arbitrary function of $\mathbf{v}$ to be determined by introducing appropriate boundary conditions.

The simplest assumption for the boundary conditions is to suppose that the scattering at the surfaces is entirely diffuse. The distribution function of the electrons **leaving** a surface must be independent of the direction of $\mathbf{v}$. Eq.19 shows that this can be satisfied only if we choose $F(\mathbf{v})$ so that $g(\mathbf{v},0) = 0$ for all $\mathbf{v}$ having $v_z>0$ and $g(\mathbf{v},d) = 0$ for all $\mathbf{v}$ having $v_z<0$. There are therefore two functions g: g^+ for electrons with $v_z>0$ and g^- for electrons with $v_z<0$:

$$g^+(\mathbf{v},z) = e\tau \mathbf{E}.\mathbf{v}\,\frac{\partial f^{(0)}(v)}{\partial \varepsilon_v}\left[1 - \exp(\frac{-z}{\tau v_z})\right] \tag{20}$$

$$g^-(\mathbf{v},z) = e\tau \mathbf{E}.\mathbf{v}\,\frac{\partial f^{(0)}(v)}{\partial \varepsilon_v}\left[1 - \exp(\frac{d-z}{\tau v_z})\right]$$

The variation of g^+ and g^- as a function of z, for a given $\mathbf{v}$ is shown in Fig.27

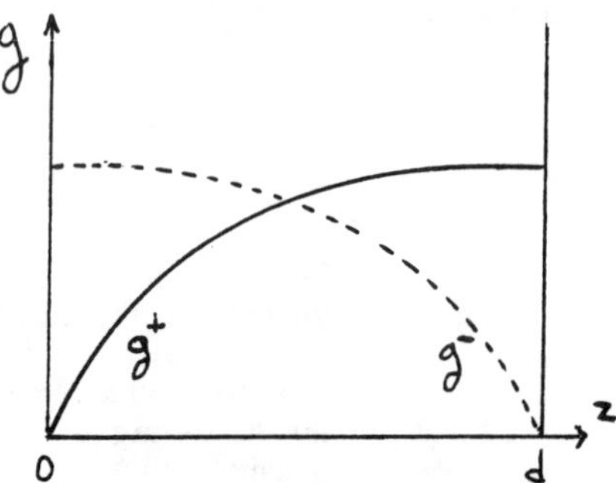

Fig.27 Variation of g^+ and g^- as a function of z for a fixed **v**.

The current carried by the film is given by an expression of the form

$$J \sim \int e\, v_x\, g(\mathbf{v},z)\, d^3v\, dz \tag{21}$$

to be compared to the current in an equivalent slab of the bulk metal:

$$J_0 \sim \int e\, v_x\, g_B(\mathbf{v})\, d^3v\, dz \tag{22}$$

where g_B is expressed by Eq.17.
Resorting to polar coordinates

$$v_z = v\cos\Theta\ ,\ v_x = v\sin\Theta\cos\phi$$
$$d^3v = \frac{v}{m}\, d\varepsilon_v \sin\Theta\, d\Theta d\phi$$

one derives the ratio of conductivity of the film to that of the bulk metal

$$\frac{\sigma}{\sigma_0} = \frac{J}{J_0} = \frac{\int_0^d dz \int_0^{\frac{\pi}{2}} d\Theta \int_0^{2\pi} d\phi\, \sin^3\Theta\, \cos^2\phi \left[1 - \exp(\frac{-d}{2\lambda\cos\Theta})\cosh(\frac{d-2z}{2\lambda\cos\Theta})\right]}{\int_0^d dz \int_0^{\frac{\pi}{2}} d\Theta \int_0^{2\pi} d\phi\, \sin^2\Theta\, \cos^2\phi} \tag{23}$$

where $\lambda = \tau v_F$ is the mean free path of the electrons in the bulk metal. After integrating over z and ϕ one obtains

$$\frac{\sigma}{\sigma_0} = 1 - \frac{3}{2k}\int_1^{\infty} (\frac{1}{t^3} - \frac{1}{t^5})\left[1 - \exp(-kt)\right] dt \tag{24}$$

where $t = 1/\cos\Theta, k = d/\lambda$ = thickness/mean free path. Approximations can be made for large and small k

$$\frac{\sigma}{\sigma_0} = 1 - \frac{3}{8k} \qquad \text{for } k = \frac{d}{\lambda} >> 1 \tag{25}$$

$$\frac{\sigma}{\sigma_0} = \frac{3k}{4}\left(\ln\frac{1}{k} + 0.423\right) \qquad \text{for } k = \frac{d}{\lambda} << 1 \tag{26}$$

which gives the main features shown in Fig.28 for the plot of $\rho/\rho_0 = \sigma_0/\sigma$ as a function of k.

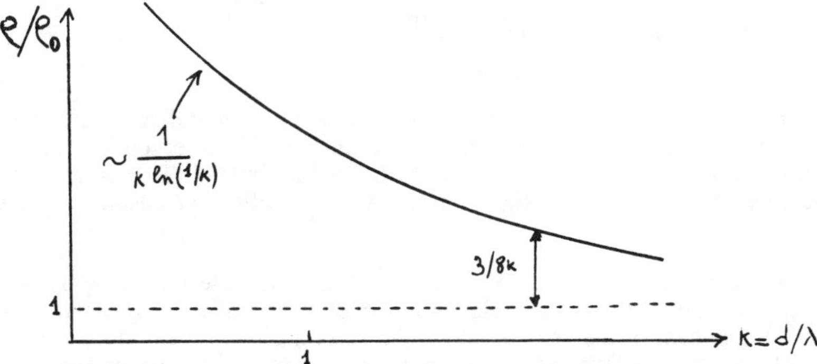

Fig. 28 Ratio of the resistivity of a thin film to the resistivity of the bulk metal versus k = d/λ (d = thickness, λ = mean free path).

It is also useful to derive expressions of the resistivity as a function of d and λ. With $\rho_0 = mv_F/ne^2\lambda$, Eq 25-26 are written as

$$\rho = \frac{mv_F}{ne^2}\left[\frac{1}{\lambda} + \frac{3}{8d}\right] \qquad \text{for } \lambda << d \tag{27}$$

$$\rho = \frac{4mv_F}{3ne^2}\left[\frac{1}{d\left(\ln\frac{\lambda}{d} + 0.423\right)}\right] \qquad \text{for } \lambda >> d \tag{28}$$

which gives the behavior shown in Fig.29.

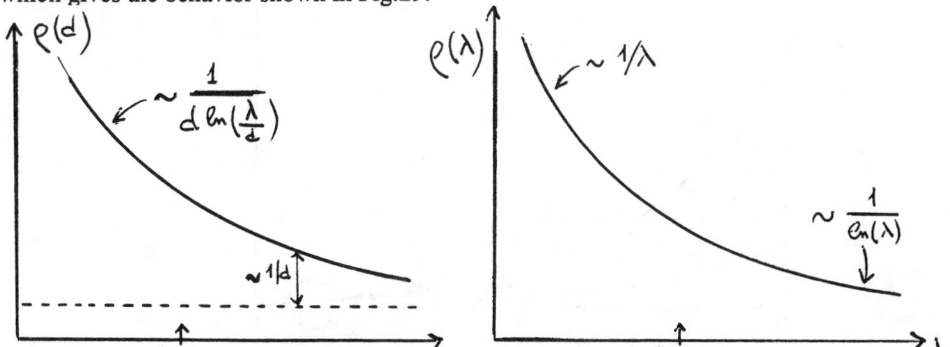

Fig.29 The Fuchs-Sondheimer resistivity as a function of the film thickness d (at fixed λ) and as a function of the mean free path λ (at fixed d).

The limit $\lambda<<d$ is clearly understood: there is a simple addition of scattering rates proportional to $1/\lambda$ and to $1/d$ arising from background and surface scattering,respectively. In the limit $\lambda>>d$ the result of the Fuchs-Sondeimer theory is puzzling as ρ tends to zero when the background scattering vanishes (i.e when $\lambda\to\infty$) at fixed d. This seems physical: for d fixed at a finite value one would rather expect a finite surface scattering and therefore a finite residual surface resistivity. As a matter of fact the assumption of classical point electrons makes the Fuchs-Sondheimer theory incorrect in the limit $\lambda\to\infty$. Electrons with velocities almost parallel to the surfaces ($v_z/v<<1$) can propagate with a vanishingly small surface scattering; more precisely, for electrons with $d/\lambda < v_z/v <<1$, Eq.20 indicates that the function g is proportional to

$$1-\exp\left(\frac{-z}{\tau v_z}\right)=1-\exp\left(\frac{z\,v}{\lambda v_z}\right)\simeq\frac{dv}{\lambda v_z} \qquad (29)$$

By integrating g over $\alpha=v_z/v$ down to d/λ we do obtain a current density diverging as $\ln(\lambda)$. We see in paragraph 3-3 that, in quantum mechanical models, this divergence of the conductivity for $\lambda\to\infty$ disappears and the resistivity remains finite for finite d. This is because the quantum mechanical zero point motion excludes momentum states confined in a plane parallel to the surfaces.

Up to now we have considered the case with only diffuse scattering by the surfaces. The opposite case is that with only specular reflections at the surfaces. Specular reflections do not change the component of the velocity along the current direction and do not contribute to the resistivity. In the intermediate case, if one assumes that a fraction p of the electrons are speculary reflected while the fraction (1-p) is scattered, calculations similar to those presented above for the case p=0 lead to

$$\frac{\sigma}{\sigma_0}=1-\frac{3\,(1-p)}{8k} \qquad \text{for } k=\frac{d}{\lambda}>>1 \qquad (30)$$

instead of Eq.25 and

$$\frac{\sigma}{\sigma_0}=\frac{3k}{4}(1+2p)\left(\ln\frac{1}{k}+0.423\right) \qquad \text{for } k=\frac{d}{\lambda},\ p<<1 \qquad (31)$$

instead of Eq.26. We see that the contribution from the surfaces decreases progressively to zero as p increases. Fig.5 shows a plot of ρ/ρ_0 versus k for several values of p.

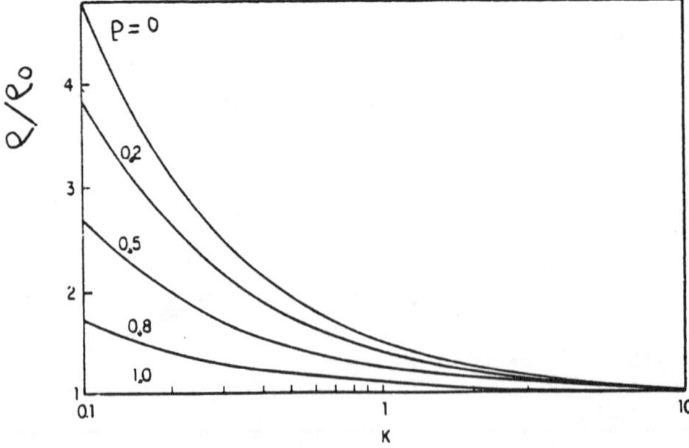

Fig.30 Ratio of the resistivity to the bulk metal resistivity as a function of $k=d/\lambda$ for several values of the proportion of specular reflections . From Campbell.[83]

Finally we summarize the results of the Fuchs-Sondheimer theory for the temperature dependence of the resistivity of thin films. The temperature dependent phonon scattering enhances the background scattering (and therefore reduces λ), which finally reduces the coefficient $\alpha = 1/\rho\, \partial\rho/\partial T$ with respect to its value in the bulk material, α_0.

$$\frac{\alpha}{\alpha_0} = 1 - \frac{3(1-p)}{8k} \qquad \text{for } k >> 1 \tag{32}$$

$$\frac{\alpha}{\alpha_0} = \frac{1}{\ln(\frac{1}{k}) + 0.423} \qquad \text{for } k << 1 \tag{33}$$

3-2 Application of the Fuchs-Sondheimer theory to thin films

The experimental verification of the predictions of the Fuchs-Sondheimer theory is not as straightforward as might be expected. This is due to the difficulties in preparing clean films with well characterized surfaces and to additional effects we will discuss later.

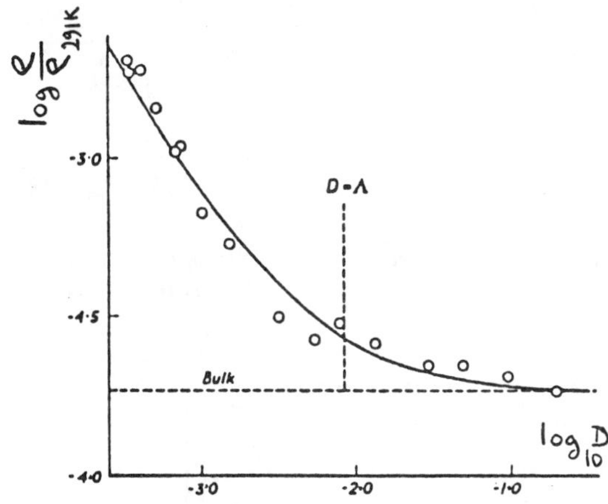

Fig.31 Log-log plot of the resistivity of Sn films at 38K versus the film thickness. The solid line is the variation predicted by the Fuchs-Sondheimer theory with p=0. From Andrew [84].

Fig. 31 shows an example of a good fit between the theoretical predictions and experimental results on a Sn film [84]. A good agreement has also been obtained by Mayer [85] in extensive studies on alkali metal films. However, even when there is a good agreement for relatively thick films, it often deteriorates at small values of $k = d/\lambda$ (say below $k \sim 10^{-1}$), as will be discussed later. In most films the best agreement is generally obtained for p = 0, that is for completely diffuse scattering at the surfaces. However certain metals such as Pb, Ag and Au have values of p very close to 1, provided they have received a suitable annealing treatment. An example is shown in Fig. 32 : the thickness dependence of the resistivity almost vanishes after annealing. Resistivity measurements during layer by layer epitaxial growth have also shown that p is maximum when a layer has just been completed. Fig. 33 shows the resulting oscillations of the resistivity and their correlation with the oscillations of the RHEED intensity [87].

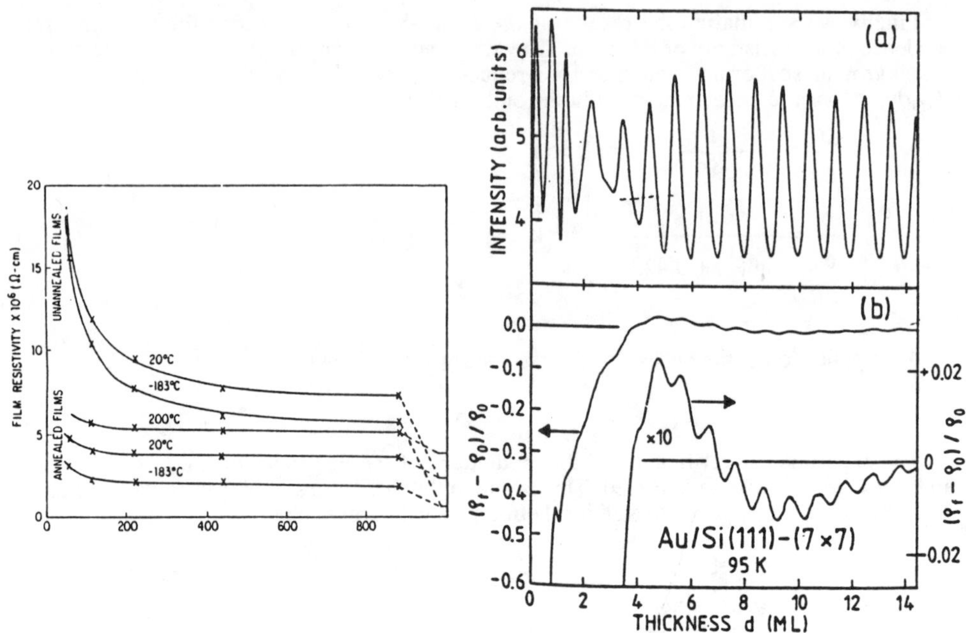

Fig.32 (left) Resistivity versus thickness for unannealed and annealed gold films at several temperatures.From Gillham, Preston and Williams [86].

Fig.33 (right) RHEED specular beam intensity (a) and resistivity oscillations (b) during growth of Au films on a Si(111)-(7x7)surface at 95K. From Jalochowski and Bauer [87].

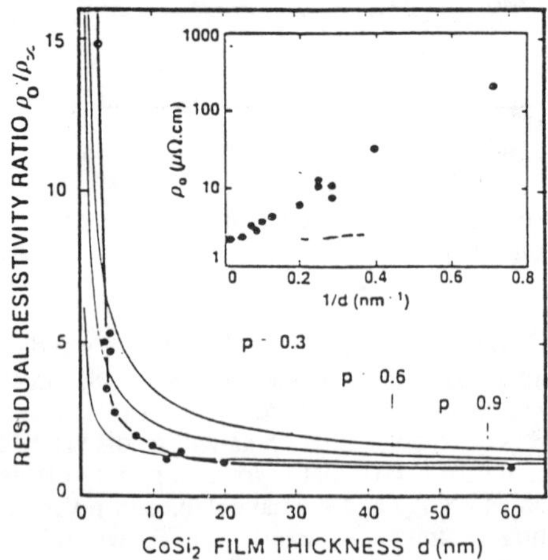

Fig.34 Low temperature residual resistivity of $CoSi_2$ single crystal films grown on Si(111) as a function of the film thickness. The thin lines refer to the Fuchs-Sondheimer calculations for different values of the parameter p. From Badoz et al. [89].

There are several additional effects which also contribute to the thickness dependence of the resistivity. In polycrystalline films the most important of these effects arises from the thickness dependence of the gain size. An extension of the Fuchs-Sondheimer theory taking into account the grain boundary scattering (with some assumptions on the thickness dependence of the grain size) has been worked out by Mayadas and Shatzkes [88] and applied to many experimental results.

Neverthless, even when these additional effects are taken into account, there are systematic departures from the theoretical predictions at small values of the parameter $k=d/\lambda$. In recent years the progress of the epitaxy techniques have enabled very thin single crystal films with long mean free path to be prepared and the departure from the Fuchs-Sondheimer theory in the regime $d/\lambda<<1$ to be studied. An example of experimental results obtained by Badoz et al. [89] on $CoSi_2$ is shown in Fig.34.The sharp upturn of the resistivity at small thickness cannot be fitted with the Fuchs-Sondheimer expressions and has been accounted for by recently developed quantum mechanical models. These models are described in the next paragraph.

3-3 Quantum mechanical models of the resistivity of thin films

Full quantum mechanical treatments of conduction in thin films have recently been worked out by Tesanovic et al. [90] and Fishman and Calecki [91]. We describe the model of Fishman and Calecki. Its starting point is the following Hamiltonian for electrons (or holes) in an ideal film with perfectly plane surfaces.

$$H_0 = \frac{p^2}{2m} + V\left[Y(z - \frac{1}{2}d) + Y(-z - \frac{1}{2}d)\right] \tag{34}$$

where Y(z) is the step function and V is the potential height outside the well for $z>d/2$ and $z<-d/2$. The quantization of the energy associated with the motion along the z axis gives rise to sub-bands and the energy in the sub-band ν is written as

$$E_{\nu k} = E_\nu + \hbar^2k^2/2m \tag{35}$$

where k is a two-dimensional wave vector in the plane parallel to the surfaces. If the surface at $z=d/2$, for example, is rough, its equation becomes

$$z = d/2 + f(\rho) \tag{36}$$

where ρ is a two-dimensional vector in the (x,y) plane ($f(\rho)<<d$) and the scattering potential induced by the roughness is written as

$$U = V\left\{Y\left[z - \frac{d}{2} - f(\rho)\right] - Y(z - \frac{d}{2})\right\} \simeq - V f(\rho)\ \delta(z - \frac{d}{2}) \tag{37}$$

U is the only scattering potential in the model. There is no additional background scattering (in the notation of Fuchs-Sondheimer $\lambda=\infty$), so that the model is implicitely worked out for the case $\lambda/d>>1$. In particular the model cannot describe the situation with the non-uniform current density occuring in the Fuchs-Sondheimer theory for $\lambda/d\leq1$. Thus, whereas the Fuchs-Sondheimer theory is relevant for relatively dirty systems, the Fishman-Calecki model is appropriate to describe very thin-high purity samples for which the Fuchs-Sondheimer theory is inadequate.

We now summarize the results obtained by calculating the scattering of the eigenstates of H_0, Eq.34, by the roughness potential U, Eq.37. The simplest results are obtained in the limit $\xi k_1 \leq 1$ where k_1 is of the order of magnitude of the wave vector at the Fermi level and ξ is the correlation length of the roughness function f(ρ) (for metallic films the condition $\xi k_1 \leq 1$ is obeyed if the surface profile is rough at the atomic scale). We first consider the case of a metal for which the number of occupied sub-bands N is very large. In this case the thickness dependence of the resistivity calculated by Fishman and Calecki [91] is written:

$$\rho \sim \frac{1}{d^2 \xi^2 \Delta^2} \left[1 - \frac{6}{(3 n \pi^5)^{\frac{1}{3}}} \times \frac{1}{d} \right]^{-1} \tag{38}$$

This expression, for 10Å<d<200Å and $n = 3 \times 10^{22} cm^{-3}$ is equivalent to $\rho \sim d^{-2.1}$. The experimental thickness dependence of the resistivity for the metallic films in the limit $d \ll \lambda$ is in much better agreement with $\rho \sim d^{-\alpha}$, $\alpha \gtrsim 2$ predicted by the quantum mechanical models than with the variation $\rho \sim d^{-1}$ of the Fuchs-Sondheimer theory. For Co Si_2 thin films Badoz et al [89] derive $\alpha = 2.3$. Fig. 35 shows the good agreement obtained by Fishman and Calecki between their calculation (solid line) and the experimental results of Badoz et al on $CoSi_2$. The best fit is obtained with $\xi = 2$ Å for the correlation length of the roughness function, $\Delta = 4$ Å for the root mean square of the height of the bumps on the surface and $n = 3 \times 10^{22} cm^{-3}$ for the electron density.

In the opposite limit, for N = 1 (one occupied subband), a variation of ρ as d^{-6} is found in the model of Fishman and Calecki. This is a well known limit for bi-dimensional semiconductors. For metals, N can be a relatively small integer only in films composed of a few ML. In this thickness range the variation of the resistivity as a function of the thickness is expected to show oscillations correlated with the change of N between integer values and superimposed to a crossover from $\alpha = 2.1$ to $\alpha = 6$. To our knowledge such quantum size effects have never been observed up to now.

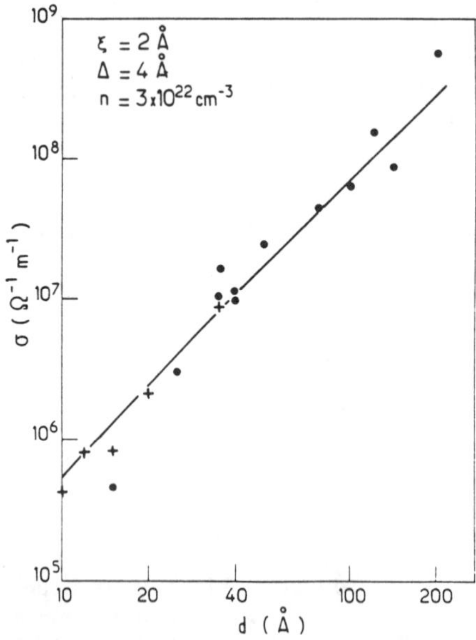

Fig. 35 Low temperature inverse residual resistivity $\sigma(d) = [\rho(d) - \rho(\infty)]^{-1}$ as a function of the film thickness d for two $CoSi_2$ samples. Symbols : experimental results of Badoz et al [89]. Solid line : best fit between the data and the Fishman-Calecki model [91].

3-4 Electron transport in multilayers

The theoretical models developed to describe the resistivity of multilayers are extensions of the Fuchs-Sondheimer theory for thin films. The simplest and most convenient model has been worked out by Carcia and Suna [92]. These authors introduce a probability p of coherent passage across an interface which is equivalent to the probability p of specular reflection in the Fuchs-Sondheimer theory. The probability of diffuse scattering at the interface is 1-p (this implies that the probability of specular reflection is zero). The calculation is similar to that described in 3-1: the functions $g^+(z)$ and $g^-(z)$, because of the interface scattering, have a minimum at the interfaces (g=0 at the interface if p=0) and vary exponentially inside the layers. The total current is obtained by integrating g over z, and Carcia and Suna [92] find for the conductivity of the multilayer:

$$\sigma(d_1+d_2)=d_1\sigma_1+d_2\sigma_2+[(\lambda_2\sigma_1+\lambda_1\sigma_2)p-\lambda_1\sigma_1-\lambda_2\sigma_2]\,I-(1-p^2)[\sigma_1\lambda_1J_1+\sigma_2\lambda_2J_2] \quad (39)$$

where d_1 and d_2 are the layer thiknesses, σ_1, σ_2 and λ_1, λ_2 are the conductivities and mean free paths of the two metals, and

$$I=\frac{3}{2}\int_0^1 d\mu\,\frac{\mu\,(1-\mu^2)\,(1-e_1)\,(1-e_2)}{(1-p^2e_1e_2)}$$

$$J_1=\frac{3}{2}\int_0^1 d\mu\,\frac{\mu\,(1-\mu^2)\,(1-e_1)\,e_2}{(1-p^2e_1e_2)} \quad (40)$$

$$J_2=\frac{3}{2}\int_0^1 d\mu\,\frac{\mu\,(1-\mu^2)\,(1-e_2)e_1}{(1-p^2e_1e_2)}$$

with $e_i = \exp(-d_i/\lambda_i\mu)$

It is easy to check that these expressions of the conductivity bears a resemblance with the expressions of the Fuchs-Sondheimer theory. For example, for $\sigma_1=\sigma_2=\sigma_0$, $\lambda_1=\lambda_2=\lambda$, p=0 one obtains for s exactly Eq.24, as expected for the situation of independent thin films of thickness d and mean free path λ. The limit $k=d/\lambda>>1$ again gives the typical variation of the resistivity as d^{-1}, Eq.27

$$\rho \sim [\,1/\lambda + 3/8d\,] \quad (41)$$

When p is different from 1, the contribution from the interfaces to the resistivity decreases and, in addition, there is some averaging of the relaxation rates in the two metals. In the limit $\lambda_1,\lambda_2>>d_1,d_2$ the current is uniform and there is complete averaging of the relaxation rates due to interface scattering and to background scattering in the metals 1 and 2.

A more sophisticated model has been developed by Dimmich [93]. This model introduces different parameters p_1 and p_2, where p_1 (p_2) is for the passage from metal 1 (2) to metal 2 (1), and takes into account the scattering by grain boundaries in polycrystalline films.

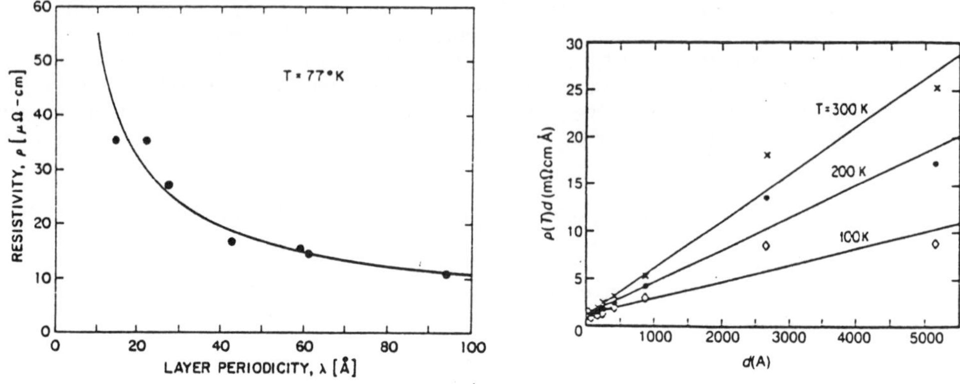

Fig. 36 (left) Resistivity at 77 K of Pd/Au multilayers as a function of the multilayer periodicity. The solid case is calculated from Eq 39 and 40 with p=0. From Carcia and Suna [92].

Fig. 37 (rigth) Resistivity times layer thickness against layer thickness for a series of Nb/Cu multilayers at three temperatures. From Falco [94].

There are few experimental results on the resistivity of metallic multilayers. In Fig. 36 we show the thickness dependence of the resistivity of Pd/Au multilayers at 77 K; a satisfactory fit is obtained by calculating the resistivity from Eq. 39 and 40 with p = 0. At room temperature it is necessary to introduce some coherent transmission (p ≠ 0) to account for the experimental results [92]. In Fig. 37 the plot of ρ versus d for Nb/Cu superlattices [94] shows that for relatively thick layers, a variation of the form of Eq. 41 is well obeyed. The experimental results are too few and more extensive data are needed for a better insight into the physics of the electron transport in metallic multilayers.

Finally, as in thin films, quantum size effects are expected in multilayers as well. In a superlattice with a periodicity of n atomic planes, the Brillouin zone is reduced by a factor n in the perpendicular direction to the layers. This leads to folded dispersion curves of the electrons and to (small) gaps in the conduction band. For metallic multilayers with layer thicknesses in the ML range (and very well defined interfaces) the resulting oscillations of the resistivity as a function of the layer thickness could possibly be detected, especially in the perpendicular direction.

3-5 Transport properties of magnetic films and multilayers

The study of the magnetotransport properties is an interesting complement of the magnetic measurements for magnetic thin films and multilayers. First the transport measurements represent a convenient tool to follow the field dependence of the magnetization, through the Hall effect and the magnetoresistance [95]. In addition recent experiments of Velu et al. [96] on Au/Co/Au sandwiches, of Baibich et al. [70] on Fe/Cr superlattices and Binasch et al. [97] on Fe/Cr/Fe sandwiches have shown a surprisingly high magnetoresistance which, by itself, could be an interesting property of the magnetic multilayers. Experimental curves are shown in figs.38 and 39.

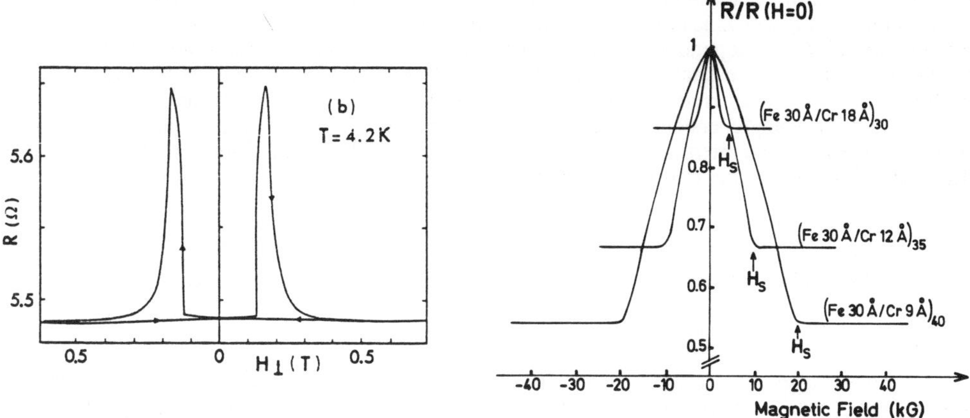

Fig. 38 (left) Magnetoresistance of a Au 250 Å /Co 7.5 Å/Au 30 Å/Co 7.5 Å/ Au 250 Å multilayer at 4.2 K in perpendical field. The maximum magnetoresistance is 3 %. Fromp Velu et al [96].

Fig.39 (right) Magnetoresistance of Fe/Cr superlattices at 4.2K. H_S is the saturation field needed to align the magnetic moments of the iron layers. From Baibich et al [70].

We describe the magnetoresitance data on bcc (001)Fe/Cr superlattices. We recall, see 2.5, that the Fe/Cr superlattices exhibit an antiferromagnetic ordering of the magnetic moments of the Fe layers ; an applied field H_S is needed to overcome the interlayer coupling and to align the moments. Fig. 39 shows that the resistivity strongly decreases during the magnetization process and then, for $H > H_S$, is practically constant. In Fe 30 Å/Cr 9 Å at 4.2 K the resistivity is reduced by about a factor of two between H = 0 and $H > H_S$. The magnetoresistance is smaller at higher temperatures and for thicker Cr layers. This giant magnetoresistance - which should be of interest for applications in magnetoresistive sensors - has been ascribed by Baibich et al [70] to spin dependent interface scattering, an effect related to the spin dependent scattering by impurities in ferromagnetic transition metals [98]. Camley and Barnas [99] and Levy and Fert [100] have worked out theoretical models in which they introduce different interface scattering rates for the spin ↑ (parallel to the majority spins of the Fe layer) and spin ↓ conduction electrons. The mechanism of the magnetoresistance is easy to describe for $\lambda >> t_{Cr}$, t_{Fe}. In this limit, as shown in 3-4, the scattering rates are averaged over many layers, which gives :

- at $H>H_S$, the magnetic moments of all the Fe layers are parallel, and there are different scattering rates and different resistivities $\rho\uparrow$ and $\rho\downarrow$ for the spin↓ and spin↑ currents. The final resistivity is $\rho_1 = (\rho^{-1}{}_{\downarrow}+\rho^{-1}{}_{\uparrow})^{-1}$, i.e. $\rho_1 \sim \rho\downarrow$ if, for example, $\rho\downarrow << \rho_{\uparrow}$ (there is a short circuit by the less scattered electrons).

- at H=0, an electron is alternately spin↓ and spin↑ so that $\rho^{eff}{}_{\uparrow(\downarrow)} = (\rho\downarrow + \rho\uparrow)/2$ and the final resistivity is $\rho_2 = (\rho\downarrow + \rho\uparrow)/4$.

In this picture an applied field lowers the resistivity from ρ_1 to ρ_2. With the values of $\rho\downarrow/\rho\uparrow$

usually observed in ferromagnets [98], the reduction factor can be very large.

When the Cr layers are thicker, the spin-dependent interface scattering affects the electron distribution function only in a layer ~ λ (mean free path) near the interface. The conductance of this layer, in the limit $t_{Cr} >> \lambda$, is completely independent of the scattering processes outside the layer and, in particular, is independent of the orientation of the magnetization at other interfaces. This means that the magnetoresistance will vanish for $t_{Cr} > \lambda$, even if the moments of Fe layers are still antiferromagnetically ordered. The magnetoresistance of the Fe/Cr superlattices thus appears to illustrate both the spin dependence of the interface scattering in transition metal multilayers and the Fuchs-Sondheimer-type interplay between thickness and mean free path. Similar magnetoresistance effects are expected in other transition metal multilayers.

3-6 Electron localization

In the presence of an arbitrary amount of disorder a bidimensional metal becomes insulating at zero Kelvin. This has been shown from scaling arguments in a famous article by Abrahams, Anderson, Licciardello and Ramakrishnan in 1979 [101]. Whereas conventional models based on the Boltzmann equation and scattering by independent defects or impurities predict a finite "residual resistivity" in the low temperature limit, the localization theories show that the resistivity diverges as the temperature T tends to zero. The departure from the conventional "Boltzmann behavior" as the temperature decreases is described in the "weak localisation "perturbation models by the onset of diverging multiple scattering terms [102]

$$\rho = \rho_B + \Delta_\rho$$

where ρ_B is the "Boltzmann resistivity" and Δ_ρ is the localization term expressed as :

$$\frac{\Delta \rho}{\rho^2} = \frac{e^2}{2\pi^2 \hbar} \ln(\frac{\tau_i}{\tau}) \qquad (42)$$

Here τ^{-1} is the total scattering rate and $\tau_i^{-1}$is the inelastic scattering rate. Eq.42 is obtained by taking into account multiple scattering in the region in which the electron wave propagates coherently. Because the phase of the waves is broken by inelastic scattering, the size of the coherent region is the coherent lenght $l_i = \sqrt{D\tau_i}$ (where D is the diffusion constant $D = v_F{}^2\tau/2$ in 2-d) and the resistivity diverges as $\ln(l_i) \sim \ln(\tau_i)$. For $\tau_i^{-1} \sim T^p$, the localization term $\Delta\rho$ diverges as ln T. We note that $\Delta\rho$ is proportional to ρ^2, so that the departure from the conventional metallic behavior is large for strongly disordered metals having a high resistivity. On the contrary, for very pure and ordered metals it would be necessary to go to considerably low temperatures (long τ_i) to detect localization effects.

When the spin-orbit coupling and the electron-electron interactions are taken into account, the weak localization effects are a little more complex than described above. For strong spin-orbit scattering (much stronger than the inelastic scattering) Eq.42 is replaced by :

$$\frac{\Delta \rho}{\rho^2} = - \frac{e^2}{4\pi^2 \hbar} \ln(\frac{\tau_i}{\tau}) \qquad (43)$$

with a different sign and an additional factor 1/2.

The electron-electron interactions give rise to an additional contribution

$$\frac{\Delta\rho^{ee}}{\rho^2} = \frac{e^2}{4\pi^2\hbar}(2 - \frac{2}{3}F)\ln(\frac{k_0T\tau}{h}) \tag{44}$$

where F is a numerical factor between 0 and 1.

All the above expressions are for a 2d-behavior, which is observed for a film or layer thickness d much shorter than l_i. At 3 dimensions, for d >> l_i, there still exists a departure from the Bolzmann behavior but the 3d localization terms are smaller than the 2d-terms and are proportional to $\sqrt{\tau_i}$ instead of ln (τ_i). Typical values for l_i in the helium range are between 500 Å for a strongly disordered metal (amorphous alloys with typically $\rho \approx 200$ μΩcm) to a few microns for weakly disordered metal films.

The weak localization effects have been extensively studied in metallic films [102]. They have been observed as well in multilayers composed of thin metallic layers separated by insulating layers [103 - 105]. Fig. 39 shows an example of the logarithmic temperature dependence of the resistivity in Au/Si multilayers [104]. The logarithmic term is the largest in the multilayer with 91 Å thick Si layers. Its decrease for t_{Si} < 50 Å is due to the onset of a coupling between the electrons of the gold layers through the silicon layers(crossover to 3-d).

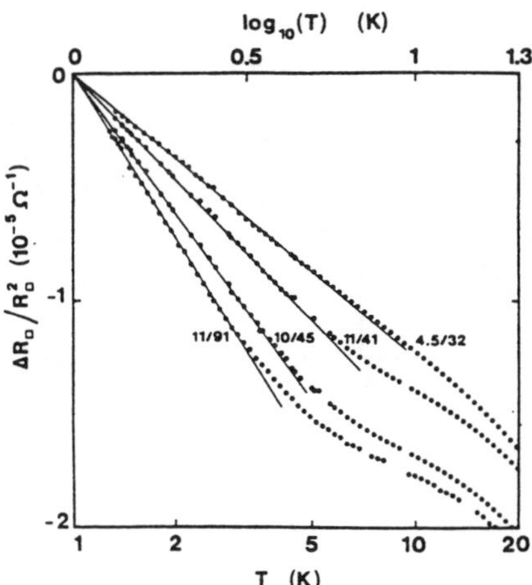

Fig. 39 Resistivity versus logarithm of temperature for Au/Si multilayers. The numbers near the curves are the thicknesses of the Au and Si layers respectively (example : Au 11 Å/Si 91 Å). From Cherradi et al [104].

There are typical magnetoresistance effects associated with the localization phenomena. For example, in the limit of strong spin-orbit coupling, the magnetoresistance is expressed as [106] :

$$\frac{\rho(H) - \rho(H=0)}{\rho^2} = \frac{e^2}{4\pi^2\hbar}\left[\phi\left(\frac{1}{2} + \frac{H_i}{H}\right) + \ln\left(\frac{H_i}{H}\right)\right] \tag{45}$$

where ϕ is the digamma function and

$$H_i = \frac{\hbar}{4eD}\tau_i^{-1} \tag{46}$$

This gives a magnetoresistance increasing as H^2 and then as lnH (positive for strong spin-orbit coupling, while it would be negative for zero spin-orbit). An experimental example obtained on a Au 13Å /Si 48 Å multilayer is shown in Fig.40. When the thickness of the insulating layers decreases, the electron coupling through Si induces a crossover to a 3d-behavior (lnH replaced by $\sqrt{H}$), as shown in Fig. 41.

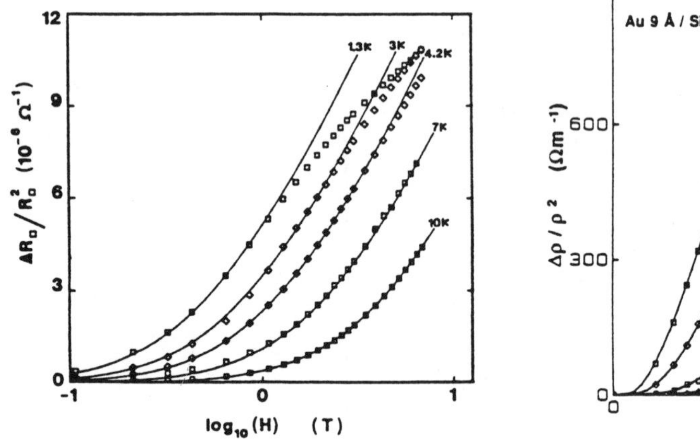

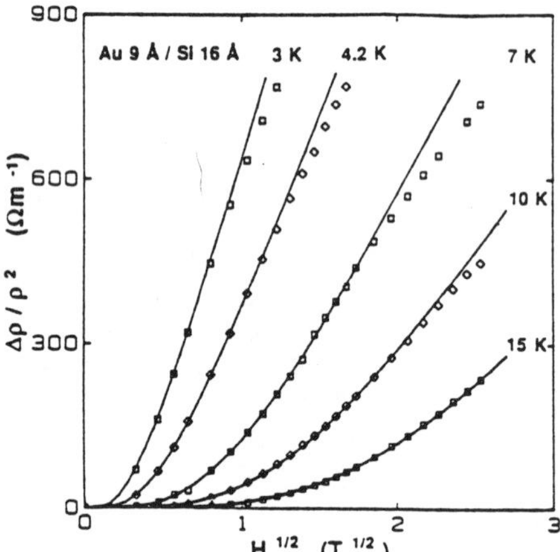

Fig. 40 (left) Perpendicular magnetoresistance versus ln H for a multilayer Au 13 Å/Si 48 Å . The solid curves are fits with the theoretical expression for 2d-localization effects, Eq(45). From Cherradi et al [104].

Fig. 41 (right) Magnetoresistance versus $\sqrt{H}$ for a Au 6.5 Å /Si 16 Å multilayer.The solid curves are fits with the theoretical expression for 3d localization effects. From Audouard et al. [105].

ACKNOWLEDGEMENTS:

I would like to thank Peter Levy and Jean-Pierre Renard for many fruitful discussions (and Jean-Pierre for several preprints). I thank my thesis student Agnès Barthélémy for her fantastic help in the preparation of the manuscript.

REFERENCES:

1) Lecture of F.Gautier in these proceedings
2) C.L.Fu, A.J.Freeman and T.Oguchi: Phys Rev Letters, 1985, 54, 2700. A.J.Freeman, C.L.Fu, S.Ohnishi and T.M.Weinert: Polarized Electrons in Surface Physics, edited by R.Feder, Advanced Series in Surface Physics, Vol 1 (World Scientific Publ., Singapore), 1985. L.M. Falicov, R.H. Victora and J.Tersoff: The Structure of Surfaces, edited by M.A. van Hove and S.Y.Tong, Springer Series in Surface Science, Vol. 2 (Springer Verlag, Berlin), 1985. S. Blügel, M. Weinert and P.H. Dederich: Phys Rev Letters, 1988, 60, 1077. F. Herman, P. Lambin, O.Jepsen: Phys Rev B, 1989, 31,4394.
3) U. Gradmann: J. Mag. Mag. Mat., 1986, 54-57, 733.
4) J.P. Renard: Thin Film Growth Techniques for Low-Dimensional Structures, edited by R.F.C. Farrow, S.S.P. Parkin, P.J. Dobson, J.H. Neave and A.S. Arrott (Plenum), 1987, 361.
5) U. Gradmann, H.J. Elmers and M. Przybylski: J. Phys. (Paris), 1989, 49, C8-1665.
6) S.P. Hong, A.J. Freeman, C.L. Fu: J. Phys. (Paris),1989, 49, C8-1683.
7) A.J. Freeman and C.L. Fu: J. Appl. Phys., 1987,61,3356.
8) S. Blügel and P.H. Dederich: Europhys. Letters, 1989, 9, 597.
9) M. Brodsky, L.R. Sill and C.H. Sowers: J. Mag. Mag. Mat., 1986, 54-57, 779.
10) A.D. Johnson, J.A.C. Bland, C. Norris and H. Lauter: J. Phys. C, 1988, 21, L 899.
11) M.C. Hanf, L. Haderbache, P. Wetzel, C. Pirri, J.C.Peruchetti, D. Bolmont and G.Gewinner: Solid State Com., 1988, 68, 113. M.C. Hanf, C. Pirri, J.C. Peruchetti, D. Bolmont and G. Gewinner: Phys.Rev. B, 1989, 39, 1546.
12) C. Rau, G. Xing, M. Robert: J. Vac. Sci. Technol. A, 1988, 6, 759.
13) W. Drube and F.J. Himpsel: Phys. Rev. B, 1987, 35, 4131.
14) M. Spampanoni, A. Vaterlaus, D. Pescia, M. Aeschlimann, F.Meier, W.Dürr and S.Blügel: Phys. Rev. B, 1988, 37, 10380.
15) M.T. Hutchings: Solid State Physics, edited by F. Seitz and D.Turnbull (Academic Press, New-York), 1964, Vol. 16.
16) P.M. Levy, private communication.
17) I. Dzyaloshinski: J. Phys. Chem. Solids, 1958, 4, 241. T. Moriya: Phys. Rev, 1960, P20, 91. A. Arrot: J. Appl. Phys., 1963, 34, 1108.
18) A. Fert and P.M. Levy: Phys. Rev. Letters, 1980, 44, 1538. P.M. Levy, C.Morgan-Pond and A. Fert, J. Appl. Phys., 1982, 53, 2168.
19) A. Fert and P.M. Levy: to be published.
20) M. Piecuch, L.T. Baczewski, J. Durand, G. Marchal, P. Delcroix, H. Nabli: J. Phys (Paris), 1989, 49, C8-1755 and to be published.
21) Z.S. Shan, D.J. Sellmyer, S.S. Jaswal, Y.J. Wang and J.X. Shen: Phys. Rev. Letters, 1989, 64, 449.
22) N. Hosoito, K. Yoden, K. Mibu and T. Shinjo: J. Phys. (Paris), 1989, 49, C8-1777.
23) M. Farle, A. Berghauss, K. Baberschke: Phys. Rev. B, 1989, 39, 4838.
24) H.J.G. Draaisma, W.J.M. de Jonge, F.J.A. den Broeder: J. Mag. Mag. Mat, 1987, 66, 351. F.J.A. den Broeder, D. Kuiper, A.A. van de Mosselaer and W. Hoving: Phys. Rev. Letters, 1980, 60, 2769. H.J.G. Draaisma: Thesis (Eindhoven 1988), unpublished.
25) M. Stampanoni, A. Vaterlaus, M. Aeschlimann, F. Meier,D. Pescia: J. Appl. Phys., 1988, 64, 5321.

26) M. Stampanoni, A. Vaterlaus, M. Aeschlimann and F. Meier: Phys. Rev. Letters, 1987, 59, 2483.
27) N.C. Koon, B.T. Jonker, F.A. Volkening, J.J. Krebs and G.A. Prinz: Phys. Rev. Letters, 1987, 59, 2463.
28) J. Araya-Pochet, C.A. Ballentine and J.L. Erskine: Phys. Rev. B, 1988, 38, 7846.
29) K.B. Urquhart, B. Heinrich, J.F. Cochran, A.S. Arrott and K.Myrtle: J. Appl. Phys., 1988, 64, 5334. S.T. Purcell, B. Heinrich and A.S. Arrot: J. Appl. Phys., 1988, 64, 5337.
30) C. Liu, E.R. Moog, S.D. Bader: Phys. Rev. Letters, 1988, 60, 2422. S.D. Bader, E.R.Moog: J. Appl. Phys., 1987, 61, 3789.
31) P. Bruno and J.P. Renard: to appear in Appl. Phys. A.
32) H.J. Elmert and U. Gradmann: Surf. Sci., 1988, 193, 94.
33) W. Dürr, M. Taborelli, O. Paul, R. Germa, W. Gudat, D. Pescia, M.Landolt: Phys. Rev. Letters, 1989, 62, 206.
34) U. Gradmann, J. Korecki, G. Waller: Appl. Phys. A, 1986, 39, 101.
35) G. Lugert, G. Bayreuther: Phys. Rev. B, 1988, 38, 11068; and J.P.Renard private communication
36) J.R. Dutcher, B. Heinrich, J.F. Cochran, D.A. Steigerwald,W.F.Egelhoff: J. Appl. Phys., 1988, 63, 3464.
37) C. Chappert, K. Le Dang, P. Beauvillain, H. Hurdequint, D. Renard: Phys. Rev. B, 1986, 34, 3192.
38) D. Pescia, G. Zampieri, M. Stampanoni, G.L. Bona, R.F. Willis, F.Meier: Phys. Rev. Letters, 1987, 58, 933.
39) J.G. Gay and R. Richter: Phys. Rev. Letters, 1986, 56, 2287. J.G.Gay and R. Richter: J.Appl. Phys. A, 1987, 6; 3362. W.Karas, J. Noffke, L. Fritsche: J. Phys. (Paris), to be published.
40) C.S. Wang, B.M. Klein and H. Krakauer: Phys. Rev. Letters, 1985, 54, 1852.
41) C. Fu and A.J. Freeman: Phys. Rev. B, 1987, 35,
42) W.A. Macedo and W. Keune: Phys. Rev. Letters, 1988, 61, 475. W. Keune: MRS Spring Meeting (San Diego, 1989).
43) M. Maurer, J.C. Ousset, M.F. Ravet and M. Piecuch: Europhys. Letters, 1989, 9, 803.
44) G.A. Prinz: Phys. Rev. Letters, 1985, 54, 1051.
45) P. Bruno: Thesis (Orsay, 1989), unpublished.
46) F.J. Lamelas, C.H. Lee, Hui He, W. Vavra and Ray Clarcke: MRS Spring Meeting (San Diego, 1989), to be published
47) B.T. Jonker, J.J Krebs and G.A. Prinz: Phys. Rev. B, 1989, 39, 1399.
48) N.D. Mermin and H. Wagner: Phys. Rev. Letters, 1966, 17, 1133.
49) Myron Bander and D.L. Mills: Phys. Rev. B, 1988, 38, 12015.
50) H. Lutz, J.D. Grunton,H.K. Schurmann,J.E. Crow, T. Mihalisin: Solid State Commun.,1974, 14, 1075. A. Bergholz and U. Gradmann: J.Mag. Mag. Mat., 1984, 45, 389.
51) J.P. Renard: Proceeding of the School" Rayonnement Synchrotron Polarisé", Mittelwhir (1989), to be published.
52) J. Korecki, U. Gradmann: Phys. Rev. Letters, 1985, 55, 2491.
53) U. Gradmann: Appl. Phys., 1974, 3, 161.
54) D. Pescia, M. Stampanoni, G.L. Bona, A. Vaterlaus, R.F. Willis, F.Meier: Phys. Rev. Letters, 1987, 58, 2126.
55) C.M. Schneider, J.J. de Miguel, P. Bressler, J. Garbe, S. Ferrer, R.Miranda, J.Kirschner: J. Phys. (Paris), 1988, 49, C8-1657.
56) C. Chappert, P. Bruno: J. Appl. Phys., 1988, 64, 5736.
57) H.K. Wong, B.Y. Jin, H.Q. Yang, J.E. Hilliart and J.B. Ketterson: J.Low Temp. Phys., 1986, 63, 307.
58) M. Taborelli, O. Paul, O. Züger, M. Landolt: J. Phys. (Paris), 1988, 49, C8-1659.
59) D. Mauri, D. Scholl, H.C. Siegmann, E. Kay: Phys. Rev. Letters, 1989, 62, 1900.
60) M. Farle and K. Baberschke: Phys. Rev. Letters, 1987, 51, 511.
61) W. Dürr, M. Taborelli, O. Paul, R. Germa, W. Gudat, D. Pescia and M.Landolt: Phys. Rev. Letters, 1989, 62, 206.

62) L.L. Hinchey and D.L. Mills: Phys. Rev. B, 1986, 33, 3329.
63) C.F. Majkrzak, J.W. Cable, J. Kwo, M. Hong, B.B. Mc Whan, Y. Yafet, J.V.Wazzczak, C. Vettier: Phys. Rev. Letters, 1986, 56, 2700. J.Kwo, M. Hong, F.J. Di Salvo, J.V. Wazzczak, C.F. Majkrzak: Phys. Rev. B, 1987, 35, 1987. J. Appl. Phys. 1988, 63, 3447.
64) Y. Yafet: J. Appl. Phys., 1987, 61, 4058.
65) M.B. Salomon, S. Sinha, J.J. Rhyne, J.E. Cunningham, W. Erwin, J.Borchers, C.P.Flyn: Phys. Rev. Letters, 1986, 56, 259.
66) J.A. Borchers, G. Nieuwenhuys, M.B. Salomon, C.P. Flyn, R. Du, W.Erwin: J. Phys. (Paris), 1989, 49, C8-1685.
67) P. Grünberg, R. Shreiber, Y. Pang, M.B. Brodsky, H. Sowers: Phys. Rev. Letters, 1986, 57, 2442.
68) C. Carbone and S.F. Alvarado: Phys. Rev. B, 1987, 36, 2433.
69) F. Nguyen Van Dau, A. Fert, P. Etienne, M.N. Baibich, J.M. Broto, G.Creuzet, A.Friederich, S. Hadjoudj, J. Massies: J. Phys (Paris), 1988, 49, C8-1633.
70) M.N. Baibich, J.M. Broto, A. Fert, F. Nguyen Van Dau, F. Petroff, P.Etienne, G.Creuzet, A. Friederich, J. Chazelas: Phys. Rev. Letters, 1988, 61, 2472.
71) S. Lequien, C. Vettier, M. Hennion, to be published.
72) K. Ounadjela, C. Sommer, P.M. Levy, A. Fert, to be published.
73) A. Cebollada, J.L. Martinez, J.M. Gallego, J.J. de Miguel, R. Miranda, S. Ferrer, F.Batallan, G. Fillion, J.P. Rebouillat, Phys. Rev. B 1989, 39, 9726.
74) M. Maurer, J.C. Ousset, C. Piecuch, private communication.
75) A.P. Malozemoff, J. Appl. Phys. 1988, 63, 3874.
76) C. Schlenker, S.S.P. Parkin, J.C. Scott and K. Howard, J. Mag.Mag. Mat 1986, 54-57, 802.
77) W.H. Meiklejohn and C.P. Bean, Phys. Rev. 1956, 102, 1413.
78) A.P. Malozenoff, Phys. Rev. B 1977, 35, 3679.
79) L.I. Maisel, Handbook of Thin Film Technology, edited by L.I. Maisel and R. Glang (Mc Graw Hill, New York) 1970, Chapter 13, 1.
80) K. Fuchs, Proc. Camb. Phil. Soc. 1938, 34, 100.
81) E.H. Sondheimer, Ad. Phys. 1952, 1, 1.
82) J.M. Ziman, Electrons and Phonons (Clarendon Press, Oxford) 1960, 267.
83) D.S. Campbell, The Use of Thin Films in Physical Investigations (Academic Press, New York) 1966, p. 299.
84) E.R. Andrew, Proc. Phys. Soc. A, 1949, 62, 77.
85) H. Mayer "Structure and Properties of Thin Films "(John Wiley Inc., New York) 1959, p. 225.
86) E.J. Gillham, J. Preston, B. Williams, Phil. Mag. 1955, 46, 1051.
87) M. Jalochowski and E. Bauer, Phys. Rev. 1988, 37, 8622.
88) A.F. Mayadas and M. Shatzes, Phys. Rev. B 1970, 1, 1382.
89) P.A. Badoz, A. Briggs, E. Rosencher, F. Arnaud d'Avitaya and C. d'Anterroches, Appl. Phys. Lett. 1987, 51, 169.
90) Z. Tesanovic, M.V. Jaric, S. Maekawa, Phys. Rev. Lett. 1986, 57, 2760.
91) G. Fischman, D. Calecki, Phys. Rev. Lett. 1989, 62, 1302.
92) P.F. Garcia and A. Suna, J. Appl. Phys. 1983, 54, 2000.
93) R. Dimmich, J. Phys. F. 1985, 15, 2477.
94) C.M. Falco, J. Phys. (Paris) 1984, 45, C5-499.
95) E.D. Dahlberg, K. Riggs, G.A. Prinz, J. Appl. Phys. 1988, 63, 4270.
96) E. Velu, C. Dupas, D. Renard, J.P. Renard, J. Seiden, Phys. Rev. B 1988, 37, 668.
97) G. Binasch, P. Grünberg, F. Sanrenbach, W. Zinn, Phys. Rev. B 1989, 39, 4828.
98) I.A. Campbell and A. Fert, Ferromagnetic Materials (North Holland, Amsterdam) 1982), p. 769.
99) R.E. Camley and J. Barnas, to be published.
100) P.M. Levy and A. Fert, to be published.
101) E. Abrahams, P.W. Anderson, D.C. Licciardello and T.V. Ramakrishnan, Phys. Rev. Lett. 1979, 42, 673.

102) For a review on localization see G. Bergmann, Phys. Reports 1984, 107, 1 or P.A. Lee and T.V. Ramakrishnan, Rev. of Modern Physics 1985, 57, 127.
103) B.Y. Jin and J.B. Ketterson, Phys. Rev. B 1986, 33, 8797.
104) N. Cherradi; A. Audouard, G. Marchal, J.M. Broto, A. Fert, Phys. Rev. B 1989, 39, 7424.
105) A. Audouard, N. Cherradi, J.M. Broto, G. Marchal, A. Fert, to be published.
106) B.L. Altshuler, D.E. Khemelnitzkii, A.I. Larkin, Phys. Rev. B 1980, 22, 5142.

Materials Science Forum Vols. 59 & 60 (1990) pp. 481-534

MECHANICAL PROPERTIES OF THIN FILMS AND COATINGS

J. Grilhé
Laboratoire de Métallurgie Physique
U.R.A. 131 du CNRS
40 avenue du Recteur Pineau, F-86022 Poitiers, France

1 INTRODUCTION

Coatings and surface treatments are widely used to improve the surface properties of materials: increase of the resistance to wear, abrasion, shocks and corrosion; reduction of friction; modification of optical, electric and magnetic characteristics, etc. The first coatings, deposited through chemical processes, were fairly thick (a few millimeters), but the progress of physical methods led to the development of thinner films, about a few micrometers thick and, in some cases, a few hundred nanometers. They are often produced in a metastable state, amorphous or with a small grain size, and present specific properties.

Metallurgists soon became aware of the advantages of these thin layers for structural, load bearing applications and they have devoted a large amount of research to determine their intrinsic mechanical properties, alone or linked with a substrate, and to understand deformation mechanisms of material covered with a coating. They have been recently joined by scientists working in the fields of materials for electronic devices and magnetic registration [32].

Microelectronic integrated circuits are made of thin layers of various materials (single or polycrystalline semiconductors, oxides, metals and passivation glasses) with complex shapes and very different mechanical and thermal properties. They have to offer adequate mechanical resistance over their life time, but they must retain their electric properties too; stresses and defects occurring during their manufacture often lead to aging and electrical failure.

Magnetic hard disks used for recording, composed until now of fine magnetic particles in a resin matrix, appeal more and more to thin film technology, the magnetic material being deposited in a thin layer covered with a passivation glass. The disk rotates at very high speed and the head, which writes and reads information by changing the magnetization of small areas, must fly at very short distances (about 200 nm) above the surface. Shocks between the head and the disk cannot be avoided and the materials must be able to resist.

The good crystallographic qualities of materials used for these technologies have brought new physical problems, never encountered in coating research before, such as the appearance of stresses and misfit dislocations in epitaxial structures. New investigation methods, coming from the microelectronic industry, have been also developed (for example, the use of very small cantilever beams).

The recent studies of metallic epitaxial multilayers will take advantage of all this progress.

This paper is not a complete bibliography of all the work done in these different fields. The proceedings of several annual symposia (M.R.S. BOSTON, M.R.S STRASBOURG, I.C.M.C. SAN-DIEGO,etc) give a good idea of their development. Our viewpoint has been to point out some fundamental characteristics and some new experimental methods leading to a best understanding of the mechanical behaviour of thin films.

We begin with the elastic properties measurement of elastic moduli and supermodulus effect in multilayers. The internal stresses and the misfit dislocations are discussed in the third part and some plastic properties (hardness, deformations induced by thermal cycling, decohesion) are briefly reported in the last part.

The main source materials used for this lecture are review papers and differents symposium proceedings quoted before, reports and PhD dissertations of Stanford University (Nix et al.) for internal stresses, and from Northwestern University (Hilliard et al.) for elastic properties.

2 ELASTIC PROPERTIES OF THIN FILMS

2-1 Introduction

The knowledge of elastic constants is fundamental to understand the mechanical properties of thin films, multilayers and coatings. For a few years their determination has been the purpose of much research and many new techniques, more appropriate for thin foils, have been recently developed: Determination of Young's, biaxial, shear and flexural moduli, surface acoustic waves velocities and Brillouin scattering measurements, etc. For single crystals and for thin foils having a strong texture, attention must be paid to the crystalline anisotropy.

In this chapter the principal moduli definition and, in each case, a determination method will be reviewed. Then the way to deduce the elastic constants from these measurements will be developed. These steps are rather long and boring, but the following example will show that it is difficult to save time on that point.

Some metallic multilayers display strong elastic moduli anomalies. We shall come back later to this "supermodulus effect" (chapter 2-4) but shall now introduce some results of Baral and al. [1,2] concerning modulated structures Cu/Ni of (111) texture. The different modulus variations as a function of the concentration modulation wavelength λ, in a direction normal to the foil, are displayed on the figures 1a,1b,1c and 1d. It can be surprising to see that the curves giving the biaxial and Young's modulus have two maxima and the curves giving the flexural and the shear modulus only one. But a rigorous analysis made by the authors shows the coherency of these variations.

One of the purposes of this chapter is to recall the basic principles needed to be able to understand and to do this kind of analysis. Afterwards we will shortly review the different aspects of the "supermodulus effect".

2-2 Elastic constants of crystals [3,4]

In linear elasticity the stiffnesses c_{ijkl} connect the components of the strain tensor e_{kl} to the components of the stress tensor σ_{ij}, following Hooke's law

$$\sigma_{ij} = c_{ijkl} \cdot e_{kl}$$

where the indices i,j,k and l take values from 1 to 3 and, with the usual convention, a summation is made over repeated subscripts.

The compliances s_{ijkl} are defined by the inverse relationship

$$e_{ij} = s_{ijkl} \cdot \sigma_{kl}$$

c_{ijkl} and s_{ijkl} are 4-rank tensors. When the axes of reference $(0x_1x_2x_3)$ are rotated and if A_{ij} are the elements of the 3x3 rotation matrix, the strains and stresses in the new axes $(0x_1'x_2'x_3')$ are obtained by the tensor transformation equations

$$e'_{ij} = A_{im}.A_{jn}.e_{mn}$$

$$\sigma'_{ij} = A_{im}.A_{jn}.\sigma_{mn}$$

and stiffnesses and compliances by the equations:

$$c'_{ijkl} = A_{im}.A_{jn}.A_{kp}.A_{lq}.c_{mnpq}$$

$$s'_{ijkl} = A_{im}.A_{jn}.A_{kp}.A_{lq}.s_{mnpq}$$

The tensors $\{\sigma_{ij}\}$ and $\{e_{ij}\}$ are symmetric; they have only six independent elements and are often converted in 6 component vectors σ_i and e_i by the following relations (Voigt notation)

$$\sigma_{11}=\sigma_1 \;,\; \sigma_{22}=\sigma_2 \;,\; \sigma_{33}=\sigma_3 \;,\; \sigma_{23}=\sigma_4 \;,\; \sigma_{31}=\sigma_5 \;,\; \sigma_{12}=\sigma_6$$

and

$$e_{11}=e_1 \;,\; e_{22}=e_2 \;,\; e_{33}=e_3 \;,\; 2.e_{23}=e_4 \;,\; 2.e_{31}=e_5 \;,\; 2.e_{12}=e_6$$

The stiffness and compliance tensors are compressed to 6x6 matrix $\{C_{ij}\}$ and $\{S_{ij}\}$ according to the relations

$$\sigma_i = C_{ij}.e_j \qquad e_m = S_{mn}.\sigma_n$$

with $\{S_{ij}\} = \{C_{ij}\}^{-1}$

Some straightforward but tedious calculations reveal that the new coefficients as a function of the old ones are given by

$$c_{ijkl} = C_{mn}$$

where Voigt notations are used for the correspondence between the subscripts ijkl and mn.

$s_{ijkl} = S_{mn}$ when m and n take the value 1,2 or 3.
$2.s_{ijkl} = S_{mn}$ when m and n take the value 4,5 or 6.
$4.s_{ijkl} = S_{mn}$ when m and n take the value 4,5 or 6.

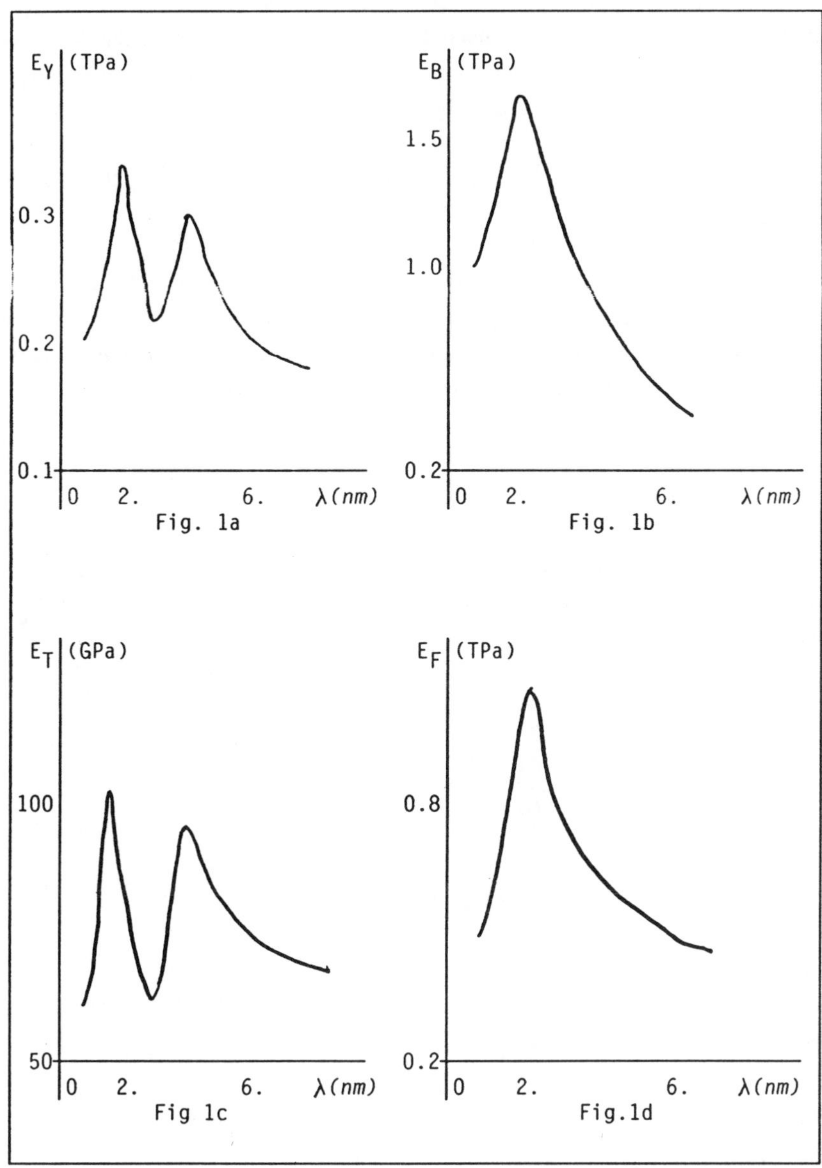

Fig 1a, 1b, 1c, 1d Variations of Young's modulus E_Y, biaxial modulus E_B, shear modulus E_T and flexural modulus E_F, with the wavelength λ of modulation for Cu/Ni multilayers of constant composition (≃ 50% Cu) and (111) texture (From ref. [1]).

The transformations, during a rotation of the axes, of these $\{C_{ij}\}$ and $\{S_{mn}\}$ 6x6 matrices are not straightforward. It is easier to come back to the 4 subscript tensor components c_{ijkl} and s_{ijkl}, to find their new values c'_{ijkl} and s'_{ijkl} with the tensor transformation equation given before and to deduce the new matrix $\{C'_{mn}\}$ and $\{S'_{mn}\}$.
Some examples are given by Nye [3] and will be widely used in the following section.

Considerations about elastic stored energy show that the $\{S_{ij}\}$ and $\{C_{ij}\}$ matrices are symmetric and have only 21 independent elements. This number is again strongly reduced by using the energy invariability through axes of rotation corresponding to the symmetry group of the crystal.

i) Elastic constants for Cubic crystals

In cubic crystals only three elements of the stiffness and compliance matrix are independent. When the coordinate axes coincide with the cubic axes, $\{C_{ij}\}$ becomes

$$\begin{vmatrix} C_{11} & C_{12} & C_{12} & 0 & 0 & 0 \\ C_{12} & C_{11} & C_{12} & 0 & 0 & 0 \\ C_{12} & C_{12} & C_{11} & 0 & 0 & 0 \\ 0 & 0 & 0 & C_{44} & 0 & 0 \\ 0 & 0 & 0 & 0 & C_{44} & 0 \\ 0 & 0 & 0 & 0 & 0 & C_{44} \end{vmatrix}$$

The three coefficients C_{11}, C_{12} and C_{44} are called the primary elastic stiffnesses. For other axes, the stiffness matrix, quoted $\{C'_{mn}\}$ in the next sections, takes the more complicated form

$$\begin{vmatrix} C'_{11} & C'_{12} & C'_{13} & 0 & 0 & 0 \\ C'_{12} & C'_{22} & C'_{23} & 0 & 0 & 0 \\ C'_{13} & C'_{23} & C'_{33} & 0 & 0 & 0 \\ 0 & 0 & 0 & C'_{44} & 0 & 0 \\ 0 & 0 & 0 & 0 & C'_{55} & 0 \\ 0 & 0 & 0 & 0 & 0 & C'_{66} \end{vmatrix}$$

the non independent C'_{ij}'s being related to the three primary elastic stiffnesses by the transformation relation.

The $\{S'_{ij}\}$ compliance matrix, equal to $\{C'_{mn}\}^{-1}$, has the same form and its elements can be expressed in terms of the primary elastic compliances S_{11}, S_{12} and S_{44}.

ii) Elastic constants for isotropic solids

A cubic crystal is isotropic if its primary stiffnesses verify the equation

$$C_{44} = (C_{11}-C_{12})/2.$$

Table I gives, for some cubic crystals, the values of the *anisotropy ratio* A defined by

$$A = 2.C_{44}/(C_{11}-C_{12})$$

Crystal	Nb	Cr	W	Al	Fe(α)	Ag	Cu	Li
A	0,5	0,7	1,0	1,2	2,5	3,0	3,4	9,0

Table I(From ref. [4])

An isotropic solid has only two independent stiffnesses. They are often expressed in terms of the two Lamé coefficients λ and μ, μ being the shear modulus

$$C_{12} = \lambda$$
$$C_{44} = \mu$$
$$\text{et} \quad C_{11} = \lambda+2.\mu$$

and

$$\sigma_{ij} = \lambda.\delta_{ij}.e_{kk} + 2.\mu.e_{ij}$$

The compliance tensor is more often given in terms of Young's modulus E_Y and Poisson's coefficient v defined by the relations:

$$E_Y = \mu.(3.\lambda+2.\mu)/(\lambda+\mu) \quad \text{et} \quad v = \lambda./2.(\lambda+\mu)$$

One obtains:

$$S_{11} = 1/E_Y \qquad S_{12} = -v/E_Y \qquad \text{et} \qquad S_{44} = (1+v)/E_Y$$

and

$$e_{11} = \sigma_{11}/E_Y-(\sigma_{22}+\sigma_{33}).v/E_Y$$
$$e_{22} = \sigma_{22}/E_Y-(\sigma_{11}+\sigma_{33}).v/E_Y$$
$$e_{33} = \sigma_{33}/E_Y-(\sigma_{11}+\sigma_{22}).v/E_Y$$
$$e_{23} = \sigma_{23}.(1+v)/E_Y$$
$$e_{31} = \sigma_{31}.(1+v)/E_Y$$
$$e_{12} = \sigma_{12}.(1+v)/E_Y$$

2-3 Elastic moduli of thin films.

Several elastic moduli are used to characterize thin films [1,2,5]. We shall give the definition of some of them and, in each case, a method for measuring it. The x_3-axis will always be chosen normal to the foil.

2-3-1 Young's modulus E_Y.

The definition and the ways to measure the Young's modulus E_Y are roughly the same as for bulk crystals. Several specific tensile testers have been devised for thin films. A tensile stress σ is applied along a direction in the plane of the foil chosen as the x_1-axis. The unit elongation in this direction $\epsilon=e_{11}$ is measured and Young's modulus is given by the relation

$$\sigma/\epsilon = E_Y.$$

i) Young modulus E_Y in terms of the C_{ij} 's

Only one component of the stress tensor σ_{11}, equal to σ, is different from zero. So it is more convenient to use the elastic compliance tensor to calculate e_{11}

$$e_{11} = S'_{11}.\sigma_{11}$$

Young's modulus is then given by

$$E_Y = 1/S'_{11}$$

To obtain E_Y in terms of the primary stiffnesses, we must (i) find the rotation matrix A_{mn} from the crystallographic axes to the axes chosen for the foil, (ii) express S'_{11} in terms of the primary compliances S_{ij} by using the procedure quoted before for the components of a 4-rank tensor, then replace the S_{ij} 's by their expression in terms of the C_{ij} 's.

For isotropic solids, E_Y is the same for all directions and is often used as a primary stiffness.

For a cubic crystal, Nye [3] obtains the general relation

$$\frac{1}{E_Y} = \frac{(C_{11}+C_{12})}{(C_{11}-C_{12})(C_{11}+2.C_{12})} + K\cdot\left(\frac{1}{C_{44}} - \frac{2}{C_{11}+C_{12}}\right)$$

with $K = l_1^2.l_2^2 + l_2^2.l_3^2 + l_3^2.l_1^2$

l_1, l_2 and l_3 being the direction of the tensile axis in the cubic coordinates.

For cubic metals with a large anisotropy ratio, E_Y is strongly dependent on the tensile direction. We give as example the Young's modulus of copper, expressed in TPa (10^{12} Newton/m^2) for the <100>, <111> et <110> directions

$E_Y[100] = 0,067$ $E_Y[111] = 0,19$ $E_Y[110] = 0,13$

2-3-2 Biaxial modulus E_B

The biaxial modulus, or bimodulus, is very often employed to characterize thin films, particularly for plane isotropic layers. The isotropic condition is given by the relations

$$C_{11} = C_{22} \qquad \text{or} \qquad S_{11} = S_{22}$$

This is the case for single crystalline foils with rotation axis perpendicular to the film, as for example the {100} and {111} planes of cubic crystals. This relationship is also verified in solids with average isotropy, as amorphous materials, films with small grain size and, in some cases that we shall see later, films with a strong texture.

For the determination of the bimodulus of an isotropic film, a biaxial stress is applied to the foil

$$\sigma_{11} = \sigma_{22} = \sigma$$

the other components of the stress tensor being equal to zero. The unit elongation ϵ in the plane is measured:

$$e_{11} = e_{22} = \epsilon$$

and the bimodulus E_B is defined by the relation

$$E_B = \sigma/\epsilon$$

i) Bimodulus E_B in terms of the C_{ij} 's

The unit elongation ϵ is easily related to the stress σ by the compliance tensor

$$\epsilon = e_{11} = (S'_{11}+S'_{12}).\sigma$$

and then

$$E_B = 1/(S'_{11}+S'_{12})$$

The procedure already used for Young's modulus must be applied to deduce E_B in terms of the C_{IJ} 's.

For isotropic solids the bimodulus is merely given by

$$E_B = E_Y/(1-v)$$

For cubic films for which the x_3-axis is parallel to the [100] or [111] cubic directions, we obtain:

$$E_B[100] = C_{11} + C_{12} - 2\cdot\frac{C_{12}^2}{C_{11}}$$

and

$$E_B[111] = 6\cdot C_{44}\cdot\frac{(C_{11}+2.C_{12})}{(C_{11}+2.C_{12}+4.C_{44})}$$

ii) Bimodulus measurement

This measurement is usually performed with a bulge tester. The film is carefully fastened by an O-ring to the circular hole of a specially designed container (Fig 2) full of gas or liquid. A pump is needed to vary the pressure P in the device. Under this pressure, the foil, of thickness t, takes approximately a spherical shape of radius R given by

$$P = 2.\sigma_{11}.t/R$$

The film is subjected to a biaxial tensile stress $\sigma_{11} = \sigma_{22}$ inducing a biaxial strain

$$e_{11} = e_{22} = a^2/6.R^2$$

where a is the hole radius of the container. Then the bimodulus is equal to

$$E_B = \frac{\sigma_{11}}{e_{11}} = \frac{3.P.R^3}{t.a^2}$$

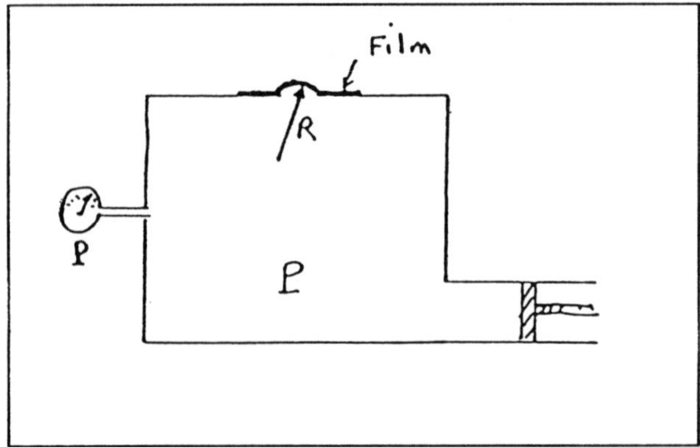

Fig. 2 Schematic diagram of a bulge testing apparatus for bimodulus measurement.

It is possible to deduce R from precise volume measurements of an incompressible liquid contained in the reservoir during the pressure variations [5]. It is also possible to measure the Z deflection [6] at the centre of the film with an interferometer or a high sensitive displacement detector. The deflection is generally small and is related to R by the straightforward relation

$$a^2 = Z.(2.R-Z) \simeq 2.Z.R$$

NOTE : The bulge test can also be used for anisotropic films but the shape of the foil under pressure is more complicated and the analysis of experimental results is more difficult. We shall mention it in chapter 3.

2-3-3 Shear modulus E_T

A torque $\sigma_{23} = \sigma_{32}$ is applied to the foil, the other components of the stress being equal to zero. The strain e_4, equal to $2.e_{23}$, is given by

$$2.e_{23} = S'_{44}.\sigma_{23}$$

The shear modulus is defined by

$$E_T = \sigma_{23}/(2.e_{23}) = 1/S'_{44}$$

and S'_{44} is expressed in terms of the C_{ij} 's by the usual procedure. For isotropic foils. S'_{44} is equal to S'_{55} and E_T does not depend of the torque axis direction, chosen as the x_1-axis.

For an isotropic solid, E_T is equal to μ

$$E_T = \mu = E_Y/(1+\nu)$$

For a cubic film with the x_3-axis parallel to a [111] direction, E_T is given by

$$E_T = (C_{11}-C_{12}+C_{44})/3$$

E_T is generally measured with a torsion pendulum, similar to the apparatus used for internal friction studies. His value is deduced from the time period of oscillation of the foil under an applied torque.

2-3-4 Flexural modulus E_F.

The flexural modulus can be determined from the deflection of a rectangular cantilever beam shaped in a film of thickness t (Fig 3). L and b are the useful length and width of the beam. A force F normal to the surface is applied to the free end and the deflection Z is measured.

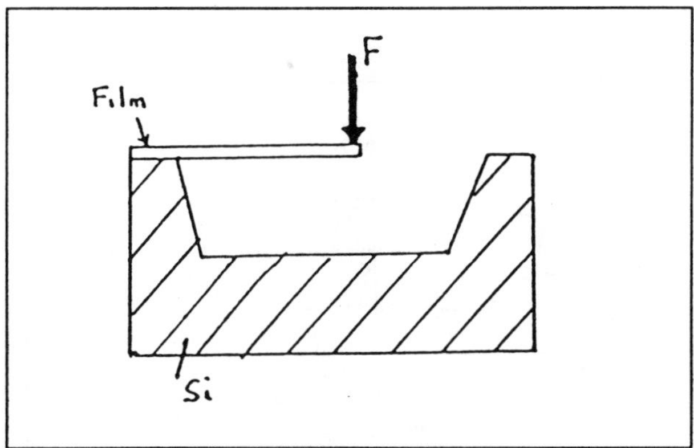

Fig. 3 Deflection of a cantilever beam used for the determination of the flexural modulus E_F.

The expression to be used to relate E_F to F and Z is not simple and needs some approximations [7], discussed by Cammarata [28].

Solving the elastic equilibrium equation shows that, (i) if the condition $t \ll b \ll L$ is fulfilled, only the strains e_{11} and e_{33} take significant values (ii) the stress component σ_{33}, equal to zero at the film surfaces, takes very small values inside and can be neglected and (iii) finally that for low forces F, the radius of curvature R is about the same all along the beam.

A fibre length dl, at a distance z from the neutral axis (Fig 4), is subjected to an elongation

$$e_{11}(z).dl=(R+z).d\Theta$$

and a strain

$$e_{11}(z) = z/R$$

The strain e_{33} can be obtained from the condition $\sigma_{33} \simeq 0$

$$\sigma_{33}= C'_{31}.e_{11} + C'_{33}.e_{33} = 0$$

and

$$e_{33} = - e_{11}.C'_{31}/C'_{33}$$

For a homogeneous film the neutral axis is situated in the middle of the foil, otherwise it must be deduced from the equation

$$F1 = \int \sigma_{11}(z).dy.dz = 0$$

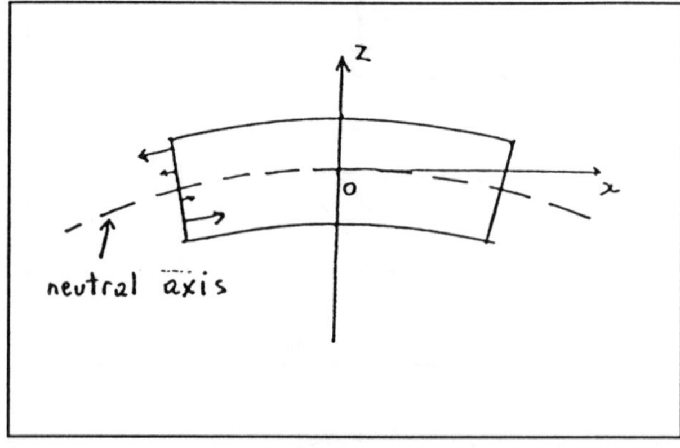

Fig. 4 Flexural deformations of a cantilever beam.

meaning that the sum of the forces acting in a section of the beam in the x_1-axis direction must vanish.

The deflection Z is obtained by minimizing the sum of the elastic energy W stored in the film and the work done by the applied force F

$$F = dW/dZ$$

W is given by

$$W = 1/2.\int (C'_{11}.e_{11}^2 + C'_{33}.e_{33}^2 + 2.C'_{13}.e_{11}.e_{33})dx.dy.dz$$

Using the values of e_{11} and e_{33} calculated before and performing an integration from z=-t/2 to z=t/2, one obtains

$$W = (C'_{11} - \frac{C'^2_{13}}{C'_{33}}).b.L.\frac{t_3}{24.R_2}$$

The flexural modulus is defined by

$$E_F = C'_{11} - \frac{C'^2_{13}}{C'_{33}}$$

The deflection Z and the radius of curvature are connected by the relation

$$L^2 = Z(2.R-Z) \simeq 2.R.Z$$

then

$$F = dW/dZ = E_F.b.t^3.Z/(3.L^3)$$

and the flexural modulus is obtained as a function of Z

$$E_F = 3.L^3.\frac{F}{b.t^3.Z}$$

For an isotropic solid, E_F is given by

$$E_F = \frac{E_Y}{1-v^2}$$

and for cubic films for which the x_3-axis is parallel to the [100] or [111] cubic directions, by

$$E_F[100] = C_{11} - \frac{C_{12}^2}{C_{11}}$$

and

$$E_F[111] = 3 \cdot C_{44} \cdot \frac{C_{11}+2.C_{12}}{C_{11}+2.C_{12}+4.C_{44}} + \frac{C_{11}-C_{12}+4.C_{44}}{6}$$

i) Measruement of flexural modulus

E_F can be measured by a vibrating reed technique. The specimen is clamped at one end and flexural vibrations are generated by an electrostatic or magnetic oscillating field [8]. A simpler method has been set up in the Department of materials Science and Engineering of Stanford [9,10]. The deflection Z of a cantilever beam and the applied force F are directly measured by using a nanoindenter. The beams have dimensions of about 1 µm x 20 µm x 100 µm. and are prepared with techniques coming from the microelectronics industry. The film is deposited on a silicon substrate and lithographic patterning and anisotropic etching techniques are used to make the structure (Fig 3)

2-3-5 Other techniques used for moduli measurements

Various techniques can be employed to determine the elastic moduli of thin films. Among the more often used processes, we can note the velocity measurements of surface acoustic waves [26,30] and Brillouin scattering of light [31]. The new nanoindentation methods, described in a next section (ch. 3), offer also very interesting possibilities.

These three techniques do not require the removal of the film from the substrate, which in some cases is a great advantage.

2-3-6 Relation between the different moduli

The primary elastic constants can be determined from different measurements of an elastic modulus along several directions, or from the measurements of different moduli for the same film orientation. However the relations between the moduli and the elastic constants are non linear and are generally difficult to reverse. It is often easier to compute the values of the moduli for several values of the elastic constants and to compare these with the experimental results. To prove the coherence of the variations in the different moduli of Cu-Ni multilayers (fig 1), Tsakalakos [2] uses the following relations demonstrated by Baral et al. [1,8] for cubic films with a [111] cubic direction normal to the foil

If A, B, and C are defined by

$$A = C_{11}+2.C_{12} \qquad B = C_{11}-C_{12} \qquad C = C_{44}$$

One obtains

$$E_B = 6.A.C/(A+4.C)$$

$$E_F = E_B/2 + (4.C+B)/6$$

$$E_Y = (4.C+B).E_B/(3.E_F)$$

$$E_T = (B+C)/3$$

In this particular case, A, B and C are readily obtained from the moduli measurements and the primary elastic constants C_{11}, C_{12} and C_{44} can be deduced.

2-3-7 Moduli of non homogeneous films

The precise moduli calculation of materials composed of a large number of small crystals $X_1.X_2...X_n$, with different orientations, different shapes, different sizes or different compositions is very complex and generally not possible. But, as for the calculation of electric resistivity, it becomes simpler for some particular geometries like the "bambou" structures and the multilayers.

i) The Moduli of multilayers

For example, let us calculate the Young's modulus of a multilayer of total thickness $t = t_1+t_2+...+t_n$ (t_i being the thickness of the X_i layer).

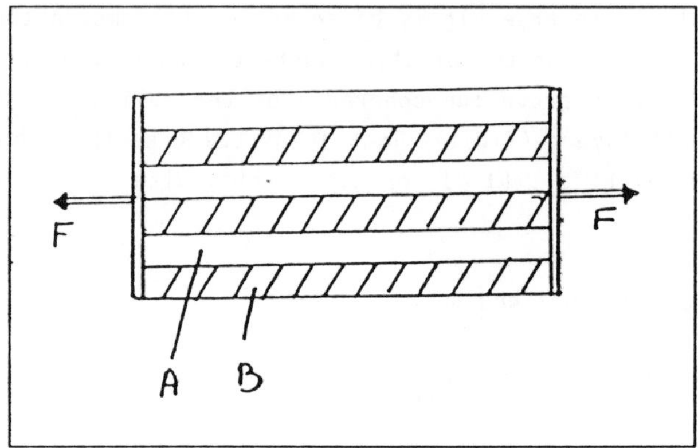

Fig. 5 Young's modulus of a multilayer composed of alternate layers of alloys A and B.

The unit elongation e_{11} is the same for each layer. The total force which must be applied at the ends of the specimen is equal to

$$F/b = \sigma_{11}.t = \sigma_{11}(X_1).t_1 + \sigma_{11}(X_2).t_2+...+\sigma_{11}(X_n).t_n$$

b is the width dimension and the stresses $\sigma_{11}(X_i)$ are given by

$$\sigma_{11}(X_i) = E_Y(X_i).e_{11}$$

The Young modulus of the multilayer is then defined by the relation

$$E_Y = E_Y(X_1).t_1/t + E_Y(X_2).t_2/t ...+ E_Y(X_n).t_n/t$$

The same form is obtained for the bimodulus E_B and the shear modulus E_T. The calculation of the flexural modulus of a multilayer is not so simple. The neutral axis position must be previously determined to continue the calculation as in ch.2-3-4. A more rigorous expression has been obtained by Townsend et al.[10,11].

NOTE: It is very important to note that the surface energy variations during the deformation are neglected to obtain these relations. It is possible, as a first approximation, to consider each interface as a layer with Young's modulus E and thickness t_s (t_s being the thickness of the perturbed crystal around the interface, roughly some interatomic distances) and to introduce them in the last formula giving E_Y. A great difficulty is that $E_Y(s)$ and t_s are not known yet. However it is easy to understand that it would not be realistic to neglect the interface effects when t_s is of the same order of magnitude as the thickness t_i of each individual layer.

ii) Moduli of films with a strong texture

We shall consider here the foils with a "pencil" structure which consists of many grains having a common crystallographic direction normal to the surface but randomly oriented in the plane of the foil. Baral *et al.* [8] have shown that for such structures it is possible to deduce the primary elastic constants C_{ij} from the different moduli values. To do that, they have expressed the moduli in terms of appropriate stiffnesses which are calculated by averaging the crystal stiffnesses for all the grain orientations in the foil. In some cases, the grains are isotropic in the plane (for example, for [111] et [100] directions of cubic crystals parallel to the x_3-axis) and the averages are equal to the values of a single crystal with the same orientation.

2-4 Elastic-constant anomalies in metallic superlattices

These anomalies, sometimes called "supermodulus effect", have been discovered by Hilliard *et al.* [6,12,13,14], at Northwestern University, in foils with a one-dimensional composition modulation. These superlattices have been originally produced to verify the Cahn's diffusion equation concerning the spinodal decomposition. Hilliard *et al.* have shown that the biaxial modulus of the modulated structures Au/Ni, Cu/Pd, Cu/Ni et Ag/Pd increases by several hundred percent for some critical values λ_c,in a 1,5 to 3 nm range, of the wavelength composition λ (Fig 1). Out of this domain E_B rapidly takes more classical lower values. This strong variation, compared with the few per cent variations generally observed between classical metals and alloys, is very surprising. This effect was also obtained with other elastic moduli and other structures (coherent superlattices). A lot of metallic couples have been tested with very different results.

The coherence seems to be a required condition to observe the supermodulus effect. The elastic moduli of some systems, such as Cu/Au and Ag/Au, do not present anomaly. For some others, such as Nb/Cu, Mo/Ni, Ti/Ni, etc, a strong softening is always observed for the same range of λ values (between 1.5 and 3 nm). A review of the main experimental results is given by Schuller [16].

This "supermodulus effect" is always surprising and its origin has not been clearly established yet. Two main hypotheses have been proposed and widely developed.

-In the first one, the new Bragg diffractions produced by the superperiodicity of the superlattice modify strongly the energy and the elastic coefficients of the alloy.

-In the second one, the very high stresses induced by the coherence between the layers are responsible for these anomalies.

-Some observations have recently revealed correlations between these anomalies and relaxations occurring at the interfaces and this third hypothesis must also be studied.

2.4.1.Modifications of the Fermi surface

Compared to an A-B homogeneous alloy taken as a reference, a concentration modulation gives rise to new Brillouin zone boundaries located at $q= \pm n.\pi/\Lambda$, q being a reciprocal space vector normal to the foil. In this space, the constant energy surfaces, and especially the Fermi surface, are perturbed in the neighbourhood of these zone boundaries and new singularities appear on the density of states curve. When one of these singularities occurs near the Fermi level, the energy of the electron gas and its variations during deformation can be strongly modified. Pickett [17] has shown that the crystal energy is particularly lowered when a superlattice zone boundary becomes tangent to the Fermi surface, inducing particular deformations of that surface called "Fermi-surface necking". Using coherent potential approximation calculations (C.P.A.), he has shown that this F.S. necking occurs for values of the wavelength λ in the range 1.5-3 nm for the superlattices Cu/Ni and Ag/Pd. Wu [18] comes to the same conclusions by computations using pair potentials taking into account the Fermi-surface modifications.

Transport properties, sensitive to the density of states at the Fermi level, should be affected too for the same λ values. Such effects have been effectively observed in Cu/Ni multilayers by Baral et Hilliard [19], Schuller et al. [20]; and in Mo/Ni superlattices, by Khan et al. [15]. The energy decrease for concentration modulations with $\Lambda \simeq \Lambda_c$ ($\simeq$ 2 nm) should promote spinodal decomposition of homogeneous alloys with the same average concentration [21]. This is not generally observed. However some concentration fluctuations in Cu-Ni alloys were revealed by positron annihilation experiments [14].

A precise calculation of elastic constants including these Fermi-surface effects is at the moment difficult, but it seems essential to perform it in the future to test correctly this hypothesis.

Coherence stresses

These stresses are studied in more detail in the following section. In heteroepitaxial layers of small thickness, the lattice mismatch at the interfaces is accommodated by uniform strains in each layer. This leads to very large biaxial stresses, larger that in the bulk materials because of the lack of dislocation sources or other relaxing mechanisms. Hilliard [12], Tsakalakos and Jankowski [22] have shown that such high stresses can lead to a 100 % increase in copper elastic constants and explain the supermodulus effect observed in Cu/Ni and others superlattices. The disappearance of the effect for larger values of λ would then be caused by the stress relaxation produced by the development of misfit dislocations. For smaller values of λ, diffusion would be responsible for interface mixing which would relax the high stresses. This model is supported by several results obtained by computations with different kinds of interatomic potentials. For example, Shuller et Rahman [23] have found, with molecular dynamic calculations, a 35 % decrease of C_{44} in nickel for a 5 % biaxial stress which would well explain the softening observed in Ni/Mo multilayers.

2-4-3 - Interface Relaxations

The biaxial deformation caused by the epitaxy must lead to a modification δa_3 of the lattice spacing a_3 between planes parallel to the surface equal to

$$a_3/a_3 = e_{33} = -e_{11}.(C'_{31}+C'_{32})/C'_{33}$$

In several cases large deformations have been observed by X-ray scattering measurements and correlated with the modulus anomaly [15,24]. However some more precise measurements have recently shown that in some cases this deformation was not homogeneous but mainly concentrated at the interfaces [25,26]. For example in Mo/Ta, Mo/Ni, Pt/Ni and Ti/Ni multilayers, a strong dilatation normal to the surface has been observed at the interfaces. A careful analysis by Clemens and

Eesley [26] shows that these interface relaxations are sufficient to explain the elastic constant softening of these superlattices. Some numerical simulations confirm these relaxations and their effect on the elastic moduli. With molecular dynamic calculations and by using Morse potentials, Imafuku [28] et al. have found strong relaxations at the interfaces of Au/Ni and Cu/Ni superlattices in close correlation with elastic modulus increases.

To conclude, the understanding of the "supermodulus effect" is still a very open problem [28]. More precise determinations of interface atomic structures and of stress distributions in the layers, and also more accurate calculations of interatomic pair potentials seem important for any further progress.

3 INTERNAL STRESSES IN THIN FILMS ON SUBSTRATES

3-1 Introduction

Thin films on substrates are very often subject to high biaxial stresses which are generally developed during their growth and subsequent thermal treatments [32,33]. The absence of dislocations or other defects prevents their relaxation and explains the high observed values, compared with the bulk materials. These stresses can be detrimental to the qualities of coatings or to the reliability of thin films for electronic or magnetic recording devices. They can lead to cracks and film decohesions which favor corrosion and mechanical failures. They can also induce the formation of defects such as, for example, misfit dislocations which can modify the electric and magnetic properties of microelectronic devices.

A simple method to check these stresses is to remove the film from its substrate : The film increases or decreases in size to take a free stress state (Fig 6). To re-attach it onto the substrate, edge tractions σ_{11} and σ_{22} must be applied to stretch it to substrate dimensions. This introduces biaxial deformations given by

$$e_{11} = S'_{11}\cdot\sigma_{11} + S'_{12}\cdot\sigma_{22}$$

$$e_{22} = S'_{21}\cdot\sigma_{11} + S'_{22}\cdot\sigma_{22}$$

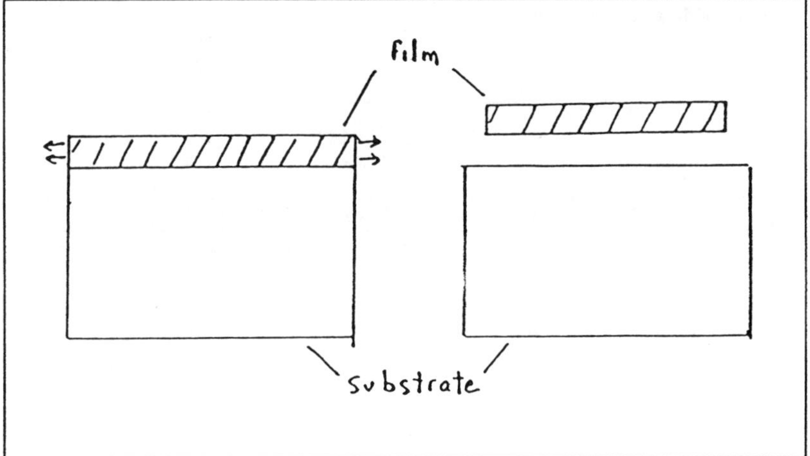

Fig. 6 (left) Film attached to its substrate and subjected to tensile stress, (right) Film detached from its substrate. The stress is relaxed and the dimensions of the film are changed.

For isotropic deformations ($e_{11} = e_{22} = \epsilon$) and isotropic films, one obtains

$$\sigma_{11} = \sigma_{22} = E_B.\epsilon$$

E_B is the biaxial elastic modulus. The density of elastic energy stored in the film, equal to the work done by the applied stresses, is given by

$$w = E_B.\epsilon^2$$

For anisotropic films or anisotropic deformations, stresses σ_{11} and σ_{22} do not take the same value. They can be calculated with the two previous equations. For anisotropic films a bimodulus is sometimes defined for two perpendicular directions parallel to x_1 and x_2 axes in the following way applied stresses σ_{11} and σ_{22} are chosen to produce the same deformation ϵ in the x_1 and x_2 axes directions ($\epsilon = e_{11} = e_{22}$). Then the two biaxial moduli are defined by:

$$E_B(1) = \sigma_{11}/\epsilon \qquad \text{et} \qquad E_B(2) = \sigma_{22}/\epsilon$$

For cubic films with direction [011] parallel to the x_3-axis, one obtains [32] for the [100] and [011] directions

$$E_B[100] = C_{11}+C_{12} - \frac{C_{12}.(C_{11}+3.C_{12}-2.C_{44})}{(C_{11}+C_{12}+2.C_{44})}$$

$$E_B[011] = \frac{(C_{11}+3.C_{12}+2.C_{44})}{2} - \frac{(C_{11}+C_{12}-2.C_{44}).(C_{11}+3.C_{12}-2C_{44})}{2.(C_{11}+C_{12}+2.C_{44})}$$

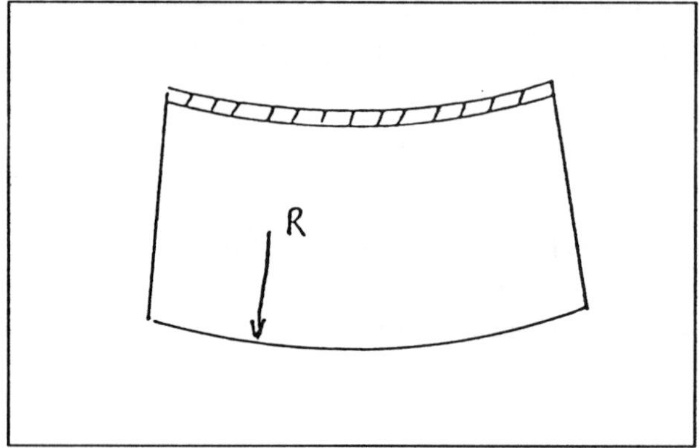

Fig. 7 Film under biaxial tensile stress leading to a bending of the substrate.

Biaxial stress in the film leads also to a bending of the substrate which can be easily observed if the substrate thickness h_s is not too large (Fig 7). For isotropic conditions, the radius of curvature R is given by [51,34]:

$$R = E_{Bs}\cdot\frac{h_s^2}{6.\sigma_f.h_f}$$

E_{Bs} is the subtrate bimodulus, h_f the film thickness and σ_f ($\sigma_f = \sigma_{11} = \sigma_{22}$) the biaxial stress in the film.

To obtain this result, h_f is assumed to be very small compared with h_s, otherwise R depends on the film bimodulus. For an anisotropic biaxial stress, two curvatures are observed and the radii formulae are the same as in the isotropic case.

3-2 Measurement of internal stresses

Measurements of substrate curvature have often been used to determine internal stresses in thin films (X-ray diffraction measurements [35,36], developed in chapter II, also provide a good method). Some techniques use circular shape substrates a few hundred µm thick for a 1 µm film thickness. The curvature is measured by optical, electromechanical or capacitive methods. An accurate laser scanning technique was recently developed by Nix et al. [32,34]. Films deposited on cantilever microbeams have also been frequently used [10,40-43]. The stress in the film is related to the measured deflection at the free end by the relation

$$\sigma_f = E_{Bs} \cdot \frac{h_s^2}{3.L^2.h_f} \cdot \delta$$

All these techniques, and particularly the last one, have been used "*in situ*" to follow the stress evolution during the growth of the film or during thermal cycling. Thermal stresses $\sigma(T)$ are then produced by the difference between the thermal expansion of the film α_f and the substrate α_s

$$\sigma(T) = E_{Bf}.(\alpha_s - \alpha_f)(T - T_0)$$

T is the temperature, T_0 the temperature of the free stress state and E_{Bf} the elastic bimodulus of the film.

Stress induced in the substrate stays at a very low level when h_s is large compared with h_s. Its maximum value, given by Nix [32], is equal to

$$\sigma_{ind.max} = -3.\sigma_f.h_f/h_s$$

For the thickness values generally employed ($h_f \simeq 1$ µm and $h_s \simeq$ a few 10^2 µm), stress in the substrate is then less than 1 % of the stress in the film.

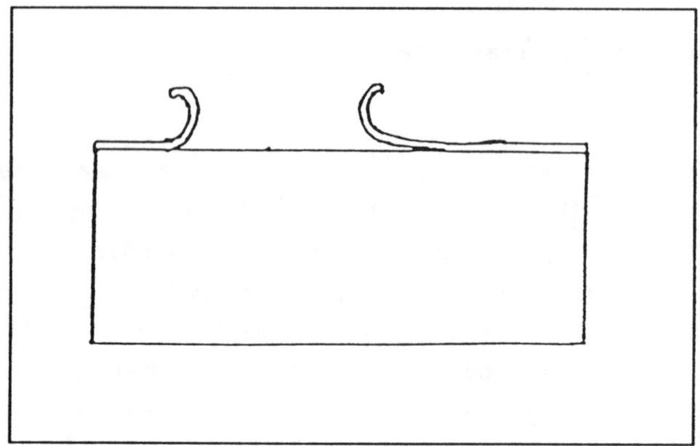

Fig. 8 Curling of a delaminated film near a crack.

In some cases stress in the film is not homogeneous. This can be observed by X-ray diffraction [33,37]. Several reasons have been put forward : temperature gradient during the growth ; diffusion and relaxation by defects (dislocations, point defects) at the interface or at the free surface; etc...The film, detached from its substrate, tends to curl up (Fig 8). A similar effect is observed, for the same reason, when a thin part of the substrate is still attached to the delaminated film [76].

3-3 Origin of internal stresses in thin films

These stresses depend strongly upon the growing process. They often appear during the deposition but can also vary for a long time afterwards. This has been observed in films of gold, silver and copper on a substrate [40]. In Cr thin films, the internal stress continues to increase after deposition for up to 200 hours.

Thermally activated mechanisms generally play a key role for the developement and relaxation of internal stresses. The thermal history during and after the growth is very essential to control the stress state of the film. Doerner and Nix have reviewed in great detail different phenomena leading to internal stresses in thin films [33]. We shall only give here a brief survey but shall develop more extensively the role of misfit dislocations.

i) Coherence stresses in epitaxial layers

These stresses are induced, when the film grows with an epitaxial relationship to the substrate, by the mismatch $\delta a_f = a_s - a_f$. between the lattice spacings of the two materials. This usually happens when films are grown using Molecular Beam Epitaxy (M.B.E.). Coherence is obtained with a biaxial strain of the film:

$$e = e_{11} = e_{22} = \delta a_f / a_f$$

and induces a biaxial stress:

$$\sigma = E_B . e$$

When the film thickness h_f becomes larger than a critical thickness h_c, this stress can be partially relaxed by misfit dislocations (cf. chapter 3-3).

ii) Stresses induced by recrystallisation and phase transformation

With several growth processes, films are usually far from a thermal equilibrium state during and just after their deposit. The projected material is often strongly quenched and non-equilibrium structures can also be induced by the substrate. The material is sometimes amorphous and sometimes crystallised in a non standard system. If the temperature is high enough, it can evolve more or less rapidly towards its equilibrium state. This is generally accompanied by dilatation and shear strains which induce stress development. For example, amorphous to crystal transformations have been observed in thin Sb layers [38] and in metallic silicide films [39]. The same effect is obtained with precipitation or spinodal decomposition.

Stress development is also often due to grain boundary relaxation and recrystallisation. The deposited thin films are often made of small grains, sometimes with a columnar structure, sometimes badly crystallized at the grain boundaries (as nanocrystals) and, if the temperature is high enough, recrystallisation takes place to decrease the large stored energy. These relaxations generally lead to high *tensile* stresses. In some cases, the film is not fully dense and grains are not closely joined together. These boundary voids can be metastable or can grow or shrink by vacancy diffusion. Several models [33], taking into account these different effects, have been developed to explain the high values of the observed stresses.

Some attention must also be paid to point defects and clusters: they would be responsible for the high *compressive* stresses observed in sputtered films and in thin films obtained with the help of ion implantation [33].

iii) Stress induced by impurity effects

Several effects on the stress in thin films can be induced by impurities. The diffusion of atoms between film and substrate has been already invoked (chapter 2-7) to explain stress relaxation in very thin films. Absorption and desorption of volatile species can also occur on the free surface during or after the growth, especially in porous and badly crystallized films. Some deposition processes do not offer a sufficient high vacuum to avoid contamination. Others requise a neutral or reactive gas (some C.V.D. and P.V.D methods,for example). The influence of O_2, H_2, N_2, CO and H_2O gas on the stress in thin films of gold, silver and copper has been studied by ABERMANN et al. [40-43]. They have found that silver deposition is very sensitive to oxygen which decreases the grain size and increases the internal stress. Haite et al. [45], and Lopata et al. [46] have shown that it is possible to obtain silicon nitride thin films with a very large range of internal stresses by varying the composition of the $N_2/NH_3/SiH_4$ mixture used in the PECVD (Plasma Enhanced Chemical Vapor Deposition) process. High tensile stresses are obtained for low N_2 gas concentrations but the stress becomes compressive for large N_2 concentration values. This effect is attributed to N_2^+ ion implantation in the film during the growth. But the nitrogen rate is too low to explain the observed stress which seems to be produced by clusters and irradiation defects. This effect is sometimes described as an "atomic peening" action [33].

3-4 Misfit dislocations

These dislocations, located at the film-substrate interface, allow a partial stress relaxation to occur in epitaxial films. They have been first theoretically introduced by Frank and Van Der Merve [47,48] and are now commonly observed by T.E.M (Transmission Electron Microscopy) or by X-ray topography. Their development can also be followed by measuring the stress relaxation that they induce in the film. Matthews et al. [49] have shown that this relaxation can only take place when the film reaches a critical thickness given by the equation[32]

$$\frac{h_c}{\ln(\beta . h_c/b)} \simeq \frac{b}{4\pi . \epsilon}$$

b is the dislocation Burgers vector, β a numerical constant of the order of unity and ϵ the biaxial strain in the epitaxial film without dislocation.

In this part, we will focus our attention on the relaxing role of these misfit dislocations and on the problem of their nucleation. We shall begin by recalling some particular points of dislocation theory.

3-4-1 Elements of dislocation theory

i) Definition

Let us consider in a solid a closed curve C, bounding a surface S. A sense is ascribed to C and the right hand rule is used to choose the orientation of the S-normal direction. Then a cut of the crystal along S defines a S^+ and a S^- surfaces. A dislocation of Burgers vector **b** is formed along C if the S^- surface is displaced by a vector **b** relative to S^+ (some material must be inserted or removed for this operation) and then the two surfaces are pasted together.In a crystalline solid, C is called a perfect dislocation when **b** belongs to the Bravais lattice and an imperfect one in the other cases.

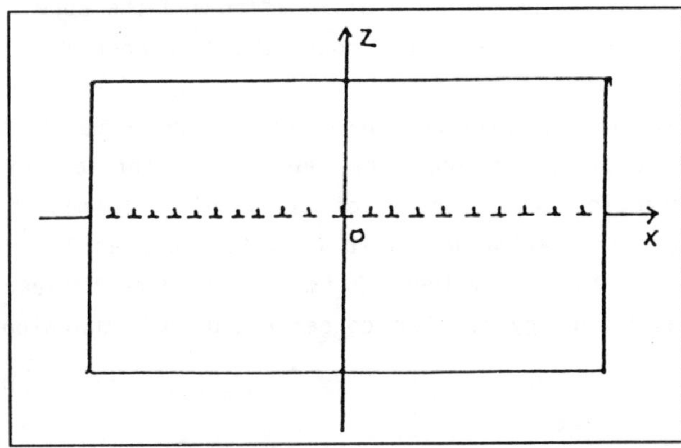

Fig. 9 Distribution of dislocations with infinitesimal BURGERS vector δb. The compressive and tensile stresses above and below the cut plane are constant for a continuous distribution.

This definition can be used to describe a continuous distribution of dislocations (Fig 9) of Burgers vector density δb/a, as the limit of a set of parallel and straight dislocations, at a distance a from one another and of Burgers vector δb. Note that the different operations which create these dislocations are the same as those needed to dilate a thin film for re-attaching it on its substrate, as described in chapter 3-1 (Fig 6). This analogy will be used later to represent and calculate the biaxial stress in epitaxial layers.

ii) Stress and energy of dislocations

For a dislocation along the x_3-axis, strains and stresses, at a point M(r,Θ) (fig 10), vary as b.f(Θ)/r. For example, the σ_{11} stress component of an edge dislocation with a Burgers vector b parallel to x_1 is given by [50]

$$\sigma_{11} = \frac{\mu.b}{2\pi.(1-\nu)} \cdot \frac{y.(3.x^2+y^2)}{(x^2+y^2)^2}$$

At large distances, these stresses must not be neglected and the elastic energy depends on the size of the solid. In a cylinder of radius R centred on the x_3-axis, its value is given by

$$W = \frac{\mu.b^2}{4\pi.K} \cdot \ln(R/r_0)$$

K is equal to (1-ν) for an edge dislocation and the core radius r_0, of the order of b, is adjusted to take into account the core energy.

At large distances, dislocation stresses are usually screened by the stresses of dislocations of opposite Burgers vector or by relaxation on surfaces. For example, the stress of a dipole, formed by two parallel dislocations of Burgers vector b and -b (Fig 10) vary as $1/r^3$ when r >> D and can be neglected for these r values (D is the distance between the two dislocations). The elastic energy is then concentrated near the dipole and is given by

$$W = 2 \cdot \frac{\mu.b^2}{4\pi.K} \cdot \ln(D/r_0)$$

We can be calculated as the sum of the two self-energies and the interaction energy. Similar formulae are obtained for multipoles (Fig 10) when the vectorial sum of the Burgers vectors is equal to zero. D is then the extension of the multipole.

The relaxation effect of a plane surface parallel to the dislocation, to a first approximation [50,52], can be taken into account by introducing a symmetric image dislocation of opposite Burgers vector. The stress is given by this dipole and the elastic energy is a function of the distance dislocation-surface D

$$W = \frac{\mu.^2}{4\pi.K} \cdot \ln(2.D/r_0)$$

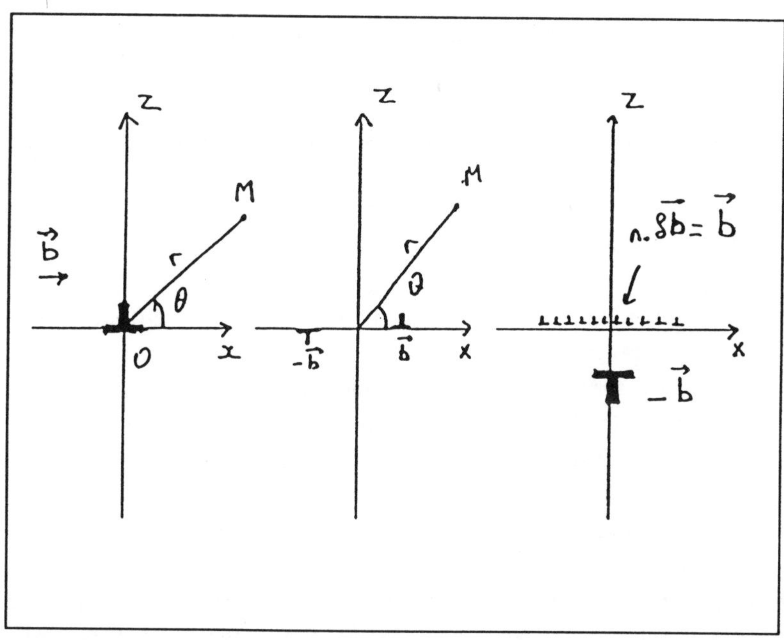

Fig. 10 (left) Edge dislocation, (center) Dipole of dislocation (right) Multipole formed with a dislocation of Burgers vector -b and a continuous distribution of infinitesimal dislocations (Σ δb = b).

iii) Array of dislocations

Some analytic formulae [52] are available to calculate the stress and energy of parallel dislocation arrays. At distances larger than d, d being the distance between two neighbouring dislocations, strains and stresses become quickly constant. For example, the strength e_{11} of a set of edge dislocations parallel to x_2 in the (x_1,x_2) plane with Burgers vector parallel to x_1 is given by

$$e_{11} \simeq \pm \frac{b}{2d} \cdot [1 - \frac{\pi|z|}{d(1-\nu)} \cdot \exp(-2\pi|z|/d) \, . \, \cos(2\pi x/d)]$$

and, with $|z|>d$

$$e_{11} \simeq \pm\, b/(2d) \qquad \text{and} \qquad \sigma_{11} \simeq \pm\, (b/d).\mu/(1-\nu)$$

strains and stresses being of opposite sign on each side of the (x_1,x_2) plane.

3-4-2 Misfit dislocations

In the following section, for simplicity, we shall neglect the difference between elastic constants of film and substrate as well as their anisotropy. These two factors are sometimes important and can lead to particular effects not to be discussed here. It is possible to take them into account with anisotropic elastic theory for more sophisticated calculations [53,54].

To see how dislocations can relax coherent stresses in epitaxial films, let us look at figures 11-14.

On Fig 11, the film is just pasted on the substrate without strain and stress. The mistmatch $\delta a = a_f - a_s$ (with $\delta a \ll a_f$) between lattices of film and substrate is not accommodated and the interface is fully incoherent. If such an interface exists in a metastable state, its energy Γ_{inc} must be very large. We call W_a the energy of this configuration

$$W_a = L^2.\Gamma_{inc}$$

L^2 is the area of the square interface.

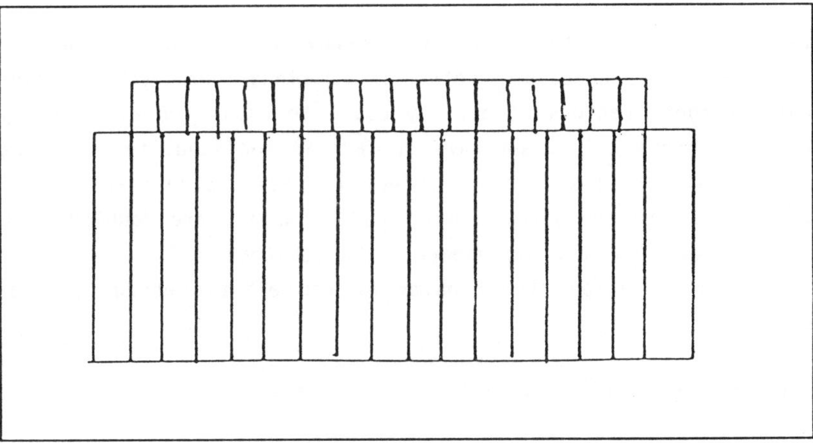

Fig. 11 Film without stress simply attached to the substrate with an incoherent interface (configuration a).

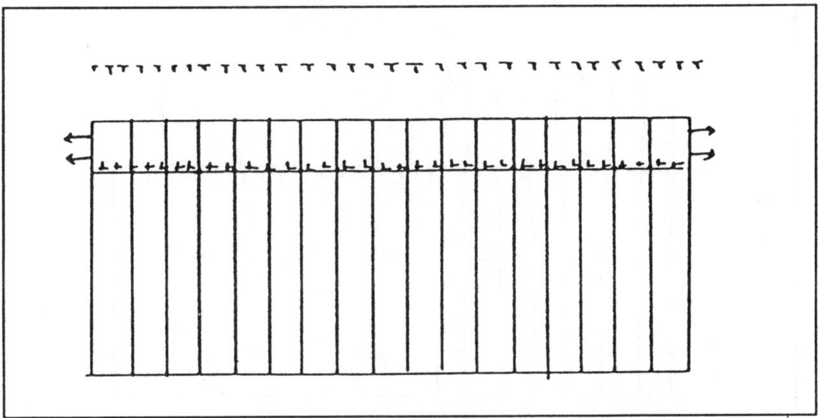

Fig. 12 Epitaxial film on substrate without misfit dislocation (configuration b).

On Fig. 12 (configuration b), the film has been homogeneously strained ($\epsilon=\delta a/a_f$) to accommodate the mistmatch between the two lattices. The interface is coherent and its energy Γ_c is much lower than in the incoherent case. This configuration is more stable than the preceding one, as long as the thickness film h_f is less than

$$h_f < (\Gamma_{inc} - \Gamma_c)/w$$

$w = E_B.\epsilon^2$ is the density of elastic energy stored in the film.

This biaxial strength and this energy can also be calculated by introducing two perpendicular sets of dislocations with infinitesimal perpendicular Burgers vectors of density equal to $\delta a/a_f$. On all the figures of this paper, only one set is drawn. The surface is replaced, for calculation, by the image dislocations. Its relaxing effect is clear. Stress and strain, which can be calculated with the already mentioned formulae, are doubled in the film by the image dislocations and disappear in the substrate.

In this approximation, the density of the elastic energy stored in the film is equal to

$$w \simeq 2\ L^2.h_f(b/d)^2.\mu/(1-\nu)$$

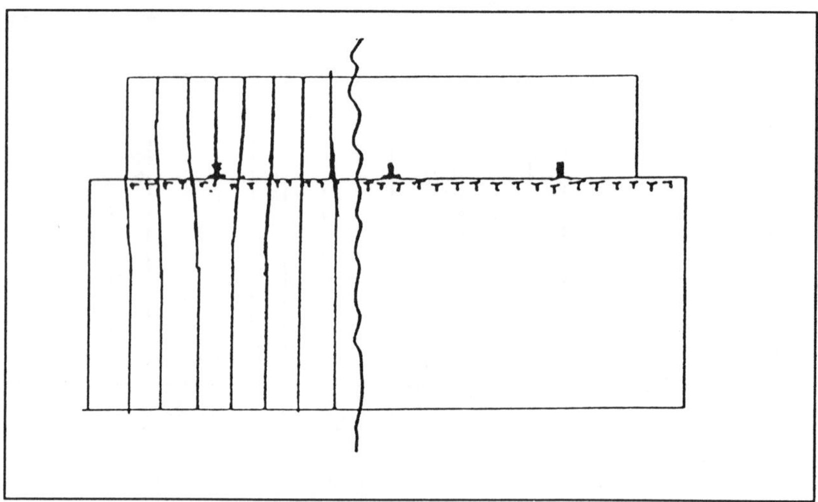

Fig. 13 Thick epitaxial film (h_f > d) on substrate with misfit dislocations (configuration c).

On Fig. 13 (configuration c), the film is not homogeneously strained,and no strain at all occurs far from the interface (z >> d) as in the case of configuration a. But some local deformations have been introduced at the interface level to obtain some coherence. In the dislocation description, this is obtained with a row of multipoles formed by adding misfit dislocations of

Burgers vector -b to the continuous distribution of configuration b. To have no stress at large distances from the interface, the Burgers vector density b/d (d is the inter-dislocation distance) of the misfit dislocation wall must be the same as for the continuous distribution

$$b/d = \delta a/a_f = e_{11}$$

With the hypothesis $d < h_f$, stress of the multipoles on the free surface can be neglected and the elastic energy is roughly the sum of the multipole energies, independent of h_f:

$$W_c = L^2.[\ \Gamma_{c.} + \frac{\mu.b^2}{\pi(1-\nu).d} \cdot \ln(d/r_0)\ ]$$

For $d < h_f$, the condition $W_c < W_b$ becomes

$$d.\ln(d/r_0) < 2.h_f$$

and is always fulfilled. Then misfit dislocations always decrease the energy of rather thick films on substrates.

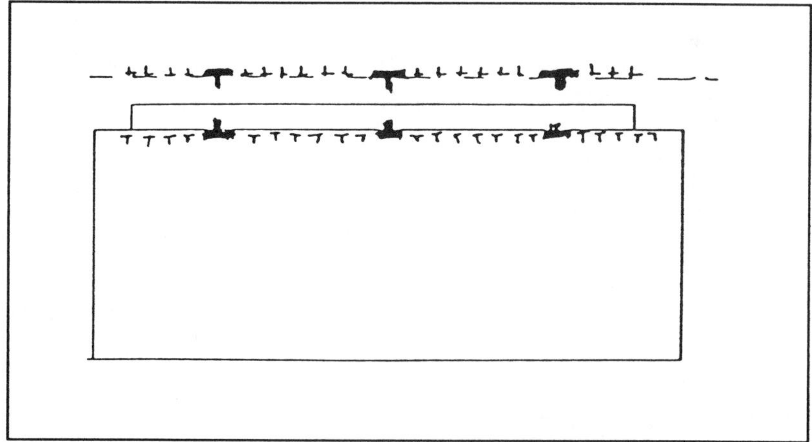

Fig.14 Thin epitaxial Film (h_f < d) on substrate with misfit dislocations.

As for the continuous distribution, two sets of misfit dislocations are needed to suppress long distance stress. For simplicity again, we shall take two families of perpendicular dislocations with perpendicular Burgers vectors. In real cases, this is not generally true and the interaction energy between the two sets must be calculated.

For very thin films ($d >> h_f$) the coherence energy W_b tends linearly to zero with h_f. This argument does not apply to misfit dislocation energy because, for low values of h_f, the dislocation core energy becomes predominant. Then for $d >> h_f$ the strained film without misfit dislocation must be the more stable configuration.

To determine the critical value h_c at which the stability change occurs, we may calculate the energy W_c for all film thicknesses. For $h_f < d$, misfit dislocations are mainly screened by their image and, to a good approximation, do not interact. W_c is then the sum of three terms: the coherence dislocation energy W_b, the misfit dislocation energy screened by the surface and the interaction energy between these two sets of dislocations.

One obtains

$$W_c = W_b + 2 \cdot L^2.[\frac{\mu.b^2}{2\pi(1-\nu).d} \cdot \ln(2h_f/r_0) - h_f \cdot \frac{2.\mu.b^2}{d^2.(1-\nu)}]$$

Misfit dislocations decrease the crystal energy when W_c becomes lower than W_b, then for $h_f > h_c$, with h_c given by the equation

$$\frac{h_c}{\ln(2.h_c/r_0)} = \frac{d}{4\pi} = \frac{b}{4\pi.e_{11}}$$

i) Misfit dislocations with Burgers Vectors non parallel to the film

Dislocations with Burgers vectors non parallel to the interface have a larger energy but are often observed because, in a lot of structures, they form more easily. The critical thickness is calculated as in the preceding case; if α is the angle between interface and Burgers vector, h_c is given by:

$$\frac{h_c}{\ln(2.h_c/r_0)} = \frac{d}{4\pi.\cos^2(\alpha)} = \frac{b}{4\pi.e_{11}.\cos(\alpha)}$$

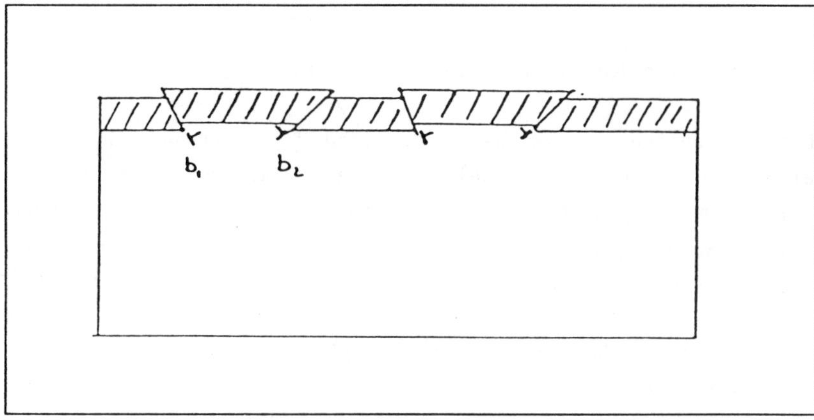

Fig.15 Misfit dislocations with Burgers vectors non parallel to the interface.

Coherence strength e_{11} is now equal to $b.\cos(\alpha)/a_f$.

It is important to note that if all dislocations have the same Burgers vector **b**, the $b_3 = b.\sin(\alpha)$ components induce an additional stress field and a misorientation between film and substrate given by $\Theta = b_3 / d$. However this misorientation is reduced or removed when the boundary is a mixture of dislocations with Burgers vectors doing a $+\alpha$ and $-\alpha$ angle with the interface. It must be also recalled that these dislocations shear the film (Fig 15) and introduce steps, b_3 high, on the free surface and on the interface. This can be detrimental to the film properties.

ii) Stacking faults and dissociated misfit dislocations

We have only considered the perfect misfit dislocations of the structure. But in some systems, as for example fcc superlattices with a [111] direction normal to the film, a stacking fault located on the interface must have a low energy and misfit dislocations with Burgers vector parallel to the interface must be dissociated. In some others cases with Burgers vector non parallel to the interface, cross slip in the interface of a partial dislocation can be favoured.

3-4-3 Mechanisms of misfit dislocation formation

i) Dislocation formation at the free surface

Let us consider an epitaxial strained film without dislocations. Fresh dislocations can appear in the bulk or on a surface by nucleation of prismatic loops for high vacancy concentrations and high enough temperature. Gliding dislocations with Burgers vectors non parallel to the film can also be introduced from the free surface. In each case, two forces are acting on the dislocation:

-the attractive force of the image dislocation varying as 1/D, D being the distance to the free surface.

-the constant force $\sigma_{11}.b_x$ normal to the surface induced by the coherence dislocation wall and its image.

The same forces are acting for the formation of a dipole in the bulk or at the interface.

The values of these two forces can be deduced from the calculation of the energy dislocation in terms of D [55]

$$W(D) = \frac{\mu.b^2}{4\pi(1-\nu)} \cdot \ln(2D/r_0) - D\cdot\frac{2.\mu.b^2}{d.(1-\nu)} \cdot \cos^2(\alpha)$$

One can see that $W(h_c) = 0$ and that the dislocation decreases the total energy for $D \geqslant h_c$, a result already found before. But the sense of the driving force on the dislocation depends on another critical value D_c defined by the equation $dW(D)/dD = 0$, which gives

$$D_c = d/(8\pi.\cos^2(\alpha))$$

or

$$D_c = \frac{b}{8\pi.e_{11}.\cos(\alpha)} = \frac{h_c}{2.\ln(h_c/r_0)} < h_c$$

When a dislocation comes from the free surface, the crystal energy always increases until $D = D_c$. As for bulk material the activation energy, equal to $W(D_c)$ per unit length of dislocation, is very large and this mechanism can only act with the help of high concentrated stresses induced by defects on the free surface [56]. The study of dipole nucleation inside the film or on the interface leads to the same conclusion.

It can also be deduced from this calculation that misfit dislocations are metastable when

$$D_c \leqslant h_f \leqslant h_c$$

ii) Misfit dislocations formed by threading dislocations

Dislocations existing at the substrate surface go through the epitaxial film during the growth and reach the free surface. When the thickness of the film becomes larger than h_c, the so called threading dislocations can be extended in the interface (Fig.16). If their Burgers vector has a component b.cos(α) parallel to the interface, then a driving force

$$f = |W(h_f)|$$

will tend to form a kink in the interface and to elongate it ($W(h_f)$ is the energy per unit length of dislocation calculated before). This mechanism has been observed in different thin films on substrates and in superlattices [58,59]. It has also been extended to explain the multiplication of misfit dislocations [57].

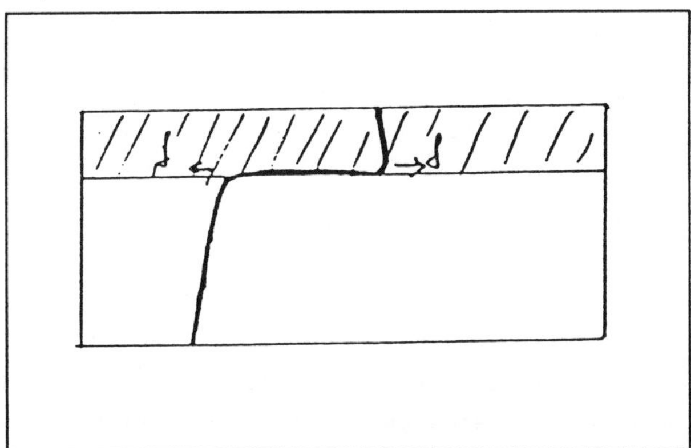

Fig.16 Threading dislocation and misfit dislocation formation.

4 PLASTIC PROPERTIES OF THIN FILMS AND COATINGS

Selecting a coating for a given material is always difficult; a compromise solution must be found to optimize different parameters: characteristics of the material and of the device to be coated, environmental constraints (mechanical stresses, shocks, high temperatures, chemical aggressions,etc), available facilities to form the deposit, cost price, etc. Among them, plastic and tribologic properties have always played a key role. The mechanical resistance and the reliability of a coating depend on its own characteristics: quality of the deposit, internal stresses, elastic modulus, hardness, etc; but they also depend on the mechanical properties of the substrate and on the adequacy of the two materials. A very hard coating on a soft substrate has generally a bad wear resistance. A good adherence between film and substrate is also fundamental: mechanical and corrosion failures often occur after decohesion of parts of the coating.

Mechanical studies are made difficult by the thinness of the film. Experimental techniques are generally adapted from bulk material testing methods: uniaxial tensile-testing apparatus especially designed for thin films, microhardness measurements of coatings on substrates, pin and disc systems for wear and friction studies. In this part, we wish to focus attention on only three points, more specific to thin films and the object of recent developments:

-the nanoindentation techniques,

-the stress measurements during thermal cycling,

-and the adherence studies.

4-1 Nanohardness measurements on thin films

4-1-1 The indentation method

Tensile-testing apparatuses are generally used for fundamental studies of plastic deformation mechanisms: the applied uniaxial and constant stress is homogeneous and very simple (at least at the beginning of the test); forces on dislocations are easily calculated and theoretical models can be developed and compared with experimental results. The mechanical resistance of the material is measured all along the deformation test.

The indentation technique for bulk material is quite simple to use. A calibrated ball or pin is applied onto the material surface with a loading force F and during a selected time. Plastic deformation is characterized by the area S of the permanent indentation. The hardness H is then defined by

$$H = F/S$$

For small loads, less than 100g, a microscope is needed to measure S. This technique is very useful to characterize and to compare different materials. But the interpretation of these tests at a microscopic level seems very difficult. The main reason is the deep complexity of the stress field and the difficulties to determine it. Only a very small area of the sample is tested; so a great number of experiments must be performed and averaged to obtain a representative value of the hardness of the material.

Stress field produced by a ball applied on a plane surface

Stresses induced by a sphere on an isotropic elastic half-space were determined by Hertz [60,61] in 1880 in the case of pure elastic deformations and frictionless contact. We shall only summarize here the main results of this calculation in order to discuss indentation experiments (fig 17):

i)-the radius a of the contact area varies as $F^{1/3}$, F being the loading force.

ii)-far from the contact zone (r >> a) stresses vary as $f(\Theta)/r^2$ (Fig.17)

iii)-on the x_3 axis, normal to the surface, stresses take their maximum values for $z \simeq 0.5.a$. Then the first plastic deformations must take place in this zone when the yield stress is overpassed.

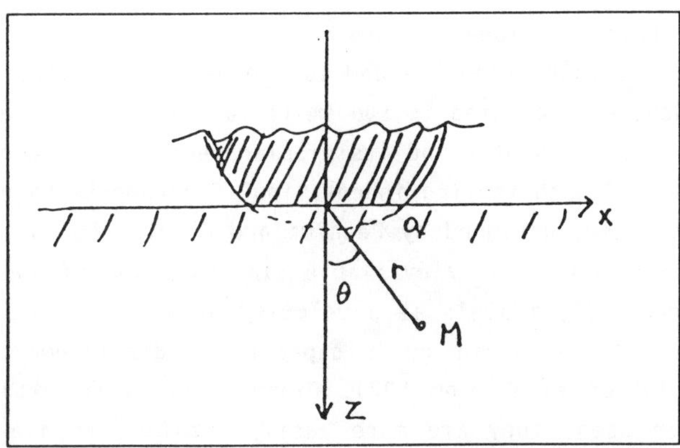

Fig. 17 Contact of a sphere on a half-space.

This elastic stress field is widely modified by plastic relaxation, friction and adhesion between the ball and the surface and these factors have been introduced in calculations [62,63] to obtain better values. It is not difficult to realize that plastic flow determination at an atomic level is very hard in such a complex stress field. The gliding systems involved have been sometimes studied in a few materials by T.E.M. Macroscopic models have been developed within the scope of the continuous solid theory and are still the aim of some research [62,64]. They are very widely used for application and research characterization.

4-1-2 Nanoindentation measurements

When the diameter of the contact area becomes less than a tenth of the film thickness ($2.a < h_f/10$), the stress induced by the ball or the pin vanishes on the interface and in the substrate. In thin films about 1 µm thick, this occurs for an area with less than a few nanometers.

When this condition is not fulfilled, empirical models can be used to deduce the intrinsic hardness H_f of the film from hardness measurement values H. These values depend on the substrate hardness H_s, on the film thickness h_f and on the indentation depth d. For example, the following formula is sometimes used

$$H_f \simeq H_s + \frac{H-H_s}{2.C.h_f/d - C^2.(h_f/d)^2}$$

where C is a constant number taken to be equal to 0.14 when the film cracks under the ball contact, and 0.07 when no crack appears.

To obtain intrinsic values of film hardness, it is however more convincing to eliminate the substrate influence and to use very low loads, in the 1 mg-1 g range. The permanent contact area is too small (a few nm) to be easily observed, so both the load on the indenter and its displacement are measured in the course of indentation. Such "depth-sensing indentation" instruments have been developed [66-68] and one of them, designed by Pethica and Olivier [67,68] and called the nanoindenter, is now commercially available. In this kind of apparatus, the tip of the indenter generally travels at a velocity of about 3 nm/s and its motion toward the surface is measured by a capacitance displacement gauge with a displacement resolution of 0.4 nm [32]. Diamond indenters with a three faced geometry are often used; they are more easily obtained from a single crystal than a four faced pin. The tip of the indenter is always rounded and its exact shape is sometimes determined with a scanning electron microscope to introduce corrections in the hardness measurements [69].

These new instruments offer more possibilities than the classical ones. As for tensile tests, the hardness can be measured as a continuous function of the deformation. Different loading and unloading programs can be used to study creep, deformation at constant velocity, etc...

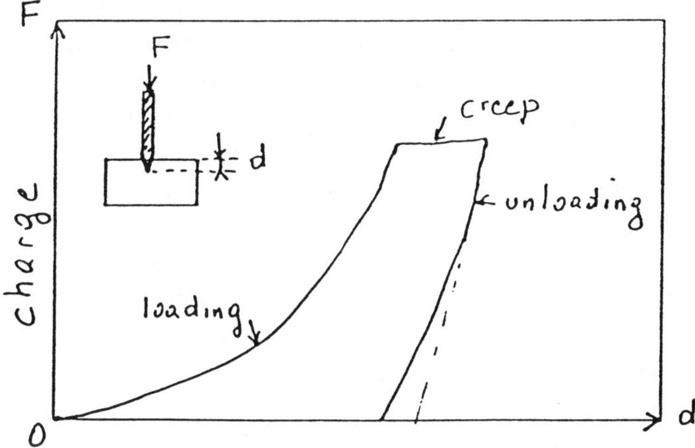

Fig. 18 Nanoindentation loading and unloading curve. From [33] and [68].

The elastic modulus $(1-v^2)/E_Y$ is obtained from the slope of the load-indentation depth curve (fig. 18) at the beginning of the unloading. This slope depends on E^* given by:

$$1/E^* = (1-v^2)/E_Y + (1-v_i^2)/E_{Yi}$$

the i subscript corresponding to the indenter material.

As already mentioned (chapter 3), a nanoindenter has been used for mechanical testing with microbeam deflection techniques [9,10].

The smallness of the tested volume can become an advantage for thin films: uniaxial tensile tests of thin films can be very sensitive to a small number of surface defects and do not reflect the intrinsic mechanical properties of the material; by averaging a great number of the measurements on very small volumes, the nanoindentation method can avoid the influence of these defects.

One of the major difficulties of indentation measurements is the wide complexity of the stress field which does not allow studies of deformation mechanisms at an atomic level.

4-2 Deformations induced by thermal cycling

Thermal cycling furnishes an elegant method to induce homogeneous biaxial stress and plastic deformation in thin films on a substrate [32-34]. Strain is imposed by the temperature and the difference between expansion coefficients of film and substrate (chapter 3-3) and stress is measured with X-ray diffraction [70] or with the substrate bending method. A stress-temperature curve is shown in Fig. 19 : the sample was an Al thin film deposited on a Si substrate at 150°C. At room temperature, the film is under a biaxial tensile stress of about 200 MPa (point A). Then, as the aluminum expansion coefficient is larger than for silicon ($\delta\alpha = \alpha_f - \alpha_s > 0$), a biaxial compressive stress of -2,3 MPa/K is applied by the increasing temperature. From point A up to point B the linearity of the curve proves that the deformation is purely elastic and the slope , equal to $-E_B.\delta\alpha$, can be used to determine the biaxial modulus. At point B, the change of slope indicates the beginning of plastic relaxation (yield stress). On cooling from 450 °C, the curve also begins with an elastic behavior, followed by a plastic deformation step until the temperature reaches room temperature. As the stress is about of the same order as at the starting point of the cycle, it can be concluded that the film has a stable structure.

More complicated curves are obtained, especially during the first thermal cycle, when the structure of the film changes on heating. Fig. 20 displays the recrystallisation curve of an amorphous film of WSi_2 deposited on a silicon substrate [71].

This technique has been widely used to study stability and mechanical properties of metallic films on silicon. The increasing rate of temperature can be modified to study the influence of the deformation velocity. But the necessity of varying the temperature makes the studies of some phenomena, such as thermally activated mechanisms, difficult.

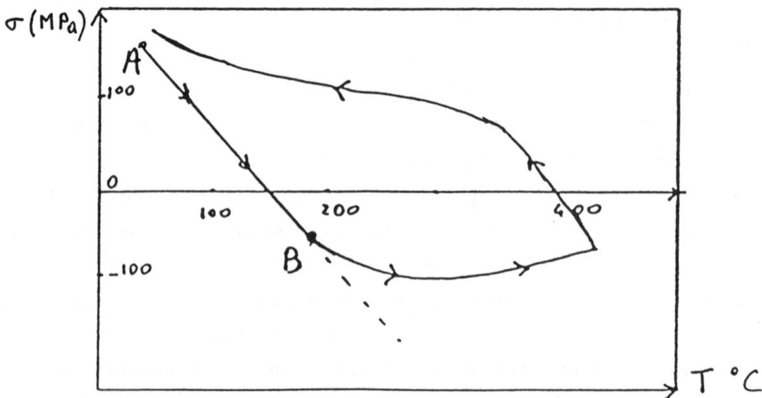

Fig. 19 Stress vs temperature for an Al film deposited on a Si substrate at 150°C from [33] and [34].

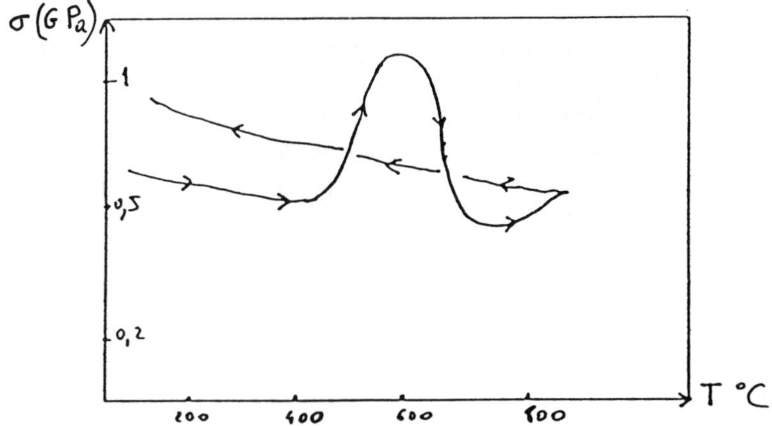

Fig. 20 Stress vs temperature for a WSi² film deposited on a Si substrate from [33] and [34].

4-3 Film-substrate adherence

4-3-1 Different factors acting on adherence

The interface between coating and substrate is generally a weak part of materials exposed to mechanical deformation. It can be sharp or lay on several planes. It is a region where defects, as cracks, dislocations, point defects, have a lower formation energy and are more easily formed and developed. It is also a boundary between two regions of different structure and different elastic constants leading sometimes to a large concentration of stresses.

The adherence of the coating, defined as the resistance to decohesion, depends on numerous factors:

- interface energy, characterizing adhesion between the two materials.
- defects (especially cracks) in the interface or its neighborhood.
- internal and applied stress.
- elastic and plastic properties of film and substrate.
- etc....

Decohesion can be likened to an intergranular failure between two different grains in a two-phase alloy. These phenomena play a key role in composite materials and are extensively studied by metallurgists. So it is not surprising that studies on decohesion of coatings have been mostly influenced by the mechanical theory of fracture. This theory supposes that microcracks are always present in the material (here, interfacial cracks) and that failure occurs when they can be extended by the applied or internal stress.

Until now, there is no wellestablished quantitative theory on adherence. We shall briefly try to characterize the role of the main parameters.

4-3-2 Adhesion energy

The energy per unit area of the interface Γ_{fs}, or adhesion energy, is one of the fundamental characteristics of coated materials. It is sometimes replaced by the Dupre energy Γ_D

$$\Gamma_D = \Gamma_f + \Gamma_s - \Gamma_{fs}$$

which appears in the calculation of the crystal energy during a crack development.

Γ_{fs} can be calculated with interatomic pair potentials or other methods used to obtain the total crystal energy [72]. Its measurement is not very simple. Some apparatuses, similar to the nanoindenter, have been especially designed [66,72]: the adhesion between the ball and the surface modifies the end of the unloading curve (Fig. 18) and the adhesion energy value between materials of the ball and of the tested sample can be deduced.

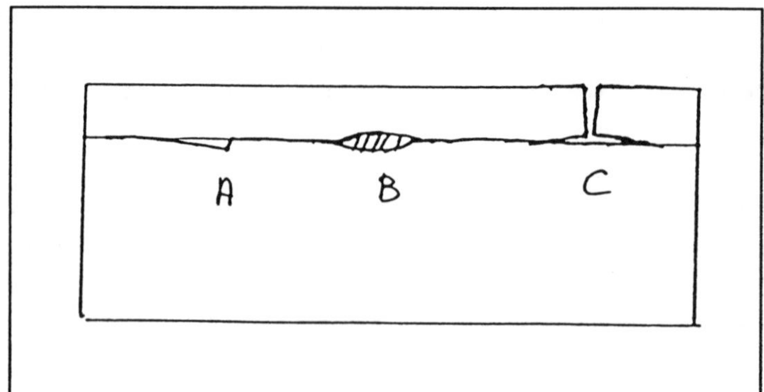

Fig. 21 Crack nucleation sites. A : Surface structural defect. B: Impurity aggregate.C: Crack normal to the interface.

Structural quality and cleanliness of the surface before the deposit are also very important. Surface irregularities and small impurity aggregates (for example, oxides) play the role of microcracks which can propagate under applied or internal stress (Fig. 21). Cracks in the film normal to the interface form also favorable sites to delamination (Fig. 21) [73].

Some techniques, such as ion implantation during the growth of the deposit, can strongly modify the interface structure and its energy. Structural defects can also be eliminated and adherence widely increased [74].

4-3-3 Crack Propagation

Three different ways of crack propagation are defined in the mechanical theory of fracture. One can give an oversimplified view of these modes by recalling that the behaviour of the moving tip of a crack has some similarities with a moving dislocation of Burgers vector b^*, b^* being the relative displacement of the two fresh surfaces near the moving crack tip (Fig. 22). We have chosen the x_2-axis along the crack tip (likened to a dislocation) and the x_1-axis in the propagating direction.

In mode I (Fig 22a), the movement of the crack tip is similar to the nonconservative *climbing motion* of an edge dislocation of Burgers vector parallel to the x_3-axis, under a tensile stress σ_{33}.

In mode II (Fig 22b), it is similar to the *glide* of an edge dislocation of Burgers vector parallel to the x_1-axis, under a shear stress σ_{31}.

The mode III (Fig 22c) is related to the glide of a screw dislocation (b^* parallel to the x_2-axis) under σ_{32}.

Elastic strains induced by a crack at a point $M(r,\Theta)$ of the crystal , calculated in the scope of linear elasticity theory and before plastic relaxation occurs, vary as $\sigma_a . f(\Theta)/\sqrt{r}$, σ_a being the stress component acting on the lips of the crack. This stress is larger than the yield stress σ_e for small values of r and plastic deformation takes place in a region around the crack tip to decrease it to about σ_e.

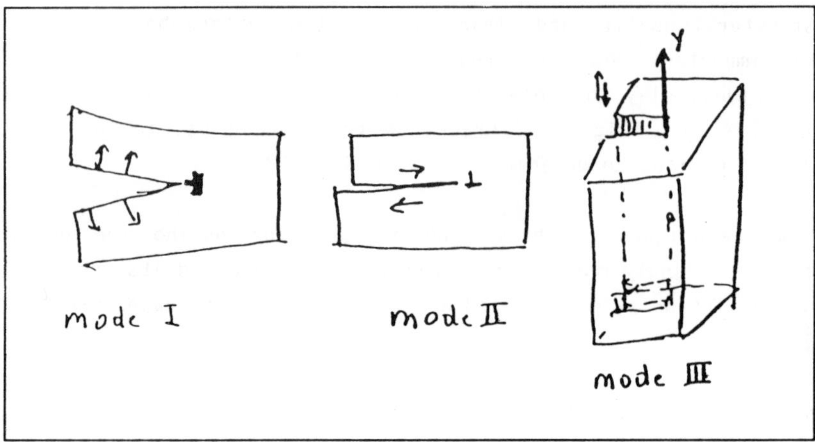

Fig. 22 a, b and c Modes I, II and III of crack propagation.

To reach a condition for crack propagation, we can write that the work done by the applied stress $\sigma_a.b^*.ds$ (ds is the area swept by the head of the crack) must be larger than the energy $\Gamma.ds$ needed to form new surfaces, plus the energy dW_p to relax by plastic deformation the region around the moving crack tip:

$$\sigma_a.b^* > \Gamma + dW_p/ds$$

When the crack develops in the interface, Γ is the DUPRE energy Γ_D. But when the crack moves in the film or in the substrate, Γ must be taken equal to $2.\Gamma_f$ or $2.\Gamma_s$ (Fig. 23).

The energy W_p needed for plastic relaxation is difficult to calculate, especially for cracks at the interface of thin films on a substrate. The stress distribution and the location of the "plastic zone" must be widely dependent on the difference between elastic constants and on the yield stress of the two materials. They must also depend on the film thickness h_f, plastic relaxation being favored in the film by image forces of the free surface.

The previous equation allows us to discuss adhesive and cohesive failures [75] (Fig. 23)

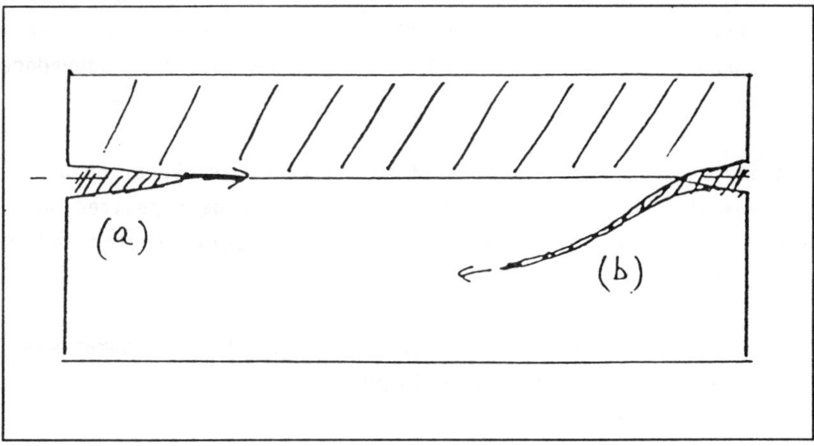

Fig. 23 a et b Adhesive failure (in the interface) and cohesive failure (in the substrate) from [75].

In mode I, the work done by the pure tensile stress does not depend on the direction of crack propagation. Then, if we neglect W_p (because we do not know how to calculate it) the driving force comes only from the surface energies and the crack runs along interface, leading to an adhesive failure (we have assumed that Γ_D is smaller than $2.\Gamma_f$ and $2.\Gamma_s$, which is generally observed).

In mode II or in a mixed mode (I and II), the maximum shear plane depends on the applied stress and the condition to observe crack propagation can be more easily satisfied in another plane. This can lead to film rupture or cohesive failure when the crack is driven into the substrate.

Delamination induced by cohesive failure has been observed by Hu [76] for Cr films deposited on a glass substrate.

4-3-4 Adherence measurement

The toughness of coatings depends on the nature of mechanical tests and no fundamental characteristic can be defined for adherence (unlike adhesion measured with DUPRE energy). But these tests can be used to compare different coatings [77]. A lot of techniques have been developed.

A sample can be prepared by pasting or depositing some material on the film and then exposed to tensile, flexural or shear stress in an especially designed apparatus. If an interfacial failure occurs, adherence can be characterized.

Coated material can be subjected to a violent acceleration or deceleration which removes the film. For example the coating can be deposited on a rotor and the rotation speed increased until decohesion occurs; coated bullets can be stopped by a steel plate.

Apparatuses using electromagnetic forces, shock waves sometimes induced by laser, etc...have been especially designed.

At last, we must mention the scratch test which becomes more and more used [78-80]. The film, displaced at constant velocity, is indented with an increasing load. The value of the load at which decohesion occurs characterizes adhesion. The beginning of decohesion can be determined by observation with a microscope or, for brittle films, by detection of acoustic emission. This technique has similarities with the indentation technique for hardness measurement, and even more, with the pin and disk method used for wear and friction characterization. The tricky problem is always the same: it is very difficult to modelize stress and deformation induced by the moving probe and the stress on the interface is not known when decohesion occurs. However, this technique, easy to use, gives fairly good results to compare materials.

REFERENCES

1) D. Baral, J.B. Ketterson and J.E. Hilliard: J. Appl. Phys. 57 (4), 1076 (1985).

2) T. Tsakalakos: J. de Phys. C5, 49, 707 (1988).

3) J.F. Nye: "Physical Properties of Crystal", Ch. 8, Clarendon Press, Oxford (1957).

4) "Elements de Metallurgie. Physique", ed. by Y. Adda, J.M. Dupouy, J. Philibert and Y. Quéré, Serv. de Doc. Saclay, Vol. 1, Ch. 5, (1976).

5) C.T. Rosenmayer, F.R. Brotzen and R.J. Gale: Mat. Res. Society, Symp. Proc, Vol 130, 77 (1989).

6) W.M.C. Yang, T. Tsakalakos and J.E. Hilliard: J. Appl. Phys. 48 (3), 876, (1977).

7) L. Landau and E. Lifchitz: "Theory of elasticity", 44 and 113, Pergamon, New York (1970).

8) D. Baral, J.E. Hilliard, J.B. Ketterson and K. Miyano: J. Appl. Phys. 53 (5), 3552 (1982).

9) T.P. Weihs, S. Hong, J.C. Bravman and W.D. Nix : Mat. Res. Society, Symp. Proc, Vol 130, 87 (1989).

10) S. Hong, T.P. Weihs, J.C. Bravman and W.D. Nix : Mat. Res. Society, Symp. Proc, Vol 130, 93 (1989).

11) P.H. Townsend, D. M. Barnett and T.A. Brunner: J. Appl. Phys., 62, 4438 (1987).

12) J.E. Hilliard: A.I.P. Conf. 53, 407 (1979).

13) G. E. Henein and J.E. Hilliard: J. Appl. Phys. 54 (2), 728, (1983).

14) T. Tsakalakos and J.E. Hilliard: J. Appl. Phys. 54 (2), 734, (1983).

15)M.R. Khan, C.S.L. Chun, G.P. Felcher, M. Gimsditch, A. Kueny, C.M. Falco and I.K. Schuller: Phys. Rev. B , 22 (12), 7186 (1983).

16) I. K. Schuller: Proc. of the IEEE Ultasonic Symposium (San Francisco), 1093 (1985).

17) W.E. Pickett: J. Phys F ; Metal Phys. 12, 2195 (1982).

18) T.B. Wu: J. Appl. Phys. 53 (7), 5265 (1982).

19) D. Baral and J.E. Hilliard: Appl. Phys. Lett. 41 (2), 156 (1982).

20) I. Schuller, C.M. Falco, J. Hilliard, J. Ketterson, B Thaler, R. Lacoe and R. Dee: A.I.P. Conf. Proc. 53, 417 (1979).

21) P.C. Clapp: "Modulated Structure Mat." ed. by T. Tsakalakos (Nato ASI series, serie E, Appl. Sc.), 83, 455 (1984).

22) A.F. Jankowski and T. Tsakalakos: J. Phys F ; Metal Phys. 15, 1279 (1985).

23) I.K. Schuller and A. Rahman: Phys. Rev. 50 (18), 1377 (1983).

24) A. Banrjea and J.R. Schmith: Phys. Rev. B, 35, 5413 (1987).

25) W.R. Bennet, J.A. Leavitt and C.M. Falco: Phys. Rev. B, 35, 4198 (1987).

26) B.M. Clemens and G.L. Eesley: Phys. Rev. Lett., **61** (20), 2356 (1988)

27)A.F. Jankowski: J. Phys F, Metal Phys. **18**, 413 (1988).

28) M. Imafuku, Y. Sasajima, R. Yamamoto and M. Doyama: J. Phys F, Metal Phys. **16**, 823 (1986).

28) R.C. Cammarata: Scripta Met. **20**, 479 (1986).

30) A. Kueny, M. Grimsditch, K. Miyano, I. Barnerjee, C. Falco and I Schuller: Phys. Rev Lett. 48 (3), 166 (1982).

31) R. Danner, R.P. Huebener, C.S.L. Chun, M. Grimsditch and I Schuller: Phys. Rev. B, **33** (6), 3696 (1986).

32) W.D. Nix :Mechanical Properties of Thin Films, Institute of Metals Lecture, Stanford University (1988).

33) M.F. Doener and W.D. Nix : Stresses and Deformation Processes in Thin Films on Substates, Stanford University (1986).

34) P.A. Flinn : Mat. Res. Society, Symp. Proc, Vol **130**, 41 (1989).

35) T. Vreeland, J.A. Dommann, C.J. Tsai and M.A Nicolet : Mat. Res. Society, Symp. Proc, Vol **130**, 3 (1989).

36) M Murakami, T.S. Kuan and I.A. Blech : Treatrise on Mat. Sc. and Tech., **24**, 163 (1982).

37) R.M. Fisher, J.Z. Duan and A.G. Fox : Mat. Res. Society, Symp. Proc, Vol **130**, 249 (1989).

38) H. Horikoshi and N. Tamura : Jpn. J. Appl. Phys. **2**, 328 (1963).

39) J.T Pan and I. Blech : J. Appl. Phys., **55**, 2874 (1984).

40) R. Abermann and R. Koch: Thin Sol. Films, **142**, 65 (1986).

41) R. Abermann and R. Koch: Thin Sol. Films, **62**, 195 (1979).

42) R. Abermann and R. Koch: Thin Sol. Films, **66**, 217 (1980).

43)R. Koch, H. Leonard and R. Abermann: Thin Sol. Films, **89**, 117 (1982).

44) M. Janda and O. Stephan: Thin Sol. Films, **112**, 127 (1984).

45) K. Haite, J. Holleman, J. Middelhoek and R. Koekoek: Mat. Res. Society, Symp. Proc, Vol **130**, 347 (1989).

46) J. Lopata, W.C. Dautremont-Smith and J.W. Lee: Mat. Res. Society, Symp. Proc, Vol **130**, 361 (1989).

47) F.C. Frank and J.H. Van Der Merve : Proc. Roy. Soc. **A198**, 205 and 216 (1949); **A200**, 125 (1949).

48) C.A.B. Ball and J.H. Van Der Merve : Dislocations in Solids, ch 27, 123, ed by F.R.N. Nabarro, North-Holland (1983).

49) J.W. Matthews, S. Mader and T.B. Light : J Appl. Phys., 41, 3800 (1970).

50) J. Friedel : Les dislocations, Gauthiers-Villars, Paris (1956).

51) G.G. Stoney: Proc. Royal Soc. London, A **82**, 172 (1909).

52) J.P. HIRTH and J. Lothe : Theory of dislocations, 2d ed., John Wiley and sons, (1982).

53) R. Bonnet : Acta Met.,**29**, 437 (1981).

54) R. Bonnet : Phil. Mag. **A51**, 51 (1985).

55) L.B. Freund, A. Bower and J.C. Ramirez : Mat. Res. Society, Symp. Proc, Vol **130**, 139 (1989).

56) S. Sharan, J Narayan, J.P. Salerno and J.C.C. Fan : Mat. Res. Society, Symp. Proc, Vol **130**, 153 (1989).

57) W. Hagen and H. Strunk : Appl. Phys., **17**, 85 (1978).

58) K. Rajan and M. Denhoff : J. Appl. Phys., **62**, 1710 (1987).

59) N.A. El-Masry, J.C.L. Tarn, N.H. Karam and S.M. Bedair : Mat. Res. Society, Symp. Proc, Vol **130**, 165 (1989).

60) H. Hertz : J. Math (Crelle's J.), 92 (1881).

61) S. Timoshenko and J.N. Goodier : Theory of Elasticity, p 372, Mc Graw-Hill (1951).

62) K.L. Johnson : Contact mechanics ,p 125, Cambridge University Press (1985).

63) D. Maugis and H.M. Pollock : Acta Met. **32**, 1323 (1984).

64) A.K. Bhattacharya and W.D. Nix : Int. J. Solids Structures, **24**, 881 (1988); **24**, 1287 (1988).

65) B. Jönsson and S. Hogmark : Thin Solids Films, **114**, 257 (1984).

66) J.L. Loubet, J.M. Georges, J.M. Marchesini and G. Meille : J. Tribology, **106**, 43 (1984).

67) J.B. Pethica, R. Hutchings and W.C. Olivier : Phil. Mag., A48, 593 (1983).

68) J.B. Pethica, and W.C. Olivier : Mater. Res. Symp. Proc., 130, 13 (1989).

69) M.F. Doerner and W.D. Nix : J. Mater. Res. 1, 601 (1986).

70) G. Bai, M.A. Nicolet, T. Vreeland JR, Q. Ye, Y.C. Kao and K.L. Wang : Mat. Res. Society, Symp. Proc, Vol **130**, 35 (1989).

71) P.H. Townsend : Ph. D., Stanford University (1987).

72) J. Ferrante, G.H. Bozzolo, C.W. Finley and A. Banerjea : Mater. Res. Symp. Proc., **119**, 3 (1989).

73) M.D. Thouless : Acta Met. **36**, 3131 (1988).

74) P. Moine, O. Popoola, J.P. Villain, N. Junqua, S. Pimbert, J. Delafond and J. Grilhé : Surface and Coatings Technology, San Diego Conf. Proc., **33**, 479 (1987).

75) M.D. Thouless : Mater. Res. Symp. Proc., **119**, 51 (1989).

76) M.S. HU : Mater. Res. Symp. Proc., **130**, 213 (1989).

77) D.S. Campbell : in Handbook of Thin Films Technology, ch 12.3, ed. by L.I. Maissel and T. Gland, Mc Graw-Hill (1970).

78) H.E. Hintermann, P. Laeng and P.A. Steinmann : ISIAT KYOTO, 1115 (1983).

79) A.J. Perry : Thin Solid Films, 107, 167 (1983).

80) J. Von Stebut, K. Anoun, J.P. Riviere and R.J. Gaboriaud : Int. Conf. on Metallurgical Coatings, San Diego, (1989).

Materials Science Forum Vols. 59 & 60 (1990) pp. 535-580

SURFACE AND INTERFACE TREATMENT OF MULTILAYERS USING ION OR LASER BEAMS

A. Pérez (a), J. Delafond (b), J.M. Pelletier (c) and B. Vannes (c)

(a) Dépt. de Physique des Matériaux, Université Claude Bernard - LYON I
43 Boulevard du 11 Novembre 1918
F-69622 Villeurbanne, France
(b) Lab. de Métallurgie Physique - Université de Poitiers
40 Avenue du Recteur Pineau
F-86022 Poitiers, France
(c) CALFETMAT-INSA
20 Avenue Albert Einstein
F-69621 Villeurbanne, France

1 INTRODUCTION

Modification of the near-surface region of materials using energetic beams (ions, electrons, photons) has been investigated extensively in recent years. The original application of ion implantation was to control the composition in semiconducting materials, and the patents dealing with this subject date from the 1950's. The nature of the process allows any element to be introduced into the near-surface region of a solid in a controlled and reproducible manner that is independent of most equilibrium constraints. Since the process is non-equilibrium in nature, compositions and structures unattainable by conventional methods may be produced. The development of high power laser applications has been achieved according to a similar process ; it was initiated by the operation of the first laser source, in the first half of the 1960's.

In ion implantation, the dopant or alloying element is the ion beam, which after an acceleration of tens to hundreds of kiloelectronvolts, impinges upon the surface of the target. The energetic ion comes to rest by displacing atoms from their normal lattice sites and by exciting the electronic system of the target, thus producing a large number of point defects. Most of the defects are annihilated by recombinations shortly (10^{-12} s) after the passage of the bombarding ion, but some survive. The resultant structure consists of the host (target) material with an impurity (alloying) addition and a defect structure characteristic of the radiation damage.

With regard to the defect creation by ion bombardment, materials can be classified into two types : (i) the materials in which defects can be created by electronic processes and (ii) materials which are sensitive to the energy deposited in nuclear collisions. Ionic crystals such as alkali halides, some oxides such as SiO_2 and organic polymers are typical of the former group in which electronic excitations lead to very efficient point defect production. In the second case we can mention the metals, semiconductors such as silicon and refractory oxides (MgO, Al_2O_3, TiO_2 ...) and ceramics. An extreme result of the damage accumulation can be the amorphization of the implanted layer and in this case the role of the implanted impurities can be to stabilize the glassy state.

The use of the ion implantation technique to modify the near-surface properties of materials is based on the following features : it allows precise control of the total number of injected ions, it permits independent control of the depth of penetration, it can use essentially all combinations of ions and targets materials, it may achieve concentrations above the equilibrium solid solubility limit, and it may be carried out at low or elevated temperatures. The concentration of dopants implanted in semiconductors is generally very low and complex radiation damage to the host lattice does not occur. In metals and structural ceramics the concentrations of alloying or impurity additions required for property alteration are in the 0.1 - 20 at. % range ; hence damage levels are high. However for doping at high fluences we must take into account that the implantation process may also remove atoms from the surface of the target by sputtering. The sputtering yield is a function of the mass and energy of the impinging ion and the mass of the target atoms. The limiting concentration (typically 10 - 75 %) is reached when the receding surface removes more impurity atoms than are being added by the impinging flux. In this last case a new technique which has been the subject of increasing interest for the past ten years : the ion beam mixing technique can be used.

Ion beam mixing utilizes the kinetic energy of ion beams to mix predeposited surface layers with substrate materials or multilayered materials. By this technique it becomes possible to introduce large non-equilibrium concentrations of alloying elements into a wide variety of host materials under carefully controlled conditions leading to many technological applications. In ion beam mixing, for every ion, several hundred atoms of the substrate and the overlayer are intermixed. Therefore the required ion doses are not so high that the sputtering limit is reached. This has stimulated much fundamental research concerning the mechanism of ion beam mixing. Originally, most of the attention was focussed on ballistic effects in the primary and secondary recoils events triggered by the ion beam. However, comparison between systems of different atomic species, that could be expected to have the same ballistic effets, surprisingly showed completely different degrees of mixing. Thus it was realized that thermodynamic and chemical effects play an important role in the ion induced collision cascades. Further experiments at widely different target temperatures revealed the importance of radiation enhanced diffusion. So, at present three different mechanisms, operating during ion beam mixing, have been identified : 1) ballistic or recoil effects, 2) cascade mixing, 3) radiation enhanced diffusion.

In the case of laser irradiations, the photon energy is mainly converted to heat in metallic systems. Consequently the interaction is nearly independent of the radiation coherency, but is a function of the number and energy of photons. So the fluence appears to be one of the predominant factors. In opposition to ions, the doping effect of the target doesn't exist directly with photon irradiation, but in the particular case of multilayers the induced melting enables the synthesis of metastable compounds and can be concurrently or complementarily used for specific applications. For these reasons, both treatment techniques are associated together in the present review.

It is the purpose of this paper to introduce the different mechanisms of particle and photon-matter interactions in order to understand the ion implantation and the ion-beam mixing as well as laser irradiation processes. Specific examples are chosen from the litterature to illustrate qualitatively, and where possible also quantitatively the damage creation and the metastable phase formation characteristics of these techniques in various types of materials. Finally the potential applications of both ion implantation, ion beam mixing and laser irradiation techniques for the surface treatment of various types of materials or new materials synthesis are mentioned in the present industrial context but also in a prospective framework.

2 INTERACTIONS OF ENERGETIC IONS IN SOLIDS

Energetic particles are slowed down in matter by momentum transfer to target atoms ("nuclear stopping"), and by exciting the electronic system of the target ("electronic stopping"). The nuclear stopping can be understood by studying the ion-atom repulsive forces, and the binary collisions. The electronic stopping is a more complex phenomenon and can be sketched only briefly in this context. Ion-slowing down, energy loss and sputtering for elemental as well as multicomponent materials have been treated comprehensively within the last twenty years. In particular, the large number of existing computer codes have turned out to be reliable in many of the predictions. The aim with the present work has not been to discuss the areas thoroughly, but largely to follow basic ideas. Readers who might want details about these subjects are referred to the many existing reviews, which are mentioned in the text in connection with each section.

The main features of the energy loss and the subsequent energy conversion are shown in figure 1. The energy loss quantities are treated in this section when the effects of the energy transfer in various types of materials will be illustrated in the next section (sect. 3).

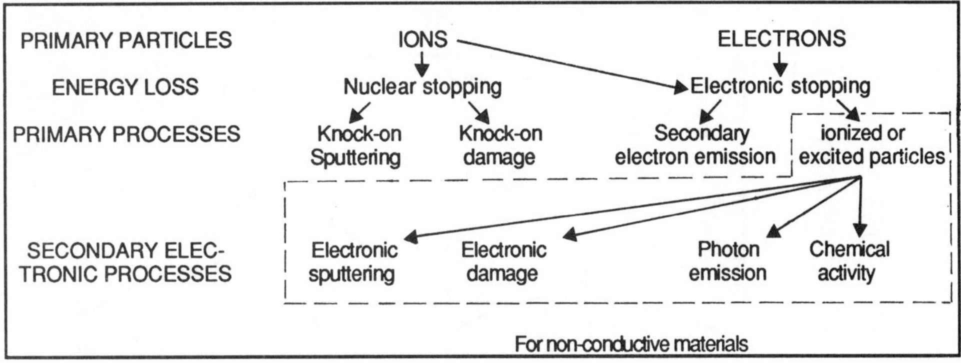

Fig. 1 A schematic survey of different processes associated with the energy loss of the primary particles, proposed by J. SCHOU [1].

The stopping power dE/dx of a medium for a particle beam is the average energy loss of the particles after passing a thin layer. It is conveniently expressed as :

$$-\frac{dE}{dx} = NS(E) \tag{1}$$

where N is the number density of atoms in the medium and S the stopping cross section. The dependence of the stopping power on the kinetic energy E of the particles has been emphasized in the stopping cross section. The stopping power is expressed in units of energy $x(\text{length})^{-1}$, and the cross section energy x area. Nevertheless, the stopping power is often tabulated in units as $eV/10^{15}$ $atoms/cm^2$ or $MeV/g/cm^2$. These units are equivalent to the stopping cross sections, since the dependence of the stopping power on the density enters only via N (apart from a density effect that is important only in the ultra-relativistic region). According to the treatment of Lindhard et al [2], the collisions are separated into those between the primary particle and the nuclei and those between the particle and the electronic system of the atoms. Consequently the stopping cross section is split into two parts :

$$S(E) = Se(E) + Sn(E) \tag{2}$$

where Se is the electronic and Sn the nuclear stopping cross section. A survey of the two stopping cross sections is shown in figure 2. The division of the energy range into four regimes is somewhat artificial, since some theories cover more than one regime. Nevertheless, the classification is convenient for the characteristics of the total stopping power. The nuclear stopping power is important only in the low-energy regime. In the intermediate, high energy and relativistic regime it is typically a factor of 2 M_p/M_e ($\simeq$ 4000) lower than the electronic stopping power [2, 3]. M_p is the proton mass and m_e the electron mass.

2-1 Electronic stopping

Historically, the first electronic stopping powers were calculated for point charges, such as protons. For particle velocities of present interest (i.e. E/M < 1 MeV/amu or energy regimes I and II in figure 2), the concept of Lindhard [4] turned out to be most fruitful. The electronic stopping of the ion in the target is treated within the local density approximation, where each infinitesimal volume element of the solid is considered to be an independant plasma ; that is, the ion-target interaction can be treated as that of a particle with a density averaged free electron gas. This model yields qualitatively the correct features, i.e. a velocity proportional stopping power at low energies, a stopping maximum around 100 keV/amu, and then - at higher energies - a gradual merge into the well-known Bethe stopping power.

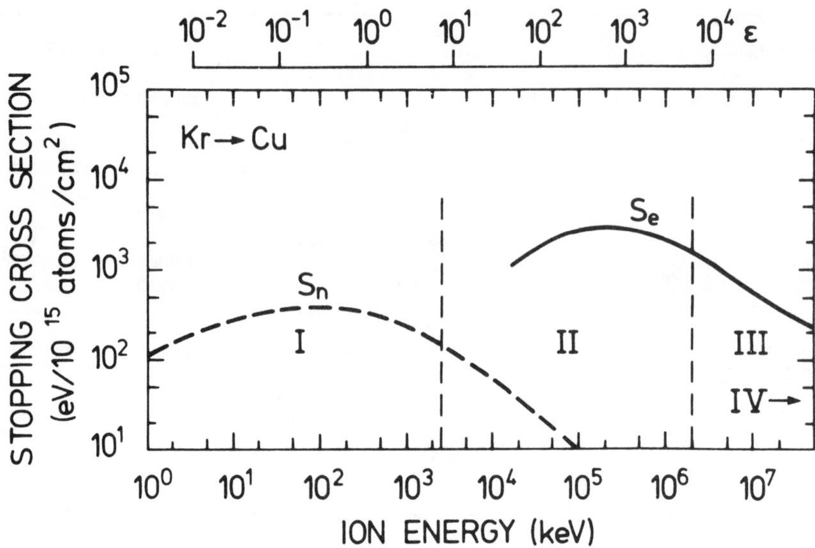

Fig. 2 The nuclear (Sn) and electronic (Se) stopping cross sections as a function of ion energy in low-energy (I), intermediate (II), high-energy (III) and ultrarelativistic (IV) regimes [1].

A relevant parameter for the interaction between collision partners in the low energy and intermediate regimes is "Lindhard's" parameter [5] :

$$\varepsilon = \frac{E_r}{Z_1 Z_2 e^2 / a_L} \tag{3}$$

Z_1, M_1 and Z_2, M_2 are the atomic numbers and masses of the primary ion and of the target material, respectively. e is the electronic charge. a_L (see below) is the screening radius and $E_r = M_2E/(M_1+M_2)$ the relative energy of the colliding particles. Lindhard et al [5] suggested that :

$$a_L = 0.8853\, a_o\, (Z_1^{2/3} + Z_2^{2/3})^{-1/2} \tag{4}$$

where a_o is the Bohr-radius (0.0529 nm) and 0.8853 a constant derived from the Thomas-Fermi potential between interacting atoms [5].

The quatity ε turned out to be a convenient scaling unit for the energy. An expression frequently used is :

$$\varepsilon = 32.53\, M_2\, E\, /\, [Z_1 Z_2\, (M_1 + M_2)\, (Z_1^{2/3} + Z_2^{2/3})^{1/2}] \tag{5}$$

where the ion energy E is in keV and the masses M_1 and M_2 in amu.

Using the energy ε expressed in Lindhard's dimensionless Thomas Fermi (TF) units, the electronic stopping power at low velocity in regimes I and II may be determined from the Lindhard-Sharff treatment [6]. The electronic stopping power is proportional to velocity and may be expressed as :

$$\sigma_e(\varepsilon) = k_L\, \varepsilon^{1/2} \tag{6}$$

ϵe is the reduced (dimensionless) electronic stopping power and :

$$k_L = \frac{0.0793\ Z_1^{2/3} Z_2^{1/2} (M_1 + M_2)^{3/2}}{(Z_1^{2/3} + Z_2^{2/3})^{3/4}\ M_2^{3/2}\ M_2^{1/2}} \tag{7}$$

Although more precise tabulations of the stopping power have emerged (Andersen and Ziegler [7], Ziegler [8, 9, 10]), the Lindhard-Scharff value is still a convenient reference standard and the predictions of the treatment are usually correct to within a factor of two.

The electronic stopping power in the region above the stopping power maximum is treated by Bethe-Bloch theory [11, 12]. A convenient expression is the velocity (v)-dependent stopping cross section :

$$Se(E) = \frac{4\,\pi\, e^4\, Z_1^{*2}\, Z_2}{m_e\, v^2}\, L(v) \tag{8}$$

Z_1^* is the effective charge of the projectile for scattering target electrons and Bethe's stopping number essentially is expressed as :

$$L(v) = \ell n\, (\frac{2\, m_e\, v^2}{I}) - \ell n\, (1 - \frac{v^2}{c^2}) - \frac{v^2}{c^2} - \frac{C}{Z_2} - \frac{\delta}{2} \tag{9}$$

I is the target mean-excitation potential, C/Z_2 the shell correction and δ an ultra-relativistic density correction [13] which otherwise will be neglected in the present context. The mean excitation potential is defined as :

$$\ell n\, I = \Sigma f n\, \ell n\, E_n \tag{10}$$

where E_n are all possible energy transitions of the target atom and fn the corresponding dipole oscillator strengths. The calculations of I are usually based on a statistical model of the target atom. The simplest result is the well-known Bloch's rule [3] :

$$I = I_o\, Z_2 \qquad \text{where } I_o \simeq 10 \text{ eV} \tag{11}$$

The shell correction accounts for the deviations from the requirement in Bethe's derivation that the projectile velocity should be much larger than that of the bound electrons. An evaluation of this term indicating a value of the shell correction which will modify stopping calculations by about 2 % has been given by Andersen and Ziegler [7].

The physical meaning of Z_1^* in equation 8 may best be described as the average projectile charge as seen by the electrons of the target, which scatter from the moving ion at various distances. Hence, the value of Z_1^* is expected to fall within the boundaries $Z_1^{**} < Z_1^* < Z_1$, where Z_1 is the nuclear charge of the ion (responsible for electron scattering close to the ion's nucleus), and Z_1^{**} is the charge state of the ion (responsible for electron scattering at large separations from the nucleus). At intermediate distances, the nuclear charge is partly screened by the non-stripped fraction of electrons. By appropriate averaging over all possible encounters (impact parameters), Brandt and Kitagawa [14] have proposed a refined equation to determine Z_1^*.

2-2 Nuclear stopping

The first step in evaluating atomic collisions and nuclear stopping always consists of determining the interaction potential. It has been shown by various authors [15, 16, 17] that the Thomas-Fermi statistical treatment [18] of the overlapping electronic clouds of colliding atoms or ions, yields quite satisfactory results, if besides of Coulomb and kinetic energy of the electrons, also exchange and correlation energies are taken into account. Comparison with more sophisticated approaches, e.g. with two-centered self-consistent field calculations (molecular Hartree-Fock-Slater, HFS), yields an excellent agreement [16, 10]. For this reason, these more computer efficient statistical methods-also known as"free electron gas" (FEG) calculations - were applied recently by Biersack and Ziegler to assess the interaction potentials of present interest [15, 16] and calculate the nuclear stopping of ions in solids (well known TRIM code [10]). Previously to these refined developments, Lindhard et al [5] showed on the basis of the Thomas-Fermi model that the nuclear stopping power $N\,S_n$ (E) for all elemental beam ion-target combinations could be expressed as :

$$N\,Sn(E) = \frac{\pi\, a_L^2\, \gamma\, N}{(\varepsilon / E)}\, s_n(\varepsilon) \qquad (12)$$

where ε and a_L are the Lindhard's parameters (dimensionless parameter for the energy and screening radius) defined in equations 5 and 4, respectively. N is the atomic density in the target and s_n the dimensionless reduced nuclear stopping.

$$\gamma = 4\,M_1 M_2 / (M_1 + M_2)^2 \qquad (13)$$

is determined by the maximum energy transfer γE from the incident particle with energy E and mass M_1 to a particle at rest in the target with mass M_2.

The important property of the relation 12 is that the nuclear stopping power for all combinations depends on the primary energy E via the dimensionless reduced nuclear stopping s_n. The maximum of $s_n(E)$ is around 0.35 for values of ε about 0.3, and the overall shape is shown by NSn in figure 2. The absolute magnitude of NSn is primarily determined by the constant ratio (ε/E) in the denominator. Heavy beam atoms incident on a heavy target lead to a combination with a small ratio, and in turn, a high nuclear stopping power. As mentionned before, recent developments in the screening determined by the Thomas-Fermi function and in particular, the combination krypton incident on carbon investigated as a representative case allowed the authors [9] to propose a simple approximation for the reduced nuclear stopping power :

$$s_n(\varepsilon) = [1/2\, \ell n\, (1 + \varepsilon)] / (\varepsilon + 0.10718\, \varepsilon^{0.37544}) \qquad (14)$$

Expression (14) shows the expected asymptotic behaviour for Rutherford scattering at $\varepsilon \gg 1$ and falls off faster than the corresponding Thomas-Fermi stopping power for $\varepsilon \ll 1$ as required.

2-3 Collision cascade effects

The displacement cascade in a solid occurs when an atom in the material is displaced from its site by an incident damaging particle and moves through the material producing additionnal displacements. This process is of central importance in the study of radiation effects in solids. Point defects and defect clusters are created and rearrangements of atomic configurations occur as a consequence of the ensuing atomic collisions in the surrounding medium. The problem of component degradation under irradiation in fission and fusion reactors provided the main technological impetus to radiation damage research for many years. Recently the growing importance of ion-implantation technology and ion-beam materials processing has generated renewed interest in cascade phenomena. Displacement cascades also pose challenging scientific problems in the areas of defect physics, kinetics, and phase stability. For example, it has been shown recently by Diaz de la Rubia et al [19] that cascades give rise to an ultrafast solidification process, with quench rates of the order of 10^{15} K/s.

Because of the inhomogeneous, non-linear and many-body character of cascades, developing a satisfactory theoretical model has been difficult. As a first approximation one may treat the short-time, ballistic phase separately from the longer-time "thermal spike" regime. One such approach would be to treat the early part of the cascade as a sequence of independent binary collisions (binary collision approximation) [20, 21] and the slowly-decaying thermal spike via the classical heat-transport equation in conjunction with chemical rate theory as performed by Vineyard [22]. Unfortunately, developing a satisfactory unified theory of cascades spanning both short - and long - time regimes has proven difficult. In recent years, however, molecular dynamics computer simulation has emerged as a practical tool for studying the entire history of cascades at energies up to multi - keV range [19]. For convenience, in this section, we will describe first the short-time ballistic regime using the models based on the binary collision approximation. This first step in the description of the collision cascade is very useful to obtain, after simple calculations, some interesting results on the cascade geometry, transferred energies, displaced atoms .... The cascade dynamics and the high density cascade effects will be introduced in a second step with some characteristic examples of bombarded solids with heavy ions.

2-3-1 ballistic phase

The theoretical description of a collision cascade is merely an extension of Lindhard's Thomas Fermi description (see section 2-2) to include the recoil atoms as well as the primary ion. Again, one separates the energy loss mechanisms into two types, the nuclear loss $\nu(E)$ and the electronic loss $\eta(E)$, with $\nu(E) + \eta(E) = E_{ion}$. Note that $\nu(E)$ is usually 20 - 30 % smaller than the total nuclear stopping power of the primary ion, because the energetic recoils created along the track of the primary ion lose some of their kinetic energy by electronic excitation. This introduces a distinction between S_n and $\nu(E)$. Winterbon et al [23] have developed analytical techniques to provide depth profiles for $\nu(E)$ and comprehensive tabulations can be found for example in references [24] and [25]. These profiles are of course closely related in shape to the corresponding ion range profiles. However the most recent and comprehensive investigations are the computer - based approaches of Parkin and Coulter [26 - 27] and computer calculations using the TRIM Monte Carlo methods [10]. Parkin and Coulter derived a set of integrodifferential equations in the binary collision approximation using the methods of Lindhard et al [2] that provide a fairly complete characterization of the nature of the displacement cascade. The TRIM code has been widely distributed by its authors and in its most recent versions can be used to calculate properties of the displacement cascade similar to those obtained by Parkin and Coulter. In addition, important spatial information about the cascade is part of the TRIM calculations. To a certain extent, the two approaches complement each other.

Generally the conversion of the "deposited energy distribution" into a damage distribution is not so easy. The usual procedure (the so-called Kinchin-Pease method [28]) is to assume that only those recoils with an energy greater than E_d become displaced, where E_d is the threshold displacement energy. This leads to a fairly simple relationship [29, 30] between $\nu(E)$ and the number N_d of displacements produced :

$$N_d = 1 \qquad \text{if } E_d < \nu(E) < 2.5\, E_d$$

$$N_d = \frac{0.8\,\nu(E)}{2\,E_d} \qquad \text{if } \nu(E) > 2.5\, E_d \tag{15}$$

How well do such predicted damage distributions compare with experiment ? Generally the position and the shape of the distribution are predicted quite accurately [31, 32]. However the absolute magnitude of the damage (i.e. the "conversion factor" from $\nu(E)$ to N_d) is usually not well predicted. The damage creation at high energy agrees quite well with equation 15 and the discrepancy is occurring at the low-energy end of each ion track. This is rather strong evidence for the existence of some sort of non-linear effect such as a "thermal spike" [31]. We shall return to a further discussion of this point in section 2-3-2. When applying cascade theory, it is extremely important to distinguish the average depth distribution of ν(E) for a large number of cascades and the distribution within an individual cascade. The latter quantity can only be obtained by Monte Carlo simulation (i.e. using the TRIM code). When the incident ion is much heavier than the substrate, the primary ion trajectory remains close to the main central axis of the ν(E) envelope and hence the damage distribution within an individual cascade is seen to be distributed fairly uniformly through most of the volume (i.e. Bi → Si) [31]. In this case the ν(E) distribution provides a fairly reasonable approximation for the individual cascade behaviour. However, at the other extreme of N^+ → Si [31], the individual cascade occupies only an extremely small fraction of the corresponding ν(E) envelope.

2-3-2 high density cascades : spike effects

Applying linear collision cascade theory to obtain the ν(E) distribution, one is assuming that, in each binary collision between a target atom and an energetic recoil atom, the target atom is initially at rest. This is probably a reasonable assumption during the early stages of cascade propagation, when only a few energetic recoils are in motion ; hence cascade theory is able to predict the approximate dimensions of the deposited energy distribution. But simple estimates show that, for most heavy ion implantations, the final deposited energy density (even if shared equally among all atoms within the cascade volume) would correspond to a very significant increase in thermal motion. Consider, for example, a 100 keV ion with a mean range of 40 nm implanted at room temperature. The total cascade volume contains less than 10^6 atoms and hence the final deposited energy density exceeds 0.1 eV/atom. Since the total cascade propagation time is too short (~ 2×10^{-13} s) for normal quenching mechanisms to remove significant energy of the substrate, it is obvious that the final stages of the cascade no longer involve stationary target atoms. Sigmund [33] has suggested that a spike mechanism might be involved to handle these high-density cascades and has proposed a model for the determination of the energy density and time constant of heavy-ion induced elastic-collision spikes in solids.

A spike is a limited volume with the majority of atoms temporarily in motion. Spike effects may be important when the spike lifetime is larger than the duration of the initiating cascade (of the order of the slowing-down time of the projectile ≃ 2.10^{-13} s). Energy densities θ_o and time constants τ $((\partial\theta/\partial t)_{\theta=\theta_o,\, t=o} = -\theta_o/\tau)$, for elastic-collision spikes have been estimated on the basis of established energy-deposition profiles and comprehensive tabulations can be found in reference [33] (note that θ_o decreases and τ increases with increasing ion energy). Very pronounced variations have been observed with the atomic numbers of the projectile ion and target and the bombarding energy. Spikes have been considered as the origin of a variety of experimental results over the years. The more compelling evidence seems to come from sputtering experiments [34, 35, 36]. The observation of a striking anomaly in the sputtering by molecular-ion bombardment was found to confirm the predicted breakdown of linear cascade theory at high energy density.

Very recently Diaz de la Rubia et al [19] have described the thermal spike behaviour from molecular dynamics (MD) simulations of energetic displacement cascades in Cu or Ni performed with primary - knock-on-atom (PKA) energies up to 5 keV. These metals have similar atomic masses but different cohesive properties (melting temperature and lattice vibrational spectra) and thus, a comparison of the cascades in these metals helps to elucidate the roles of the ballistic - and thermal spike phases of cascade dynamics. Their results indicate that the primary state of damage produced by displacement cascades is controlled basically by two phenomena : replacement collision sequences during the ballistic phase, and melting and resolidification during the thermal spike. The thermal-spike phase is of longer duration and has a more marked effect in Cu than in Ni. The local melting occurring at the core of a cascade is a key result. At the outset of the thermal spike (t = 0.25 ps), it can be shown that the temperature near the center of the cascade (r < 1.4 nm) reaches ~4000 K in both metals, and the temperature gradient outside this central core ≃ 140 K/Å. The initial cooling rate is approximately 10^{15} K/s at the center of the cascade. At a time t ≃ 1.4 ps, the temperature begins to fall below the melting temperature T_m at r ≃ 2 nm in the Cu cascade. In the case of Ni, the temperature falls below the melting point much earlier, although the absolute temperature profile remains similar to that in Cu. Also it is to be noted that the local densities in the core of the cascade are reduced by as much as 15 - 20 % at 1.14 ps in Cu and 15 % in Ni at 0.8 ps, relative to the surrounding solid. A high-density ridge, particularly evident in Cu, forms just outside the central core. This ridge is due to a shock-wave like expansion. Eventually this compression at the cascade periphery relaxes and the density returns to its equilibrium value, perturbed only slightly by the presence of self-interstitial atoms that survive recombination.

2-4 Energy loss in multicomponent materials and range

The more general case of energy losses in polyatomic materials, those consisting of two or more elements, has received much less study. These materials, which are more representative of the class of materials used in applications, introduce many new parameters into the problem including elemental composition, material stoichiometry, complex crystal structure, bonding type (i.e. metal, ionic, covalent) and additional types of defects. Usually the stopping power in multicomponent materials is determined by Bragg's rule which expresses the total stopping cross section as a weighted sum of the contributions [3] :

$$S(E) = \Sigma N_j S_j(E) \tag{16}$$

N_j is the atomic number density and S_j the stopping cross section of the j' th element.

The electronic-stopping cross-section is influenced by changes in the electronic structure as a result of chemical binding. Since only the outermost electron shell changes, the total change in the stopping cross section depends on the relative contribution of this shell to the full value. Of course, Bragg's rule has to break down for any condensed material or compound at a sufficiently high experimental accuracy. Deviation from the additivity is expected mostly for : (i) comparatively low energies where the relative contribution from valence electrons to the stopping power becomes large and (ii) for light elements where the valence electrons constitute a major fraction of the total number of electrons. For light ions at the stopping maximum the deviation rarely exceeds 10 per cent. For energies above the stopping power maximum, the changes in the binding from chemical effects or condensation enter into the mean excitation potential I alone (equation 10). A comprehensive discussion for the determination of I-values has been presented recently by Schou [1]. Therefore, the tabulations of I usually contain a value for the solid state, the gaseous state, and one or more for the atoms in a chemical compound (see i.e. the ICRU table) [37]. Occasionally, an analogous additivity rule is formulated for the mean excitation energy of a molecule [37, 38] :

$$\ell n\, I = \Sigma N_J Z_J\, \ell n\, I_J / (\Sigma N_J Z_J) \tag{17}$$

where I_j is the mean excitation energy of the j'th element. The application of this additivity requires the choice of appropriate values for the mean excitation potentials of the constituants. The simplest procedure, often used in the past, is to take the same I-values for the atomic constituents of a compound as for the corresponding elements. This introduces some error because of the neglect of molecular binding effects. An additional error may be incurred when elemental I-values for gases (i.e. oxygen or nitrogen) are applied to the constituents of solid compounds. The systematic studies on solid oxides demonstrate clearly that the stopping cross section of oxygen atoms in oxides is significantly less than that for gaseous oxygen [39]. The accuracy of the additivity rule may be improved by assigning values of I depending on the type of compound and on the chemical state. An instructive procedure has been applied by Sabin et al [40]. The mean excitation potential is decomposed into orbital mean energies (I_{core} and I_{val}) from the core electrons and the valence electrons. An analysis of carbon hydrogen molecules led the authors to the conclusion that the type of chemical bonds determined the mean excitation energy. For example, it was found that the stopping cross section of a carbon atom in a double-bond compound is larger than that in a single-bond one [40, 41]. For metallic alloy no systematic deviations from Bragg's rule have been observed. This is in agreement with the point that deviations only occur if the state of the valence electrons changes from the element to the mixture.

The nuclear-stopping cross-section of multicomponent materials behaves strictly according to Bragg's rule (at least in the usual approximation of binary collisions). As mentioned in section 2-3-1, for monoatomic materials, the modified Kinchin- Pease formula is the most widely accepted method of calculating the number of displacements. Through various approximations it is often extended to the more general case of polyatomic materials. However, as shown by Parkin et al. [26, 27], the principal characteristics of the displacement cascade in polyatomic materials can only be obtained from a model that explicitly includes the relevant material parameters.

The total pathlength (R) of incident ions in the target may easily be obtained by integration of the deduced total stopping power (-dE/dx = N S(E), see section 2) :

$$R = \int_0^E \frac{dE}{NS(E)} \qquad (18)$$

but the total pathlength is of very limited interest since what may actually be observed is the final distribution of implanted atoms. It is thus of much more interest to find the mean projected range and possibly also the range straggling. The mean ion depth (Rp) is defined as the mean value of the final depth distribution which is approximately Gaussian in shape. The longitudinal straggling (σ// or ΔRp) is the standard deviation of the final implanted depth distribution and the lateral straggling ($\sigma\perp$) is the standard deviation perpendicular to the initial ion trajectory.

In the high velocity regime, where electronic excitations are the main interaction processes of the incident ions with the target atoms (see section 2-1), the ion trajectory is quite linear and the pathlength calculated with the electronic stopping power injected in equation 18 gives a rather good value for the position of the implanted ion distribution. However, in the low velocity regime of interest in the present context (see sections 2-2 and 2-3), by each nuclear collision with a target atom, the ion loses energy and at the same time changes direction, where the so-called "nuclear" energy loss (due to momentum transfer to the target atom) is directly related to the deflection angle. Note that the direction of motion changes at random with each collision, with no correlation between successive deflections. In this case, the mean projected range of ions can be obtained by summing the mean projections of each path-length element. To accomplish this, we need to know the mean directional cosine of the ion motion during the slowing down process. This calculation method was used recently by Biersack [32, 42] and a fast algorithm was developed, which is based on the known stopping powers and energy loss stragglings. The achieved precision in stopping powers and ranges corresponds now to a standard deviation of better than 10 %, as compared to presently available experimental data. If more details of the radiation effects need to be known, such as damage or ionization distributions in 3-dimensional space (including the motion of recoils), it appears most promising to use Monte Carlo simulations (see sections 2-2 and 2-3, TRIM code [10, 32]), although other numerical methods are available [24, 25, 43].

3 CHARACTERISTIC EFFECTS OF ION BOMBARDMENTS IN VARIOUS TYPES OF MATERIALS

Radiation damage as a consequence of the energy deposition mechanisms (electronic and nuclear, see section 2) and the doping of the target materials in the implantation zone of incident ions are the characteristic effects which are observed in the different types of materials (metals, semiconductors, insulators). These effects in semiconductors have been extensively studied for applications in microelectronic devices and a large number of comprehensive review articles exist in this field. For this reason most of the effects and applications described in this section will concern mainly other classes of materials.

3-1 Radiation damage

This is the key problem area in almost all implantation studies. It involves a complex interplay between bombardment processes and subsequent annealing processes and cannot adequately be treated in a single and short section. Hence, we consider here only some simple general concepts. The first obvious sub-division of the problem proposed by Davies [31] is into bombardment and annealing processes.

- bombardment process : this involves the high excitation density or the formation of the nuclear collision cascade around each individual ion track. The main parameters involved here are Z_1, Z_2 and E.

- annealing processes : the high concentration of defects within a single cascade may cause them to recombine or unite to form larger clusters, loops, voids, etc ... ; alternatively they may migrate and interact with the surface or with other cascades or impurity atoms. This is the most complex part of the radiation damage problem, since these defect migration and interaction processes depend strongly on the substrate temperature and on the concentration of other defects. Consequently, the total dose and the dose rate can both play an important role.

Another useful subdivision of the problem proposed by Davies [31] is to consider the low-dose and high-dose limits.

- low dose : when the implant dose is sufficiently low ($\leq 10^{12}$ ions/cm^2), we have individual isolated damage regions around each ion track, with negligible probability of cascade overlap. Within such regions, the damage level is independent of both the total dose and the dose rate. Hence in this low-dose limit, the bombardment and anneal processes can be fully separated from one another. Furthermore, one can minimize and even eliminate completely the annealing processes by implanting at very low temperature. When the total dose increases sufficiently that the average separation between cascades becomes comparable to the cascade dimensions, then significant inter-cascade effects start to occur and the problem becomes more complex.

- high dose : when the implant dose is sufficiently high ($\geq 10^{14}$ ions/cm^2), one gets complete overlap of cascade and an essentially uniform lateral distribution of damage. This damage level will often exhibit a complex dose rate dependence, due to in situ annealing effects during the implantation. Again low temperature can be used to minimize this annealing, but at this high dose level, one will still have complex defect-defect intractions during the implantation, due to overlap of each new cascade with residual defects from the previous ones. At typical implantation currents of 1-10 μA/cm^2, the mean time between cascade overlap is approximately 1 second. Hence one interesting simplification is to implant at a sufficiently high temperature to anneal all mobile defects before the next cascade arrives ; this again should eliminate any dose rate dependence.

For the rest of this section, we are going to simplify the discussion by considering only the isolated tracks (i.e. the low dose limit). In most cases, we will also restrict ourselves to low temperatures in order to study the basic mechanisms of defect creation without the further complications of post-implantation annealing or defect-defect interaction.

3-1-1 Radiolysis mechanisms in ionic crystals and tracks produced by high energy heavy ions in insulators

It has long been known that defects are created when ionic cyrstals such as alkali halides (LiF, NaCl, KCl, KBr ...) are exposed to ionizing radiation. Such materials are very sensitive to electronic excitations for the creation of point defects in the anionic sublattice (colour centres). This results from the rather large amount of ionic relaxation that follows any electronic change (non-radiative decay of excitons), inducing a directed ionic motion of a halide ion. The latter is ejected from its lattice site and, via a favourable collision sequence in the [110] direction, comes to rest at a site that is well separated from the corresponding vacancy. For electrical neutrality, the anionic vacancy can trap an electron to form the so- called F-centre, whereas the interstitial atom binds strongly to a halogen anion to form an X_2^- halogen molecular species the so-called H-center. Corresponding to the electron-trapped centres (F-type centres) created in the anionic sublattice we can observe hole-trapped centers (V-type centres) in the cationic sublattice. However, the latter cannot be created by an excitonic mechanism as the F-H centre pair, but can be created by direct displacements in collision cascades for example. Complete reviews of the different types of defects in ionic crystals including alkali halides and oxides ($Mg0$, Al_20_3 ...), as well as the creation mechanisms, have been presented in reference [44] and more recently by Catlow in reference [45]. Note that in another type of insulator such as silica (amorphous $Si0_2$), the self-trapped exciton seems also to be at the origin of the point defect creation [46]. However, refractory oxides such as $Mg0$, Al_20_3 ... are not sensitive to electronic excitations and only the direct displacements are able to produce defects in both sublattices [44].

High energy heavy ion tracks are observed in various types of insulators sensitive to electronic excitations for the defect creation. As mentioned in section 2, in the high energy regime, the incident ions are slowed-down mainly by exciting the electronic system of the target. This results in a large number of energetic secondary electrons emitted, which propagate around the ion trajectory, and lose their energy also by electronic excitations of the target atoms. In this case, the defects are localized in tracks surrounding the ion trajectories in which very high-energy densities are deposited by the delta rays emitted. The model developed by Fain et al [47, 48] allows to calculate the radial energy distribution deposited by the secondary electrons in the direction perpendicular to the ion trajectory. In this model, the ion trajectory is considered as a linear source of electrons. The energy and angular distributions of secondary electrons emitted are thus calculated, and using the stopping power of electrons in the materials of the target, it is possible to obtain the radial energy distribution deposited inside of the volume around the ion trajectory irradiated by the secondary electrons (secondary track). From the experimental study performed by Perez et al. [49] on the point defect (F-centres) creation in individual isolated tracks produced in alkali halides bombarded with low doses ($\geq 10^{12}$ ions/cm^2) of high-energy heavy-ions (E > 20 MeV/amu), it has been shown that in the largest part of the tracks the energy density is considerably higher than the density for primary defect saturation. Consequently in these tracks, saturated with primary defects, some complex aggregation or annealing mechanisms have been observed. Note that in some minerals or organic materials the heavy ion tracks can be revealed by chemical etching [50]. This preferential etching of the tracks is directly related with the high level of damage as mentioned above.

The damage created by electronic excitations is at the origin of some applications in insulating materials : namely, very sensitive heavy ion track-detectors using some specific organic materials, high performance selective membrane obtained after track etching and creation of well calibrated cylindrical holes, radiation induced polymerisation of plastics, high resolution lithography in resists using ion or electron beams in microelectronic devices .... A complete and detailed review on these effects and applications in various insulators is presented in reference [51].

3-1-2 Defect creation in nuclear collision cascades

Concerning the defect production, it is known that the modified Kinchin-Pease expression (15) overestimates the number of Frenkel pairs produced in a cascade by a factor of about three in most metals [31, 52, 53]. Previous molecular dynamics (MD) simulations of cascades in W [54] suggested that defect motion and recombination during the thermal spike is responsible for the low defect production efficiencies observed experimentally. In the case of Cu and Ni (f.c.c. metals) [19], the replacement collision sequences (RCS) are mostly in [100] and [110] directions, although some sequences along [111] directions are observed in the simulations. Most of the self interstitial atoms (SIA) are found near the end of a trail of replacements associated with one or more RCS's. Those SIA's not associated with RCS trails lie very close to the perimeter of the once-melted region. These SIA's most likely originate in short RCS's that do not extend far beyond the core. Note that the RCS's propagate at supersonic speeds and are therefore able to transport interstitials beyond the core region before melting occurs. The important role of RCS's in the creation of stable defects has been amply demonstrated in simulations at energies close to the displacement threshold [55, 56]. The new feature that emerges from the higher energy simulations [19] is that only those interstitials that are transported beyond the melted core region survive recombination during the thermal spike. Thus, melting in the cascade core creates a zone ($r \leq 1.5$ nm) in which all SIA's are annihilated and is responsible for the reduced defect production efficiency. These results suggest that defect-production efficiencies are controlled by two characteristic lengths : the RCS ranges and the dimension of the melt zone. The relatively high efficiencies observed in heavier metals such as W and Pt [57, 58] would then follow naturally from the long RCS's and compact melt regions in these materials.

The degree of clustering of SIA's and vacancies in cascades is important for the kinetics of high-temperature radiation effects. Of course MD simulations correspond to a static (zero K) crystal and cannot be assumed a priori to represent the behaviour of cascades at elevated temperatures. Nevertheless, it is of some interest to ascertain the clustering behavior in low-temperature irradiations. Low-temperature irradiations avoid the possible ambiguities in separating the effects of the primary state of damage from those of thermal activation. The removal of atoms from the core of the cascade by RCS's results in a vacancy-rich the depleted zone after resolidification [19]. In situ transmission electron microscopy studies have demonstrated that the depleted zone can collapse athermally into dislocation loops and stacking-fault tetrahedra in many metals [59, 60]. The probability for collapse increases with increasing cascade energy density and with decreasing melting temperature. Wei et al. [57] have suggested that vacancy clusters precipitate in the highly supersaturated depleted zone and subsequently grow by absorbing additional vacancies during the thermal spike. Another suggested mechanism for collapse is the thermo-transport of vacancies toward the center of the cascade and SIA's away from the center driven by the steep temperature gradients in cascades [61, 62]. These mechanisms both require high vacancy mobility during the thermal spike. The creation of a melt zone in cascades suggests an alternative interpretation, based on the kinetics of resolidification [19]. As mentioned above (see section 2-3-2), the atomic density in the melt zone is lower than in the surrounding matrix, both as a result of the cascade induced shock-wave and the generation of RCS's. If the solid-melt interface moves relatively slowly, most of the vacancies will not "crystallize" until resolidification is essentially complete. This leads to a compact, highly clustered vacancy configuration. If the interface motion is rapid, on the other hand, a more dispersed vacancy distribution will occur, which hinders cascade collapse. Some additional evidence supporting this interpretation is provided by heat-spike simulations in Cu and Ni [19]. Heat spike were simulated with an energy density of 1.1 eV/atom and a concentration of 2 at. % vacancies randomly arranged in the heated region. It was found that the radius of gyration of the vacancy distribution decreased from an initial value of 1.625 nm to a final value of 0.99 nm in the case of Cu, but it was essentially unchanged in the Ni heat spike. This behaviour is a consequence of the more rapid interface motion in Ni than in Cu. This picture is in accord with experiment : both the fraction of depleted zones that collapse and the number of vacancies contained in collapsed loops are observed to be larger in Cu than in Ni [63]. Finally it seems that vacancy clustering in cascades is presumed to occur by a process akin to zone-refining [19]. The slower the motion of the melt interface, the more effective the "refining", i.e., the more clustered the vacancies. Note that the zone-refining picture has implications for impurity segregation in cascades as well as vacancy clustering. Also vacancy collapse has not been observed experimentally in Cu and Ni at energies under 10 keV. Higher energies produce more favorable conditions for collapse because (i) more vacancies are available, and (ii) the cascade cools more slowly due to the smaller surface-to-volume ratio.

3-2 Ion implantation

The doping effect of the target materials in the stopping zone of incident ions is interesting for the synthesis of new materials having some original properties. Firstly, let us remember some of the unique advantages of ion implantation :

- precise control of the total number of implanted ions and consequently of the local concentration in the implanted zone.

- Independent control of the penetration depth Rp (mean projected range ; see section 2.4.).

- Almost all combinations of ion (Z_1, M_1) and target (Z_2, M_2) may be used.

- A wide concentration range is achievable, with the upper limit generally being set by the sputtering yield rather than by equilibrium solubility.

- Independent control of the target temperature during implantation.

- It is an ideal "clean room" technique, provided one uses a good vacuum (UHV) chamber. Until recently, this feature has not been properly exploited, since at 10^{-6} torr most accelerator systems have serious (hydro) carbon contamination problems.

A second important feature of ion implantation is the superposition, in the implanted zone, of a high concentration of defects with the implanted impurity distribution. This may cause some complex structural changes and chemical effects. The solid phase which is created can incorporate implanted impurities in various states, which can be associated with lattice defects thus creating a multitude of metastable impurity-defect structures. Thermal annealing usually removes a great deal of damage and leads via nucleation and diffusion to the growth of different precipitates of various equilibrium phases, depending on the implantation dose regime (see high-dose or low-dose regimes at the beginning of section 3). These effects which involves a complex interplay between several parameters (nature of the implanted ions and the target, implantation dose and energy, target temperature during implantation and temperature and atmosphere during subsequent thermal treatments) cannot adequately be treated in a single section. On the other hand, in the context of this school on metallic multilayers, it appears more interesting to develop in more details the ion beam mixing technique (see section 4), which is a suitable method for the synthesis of new phases from samples having the multilayer geometry. However a comprehensive review on the implantation effects in metals and ceramics can be found for example in reference [64]. Except for the well known application of ion implantation for the doping of semiconductors, note that it can be used for the modification of near-surface physical and chemical properties of various other types of materials. Examples of significant modifications of electrical conductivity in metals and insulators (i.e. Mg0 [65]), of magnetic properties (i.e. bubble garnets [66]), and surface mechanical properties (hardness, friction and wear - i.e. nitrogen implanted steel or metallic ions implanted ceramics [64]), have been reported. Also protections against oxydation or corrosion have been obtained in implanted pure metals or alloys [64]. The relation between the observed near-surface property modification and the metastable phase formation (crystalline or amorphous) in the implanted zone is understood in most of the cases. These processes are also well discussed in references [51, 64].

4 ION BEAM MIXING

Ion beam mixing (IM) has been the subject of increasing interest for the past 10 years. The literature for metal-metal and metal-semiconductor systems is very large and still growing. Instead of attempting complete and detailed coverage of all work done to date, it seems more useful to concentrate on the basic concepts. A wide variety of effects are involved in IM, the most evident is the ballistic effect but thermodynamic and chemical effects play also an important role on the mixing efficiencies. Details of these mechanisms, their magnitudes and their dependence on the system and conditions of irradiations are still not completely understood. The first classification of the mechanisms is given by temperature conditions [67, 68, 69]. The sample temperature dependent mechanism is called "radiation enhanced diffusion" and the athermal mechanism is called "ballistic mixing". As mentioned in section 2-3, molecular dynamics simulations [19, 70] provide microscopic insight into the cascade describing the time-scale during the slowing-down of incident ions. Within the first $2x10^{-13}$ s, the transport is predominantly collisional, leading to a recoil mixing mechanism (temperature independent). Up to a few times 10^{-11} s, there is a "relaxation" phase, during which the kinetic energy of the recoiled particles is thermalized which induces small displacements in random directions like a diffusional motion : this is a "cascade mixing". At about 10^{-11} s, the cascade reaches its thermal equilibrium, and if the sample temperature is high enough, the radiation defects are mobile and the atomic motion occurs by thermal diffusion with the defect concentration over the equilibrium one. This is a radiation enhanced diffusion. After a brief presentation of experimental sample configurations, we will describe in more details the three different mixing mechanisms mentioned above, mainly in metallic systems [71, 72]. New developments in metal-ceramic systems and dynamic mixing will be also mentioned.

4-1 Sample configurations and mixing parameter measurements

Experimental investigations in the field of ion-mixing have been executed mainly with three sample configurations represented in figure 3. Bilayered samples consist of a thin film deposited on a substrate

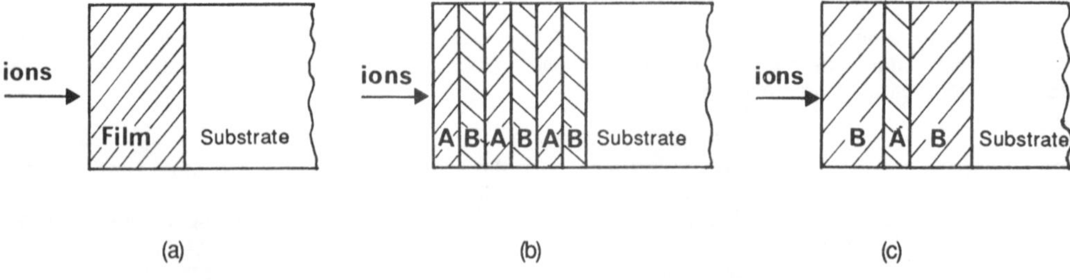

Fig.3 Sample configuration. (a) bilayered sample, (b) multilayered sample and (c) marker of element A in element B .

(Fig.3a). Multilayered samples consist of a multiple sequence of thin alternating layers of generally two elements (Fig.3b). Experiments with a thin marker of one given element embedded in a matrix (Fig.3c) can also be performed.

The mixing rate is measured by different techniques adapted to the sample configuration. In bilayered and thin marker samples, most studies have been performed using the Rutherford backscattering technique (RBS) [73] of light particles (alpha-particles). The typical depth resolution of this technique is ~ 10 - 15 nm, and the mixing rate is estimated from the broadening of the interface. If we assume that this broadening can be described as a diffusion problem, with a characteristic diffusion coefficient D, then the mixing kinetics can be analyzed by the diagram $D \times t$ versus irradiation fluence Φ. t is the irradiation time ($t = \Phi/\Phi_0$, Φ_0 being the ion flux). Every other technique with sufficient depth resolution can be used for the study of ion beam mixing : i.e. secondary ion mass spectroscopy profiling technique (SIMS), transmission electron microscopy observation (TEM) in cross section geometry .... In the case of multilayered samples, the individual layer thickness is typically 5-10 nm which is lower than the RBS resolution. In the recent years some techniques such as grazing X-ray reflectometry [74] have reached a sufficient resolution (0.5 nm) to lead to quantitative values of the mixing rate in the first stage of mixing. Electrical resistivity and TEM measurements are able to study the mixing kinetics over the total reaction when the multilayers are completely mixed and an homogeneous new compound formed.

4-2 Mixing mechanisms

4-2-1 Recoil mixing

Recoil mixing is the process which implies atomic rearrangements caused by a direct collision between an incident ion and target atom. Many theoretical treatments (see sections 2.2. and 2.3. and references [75 to 79] lead to a linear dependence on fluence for this process. The total number of these direct recoils is small compared to all atoms that are intermixed in the nearest vicinity of the interface. However, direct collision events with high energy transfer are rare, but produce recoil atoms that are projected deeply into the substrate leading to a long range mixing with a profile nearly exponential. The figure 4 shows an example of the long range mixing in the Al/Sb/Al system [80, 81]. These measurements have shown that the long range mixing is proportional to the fluence and this effect is independent of the target temperature.

4-2-2 Cascade mixing

The number of primary recoils increases rapidly as their energy decreases. This causes a short-range intermixing attributed to numerous small displacements created by low energy events in a collision sequence that involves atoms on either side of the interface. These displacements may be regarded as isotropic. Sigmund et al [79] and Littmark et al [82] have developed a collisional model of cascade mixing. Let us consider a semi-infinite target with a plane surface at x = 0, and an impurity profile Co(x). After bombardment with an ion fluence Φ, the impurity profile has changed to a profile $C(x, \Phi)$ because of atomic transport so that $C(x, 0) = Co(x)$. At depth x a relocation function $F(x, z)$ is defined such that $NC(x, \Phi) F(x,z)\, dx\, dz$ is the mean number of impurity atoms transported per incident ion from a layer at depth x of thickness dx to a layer at depth x + z of thickness dz. N is the atomic density of the target. Ignoring the sputtering effect, the balance equation becomes :

$$\frac{\partial}{\partial \Phi} C(x, \Phi) = \int dz\, [C(x - z, \Phi)\, F(x - z, z) - C(x, \Phi)\, F(x, z)] \qquad (19)$$

If we take into account the numerous, isotropic short-range displacements (z small), we can expand the function $C(x - z, \Phi) F(x - z, z)$ in Taylor expansion at the second order [83, 84]. One finds :

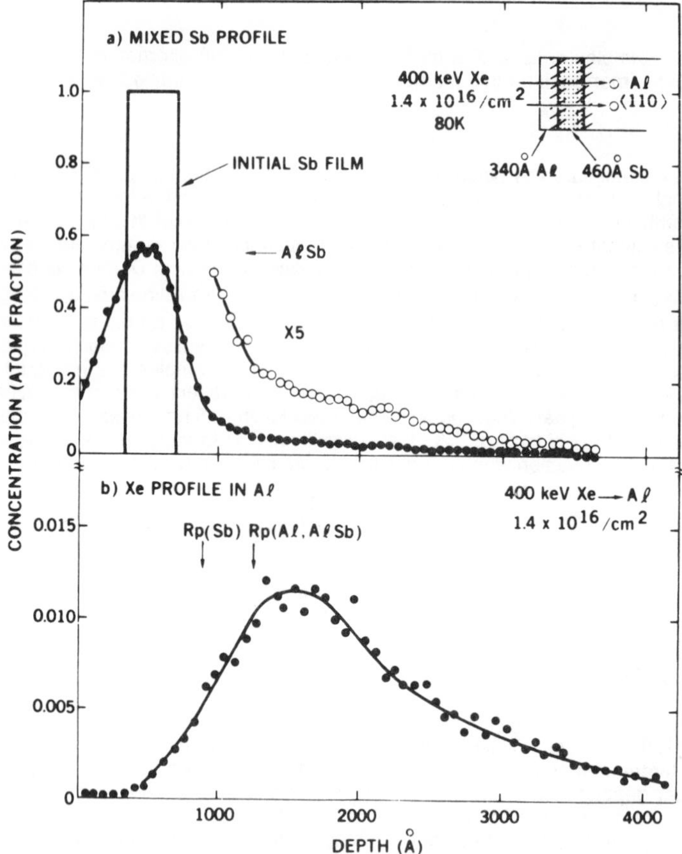

Fig. 4 Concentration vs depth profiles : (a) Sb profile for the thin-film system 340-Å Al on 460 Å Sb on Al <110> substrate after 400-keV Xe bombardment to $1.4x10^{16}/cm^2$ at 80 K ; and (b) Xe profile after same fluence as measured in pure Al [80].

$$\frac{\partial}{\partial\Phi} C(x,\Phi) = \frac{1}{2}\frac{\partial^2}{\partial x^2}\left[C(x,\Phi)\int dz \times z^2 \times F(x,z)\right] - \frac{\partial}{\partial x}\left[C(x,\Phi)\int dz \times z \times F(x,z)\right] \qquad (20)$$

This suggests the introduction of the quantities :

$$\frac{D(x)}{\Phi_o} = \frac{1}{2}\int dz \times z^2 \times F(x,z) \qquad (21)$$

$$\frac{V(x)}{\Phi_o} = \int dz \times z \times F(x,z) \qquad (22)$$

where D (x) and V (x) are the effective diffusivity and the shift rate of the profile, respectively. Φ_o is the ion flux. The balance equation 20 becomes :

$$\frac{\partial}{\partial\Phi} C(x,\Phi) = \frac{\partial^2}{\partial x^2}\left[\frac{D}{\Phi_o} C(x,\Phi)\right] - \frac{\partial}{\partial x}\left[\frac{V}{\Phi_o} C(x,\Phi)\right] \qquad (23)$$

which is a generalized diffusion equation. According to Sigmund et al [79], the D value for the recoils in the eV-region is connected intimately with displacement and replacement processes in radiation damage. Assuming that the relocalisation process is essentially equivalent to Frenkel-pair formation, they found the well know Andersen formula [85] :

$$\frac{D \times t}{\Phi} = \frac{0.42}{6} \frac{F_D \times R_c^2}{N \times E_d} \qquad (24)$$

where R_c is the distance between a stable vacancy and interstitial complex. E_d is the displacement threshold energy in radiation damage (see section 2.3.), F_D is the amount of energy deposited in nuclear processes per unit length, t the irradiation time and Φ the ion fluence. Note that all the models predict a linear dependence of Dt on Φ and F_D.

4-2-3 Radiation enhanced diffusion

The radiation enhanced diffusion is the extra-cascade transport which arises from the long-range migration of irradiation-induced defects (vacancies, interstitials and their clusters), away from their nascent cascades. At temperatures above the migration stage of defects, radiation enhanced diffusion is likely to play a role. The enhanced diffusivity D* is given by :

$$D^* = C_V D_V + C_I D_I \qquad (25)$$

where C_V and C_I are the concentrations of vacancies and interstitials respectively and D_V and D_I are the corresponding diffusivities. C_V and C_I are calculated by solving two independent equations [86] :

$$\frac{\partial C_V}{\partial t} = D_V \frac{\partial^2 C_V}{\partial x^2} + \alpha P - \frac{D_V}{\ell^2} (C_V - C_V^{th}) - 4\pi R_{iv} N D_i C_V C_i \qquad (26)$$

$$\frac{\partial C_i}{\partial t} = D_i \frac{\partial^2 C_i}{\partial x^2} + \alpha P - \frac{D_i}{\ell^2} C_i - 4\pi R_{iv} N D_i C_V C_i \qquad (27)$$

where α is the fraction of point defects which escape the cascade to migrate freely in the lattice, C_V^{th} the thermal equilibrium vacancy concentration, P the rate of displacements per target atom, ℓ a characteristic diffusion length to fixed annihilation sites, N the atomic density of the materials and R_{iv} the vacancy-interstitial radius at which spontaneous recombination occurs. Finally the four terms of the equations 26 and 27 are connected to defect diffusion, defect production, point-defect annihilation at fixed sinks and spontaneous recombinations.

The enhanced diffusivity D* has three regions of different temperature behaviour. At high temperatures, the thermal diffusivity dominates and D* is approximately equal to the normal thermal diffusion constant. At intermediate temperatures where point defects annihilate mostly at fixed sinks $D^* \simeq 2 \alpha P \ell^2$ and at still lower temperatures $D^* \simeq (\alpha P D_V / \pi N R_{iv})^{1/2}$. Finally the radiation enhanced diffusion is treated like normal thermal diffusion but with an enhanced diffusivity D*.

Radiation enhanced diffusion can lead to a second effect which consists of the coupling of the solute atoms to the local defect flux. The coupling coefficient may be positive or negative and sufficiently strong to cause flow against a concentration gradient. In such cases alloys can segregate at sinks. Not only the atom migration depends strongly on the kind and mobility of defects, also the final location of solute atoms in the ion irradiated matrix is affected by solute trapping. Unfortunately, although radiation enhanced diffusion dominates ion mixing in many systems at elevated temperatures, there are no general rules.

4-3 Chemical effects and thermal spike regimes

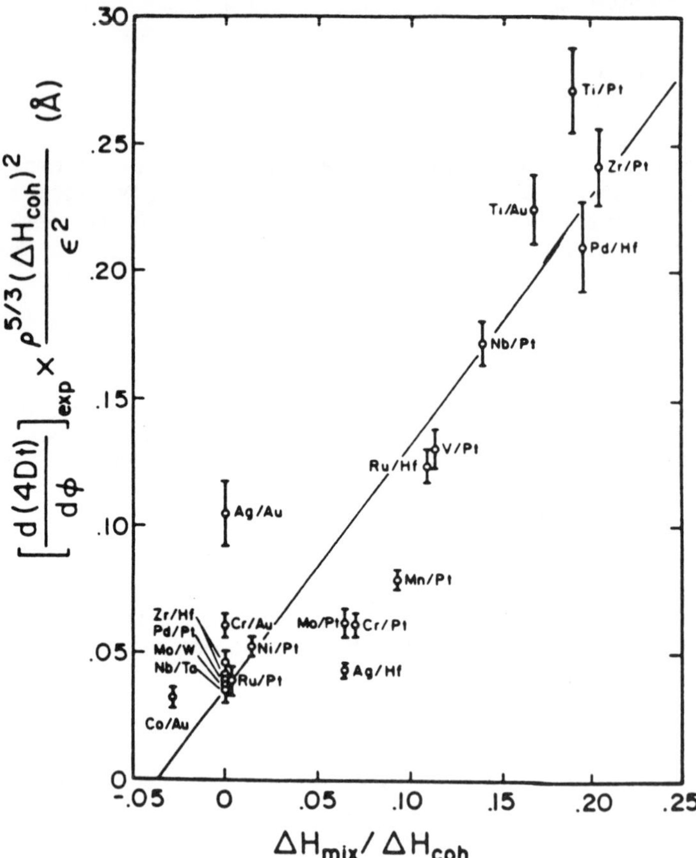

Fig. 5 Summary of results of ion mixing of metallic bilayers [90].

The temperature independent component is apparently associated with collisional effects. Thus, if the kinematics of the irradiations are similar for two different systems, the ion mixing effects should be similar. Some experimental observations agree with this point of view. For example, the mixing rate of Al-Fe, Al-Ni, Al-Co and Fe-Mg at 77 K does not depend on the system [87]. However, some other results suggest that this is not necessarily the case. For example the mixing of Cu-Au is much larger than that of Cu-W even though the kinematics are quite the same [88]. To explain these results, Cheng et al [89] suggest a mixing model based on Darken's analysis of chemical interdiffusion coefficient in random regular solutions. Taking ΔHm (heat of mixing) as the origin of the chemical driving force for diffusion, the Darken effect can be expressed as :

$$D = D_O \left(1 - \frac{2\,\Delta Hm}{RT}\right) \qquad (28)$$

where Do is the diffusion coefficient for pure random walk mixing (ballistic). T is not the bulk sample temperature but defined from the average kinetic energy per atom (θo, see section 2-3-2) involved in the mixing process by $\frac{3}{2}kT = \theta o$. If the energy density θo in the cascade is high enough, it becomes possible to use the thermal spike concept (see sections 2-3-2 and 3-1-2). Cheng et al [89], who started from Vineyard's [22] classical work concerning thermal spikes assuming that the random walk of the atoms is biased by chemical interactions, have come up with the expression :

$$\frac{D \times t}{\Phi} = \frac{1}{4} \frac{K_1 \times \theta_0^2}{N^{5/3} (\Delta H_{coh})^2} (1 + K_2 \frac{\Delta H_m}{\Delta H_{coh}}) \quad (29)$$

where ΔH_{coh} is taken as the average cohesive energy of the constituents + ΔH_m, and K_1 and K_2 two adjustable parameters. Experimental support [90] for the validity of equation 29 is shown in figure 5 for a large number of metallic systems.

4-4 Metastable phase formation

Ion beam mixing allows the formation of metastable phases, crystalline or amorphous. To illustrate these effects we have chosen to describe the phase formation in the metallic systems Ni-Al and Fe-Al [91] after mixing at low temperature (77 K) when chemical diffusion is prevented. The comparative study of equiatomic Fe-Al and Ni-Al multilayers has shown amorphous phase formation in the case of Ni-Al and a b.c.c. disordered solution for Fe-Al. Usual criteria for predicting amorphization fail : the Ni-Al compound presents a congruent

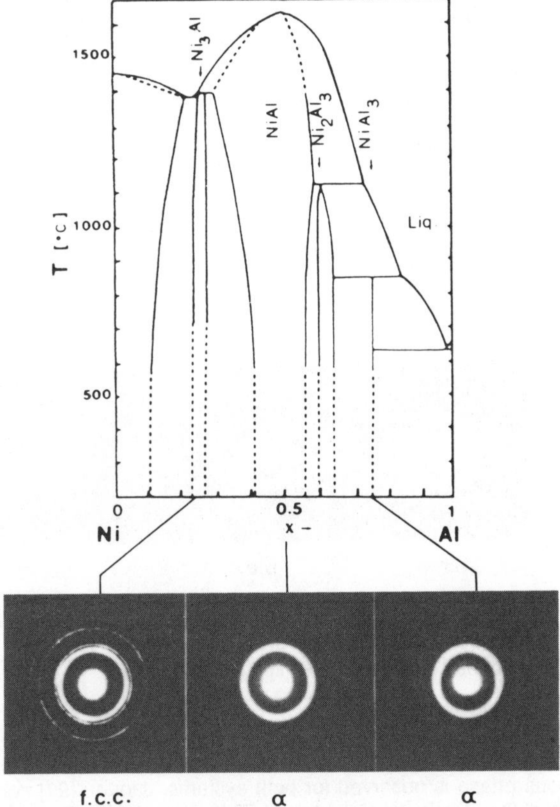

Fig.6 Equilibrium phase diagram and diffraction patterns of mixed samples at three compositions (x = 0.25, 0.50 and 0.75) for the system $Ni_{1-x}Al_x$. Ion beam mixing was performed at 77 K with 360 keV Xe ions [91].

melting, the terminal crystalline solid solutions are of identical nature and the intermetallic compound Ni-Al has a very simple structure. Moreover, the structural characteristics relatively to Fe-Al are quite comparable : the atomic radii ratio of both elements and electronegativity difference are similar. Then the question arises why the crystalline lattice collapses into topological disorder only for Ni-Al ? According to Jonhson et al [92], metastable phase formation in ion mixing may be successfully explained with the aid of calculated phase diagrams comparing the free energy differences between the amorphous phase and the metastable solid solutions. In this thermodynamic approach, we shall try to determine the form under which the energy responsible for the transformation is stored.

The results of the mixing for both Ni-Al and Fe-Al systems are shown in figures 6 and 7 respectively, in the form of diffraction patterns. Only metastable phases are observed, crystalline or amorphous solid solutions depending on the composition range and also on the system.

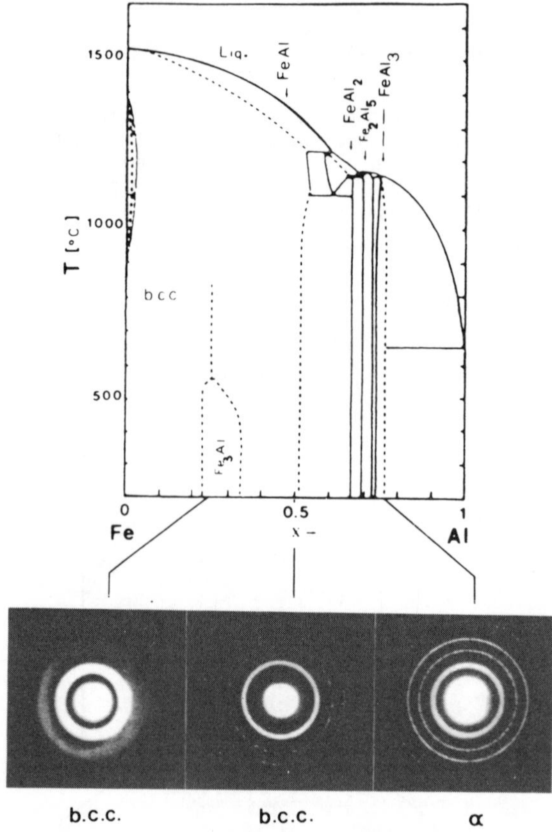

Fig. 7 Equilibrium phase diagram and diffraction patterns of mixed samples at three compositions (x = 0.25, 0.50 and 0.75) for the system $Fe_{1-x}Al_x$. Ion beam mixing was performed at 77 K with 360 keV Xe-ions [91].

In the transition metal rich concentration range, a crystalline solid solution is observed. For equiatomic composition, an amorphous state is obtained for Ni-Al and a b.c.c. solid solution for Fe-Al. On the aluminium rich side, the amorphous phase is observed for both systems. Jaouen [93] has computed free enthalpy variations as a function of the atomic concentration at 77 K using the Kaufman method [94] for the different phases of the systems. These results are reported in figures 8 for Ni-Al (8a) and Fe-Al (8b). The more stable phase without irradiation is the ordered intermetallic compound. Nevertheless under ion irradiation, chemical disorder is forced, and at low temperature long range migration necessary to allow crystalline complex ordered

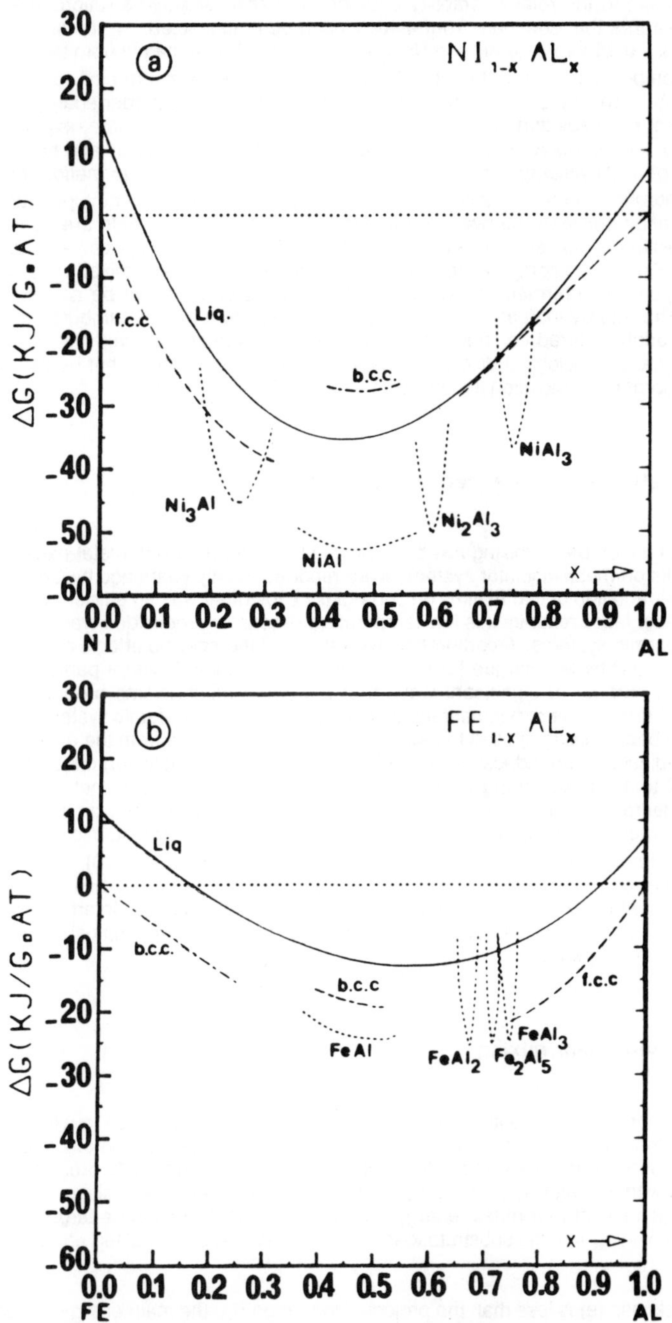

Fig. 8 Free enthalpy as a function of the atomic composition for Ni-Al (a) and Fe-Al (b) systems [91, 93]. The solid lines are the results of the calculations (extrapolated at 77 K) for the liquid state and the dotted lines for the crystalline phases.

phase is prevented, then intermetallic ordered compounds cannot exist. So the studied states can be considered as constrained equilibrium states. With such a kinetic restriction, the nature of the phase formation can be discussed with respect to the relative stability of disordered solid or liquid solutions (i.e. glassy state). Therefore the main features that dominate from analysis can be emphasized : (i) the stability of b.c.c. and f.c.c. solid solutions at x = 0.25 for $Fe_{1-x}Al_x$ and $Ni_{1-x}Al_x$ respectively is apparent from the diagrams in figure 8. (ii) At equiatomic composition x = 0.5, the stability of the disordered phase (b.c.c.) relatively to the liquid one is deduced for FeAl. In the case of NiAl, as a result of the strong order-disorder energy of the intermetallic compound, the b.c.c. solid solution is unstable relatively to the amorphous state, explaining the experimental results. (iii) At x = 0.75, equivalent values of the f.c.c. solid solution energy and of the liquid one are observed in the case of Ni-Al yielding experimentally to the amorphous phase formation. For $Fe_{25}Al_{75}$ no direct evidence of amorphous phase stability may be obtained from the free enthalpy diagram (Fig. 8b). Nevertheless, discrepancy between the two diagrams in this concentration region are questionable, since solubilities of both transition metals are very similar and at x = 0.75 the free energy curve of f.c.c. solution is a very extrapolated line. Finally, the comparative study of Ni-Al and Fe-Al systems mixed at 77 K shows that an amorphous phase appears in intermetallic systems if chemical disorder cannot be accommodated by the crystalline structure. The energy increase due to "forced" solid solution can be attributed mainly to antisite defects in the very strongly ordered compounds. Point defects, increasing partially the energy, can play a role in this collapse towards topological disorder ; however their contribution cannot be separated from the one of the chemical disorder in irradiation experiments.

4-5 Specific effects in metal-ceramic systems

Although research on ion beam mixing has been principally concerned with metal-metal or metal-silicon systems, the first results on metal insulator systems were reported twenty years ago by Collins et al [95] who have found that Ar-ion irradiation of thin Al films on soda-glass substrates was an efficient way to increase the adhesion at the interface. More recently this low temperature processing technique was applied to the particular case of metal-ceramic systems. A comprehensive review of the specific effects of ion mixing in these systems has been reported by Mc Hargue [96]. The effects of energetic incident particle in ceramics and metals differ in such a number of significant ways that the same mechanisms proposed in metal-metal systems cannot explain directly the experimental results obtained in metal-ceramic systems. One of the most interesting result in ion beam mixing applied to metal-ceramic systems arises from the experimental observation that ion beam irradiation often induces an enhancement of adhesion between the metallic film and the ceramic substrate [97 to 100]. According to these results, ion bombardment of metal films deposited on ceramics produces a decrease of the interfacial energy [99, 100] and consequently an increase of adhesion. The microstructural characterization of the irradiated interface, performed on the atomic scale in the Cu/Al_2O_3 system using the high resolution transmission electron microscopy (HRTEM) [100], exhibits an important increase of the interface roughness without any chemical reaction between these two immiscible materials. The induced roughness, which concerns few atomic planes on both sides of the interface, seems to be the result of the atomic transport during the ballistic phase only, and no long-range diffusion seems to occur in this system mixed at room temperature.

4-6 Dynamic ion beam mixing

The generic term "ion-assisted coating" (IAC) describes any coating process in which an ion beam is used to enhance the properties of the coating materials, but specific processes are often described in other terms. The term "ion beam enhanced deposition" (IBED) describes those IAC processes in which the substrate is irradiated with low-energy (< 30 keV) ion beam during coating deposition. Dynamic ion beam mixing (DIM) uses an ion beam with mass, energy and dose sufficient to cause a large number of atomic displacements to mix a coating with the substrate to form a broad interface several ten nm wide.

When the film thickness (e) is less than the projected ion range R_p, the main energy losses take place in the substrate. If $e \simeq R_p$, the mixing process at the interface occurs. When e becomes larger than R_p, all the ion energy is deposited in the volume of the film leading to the phase formation as described previously in the ion beam mixing section (section 4). Finally the results show that ion implantation into the substrate during deposition strongly influences the interfacial properties, nucleation processes, coalescence and orientation of film growth and consequently the bulk properties of the growing film (densification, homogeneity ...). The possibility to produce thick films (a few microns) by this technique is also interesting for industrial applications. For example, He^+, and Ar^+ ion implantations with energy of 100 keV into mica substrates during the deposition of silver reduces the equivalent mean thickness required for a continuous film from 7 nm (without ion bombardment) to 1 nm for Ar^+ and 2.5 nm for He^+ [101, 102]. The same authors have shown that Ag layers on KCl substrates obtained by DIM with He^+ ions are epitaxially oriented with respect to the substrate. Those formed on KCl substrates with Ar^+ ions have a larger density but, are not well oriented relatively to the substrate. In all cases, good adhesion was achieved between substrate and film.

Examples of the apparatus generally used for dynamic ion beam mixing experiments [103, 104] are shown schematically in figure 9, using two possibilities for deposition : the electron beam evaporator (9a) and the sputtering evaporator (9b).

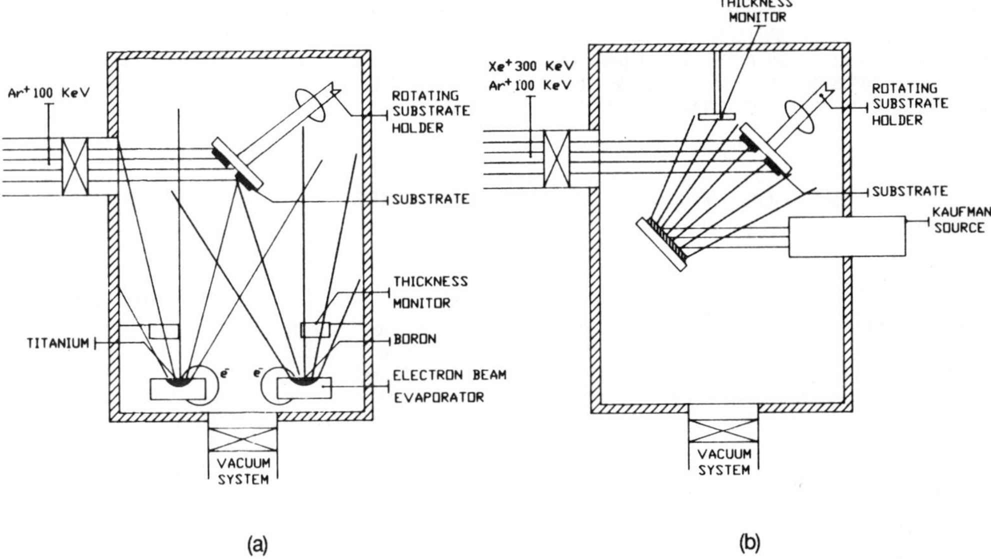

Fig. 9 Dynamic ion beam mixing setup with electron beam (a) or sputtering (b) deposition systems [103, 104].

With the usual deposition methods using ion sources of low energy (1 - 2 keV), high ion fluences are necessary to induce significant modifications of the properties of thin layers. The use of an implantor in line with the deposition chamber allows the same energy deposition with a much lower ion fluence due to the high energy of bombarding ions, thus reducing film sputtering. Moreover, in contrast with classical sputtering gazeous ion sources used for the bombardement of the film, there is no limit with the implantor on the choice of bombarding elements, preventing bubble formation.

The defects and the modifications induced by such bombardments can reach depths of 100 to 500 nm depending on the nature and energy of the ions, whereas with classical low-energy sources depths do not exceed some nm. The defect distribution obtained differs markedly in both cases and is efficient in aiding the migration of species in the case of DIM.

The evaporation rate is higher for the electron gun system than for the sputtering system. In contrast, a low deposition rate allows the energy density deposited by ion implantation to be increased. Practically, the choice of the evaporation source depends on the kind of film to be synthetized. Generally pure materials can be easily evaporated with the electron gun system whereas alloy targets can be efficiently used with the sputtering evaporator. The potential applications of DIM can be illustrated with some typical results obtained recently. Coatings can be used to improve tribological and corrosion properties of some specific substrates. In this case hard ceramic-like coatings such as various nitrides or carbides have been successfully obtained. Guzman [105] has realized a chromium nitride (CrN) with a stoichiometric composition by multiple sequential steps of deposition and irradiation. The adhesion between the CrN coatings and copper or tool steel substrates were greatly improved as well as the mechanical properties of the film. Sato et al [106] and Satou et al [107] have synthetized titanium nitride with a preferential orientation varying from <111> to <100> axis normal to the surface, as decreasing the ratio (Ti/N) of evaporated titanium over implanted nitrogen. Titanium carbides and silicon carbides have been prepared by Pimbert-Michaux et al [103, 104]. The titanium carbide was synthetized at room temperature by simultaneous evaporation of titanium and carbon assisted with Ar ion bombardment at high energy (100 keV). Microstructural analysis by TEM shows that the coatings are homogeneous with very small grains and the diffraction patterns indicate an ordered f.c.c. crystalline structure of titanium carbide whatever the composition of the layer. Strong adhesion of these films to the steel substrate is obtained and wear resistance was also considerably improved. For the synthesis of silicon carbide films at room temperature, simultaneous sputtering of silicon carbide at equiatomic composition assisted with Ar ion bombardement (100 keV) were performed. In this case the coatings presented an amorphous structure stable up to 800°C leading to a considerable improvement in the tribological behaviour. Finally, DIM which is an extension of the classical mixing technique is often capable of providing thick coatings which cannot be produced by other conventional methods. It is especially useful when substrate heating cannot be tolerated and metastable phases needed.

5 INTERACTION OF LASER BEAMS WITH METALS

A local heating is induced by laser interaction, with fluences between a few mJ/cm^2 and several J/cm^2. Heating is rapid (10^3-10^5 K/s), and surface temperatures are high enough to produce melting or even vaporization. In this last case a plasma occurs above the specimen and changes significantly the laser-material interaction. Duration of this interaction is well controlled and thus surface and depth of the treated zone may be monitored. Modifications of both composition and microstructure are observed; they depend on the nature of materials, diffusion phenomena and, in some cases, on convective motions in the melt pool. Also a relevant parameter of laser treatments is the cooling rate,up to 10^{12} K/s.

Therefore, the major part of surface treatments may be carried out by using photonic irradiation. Due to the mechanisms of energy transfer involved in the process, a minimum thickness is concerned. As it will be shown, its value is a function of the laser wavelength and material composition. For very large irradiances (> 10^8 W/cm^2) and very short pulse durations (<10^{-8} s) mechanical effects prevail over thermal effects; a shock wave is created which produces a "photonic shot peening". Finally it is the purpose of this review to illustrate the potentiality of this method, mainly in metallic materials.

5 - 1 Different kinds of laser

Whatever the energy delivered by a laser beam, the principle of a laser cavity is unchanged and well known (see for example ref.[108,109]). In the present context, we will only recall the laser characteristics which are useful for treatments. The available laser beams supply irradiation times ranging from a few picoseconds up to several seconds. Their wavelength is mainly in infrared or more recently in ultraviolet. However, for some of them, especially YAG lasers, the frequency may be multiplied by 2, 3 or even 4; so intermediate wavelengths are available for experiments. The mainly used laser beams are as follows :

- CO_2 laser, emitting at 10.6 μm, monomode (TEM_{00}) or multimode, in continuous (CW) or pulsed (TEA) regime; the peak power may reach 10 MW.
- YAG-Nd or glass-Nd lasers; their wavelength is 1.06 μm; the duration of the pulse ranges from a few nanoseconds (Q-switched mode) to a few milliseconds (relaxed mode). The Q-switched mode locked laser produces picosecond (ps) pulses. The energy delivered is between a few mJ and a few tenths of J.
- excimer lasers are constitued by diatomic molecules (ArF, KrF, XeCl...), emitting in UV and producing pulses of a few hundreds of mJ. The pulse duration is about 10^{-8}s. The technology of these sources is quickly evolving.
- in some experiments ionic lasers are used. For example argon lasers emit continous waves at 488 or 514.5 nm. However, powers are limited to about 20W.

5 - 2 Characterization of the laser beam

A precise analysis of the interaction requires the knowledge of the photon beam.The first feature of the laser beam is its wavelength (l), in µm or in nm, or its frequency (ν), in Hz, or its pulsation (ω), in rad/s. Then it is possible to show [110,111] that a beam may be defined by using three additional parameters (fig. 10), chosen among various possibilities; for example :

- output power P of the cavity (W)

- power density or irradiance I (W/cm^2 or W/m^2)

- fluence or energy density J (J/cm^2 or J/m^2)

These two last parameters may be defined either globally (I, J), or by using the average values ($\bar{I}$, $\bar{J}$), in a given direction normal to the propagation (I_x, J_x) or locally (I_{xy}, J_{xy}), by the maximum values (I_M, J_M) or by the threshold values (I_S, J_S) required for a given phenomenon.

- geometrical parameters :
  - the effective radius of the beam, defined either from an interaction threshold (I_S), or by using geometrical data; for example the integral radius (r^+), which is given, for an axisymetric beam by :

$$r^+ = \sqrt{\frac{P}{\pi\, I_M}} \qquad (30)$$

In the same way, an equivalent dimension r_x^+ may be introduced in a given direction :

$$r_x^+ = \frac{1}{2}\int_{S_1}^{S_2} \psi(x)\, dx \qquad (31)$$

where $\Psi(x)$ is the distribution function of energy in the x direction, and S_1, S_2 the efficiency thresholds. From the average value of this radius :

$$\bar{r}_x^+ = \frac{1}{2\pi}\int_0^{2\pi} d\theta \int_0^{s(\theta)} du\ \psi(u,\theta) \qquad (32)$$

one can define a spreading parameter E :

$$E = \frac{\int_0^{2\pi} d\theta \int_0^{r^+} \psi(u,\theta)\, du}{\pi\, r_+^2} \qquad (33)$$

which characterize the energy distribution within the cross-section of the laser beam.

5 - 3 Short review of the interaction

The laser-material interaction depends on different factors :
- laser radiation
- nature of the materials (in the present review only metals will be considered)
- environment (fig. 11) [112].

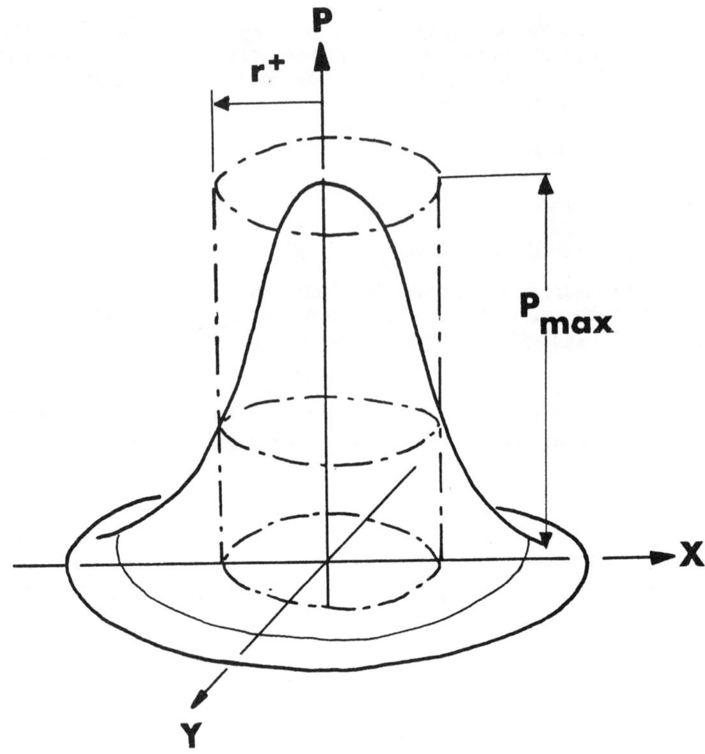

Fig. 10 schematic representation of the spatial distribution of the energy within a laser beam.

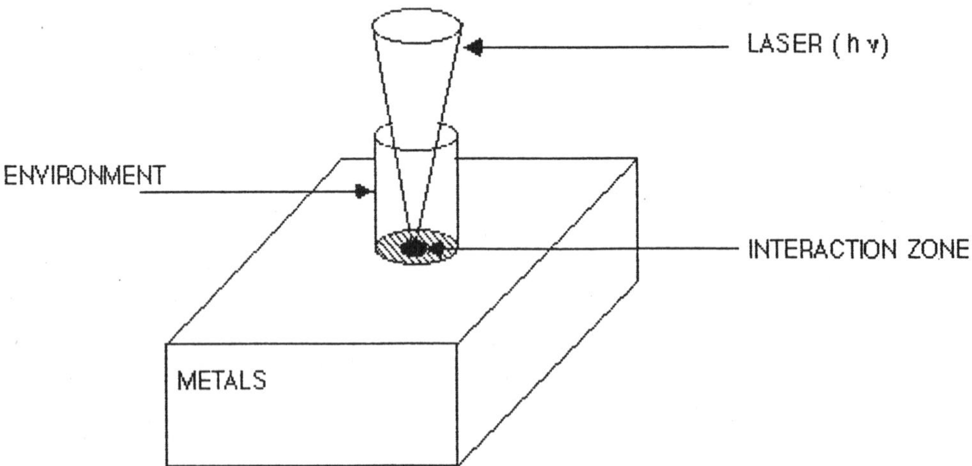

Fig. 11 schematic representation of the laser-material interaction.

In a first approximation the laser-material interaction will be considered independently from the environment. Then the main phenomenon is a photon-free electron interaction. The electrons excited by absorption of the laser radiation lose energy to return to an equilibrium state after photon excitation. This occurs by scattering with phonons and lattice defects. Equations of motion of these excited electrons have been described by Drude (see for example 111,112,113] and three ranges may be distinguished, according to the radiation pulsation ν and to the relaxation time t_c of the electrons :

- $0 < \omega < (1/t_c)$: reflection occurs
- $(1/t_c) < \omega < \omega p$: absorption occurs
- $\omega_p < \omega$: transparency occurs

ω_p is the plasma frequency, given by :

$$\omega_p^2 = (Ne^2)/(m\varepsilon_0) \tag{34}$$

where N is the number of free electrons, e is the electron charge, m is the electron mass and ε_0 is the vacuum permittivity. For materials and radiation considered in the present context, only the two first phenomena occur. For example in copper $1/t_c = 3\ 10^{13}\ s^{-1}$, while $\omega = 1.8\ 10^{15}$ s-1 for $\lambda = 1.06\ \mu m$ or $1.8\ 10^{14}\ s^{-1}$ for $\lambda = 10.6\ \mu m$. So it is evident that the incident radiation will be mainly reflected by the surface of the specimen. For normal incidence of radiation on a plane metallic surface, the reflectivity can be written as :

$$R = \frac{(n\ -1)^2 + k_e^2}{(n\ +1)^2 + k_e^2} \tag{35}$$

n is the refractive index and k_e is an extinction coefficient related to the dielectric properties ot the surface. Experimental data have been reported for the reflectivity [114]. Some examples are given in table I:

λ- R	Al	Fe	Si
0.25 μm	0.92	-	0.61
1.06 μm	0.94	0.82	0.33
10.6 μm	0.98	0.94	0.30

Table I effect of wavelength on the reflectivity of some materials.

This parameter R depends on various conditions :

- temperature T : R decreases with temperatrure for Al and Fe (and other metals), but increases with temperature for silicon (and other semi-conductors).
- surface roughening
- surface state (liquid or solid) : melting generally induces a drastic decrease of R for metals. For semi-conductors, changes are different : R increases above T_M for silicon. Values reported are also very different from one author to another one; for example in liquid Al data between 8% and 40% have been reported. In the solid state, R(T) is well approximated by the relation of Hagen-Rubens [111-113] :

$$R = 1 - \sqrt{\frac{2\ \varepsilon_0\ \omega}{\sigma}} \tag{36}$$

where σ is the electrical conductivity of the materials.

- surface homogeneity : impurities, especially oxides, induce some large local variations of reflectivity.

To summarize, the coupling coefficient for laser radiation (especially in the infrared) is fairly small for metals. Consequently, only a small fraction of the incident energy is absorpted by the specimen. Note that this coefficient depends strongly on the environment.

5 - 4 Energy transfer within the metal [114] :

Laser radiation is mainly converted to heat in metallic systems. Consequently, the interaction is nearly independent of the radiation coherency, but is a function of the number and energy of photons. So fluence appears like the predominant factor.

When a solid metallic surface is irradiated by a laser pulse, the energy is deposited in a layer of thickness a (absorption coefficient) given by :

$$a = \frac{2\,\omega\,k_e}{c} = \frac{4\,\pi\,k_e}{\lambda} \tag{37}$$

where c is the light velocity. It may be assumed that d is nearly equal to a^{-1}, where d is the absorption length. For metals d is limited, due to the large number of charge carriers; some values are gathered in table II :

	λ (μm)=0.25	1.06	10.6
Al	a^{-1} = 8 nm	10 nm	12 nm
Si	6 nm	200 μm	1 mm

Table II absorption coefficient for Al and Si, at various wavelengths.

More generally, for metals, a^{-1} is between a few nm and 100 nm, while for semi-conductors a large wavelength dependence is observed. This feature is fundamental; indeed the thickness of the surface layer affected during a laser irradiation, as short as the pulse duration may be, is not less than a^{-1}. Consequently a limitation of the thickness requires a decrease of the wavelength used. Excimer lasers are attractive in this way.

To determine the depth treated, we have to compare d with the thickness l_T of the layer heated up by thermal conduction; this thermal diffusion length is approximated by [115,116]:

$$l_T = 2\sqrt{\alpha\,\tau} \tag{38}$$

where α is the thermal diffusivity and τ is the interaction time. Experimental data for a are not accurately known and are temperature dependent. However, for metals, α is on the order of 10-100 mm²/s. Therefore, two cases have to be considered : $d \gg l_T$ and $d \ll l_T$; this condition may be expressed by means of a critical interaction time τ_c given by :

$$\tau_c = d^2 / 4a \tag{39}$$

which is in the range 2.5 10^{-11} - 2.5 10^{-10} s. So we consider two situations :

- $\tau < \tau_c$: very short pulse durations [117]. In this limit only picosecond or better femtosecond lasers may be used. Absorption of energy occurs in a layer with a thickness equal to the thermal diffusion length. An exponential temperature profile is induced and the heating rate is then :

$$\left(\frac{dT}{dt}\right)_h = \frac{1-R}{\rho C_v}\, a\, I\, e^{-\alpha z} \tag{40}$$

with z : depth, c_v : volumic thermal conductivity, ρ : mass density; the cooling rate is given by :

$$\left(\frac{dT}{dt}\right)_c = (1-R)\, a^3 \left(\frac{2\alpha}{\rho c_v}\, I\, t_p\right) \tag{41}$$

where t_p is the pulse duration.

- $\tau > \tau_C$: pulse durations are higher than in the previous case; two regimes may then be distinguished:

- irradiation by a pulsed laser : irradiation times are short, between a few ns and a few ms; energy absorption induces a temperature increase ΔT on the surface which may be written as :

$$\Delta T = \frac{(1 - R)\, I\, t_p}{\sqrt{(2\ \alpha t_p)}\ \rho c_v} \qquad (42)$$

However, this value depends on the pulse shape and latent heat corrections are required, when melting occurs. So only an order of magnitude is obtained.

Heat transfer in the sample is then governed by the heat flow equation and various models have been proposed [118,119].

- irradiation by a continous laser : interaction duration is higher than 1 ms. A first useful approach is given by a one-dimension model The treated material is assumed to occupy a semi-infinite space.and temperature is then approximated by :

$$\Delta T(z,t) = \frac{2\alpha I}{k} \sqrt{Dt}\ \mathrm{ierfc}\left(\frac{z}{2}\sqrt{Dt}\right) \qquad (43)$$

ierfc is the integral of the complementary error function. In practice different operating conditions occur [110]; but heat affected zones are often larger than a few tenths of µm.

5 - 5 Influence of the environment

The environment has a drastic influence on the interaction. This environment may be controlled and constant during the experiment or evolve during interaction To obtain a controlled environment, the experiments may be performed in vacuum or in a given atmosphere : shielding gas or active gas (for example in laser assisted chemical vapor deposition, see further). In addition, absorbing coatings are sometimes used to enhance the coupling between laser radiation and metal. In this case two situations are possible :

- the thickness of the coating is smaller than the absorption length d: under these conditions part of the radiation reaches the substrate and is then directly absorbed by the metal, but the major fraction of the energy is reflected within the coating and then absorbed ;

- the thickness of the coating is larger than d; conditions are then similar to those encountered in conventional experiments.

In given conditions, the environment is evolving during experiments and melting or vaporization of the surface layer may be achieved. For example in aluminium with R = 92% and $t_p = 2\ 10^{-9}$s, vaporization occurs for $I >= 4.2\ 10^5$ W/cm^2. This value is easily obtained and for larger specific powers, a plasma formation is observed above the sample. The interaction is then very different, since plasma is generally opaque for laser radiation [121].

Note that for intensities higher than 10^9 W/cm^2 and pulse durations smaller than 10^{-7} s, the plasma induced by irradiation grows above the target and becomes absorbing. It has been shown [122] that the plasma formation may be described by models proposed for atomic explosions. A shock wave is produced (compression wave), and induces a large deformation in the target (fig. 12). In alloys with an appropriate composition, a localized martensitic transformation has been observed (fig. 13). In fact, this transformation is not due to the initial compressive wave, but to the expansive wave reflected by the opposite face of the sample.

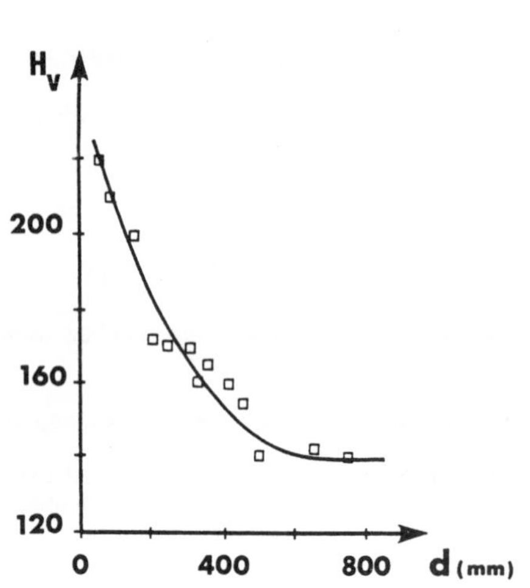

Fig. 12 shock hardening in a low carbon steel.

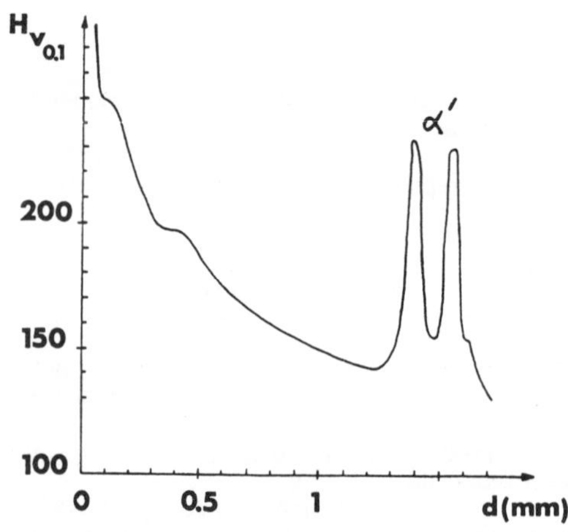

Fig. 13 Vickers microhardness profile in a laser irradiated Fe-30 at.% Ni.

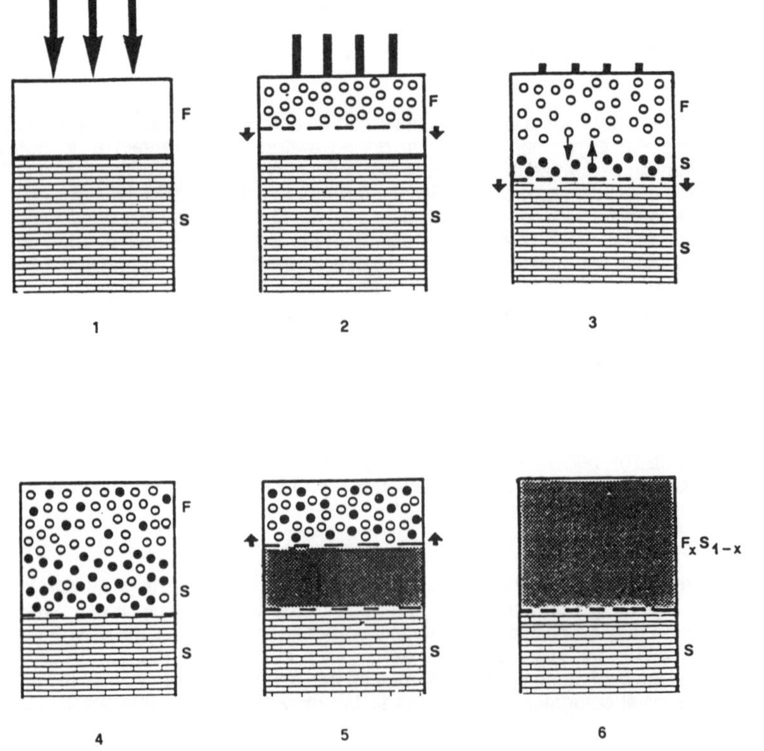

Fig. 14 sequential cross-section schematic illustration of mechanism of laser surface alloying from arrival of laser pulse to complete resolidification.

6 APPLICATIONS OF LASER BEAMS TO SURFACE TREATMENT

Surface treatment by laser was first carried out for annealing silicon; one reason was the perceived technological importance of laser processing of silicon. Furthermore many of the physical properties of this semiconductor, such as reflectivity or electrical conductivity, are ideally suited for time-resolved measurements [123]. In metallic materials, many applications were then developed. Laser induced transformations may be classified in different ways; we will use the following classification :

- treatments without changes of the chemical composition :

- in a pure metal it has been shown that some allotropic transformations occurring in Mn could be suppressed either by rapid heating or by rapid solidification from the melt [124].

- in alloys, hardening by laser annealing has been extensively studied in steel and cast iron, since technological applications are evident [125] . In this case, rapid cooling induces a martensitic transformation and consequently a large improvement of both mechanical and tribological properties is obtained.

- treatments with near-surface composition changes [126,127]: melting is then necessary; indeed diffusion phenomena require too long durations in the solid state. The main stages of the process are as follows (fig. 14):

- energy deposition in a layer of the order of the absorption depth d, which spreads by heat conduction over a layer of thickness l_T. In appropriate conditions this layer melts.

- during the melt-in period, the solid-liquid interface progresses towards the bulk. The lifetime of the melt is larger than the pulse duration, due to the large temperature gradient within the sample and to the thermal diffusion phenomena.

- when the maximum melt depth is reached, the crystal-melt interface reverses direction and solidification progresses towards the surface. It is remarkable to notice that at the inversion stage, solidification velocity v_S is theoretically zero; a progressive increase of v_S occurs during the progressive disparition of the molten layer. In experiments performed with continuous lasers, the solidification velocity cannot exceed the scanning speed of the sample under the laser beam.

In the present work, applications are concerned only with surface treatments involving a composition change.

6- 1 Different lasers used for experiments

Schematically the lasers used fall into two categories :

- short pulse lasers. Pulse duration is either on the order of a few ns (for example Q-switched YAG-Nd laser or excimer ones) or on the order of a few tenths of ps (for example mode-locked Q-switched lasers). Differents goals are investigated :

- extension of the solubility.

- formation of metastable phases or extension of present ranges of glass forming ability to new materials.

- continuous laser beams in a fast scanning mode. Interaction durations are then in the range 1ms - 1s. Let us mention that a pulsed laser operating in a relaxed mode (pulse duration : about 1 ms) produces the same metallurgical effects than a continous beam impinging on a moving sample with a high scanning rate (several m/s). Expected effects are then :

- formation of some metastable phase.

- but mainly obtention of fine microstructure with attractive properties.

Some examples will be presented to illustrate the potentiality of the method, rather than to give a comprehensive account. Classification is made as a function of irradiation duration.

6 - 2 Picosecond laser quenching

YAG-Nd infrared (λ = 1.06 µm) Q-switched mode locked lasers produce 30 ps pulses. Pulse duration is therefore larger than the time required for the transfer of the energy from the electrons which interact with the laser light to the lattice and then the process can be described as a thermal one [128]. It is only for pulse duration of less than 1 ps that l_T becomes smaller than a^{-1}. In this regime the thickness ot the initial molten layer and the thermal gradients become independent of the pulse duration. For this reason femtosecond pulses which are attractive for time-resolved analysis of interactions do not produce higher quenching rates than ps pulses. The order of magnitude can be estimated from dimensional arguments; thermal diffusion length l_T is given by :

$$l_T = \sqrt{2\ D_{th}\ \tau} \qquad (44)$$

for τ = 30 ps and D_{th} = 0.5 cm^2/s (for a metal) , we obtain l_T = 50 nm. For a typical laser fluence of 0.5 J/cm^2 and a metallic reflectivity (R = 90%), the absorbed energy leads in fact to melting to a total depth on the order of 100 nm. Heat flow calculations and direct observations of melting (in silicon) show that the total lifetime of the melt is 1 ns, for a 30 ps pulse [129]. Since temperature increase is on the order of 10^3 K, the average cooling rate during solidification can be estimated to about 10^{12} K/s, with a thermal gradient of about 10^{10} K/m. A simple linear approximation gives the isotherm velocity u_T :

$$u_T = \frac{dT}{dt} / \frac{dT}{dz} \approx 100 \text{ m/s} \qquad (45)$$

These values are gathered in table III, and compared with estimated values reached in melt spinning experiments :

		Laser Quenching	Melt Spinning
Melt temperature	T_m (K)	10^3	10^3
Melt thickness	d (m)	10^{-7}	5×10^{-5}
Temperature gradient	∇T (K/m)	10^{10}	2×10^7
Cooling rate	$\dot{T}$ (K/s)	10^{12}	4×10^6
Melt lifetime	τ (s)	10^{-9}	not applicable
Isotherm velocity	u_T (m/s)	100	0.2
Heat-flow limited crystal growth velocity	u_h (m/s)	230	0.5

table III thermal parameters in melt quenching (from ref. [128])

The solidification mechanism depends on the ratio between the isotherm velocity u_T and the solidification speed, i.e. the melt-crystal interface velocity u_C. This value is a function of the interface kinetics and of the thermodynamics driving force ΔG [128,130] :

$$u_C = u_0 \left(1 - \exp\left(\frac{-\Delta G}{kT}\right)\right) \qquad (46)$$

where ΔG is the free energy difference between crystal and melt; u_0 depends on the nature of atomic rearrangements occurring during the liquid/solid transformation. Two cases can be distinguished :

- for pure metals simple collisions of atoms in the liquid state with the crystal surface are required. Atomic jump frequency is then simply equal to the vibration frequency in the liquid and, therefore, u_C approaches the sound velocity u_S in the liquid (several thousands of m/s). Crystallisation is then partitionless.

- for alloys, crystal formation requires atomic jumps by a diffusion mechanism in order to change the nature of nearest neighbours. In such conditions u_0 is then nearly equal to D/λ, where D is the diffusivity in the liquid and λ the jump distance (i.e. interatomic distance). D depends on temperature and can be approximated by a Fülcher-Vogel law. This required diffusion slows down the crystal-melt interface velocity. To calculate or at least to estimate the driving force ΔG, different parameters are needed : concentrations of each element, interaction energy and free energy difference between liquid and crystal for the various components. Then the melt-crystal interface velocity may be deduced and compared with u_T. For metallic glass formation, the isotherm velocity must exceed u_C. Spaepen et al give an example for Fe-B alloys irradiated with a pulsed laser (t_p = 30 ps); isotherm velocity is on the order of several hundreds of m/s. Under these conditions the time to solidify a monolayer (t= l/u) is less than 1 ms and the corresponding diffusion length is less than an interatomic distance. Consequently, only a partitionless crystallisation can occur, without any composition change. A crystal is then obtained only for a pure metal and for a dilute alloy.

From these considerations Spaepen et al [130] have calculated the crystal growth speed (in m/s) as a function of boron concentration (fig. 15). A transition occurs near 5at.% B : for x > 5at.%B, amorphization is possible, since these velocities are on the order of the isotherm velocity during ps experiments. For x < 5 at.%B amorphization cannot be achieved, even with these very short laser pulses.

Two additional remarks may be made :

- a singularity is observed at 13 at.% B; this value is considered as the threshold composition for glass formation using melt spinning. This model enables an explanation of this critical value.

- only homogeneous nucleation is taken into acccunt; therefore crystal growth velocity is probably underestimated.

In these experiments a new problem appears, concerning the sample preparation. Since diffusion lengths are very short (only a few nm), due to the very short lifetime of the melt, the components must be mixed in the starting sample on a finer scale than this length, in order to ensure the homogeneity of the melt before solidification. For an initial solid solution it is easy to obtain; otherwise multilayers have to be prepared. This solution was retained by Lin and Spaepen [131] by depositing alternative layers of Ni and Nb, about 1 nm thick, to form a 100 nm compositionally modulated layer. This film is deposited on a conducting sample (Al+Cu) (fig. 16). Laser pulse with a duration of 30 ps enables the formation of an homogeneous Ni-Nb alloy, the composition of which depends on the thicknesses of the initial Ni and Nb layers. These authors have shown that laser quenching extends the glass-forming range considerably beyond that of conventional melt-quenchinq techniques. By these methods, including solid state diffusion, amorphous alloys are obtained only for nickel concentrations x between 40 and 79%; by using laser quenching, this range is extended : 23 < x < 82% Ni. For smaller values (0 < x < 18%), homogeneous supersaturated bcc solid solutions are formed; in the range 89 < x < 100%, homogeneous solid solutions are fcc and contain many twins and stacking faults.

The existence of lattice defects is a general feature for metals and metallic alloys treated by laser with short pulses. This is in opposition with observations performed in silicon, where laser annealing is used to remove irradiation defects and to obtain "defectless materials"[132]. This difference can be explained in the following way : with the same pulse duration (25 ns for example), and the same fluence, thermal gradients are much higher in a metal (10^{10} K/m) than in a semiconductor (10^8K/m), due to the difference in :

- absorption length : 0.02 µm for a metal, 1µm for Si,
- thermal conductivity : 0.2 against 2 W.cm^2.K

Furthermore,as thermal expansion is much larger in metals than in semiconductors (25 against 2 10^{-6} /K), with a smaller yield strength, formation of defects during solidification is very easy in metals, but very difficult in semiconductors.

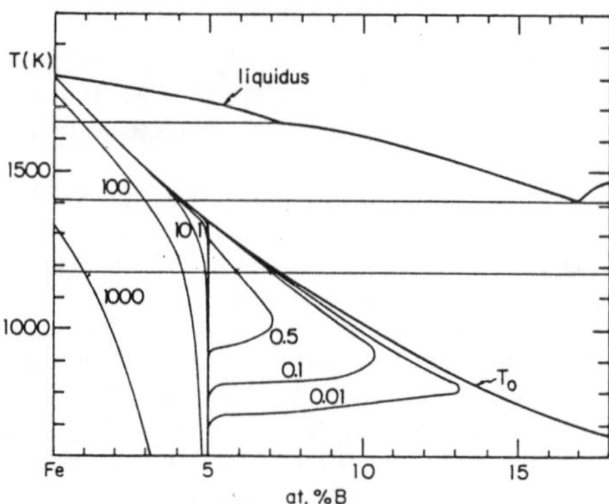

Fig. 15 the Fe-B phase diagram; the T_0 line corresponds to the liquid-bcc transition. The contours represent the crystal growth speed (in m/s) as a function of composition and interface temperature (from ref. 130).

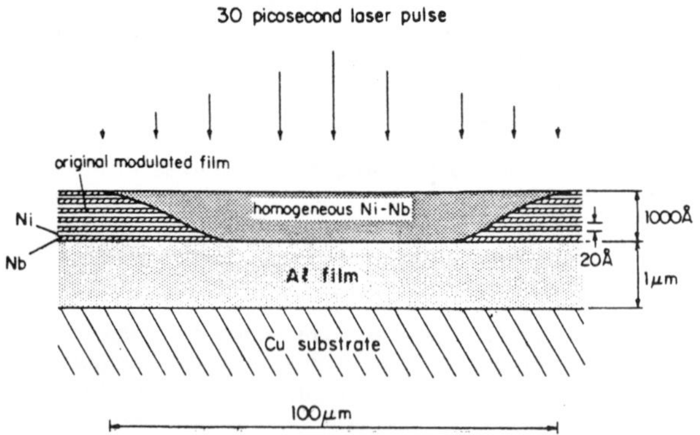

Fig. 16 schematic diagram, drawn on a variable scale, of a multilayer film after irradiation with a 30 ps pulse with a Gaussian intensity profile (from ref. 133).

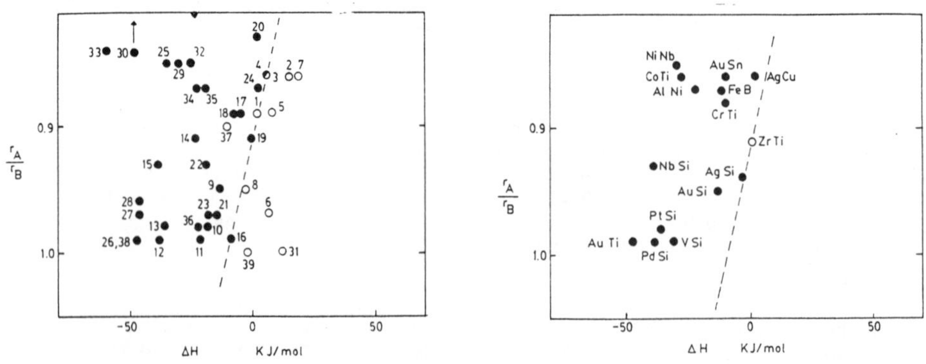

Fig. 17 glass formation map; a : ion beam mixing; b: laser quenching; full circles represent alloys forming amorphous phases; whereas open circles are for alloys not forming amorphous phases (see ref. 141 for more details).

6 - 3 Nanosecond laser quenching

The same kind of experiments performed with ps lasers are performed with ns laser, but the depth of the treated zone is larger (l_T is proportional to t_p). Two cases will be emphasized :
- obtaining glassy metals from metallic multilayers
- laser surface alloying by melting of a predeposited layer on a substrate.

6 - 3 - 1 Glassy metals

Au and Si are well known to form a metallic glass by splat cooling at composition near the eutectic (17 at% Si). In fact the Au-Si glass was the first to be obtained. By nanosecond laser treatment, Von Allmen [133] extended the range of composition that can be obtained in the amorphous state : between 9 and 91 at.%Si. Other authors made many attempts in various systems, especially prepared in the form of thin films (with a thickness of about 30µm) and by using short pulsed lasers. Let us mention for example : Fuxi et al (Al-Si)[134], Blatter et al (Au-Ti, Cr-Ti, Ag-Cr)[135], Kaufmann et al (Ta-Tr)[136], Woychik et al (Cu-Ti)[137], Alden et al (Cr-Ta, Zr-Ge)[138], Yatsuda et al (Ag-Cu-Ge)[139] and Mazzoldi et al (Al)[140]. Experimental results can be summarized as follows :

- owing to the short laser-material interaction time, and therefore to the high values of both thermal gradient and cooling rate, most of these materials can be obtained in the amorphous state, in the near surface region of the sample. Due to the experimental process, the thickness of the transformed zone is very limited, typically 100-150 nm.
- in some cases (Au_{75}-Ti_{25} or Cu_{50}-Ti_{50}), a careful observation by transmission electron microscopy reveals the existence of fine crystallites. Phases which are formed are metastable phases and not equilibrium ones. For example in Cu-Ti a fcc phase is observed, while the equilibrium phase detected during crystallization of a thin film with the same composition is the tetragonal Cu-Ti phase, with a c/a ratio very different from 1 [137].
- no glass, or only a few ones, can be formed in materials of pure ionic, metallic or covalent bonds. It can be formed only in materials in which the ionic or metallic bond is mixed with the covalent bond [134].
- glass forming ability of given materials is improved by depositing this material on a substrate with a high thermal conductivity. For example, a tungsten substrate is better than a saphir substrate in the case of Au-Ti or Cr-Ti films [135].
- some results have to be carefully analyzed. For example, the existence of a glassy surface layer on pure aluminium irradiated by a ns pulsed laser [149] is doubtful. It could be perhaps only a thin oxide layer. Due to the very small thickness of the studied film, experiments must be performed very carefully.

Several rules exist in the literature which correlate the glass forming ability with different properties of the components. Alonso and Lopez [141] have constructed glass formation diagrams from results obtained either by laser quenching or by ion beam mixing. Two parameters are used:
- atomic radius ratio : r_A/r_B
- heat of compound formation ΔH at the equiatomic composition obtained from the Miedema theory.

As shown in fig. 17, the thermochemical coordinates are successful in differentiating between glassy and crystalline alloys obtained by the fast quenching techniques.

6 -3 - 2 Laser surface alloying

Let us describe a representative experiment. A thin aluminium film (e = 100 nm) was first deposited under vacuum on a stainless steel (<0.02%C, 18%Cr, 12%Ni)[142]. This target was irradiated by a Gaussian beam (TEM_{00}), with an effective radius r^+ = 1.5 mm. Pulse duration was about 15 ns, fluence was about 5 J/cm^2, leading to an irradiance I_M on the order of 3.10^{8}W/cm^2. A melt pool is induced, which reproduces the shape of the incident beam. Consequently the melt depth was not uniform and its value in the beam axis was about 100 nm. Heat flow calculations show that the melt lifetime was about 100 ns. After solidification, the aluminium concentration profile was measured by SIMS (fig. 18). Curve 1 corresponds to the untreated material: no aluminium is detected in steel; curves 2, 3 and 4 reveal aluminium penetration into the substrate in different areas of the sample, i.e. for various fluences. From these results, we can deduce that :

- a partial vaporization of the aluminium layer has occurred (about 50% of the initial mass), in agreement with the assumptions and predictions reported above (5-3). We have shown that vaporization occurs as soon as irradiance exceeds 10^{5}W/cm^2, and in the reported work a value of 10^{8}W/cm^2 was used.

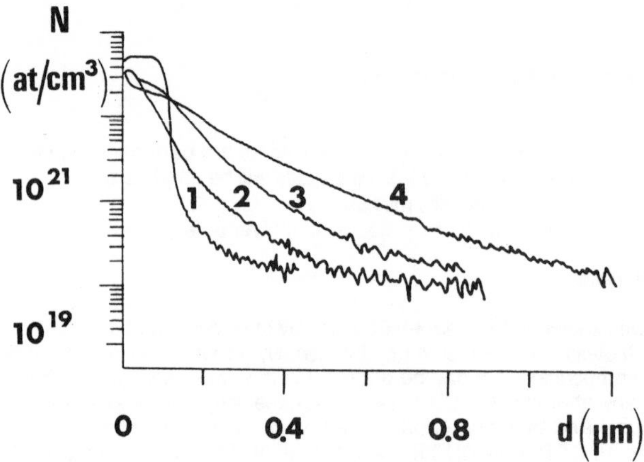

Fig. 18 SIMS aluminium concentration profiles.

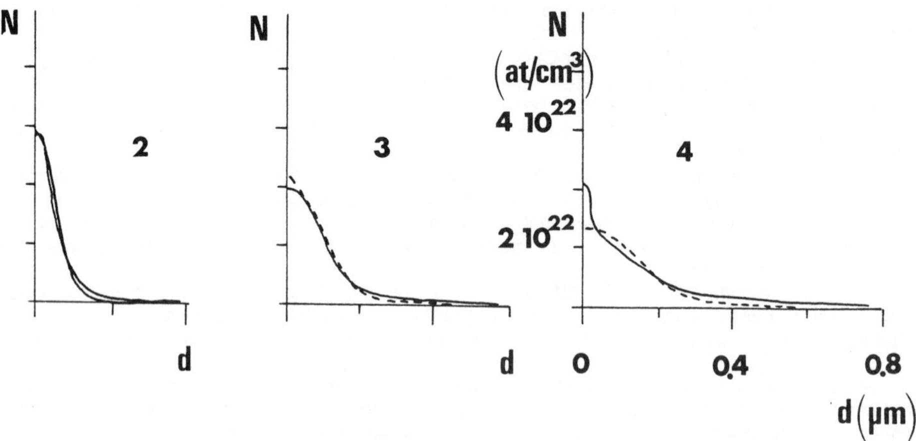

Fig. 19 fitting of the aluminium concentration profiles with diffusion laws in the liquid state.

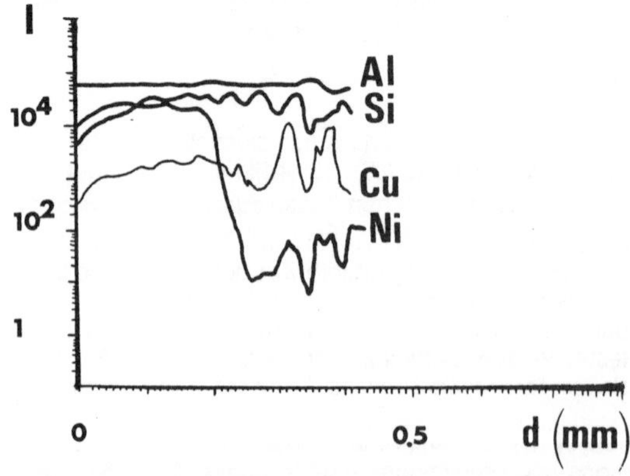

Fig. 20 various element profiles, determined by SIMS experiments on a cross-section of the sample.

- the penetration of aluminium changes in a continous way. Results may be fitted with equations which govern diffusion phenomena [142]:

$$c(z) = \frac{c_0}{2}\left(\operatorname{erf}\left(\frac{h - z}{2\sqrt{Dt_l}}\right) + \operatorname{erf}\left(\frac{h + z}{2\sqrt{Dt_l}}\right)\right) \qquad (47)$$

with : c(z): aluminium content at a depth z
c_0 : initial concentration
h : estimated value of the non-vaporized layer (45 nm)
t_l : lifetime of the melt pool
D : diffusion coefficient of aluminium in liquid iron

As shown in fig.19, good fits of the experimental curves are obtained with a diffusivity of 8 10^{-4} cm^2/s. This value agrees with the reported data. Consequently, for such short pulse durations, laser surface alloying is achieved by a simple diffusion mechanism. A metastable Al-Fe rich phase is identified, as predicted from rapid solidification models (in the present work v_S> 7m/s).

6 - 4 Continous beam in a scanning mode

The obvious advantage of using continous rather than pulsed laser radiation is that a continuous treated zone is obtained. This may be achieved in two ways : either the continous laser beam is directed onto a sample mounted on a moving table (often a rotating one), or the beam is scanned with the aid of moving mirrors across a fixed specimen. Scanning speeds between 1cm/s and several m/s are used, depending on the application. Incident irradiances are on the order of 10^5-10^6 W/cm2. Interaction time is then simply defined as the ratio of laser beam diameter and scanning speed. The two main categories of experiments will be described :
- surface amorphization of bulk specimens
- surface alloying : crystalline structures are expected, but with a very fine microstructure.

6 - 4 -1 Surface amorphization

Due to the large interaction time (τ > 1ms), the melt depth is also large, typically a few tenths of µm and cooling rates cannot exceed about 10^6K/s. Therefore only systems with a good glass-forming ability have been studied. Two main categories of experiments have been performed :

- surface vitrification of bulk materials. Attempts were successful in different systems : Pd-Si-Cu [143], Ni-Nb [144] and Fe-Cr-P-C-Si [143]. However, experiments to vitrify a large surface by overlapping of laser traces encounter two problems:
  - crystallisation of the amorphous phase in the heat affected zone of the next laser trace
  - cracking of the surface alloy : in these materials hard and brittle phases are often present in the bulk substrate (borides, sulfides..), and cracks form during rapid solidification of the near surface region..
- laser surface treatment of specimens prepared by bonding iron-base alloy ribbons to the surface of a crystalline substrate, pure nickel metal or nickel-plated mild steel [145]. The entire surface were vitrified if the metalloïd content were adequally controlled by this special technique of multilayers. However making use of this process is difficult and long.

6 - 4 - 2 Laser surface alloying (crystalline alloys)

In this case, the procedure is derived from the multilayer method : one or several metallic layers are first deposited on the specimen surface by electrolytical or chemical deposition or plasma spraying. Then the surface melting is achieved by laser treatment. In adequate conditions melting occurs both in the coating and in a fraction of the substrate. After solidification a surface alloy is formed. The treated depth concerns some tenths or hundreds of µm. As the solidification rate is smaller or equal to the scanning speed (on the order of 1 cm/s) the classical theory of constitutional supercooling can be used as an adequate guideline in determining the growth conditions that result in unstable interfaces. The absolute stability criterion is never operating. Diffusion can occur and crystallization results in the formation of various phases. Different questions are open:
- equilibrium or non-equilibrium phases?
- operating diffusion mechanism?
- type of microstructure and influence on surface properties?

a - Phases formed during cooling

Let us first mention the example of a surface alloy obtained by melting a nickel coating on an Al-Si cast alloy (Al-5at% Si)[146,147]. With an irradiance of 10^5W/cm^2 and an interaction time of about 100 ms, the melt depth is on the order of 250 μm. As the initial nickel layer thickness is 25μm, the nickel concentration within the surface alloy may be estimated to about 10%. This value agrees with microanalysis results (147). The intermetallic compound Al_3Ni is identified by X-ray diffraction experiments; for higher nickel contents, both Al_3Ni and Al_3Ni_2 are observed; these phases are both predicted by the equilibrium phase diagram. Furthermore laser melting does not induce an extension of solid solubility of nickel in aluminium. These results are at variance with those reported in the same system, but with different experimental conditions [146] : in that case rapid solidification (v_S = 7m/s) suppressed the formation of the equilibrium phase and only one metastable compound was observed (AlNi). Formation of Al-rich phases with a more complex crystalline structure was hindered. With even more drastic quenching rates, achieved by ion implantation of Ni in Al [148], the formation of an amorphous phase was reported.

This example clearly evidences the major influence of the solidification rate on the nature of the phases which are observed after melting metallic multilayers.

b - Diffusion mechanism

We have shown above that for short laser pulses only conventional diffusion mechanisms have to be taken into account to fit solute profiles. Results are different for experiments using continous beams in a scanning mode. Fig. 20 shows the concentration profile of Ni in Al-Si after irradiation with the parameters given above (10^5W/cm^2, 100ms). Analysis has been carried out by SIMS on a cross-section of the sample. We observe that the nickel content is nearly constant in the whole near-surface layer; this result is confirmed, in an indirect way, by a constant hardness value in this surface alloy [146]. A fit of the experimental Ni profile cannot be achieved by considering only the diffusion equation; convection movements have to be taken into account.

Different workers [149,150] have shown that the laser surface alloying ability may be characterized by a dimensionless parameter S given by :

$$S = \frac{M_r}{N_u\, R_e} = \frac{\sigma'\, I\, \tau}{\mu\, k} \qquad (48)$$

with M_r : Marangoni number
N_u : Nusselt number
R_e : Reynolds number
and I : irradiance
τ : interaction time
σ' = ds/dT; σ : surface tension
μ : dynamic viscosity of the melt pool
k : thermal conductivity

The convective motions may be considered as due to a coupling of both natural (buoyant) and Marangoni (surface tension driven convection) phenomena. Their origin is probably the occurrence of small surface perturbations, which may be induced either by laser beam heterogeneities (for example for a TEM_{00}, energy is not spatially constant), or to chemical surface heterogeneities (oxydes, segregations...). This convection phenomenon increases with S and contributes to the homogeneization of the melt pool and therefore to the final homogeneity of the surface alloy.

c - Microstructure and influence on the properties

As mentioned previously, solidification speed in a treatment involving a continous laser depends on the depth : it starts from zero at the bottom of the melt layer and reaches a maximum value at the top. This extremum cannot exceed the scanning speed of the specimen under the fixed beam. Due to the small value of this speed, typically 1cm/s, the solidification rate is then around the constitutional supercooling range [151]. So different microstructures can be observed :

- a planar interface (no segregations) at the interface with the substrate
- cellular or dendritic solidification (involving segregation) in the bulk of the surface alloy.

Occurrence of both kinds of microstructure has been reported [152]; a nickel coating was deposited on a mild steel before laser treatement.

However, quenching rate is fairly high compared to values induced by conventional metallurgical methods (for example 10^4°C/s). Consequently interdendritic distances are very low (a few µm) and a large enhancement of superficial properties is induced, especially for mechanical and electrochemical properties. In addition the modification of the surface composition is a favourable factor. Let us for example mention two works :

- in (Al-Si) + Ni, Young's modulus is only 72 GPa for the substrate, while a value of 130 Gpa is achieved for the surface alloy [146]. Hardness is also greatly improved : up to 600 $Hv_{0.1}$ against only about 95$Hv_{0.1}$
- surface alloying of low-carbon steel with either boron or aluminium can drastically inhibit the iron oxydation at high temperature [150].

These examples of the preparation of surface alloys by the melting of predeposited layers appear like borderline cases of metallic multilayers, since in these experiments a single layer is used. However it is obvious that when several elements have to be added, both nickel and chromium for example, a multilayer coating is achieved before laser irradiation.

6 - 5 Laser-assisted treatments : thermochemical and photochemical effects

Laser-assisted processes are currently being investigated for deposition or etching of thin-film of electronic materials and for various applications (welding, machining...) in the field of advanced materials, laser irradiation can induce locally high surface temperature and therefore assist a chemical reaction, either by photodecomposition or by photoexcitation. Until recently, this last effect was mainly used for metals, especially for laser assisted chemical vapor deposition (LCVD)[153-156]. Depending on the nature of the ambient reactive gas, deposition of the material can occur in the laser spot area and thus only a local heating of the sample is performed, while in conventional technique the whole specimen is maintained at high temperature. Otherwise the phenomenon is basically the same.

For example the LCVD process has been used for cladding titanium and titanium alloys (TA6V) with ceramics, in order to increase the surface hardness [154]. Deposition of carbides and borides from BCl_3-CH_4-H_2 has been studied. The LCVD rates are 1 to 3 orders of magnitude higher than conventional CVD rates. For example, the deposition rates varies from 8 µm/s to 50 µm/s in the (1000-1300°C) range, compared to (0.01-1 µm/s). The apparent activation energy is unchanged, but LCVD microstructures are smaller than those observed with conventional CVD. Consequently, hardness and wear resistance improvements are observed.

Fig. 24 shows the evolution of a Cr_2O_3 layer as a function of the number of laser pulses [157]. Growth is in agreement with oxidation laws, but the kinetics is accelerated.

When a photochemical effect is involved, the laser irradiation wavelength is a major parameter. This value must be adapted to the nature of the gas in order to induce the desired photochemical effects. Short wavelengths, especially those delivered by UV lasers (excimer lasers for example) are very suitable for such experiments.

7 - CONCLUSION

To summarize this short and non-exhaustive review of the potentiality of laser irradiation for surface treatment, several points are emerging that would permit developments of guidelines for alloy design and process control to achieve the desired microstructures and properties :

- treated depth is at least equal to the absorption length, i.e. between a few nm and a few tenths of nm for metals,

- with short pulses ($<10^{-8}$s), surface alloying is mainly governed by classical diffusion laws,

-for longer pulse durations, convection phenomena occur and have a large effect on solute distribution,

- various kinds of microstructures can be achieved :

- in metallic alloys with an appropriate composition irradiated with short laser pulses which induce high cooling rates (10^6 - 10^{12}K/s), metastable phases are observed, especially amorphous ones. Only limited thicknesses are concerned;

- with continous lasers in a scanning mode, solidification speeds are generally small (a few cm/s). Fine microstructures are produced which lead to a large improvement of mechanical and physicochemical properties of surfaces.

The development of lasers emitting in the UV range offers attractive possibilities, especially for laser assisted processes (LCVD for example). In this case laser irradiation is only an element of the treatment which produces a photodecomposition.

REFERENCES

1) Schou J. : in structure-property relationships in surface - modified ceramics, Eds C.J. Mc Hargue, R. Kossowsky and W.O. Hofer, NATO ASI Series E : Appl. Sci., Kluwer Academic Pub., London, 1989, vol. 170, p. 61.

2) Lindhard J., Nielsen V., Scharff M. and Thomsen P.V. : Mat. Fys. Medd. Dan. Vid. Selsk., 1963, 33, N° 10.

3) Sigmund P. : in Radation damage processes in materials, Ed. C.H.S. Dupuy, NATO-ASI Series, Noordhoff, Leyden, 1975, p. 3.

4) Lindhard J. : Mat. Fys. Medd. Dan. Vid. Selsk., 1954, 28, N° 8.

5) Lindhard J., Nielsen V. and Scharff M. : Mat. Fys. Medd. Dan. Vid. Selsk., 1968, 36, N° 10.

6) Lindhard J. and Scharff M. : Phys. Rev., 1961, 124, 128.

7) Andersen H.H. and Ziegler J.F. : Hydrogen stopping powers and ranges in all elements, Pergamon, New York, 1977.

8) Ziegler J.F. : Helium stopping powers and ranges in all elemental matter, Pergamon, New York, 1977.

9) Ziegler J.F. : Handbook of stopping cross section for energetic ions in all elements, Pergamon, New York, 1980.

10) Ziegler J.F., Biersack J.P. and Littmark U. : The stopping and range of ions in solids, Pergamon, New York, 1985.

11) Fano U. : Ann. Rev. Nucl. Sci., 1963, 13, 1.

12) Ahlen S.P. : Rev. Mod. Phys., 1980, 52, 121.

13) Sternheimer R.M. and Peierls R.F. : Phys. Rev., 1971, B3, 3681.

14) Brandt W. and Kitagawa M. : Phys. Rev., 1982, B25, 5631.

15) Wilson W.D., Haggmark L.G., Biersack J.P. : Phys. Rev., 1977, B15, 2458.

16) Biersack J.P. and Ziegler J.F. : Nucl. Instr. Meth., 1982, 194, 93.

17) Biersack J.P. and Ziegler J.F. : in Springer Series in Electrophysics, vol. 10, Springer, Berlin, 1982, P. 122.

18) Gombas P. : "Die statistische Theorie das Atoms", Springer-Verlag, Wien, 1949.

19) Diaz de la Rubia T., Averback R.S. and Horngming Hsieh : J. Mater. Res., 1989, 4, 579.

20) Beeler J.R. : Phys. Rev., 1966, 150, 470.

21) Robinson M.T. and Torrens I.M. : Phys. Rev., 1974, B9, 5008.

22) Vineyard G.H. : Rad. Eff., 1976, 29, 245.

23) Winterbon K.B., Sigmund P. and Sanders J.B. : Kgl Danske Vid. Selsk. Mat. Fys. Medd., 1970, 37, N° 14.

24) Brice D. : Ion implantation range and energy deposition distributions, vol. 1 - high energy, Plenum Press, New York, 1975.

25) Winterbon K.B. : Ion implantation range and energy deposition distributions, vol. 2 - low energy, Plenum Press, New York, 1975.

26) Coulter C.A. and Parkin D.M. : J. Nucl. Mater., 1980, 88, 249.

27) Parkin D.M. and Coulter C.A. : J. Nucl. Mater, 1979 85/86, 611.

28) Kinchin G.H. and Pease R.S. : Report Prog. Phys., 1955, 18, 1.

29) Sigmund P. : Rad. Eff., 1969, 1, 15.

30) Norgett M.H., Robinson M.T. and Torrens I.M. : I.M. Nucl. Eng. Design, 1975, 33, 50 and 1975, 33, 91.

31) Davies J.A. and Howe L.M. : in Site characterization and aggregation of implanted atoms in materials, Eds. Perez A. and Coussement R., NATO-ASI series B : Physics, Plenum, New York, 1980, vol. 47, p. 7.

32) Biersack J.P. : in Ion beam modification of insulators, Eds. Mazzoldi P. and Arnold G.W., Elsevier, Amsterdam, 1987, Vol. 2, p.1.

33) Sigmund P. : Appl. Phys. Lett., 1974, 25, 169.

34) Chapman G.E., Farmery B.W. and Thompson M.W : Rad. Eff., 1972 13, 121.

35) Andersen H.H. and Bay H. : Rad. Eff., 1973, 19, 139.

36) Thompson and Johar : Appl. Phys. Lett., 1979, 34, 342.

37) ICRU Stopping powers for electrons and positrons, Report 37, ICRU-publications, 7910 Woodmont Avenue, Bethesda, MD-20814 USA, 1984.

38) Twaites D.I. : Nucl. Inst. Meth., 1987, B27, 293.

39) Ziegler J.F; Chu W.K. and Feng J.S.Y. : Appl. Phys. Lett., 1975, 27, 387.

40) Sabin J.R. and Oddershede J. : Nucl. Instr. and Meth., 1987, B27, 280.

41) Powers D., Chu W.K., Robinson R.J. and Lodhi A.S. : Phys. Rev., 1972, A6, 1425.

42) Biersack J.P. : Nucl. Instr. and Meth., 1981, 182/183, 199.

43) Manning I. and Mueller G.P. : Comp. Phys. Comm., 1974, 7, 85.

44) Sonder E. and Sibley W.A. : in Point defects in solids, vol. 1, General and ionic crystals, Eds. J.H. Crawford and Slifkin L.M., Plenum, New York, 1972, p. 201.

45) Catlow C.R.A. : Structure property relationships in surface - Modified ceramics, Eds. Mc Hargue C.J., Kossowsky R. and Hofer W.O., NATO-ASI Series E : Appl. Sci., Kluwer Academic Pub., London, 1989, vol. 170, p.1.

46) Itoh C., Tanimura K. and Itoh N. : proceedings of the 5th Int. Conf. on Radiation Effects in Insulators, Hamilton, Canada, 19-23 June 1989, Nucl. Instr. and Meth., to be published.

47) Fain J., Monnin M. and Montret M. : Rad. Res., 1974, 57, 379.

48) Perez A., Davenas J. and Dupuy C.H.S. Nucl. Instr. and Meth., 1976, 132, 219.

49) Perez A., Balanzat E. and Dural J. : Phys. Rev. B., 1989, to be published.

50) Dran J.C., Langevin Y. and Petit J.C. : Nucl. Instr. and Meth. in Phys. Res., 1984, B1, 557.

51) Ion Beam Modification of Insulators, Eds. Mazzoldi P. and Arnold G.W. : Elsevier, Amsterdam, 1987, vol.. 2.

52) Averbach R.S., Benedek R. and Merkle K.L. : Phys. Rev., 1978, B18, 4156.

53) Jung P. : J. Nucl. Mater., 1983, 117, 70.

54) Guinan M.W. and Kinney J.H. : J. Nucl. Mater., 1981, 103/104, 1319.

55) Gibson J.B., Goland A.N., Milgram M. and Vineyard G.H. : Phys. Rev., 1960, 120, 1229.

56) King W.E. and Benedek R. : J. Nucl. Mater., 1983, 117, 26 and Phys. Rev., 1981, B23, 6335.

57) Wei C.Y., Current M.I. and Seidman D.N. : Phil. Mag., 1981, 44, 459.

58) Pramanik D. and Seidman D.N. : Nucl. Instr. and Meth., 1983, 209/210, 453.

59) English C.A. and Jenkins M.L. : Mats. Science Forum, 1987, 15/18, 1003.

60) Kirk M.A., Robertson I.M., Jenkins M.L., English C.A., Black T.J. and Vetrano J.S. : J. Nucl. Mater., 1987, 149, 21.

61) Protasov V.I. and Chudinov V.G. : Rad. Eff., 1982, 66, 1.

62) Kapinos V.G. and Platonov P.A. : Rad. Eff., 1987, 103, 45.

63) Smalinskas K. : Master's thesis, Univ. of Illinois at Urbana, 1988.

64) Mc Hargue C.J. : Int. Metals. Rev., 1986, 31, 49.

65) Meaudre R. and Perez A. : Nucl. Instr. and Meth. in Phys. Res., 1988, B32, 75.

66) Gerard P. : in Ion beam modification of insultators, Eds. Mazzoldi P. and Arnold G.W., Elsevier, Amsterdam, 1987, vol. 2, p. 449.

67) Dearnaley G. : Rad. Eff., 1982, 63, P. 25.

68) Nicolet M.A., Banwell T.C. and Paine B.M. : Mat. Res. Soc. Symp. Proc., 1984, vol. 27.

69) Averback R.S. : Nucl. Instr. and Meth., 1986, B15, 675.

70) King W;E. and Benedek R. : J. Nucl. Mat. 1983, 117, 26 and Phys. Rev., 1981, B23, 6335.

71) Saris F.W. : in Structure-property relationships in surface modified ceramics, Eds. Mc Hargue C.J., Kossowsky R. and Hofer W.O., NATO-ASI series E : Appl. Sci., Kluwer Academic Pub., London, 1989, vol. 170, p. 103.

72) Collins R. : Proceedings of the 3rd Int. Conf. on Radiation Effects in Insulators, Rad. Eff. 3, Eds. Wilson I.H. and Webb R.P., 1986, vol. 1, p. 167.

73) Chu W.K., Mayer J.W. and Nicolet M.A. : Backscattering Spectrometry, Academic Press, New York, 1978.

74) Traverse A. : J. of the less common metals, 1988, 145, 451.

75) Kelly R. and Sanders T.B. : Surf. Sci., 1976, 57, 143.

76) Dzioba S. and Kelly R. : J. Nucl. Mat., 1978, 76, 175.

77) Christel L.A., Gibbons J.F. and Mylroie S. : Nucl. Instr. and Meth., 1981, 182/183, 187.

78) Nelson R.S. : Rad. Eff., 1969, 2, 47.

79) Sigmund P. and Gras-Marti A. : Nucl. Instr. and Meth., 1981, 182/183, 25.

80) Delafond J., Picraux S.T. and Knapp J.A. : Appl. Phys. Lett., 1981, 38, 237.

81) Picraux S.T., Follstaedt D.M. and J. Delafond : Mat. Res. Soc. Symp. Proceedings, 1982, 7, 71.

82) Littmark U. and Hofer W.O. : Nucl. Instr. and Meth., 1980, 168, 329.

83) Collins R. and Jimenez-Rodriguez J.J. : Rad Eff. Lett., 1982, 68, 19.

84) Collins R., Marsh T. and Jimenez-Rodriguez J.J. : Nucl. Instr. and Meth., 1983, 209/210, 147.

85) Andersen H.H. : Appl. Phys., 1979, 18, 131.

86) Dienes G.J. and Damask A.C. : J. Appl. Phys., 1958, 29, 1713.

87) Jaouen C., Delafond J., Junqua N. and Goudeau P. : Nucl. Instr. and Meth. in Phys. Res., 1989, B43, 34.

88) Saris F.W. : in Structure-property relationships in surface modified ceramics, Eds. Mc Hargue C.J., Kossowsky R. and W.O. Hofer : NATO-ASI series E : Appl. Sci., Kluwer Academic Pub., London, 1989, vol. 170, p. 103.

89) Cheng Y.T., Van Rossum M., Nicolet M.A. and Johnson W.L. : Appl. Phys. Lett., 1984, 45, 185.

90) Workman T.W., Cheng Y.T., Johnson W.L. and Nicolet M.A. : Appl. Phys. Lett., 1987, 50, 1485.

91) Jaouen C., Riviere J.P. and Delafond J. : Nucl. Instr. and Meth. in Phys. Res., 1987, B19/20, 549.

92) Johnson W.L., Cheng Y.T., Van-Rossum M. and Nicolet M.A. : Nucl. Instr. and Meth., 1985, B718, 657.

93) Jaouen C. : Thesis , Univ. Poitiers, France, 1987.

94) Kaufman L. and Bernstein H. : in Computer calculations of phase diagram, Academic Press, New York, 1970.

95) Collins L.E., Perkins J.G. and Stroud P.T. : Thin Solid Films, 1969, 4, 41.

96) Mc Hargue C.J. : in Structure-property relationships in surface-modified ceramics, Eds. Mc Hargue C.J., Kossowsky R. and Hofer W.O. ; NATO-ASI series E : Appl. Sci., Kluwer Academic Pub., 1989, vol. 170, p.117.

97) Gazeki J., Sai-Halasz G.A., Elliman R.G., Kellock A., Nyberg G.L. and Williams J.S. : Appl. Surf. Sci., 1985, 22/23, 1034.

98) Battaglin G., Carbucicchio M., Dal Maschio R., Machetti F., Mazzoldi P. and Valenti A. : XIV Int.Conf. on Glass, New Delhi, India, March 1986.

99) Baglin J.E.E. and Clark G.J. : Nucl. Instr. and Meth. in Phys. Res., 1985, B7/8, 881.

100) Abonneau E., Fuchs G., Treilleux M. and Perez A. : Proc. of the Vth Int. Conf. on Radiation Effects in Insulators, Hamilton, Canada, 19-23 June 1989, to be published.

101) Pranevicius C. : Nucl. Instr. and Meth., 1981, 182/183, 251.

102) Legg K.O. : Nucl. Instr. and Meth., 1987, B24/25, 565.

103) Pimbert-Michaux S., Chabrol C., Denanot M.F. and Delafond J. : Mat. Sci. and Engin., 1989, A115, 209.

104) Jaulin M., Laplanche G., Delafond J. and Pimbert-Michaux S. : Surf. and Coat. Tech.,1989, 37, 225.

105) Guzman L. : Adv. Mat. and Man. Processes : 1988, 3(2), 279.

106) Sato T., Okata K., Asahi N., Ono Y., Oka Y. and Hashimoto I. : J. Vac. Sci. Tech. : 1986, A4(3).

107) Satou M., Fujii K., Kiuchi M. and Fusimoto F. : Nucl. Instr. and Meth., to be published.

108) Maillet H.: "Le laser, principes et techniques d'applications", Tec. et Doc., Lavoisier, Paris 1986.

109) Boulon G. : "Lasers et industries de transformation", A.B. Vannes coor.,Paris -1986 Tec. et Doc., Lavoisier, p 1.

110) Dietz J. : Thesis, Lyon, France, 1987.

111) Spalding I.S.: "Physical processes in laser-material interactions", Bertolotti edit., Plenum Press, New-York, 1983, p 1.

112) Girardeau-Montaut J.P. : cf ref. 109, p. 98.

113) Duley D.M. : "Laser treatment of metals", Drapper C.W. and Mazzoldi P. edit., NATO AISI Series, Martinus Nijhoff publishers, Dordrecht ,1986.

114) Von Allmen M. : cf ref. 111, p 49.

115) Fenk A. : Metall. Trans. , to be published.

116) Touloukian M. : TPRC Data Series, LFI / Plenum Press, New York, 1973.

117) Root R.G. : J. de Phys.,1980, C9, 41.

118) Carslaw H.S. and Jaeger J.C. : "Conduction of heat in solids", Oxford Clarendon Press, 1959

119) Laurent M. : cf ref. 109, p 146.

120) Mac Lachlan A.D. and Witbourn L.B. : J. Appl. Phys.,1982, 53, 4038.

121) Girault G., Damiani D., Aubreton J. and Catherinot A. : Appl. Phys. Letters, 1989, 43, 128.

122) Grevey D. : Thesis, Lyon, France, 1988.

123) Hart J.M. and EVANS A.G.R. : Semicond. Sci. Tech.,1988, 3, 421.

124) Peercy P.S. : cf Ref. 119, p 57.

125) Bamberger M., Boas M. and Akin O. : Z. Metalkde,1988, 79, 806.

126) Draper C.W. and Ewing C.A. : J. Mat. Sci.,1984, 19, 3815.

127) Galerie A., Pons M. and Caillet M. : Mat. Sci. Eng.,1987, 88, 127.

128) Spaepen F. : cf Ref. 113, p 79.

129) Lin C.J., Spaepen F. and Turnbull D. : J. Non Cryst. Solids, 1984, 61-62, 767.

130) Spaepen F. and Lin J.C. : Proc. MRS Congress, Editions de Physique, Paris 1984, p 65.

131) Lin J.C. and Spaepen F. : Acta Metall. ,1986, 34, 1367.

132) Peercy P.S. and Wampler W.R. : Appl. Phys. Letters,1982, 40, 768.

133) Von Allmen M. : Top Appl. Phys. ,1983, 53, 261.

134) Fuxi G., Baorong S. and Wang H. : J. Non Cryst. Solids,1983, 56, 201.

135) Blatter A. and Von Allmen M. : Phys. Rev. Letters,1985, 54, 210.

136) Kaufmann E.N., Wallace R.J.,Mahin K.W., Echer C.J., HuegelF.J. and Draper C.W. : cf Ref. 23, p 59.

137) Woychik G., Lowndes D.H. and Massalski T.B. : Acta Metall.,1985, 33, 1861.

138) Alden P.A., Massalski T.B., Lowndes D.H. and Kaufmann E.N. : Scripta Metall., 1987, 19, 67.

139) Tatsuda G. and Massalski T.B. : Proc. RQM III, Sandaï - 1981, Jap. J. Metals, 161.

140) Mazzoldi P., Della Mea G., Battaglin G., Miotello M., Servidori M., Bacci D. and Janitti E : Phys. Rev. Letters,1980, 44, 88.

141) Alonso J.A. and Lopez J.M. : Materials Letters,1986, 4, 313.

142) Juhel M. :Thesis, Grenoble, France, to be published.

143) Pergue D. : Thesis, Lyon, France , 1987.

144) Bergmann H. and Mordike B.L. : cf ref 139, 181.

145) Yoshioka H, Asami K., Kawashima A. and Hashimoto K. : RQM IV, Wurtzburg - 1985, Elsevier Science Publishers, p 123.

146) Bonnet-Jobez S. : Thesis, Lyon, France, 1989.

147) Gaffet E., Pelletier J.M. and Bonnet-Jobez S. : Acta Metall, 1989, 37, 3205.

148) Follstaedt D.M. and Romig A.M. : Proc. 20th Annual Meeting of the Microbeam Analysis Society, Louisville - 1985.

149) Chan C, Mazumder J. and Chen M.H. : Metall. Trans. A,1984, 15, 2175.

150) Pons M., Caillet M; and Galerie A : J. Mat. Sci.,1986, 21, 4101.

151) Kurz W.and Ficher K.: Fundamentals of solidification, Lavoisier ed., 1988, Paris.

152) Renaud L.,Chabaud B., Fouquet F., Mazille H.and Crollet J.L. : "Surface Engineering with high energy beams", Proc. 2nd Seminar, Lisbon, September 25-27, 1989, 305.

153) Allen S.D. : "Physical Process in laser-material interactions", Bertolloti M. edit., Nato Adv. Series B : Physics 1983, Plenum press, p 455.

154) Wehr M. and Matsuwana A. : Proc of 4th CISFEL, Cannes, France, 26-30 Sept. 1988, p 573.

155) Mewow S.K. and Jervis T.R. : Scripta Metall.,1986, 20, 1519.

156) Tonneau D.S., Pauleau Y. and Auvert G. : J. Appl. Phys., 1989, 65, 4410.

157) Metev S.M.,Savtchenko S.K.,Statenov K.V.,Veiko V.I., Kotov G.A. and Shandibina D.G. : IEEE, J. Quant. Electr., 1981, QE 17, 2004.

Materials Science Forum Vols. 59 & 60 (1990) pp. 581-616

METALLIC SUPERLATTICES: APPLICATIONS, MANUFACTURE AND CHARACTERIZATION

Ph. Houdy
Laboratoires d'Electronique Philips
3, avenue Descartes, B.P. 15, F-94451 Limeil-Brevannes, France

Abstract

This is a review of the application, the preparation and the various characterization methods of metallic layer stacks on the atomic or nanometric scale. Superconductors, semiconductors, magnetic and X-ray optical applications are considered. Sputtering, evaporation and molecular beam epitaxy fabrication processes are described with in-situ characterization methods such as ellipsometry, X-ray reflectometry and RHEED. A posteriori characterization methods such as XRD, TEM, AES, GXR and RBS are presented, XPS, EXAFS, SIMS, STM and others are also summarily presented.

1 INTRODUCTION

For a few years the capacity to deposit thin films on the atomic scale has allowed new structures to be made for different domains of application [1,2,3,4,5]. Recently for semiconductor applications, quantum wells [66] have been made consisting of alternating nanometric scale semiconducting layers. Also recently, superconductors films have been made of alternating monolayers of rare earth and metallic materials giving a superconducting effect at nitrogen temperature [6, 65]. For magnetic applications [7,8,9], metallic superlattices have been grown: ultra-thin layers show increases of magnetoresistance, of Kerr effect and strong magnetization anisotropy which is very interesting for future applications. Finally, over the last few years multilayers have been made as soft X-ray optics in the range 1 nm-100 nm. These mirrors are made of a few tens of layers alternating light and heavy materials [10,11,12,13,14].

For all these applications, three deposition methods have been particularly used : sputtering in its different forms (magnetic, RF diode, triode) evaporation and molecular beam epitaxy. For semiconductor applications, MBE is the most used method. For superconductors and magnetic superlattices, the three methods have been used and for soft X-ray applications, sputtering gives the best results. In situ growth control methods can be adapted to each method : RHEED in an MBE system allows the control of the growth monolayer by monolayer, X-ray reflectometry in the evaporation system and ellipsometry in the sputtering chamber give information in real time on the composition and the thickness of the polycrystalline or amorphous layers grown.

In all cases, to understand the properties we can expect from the stacks for all these kinds of applications, it is important to know the structures of the layers : are they both amorphous, amorphous and crystalline or both crystalline (Fig. 1).

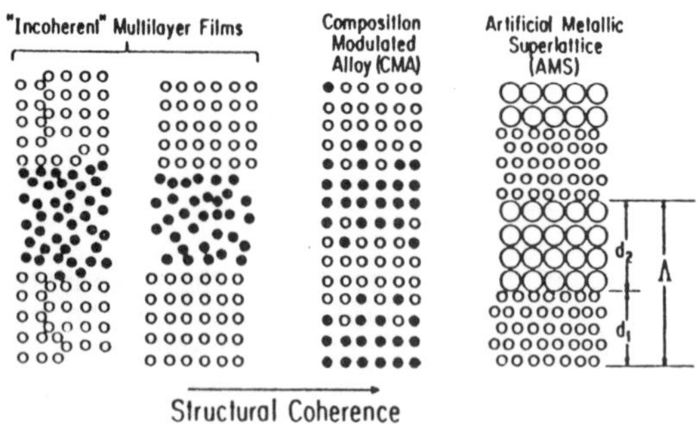

Fig. 1 Diagram illustrating the degrees of structural coherence obtained in metallic multilayers and superlattices (from ref. 1).

The most used a posteriori characterization methods are X-ray diffraction, transmission electron microscopy, Auger electron microscopy, grazing X-ray reflection and Rutherford backscattering. These methods give information on the crystalline structure of the layers and on the interfaces [15], information on the thicknesses, roughnesses and the interdiffusion. Less used up to now for metallic superlattices are XPS, EXAFS, and SIMS. These methods give chemical information on the layers and the chemical bonding at the interface. Finally STM and other methods such as neutron scattering have also been used recently by some laboratories. We will present here how all these methods, well known for bulk or surface characterization, have been used for nanometric metallic superlattice analysis.

2 APPLICATIONS

2.1 Microelectronics

For applications in microelectronics different stacks have been made : alternating two semiconductors such as GaAs/GaAlAs for nanometric ultra high speed devices (quantum well [66] : fig. 2) ; metal/metal nitride multilayers such as W/WN for diffusion barrier [72] on semiconductor materials ; metal/metal stacks such as Cu/Mo for special electrical resistivity behaviours [16,63,75].

For quantum well applications, a resonance phenomena is used, the device size (a few ten nanometers) being of the order of the electron wavelength. This resonance phenomemon increases the transistor intensity and the inversion speed (1000 times faster than a classical transistor) for a basic device size 100 times smaller than actual devices.

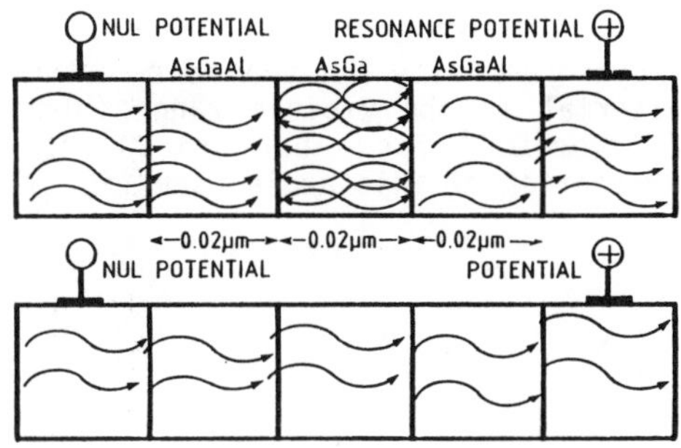

Fig. 2 Tunnel effect and quantum well (from ref. 66).

For the second type, (tungsten/tungsten nitride) alternating metal and metal nitride avoids the diffusion of the metal into the semiconductor during device annealing, conduction always being carried out by the metallic layer (the metal nitride bulk only is an insulator with respect to the metal layer).

In the third case (Cu/Mo), the variation of resistivity with the multilayer period allows the resistivity of the deposited film to be ajusted.

2.2 Magnetic superlattices

Many experiments have been done in the magnetic domain especially on bulk materials for magnetic head or recording media. For a few years, new magnetic properties have been observed using superlattices [64,73]. Indeed different authors have observed an enhancement of magnetoresistance or of the Kerr effect for magneto-optical applications [8,9] (see fig. 3) or a change of magnetization direction for ultra thin layers [7]. Cobalt has been particularly studied and it has been shown that down to 1.2 nm for one or few monolayers cobalt sandwiched in crystallized gold presents a perpendicular aimantation. Elsewhere it has been shown in cobalt/iron multilayers that the coercitivity decreases with the cobalt layer thickness due to the cobalt structure being induced by iron structure and that the magnetization is partially due to an interface effect.

As well as cobalt and its aimantation anisotropy interesting for perpendicular recording, other materials have been studied such as rare earth (Terbium) for magneto-optical interest.

Terbium iron alloys give rise to the Kerr effect. Alternating this alloy with transparent material in the visible range (such as SiO_2 or Si_3N_4) increases the Kerr angle (from 0.1° for a bulk to 10° for a multilayer) thus enabling a very easy determination of this angle (see fig. 3).

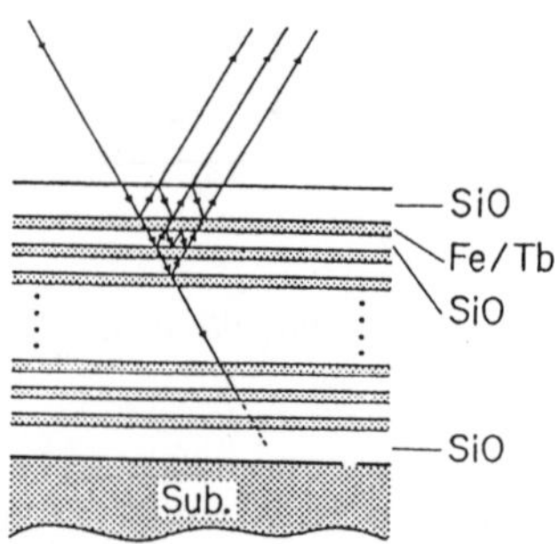

Fig. 3 Multilayers for magneto-optical applications (from ref. 9).

For all these reasons, magnetic monolayer behaviour and high density recording or magneto-optical reading, metallic superlattices are now very intensively investigated.

2.3 Superconductors

Very recently it has been shown that special crystalline structures with alternating metal (copper, barium) and rare earth (yttrium) oxides have a superconducting behaviour at nitrogen temperature [6,66].

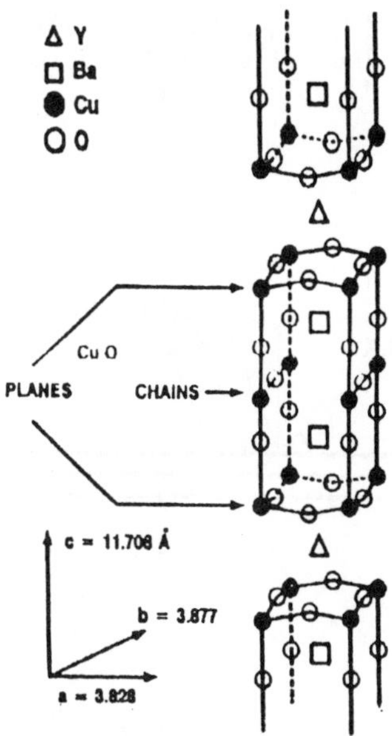

Fig. 4 Crystal structure of $YBa_2Cu_3O_{7-x}$ *(from ref. 65).*

The interests of such structures are : no electrical loss, high speed possibilities and high sensitivity to the magnetic fields.

The absence of electrical loss due to a zero resistance allows electrical transport without loss. This property can be used in microelectronics for interconnections. Another advantage is the capacity to create very intense magnetic fields. The high speed also due to zero resistance enables rapid commutation in Josephson junctions for electronic devices. High sensitivity allows the tension, magnetic field and infra-red radiation to be measured.

The greatest domains of application of these kinds of structure are in microelectronics and in electrical engineering (alternators, transformators).

2.4 Soft X-ray mirrors

For applications in the soft X-ray range (1 nm - 100 nm) superlattices have been made for some years since E. Spiller proposed to use artificial "crystals" in the X-UV range. At the beginning of the eighties E. Spiller and T. Barbee demonstrated the feasibility of such structures [10,11,12,13,14].It consists of a strictly regular alternating arrangement of heavy materials (W, Mo, Ni) and light materials (C, Si, B_4C) (fig. 5)[69, 77, 78]. These mirrors use Bragg's law $2d \sin\theta = k\lambda$. The reflectivity of such systems can be increased by a factor of 10^3 with regard to bulk materials.

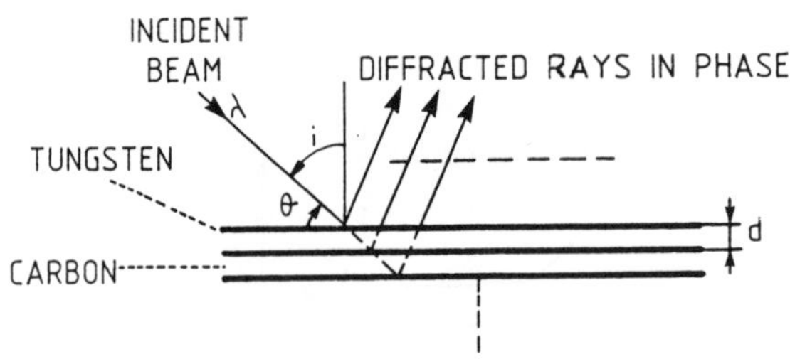

Fig. 5 Schematic diagram for Soft X-ray reflector using Bragg diffraction ($2d \sin\theta = k\lambda$).

The main applications are in the astrophysic, biology, microelectronics and X-ray instrumentation domains. The authors have obtained sun coronar micrographs in the V-UV range to understand sun behavior. In few years, a sky map in the soft X-ray range will be made. Soft X-ray mirrors will allow an X-ray microscope to study living biological cells with great resolution (50 nm) to be built. The semiconductor industry is interested in X-ray lithography to decrease the size of the integrated circuits. Industry and research are also interested in X-ray multilayers to be used as optics for different analyses (metallurgy, plasma) or in different systems such as synchrotron rings.

3 MANUFACTURE

Two kinds of methods can be used to produce metallic superlattices. One is the sputtering in various forms (magnetron, RF diodes and triode). The other is the evaporation where we will distinguish electron beam evaporation and molecular beam epitaxy.

3.1 Sputtering

Sputtering [17] consists of bombarding a material target with plasma ions (typically argon) in a 1 Pa vacuum (fig. 6).

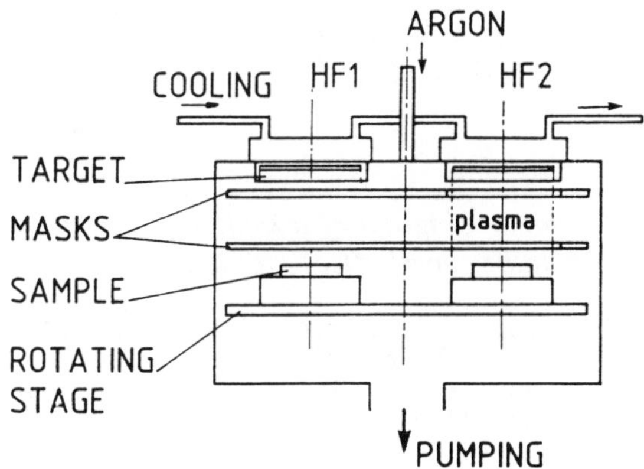

Fig. 6 Schematic view of a R.F. diode sputtering system (from ref. 70).

This process implies that there is a light bombardment of the substrate where atoms of the target material have to be deposited. In this case, target atoms reach the sample with energy in the range of 10 eV and the argon ions reach the sample typically with 50 eV energy. Thus, we observe argon inclusion in the films and it is very difficult to obtain monocrystalline films. In general, at room temperature, layers are amorphous or polycrystalline with slight argon inclusions. The various processes of sputtering give a partial solution to this problem. Magnetron sputtering uses magnets to create a magnetic field near the target which confines the plasma near the target far from the sample, decreasing the sample bombardment by the argon and the amount of argon included in the deposited film. Triode sputtering [18] (fig. 7) consists in creating the plasma between two anodes which are neither the sample nor the target. So the sample is not located in the plasma contrary to the diode system. Ion beam sputtering [19,76] also decreases argon pollution for similar reasons. For RF diode sputtering at low RF power ($1W/cm^2$), low plasma pressure (few 10^{-1} Pa) and large target-sample distance (10 cm), it is possible to obtain perfect compact material with a smooth surface on the atomic scale. Thus the deposition rates are in the range of a few 10^{-2} nm/s.

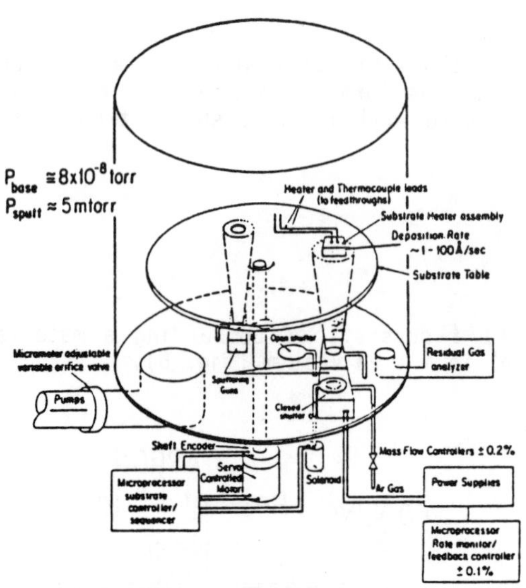

Fig. 7 Schematic diagram of a sputtering system used for depositing metallic superlattices (from ref. 1).

Owing to all the sputtering characteristics and using a sample heater in special conditions for materials as silver and palladium it is possible to produce monocrystalline layers [79]. However, sputtering is mostly used to deposit amorphous or polycrystalline layers.

3.2 Electron beam evaporation

Electron beam evaporation [20,21] consists in heating a material held in a crucible with an electron gun (fig. 8). The heated particles are then evaporated throughout the chamber and especially on the sample. The vacuum during the deposition is in the range 10^{-4} Pa. For a 20 cm crucible-sample distance and a 1A, 10 kV electron gun bias and voltage, deposition rates of 10^{-4} to 1 nm/s are achieved.

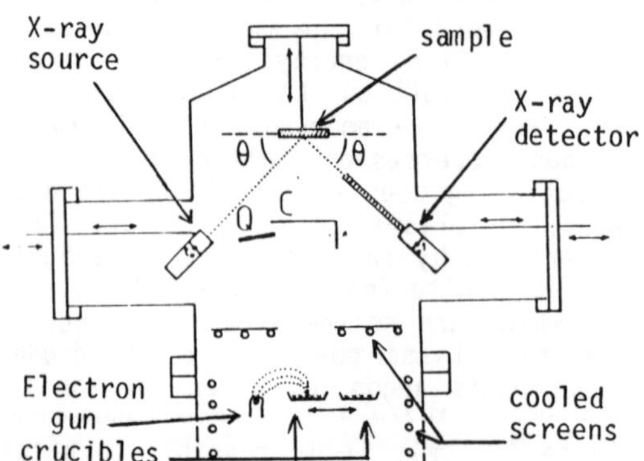

Fig. 8 Electron beam evaporator (from ref. 11, 20).

The emission cone, the difficulties to have a very accurate control on the electron beam and the standard basic vacuum of the chamber (10^{-3} Pa) are suitable for the production of true epitaxy. With regard to sputtering, the advantage is that no plasma is used, but evaporated atoms reach the sample without energy and an interface roughness increase is often observed during the growth. As for sputtering, various adaptations (sample heater, ultra-high basic vacuum) can be used to obtain epitaxial growth.

3.3 Molecular beam epitaxy

Molecular beam epitaxy [22, 23, 24] consists of thermal heating by the Joule effect of the material held in a special effusion cell designed to provide an uniform stable high temperature uniformity and stability (fig. 9). This method applied in ultra-high vacuum chamber (10^{-9} Pa) enables epitaxial growth of materials and real monolayer by monolayer growth to be achieved. For a 25 cm effusion cell-sample distance, 10A, 10V heating resistance bias and voltage and 500 °C to 1200 °C sample temperature (respectively for gallium arsenide and strontium fluoride) growth rates of a few 10^{-1} nm/s are obtained. The uniformity of the heating and the capability to have an accurate regulation of the material evaporation enable a very good control of the growth rate on large surfaces to be made. But for high evaporation temperature materials, for oxides or complex molecules, electron beam evaporation has to be used instead of MBE.

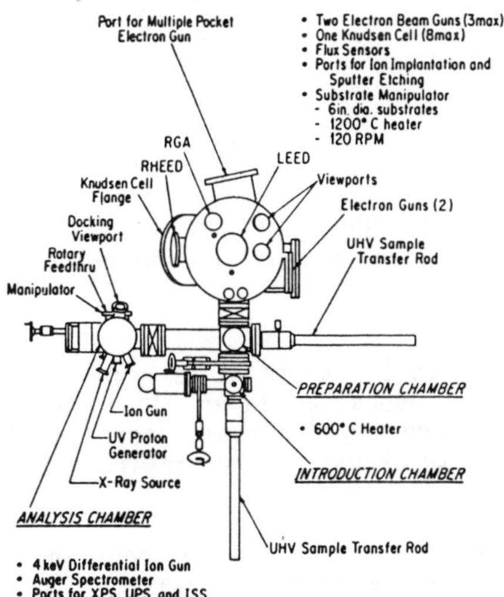

Fig. 9 Schematic diagram of a MBE system for producing metallic superlattices (from ref. 1).

In other respects chemical vapor deposition or liquid phase epitaxy have also been used to produce multilayers but especially for semiconductor sandwiches and for thick (micrometric) layers [25].

4 IN SITU CONTROL METHODS

Three characterization methods have been used to control the multilayer growth. Each method has been adapted to one kind of growth : ellipsometry to diode RF sputtering, soft X-ray reflectometry to electron beam evaporation and RHEED to molecular beam epitaxy.

4.1 Ellipsometry

Ellipsometry [26, 27] consists of the analysis of the polarization variation after reflection from a sample of a polarized beam (632,8 nm) (fig.10).This variation is expressed by two parameters tg Ψ and cos Δ where tg Ψ is the ratio of the reflection coefficients of the p wave on the S wave and where Δ is the difference in phase of the p wave and S wave ($\Delta = \delta_p - \delta_s$).

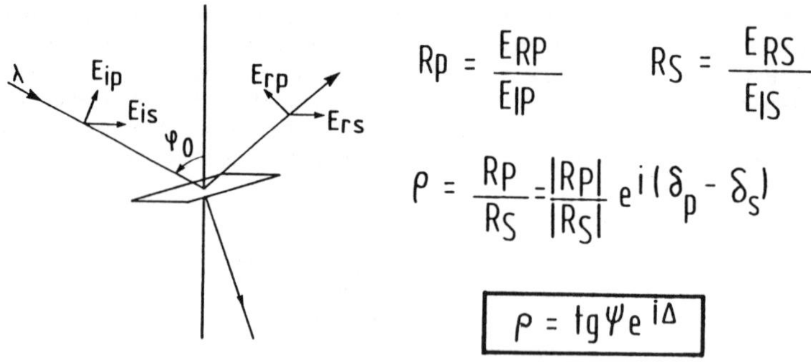

Fig. 10 Ellipsometry principle. Polarization before and after reflection on a surface (S and P compounds of the electric field) (from ref. 26).

In the plane tg Ψ , cos Δ each point is representative of the composition and the thickness of the deposited layer. Then during the growth we can observe trajectories in the plane (tg Ψ , cos Δ) corresponding to the evolution of the composition and of the thickness of the growing film.

This method is very accurate (a few per cent in composition, 0.1 Å in thickness) and can be used in gas atmosphere (plasma). It is very well adapted to sputtering but can also be used in cases of evaporation or MBE.

The light coming from a helium-neon laser (632,8 nm) is polarized by a fixed polarizer before being reflected from the sample. The signal is analyzed, after reflection, by a rotating polarizer (analyzer) and detected by a photomultiplier (fig. 11).

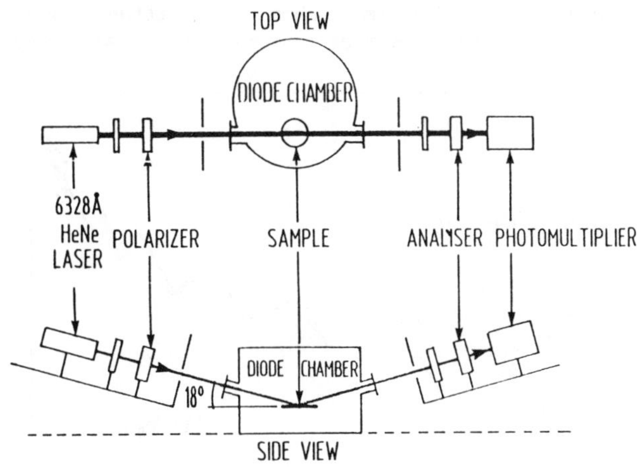

Fig. 11 Schematic drawing of an in situ kinetic ellipsometer adapted to a deposition chamber (from ref. 26).

The Hadamar method enables the calculation of tg Ψ and cos Δ . We can observe on figure 12 typical trajectories of "thick" layer growth : 35 nm tungsten followed by 150 nm carbon. Simulation shows the determination of the optical index using a simple model of iso-optical index all along the trajectories.

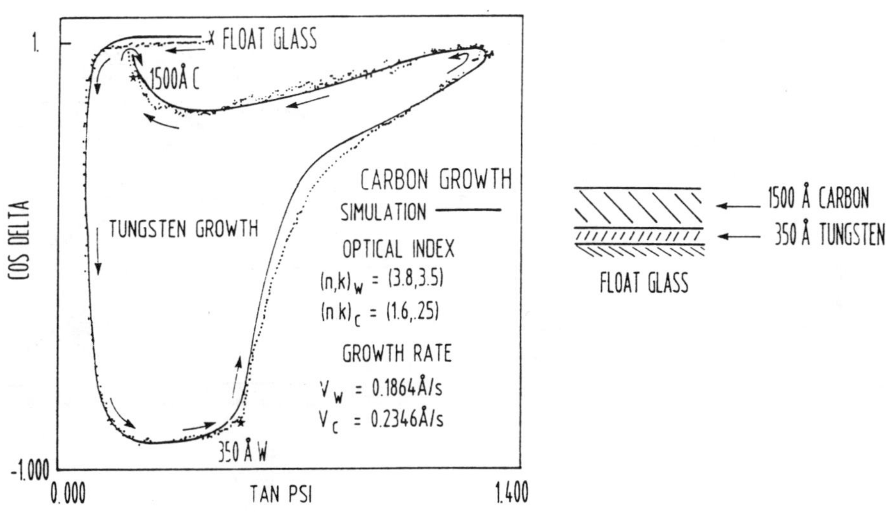

Fig. 12 Ellipsometric trajectories of the tungsten growth (35 nm) followed by a carbon growth (150 nm).

Figure 13 shows the trajectory corresponding to a 10 period stack of alternating 1.5 nm tungsten and 2.5 nm carbon. Simulation shows that the iso-optical index model is valid here and that the interfaces are very sharp.

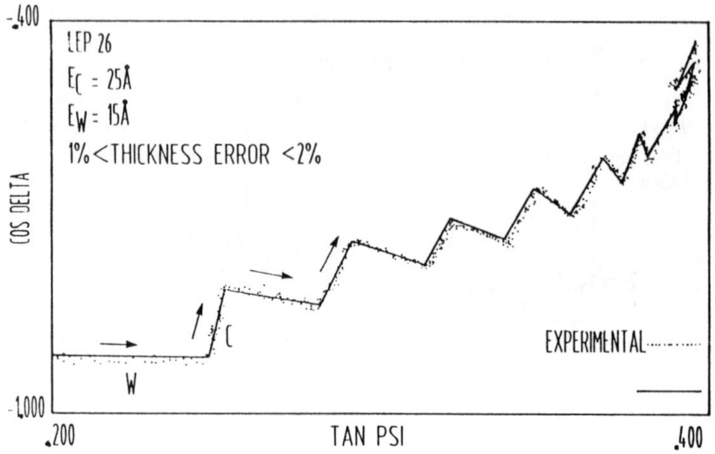

Fig. 13 Ellipsometric trajectories of a carbon (2-5 nm)/-tungsten (1.5 nm) multilayer growth (from ref. 26).

We can compare in figure 14 and figure 15 two different interfaces. The first one (fig. 14) is very abrupt corresponding to the growth of tungsten on a compact flat carbon surface without any diffusion. The second one shows three steps : first the etching of the silicon roughness by the tungsten atoms, second the diffusion-reaction (WSi_x) between tungsten and silicon and then third the growth of tungsten.

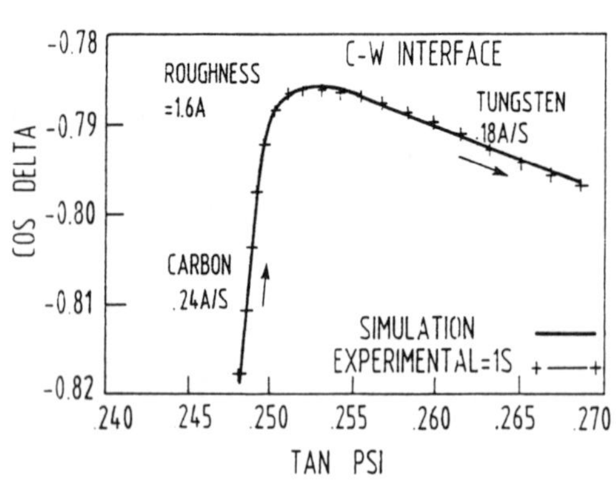

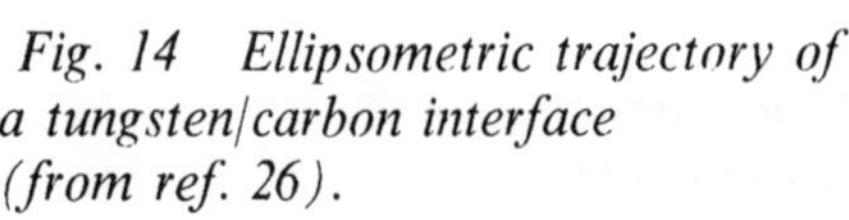
Fig. 14 Ellipsometric trajectory of a tungsten/carbon interface (from ref. 26).

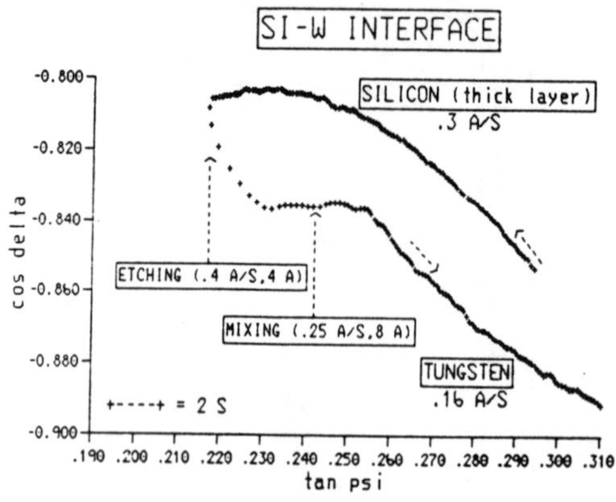

Fig. 15 Ellipsometric trajectory of a tungsten/silicon interface (from ref. 26).

These figures (12 to 15) illustrate the capacity of ellipsometry to control growth of ultra thin layers and particularly the capacity to give informations on layer composition, layer thicknesses and interface formation processes.

4.2 Soft X-ray reflectometry

In situ X-ray reflectometry [20, 28, 29] consists in measuring the reflectivity of a sample in the soft X-ray range (typically 4.47 nm or 6.78 nm with incidence angle of 64° or 50°). This can be done in a vacuum chamber and it has been observed during electron beam evaporation. The reflectivity is recorded against the time and we can observe oscillations corresponding to the thickness growth (fig. 16). Maxima and minima give an accurate value (proportionnal to the wave-length) of the thickness of the layer. The same analysis can be done depositing a multilayer. Oscillations give information on all thicknesses. This method is very well adapted to a small number of layers : the signal being absorbed for high thicknesses, the oscillation variations are not accurate enough to control the growth.

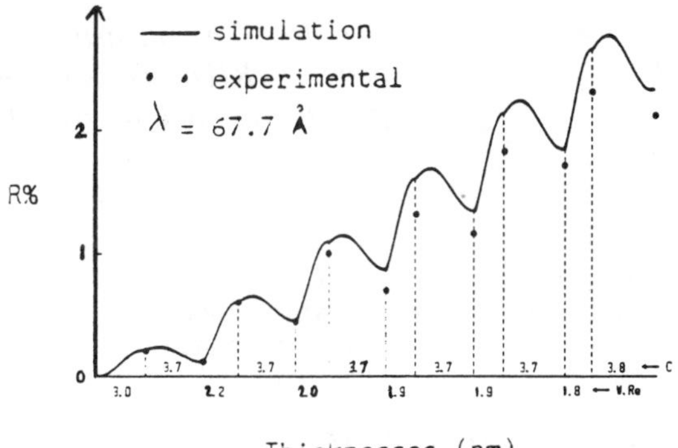

Fig. 16 Evolution of the reflectivity during WRe/C multilayer growth (from ref. 11, 20).

4.3 RHEED

Reflection high energy electron diffraction [30,31,32] consists in measuring the electron beam intensity diffracted in a particular direction by a crystalline surface. In this direction the signal is maximum when the surface is a perfect monolayer. When the growth starts, the monocrystalline surface is partially covered with atoms and then the signal decreases quickly as the surface roughness increases.

When the surface roughness decreases, that is when the atoms fill up the layers, the signal increases to reach the initial value when the film becomes again a perfect crystalline monolayer. Then during an epitaxial growth, monolayer by monolayer, we observe oscillations as simulated fig. 17. A typical example is given in fig. 18 for a metallic superlattice consisting in copper molybdenum layers : oscillations are visible during copper growth indicating clear crystalline growth (fig. 18).

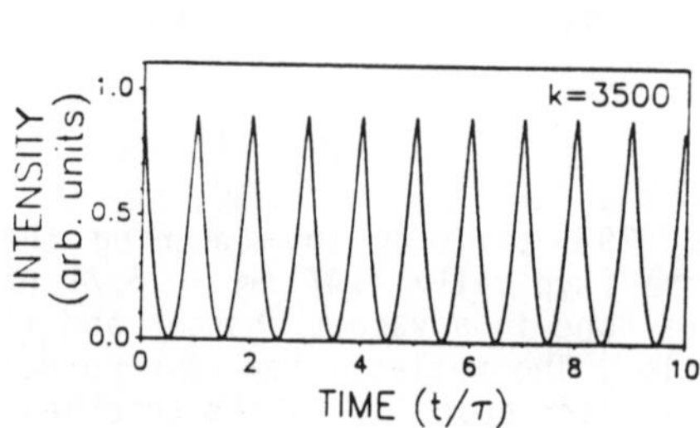

Fig. 17 Calculated Rheed oscillations (from ref. 31).

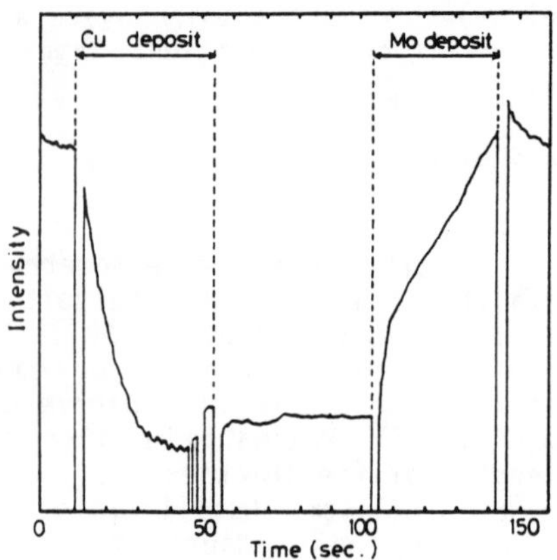

Fig. 18 The intensity of the specular beam of RHEED as a function of time function of time of growth (from ref. 30).

RHEED is a very powerful method to control epitaxial growth but it cannot be used with sputtering or evaporation because the electron beam is disturbed both by the plasma and by the electron gun.

5 A POSTERIORI CHARACTERIZATION METHODS

5.1 Current methods

5.1.1 X-Ray diffraction (XRD)

X-ray diffraction [33,34,35,36,37,38] consists in the analysis of the diffraction of an X-ray beam (0.154 nm for instance) by the crystallites present in the layer.

The analysis of the diffracted intensity versus the angle (10° to 40°) gives peaks in which the angular positions are related to the inter-reticular distance of the crystallite by the Bragg law ($2d \sin\theta = k\lambda$). The apparatus can be the same as for grazing X-ray reflection (see above fig. 22) but used at high angle.

We can see fig. 19 a XRD spectrum from 5° to 40° for a rhodium/boron carbide multilayer. The left figure corresponds to an unannealed sample and the right one corresponds to a sample annealed at 400 °C. We can observe the evolution of the material structure, different phases disappearing. At room temperature Rh_7B_3 crystallites are observed with also some B_4C crystallites. When we increase the temperature there is a strong diffusion of boron and carbon in the rhodium layers destroying the crystalline interfaces made of Rh_7B_3. Then the Rh_7B_3 peaks disappear when the multilayer is completely melted.

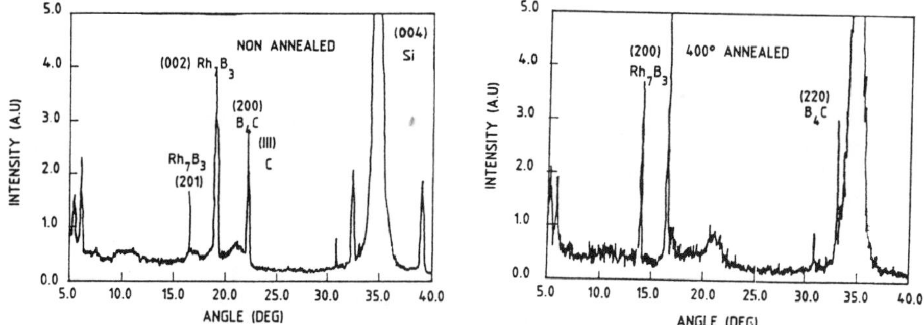

Fig. 19 X-ray diffraction of rhodium/boron carbide multilayer at room temperature and at 400°C (from ref. 68).

Another way is to observe a particular diffraction peak versus a parameter. We can see (fig. 20), the analysis of the crystallization of rhodium and nickel in rhodium/carbon and nickel/carbon multilayers. We can observe that crystallization of nickel appears very early (1.3 nm) in comparison with the rhodium one. That can be connected with other measurements (ellipsometry, Auger electron profil) which show a strong diffusion of carbon on nickel and a high interface roughness value (1.6 nm), a slight diffusion of carbon into rhodium with a middle roughness and no diffusion of carbon in tungsten with very low interface roughness (0.2 nm) (tungsten layers down to 4 nm are amorphous).

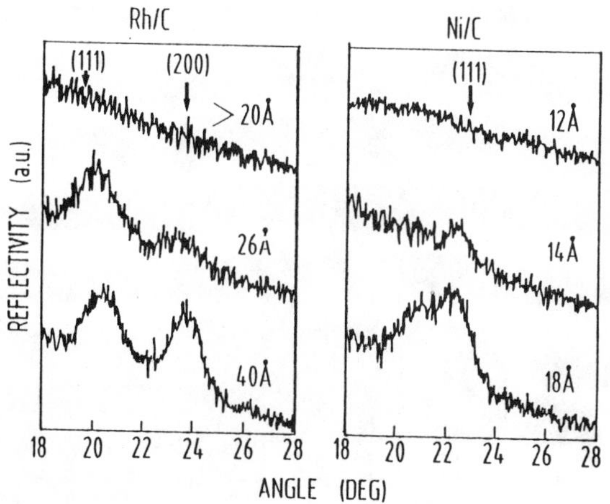

Fig. 20 X-ray diffraction of rhodium/carbon and nickel/ carbon multilayers. An evolution of layer crystallization is observed versus the metallic layer thickness (from ref. 67).

In conclusion we can say that XRD is one of the more important characterization methods to understand the ultra-thin layer superlattice structure, the diffusion processes and the interface interactions.

5.1.2 Transmission electron microscopy (TEM)

To observe superlattices, using TEM [39,40,41,42 , a very precise sample preparation method is needed. Indeed, to observe very accurately the stack cross section, preparation is needed to reach sample thickness down to values of the order of a few tens of nanometers. Three principal techniques exist to prepare such samples : ion beam, microtome and cleavage.

Ion beam preparation method consists in making a hole in the sample enclosed in glue.This technique works very well, but during this kind of preparation, the sample is bombarded with an ion beam which has a strong interaction with the sample. So the characterized sample can be different from the as-deposited sample especially concerning the presence of crystallites. Microtome preparation has limits on the thicknesses we can reach and a mechanical interaction is observed with the sample. Actually the cleavage preparation seems to be the best one to be sure of the observations done on the sample. We can summarize these observations in : roughnesses, thicknesses and crystallite formation and their evolution.

Figure 21 shows the evolution of the roughness all along the sample from the substrate to the top. It has been observed that in the sputtering process (on the contrary to evaporation) the carbon/tungsten multilayer smooths substrate defects on the nanometric scale.

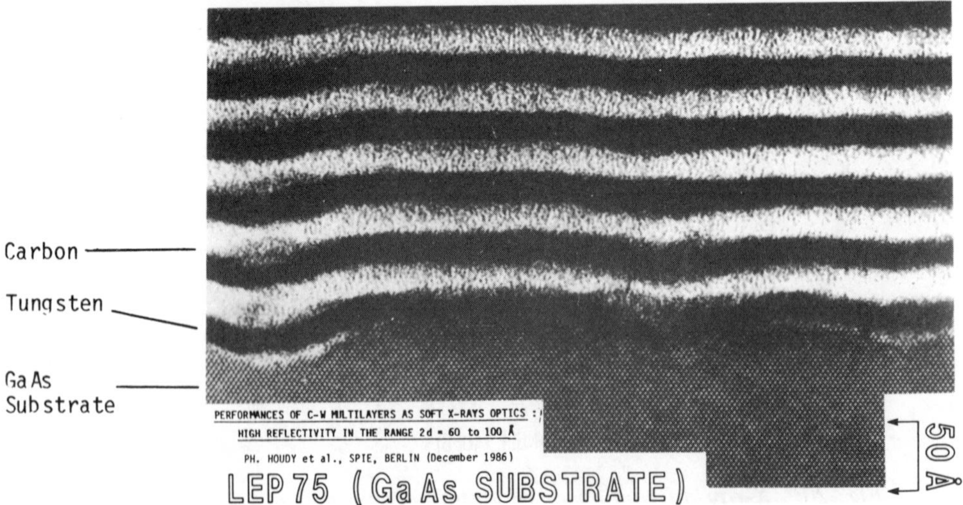

Fig. 21 Smoothing effect observed on a carbon/tungsten sputtered multilayer (from ref. 42).

We can see in fig. 22 what can be observed on an X-ray mirror. From the substrate crystalline structure we can deduce a thickness scale for the micrograph and then measure the superlattice thickness.

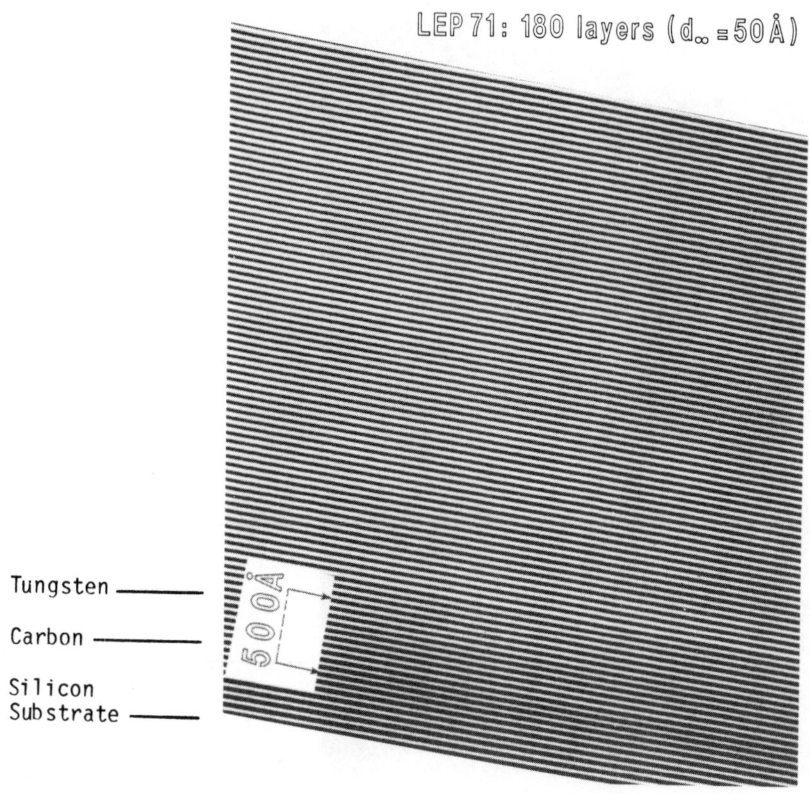

Fig. 22 T.E.M. micrograph of a 180 layers X-ray mirror (from ref. 77).

Another important information one can get is the structure of the layers (amorphous, polycrystalline, monocrystalline) and the evolution of this structure. The evolution of the crystallization of the rhodium in a carbon/rhodium multilayer can be see on fig. 23. The first rhodium layer (2 nm directly on the desoxidized silicon substrate is completely amorphous and the interface roughness with the first carbon layer is very low. When we increase the rhodium deposition time we observe a crystallization of the rhodium accompanied by an interface roughness increase. For the fifth rhodium layer, the roughness is very high and rhodium crystallites can be observed up to the carbon layer.

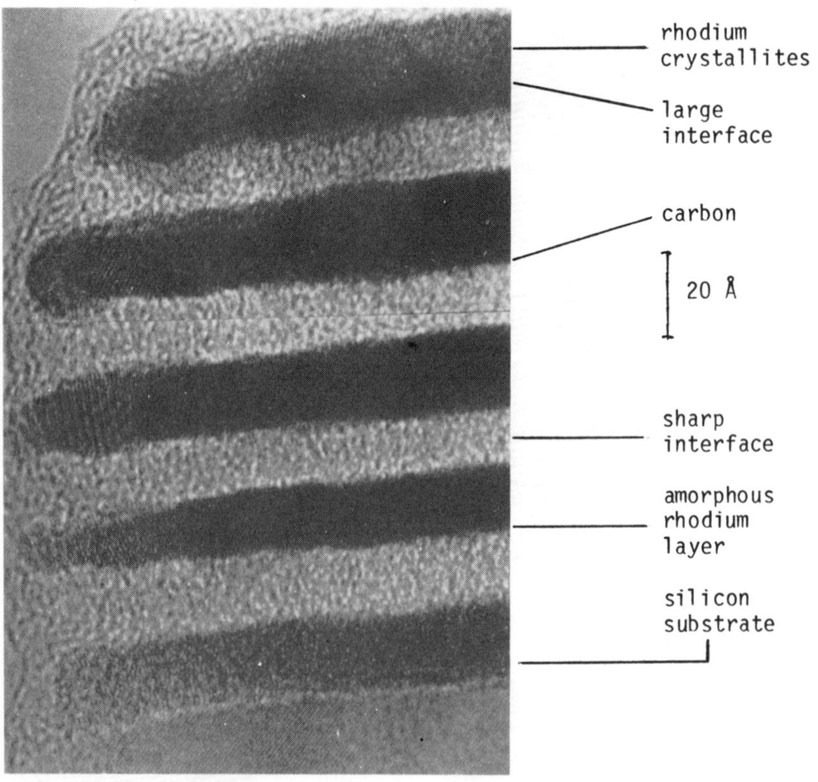

Fig. 23 Crystallinity and roughness increase in a rhodium/carbon multilayer.

This figure illustrates all the information we can get from TEM : thickness, roughness, crystallization and their evolution. From electronic diffraction of such a prepared sample we can also identify the kind of crystallites present in the layers.

5.1.3 Auger electron spectroscopy (AES)

In AES [43,44,45,46,47,48] we analyze the electron emission of a sample bombarded by an incident electron beam (electron energy of a few keV). The Auger process corresponds to a non radiative desexcitation of an atom where the relaxation energy is given as kinetic energy of an electron called Auger electron.

Then AES gives information on the chemical composition of a surface, looking at the energy levels of all its constituants.

To investigate a multilayer it is possible to use AES coupled with an etching ion beam. We can look at the evolution of the intensity corresponding to a special energy level (that is corresponding to a special kind of atom) during the etching.

We can see such an evolution on fig. 24 for a sputtered nickel-carbon multilayer (2 nm/4 nm). We observe oscillations corresponding to the penetration of the analysis all along the stack. Then we start at the surface with a carbon layer. The carbon signal decreases from the surface while the nickel one increases. We can observe a small shift between the carbon minima and the nickel maxima corresponding to the diffusion of the carbon in the nickel at the interface carbon on nickel. Concerning argon inclusions, we can observe that the maxima correspond to the beginning of the nickel layers. The argon is trapped in the amorphous part of the nickel. As soon as the nickel crystallizes, the argon is rejected out of the layer and the argon signal decreases.

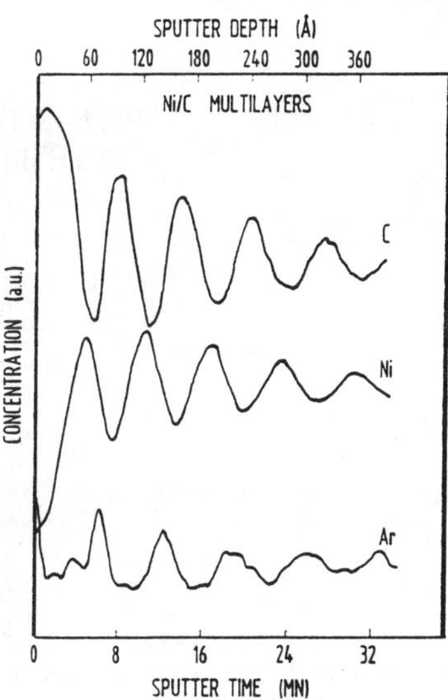

Fig. 24 Carbon, nickel and argon concentrations in a sputtered carbon-nickel multilayer versus the depth from the surface (from ref. 67).

Thus, using AES with etching we can make concentration profiles in a multilayer and have information on the chemical composition of the layers and of their interfaces and detect the diffusion process if it occurs.

5.1.4 Grazing X-ray reflection (GXR)

The GXR [49, 70] consists of the analysis of the reflection of a sample in the X-ray range at low angle (inferior to 5°). Copper K_{α} radiation (0.154 nm) is the typical wavelength used for GXR measurements. It allows the characterization of single layers from 10 nm to few hundred nanometers and the characterization of stacks with periods in the range 1 nm to a few ten nanometers.

Fig. 25 presents a schematic drawing of a grazing X-ray reflectometer. The beam emitted by an X-ray tube is monochromatized, passes through slits, is reflected on a $\theta, 2\theta$ goniometer and is detected by a proportional counter. Then the signal is recorded through a microcomputer in order to be analyzed.

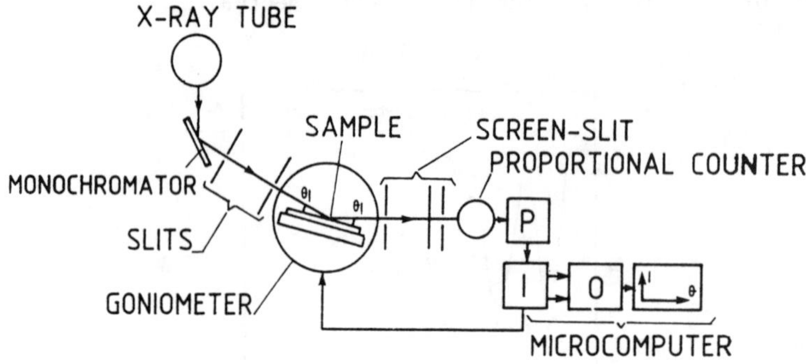

Fig. 25 Schematic view of a grazing X-ray reflectometer (1.54Å) (from ref. 70).

We can see fig. 26 a typical simulated curve of reflectivity versus the angle for a regular stack. From the total reflection limit angle ($\theta_c = \sqrt{2\delta}$) we can deduce the average density of the stacks through the real part of the optical index $n = (1-\delta) + i\beta$.

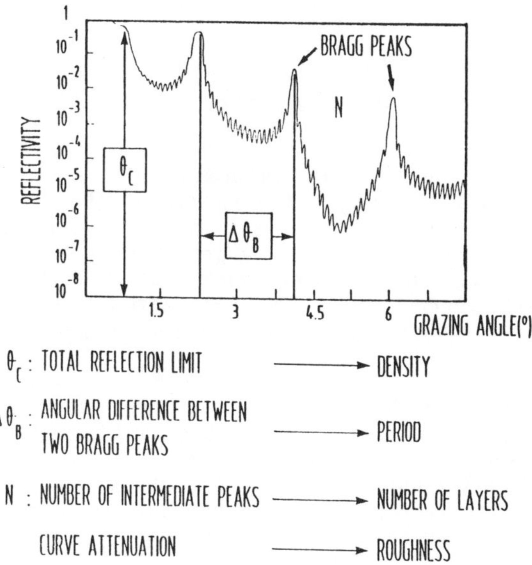

Fig. 26 Typical calculated grazing X-ray reflection curve of a multilayer and information given by the analysis of the curve (from ref. 70).

From the angular difference between bragg peaks ($2d \sin\theta = k\lambda$) we can measure the period of the multilayer and from the curve attenuation we can calculate the average roughness (σ) of the stack using a Debye-Waller factor multiplying the ideal reflectivity

$$R(\theta) = R_{id}(\theta) \exp\left[\frac{-16\pi^2\sigma^2 \sin^2\theta}{\lambda^2} \right]$$

Other models can be introduced, to take into account the roughness ( ), such as the error function where the optical index is given at the interface by the relation

$$n(z) = n_1 + \frac{n_2 - n_1}{\sigma\sqrt{2\pi}} \int_{-\infty}^{z} \exp\left(\frac{-x^2}{2\sigma^2} \right) dx$$

Finally, information on thickness regularity can be deduced from the Bragg peak broadening. In conclusion GXR is a powerful method to give average values of density, thickness and roughness in the stack and to give information on the bottom layer (interaction with substrate) and on the top layer (oxidation in case of metal surface material).

5.1.5 Rutherford backscattering (RBS)

RBS [50, 57] consists in bombarding the sample with an ion beam (ion being typically He) and to analyze the energy of the backscattered ions after interaction with the sample. The incident ion energy is very high (of the order of one MeV) and the interaction can approximately be described by a simple "billiard ball" model. Then we observe peaks due to each element present in the sample, high energy corresponding to heavy atoms. With such a method it is possible to calculate accurately the composition of a multilayer.

We can observe fig. 27 the RBS measurements of a nickel-carbon multilayer deposited by sputtering with the plasma lighted on nickel and carbon targets, but also working on a tungsten target set aside in the chamber. So we observe carbon and nickel but also silicon which is the substrate, argon (2%)and tungsten (0.5 %). This method is very important to quantitatively measure the constituants of the sample analyzed.

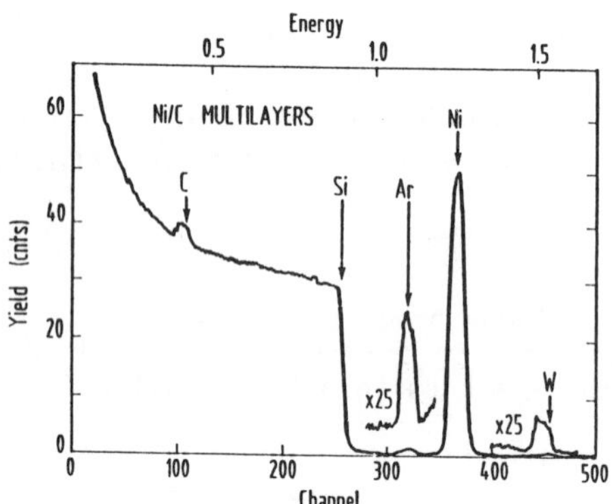

Fig. 27 R.B.S. analysis of a carbon/nickel multilayer sputtered on a silicon substrate with plasma working on an aside tungsten target (from ref. 67).

These five characterization methods (XRD, TEM, AES, GXR and RBS) are most frequently used for superlattice structure analysis and are generally sufficient to provide a first simple model of layer structure and interface state. However, typical surface chemical analysis methods as XPS, EXAFS and STM are beginning to be used for superlattice analysis and especially for interface chemical binding characterization.

5.2 Other methods

5.2.1 X-ray photoemission spectra (XPS)

The X-ray photoemission spectra[method] 51,52,53 consists in the analysis of the energy of the photoelectron emitted by a sample bombarded by an X-ray beam. XPS is a surface analysis method. As a matter of fact for 1486,7 eV incident photons (Al K_α X-ray radiation) with an incident angle of 60° we collect photoelectrons coming from the surface up to 20 nm below the surface, the electron mean free path being of two to one hundred monolayers, and their energy being in the range 1 eV to 1000 eV. The accuracy can be about 100 meV and allows atom emission peaks and the chemical binding of materials present in the sample to be observed. By tilting the sample it is possible to get information on the photoelectron emitted deeply or at the extreme surface.

We can observe fig. 28 photoemission spectra of the valence band for Cu/Hf and Cu/Nb multilayers for different copper thickness. The evolution of the peak from Hf or Nb to copper gives information on the way of growth of copper on these materials : for Hf Cu layers are disordered and contain Hf ; for Nb, interfaces are abrupt and copper layers don't present impurities.

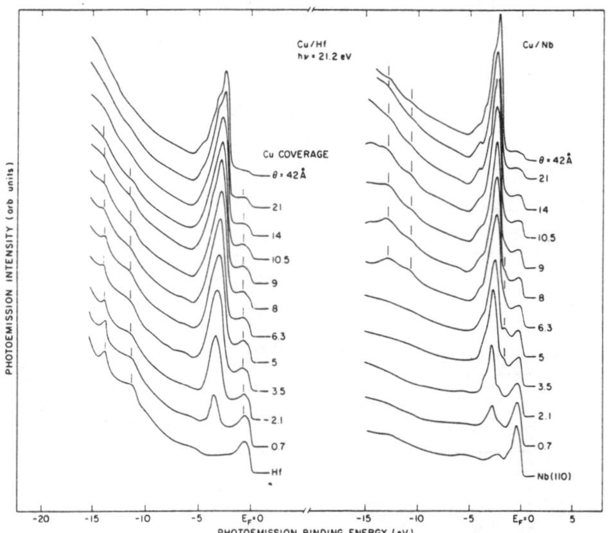

Fig. 28 Valence band photoemission spectra taken at 21.2 eV photon energy with a pass energy of 10 eV are shown for increasing Cu coverages on polycrystalline Hf (left) and Nb (right) substrates (from ref. 51).

This method is very useful to describe the state of the interface especially concerning diffusion and chemical binding.

5.2.2 Extended X-ray absorption fine structure (EXAFS)

EXAFS [54, 55, 56] consists of the analysis of the absorption of X-ray photons by a sample. The absorption energy analysis in the range 30 eV to a few hundreds eV gives information on the position, the distance and the number of first neighbour atoms.

Then it is possible to analyse very accurately the structure of the interface by studying multilayers with different thicknesses.

An example of such a study is given in figures 29 a and b for copper/ thallium superlattices 54 . Comparing the EXAFS signal for various stacks alternating from half monolayer to twenty monolayers allows the bulk contribution to be separated from the interface contribution. Thus it is possible to know the exact structure of the interface. In the present case the interface is not diffused as one might expect (from other considerations) but is a thin layer of amorphous material (more copper than hafnium) sandwiched between two metallic bulk layers.

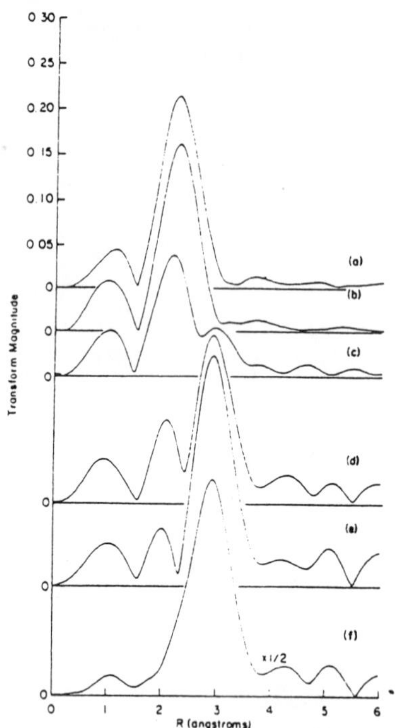

Fig. 29-a Fourier transforms of EXAFS data of Cu-Hf multilayer samples ;
a) 1/2 monolayers, b) 5 monolayers,
b) 10 monolayers, d) 14 monolayers,
e) 20 monolayers, f) hafnium metal
(from ref. 54).

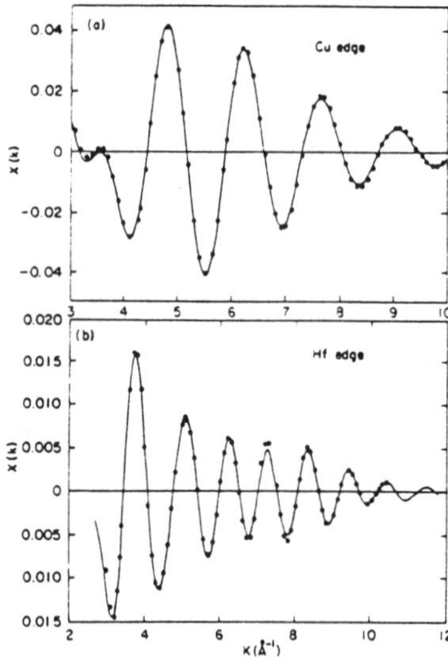

Fig. 29-b Filtered, back transformed EXAFS data (line) and fit (points) for a 20 monolayers Cu-Hf multilayer (from ref. 54).

5.2.3 Secondary ion mass spectroscopy (SIMS)

SIMS [57] consists in bombarding a sample with energetic ions and to analyse the ejected substrate atoms and molecules in charged states (secondary ions) with mass spectrometer.

The analysis depth is around 5 nm and the accuracy on the determination of the number of the atoms of one kind can reach 1 ppm to a thousandth of ppm. Changing the incident ion energy allows different parts of the sample to be probed : high energy incident ions give information on a scale of a few nanometers (up to 5 nm), low energy incident ions give information on the extreme surface (down to 1 nm). In other respects, for a superlattice, etching the sample allows the composition profile all along the stack to be determined.

We can observe such experiments carried out on $Ge/Ge_{1-x}\,Si_x$ multilayers with thicknesses in the order of 40 nm in fig. 30. Interfaces are assumed to be abrupt at this scale and we can observe that the boron profile follows the silicon as expected.

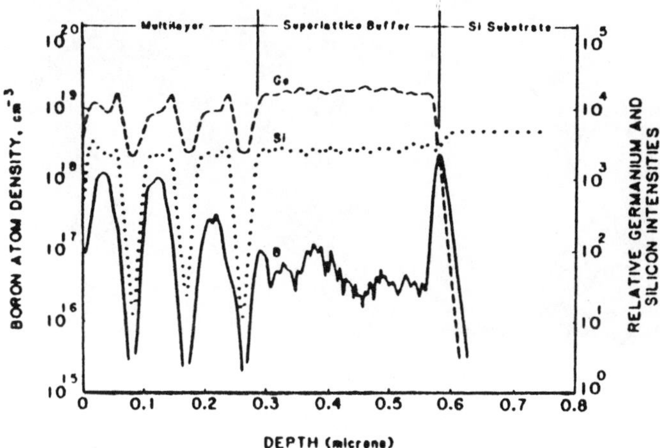

Fig. 30 SIMS profile of $Si_{0.4}Ga_{0.6}$-Ge multilayer on si (from ref. 57).

5.2.4 Scanning tunneling microscopy (STM)

S.T.M. [58, 70] is a surface analysis method which gives information on the atomic structure and the electronic properties of the top layer of a sample. A metallic tip is moved in front of the sample at distances down to 1 nm, over the surface plane. A tunneling current is established between the tip and the sample. To maintain a constant current, the tip moves up and down, following the surface. The recording of the tip displacement gives a very accurate mapping at atomic scale of the sample surface.

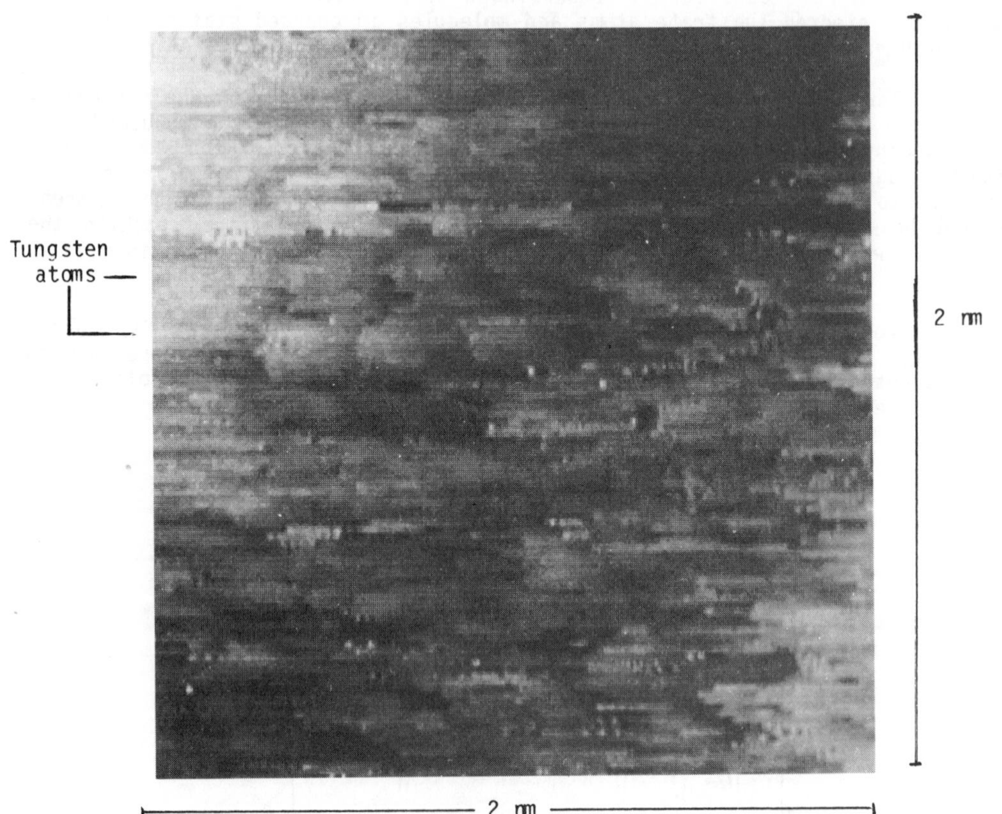

Fig. 31 S.T.M. micrograph of a sputtered tungsten surface (scale : 2 nm x 2 nm) : tungsten atoms are visible (from ref. 70).

We can observe fig. 31 a STM micrograph of a 30 nm tungsten layer surface showing tungsten atoms indicating a polycrystalline surface. It should be noticed that in the case of tungsten, the surface is very flat and no islands are observed.

On the other hand, we can observe (fig. 32) in the case of rhodium, that the surface is formed by rhodium islands with diameter in the order of 7 nm and height of about 1,5 nm. The tungsten surface micrograph has been recorded in air under oil to avoid tungsten oxidation which forms an insulated layer at the top preventing the tunnel current from passing. The rhodium surface micrograph has been recorded directly in air, rhodium being not easily oxidized.

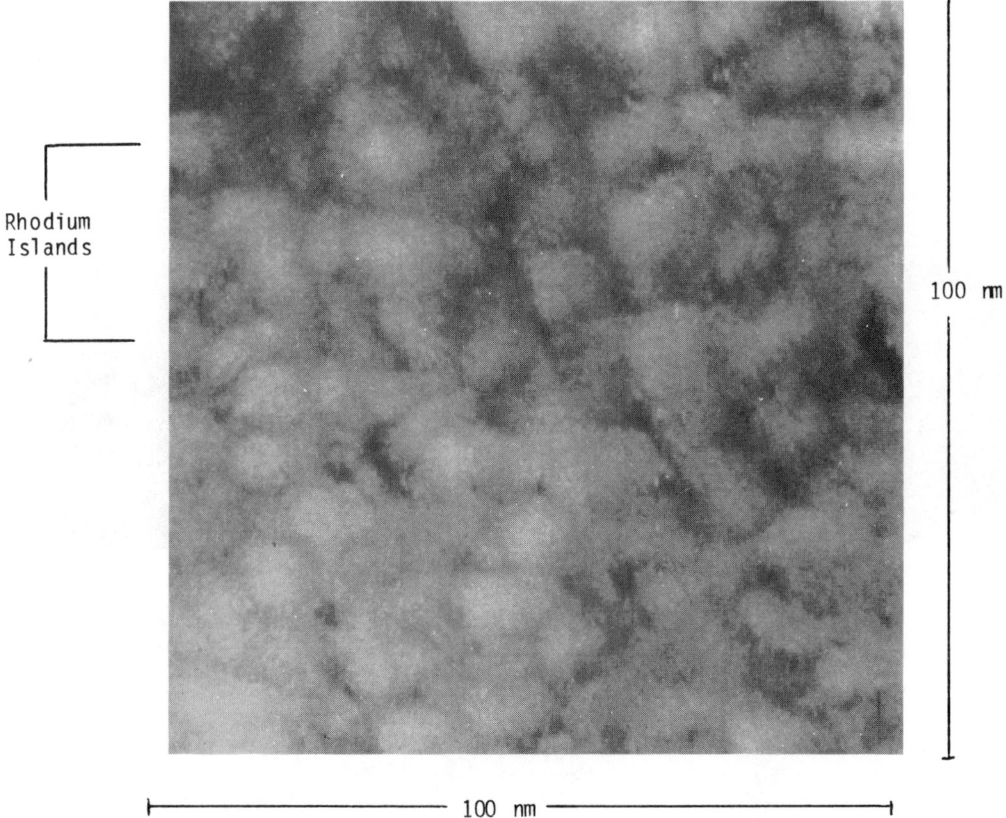

Fig. 32 S.T.M. micrograph of a sputtered rhodium surface (scale : 100 nm x 100 nm, contrast 2 nm) : rhodium islands (ϕ = 7 nm) are visible (from ref. 70).

We can imagine using this method in cross section on cleaved superlattices to get very precise information on the layer structure and on their interfaces at the atomic scale.

5.2.5 Scanning electron microscopy (SEM)

SEM [59] consists in obtaining on a submicronic scale a micrograph made from the secondary electrons emitted by a sample probed by a primary incident beam. This method enables micrographs with resolution down to 10 nm to be obtained. This method can be used for superlattices with thick layers with thicknesses of the order of 100 nm.

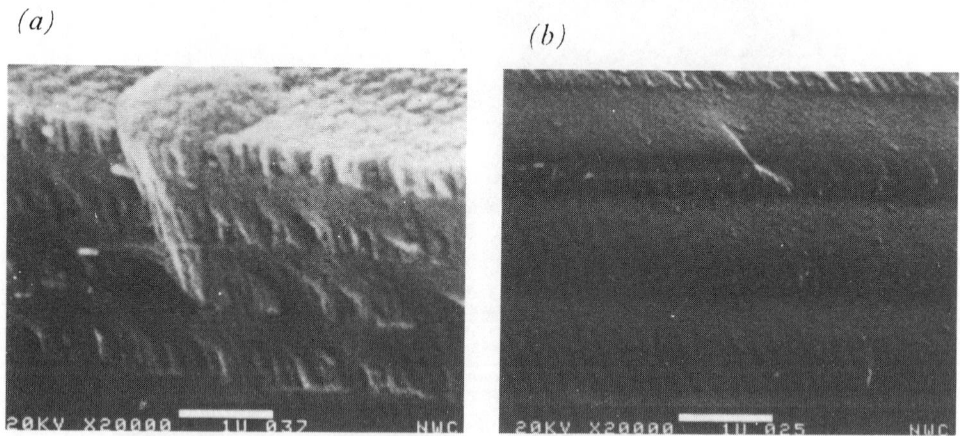

Fig. 33 a&b Scanning electron micrographs showing a cross section of SiH/SiO_2 multilayer with (a) and without (b) modular growth defect (from ref. 59).

We can see in fig. 33 (a&b) SH/SiO_2 multilayer micrographs of the cross section profile. In the first case (fig. a : low sputtering power on SiH) nodules are observed in SiH layers inducing nodular growth through the layers and then important defects in the stack. In the second case (fig. b : high sputtering power on SiH) the micrograph shows that the multilayer is nodule-free and that their is no columnar growth all along the sample.

This method is interesting for the observation of material growth on a scale of 100 nm.

5.2.6 Raman scattering (RS)

Raman scattering [66, 61] consists of the analysis of the scattering of a light beam by the optical phonons or other elementary excitations with frequency in the range 10 - 5000 cm^{-1}. The light scattered by the sample is analyzed by a monochromator and the intensity is plotted versus the frequency. The spectrum obtained is then characteristic of the structure and of the sample composition.

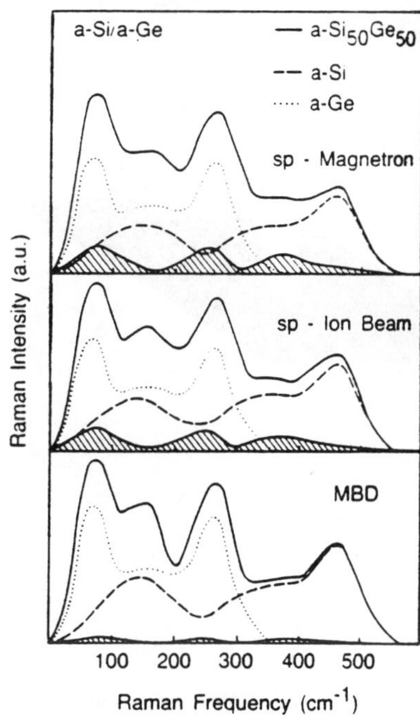

Fig. 34 Raman spectra for a-Si/a-Ge multilayers prepared by magnetron sputtering (top), ion beam sputtering (middle), and MBD (bottom). Each spectrum has been decomposed into three components, s-Si (dashed), a-Ge (dotted), and a-$Si_{1-x}Ge_x$ (continuous) (from ref. 60).

Raman scattering has been used to characterize the structure of amorphous Si/Ge superlattices. We can observe fig. 34 the Raman spectra obtained for three samples prepared by different methods. Simulations indicate that the interface (a-Si_{50} Ge_{50}) decreases in going from the magnetron method to the ultra-high vacuum evaporation one. The proportion of a-Si_{50} Ge_{50} has almost disappeared in the case of evaporation, indicating very smooth and abrupt interfaces. Thus, using RS, it is possible to characterize the structure (amorphous, crystalline) and the interface of superlattices at nanometric scale.

5.2.7 Neutron scattering (NS)

It is possible from the analysis of the neutron beam reflection peak intensity to deduce the interdiffusion coefficient in superlattices [62]. This has been done for amorphous Si/Ge multilayers. One can see on fig. 35 the intensity around the first order peak reflection versus the angle and the corresponding simulation. From the analysis of such curves for different temperatures it has been possible to determine the diffusion processes during annealing : saturation of the diffusion, stopping of the diffusion due to Ge crystallization, diffusion law as in crystalline silicon.

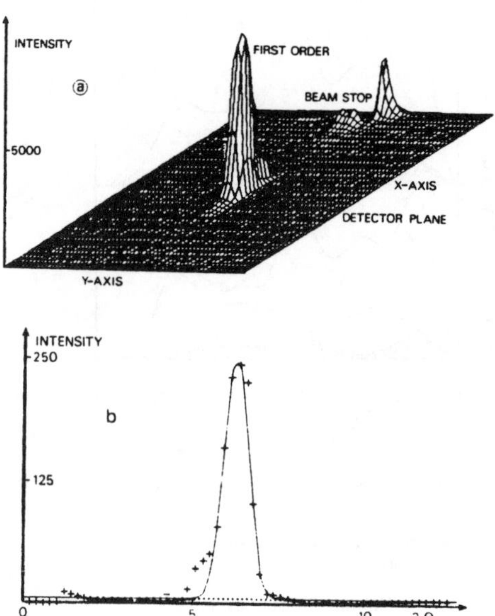

Fig. 35 Typical $\theta - 2\ \theta$ neutron scan of the first order reflection on a Si/Ge amorphous multilayer sample. a) 3D representation of intensities measured on the 2d multidetector of D17 (ILL). b) Regrouped data and Gaussian fit used to calculate the integrated intensity (from ref. 62).

This method gives results similar to grazing X-ray reflection but is not so easy to perform in standard laboratories.

6 CONCLUSION

For new applications, great interest has been shown in recent years, in devices structured on the atomic scale. New properties have been observed such as electronic resonance in quantum wells, superconductivity at nitrogen temperature in ceramics, magnetic anisotropy in magnetic superlattices and high reflectivity in soft X-ray multilayers. This has been made possible due to improvements in deposition processes and in characterization methods.

Indeed it is now possible to control the deposition of materials on the atomic scale using sputtering, evaporation or MBE. Characterization methods such as XRD, TEM, AES, GXR and RBS are used daily to determine the typical parameters of superlattices : thickness, composition and crystalline structure of the layers and their interfaces, interaction with the substrate and with the external medium.

Other methods well known for semiconductors analysis, bulk analysis, or micronic device characterization are now investigated to characterize nanometric superlattices. These methods are especially XPS, EXAFS, SIMS and STM.

In any case one method is not enough to have a unique model of the superlattice structure and several characterizations have to be done to be sure of the deduced stack structure model. Thus it is possible to connect this structural model to specific electronic, magnetic or optical properties of each kind of superlattice. So metallic superlattices are part of a new technology for future devices on the atomic scale.

REFERENCES

1 C.M. Falco
J. Phys. Colloq., Vol. 48, n° C-5, 1987, pp. 57-74

2 I. K. Schuller, C.M. Falco
Microstructure Science and Engineering/VLSI, Vol. 4, Norman G. Einspruch, editor

3 L.M. Williams, P.Y. Lu, S.N.G. Chu, C.H. Wang
J. Appl. Phys., Vol. 62, n° 1, 1987, p. 295-7

4 T. Kaneko, T. Sasaki, M. Sakudat, R. Yamamoto, T. Nakamura, H. Yamamoto and S. Tanaka
J. Phys. F:Met. Phys. 18 (1988) pp. 2053-2060

5 R.T. Tung, A.F.J. Levi, J.M. Gibson
J. Vac. Sci and Technol. B, Vol. 4, n° 6, 1986, pp. 1435-43

6 E.M. Logothetis, R.E. Soltis, R.M. Ager, W. Win, C.J. McEwan, K. Chang, J.T. Chen, T. Kushida, L.E. Wenger
Physica C, Vol. 153-155, pt.2, 1988, pp. 1439-40

7 P. Beauvillain, P. Bruno, C. Chappert, C. Dupas, F. Trigui, E. Vélu, D. Renard
To be published in Proc. of ICM 88 (Journal de Physique)

8 M. Komuro, Y. Kozono, S. Narishige, M. Hanazono and Y. Sugita
Japan. Journ. of Applied Physics, Vol. 27, N° 11, 1988, pp.L2105-L2107

9 T. Morishita, R. Sato, K. Sato and H. Kida
To be published in Proc. of ICM'88 (J. de Physique)

10 F.E. Fernandez, C.M. Falco
IEEE Journal of Quantum Electronics, Vol. 24, n° 8, 1988, pp. 1758-62

11 Ph. Houdy, J.P. Chauvineau
Le Vide, Les Couches Minces, n° 245, 1989, pp. 69-86

12 Many articles in SPIE, Vol. 563, Aug. 20-22, 1985, San Diego, California

13 Many articles in SPIE, Vol. 689, Aug. 19-20, 1986, San Diego, California

14 Many articles in SPIE, Vol. 984, Aug. 17-19, 1988, San Diego, California

15 D.B. McWhan, C. Vettier
AIP Conf. Proc., n° 138, 1986, pp. 80-92

16 H. Hoffmann, P. Kucher
Thin Solid Films, Vol. 146, n° 2, 1987, pp. 155-64

17 R.E. Somekh, C.S. Baxter
J. Cryst. Growth, Vol. 79, n° 1-3, pt.1, 1986, pp. 119-25

18 C. Sella, K.B. Youn, R. Barchewitz, M. Arbaoui
Vacuum, Vol. 16, n° 1-3, 1986, pp. 121-123

19 T. Suzuki, T. Yamazaki, K. Takahashi, T. Kageyama, H. Oda
Journ. Materials Science Letters, 7, 1988, pp. 79-80

20 J.P. Chauvineau, J. Corno, D. Naccache, L. Nevot, B. Pardo, L. Valiergue
J. Optics, Vol. 15, n° 4bis, 1984, pp. 265-269

21 S.F. Alvarado, C. Carbone
Physica B & C, Vol. 149 B+C, N° 1-3, 1988, pp. 43-8

22 P. Etienne, J. Chazelas, G. Creuzet, A. Friederich, J. Massies, F. Nguyen-Van-Dau, A. Fert
J. of Crystal Growth, Vol. 95, 1989, pp. 410-414

23 J.M. Baribeau, D.C. Hougton, D.J. Lockwood, M.W.C. Dharma-wardana, G.C.Aers
J. of Cryst. Growth, Vol. 95, 1985, pp. 447-450

24 G.M. Williams, C.R. Whitehouse, A.G. Cullis, N.G. Chew, G.W. Blackmore
Applied Physics Letters, Vol. 53, n° 19, 1988, pp. 1847-9

25 D. Käss, M. Warth, H.P. Strunk, E. Bauser
Physica 129 B, 1985, pp. 161-165

26 Ph. Houdy
Revue Phys. Appl. 23, 1988, pp. 1653-1659

27 T.E. Tiwald, J.A. Woollam, D.J. Sellmyer
J. of Appl. Phys., Vol. 63, n° 8, pt.2A, 1988, pp. 3215-17

28 R. Barchewitz, R. Marmoret
Revue de Physique Appliquée, Vol. 23, n° 10, 1988, pp. 1661-74

29 H.U. Krebs, D.J. Webb, A.F. Marshall
J. Less-Common Met., Vol. 140, 1988, pp. 17-24

30 M. Doyama, R. Yamamoto, T. Kaneko, M. Imafuku, C. Kokubu, T. Izumiya, T. Hanamure
Vacuum, Vol. 36, n° 11-12, 1986, pp. 909-11

31 G.S. Petrich, P.R. Pukite, A.M. Wowchak, G.J. Whaley, P.I. Cohen, A.S. Arrott
Journal of Crystal Growth 95, 1989, pp. 23-27

32 G.W. Jones, J.A. Venables
Ultramicroscopy, Vol. 18, n° 1-4, 1985, pp. 439-44

33 W.H. Goldstein, C.J. Hailey, J.H. Lupton
Opt. Commun., Vol. 62, n° 4, 1987, pp. 259-64

34 W.J. Bartels
Microscopy of Semiconducting Materials, Prooceedings of the Institute of Physics Conference, Oxford, 1987, pp. 599-608

35 W.J. Bartels, J. Hornstra, D.J.W. Lobeek
Acta Crystallogr., Vol. A42, pt.6, 1986, pp. 539-45

36 A. Cavalleri, M. Dapor, F. Giacomozzi, L. Guzman, P.M. Ossi, M. Scotoni
Mat. Sci. Eng., Vol. 99, 1988, pp. 201-5

37 Z.S. Shan, Z.R. Zhao, J.G. Zhao, D.J. Sellmyer
J. Appl. Phys., Vol. 61, n° 8, pt. 2B, 1987, pp. 4320-2

38 O.B. Loopstra, W.G. Sloof, T.H. de Keijser, E.J. Mittemeijer, S. Radelaar, A.E.T. Kuiper, R.A.M. Wolters
J. Appl. Phys., Vol. 63, n° 10, 1988, pp. 4960-9

39 A.K. Petford-Long, M.B. Stearns, C.H. Chang, S.R. Nutt, D.G. Stearns, N.M. Ceglio, A.M. Hawryluk
J. Appl. Phys., Vol. 61, n° 4, 1987, pp. 1422-8

40 H. Oppolzer
J. Phys. Colloq., Vol. 48, n° C-5, 1987, pp. 65-74

41 Y. Lepetre, Ivan K. Schuller, G. Rasigni, R. Rivoira, R. Philip, P. Dhez
SPIE, Vol. 563, Applications of Thin Film Multilayered Structures to Figured X-Ray Optics, 1985, pp. 258-263

42 P. Ruterana, J.P. Chevalier, Ph. Houdy
J. Appl. Phys., 65 (10), 1989, pp. 3907-3913

43 J.P. Petrakian, P. Renucci
Surf. Sci., Vol. 186, n° 3, 1987, pp. 447-59

44 S. Hofmann, A. Zalar
Surf. & Interface Anal., Vol. 10, n° 1, 1987, pp. 7-12

45 R. Pantel, D. Levy, D. Nicolas
J. Vac. Sci. Technol., Vol. 6, n° 5, 1988, pp. 2953-6

46 T. Kobayashi, R. Nakatani, S. Ootomo, N. Kumasaka
J. Appl. Phys., Vol. 63, n° 8, pt.2A, 1988, pp. 3203-5

47 R.W. Judd, M.A. Reichelt, E.G. Scott, R.M. Lambert
Surf. Sci., Vol. 185, n° 3, 1987, pp. 515-28

48 G. Zajac, S.D. Bader, R.J. Friddle
Phys. Rev. B, Vol. 31, n° 8, 1985, pp. 4947-53

49 Y. Utsumi, H. Kyuragi, T. Urisu, H. Maezawa
Appl. Opt., Vol. 27, n° 18, 1988, pp. 3933-6

50 J. Eridon, G. Was
Thin Films, Interfaces and Phenomena, Part of the Fall 1985 Meeting of the Materials Research Society, Boston, MRS, 1986, pp. 693-8

51 M.W. Ruckman, Li-Qiang Jiang, M. Strongin
Mater. Lett., Vol. 5, n° 4, 1987, pp. 147-52

52 R.J. Gorte, E. Altman, G.R. Corallo, M.R. Davidson, D.A. Asbury, G.B. Hoflund
Surf. Sci., Vol. 188, n° 3, 1987, pp. 327-34

53 J.W. Erickson, T.B. Fryberger, S. Semancik
J. Vac. Sci. Technol. A, Vol. 6, n° 3, pt.2, 1988, pp. 1593-8

54 S.M. Heald, J.M. Tranquada, B.M. Clemens, J.P. Stec
J. Phys. Colloq., Vol. 47, n° C-8, pt.2, 1986, pp. 1061-4

55 C. Brouder, G. Krill, P. Guilmin, G. Marchal, E. Dartyge, A. Fontaine, G. Tourillon
Phys. Rev. B, Vol. 37, n° 5, 1988, pp. 2433-9

56 H. Van Brug, M.J. van der Wiel, M.P. Bruijn, J. Verhoeven
J. Vac. Sci. Technol. A, Vol. 5, n° 4, pt.3, 1987, pp. 2028-9

57 R.M. Ostrom, F.G. Allen, P.K. Vasudev
Proceedings of the Second International Symposium on Silicon Molecular Beam Epitaxy, Honolulu, Elec. Soc. 1988, pp. 85-94

58 M. Green, M. Richter, J. Kortright, T. Barbee, R. Carr, I. Lindau
J. Vac. Sci. Technol. A, Vol. 6, n° 2, 1988, pp. 428-31

59 L.F. Johnson, E.J. Ashley, T.M. Donovan, J.B. Franck, R.W. Woolever, R.Z. Dalbey
SPIE Conference, Los Angeles, Vol. 525, 1985, pp. 127-39

60 D.D. Allred, J. Gonzalez-Hernandez, O.V. Nguyen, D. Martin, D. Pawlik
J. Mater. Res., Vol. 1, n° 3, 1986, pp. 468-75

61 J. Gonzalez-Hernandez, D.D. Allred, O.V. Nguyen, D. Martin, D. Pawlik
Layered Structures and Epitaxy Symposium, Boston, MRS, 1985, pp. 389-94

62 C. Janot, A. Bruson, G. Marchal
J. Phys., Vol. 47, n° 10, 1986, pp. 1751-6

63 S.M. Vaezi-Nejad, C. Juhasz
Semicond. Sci. Technol., Vol. 3, n° 7, 1988, pp. 664-74

64 S. Mantl, C. Buchal, B. Stritzker, B. Saftic
Layered Structures and Epitaxy Symposium, Boston, MRS, 1985, pp. 195-200

65 Superconductivity News, p. 7, Feb. 1988

66 L'Usine Nouvelle, n° 51/52, Dec. 1988

67 P. Boher, Ph. Houdy, R. Barchewitz, J.C. Joud, L.J. Van Ijzendoorn
12 IXCOM Conference Cracovie, August 1989

68 N. Nacitas
Rapport d'IUT Créteil, June 1989

69 P. Boher, Ph. Houdy, P. Kaikati, L.J. Vàn Ijzendoorn
TATF 89 2nd Int. Symp. Regensburg, 27 Feb-3 March 1989

70 Ph. Houdy, Habilitation à diriger des recherches, Univ. Paris 7, 22 June 1989

71 M. Erman, Thèse 3ème cycle, Univ. Paris 11, 26 Mars 1982

72 P. Boher & al.
Journal of Vacuum Science (in press)

73 R. Krishnan, M. Porte, M. Tessier, J.P. Vitton, Y. Le Cars
Journal of the Less-Common Metals, Vol. 145, 1988, pp. 613-19

74 C.D. England, W.R. Bennett, C.M. Falco
J. Appl. Phys., Vol. 64, n° 10, pt. 2, 1988, pp. 5757-9

75 T. Kaneto, T. Sasaki, M. Sakuda, R. Yamamoto, T. Nakamura, H. Yamamoto, S. Tanaka
J. Phys. F. Met. Phys., Vol. n° 18, n° 9, 1988, pp. 2053-60

76 T. Suzuki, T. Yamazaki, K. Takahashi, T. Kageyama, H. Oda
Dept. of Ind. Chem., Tokyo Univ. of Agric. & Technol., Japan, J. Mater. Sci. Lett., Vol. 7, n° 1, 1988, pp. 79-80

77 Ph. Houdy, V. Bodart, C. Hily, P. Ruterana, L. Névot, M. Arbaoui, N. Alehyane, R. Barchewitz
Proc. SPIE, Vol. 733, 1986, pp. 389-97

78 C.M. Falco, F.E. Fernandez, P. Dhez, A. Khandar-Shahabad, L. Névot, B. Pardo, J. Corno
Proc. SPIE, Vol. 733, 1987, pp. 343-52

79 F. Reniers, P. Delcombe, L. Binst, M. Jardinier-Offergeld and F. Bouillon
to be published in Thin Solid Films.

Acknowledgments

I want to thank Mr. P. Ruterana (Ecole Polytechnique de Lausanne) for the T.E.M. micrographs and Mr. O. Siboulet (Laboratoire GPS ENS Univ. Paris 6) for the S.T.M. micrographs.

Special thanks to all the authors from whom I have used results to illustrate this review paper.

Materials Science Forum Vols. 59 & 60 (1990) pp. 617-632

MULTILAYERS AND HIGH-T_C SUPERCONDUCTIVITY

G. Creuzet, R. Cabanel and A. Schuhl
Laboratoire Central de Recherches Thomson-CSF
F-91404 Orsay Cédex, France

1 INTRODUCTION

The discovery of the first high-T_c superconducting oxide compound at the end of 1986 [1] was the starting point for a huge activity and hence a great number of new materials were identified and studied. Despite an intense theoretical activity, the mechanisms which lead to superconductivity in these compounds are still unknown.

Moreover, on the more technical side some problems have been found to be particularly difficult to solve. This is the case for the preparation of high quality thin films and junctions, which is of great importance as it is a critical step in the application for these new compounds to electronics. Furthermore, it is clear that the present state of the art is far from the moment when controlled superlattices involving high-T_c superconducting oxides (HTSC) will be available.

In this context, the present review will try to discuss the different aspects of high-T_c superconductivity where usual multilayer concepts may be relevant. It is primarily concerned with the preparation of thin films and junctions. The basic features of superconducting superlattices will then be described, as well as the main results on low-T_c superconducting materials. Finally, some preliminary results and ideas for superlattices involving high-T_c superconducting oxides will be discussed.

2 HTSC STRUCTURE

From a microscopic point of view, the high-T_c superconducting oxides appear as a "super" lattice, in the sense that the structure is made up from the superposition of planes along the c direction. Figure 1 shows the structure of the model compound $YBa_2Cu_3O_7$ (1:2:3). Here the basic alternance is $Y_{1/2}$-CuO_2-BaO-CuO-BaO-CuO_2-$Y_{1/2}$. Such a description is supported in this case by the fact that the

CuO_2 planes play the major role in the conduction process.

This situation is favourable to a layer-by-layer growth of HTSC thin films, as discussed later. The main consequence of this structure is the strong anisotropy, as shown by the superconducting parameters, which have been measured on single crystals [3, 4]. In particular, the coherence length is around 30 Å in the a-b plane but only 6 Å along the c direction (less than half the c parameter). As we will see, this extremely low value leads to strong constraints for the elaboration of junctions.

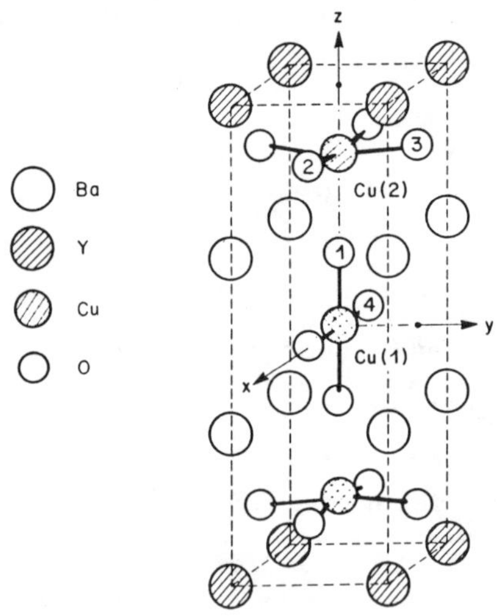

Fig.1 Crystal structure of $YBa_2Cu_3O_7$ (from ref.2).

Moreover, the different HTSC families can be classified using two numbers n and m via the general formula $(AO)_n(A'CuO_{3-x})_m$. This expression supports the description of HTSC as interlayer growth of n layers with NaCl structure $(AO)_n$ and m oxygen-deficient perovskite layers $(A'CuO_{3-x})_m$. A and A' can be Y, a rare earth, Bi, Tl, Pb, Sr, Ba or Ca. Let us note that, within a given family, T_c often appears to be strongly correlated to the parameter m, i.e. the number of Cu-O planes. This is the case for the n = 2 and n = 3 Tl oxides [5], up to the established T_c record at 125 K [6] in the $Tl_2Ba_2Ca_2Cu_3O_{10}$ compound (n = 3, m = 3). This observation was at the origin of attempts to increase T_c further by increasing the number of Cu-O planes beyond 3, as suggested by theoretical predictions [7, 8]. Unfortunately, the m = 4 compounds show lower T_c values [9]. Then the crossover between m = 3 and m = 4 can be regarded as the transition between 2D and 3D dimensionality inside the Cu-O sheets. The first HTSC family, where $La_{2-x}Sr_xCuO_4$ is the typical compound, corresponds to n = m = 1. The well known $YBa_2Cu_3O_7$ is the only n = 0 compound (with m = 3). Presently, almost all the families have been isolated, from n = 1 to 3 and m = 1 to 4, except the n = 1, m = 4 compounds.

3 FROM MULTILAYERS TO THIN FILMS

3.1 HTSC thin film growth: general features

All the available techniques normally used for other materials have been tested for the preparation of HTSC thin films. Superconducting films have been obtained in all cases more or less, but the film quality is often very bad and still unsatisfactory for realistic applications. Schematically, the results can be classified in three categories: (A) unoriented grains with other phases inside the material, especially in the grain boundaries, and rather bad surface roughness (several hundred Å); (B) juxtaposition of oriented grains without extra phases and better surface roughness than in A (100 Å or less); (C) single crystal films. Most of the available films are in the A category, and only a few groups present reproducible results of the B type. The best critical current densities ever obtained with HTSC (typically a few 10^6 A cm^{-2} at 77 K and around 10^7 A cm^{-2} at 4.2 K) have been obtained on B type thin films. A crucial point is the necessity, or not, of a post-annealing in the fabrication process. Such a step is very often required in order to complete the oxygen stoichiometry, and also to build up the structure if the as-deposited film is amorphous. For the standard compound $YBa_2Cu_3O_7$, "classical" annealing consists of warming in oxygen or inert gas up to 800°C to 900°C, then a plateau in that state (for the structure), followed by a slow cooling in oxygen with or without another plateau around 500°C (for the oxygen content). This kind of annealing usually leads to A type films, with very bad surface roughness, and also interdiffusion between the film and the substrate. B type films often require a short post-annealing at moderate temperature in order to complete the oxygen stoichiometry. It is expected that "perfect" C type films need to be obtained as-deposited, without any kind of annealing. This is to avoid any interdiffusion with the substrate and hence any defect at the interface, but mainly because, as in single crystals, a perfect lattice does not allow oxygen diffusion into the structure.

3.2 HTSC preparation by multilayer evaporation

Some groups present results using a method derived from multilayer preparation. It consists of two steps: (i) sequential deposition of the constituants by electron beam evaporation or any other appropriate technique; (ii) post-annealing at high temperature (800-900°C) in oxygen. The supercell in the intermediate state is made up of layers with thicknesses of the order of 100 Å, the ratio being adjusted in order to obtain finally the correct stoichiometry. The annealing is required for three reasons: interdiffusion of the species, structure building and oxygen introduction up to the desired value. The main advantages of this technique are the possibility of preparing films with thicknesses well above 1 μm, and the precise adjustment of the relative quantities of the constituent atoms. $YBa_2Cu_3O_7$ thin films were prepared in such a way [10] rather soon after the isolation of the compound in ceramic form but up to now the results are always of A type. This is mainly due to inhomogeneities in the concentrations of the species in the film and interdiffusion with the substrate, as revealed by AES depth profile measurements (see Fig. 2) [11].

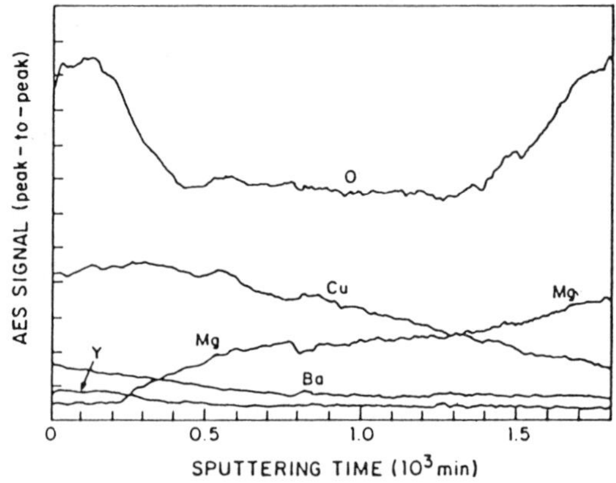

Fig.2 AES depth profile of an YBaCuO thin film on MgO prepared by sequential deposition and rapid thermal annealing (from ref.11).

Consequently, this kind of preparation technique appears to be restricted to specific applications where rather thick films are required with small constraints on the microstructure and superconducting characteristics.

3.3 HTSC preparation by Atomic Layer Epitaxy

Starting from the multilayer evaporation process for HTSC thin film preparation, it is possible to reduce the thickness of each individual layer to very low values. The ultimate step is the layer-by-layer deposition of the compound, where the structure is directly built according to the appropriate sequence, i.e. Y-CuO_2-BaO-CuO-BaO-CuO_2. This technique is known as Atomic Layer Epitaxy (ALE). It is derived from the Molecular Beam Epitaxy (MBE) by computer controlled flux modulation. In the particular case of oxides, one should note that the oxygen flux is not modulated. However we will speak here of ALE.

Only C type films will provide high critical-current SNS or SIS junctions. MBE appears in this context to be very promising [12-14]. Good quality B type films have been obtained by standard MBE growth, followed by a short annealing at 550°C [12]. Recently, the same group also managed to eliminate this final post-treatment [13]. The first results using a modulation of the Ba and Dy flux were reported by the Stanford group [14], are shown on Fig. 3. The method needs to be optimized, but they obtained as-deposited superconducting films.

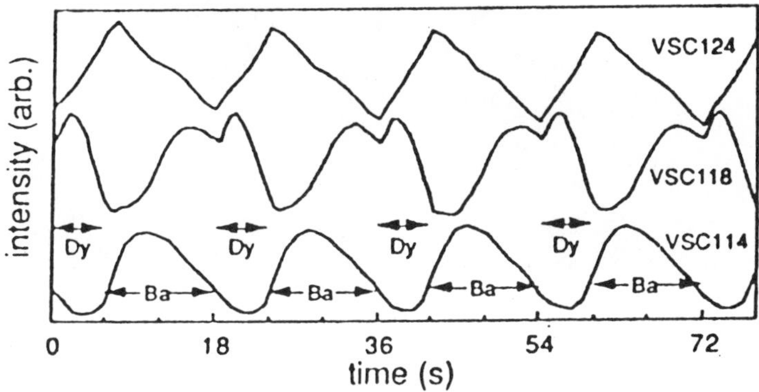

Fig.3 Intensity of the central RHEED streak as a function of time during three different "shuttered " growths (from ref.14).

Let us now describe the ALE process, as developed and used in our laboratory. Thin films of the DyBaCuO system with typical thicknesses of 300 Å are grown in a modified Riber machine specially designed for superconductors. The cations (Dy, Cu and Ba) are successively sent to the sample while a relatively high oxygen pressure (10^{-5} Torr) is maintained in the chamber. Growth rates vary between 0.1 and 1Å/s. Standard (001) $SrTiO_3$ wafers are prepared in order to present a very good crystalline quality, as checked by Rutherford Backscattering (RBS) experiments [15]. An oxygen cell using a dc plasma source delivers a ratio of active oxygen over molecular oxygen of about 10 %. The oxygen flux during deposition is around 2.10^{16} atoms per sec. per cm^2, which is two orders of magnitude above the cation flux. The stoichiometry adjustments are achieved inside a loop where the RBS measurements take place [15]. Moreover, we did not observe any interdiffusion between the film and the substrate for growth temperatures below 550°C, which is of great importance to obtain well controlled Josephson junctions.

In the best experimental conditions, namely from 500°C to 600°C for substrate temperature and high oxygen flux, RHEED patterns indicate good epitaxial growth, as shown in Fig.4. Up to 400 Å, RHEED patterns remain more or less identical. Due to the complexity of the process, and the requirement for very thin films to perform RBS analysis, we do not intend to grow thicker layers.

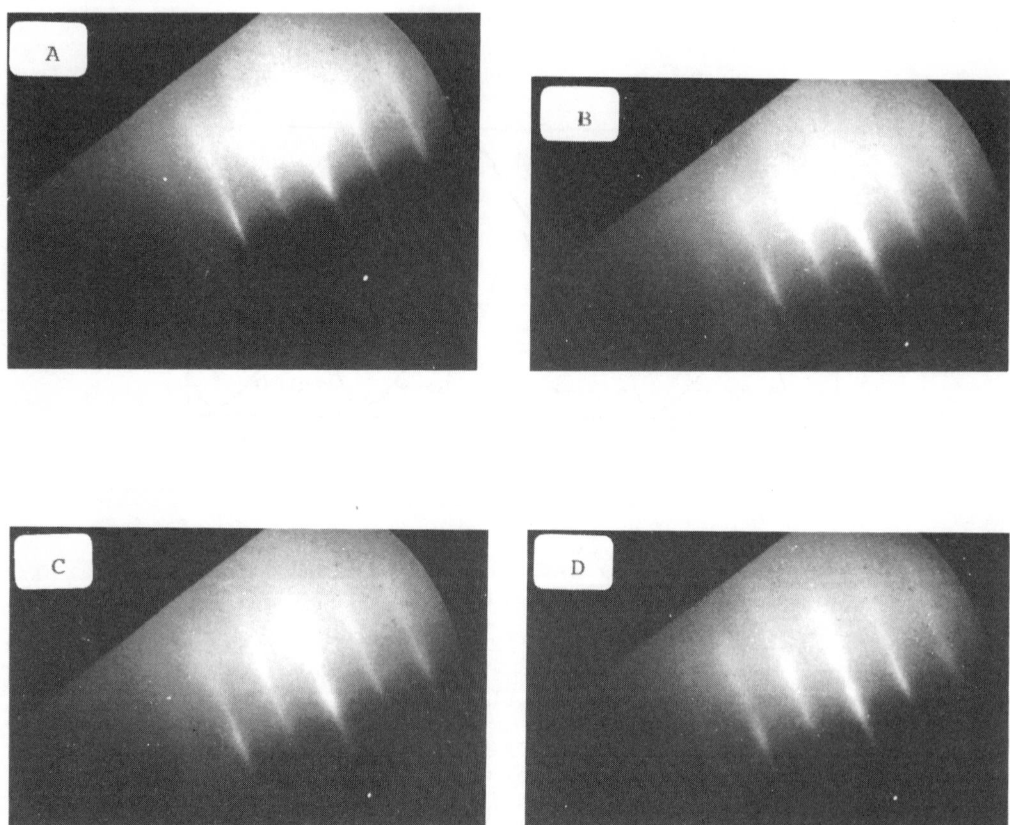

Fig.4 RHEED patterns of DyBaCuO film during deposition at 550°C in the (100) direction: a, after 3 atomic cells (36 Å); b, after 20 cells and a Ba atomic layer; c, after the following Cu layer; d, after the following Dy layer (from ref.15).

X-Ray measurements reveal the presence of the 1:2:3 phase, which is oriented with the c-axis perpendicular to the substrate, as expected from the ALE process. The best films are still poor in oxygen content (6.2-6.3), as shown by the c parameter determination and checked by transport properties. Nevertheless, the present situation is very promising, as layer-by-layer growth has been achieved up to reasonable thicknesses. Improvements of the oxygen cell efficiency will give better chances of obtaining as-deposited ALE films with oxygen contents close to 7. This prediction is supported by the fact that we have obtained superconducting thin films without any annealing with a co-deposition process at 600°C. The transition onset takes place at 92 K, which proves that the 1:2:3 phase present with an oxygen content close to 7 is stable under these experimental pressure and temperature conditions. Other improvements are needed, like the transitory regime occurring at each shutter opening, and the flux control which must be better than 1%. Under these conditions, the ALE process will be at the forefront of the preparation techniques for HTSC films and junctions.

4 FROM THIN FILMS TO MULTILAYERS

4.1 Junctions by Atomic substitution

Assuming that C type films are available, we can then imagine growing superlattices. In this context, the first step will be the sandwich structure which is precisely a controlled-layer SNS or SIS junction, depending on the nature of the material in the middle. Considering the fact that, for example, the $PrBa_2Cu_3O_7$ compound presents a semiconducting-like behaviour, but is isostructural with any other 1:2:3 material, we can envisage a sandwich of the type $YBa_2Cu_3O_7$ -$PrBa_2Cu_3O_7$ -$YBa_2Cu_3O_7$ grown in epitaxial conditions, with c-axis perpendicular to the substrate. As the thickness of the barrier in SNS or SIS junctions must be of the order of magnitude of the coherence length, we deduce that the praseodynium substitution must be limited to a few unit cells. This requirement supports the use of the ALE technique for thin film deposition.

Recently, preliminary results were reported on such sandwiches [16,17]. For the Jülich group using a sputtering technique [16], the growth was not fully epitaxial, but the lattice orientation was not affected by the substitution of Pr to Y. However, only very small deterioration of the transport properties of the final heterostructure was observed, which is probably due to large uncertainties in the microstructure. On the other hand, the Bellcore work using laser deposition [17] obtained current-voltage characteristics similar to those of traditional SNS components. However, these are preliminary results as the barrier is not fully controlled and is too thick.

Nevertheless, the $YBa_2Cu_3O_7$-$PrBa_2Cu_3O_7$-$YBa_2Cu_3O_7$ structure appears to be very promising for HTSC devices. In future, multilayered structures from this fundamental building block may be also of interest for advanced superconducting components.

4.2 Superconducting superlattices

4.2.1 The situation of low-T_c materials superlattices

The superlattices built from low-T_c materials (LTSC) have been extensively studied for many years. Recent reviews gave most of the theoretical background and experimental results [18, 19]. The more interesting physical effects can be classified in three major domains: dimensional effects (from 2D when the thickness of individual layers is less than the coherence length to 3D in the opposite case), proximity effects and magnetic effects. In addition, attempts have been made to test the different theoretical mechanisms exploring the same family with different superstructure parameters. Then the studied superlattices belong to three categories: LTSC - insulator, LTSC - normal metal and LTSC - magnetic metal.

The crossover between the 2D and 3D regimes is well illustrated in the Nb/Cu system. The superlattices are symmetric ($d_{Nb} = d_{Cu} = d$) and the transition temperature increases regularly between d = 10 Å (limit of the 2D region) to around a few thousand Å [20], as shown on Fig. 5. In this case, Cu is playing a role of inert spacer between superconducting Nb layers.

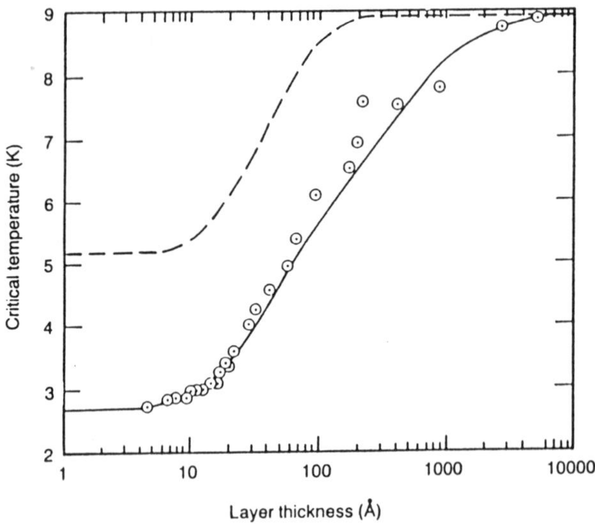

Fig.5 Dependence of the transition temperature of symmetric Nb/Cu superlattices on layer thickness (from ref.20). o is for experiment, — and -- are theoretical fits.

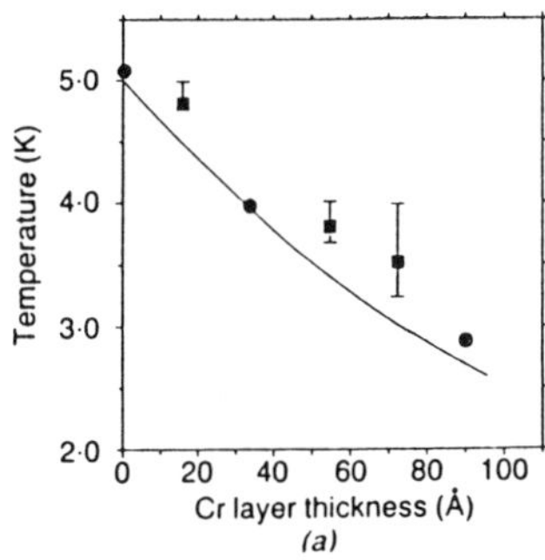

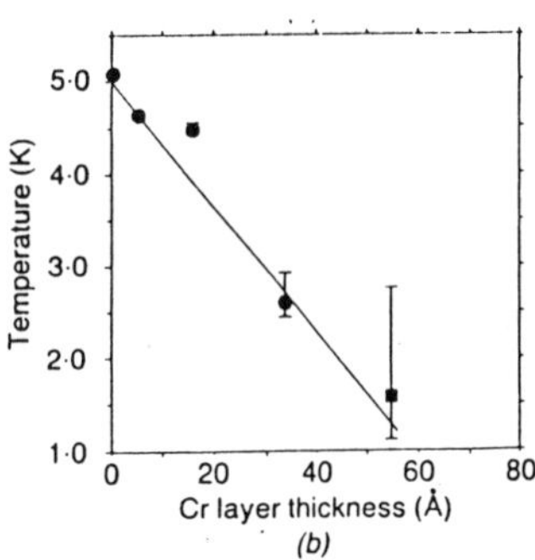

Fig.6 The transition temperature as a function of the Cr layer thickness for several V/Cr superlattices with fixed V layer thickness (a) d_V = 374 Å and (b) d_V = 209 Å (from ref.21). • and • are for experiment, – is a theoretical fit.

The influence of magnetic layers on the superconductive properties of superlattices has been studied extensively in a large variety of systems. Let us illustrate the problem by the influence of the antiferromagnetic order of Cr layers on V/Cr superlattices [21]. Fig. 6 shows a rapid decrease of the transition temperature with the Cr layer thickness. This effect is more pronounced for the thinner V layers, as expected.

A general discussion of the literature would be out of place here, so we refer for more details to the recent reviews already mentioned [18, 19].

4.2.2 Future high-T_c and mixed superlattices

As mentioned above, the state-of-the-art on HTSC thin film deposition is such that well defined superlattices involving HTSC materials are not yet available. Nevertheless, a group recently claimed to have succeeded in growing $YBa_2Cu_3O_7$-$DyBa_2Cu_3O_7$ superlattices by dc magnetron sputtering on $SrTiO_3$ and MgO substrates [22]. X-Ray diffraction curves present satellite peaks which are clearly related to the modulation, as compared to pure $YBa_2Cu_3O_7$ and $PrBa_2Cu_3O_7$. The superlattices are certainly not perfect, but at least there is an evidence for a modulation along the c direction in the Y + Dy concentration. The superconductive properties are very similar to those of pure $YBa_2Cu_3O_7$ and $PrBa_2Cu_3O_7$, which are basically very close to each other.

Furthermore, one can imagine to introduce HTSC in superlattices with either insulators, normal metals or magnetic metals as for LTSC. The main resulting properties can be expected from the well known results with LTSC, but specific effects may arise from particular characteristics of HTSC like very high critical fields. Provided that crystal growth problems can be solved, the mixed

LTSC-HTSC superlattices might be of particular interest. Due to proximity effects, on can imagine to combine high critical temperature (from HTSC) with high coherence length (from LTSC). This is a pure conjecture, as no theoretical background supports this hypothesis. Anyway, the study of such systems will certainly show new and interesting properties.

REFERENCES

[1] J.G. Bednorz and K.A. Muller, Z.Phys. B 64, 189 (1986).

[2] M. Stavola, D.M. Krol, W. Veber, S.A. Sunshine, A. Jayaraman, G.A. Kourouklis, R.J. Cava and E.A. Rietman, Phys. Rev. B 36, 850 (1987).

[3] T.K. Worthington, W.J. Gallagher and T.R. Dinger, Phys. Rev. Lett. 59, 1160 (1987).

[4] A. Umezawa, G.W. Crabtree, J.Z. Liu, T.J. Moran, S.K. Malik, L.H. Nunez, W.L. Kwok and C.H. Sowers, Phys. Rev. B 38, 2843 (1988).

[5] M.A. Subramanian, J. Gopalakrishnan, C.C. Torardi, J.C. Calabrese, K.J. Morrissey, J. Parise, P.L. Gai and A.W. Sleight, Advances in Superconductivity, K. Kitizawa and T. Ishiguro Eds., Springer Verlag 1989, p763 and references therein.

[6] S.S.P. Parkin, V.Y. Lee, E.M. Engler, A.I. Nazzal, T.C. Huang, G. Gorman, R. Savoy and R. Beyers, Phys. Rev. Lett. 60, 2539 (1988).

[7] P.M. Grant, unpublished work.

[8] J. Bok, Solid State Comm. 67, 251 (1988).

[9] P.T. Wu, R.S. Liu, J.M. Liang and L.J. Chen, Advances in Superconductivity, K. Kitizawa and T. Ishiguro Eds., Springer Verlag 1989, p799.

[10] B.Y. Tsaur, M.S. Dilorio and A.J. Strauss, Appl. Phys. Lett. 51, 858 (1987).

[11] Q.Y. Ma, T.J. Licata, X. Wu, E.S. Yang and C.A. Chang, Appl. Phys. Lett. 53, 2229 (1988).

[12] J. Kwo, M. Hong, D.J. Trevor, R.M. Fleming, A.E. White, R.C. Farrow, A.R. Kortan and K.T. Short, Appl. Phys. Lett. 53, 2683 (1988).

[13] J. Kwo, M. Hong, D.J. Trevor, R.M. Fleming, A.E. White, J.P. Mannaerts, R.C. Farrow, A.R. Kortan and K.T. Short, M^2S-HTSC 89 Proceedings, Physica C 162-164, 623 (1989).

[14] D.G. Schlom, J.N. Eckstein, E.S. Hellman, S.K. Strieffer, J.S. Harris, Jr., M.R. Beasley, J.C. Bravman, T.H. Geballe, C. Webb, K.E. von Dessonneck and F. Turner, Appl. Phys. Lett. 53, 1660 (1988).

[15] A. Schuhl, R. Cabanel, S. Koch, J. Siejka, M. Touzeau, J.P. Hirtz and G. Creuzet, M^2S-HTSC 89 Proceedings, Physica C 162-164, 627 (1989).

[16] U. Poppe, P. Prieto, J. Schubert, H. Soltner, K. Urban and Ch. Buchal, submitted to Solid State Comm.

[17] C.T. Rogers, A. Inam, M.S. Hedge, D. Dutta, X.D. Wu and T. Venkatesan, submitted to Appl. Phys. Lett.

[18] I.K. Schuller, Physics, fabrication and applications of multilayered structures, P. Dhez and C. Weisbuch Eds., NATO ASI Series, Plenum 1988, p139.

[19] B.Y. Jin and J.B. Ketterson, Advances in Physics 38, 189 (1989).

[20] I. Banerjee, Q.S Yang, C.M. Falco and I.K. Schuller, Solid. State Com., 41, 805 (1982).

[21] B. Davis, J.Q. Zheng, P.R. Auvil, J.K. Ketterson and J.E. Hilliard, Superlattices Microstructures 4, 465 (1988).

[22] J-M. Triscone, M.G. Karkut, L. Antagnozza, O. Brunner and O. Fisher, submitted to Phys. Rev. Lett.

SUBJECT INDEX